New Perspectives in Stochastic Geometry

New Perspectives in Stochastic Geometry

Edited by

Wilfrid S. Kendall
University of Warwick

Ilya Molchanov
University of Bern

OXFORD
UNIVERSITY PRESS

has been printed digitally and produced in a standard specification in order to ensure its continuing availability

OXFORD
UNIVERSITY PRESS

Great Clarendon Street, Oxford OX2 6DP
United Kingdom

Oxford University Press is a department of the University of Oxford.
It furthers the University's objective of excellence in research, scholarship,
and education by publishing worldwide. Oxford is a registered trade mark of
Oxford University Press in the UK and in certain other countries

Reprinted 2012

British Library Cataloguing in Publication Data
Data available

Library of Congress Cataloging in Publication Data
Data available

ISBN 978-0-19-923257-4

In memory of David Kendall,
15 January 1918–23 October 2007

PREFACE

The roots of stochastic geometry derive from classical highlights of geometric probability. Examples include concepts arising from explorations of the notion of 'natural' probability, questions raised in nineteenth-century UK recreational mathematics, and the famous problem of the eighteenth-century French polymath Georges-Louis Leclerc, Comte de Buffon. The actual phrase 'stochastic geometry' appears first to have been employed in its current sense by D.G. Kendall and K. Krickeberg in 1969, arising from their planning for an Oberwolfach workshop (Stoyan, Kendall and Mecke, 1987, foreword), though Andrew Wade has pointed out its use by Frisch and Hammersley (1963, p. 895) as one of two possible terms to describe the study of 'random irregular structures' motivated by percolation problems. It is now plain that stochastic geometry has woven together various strands from a wide-ranging variety of sources, of which we mention just four representative examples:

- impetus arising from the creation and development of geostatistics and mathematical morphology by G. Matheron and J. Serra and other French workers;
- the German school of point process and queueing theory, involving such names as J. Kerstan, K. Matthes, J. Mecke and K. Krickeberg;
- British and Australian research, growing out of the study of classical geometric probabilities, and developing relationships with stochastic analysis;
- the Armenian school of combinatorial geometry led by R.V. Ambartzumian.

After 40 years stochastic geometry is now an established part of probability theory, expounded in several well-established monographs, and finding vital employment in numerous application areas such as spatial statistics, image analysis, materials science, and even finance. In common with many areas of applied mathematics, the subject has been heavily influenced by the availability of cheap and powerful computing, justifying deeper study of models which are flexible but computationally demanding, and providing opportunities for simulation study of systems which would otherwise be inaccessible. A subtle and expressive mathematical vocabulary has been developed, allowing us to describe and investigate random patterns of points, lines, planes, fibres, surfaces, tessellations and sets. A wealth of hard theoretical problems can be found in the wide variety of application areas, and it is an ongoing challenge for stochastic geometry to contribute effectively to the intense and powerful scientific effort devoted to dealing with these problems.

The present volume has been motivated by the sense that stochastic geometry is now poised for a further phase of development, acquiring new perspectives from revival of its classical roots, from new connections with recent developments in the main body of probability, and stimulated by the plethora of rapidly developing applications. The purpose of the volume is therefore to present recent developments so as to form an entry point into stochastic geometry for mathematical scientists from other areas who are curious about what it might offer, and to provide a resource for young researchers who wish to engage with the subject and seek out areas in which they might make an original contribution. The editors gathered together a group of experts, each of whom was tasked with writing a chapter to cover a particular aspect of stochastic geometry, reflecting the current state of the art but also commenting on and providing new developments. To this end, the volume is divided into a chapter on the fundamentals, and then into four parts. The initial Chapter 1 surveys what can be described as classical stochastic geometry, summarizing its historical development and clarifying the core mathematical concepts which arise throughout stochastic geometry as a whole. Classical stochastic geometry has itself recently seen a revival; Part I ('New Developments in Classical Stochastic Geometry') contains four chapters describing new developments in random polytopes, random measures, limit theory and tessellations (Chapters 2, 3, 4, 5). In addition, recently there has arisen a number of strong connections with other areas of mathematics such as percolation, random network theory and fractals, forming the three Chapters 6, 7 and 8 of Part II 'Stochastic Geometry and Modern Probability'. There has always been a strong statistical theme in stochastic geometry, represented here by Part III 'Statistics and Stochastic Geometry'. This commences with Chapter 9, discussing inferential issues arising when one seeks to estimate parameters of random pattern models, and continues with three further Chapters 10, 11 and 12, respectively covering statistical shape, estimation of sets, and notions of data depth. Finally Part IV 'Applications' contains five chapters surveying applications in image analysis (Chapter 13), stereology (Chapter 14), materials science (Chapter 15), telecommunications (Chapter 16) and finance (Chapter 17); the growth of stochastic geometry has always been fuelled by ever-changing demands from application areas, and these chapters demonstrate strong possibilities for future growth.

The authors and editors offer this volume to the mathematical community as a stepping stone towards the new perspectives which stochastic geometry now offers.

References in the Preface

Frisch, H. L. and Hammersley, J. M. (1963). Percolation processes and related topics. *J. Soc. Indust. Appl. Math.*, **11**, 894–918.

Stoyan, D., Kendall, W. S., and Mecke, J. (1987). *Stochastic Geometry and its Applications*. Wiley Series in Probability and Mathematical Statistics: Applied Probability and Statistics. John Wiley & Sons Ltd., Chichester. With a foreword by D. G. Kendall.

ACKNOWLEDGEMENTS

The editors and authors would like to thank the Oxford University Press staff for help with the production of this volume.

They are also grateful to Mathematisches Forschungsinstitut Oberwolfach for the opportunity to organize a workshop under the same name, which greatly aided the process of preparation of this volume, nearly 40 years after the Oberwolfach workshop at which the phrase 'Stochastic Geometry' first saw the light of day.

CONTENTS

LIST OF CONTRIBUTORS

Dr. Pierre Calka
(pierre.calka@math-info.univ-paris5.fr)
MAP5, Université Paris Descartes, 45,
rue des Saints-Pères, 75270 Paris Cedex 06, France.

Dr. Ignacio Cascos
(ignacio.cascos@uc3m.es)
Department of Statistics,
Universidad Carlos III de Madrid,
Av. Universidad 30, E-28911 Leganés (Madrid), Spain.

Prof. Dr. Antonio Cuevas
(antonio.cuevas@uam.es)
Department of Mathematics,
Facultad de Ciencias,
Universidad Autonoma de Madrid,
28049 Madrid, Spain.

Prof. Dr. R. Fraiman
(rfraiman@cmat.edu.uy)
Centro de Matematica,
Universidad de la Republica,
Eduardo Acevedo, 1139,
Montevideo, Uruguay.

Prof. Dr. Remco van der Hofstad
(rhofstad@win.tue.nl)
Eindhoven University of Technology
Department of Mathematics and Computer Science, HG 9.04
P.O. Box 513, 5600 MB Eindhoven, the Netherlands.

Prof. Dr. W.S. Kendall
(w.s.kendall@warwick.ac.uk)
Dept of Statistics,
University of Warwick
Coventry CV4 7AL, United Kingdom.

Prof. Dr. Günter Last
(last@math.uni-karlsruhe.de)
Institut für Stochastik,
Universität Karlsruhe, D-76128 Karlsruhe, Germany.

Dr. Huiling Le
(huiling.le@nottingham.ac.uk)
School of Mathematical Sciences,
University of Nottingham
Nottingham NG7 2RD, United Kingdom.

Dr. M.N.M. van Lieshout
(colette@cwi.nl)
CWI, Science Park 123, NL-1098, XG Amsterdam, The Netherlands.

Prof. Dr. Klaus Mecke
(klaus.mecke@physik.uni-erlangen.de)
Institut für Theoretische Physik,
Universität Erlangen-Nürnberg,
Staudtstrasse 7, 91058 Erlangen, Germany.

Prof. Dr. Ilya Molchanov
(ilya@stat.unibe.ch)
IMSV, University of Bern
Sidlerstrasse 5, 3012 Bern, Switzerland.

Prof. Dr. Jesper Møller
(jm@math.aau.dk)
Department of Mathematical Sciences, Aalborg University,
F. Bajers Vej 7G,
DK-9220 Aalborg Øst, Denmark.

Prof. Dr. Peter Mörters
(maspm@bath.ac.uk)
Department of Mathematical Sciences, University of Bath
Claverton Down, Bath BA2 7AY,
United Kingdom.

Dr. Werner Nagel
(werner.nagel@uni-jena.de)
Friedrich-Schiller-Universität Jena,
Fakultät für Mathematik und Informatik
Institut für Stochastik, D-07737 Jena, Germany.

Prof. Dr. Mathew D. Penrose
(m.d.penrose@bath.ac.uk)
Department of Mathematical Sciences, University of Bath
Claverton Down, Bath BA2 7AY,
United Kingdom.

Prof. Dr. Matthias Reitzner
(mreitzner@uni-osnabruck.de)
Mathematik/Informatik
Universität Osnabrück,
Albrechtstraße 28a
49706 Osnabrück, Germany

Prof. Dr. Rolf Schneider
(rolf.schneider@math.uni-freiburg.de)
Mathematisches Institut,
Albert-Ludwigs-Universität,
Eckerstr. 1, D-79104, Freiburg i. Br., Germany.

Dr. Tomasz Schreiber
(tomeks@mat.uni.torun.pl)
Faculty of Mathematics and Computer Science
Nicolaus Copernicus University,
Chopina 12/18, 87-100 Torun,
Poland.

Dr. Andrew R. Wade
(Andrew.Wade@bris.ac.uk)
Department of Mathematics,
University of Bristol
University Walk, Bristol BS8 1TW,
United Kingdom.

Prof. Dr. Wolfgang Weil
(weil@math.uni-karlsruhe.de)
Institut für Algebra und Geometrie,
Universität Karlsruhe (TH),
D-76128 Karlsruhe, Germany.

Prof. Dr. Sergei Zuyev
(sergei.zuyev@chalmers.se)
Department of Mathematical Sciences, Chalmers University of Technology, 412 96 Gothenburg,
Sweden

1

CLASSICAL STOCHASTIC GEOMETRY

Rolf Schneider and Wolfgang Weil

Abstract

The aim of this chapter is to introduce the basic tools and structures of stochastic geometry and thus to lay the foundations for much of the book. Before this, a brief historic account will reflect the development from elementary geometric probabilities over heuristic principles in applications to the advanced models employed in modern stochastic geometry. After the basic geometric and stochastic concepts have been presented, their interplay will be demonstrated by typical examples.

1.1 From geometric probabilities to stochastic geometry – a look at the origins

The origins of stochastic geometry can be traced back to two different sources. These are, on one hand, geometric probabilities and integral geometry, with their intuitive problems and imagined experiments, and on the other hand the investigation of real-world materials by stochastic-geometric methods, which in the beginning were often heuristic and required sound mathematical foundations. We illustrate these two aspects by describing a few landmarks.

The birth of geometric probability can be attributed to a game of chance, in a geometric version, due to Georges–Louis Leclerc, Comte de Buffon. In 1733 he considered the chances that a randomly thrown coin hits an edge of a regular mosaic paving on the floor. His results were only published much later, as part of a longer essay, in 1777. A simplified version of such geometric games is Buffon's needle problem, where the mosaic is given by parallel lines of distance D and the coin is replaced by a needle of length $L < D$ (Buffon at first spoke of a rod, a *baguette* in French, and then suggested to play the game with a needle). Considering the position of the midpoint of the needle and the angle between the needle and the lines, and using integration (apparently, for the first time in a probabilistic problem), Buffon calculated the probability p for the needle to hit a line as

$$p = \frac{2L}{\pi D}.$$

The appearance of π in the formula prompted later experiments, and probably added to the lasting popularity of Buffon's needle problem. Buffon's calculation rested on the assumption that the distance of the midpoint of the needle from the nearest line and the angle between needle and lines, in modern terminology, were independent and uniformly distributed in their respective range.

More problematic was another historical question, Sylvester's four-point problem of 1864. He asked for the probability that four points taken at random in the plane are the vertices of a 're-entrant quadrilateral' (that is, their convex hull is a triangle). Several contradictory answers were received. Only later was the problem given a precise version, by specifying that the four random points should be independent and uniform in a given convex domain.

The ambiguous nature of such intuitive assumptions in geometric probability problems was made evident in the book *Calcul des probabilités* published in 1889 by J. Bertrand. He described several situations where random geometric objects were parametrized in different ways and the natural uniform distributions in the parameter spaces resulted in different distributions of the objects themselves.

From a purely mathematical point of view, this dilemma can be overcome by a principle that took its origin in a paper by M.W. Crofton in 1868 and and was further developed by H. Poincaré. An extended version of this principle may be formulated as follows. *If a probabilistic problem on geometric objects is invariant under geometric transformations of a certain kind, a natural distribution of the objects can be obtained from a measure which is invariant under these transformations.* Therefore, the use of Haar measures on topological groups and homogeneous spaces clarifies in many cases the definition of a canonical probability measure for geometric problems. It also opens the way to a unified treatment of large classes of problems, by establishing and applying formulae from integral geometry, which deals with invariant integrals involving functions of geometric objects undergoing transformations. Of course, for modelling real-life situations, such invariance assumptions on the distribution may be too restrictive. They will often be convenient approximations and tentative working assumptions only. Nevertheless, the use of invariant integral geometry is a first step to obtain explicit results, and it often gives hints to the necessary extensions, as partly explained below.

Symbolically, the mentioned formulae of integral geometry have the form

$$\int_{\mathcal{B}} f(A \odot \tilde{B})\, \rho(\mathrm{d}\tilde{B}) = F(A, B) \tag{1.1}$$

where A, B are sets, say in $\mathbb{R}^d$, $\odot$ denotes a geometric operation (this could be intersection, sum, projection, etc.), f is a geometric functional (volume, surface area, integral mean curvature, Euler characteristic, etc.) and the integration is with respect to an invariant measure ρ over a class $\mathcal{B}$ of congruent copies of B. The challenge is to express this integral in a simple way in terms of geometric functionals applied to A and B separately, if possible. An important example is the principal kinematic formula

$$\int_{G_d} V_j(K \cap gM)\, \mu(\mathrm{d}g) = \sum_{k=j}^{d} c(d,j,k) V_k(K) V_{d-k+j}(M), \quad j = 0, \ldots, d. \tag{1.2}$$

Here G_d is the group of rigid motions, with Haar measure μ (unique up to a factor and bi-invariant, due to the unimodularity of G_d), and K, M are convex bodies (nonempty, compact, convex sets), for example. The functionals $V_0, \ldots, V_d$ are the intrinsic volumes, and the constants are given by (1.13) below. The formula holds for much more general set classes (finite unions of convex bodies, sets of positive reach, etc.). For convex bodies K, M, the case $j = 0$ of (1.2) yields the measure for the event $G_d(K, M)$ of nonempty intersection,

$$\mu(G_d(K, M)) = \mu(\{g \in G_d : K \cap gM \neq \emptyset\}) = \sum_{k=0}^{d} c(d, 0, k) V_k(K) V_{d-k}(M)$$

(this follows since $V_0(K \cap gM) \in \{0, 1\}$, the Euler characteristic of the intersection, is equal to one precisely if the intersection is nonempty). It is now evident how formulae from integral geometry can be used in problems of geometric probability. A typical result of this transfer concerns the situation of convex bodies K, L, M with $L \subset K$, then

$$\frac{\mu(G_d(L, M))}{\mu(G_d(K, M))} = \frac{\sum_{k=0}^{d} c(d, 0, k) V_k(L) V_{d-k}(M)}{\sum_{k=0}^{d} c(d, 0, k) V_k(K) V_{d-k}(M)}.$$

The left side can be interpreted as the probability that a randomly moving body M which hits K, also hits the smaller body L. This result is expressed solely in terms of intrinsic volumes of the three bodies K, L, M. As in this example, geometric probabilities are frequently of a conditional type: since the underlying measure μ is infinite, a randomly moving body M makes only sense under some restriction, such as to hit K, since the restriction of the measure μ to $G_d(K, M)$ is finite and can therefore be normalized. A second typical aspect is that the random object, here the randomly moving body M, has a fixed shape, only its position and orientation are random (given by a random motion g applied to M).

The roughly 200 years from the publication of Buffon's essay in 1777 to that of the book by Santaló (1976) on *Integral Geometry and Geometric Probability* can be structured by a few more dates. In the nineteenth century, various elementary geometric probability questions were considered, many of them asked and answered in *The Educational Times*. First accounts of the field were given in Crofton's article on *Probability* in the Encyclopaedia Britannica in 1885 and in the book by Czuber (1884) on *Geometrische Wahrscheinlichkeiten und Mittelwerte*, which collected 206 problems and their solutions. A more systematic treatment was presented by Deltheil (1926) in his book *Probabilités géométriques*. During the next decades, an increasing number of geometric probability questions came from various sciences, so that Kendall and Moran (1963) in their booklet on *Geometrical Probability* listed the following fields of current applications in their preface: astronomy, atomic physics, biology, crystallography, petrography, sampling theory, sylviculture. The particular role of invariant integrals for geometric probability was emphasized by G. Herglotz in a course on *Geometrische*

Wahrscheinlichkeiten that he gave in Göttingen in 1933 (and of which there exist mimeographed notes). W. Blaschke mentioned that he was much inspired by Herglotz when, in the mid-1930s, he developed his Integral Geometry, which included first versions of kinematic formulae. The great geometers S.S. Chern, H. Hadwiger and L.A. Santaló all worked with Blaschke in Hamburg for some time, and each of them contributed substantially, in his own personal style, to the further development of integral geometry. The work of Santaló has the closest ties with geometric probability, culminating in his already mentioned fundamental monograph of 1976.

Parallel to the establishment of integral geometry and geometric probability, scientists working in different applied fields such as geology, medicine, biology, mineralogy and others, used stochastic-geometric approaches and methods. Theoretical justification and further development later became an essential part of stochastic geometry and provided much motivation for mathematical research. An example is given by the history of stereology. In 1847, the geologist A. Delesse suggested that the usual procedure to estimate the amount of mineral in a solid piece of rock, namely crushing the rock into small pieces in order to separate rock and mineral, could be simplified substantially by investigating a polished planar section of the rock and investigating the area fraction of the mineral in the section. He made it plausible that the area fraction A_A in the planar section and the volume fraction V_V in the whole material are related by the simple equation $A_A = V_V$, provided the distribution of the mineral in the rock is sufficiently homogeneous. It took some time until it was realized that the same principle could be applied in the planar section again. A. Rosiwal (1898) showed that the area fraction can be replaced by the length fraction L_L of the mineral part along a grid of lines laid out in the sectioning plane. A.A. Glagolev (1933) and E. Thomson (1930) finally introduced the simplest estimation method, namely superimposing a grid of points onto the plane and counting the fraction P_P of points covered by the mineral. The resulting formulae

$$V_V = A_A = L_L = P_P \tag{1.3}$$

marked the first set of basic relations in stereology. A next step was undertaken by S.A. Saltykov (1952), H.W. Chalkley (1949), S.T. Tomkeieff (1945) and others by considering the surface area per unit volume S_V of an embedded surface in three-space and estimating it by the boundary length B_A per unit area in a planar section or the number I_L of intersection points in a grid of lines. Here, the formulae read

$$S_V = \frac{4}{\pi} B_A = 2I_L. \tag{1.4}$$

It was clear that such formulae required some isotropy (rotational invariance) of the material under investigation and it also became apparent that there must be a common background from mathematics for these results. Scientists who developed such formulae and applied them, met occasionally at conferences, although working in quite different fields. In 1961 the *International Society for Stereology* was founded in the Black Forest. Only then, mathematicians pointed out that these fundamental formulae of stereology are applications and variants of the classical Crofton formulae in integral geometry. R.E. Miles and P. Davy, in a series of papers starting in 1976, analysed the stereological situation carefully and provided the correct assumptions for their validity. In particular, for the 'design-based approach', where the material is deterministic and the sectioning is random, it was clarified how the random elements had to be weighted in order that the estimators corresponding to (1.3), (1.4) become unbiased. In contrast to this, the 'model-based approach' assumes that the structure F under consideration is the realization of a stationary and isotropic random (closed) set $Z \subset \mathbb{R}^3$. More precisely, since such a random set is unbounded with probability one and the structure F is bounded, one assumes that F is the realization of Z in a sampling window W. In that case, the fundamental formulae of stereology, (1.3) and (1.4), can be verified as expectation formulae for random sets.

There are many other instances where material structures have been described by stochastic-geometric models, for example, fibres in paper by line processes, foams by random tessellations, components of two-phase materials by random sets, porous media by processes of particles. The general introduction and systematic study of the basic models of stochastic geometry began in the 1960s. Apart from the theory of point processes, which was applied and extended to spaces of geometric objects, such as lines, flats, curves, or compact sets, a general notion of random sets was established. An essential feature in the definition of the latter is the reduction to hit-or-miss events, which leads to a natural σ-algebra and the useful tool of the capacity functional. The thus established new field of stochastic geometry had its early development at different places, of which we mention the following. In Cambridge, the thesis of R.E. Miles in 1961 on Poisson flats, the work of R. Davidson on line and flat processes, and D.G. Kendall's foundations of a theory of random sets were fundamental. Independently, in Fontainebleau, G. Matheron, motivated by geostatistics, developed his theory of random closed sets, as it is much used today, and combined it with mathematical morphology, developed by J. Serra and others. Both Kendall and Matheron give credit to preceding work of G. Choquet. Important for the further development of stochastic geometry were R.V. Ambartzumian and his school in Yerevan (see the books by Ambartzumian (1982, 1990)) and the East-German school of J. Mecke and D. Stoyan. The appearance of the collection on *Stochastic Geometry* by Harding and Kendall (1974) and of the book by Matheron (1975) marks the establishment of the new field. Its origins were remembered at the Buffon Bicentenary Symposium held in Paris in 1977, the proceedings of which were edited by Miles and Serra (1978). The rapid development of stochastic geometry is demonstrated by the volume by Stoyan, Kendall and Mecke (1995, first edition 1987).

The newly created stochastic geometry linked well with the existing integral geometry, as long as models with appropriate group invariance were considered.

The conditions of stationarity and isotropy of random sets and processes of flats or compact sets allowed to transfer kinematic formulae from integral geometry to the new random setting. From a practical point of view, such assumptions are often too restrictive. Consequently, also the scope of integral geometry had to be widened. The need to study stationary, but not necessarily isotropic models in stochastic geometry put translative integral geometry into the focus, a topic started by W. Blaschke and others in 1937 but then nearly forgotten for a long time. New stereological principles required the extension of integral geometry in a different direction. For example, local stereology, developed in Aarhus by E.B.V. Jensen and her group, used rotational formulae (without translations) and new Blaschke–Petkantschin formulae; see the books by Jensen (1998) and, more generally, Baddeley and Jensen (2005). Also of importance, for example, for the introduction of densities, were local versions of the classical functionals of integral geometry in the form of curvature measures. These were already introduced by H. Federer in 1959, and the local kinematic formulae that he proved for these measures found direct applications in stochastic geometry. Surprisingly, translative integral formulae for curvature measures, as they have been established in the last 20 years, even turned out to be essential in the study of non-stationary random sets and geometric point processes. This is still one of the current areas of research.

1.2 Geometric tools

In this section, we introduce general geometric notation and provide the geometric tools that are used later in this chapter and in many parts of stochastic geometry.

Much of this geometry takes place in d-dimensional Euclidean space $\mathbb{R}^d$ ($d \geq 2$), with standard scalar product $\langle \cdot, \cdot \rangle$ and induced norm $\| \cdot \|$. We use o for the origin of $\mathbb{R}^d$. The Euclidean metric is denoted by ρ, thus $\rho(x,y) = \|x-y\|$ for $x, y \in \mathbb{R}^d$, and $\rho(x, A) = \inf_{a \in A} \rho(x, a)$ is the distance of a point x from a nonempty set $A \subset \mathbb{R}^d$ (if A is closed, then $\rho(x, A) = \min_{a \in A} \rho(x, a)$). The set $B^d = \{x \in \mathbb{R}^d : \|x\| \leq 1\}$ is the *unit ball*, and $\mathbb{S}^{d-1} = \{x \in \mathbb{R}^d : \|x\| = 1\}$ is the *unit sphere*. A linear map of $\mathbb{R}^d$ into itself that preserves the scalar product is called an *orthogonal map*, and a *rotation* if it in addition preserves the orientation (has positive determinant). The group SO_d of all rotations, with its usual topology, is compact. A map of $\mathbb{R}^d$ into itself that preserves the metric is called an *isometry*. Every isometry is the composition of an orthogonal map and a translation. If it preserves the orientation, it is called a *rigid motion*. The group G_d of all rigid motions of $\mathbb{R}^d$, with its usual topology, is locally compact.

Lebesgue measure on $\mathbb{R}^d$ is denoted by λ_d, and spherical Lebesgue measure on $\mathbb{S}^{d-1}$ by σ_{d-1}. In particular, we have

$$\kappa_d = \lambda_d(B^d) = \frac{\pi^{\frac{d}{2}}}{\Gamma\left(1+\frac{d}{2}\right)}, \quad \omega_d = \sigma_{d-1}(\mathbb{S}^{d-1}) = d\kappa_d = \frac{2\pi^{\frac{d}{2}}}{\Gamma\left(\frac{d}{2}\right)}.$$

By $\mathcal{H}_k$ we denote k-dimensional Hausdorff (outer) measure ($k \geq 0$). Restricted to the Borel sets, it is a measure. The σ-algebra of Borel sets of a topological space E is denoted by $\mathcal{B}(E)$.

Let $\mathcal{C}$ denote the system of compact subsets of $\mathbb{R}^d$, and $\mathcal{K}$ the subsystem of convex compact sets (thus, the dimension d is suppressed in this notation, but should be clear from the context). We write $\mathcal{C}'$, $\mathcal{K}'$ for the subsystems of nonempty sets, in each case. The elements of $\mathcal{K}'$ are called *convex bodies*. Thus, in our terminology (which follows Schneider 1993), a convex body need not have interior points. This is convenient but, as the reader is warned, different from the usage in part of the literature. A set $A \subset \mathbb{R}^d$ is *polyconvex* if it is the union of finitely many convex bodies, and *locally polyconvex* if $A \cap K$ is polyconvex for every convex body K. We denote the system of polyconvex sets in $\mathbb{R}^d$ by $\mathcal{R}$ and call it the *convex ring*, and the system of locally polyconvex sets is denoted by $\mathcal{S}$. For a nonempty set $A \subset \mathbb{R}^d$, the *convex hull* of A, denoted by $\operatorname{conv} A$, is the set of all convex combinations of finitely many points from A, and also the intersection of all convex sets containing A. If A is compact, then $\operatorname{conv} A$ is a convex body. The convex hull of a finite set is a *polytope*. The polytopes are also the bounded intersections of finitely many closed halfspaces. The system of polytopes in $\mathbb{R}^d$ is denoted by $\mathcal{P}$.

On $\mathcal{C}'$, the *Hausdorff metric* is defined by

$$\rho_H(A, B) = \max\{\max_{x \in A} \rho(x, B),\ \max_{y \in B} \rho(y, A)\}, \qquad A, B \in \mathcal{C}'.$$

In the following, $\mathcal{C}'$ and its subspaces are always equipped with the Hausdorff metric and the induced topology. The space $\mathcal{C}'$ is locally compact and has a countable base (in this terminology, 'locally compact' includes the Hausdorff separation property). Every bounded infinite sequence in this space has a convergent subsequence. The subspace $\mathcal{K}'$ is closed.

The *Minkowski addition* on $\mathcal{C}'$ is defined by the vector sum, thus

$$A + B = \{a + b : a \in A,\ b \in B\}, \qquad A, B \in \mathcal{C}'.$$

The sum $A + B$ is again compact, and if A and B are convex, then $A + B$ is convex. For $x \in \mathbb{R}^d$, one writes $A + x = A + \{x\}$ for the image of A under the translation by the vector x. The dilatation by the number $r \geq 0$ is defined by $rA = \{ra : a \in A\}$. If $A_1, \dots, A_k \in \mathcal{C}'$ and $A_i \subset RB^d$ for $i = 1, \dots, k$, with some number R, then

$$\rho_H\left(\frac{1}{k}(A_1 + \cdots + A_k), \frac{1}{k}\operatorname{conv}(A_1 + \cdots + A_k)\right) \leq \frac{dR}{k} \tag{1.5}$$

(see, for example, Schneider 1993, Corollary 3.1.3); thus, Minkowski averaging has a convexifying effect.

Convex bodies have useful descriptions by functions or measures. For $K \in \mathcal{C}'$, the *support function* is defined by

$$h_K(x) = \max\{\langle x, y\rangle : y \in K\}, \quad x \in \mathbb{R}^d.$$

The support function h_K is *sublinear*, namely positively homogeneous, satisfying $h_K(rx) = rh_K(x)$ for $r \geq 0$ and $x \in \mathbb{R}^d$, and subadditive, which means that $h_K(x+y) \leq h_K(x) + h_K(y)$ for $x, y \in \mathbb{R}^d$. In particular, support functions are convex functions. Below, we do not distinguish between h_K and its restriction to the unit sphere. The following fact is very useful.

Theorem 1.1 *Every sublinear function on $\mathbb{R}^d$ is the support function of a convex body. This body is uniquely determined.*

In terms of the support function, the Hausdorff distance of $K, M \in \mathcal{K}'$ is expressed by

$$\rho_H(K, M) = \max\{|h_K(u) - h_M(u)| : u \in \mathbb{S}^{d-1}\} = \|h_K - h_M\|_\infty,$$

where $\|\cdot\|_\infty$ denotes the maximum norm on the space $\mathbf{C}(\mathbb{S}^{d-1})$ of continuous real functions on $\mathbb{S}^{d-1}$. Moreover, for $K, M \in \mathcal{K}'$ one has $h_{K+M} = h_K + h_M$, $h_{rK} = rh_K$ for $r \geq 0$, $h_{\vartheta K}(x) = h_K(\vartheta^{-1}x)$ for $x \in \mathbb{R}^d$ and $\vartheta \in SO_d$, and $h_{K+z} = h_K + \langle z, \cdot\rangle$ for $z \in \mathbb{R}^d$. The inclusion $K \subset M$ is equivalent to $h_K \leq h_M$.

The translation invariant version of the support function is the *centred support function*, defined by

$$h_K^* = h_{K-s(K)} = h_K - \langle s(K), \cdot\rangle, \quad K \in \mathcal{K}',$$

where

$$s(K) = \frac{1}{\kappa_d}\int_{\mathbb{S}^{d-1}} h_K(u)u\,\sigma_{d-1}(\mathrm{d}u)$$

is the *Steiner point* of K.

For a convex body K and a subset $A \subset \mathbb{S}^{d-1}$, let $\tau(K, A)$ be the set of boundary points of K at which there exists an outer normal vector u of K with $u \in A$. Defining

$$S_{d-1}(K, A) = \mathcal{H}_{d-1}(\tau(K, A))$$

for $A \in \mathcal{B}(\mathbb{S}^{d-1})$, we obtain a finite Borel measure $S_{d-1}(K, \cdot)$ on $\mathbb{S}^{d-1}$. It is called the *surface area measure* of K. The one-to-one correspondence between convex bodies and sublinear functions on $\mathbb{R}^d$ is paralleled by the one-to-one correspondence between translation classes of full-dimensional convex bodies and a class of measures on $\mathbb{S}^{d-1}$. This is the content of *Minkowski's existence and uniqueness theorem.*

Theorem 1.2 *Let φ be a finite Borel measure on the unit sphere $\mathbb{S}^{d-1}$, which satisfies $\int_{\mathbb{S}^{d-1}} u\,\varphi(\mathrm{d}u) = o$ and is not concentrated on a great subsphere. Then there exists a convex body K with interior points that has surface area measure φ. The body K is uniquely determined up to a translation.*

The surface area measure of a d-dimensional convex body K, which is a measure on the unit sphere, must be well distinguished from the *boundary measure* of K. This is the Borel measure concentrated on ∂K, the boundary of K, which is defined by

$$C_{d-1}(K, A) = \mathcal{H}_{d-1}(A \cap \partial K), \quad A \in \mathcal{B}(\mathbb{R}^d).$$

(The notation comes from the fact that this is one in a series of curvature measures.)

The following special convex bodies appear in several applications. A Minkowski sum of finitely many closed line segments is called a *zonotope*. All the faces of a zonotope (including the zonotope itself) are centrally symmetric. Conversely, if all the two-dimensional faces of a polytope P are centrally symmetric, then P is a zonotope. A *zonoid* is a convex body that can be approximated, in the Hausdorff metric, by a sequence of zonotopes. Every zonoid has a centre of symmetry. A convex body K is a zonoid with centre o if and only if its support function has the representation

$$h_K(x) = \int_{\mathbb{S}^{d-1}} |\langle x, u\rangle| \, \varphi(\mathrm{d}u), \quad x \in \mathbb{R}^d, \tag{1.6}$$

with a finite Borel measure φ on the sphere $\mathbb{S}^{d-1}$. This measure can be assumed to be even (that is, satisfy $\varphi(A) = \varphi(-A)$ for all $A \in \mathcal{B}(\mathbb{S}^{d-1})$); it is then uniquely determined and called the *generating measure* of the zonoid K.

One can associate zonoids with more general measures. Let μ be a Borel measure on $\mathbb{R}^d$ satisfying $\int_{\mathbb{R}^d} \|x\| \, \mu(\mathrm{d}x) < \infty$. Then

$$h_{Z(\mu)}(x) = \int_{\mathbb{R}^d} \langle x, y\rangle_+ \, \mu(\mathrm{d}y), \quad x \in \mathbb{R}^d, \tag{1.7}$$

(where $a_+ = \max\{0, a\}$ denotes the positive part of a) defines the support function of a zonoid $Z(\mu)$, which has centre $\frac{1}{2}\int_{\mathbb{R}^d} x \, \mu(\mathrm{d}x)$. However, $Z(\mu)$ does not determine the measure μ uniquely. To restore a one-to-one correspondence, one 'lifts' the measure μ to the product space $\mathbb{R} \times \mathbb{R}^d$, by defining $\widehat{\mu} = \delta_1 \otimes \mu$, where δ_1 is the Dirac measure at 1. Then $\widehat{Z}(\mu) = Z(\widehat{\mu})$ (in $\mathbb{R} \times \mathbb{R}^d$) is a zonoid, called the *lift zonoid* of μ. Its support function is given by

$$h_{\widehat{Z}(\mu)}(\xi, x) = \int_{\mathbb{R}^d} (\xi + \langle x, y\rangle)_+ \, \mu(\mathrm{d}y), \quad (\xi, x) \in \mathbb{R} \times \mathbb{R}^d. \tag{1.8}$$

The measure μ is uniquely determined by the zonoid $\widehat{Z}(\mu)$.

Now we turn to volume, surface area and similar functions, measuring the size of a convex body, and to their extensions to polyconvex sets. For $K \in \mathcal{K}'$, the *parallel body* at distance $r > 0$ is defined by

$$K^r = \{x \in \mathbb{R}^d : \rho(x, K) \le r\} = K + rB^d.$$

Its volume is a polynomial in r, which can be written as

$$\lambda_d(K^r) = \sum_{j=0}^{d} r^{d-j} \kappa_{d-j} V_j(K). \tag{1.9}$$

This *Steiner formula* defines functions $V_j : \mathcal{K}' \to \mathbb{R}$, $j = 0, \ldots, d$, which are called the *intrinsic volumes*. The normalization they obtain from (1.9) is convenient, but it must be pointed out that these functions appear in the literature also with different normalizations and indexing, and then are called *quermassintegrals* or *Minkowski functionals*.

Clearly, $V_d(K) = \lambda_d(K)$ is the *volume* of K. Due to the chosen normalization, the value $V_j(K)$ does not depend on the dimension of the surrounding space in which it is computed. In particular, if the convex body K has dimension j, then $V_j(K) = \mathcal{H}_j(K)$. If K has interior points, then $2V_{d-1}(K) = \mathcal{H}_{d-1}(\partial K)$ is the total boundary measure or *surface area* of K. Trivially, $V_0(K) = 1$ for every $K \in \mathcal{K}'$. Further intuitive interpretations of the intrinsic volumes are given below.

Each function V_j is nonnegative, continuous, and invariant under rigid motions. Since the intrinsic volumes are derived from a measure, they inherit an additivity property, in the following sense. Let ϕ be a function on an intersectional family $\mathcal{M}$ of sets with values in an abelian group. It is called *additive* or a *valuation* if

$$\phi(K \cup L) + \phi(K \cap L) = \phi(K) + \phi(L)$$

for all $K, L \in \mathcal{M}$ with $K \cup L \in \mathcal{M}$. Without loss of generality, we may always assume that $\emptyset \in \mathcal{M}$ and $\phi(\emptyset) = 0$. With this definition, the intrinsic volumes are additive on $\mathcal{K}$. Their predominant role in the theory of convex bodies is illuminated by the following fundamental result, known as *Hadwiger's characterization theorem*.

Theorem 1.3 *Every rigid motion invariant, continuous real valuation on $\mathcal{K}$ is a linear combination, with constant coefficients, of the intrinsic volumes.*

For $j = 0, \ldots, d$, the intrinsic volume V_j has an additive extension, also denoted by V_j, to the convex ring $\mathcal{R}$. This follows, for example, from *Groemer's extension theorem*.

Theorem 1.4 *Every continuous valuation on $\mathcal{K}'$ with values in a topological vector space has a unique additive extension to the convex ring $\mathcal{R}$.*

The extended function V_d coincides, of course, with the Lebesgue measure on $\mathcal{R}$. For a polyconvex set K which is the closure of its interior, $2V_{d-1}(K) = \mathcal{H}_{d-1}(\partial K)$ is still the surface area. The remaining intrinsic volumes can attain negative values on $\mathcal{R}$. Particularly important is the function V_0, which is called the *Euler characteristic* and denoted by χ. This is the unique additive function on $\mathcal{R}$ which satisfies $\chi(K) = 1$ for $K \in \mathcal{K}'$ and $\chi(\emptyset) = 0$.

The intrinsic volumes have local versions, in the form of measures, which can be introduced by means of a local Steiner formula. To obtain it, we use the *nearest-point map* $p(K,\cdot):\mathbb{R}^d\to K$, for $K\in\mathcal{K}'$. For $x\in\mathbb{R}^d$, the point $p(K,x)$ is, by definition, the unique point $p\in K$ with $\rho(x,K)=\rho(x,p)$. If $x\in\mathbb{R}^d\setminus K$, then the vector $u(K,x)=(x-p(K,x))/\rho(x,K)$ is an outer unit normal vector to K at the point $p(K,x)$, and the pair $(p(K,x),u(K,x))$ belongs to the *generalized normal bundle* $\operatorname{Nor} K$ of K. Here, $\operatorname{Nor} K$ is the set of all pairs (p,u) where $p\in\partial K$ and u is an outer unit normal vector to K at p. It is a closed subspace of the product space $\Sigma=\mathbb{R}^d\times\mathbb{S}^{d-1}$. Now the *local parallel set* of K at distance $r>0$ corresponding to a Borel set $A\in\mathcal{B}(\Sigma)$ is defined by

$$M_r(K,A)=\{x\in K^r\setminus K:(p(K,x),u(K,x))\in A\}.$$

The *local Steiner formula* says that

$$\lambda_d(M_r(K,A))=\sum_{j=0}^{d-1} r^{d-j}\kappa_{d-j}\,\Xi_j(K,A),\quad A\in\mathcal{B}(\Sigma), \tag{1.10}$$

with finite measures $\Xi_0(K,\cdot),\dots,\Xi_{d-1}(K,\cdot)$ on $\mathcal{B}(\Sigma)$, which are concentrated on $\operatorname{Nor} K$. The measure $\Xi_j(K,\cdot)$ is called the jth *support measure* or *generalized curvature measure* of K. Of particular importance are the marginal measures, for which we use the notation

$$\Phi_j(K,A)=\Xi_j(K,A\times\mathbb{S}^{d-1}),\quad A\in\mathcal{B}(\mathbb{R}^d),$$
$$\Psi_j(K,B)=\Xi_j(K,\mathbb{R}^d\times B),\quad B\in\mathcal{B}(\mathbb{S}^{d-1}).$$

One calls $\Phi_j(K,\cdot)$ the jth *curvature measure* of K, and $\Psi_j(K,\cdot)$ the jth *area measure* of K. The first series of measures is supplemented by putting

$$\Phi_d(K,\cdot)=\lambda_d(K\cap\cdot).$$

Also here, other notation and normalizations are used in the literature. We mention only the connection with the boundary measure and the surface area measure introduced earlier, namely

$$2\Phi_{d-1}(K,\cdot)=C_{d-1}(K,\cdot),\quad 2\Psi_{d-1}(K,\cdot)=S_{d-1}(K,\cdot).$$

Clearly,

$$\Xi_j(K,\Sigma)=\Phi_j(K,\mathbb{R}^d)=\Psi_j(K,\mathbb{S}^{d-1})=V_j(K),\quad j=0,\dots,d-1.$$

To explain the name 'curvature measure', we mention that for a convex body K with a sufficiently smooth boundary, the jth curvature measure can be represented by

$$\Phi_j(K,A)=\frac{\binom{d}{j}}{d\kappa_{d-j}}\int_{A\cap\partial K} H_{d-1-j}\,\mathrm{d}\mathcal{H}_{d-1}$$

for $A \in \mathcal{B}(\mathbb{R}^d)$, $j = 0, \ldots, d-1$. Here, H_k denotes the kth normalized elementary symmetric function of the principal curvatures at points of ∂K. A similar representation exists for $\Psi_j(K, \cdot)$, involving principal radii of curvature, as functions of the outer unit normal vector.

The jth support measure has the following properties. It is covariant under rigid motions, that is $\Xi_j(gK, g{\cdot}A) = \Xi_j(K, A)$ for $g \in G_d$, where $g{\cdot}A = \{(gx, g_0 u) : (x, u) \in A\}$ and g_0 denotes the rotation part of g. It is homogeneous of degree j, satisfying $\Xi_j(rK, r \cdot A) = \Xi_j(K, A)$ for $r \geq 0$, where $r \cdot A = \{(rx, u) : (x, u) \in A\}$. It is continuous with respect to the weak topology on the space of finite Borel measures on Σ. For each fixed $A \in \mathcal{B}(\Sigma)$, the function $\Xi_j(\cdot, A)$ on $\mathcal{K}'$ is measurable and additive. Corresponding properties are shared by the curvature measures and the area measures. By Groemer's extension theorem, the support measures, curvature measures and area measures have additive extensions, in their first argument, to the convex ring $\mathcal{R}$. The extensions are denoted by the same symbols. Note that $\Xi_j(\emptyset, \cdot) = 0$ for $j = 0, \ldots, d-1$.

The curvature measures, and hence also the intrinsic volumes, satisfy a series of integral geometric mean value formulae. They refer to invariant measures on the motion group and on Grassmannians. The rotation group SO_d, being a compact topological group, carries a unique bi-invariant (Borel) probability measure. We denote it by ν (again suppressing the dimension d, which should be clear from the context). From this, an invariant measure μ on the motion group G_d is obtained as the image measure of $\lambda_d \otimes \nu$ under the map from $\mathbb{R}^d \times SO_d$ to G_d that associates with (x, ϑ) the rotation ϑ followed by the translation by x. Thus, for any nonnegative, measurable function f on $\mathcal{K}'$ we have

$$\int_{G_d} f(gK)\, \mu(\mathrm{d}g) = \int_{SO_d} \int_{\mathbb{R}^d} f(\vartheta K + x)\, \lambda_d(\mathrm{d}x)\, \nu(\mathrm{d}\vartheta). \tag{1.11}$$

The integral geometric formulae to be considered concern mean values, formed with invariant measures, involving the intersection of a fixed and a moving set in $\mathbb{R}^d$. These sets will be polyconvex sets or flats. The *local principal kinematic formula* holds for polyconvex sets $K, M \in \mathcal{R}$ and Borel sets $A, B \in \mathcal{B}(\mathbb{R}^d)$ and says that

$$\int_{G_d} \Phi_j(K \cap gM, A \cap gB)\, \mu(\mathrm{d}g) = \sum_{k=j}^{d} c(d, j, k) \Phi_k(K, A) \Phi_{d-k+j}(M, B) \tag{1.12}$$

for $j = 0, \ldots, d$, where the coefficients are given by

$$c(d, j, k) = \frac{k! \kappa_k (d-k+j)! \kappa_{d-k+j}}{j! \kappa_j d! \kappa_d}. \tag{1.13}$$

The global case, known as the *principal kinematic formula*, reads

$$\int_{G_d} V_j(K \cap gM)\, \mu(\mathrm{d}g) = \sum_{k=j}^{d} c(d, j, k) V_k(K) V_{d-k+j}(M). \tag{1.14}$$

For $j = 0$ and for convex bodies K, M, we obtain a formula for the total measure of the set of rigid motions bringing M into a hitting position with K, namely

$$\int_{G_d} \chi(K \cap gM)\, \mu(\mathrm{d}g) = \sum_{k=0}^{d} c(d, 0, k) V_k(K) V_{d-k}(M). \tag{1.15}$$

The case $M = rB^d$ reproduces the Steiner formula.

Let $q \in \{1, \dots, d-1\}$. By $A(d,q)$ we denote the *affine Grassmannian* of q-flats (q-dimensional affine subspaces), with its usual topology, making it a locally compact space. This space carries a rigid motion invariant Borel measure μ_q, which is unique up to a constant factor. Choosing a convenient normalization, we can assume that, for any q-dimensional linear subspace L_q of $\mathbb{R}^d$,

$$\int_{A(d,q)} f \,\mathrm{d}\mu_q = \int_{SO_d} \int_{L_q^\perp} f(\vartheta(L_q + x))\, \lambda_{d-q}(\mathrm{d}x)\, \nu(\mathrm{d}\vartheta) \tag{1.16}$$

for every nonnegative, measurable function f on $A(d,q)$. With this measure, the *local Crofton formula*

$$\int_{A(d,q)} \Phi_j(K \cap E, A \cap E)\, \mu_q(\mathrm{d}E) = c(d,j,q) \Phi_{d-q+j}(K, A), \quad j = 0, \dots, q, \tag{1.17}$$

holds for polyconvex sets $K \in \mathcal{R}$ and Borel sets $A \in \mathcal{B}(\mathbb{R}^d)$. The *Crofton formula* is the global version,

$$\int_{A(d,q)} V_j(K \cap E)\, \mu_q(\mathrm{d}E) = c(d,j,q) V_{d-q+j}(K), \quad j = 0, \dots, q. \tag{1.18}$$

For $j = 0$ and a convex body K, we get

$$\int_{A(d,q)} \chi(K \cap E)\, \mu_q(\mathrm{d}E) = c(d,0,q) V_{d-q}(K), \tag{1.19}$$

which interprets the intrinsic volume $V_{d-q}(K)$, up to a normalizing factor, as the total invariant measure of the set of q-flats hitting K.

An important feature of the kinematic formula (1.12) is the fact that the convex bodies K and M on the right side are separated. This is due to the integration over all rotations. For this reason, the formula can easily be iterated, that is, applied to $K_1 \cap g_2 K_2 \cap \cdots \cap g_k K_k$. The corresponding formulae of integral geometry with respect to the translation group, which are useful in stochastic geometry for the treatment of stationary, non-isotropic structures (and even of non-stationary models), are necessarily more complicated. We formulate here only the *iterated local translative formula* for curvature measures. Let $k \in \mathbb{N}$, $j \in \{0, \dots, d\}$, and let $m_1, \dots, m_k \in \{j, \dots, d\}$ be numbers satisfying $m_1 + \cdots +$

$m_k = (k-1)d + j$. For convex bodies $K_1, \dots, K_k \in \mathcal{K}'$, there exists a finite measure $\Phi^{(j)}_{m_1,\dots,m_k}(K_1, \dots, K_k; \cdot)$ on $\mathcal{B}((\mathbb{R}^d)^k)$ such that the following holds. If $A_1, \dots, A_k \in \mathcal{B}(\mathbb{R}^d)$, then

$$\begin{aligned}\int_{(\mathbb{R}^d)^{k-1}} & \Phi_j(K_1 \cap (K_2 + x_2) \cap \cdots \cap (K_k + x_k), \\ & \qquad A_1 \cap (A_2 + x_2) \cap \cdots \cap (A_k + x_k))\, \lambda_d^{k-1}(\mathrm{d}(x_2, \dots, x_k)) \\ = & \sum_{\substack{m_1,\dots,m_k=j \\ m_1+\cdots+m_k=(k-1)d+j}}^{d} \Phi^{(j)}_{m_1,\dots,m_k}(K_1, \dots, K_k; A_1 \times \cdots \times A_k). \end{aligned} \tag{1.20}$$

The measures $\Phi^{(j)}_{m_1,\dots,m_k}(K_1, \dots, K_k; \cdot)$ are called the *mixed measures*, and their global versions, $V^{(j)}_{m_1,\dots,m_k}(K_1, \dots, K_k) = \Phi^{(j)}_{m_1,\dots,m_k}(K_1, \dots, K_k; (\mathbb{R}^d)^k)$, are known as the *mixed functionals*. The mixed measures are additive and weakly continuous in each of their arguments from $\mathcal{K}'$, hence they have additive extensions to the convex ring. Formula (1.20) then extends to polyconvex sets $K_1, \dots, K_k$.

Hints to the literature For a detailed treatment of the last result, including properties of the mixed measures, we refer to Schneider and Weil (2008). In the Appendix of that book, proofs of Hadwiger's characterization theorem and of Groemer's extension theorem are reproduced. For lift zonoids, see Mosler (2002). All other facts stated here without proof can be found in the book by Schneider (1993).

1.3 Point processes

Point processes are models for random collections of points in a space E. Originating from stochastic processes on the real line (modelling, for example, the times where certain events occur), the classical extension is to spatial processes in $\mathbb{R}^d$. Since, in stochastic geometry, point processes are used to model random collections of sets (such as balls, lines, planes, fibres), a more general setting is required. For our purposes and the later applications, it is convenient to consider, as the basic space, a locally compact space E with a countable base. Most of the results in this section hold under more general assumptions (for example, for Polish spaces, or even for measurable spaces with suitable additional structure).

We shall introduce point processes as random locally finite counting measures on E. Without much extra effort, general (locally finite) random measures on E can be introduced and so we will do that, although we shall soon concentrate on the subclass of counting measures. The space E is supplied with its Borel σ-algebra $\mathcal{B}(E)$. Further, $\mathcal{F}(E)$ and $\mathcal{C}(E)$ denote the classes of closed, respectively compact, subsets of E, and $\mathcal{F}'(E)$, $\mathcal{C}'(E)$ are the corresponding classes of nonempty sets.

Let $\mathsf{M}(E)$ be the set of all Borel measures η on E which are *locally finite*, that is, satisfy $\eta(C) < \infty$ for all $C \in \mathcal{C}(E)$, and let $\mathsf{N}(E)$ be the subset of all counting measures. Here, a measure $\eta \in \mathsf{M}(E)$ is a *counting measure* if $\eta(A) \in \mathbb{N}_0 \cup \{\infty\}$, for all $A \in \mathcal{B}(E)$. If, in addition, $\eta(\{x\}) \in \{0, 1\}$ for all $x \in E$, then the counting measure η is called *simple*. Let $\mathsf{N}_s(E)$ be the corresponding class of simple counting measures. We supply $\mathsf{M}(E)$ with the σ-algebra $\mathcal{M}(E)$ generated by the evaluation maps

$$\begin{array}{rcl} \Phi_A : \mathsf{M}(E) & \to & \mathbb{R} \cup \{\infty\}, \\ \eta & \mapsto & \eta(A) \end{array} \qquad A \in \mathcal{B}(E).$$

The subsets $\mathsf{N}(E)$ and $\mathsf{N}_s(E)$ carry the induced σ-algebras $\mathcal{N}(E)$ and $\mathcal{N}_s(E)$. A convenient generating system of $\mathcal{M}(E)$ is $\{\mathsf{M}_{G,r}\}$, where $r \geq 0$ and G varies through the open, relatively compact subsets of E. Here,

$$\mathsf{M}_{A,r} = \{\eta \in \mathsf{M}(E) : \eta(A) \leq r\}$$

for $A \in \mathcal{B}(E)$ and $r \geq 0$.

A counting measure η is a locally finite sum of Dirac measures,

$$\eta = \sum_{i=1}^{k} \delta_{x_i}, \quad k \in \mathbb{N}_0 \cup \{\infty\},$$

with $x_i \in E$. More precisely, it can be enumerated in a measurable way, that is, there exist measurable mappings $\zeta_i : \mathsf{N}(E) \to E$ such that

$$\eta = \sum_{i=1}^{\eta(E)} \delta_{\zeta_i(\eta)} \quad \text{for } \eta \in \mathsf{N}(E).$$

A simple counting measure η can be identified with its support $\{\zeta_1(\eta), \zeta_2(\eta), \ldots\}$, and so we can imagine a simple counting measure η also as a locally finite set in E. This interpretation will often be used, in the following. For example, it allows us to write $x \in \eta$ instead of $\eta(\{x\}) > 0$. This identification also shows that $\mathcal{N}_s(E)$ is generated by the simpler system $\{\mathsf{N}_{s,G}\}$, with G varying through the open, relatively compact subsets of E. Here,

$$\mathsf{N}_{s,A} = \{\eta \in \mathsf{N}_s(E) : \eta(A) = 0\}$$

for $A \in \mathcal{B}(E)$.

In the following, we assume a basic probability space $(\Omega, \mathcal{A}, \mathbf{P})$ to be given. Measurability then always refers to the corresponding σ-algebras.

Definition 1.5 *A* random measure *on E is a measurable mapping $M : \Omega \to \mathsf{M}(E)$. If $M \in \mathsf{N}(E)$ a.s. (respectively $M \in \mathsf{N}_s(E)$ a.s.), the random measure M is called a* point process *(respectively a* simple point process*) in E.*

Standard notions such as distribution, independence, equality in distribution (denoted by $\stackrel{d}{=}$), expectation, weak or vague convergence, etc., are used now for random measures and point processes without further explanation. The simple structure of the generating system of $\mathcal{N}_s(E)$ mentioned above yields the following quite useful result.

Lemma 1.6 *If N, N' are simple point processes in E with*

$$\mathbf{P}\{N(C) = 0\} = \mathbf{P}\{N'(C) = 0\}$$

for all $C \in \mathcal{C}(E)$, then N and N' are equal in distribution.

For random measures M, M', the sum $M + M'$ and the restriction $M \llcorner A$ to a set $A \in \mathcal{B}(E)$ are random measures again. For simple point processes N, N', these operations correspond to taking the union $N \cup N'$, respectively the intersection $N \cap A$.

If G is a group operating on E in a measurable way, then G acts also on M, in a canonical (and measurable) way, by letting $g\eta$ for $g \in G$ and $\eta \in \mathsf{M}$ be the image measure of η under g,

$$g\eta(B) = \eta(g^{-1}B) \quad \text{for } B \in \mathcal{B}(E).$$

Hence, for a random measure M (or a point process N) on E and for $g \in G$, also gM is a random measure (and gN is a point process) on E. This will be used later, mainly for the two spaces $E = \mathbb{R}^d$ or $E = \mathcal{F}'(\mathbb{R}^d)$, where G is the group G_d of rigid motions of $\mathbb{R}^d$, or one of its subgroups, SO_d (the group of rotations) or $T_d = \mathbb{R}^d$ (the group of translations t_x, where t_x is identified with the point $x \in \mathbb{R}^d$). In all these cases, G even acts continuously on E. Instead of $t_x\eta$, for $\eta \in \mathsf{M}(E)$ and $x \in \mathbb{R}^d$, we write $\eta + x$ (and we use similar notations for random measures, sets of measures, etc.). We call a random measure M on $E = \mathbb{R}^d$ or $E = \mathcal{F}'(\mathbb{R}^d)$ *stationary* if $M \stackrel{d}{=} M + x$ for all $x \in \mathbb{R}^d$. M is *isotropic* if $M \stackrel{d}{=} \vartheta M$ for all rotations $\vartheta \in SO_d$.

We return to the general situation and introduce, for a random measure M on E, the *intensity measure* $\Theta = \Theta_M$ by

$$\Theta(A) = \mathbf{E}M(A) \quad \text{for } A \in \mathcal{B}(E).$$

If N is a simple point process, then $\Theta(A)$ is the mean number of points of N lying in A. Although the random measure M is locally finite a.s., the intensity measure Θ need not have this property. We will later require this, as an additional assumption, in order to simplify some of the formulae.

If M is a stationary random measure on $\mathbb{R}^d$, its intensity measure Θ, which is now a measure on $\mathbb{R}^d$, is invariant under translations. The only translation invariant, locally finite measure on $\mathbb{R}^d$ is, up to a constant factor, the Lebesgue measure λ_d. Hence, if Θ is locally finite, then

$$\Theta = \gamma\lambda_d$$

with a constant $\gamma \in [0, \infty)$. The number γ is called the *intensity* of the (stationary) random measure M. We often exclude the case $\gamma = 0$, since it corresponds to the trivial situation where $M = 0$ almost surely.

For a stationary random measure M on $\mathcal{F}'(\mathbb{R}^d)$, the intensity measure Θ is a translation invariant measure on $\mathcal{F}'(\mathbb{R}^d)$. If M (and hence Θ) is supported by certain subclasses of $\mathcal{F}'(\mathbb{R}^d)$, the class $\mathcal{C}'(\mathbb{R}^d)$ of compact sets (particles) or the class $A(d,k)$ of k-dimensional affine flats, the translation invariance of Θ will lead to basic decomposition results, as we shall see in Section 1.5.

The following simple observation (the *Campbell theorem*) is quite useful. It is a direct consequence of the definition of Θ and the usual extension arguments from indicator functions to (nonnegative) measurable functions.

Theorem 1.7 *Let M be a random measure on E with intensity measure Θ, and let $f : E \to \mathbb{R}$ be a nonnegative, measurable function. Then $\int_E f \,\mathrm{d}M$ is measurable, and*

$$\mathbf{E} \int_E f \,\mathrm{d}M = \int_E f \,\mathrm{d}\Theta.$$

Clearly this result holds for Θ-integrable functions, as do its relatives to be discussed below.

For simple point processes N, it is convenient to use the identification with their supports and to write Campbell's theorem in the form

$$\mathbf{E} \sum_{x \in N} f(x) = \int_E f \,\mathrm{d}\Theta.$$

Let M be a random measure on E. In generalization of the intensity measure Θ, also called the *first moment measure*, one defines the mth *moment measure* $\Theta^{(m)}$ of M as the Borel measure on E^m with

$$\Theta^{(m)}(A_1 \times \cdots \times A_m) = \mathbf{E}\, M^m(A_1 \times \cdots \times A_m) = \mathbf{E}\, M(A_1) \cdots M(A_m)$$

for $A_1, \dots, A_m \in \mathcal{B}(E)$. Since the product measure M^m is a random measure on the locally compact product space E^m, the mth moment measure $\Theta^{(m)}$ is nothing but the intensity measure of M^m. For each $m \in \mathbb{N}$, the set

$$E^m_{\neq} = \{(x_1, \dots, x_m) \in E^m : x_i \text{ pairwise distinct}\}$$

is an open subset of E^m. The mth *factorial moment measure* of M is the Borel measure $\Lambda^{(m)}$ on E^m with

$$\Lambda^{(m)}(A_1 \times \cdots \times A_m) = \mathbf{E}\, M^m(A_1 \times \cdots \times A_m \cap E^m_{\neq})$$

for $A_1, \ldots, A_m \in \mathcal{B}(E)$. In particular, for a simple point process N and for $A \in \mathcal{B}(E)$,

$$\begin{aligned}\Lambda^{(m)}(A^m) &= \mathbf{E} \sum_{x_1 \in N \cap A} \; \sum_{x_2 \in N \cap A,\, x_2 \neq x_1} \cdots \sum_{x_m \in N \cap A,\, x_m \neq x_1, \ldots, x_{m-1}} 1 \\ &= \mathbf{E}[N(A)(N(A)-1)\cdots(N(A)-m+1)]\end{aligned}$$

is the mth factorial moment of the random variable $N(A)$; this explains the name. Note that $\Lambda^{(m)}$ is the intensity measure of the random measure

$$M^m_{\neq} = M^m \llcorner E^m_{\neq},$$

which is, in general, different from M^m. It is clear that also the mth moment measure $\Theta^{(m)}$ and the mth factorial moment measure $\Lambda^{(m)}$ satisfy Campbell type theorems. We formulate them only for simple point processes.

Corollary 1.8 *Let N be a simple point process in E, let $f : E^m \to \mathbb{R}$ be a nonnegative measurable function ($m \in \mathbb{N}$). Then $\sum_{(x_1,\ldots,x_m) \in N^m} f(x_1, \ldots, x_m)$ and $\sum_{(x_1,\ldots,x_m) \in N^m_{\neq}} f(x_1, \ldots, x_m)$ are measurable, and*

$$\mathbf{E} \sum_{(x_1,\ldots,x_m) \in N^m} f(x_1, \ldots, x_m) = \mathbf{E} \int_{E^m} f \, \mathrm{d}N^m = \int_{E^m} f \, \mathrm{d}\Theta^{(m)}$$

and

$$\mathbf{E} \sum_{(x_1,\ldots,x_m) \in N^m_{\neq}} f(x_1, \ldots, x_m) = \mathbf{E} \int_{E^m} f \, \mathrm{d}N^m_{\neq} = \int_{E^m} f \, \mathrm{d}\Lambda^{(m)}.$$

Very useful for the study of random measures and point processes is the notion of Palm measure and its normalized version, the Palm distribution. Since Palm measures are treated, in greater generality, in Section 2.2, we give here only a short introduction and we concentrate on stationary random measures M on $\mathbb{R}^d$. We assume that the intensity γ of M is positive and finite. Then we define the *Palm distribution* $\mathbf{P}^o$ of M by

$$\mathbf{P}^o(A) = \gamma^{-1} \mathbf{E} \int_{\mathbb{R}^d} \mathbf{1}_B(x) \mathbf{1}_A(M - x)\, M(\mathrm{d}x), \quad A \in \mathcal{M}.$$

Here, $B \subset \mathbb{R}^d$ is an arbitrary Borel set with $\lambda_d(B) = 1$. If $M = N$ is a stationary point process in $\mathbb{R}^d$, then $\mathbf{P}^o$ can be considered as the (regular version of the) conditional distribution of N given that N has a point at the origin o.

We mention two important results on Palm distributions. The first one is the *refined Campbell theorem*.

Theorem 1.9 *Let M be a stationary random measure on $\mathbb{R}^d$ with intensity $\gamma \in (0,\infty)$, and let $f : \mathbb{R}^d \times \mathsf{M} \to \mathbb{R}$ be a nonnegative measurable function. Then $\omega \to \int_{\mathbb{R}^d} f(x, M(\omega))\, M(\omega, \mathrm{d}x)$ is measurable, and*

$$\mathbf{E} \int_{\mathbb{R}^d} f(x,M)\, M(\mathrm{d}x) = \gamma \int_{\mathbb{R}^d} \int_{\mathsf{M}} f(x, \eta + x)\, \mathbf{P}^o(\mathrm{d}\eta)\, \lambda_d(\mathrm{d}x).$$

The following is known as the *exchange formula of Neveu.*

Theorem 1.10 *Let M_1, M_2 be stationary random measures on $\mathbb{R}^d$ with intensities γ_1, γ_2 and Palm distributions $\mathbf{P}_1^o, \mathbf{P}_2^o$, respectively. Let $f : \mathbb{R}^d \times \mathsf{M} \to \mathbb{R}$ be a nonnegative measurable function. Then*

$$\gamma_1 \int_{\mathsf{M}} \int_{\mathbb{R}^d} f(y, \eta - y)\, \eta(\mathrm{d}y)\, \mathbf{P}_1^o(\mathrm{d}\eta) = \gamma_2 \int_{\mathsf{M}} \int_{\mathbb{R}^d} f(-x, \eta)\, \eta(\mathrm{d}x)\, \mathbf{P}_2^o(\mathrm{d}\eta).$$

Assumption From now on we assume that all point processes occurring in this chapter have *locally finite* intensity measures.

Marked point processes Now we leave the general framework and study the notion of marked point processes in $\mathbb{R}^d$. These are point processes in $E = \mathbb{R}^d \times Q$, where Q, the *mark space*, is supposed to be a locally compact space with countable base. A simple point process N in E is called a *marked point process* if

$$\Theta(C \times Q) < \infty \qquad \text{for all } C \in \mathcal{C}.$$

The image process $N^0 = \pi N$ under the projection $\pi : (x, m) \mapsto x$ is called the *unmarked process* or *ground process.*

We define

$$t_x(y, q) = (y + x, q)$$

for $x, y \in \mathbb{R}^d$ and $q \in Q$, thus letting translations work on the first component only. The image of N under t_x is again denoted by $N+x$. Stationarity of a marked point process N then implies a basic decomposition of the intensity measure Θ.

Theorem 1.11 *If N is a stationary marked point process in $\mathbb{R}^d$ with mark space Q and intensity measure $\Theta \neq 0$, then*

$$\Theta = \gamma \lambda_d \otimes \mathbf{Q}$$

with a number $0 < \gamma < \infty$ and a (uniquely determined) probability measure $\mathbf{Q}$ on Q.

We call γ the *intensity* and $\mathbf{Q}$ the *mark distribution* of N. Obviously, γ is also the intensity of the (stationary) ground process N^0.

In analogy to the construction above, we define, for stationary N, the *Palm distribution* $\mathbf{P}^o$ of the marked point process N as a probability measure on $Q \times \mathsf{N}_s(\mathbb{R}^d \times Q)$ by

$$\mathbf{P}^o(A) = \gamma^{-1}\mathbf{E} \sum_{(x,q)\in N} \mathbf{1}_B(x)\mathbf{1}_A(q, N - x), \quad A \in \mathcal{B}(Q) \otimes \mathcal{N}_s(\mathbb{R}^d \times Q).$$

Again, $B \subset \mathbb{R}^d$ is an arbitrary Borel set with $\lambda_d(B) = 1$. We may interpret $\mathbf{P}^o(A\times \cdot)$, for $A \in \mathcal{B}(Q)$, as the conditional distribution of N under the condition that there is a point of N^0 at the origin o with mark in A.

Using the fact that $Q \times \mathsf{N}_s(\mathbb{R}^d \times Q)$ is a locally compact space with countable base (where the topology on $\mathsf{N}_s(\mathbb{R}^d \times Q)$ is induced by the hit-or-miss topology on $\mathcal{F}'(\mathbb{R}^d \times Q)$, see the next section), we can go one step further and disintegrate $\mathbf{P}^o$ with respect to the mark distribution $\mathbf{Q}$. The result is a regular family $(\mathbf{P}^{o,q})_{q\in Q}$ of conditional distributions $\mathbf{P}^{o,q}$ on $\mathsf{N}_s(\mathbb{R}^d \times Q)$ with

$$\mathbf{P}^o(A \times B) = \int_A \mathbf{P}^{o,q}(B)\,\mathbf{Q}(\mathrm{d}q)$$

for $A \in \mathcal{B}(Q)$ and $B \in \mathcal{N}_s(\mathbb{R}^d \times Q)$. We also call $\mathbf{P}^{o,q}$ a *Palm distribution* and interpret it as the distribution of N under the condition $(o, q) \in N$, that is, under the condition that N^0 has a point at the origin o with mark q.

We state the corresponding *refined Campbell theorem* for $\mathbf{P}^{o,q}$.

Theorem 1.12 *Let N be a stationary marked point process in $\mathbb{R}^d$ with mark space Q and intensity $\gamma > 0$. Let $f : \mathbb{R}^d \times Q \times \mathsf{N}_s(\mathbb{R}^d \times Q) \to \mathbb{R}$ be a nonnegative measurable function. Then $\sum_{(x,q)\in N} f(x, q, N)$ is measurable, and*

$$\begin{aligned}&\mathbf{E} \sum_{(x,q)\in N} f(x,q,N)\\ &\quad = \gamma \int_{\mathbb{R}^d}\int_Q \int_{\mathsf{N}_s(\mathbb{R}^d\times Q)} f(x,q,\eta + x)\,\mathbf{P}^{o,q}(\mathrm{d}\eta)\,\mathbf{Q}(\mathrm{d}q)\,\lambda_d(\mathrm{d}x).\end{aligned}$$

Poisson processes A class of point processes which is of fundamental importance in stochastic geometry is given by the Poisson processes. We assume again that E is locally compact with countable base. A *Poisson process* in E is usually defined as a point process N (with intensity measure Θ) having the two properties that

(i) the random variable $N(A)$ has a Poisson distribution, for each $A \in \mathcal{B}(E)$ with $\Theta(A) < \infty$,

(ii) for pairwise disjoint sets $A_1, \ldots, A_k \in \mathcal{B}(E)$ with $\Theta(A_i) < \infty$, the random variables $N(A_1)$, ..., $N(A_k)$ are (stochastically) independent.

If N is a Poisson process, then

$$\mathbf{P}\{N(A) = k\} = \mathrm{e}^{-\Theta(A)} \frac{\Theta(A)^k}{k!}, \quad k = 0, 1, 2, \ldots$$

for $A \in \mathcal{B}(E)$ with $\Theta(A) < \infty$. In particular, if $N(\{x\}) > 0$ with positive probability, then x must be an atom of Θ and N is not simple. Of course, for stationary Poisson processes (or stationary marked Poisson processes) in $\mathbb{R}^d$, this cannot occur (because the intensity measure is translation invariant). Since these are the main applications which we have in mind, we shall concentrate on simple Poisson processes, in the following, without further mentioning this condition. In that case, condition (i) implies condition (ii), that is, a simple point process with counting variables $N(A)$, $A \in \mathcal{B}(E)$, which are Poisson distributed has automatically independent 'increments' $N(A_1), \ldots, N(A_k)$ (A_i pairwise disjoint). This is a direct consequence of the construction underlying the following existence theorem and the corresponding uniqueness.

Theorem 1.13 *Let Θ be a locally finite measure without atoms on E. Then there exists a Poisson process in E with intensity measure Θ; it is uniquely determined (in distribution).*

To give a sketch of the proof, we start with a sequence of pairwise disjoint Borel sets $A_1, A_2, \ldots$ in E with $E = \bigcup_{i \in \mathbb{N}} A_i$, $\Theta(A_i) < \infty$, and such that each $C \in \mathcal{C}$ is covered by some finite union $\bigcup_{i=1}^k A_i$. In each A_i, we define a point process with intensity measure $\Theta \llcorner A_i$, satisfying condition (i) above, by specifying its distribution as

$$\mathbf{P}_i = \mathrm{e}^{-\Theta(A_i)} \left(\Delta_0 + \sum_{r=1}^{\infty} \frac{1}{r!} \Gamma_r \left((\Theta \llcorner A_i)^r \right) \right).$$

Here, the map $\Gamma_r : A_i^r \to \mathsf{N}(E)$ is defined by

$$\Gamma_r(x_1, \ldots, x_r) = \sum_{j=1}^{r} \delta_{x_j},$$

and Δ_0 is the Dirac measure on $\mathsf{N}(E)$ concentrated at the zero measure. Next, let $(N_1, N_2, \ldots)$ be an independent sequence of point processes in E such that N_i has distribution $\mathbf{P}_i$, for $i \in \mathbb{N}$, and put

$$N = \sum_{i=1}^{\infty} N_i.$$

Then N is a point process in E, it has Poisson counting variables and the intensity measure is Θ. This also implies that N is simple.

If N' is another point process in E with the same properties (intensity measure Θ and Poisson counting variables), we obtain $\mathbf{P}\{N(A) = 0\} =$

$\mathbf{P}\{N'(A) = 0\}$, for each $A \in \mathcal{B}(E)$. By Lemma 1.6, for the distributions we have $\mathbf{P}_N = \mathbf{P}_{N'}$. This shows uniqueness, but it also implies the independence property (ii). Namely, let pairwise disjoint sets $A_1, \dots, A_k \in \mathcal{B}(E)$ be given; we may assume that $\Theta(A_i) < \infty$. We can extend $(A_1, \dots, A_k)$ to a sequence $A_1, A_2, \dots$ satisfying the conditions underlying the construction above. Thus, we obtain a Poisson process N' in E deduced from the sequence $A_1, A_2, \dots$ The uniqueness implies $\mathbf{P}_N = \mathbf{P}_{N'}$. Since $N'(A_1), \dots, N'(A_k)$ are independent by construction, the same holds true for $N(A_1), \dots, N(A_k)$.

This proof shows a bit more, namely that, for a Poisson process N and pairwise disjoint sets $A_1, \dots, A_k \in \mathcal{B}(E)$, the induced processes $N \llcorner A_1, \dots, N \llcorner A_k$ are independent. Also, it yields a description of the conditional distribution

$$\mathbf{P}\{N \llcorner A \in \cdot \mid N(A) = k\} \quad \text{for } 0 < \Theta(A) < \infty.$$

Namely, if $N(A) = k$, the k points of $N \llcorner A$ are distributed as k independent, identically distributed random points $\xi_1, \dots, \xi_k$ in E with distribution

$$\mathbf{P}_{\xi_i} = \frac{\Theta \llcorner A}{\Theta(A)}, \quad i = 1, \dots, k.$$

The latter result is important for simulating a Poisson process in a given window $A \subset E$. It also implies that

$$\begin{aligned} &\mathbf{E} f(N \llcorner A) \\ &= \mathrm{e}^{-\Theta(A)} \left(f(0) + \sum_{k=1}^{\infty} \frac{1}{k!} \int_A \cdots \int_A f\left(\sum_{i=1}^{k} \delta_{x_i} \right) \Theta(\mathrm{d}x_1) \cdots \Theta(\mathrm{d}x_k) \right) \end{aligned}$$

for a Borel set A with $\Theta(A) < \infty$ and a nonnegative measurable function $f : \mathsf{N}(E) \to \mathbb{R}$.

We mention two important characterizations of Poisson processes. Due to our limitation to simple processes, we have to assume that the intensity measure of the given point process N is atom-free. In general, the results hold without this condition.

Theorem 1.14 *Let N be a point process in E, the intensity measure Θ of which has no atoms. Then N is a Poisson process if and only if*

$$\mathbf{E} \prod_{x \in N} f(x) = \exp \left(\int_E (f - 1) \, \mathrm{d}\Theta \right) \tag{1.21}$$

holds for all measurable functions $f : E \to [0, 1]$.

Theorem 1.15 *Let N be a point process in E, the intensity measure Θ of which has no atoms. Then N is a Poisson process if and only if*

$$\mathbf{E} \sum_{x \in N} g(N, x) = \int_E \mathbf{E}\, g(N + \delta_x, x) \, \Theta(\mathrm{d}x) \tag{1.22}$$

holds for all nonnegative measurable functions g on $\mathsf{N}(E) \times E$.

By iteration of (1.22), we obtain the *Slivnyak–Mecke formula*.

Corollary 1.16 *Let N be a Poisson process in E with intensity measure Θ, let $m \in \mathbb{N}$, and let $f : \mathsf{N}(E) \times E^m \to \mathbb{R}$ be a nonnegative measurable function. Then*

$$\mathbf{E} \sum_{(x_1,\ldots,x_m)\in N^m_{\neq}} f(N, x_1, \ldots, x_m)$$

$$= \int_E \cdots \int_E \mathbf{E} f\left(N + \sum_{i=1}^{m} \delta_{x_i}, x_1, \ldots, x_m\right) \Theta(\mathrm{d}x_1) \cdots \Theta(\mathrm{d}x_m).$$

This corollary implies that, for a Poisson process N in E and for $m \in \mathbb{N}$,

$$\Lambda^{(m)} = \Theta^m. \tag{1.23}$$

Now, we turn to the case $E = \mathbb{R}^d$ and to stationary processes. A stationary Poisson process N in $\mathbb{R}^d$ is uniquely determined (in distribution) by its intensity γ and is automatically isotropic. The Poisson property can be characterized by the following theorem. In its formulation, we interpret a simple counting measure again as a locally finite set in $\mathbb{R}^d$ (hence as an element of $\mathcal{F}$). The following is known as the *theorem of Slivnyak.*

Theorem 1.17 *Let $\mathbf{P}^o$ be the Palm distribution of a stationary simple point process N in $\mathbb{R}^d$ with intensity $\gamma > 0$. Then N is a Poisson process if and only if*

$$\mathbf{P}^o(A) = \mathbf{P}\{N \cup \{o\} \in A\} \tag{1.24}$$

holds for all $A \in \mathcal{B}(\mathcal{F})$.

The proof uses Theorem 1.15 and the refined Campbell Theorem 1.12.

If N^0 is a Poisson process in $\mathbb{R}^d$ (not necessarily stationary) and if $N^0 = \{\xi_1, \xi_2, \ldots\}$ is a measurable enumeration of N^0, we can define a marked point process N with ground process N^0 by choosing identically distributed marks $\kappa_1, \kappa_2, \ldots$ (in a mark space Q) with distribution $\mathbf{Q}$, which are independent (and independent of N^0), and putting

$$N = \sum_{i=1}^{\infty} \delta_{(\xi_i,\kappa_i)}.$$

Then, N is a Poisson process in $\mathbb{R}^d \times Q$ which we call *independently marked.* It is easy to see that not every (marked) Poisson process N in $\mathbb{R}^d \times Q$ is independently marked, however the latter is true if N is stationary.

For stationary marked point processes, there is also a version of Slivnyak's theorem.

Theorem 1.18 *Let N be a stationary marked point process in $\mathbb{R}^d$ with intensity $\gamma > 0$ and with mark space Q and mark distribution $\mathbf{Q}$ and let $(\mathbf{P}^{o,q})_{q\in Q}$ be the family of Palm distributions of N. Then, N is a Poisson process if and only if, for $\mathbf{Q}$-almost all $q \in Q$, we have*

$$\mathbf{P}^{o,q}(A) = \mathbf{P}\{N \cup \{(o, q)\} \in A\}$$

for all $A \in \mathcal{B}(\mathcal{F}(\mathbb{R}^d \times Q))$.

Starting from Poisson processes, one can construct useful classes of more general point processes. A *Cox process* (or *doubly stochastic Poisson process*) N (directed by a random measure M on E) can be considered as a Poisson process in E with random intensity measure M. More precisely, given a locally finite random measure M on E, the distribution of N is specified by

$$\mathbf{P}\{N(A) = k\} = \int_{\mathsf{M}(E)} \mathrm{e}^{-\eta(A)} \frac{\eta(A)^k}{k!} \mathbf{P}_M(\mathrm{d}\eta)$$

for $k \in \mathbb{N}_0$ and $A \in \mathcal{G}_c$ (the system of open sets with compact closure in E). The Cox process N exists if M is not identically 0 and has a.s. no atoms. The intensity measure of N is equal to the intensity measure of M.

A *cluster process* N in $\mathbb{R}^d$ is defined by a marked point process $\tilde{N}$, where the mark space Q is the subset $\mathsf{N}_{sf} \subset \mathsf{N}_s$ of simple finite counting measures in $\mathbb{R}^d$. The cluster process is obtained by superposition of the translated marks, that is, by

$$N = \sum_{(x,\eta)\in\tilde{N}} \sum_{y\in\eta} \delta_{x+y} = \sum_{(x,\eta)\in\tilde{N}} (\eta + x).$$

This is a point process if we assume that the clusters are uniformly bounded. For $(x, \eta) \in \tilde{N}$, one calls x a parent point of N, and the points $x + y$, $y \in \eta$, are called daughter points. They form a 'cluster' around the 'centre' x. If $o \in \eta$, the parent points appear in the cluster process, but this is not required by the definition. That the cluster process has locally finite intensity measure has to be guaranteed by an additional assumption. If the marked point process $\tilde{N}$ is stationary (with intensity $\tilde{\gamma} > 0$), the cluster process N is stationary and has intensity $\tilde{\gamma} n_c$, where n_c is the mean number of points in the typical cluster. If $\tilde{N}$ is an independently marked Poisson process, the cluster process N is called a *Neyman–Scott process*. A special Neyman–Scott process is the *Matérn cluster process*. It is obtained if the mark distribution is the distribution of a second stationary Poisson process Y, restricted to the ball RB^d (thus, a Matérn cluster process has Poisson clusters). The intensity μ of Y and the cluster radius $R > 0$ are additional parameters. A Neyman–Scott process is simple, but the definition of a general cluster process allows multiple points. If we identify $\eta \in \mathsf{N}_{sf}$ with its support, then

$$\hat{N} = \bigcup_{(x,\eta)\in\tilde{N}} (\eta + x)$$

defines a simple point process which is also called a cluster process. For Neyman–Scott processes, both definitions coincide. In this latter interpretation, clusters are finite sets, thus cluster processes appear as union sets of special particle processes, as they will be discussed in Section 1.5. Neyman–Scott processes are then special cases of Boolean models.

A *hard core process* N can be obtained from a stationary Poisson process $\tilde{N}$ in $\mathbb{R}^d$ by deleting some points of $\tilde{N}$, so that the distances between the remaining points in N have a given positive lower bound, the hard-core distance $c > 0$. Several methods of thinning are popular. The simplest one consists in deleting all pairs of points $x, y \in \tilde{N}$, $x \neq y$, with distance $\rho(x,y) < c$. The resulting process is called the *Matérn process (first kind)*. For the *Matérn process (second kind)* N, we start with a stationary marked Poisson process $\tilde{N}$ with intensity $\tilde{\gamma}$ and mark space $[0,1]$ (with uniform mark distribution). For each pair (x_1, w_1), $(x_2, w_2) \in \tilde{N}^2_{\neq}$ with $\rho(x_1, x_2) < c$, we delete the point $x_i \in \tilde{N}^0$ with the higher weight w_i. The undeleted points then form the point process N. For the intensity γ of N, one can obtain from Theorems 1.15 and 1.11 that

$$\gamma = \frac{1}{c^d \kappa_d}\left(1 - \mathrm{e}^{-\tilde{\gamma} c^d \kappa_d}\right).$$

Similarly, for the intensity γ of the Matérn process (first kind) one obtains

$$\gamma = \tilde{\gamma}\, \mathrm{e}^{-\tilde{\gamma} c^d \kappa_d},$$

where $\tilde{\gamma}$ is the intensity of the original Poisson process $\tilde{N}$ before thinning.

For the statistical analysis of spatial point patterns, the model classes 'completely random (Poisson)', 'clustered' and 'hard core' yield important first distinctions. Various second order quantities can be used for classification. We mention here only the *pair-correlation function* g_2 of a point process N in $\mathbb{R}^d$. For its definition, we assume that the intensity measure Θ and the second factorial moment measure $\Lambda^{(2)}$ are both absolutely continuous (with respect to λ_d, respectively $\lambda_d \otimes \lambda_d$). Let f and f_2 be the corresponding densities. Then g_2 is defined as

$$g_2(x,y) = \frac{f_2(x,y)}{f(x)f(y)}, \quad x, y \in \mathbb{R}^d,$$

provided that $f(x), f(y) > 0$. If N is stationary and isotropic, $g_2(x,y)$ depends only on $\|x - y\|$,

$$g_2(x,y) = g(\|x - y\|).$$

For the stationary Poisson process, $g \equiv 1$. In general, $\lim_{r\to\infty} g(r) = 1$, if N satisfies a mixing condition (that is, if points in N far away are asymptotically independent). If $g > 1$ in an interval (a,b), then point pairs $x, y \in N$ with distance in (a,b) are more frequent, whereas $g < 1$ indicates that such point pairs are more rare. Therefore, large values of g near zero indicate clustering (g can even have a pole at zero), and small values indicate repulsion. In particular, for hard core processes with hard core distance $c > 0$, the pair-correlation function g vanishes in $[0, c)$.

Hints to the literature For a general introduction to the theory of point processes, we refer to the two volumes of Daley and Vere–Jones (2005, 2008). Applications of point process theory in spatial statistics are presented, for example, in the recent book by Illian, Penttinen, H. Stoyan, and D. Stoyan (2008).

1.4 Random sets

Whereas random sets can be defined in general topological spaces, we restrict ourselves here to a locally compact space E with a countable base, as this is the common situation in stochastic geometry. In order to define a random set in E, one has to specify a class of (Borel) sets in E together with a σ-algebra. The latter should not be too small, in order to allow the usual set operations to be measurable, but also not too big, such that a rich variety of random sets can be constructed. Under these aspects, the class $\mathcal{F}(E)$ of closed sets has proved quite useful. It can be supplied with a natural topology and the corresponding Borel σ-algebra. A notion of open random sets could be obtained similarly, but the random closed sets have the advantage that simple point processes are subsumed. Therefore, we shall concentrate on the latter.

The *topology of closed convergence* on $\mathcal{F}(E)$ (also called *hit-or-miss topology*) is generated by the subbasis $\{\mathcal{F}^C : C \in \mathcal{C}(E)\} \cup \{\mathcal{F}_G : G \subset E \text{ open}\}$. Here we used the notation

$$\mathcal{F}_A = \{F \in \mathcal{F}(E) : F \cap A \neq \emptyset\}$$

and

$$\mathcal{F}^A = \{F \in \mathcal{F}(E) : F \cap A = \emptyset\}$$

for $A \subset E$. Supplied with this topology, $\mathcal{F}(E)$ is a compact space with a countable base. The subspace $\mathcal{F}'(E)$ is locally compact. The Borel σ-algebra on $\mathcal{F}(E)$ or $\mathcal{F}'(E)$ is generated, for example, by the system

$$\{\mathcal{F}_C : C \in \mathcal{C}\}.$$

The class $\mathcal{C}'(E) \subset \mathcal{F}'(E)$ of nonempty compact sets is usually supplied with the Hausdorff metric, which generates a topology different from the one induced by $\mathcal{F}'(E)$. However, the Borel σ-algebras are the same. The usual set operations on $\mathcal{F}(E)$, such as union $\cup$, intersection $\cap$, or the boundary operator ∂, are continuous or semi-continuous, hence they are measurable. The same holds for transformations $(g, F) \mapsto gF$, $g \in G$, $F \in \mathcal{F}(E)$, if G is a group operating on E in a measurable way.

As in the case of random measures, we assume that an underlying probability space $(\Omega, \mathcal{A}, \mathbf{P})$ is given.

Definition 1.19 *A* random closed set *in E is a measurable mapping $Z : \Omega \to \mathcal{F}(E)$.*

Since all random sets occurring will be closed, we often just speak of a random set Z.

An important characteristic of a random closed set Z is the *capacity functional* T_Z (also known as hitting functional or Choquet functional). It is defined on $\mathcal{C}(E)$ as

$$T_Z(C) = \mathbf{P}\{Z \cap C \neq \emptyset\} = \mathbf{P}_Z(\mathcal{F}_C)$$

and replaces the distribution function of real random variables. Namely, $T = T_Z$ has the following similar properties:

(1) $0 \le T \le 1$, $T(\emptyset) = 0$,
(2) $T(C_i) \to T(C)$ for every decreasing sequence $C_i \searrow C$,
(3) T is alternating of infinite order, that is,

$$S_k(C_0; C_1, \dots, C_k) \ge 0 \quad \text{for all } C_0, C_1, \dots, C_k \in \mathcal{C}(E),\ k \in \mathbb{N}_0.$$

Here, $S_0(C_0) = 1 - T(C_0)$ and for $k \in \mathbb{N}$,

$$S_k(C_0; C_1, \dots, C_k) = S_{k-1}(C_0; C_1, \dots, C_{k-1}) \\ -S_{k-1}(C_0 \cup C_k; C_1, \dots, C_{k-1}).$$

Moreover, T_Z determines Z uniquely (in distribution). This is an easy consequence of the fact that the complements $\mathcal{F}^C$ of $\mathcal{F}_C$, $C \in \mathcal{C}(E)$, form a $\cap$-stable generating system of the σ-algebra $\mathcal{B}(\mathcal{F}(E))$.

The following *theorem of Choquet* characterizes the capacity functionals of random closed sets.

Theorem 1.20 *If a functional T on $\mathcal{C}(E)$ has properties* (1)–(3), *then there is a random closed set Z in E with $T = T_Z$.*

Since we have identified simple counting measures with locally finite (closed) sets in E, the set N_s now appears as a measurable subset of $\mathcal{F}(E)$. Hence, a simple point process N can also be interpreted as a locally finite random closed set in E. This will be pursued further in Section 1.5, when we consider point processes in $E = \mathcal{F}'$ (the space of nonempty closed sets in $\mathbb{R}^d$) or in certain subclasses.

For $E = \mathbb{R}^d$, further subclasses of $\mathcal{F}$ are of interest. We call a random closed set Z in $\mathbb{R}^d$ a random compact set, random convex body, random polyconvex set, random k-flat if $\mathbf{P}_Z$ is concentrated, respectively, on $\mathcal{C}$, $\mathcal{K}$, $\mathcal{R}$, $A(d,k)$. The random set Z is called *stationary* if $\mathbf{P}_Z$ is invariant under translations, and *isotropic* if $\mathbf{P}_Z$ is invariant under rotations.

Lemma 1.21 *A stationary random closed set Z in $\mathbb{R}^d$ is almost surely either empty or unbounded.*

For a stationary random set Z, the value $p = T_Z(\{x\})$ is independent of $x \in \mathbb{R}^d$, since

$$T_Z(\{x\}) = \mathbf{P}\{x \in Z\} = \mathbf{P}\{o \in Z - x\} = \mathbf{P}\{o \in Z\} = T_Z(\{o\}).$$

Moreover, a simple Fubini argument shows that

$$\mathbf{E}\,\lambda_d(Z \cap A) = p\,\lambda_d(A), \quad A \in \mathcal{B}. \tag{1.25}$$

The constant p is therefore called the *volume fraction* of Z; we also denote it by $\overline{V}_d(Z)$.

A further basic characteristic for random sets Z is the *covariance* $C(x,y) = \mathbf{P}\{x, y \in Z\}$, $x, y \in \mathbb{R}^d$, of Z. For stationary Z, we have

$$C(x,y) = C(o, y-x) = \overline{V}_d(Z \cap (Z + x - y)).$$

We now combine the two notions of point process and random set by considering a point process X in the space $E = \mathcal{F}'$ of nonempty closed sets in $\mathbb{R}^d$. If X is simple, it consists of a locally finite collection of random closed sets. Since local finiteness means that any compact subset of $\mathcal{F}'$ a.s. contains only finitely many sets from X, and since $\mathcal{F}_C$, $C \in \mathcal{C}'$, is compact, it follows that, with probability one, any nonempty compact set $C \subset \mathbb{R}^d$ is hit only by finitely many $F \in X$. This implies that the union set of X is closed.

Theorem 1.22 *Let X be a simple point process in $\mathcal{F}'$. Then the union set*

$$Z = Z_X = \bigcup_{F \in X} F$$

is a random closed set in $\mathbb{R}^d$. If X is stationary (isotropic), then Z_X is stationary (isotropic).

If X is a stationary Poisson process, the union set Z_X is infinitely divisible and has no fixed points. Here, a random closed set Z is called *infinitely divisible* with respect to union if to each $m \in \mathbb{N}$ there are independent, identically distributed random sets $Z_1, \dots, Z_m$ such that Z equals $Z_1 \cup \cdots \cup Z_m$ in distribution. Further, a point $x \in \mathbb{R}^d$ is a *fixed point* of Z if $\mathbf{P}\{x \in Z\} = 1$. These properties even characterize stationary random sets Z which arise from Poisson processes.

Theorem 1.23 *For a stationary random closed set Z in $\mathbb{R}^d$ satisfying $Z \neq \mathbb{R}^d$ almost surely, the following conditions* (a), (b), (c) *are equivalent:*

(a) *Z is (equivalent to) the union set of a Poisson process X in $\mathcal{F}'$.*

(b) *There is a locally finite measure Θ without atoms on $\mathcal{F}'$ with*

$$T_Z(C) = 1 - e^{-\Theta(\mathcal{F}_C)}, \quad C \in \mathcal{C}.$$

(c) *Z is infinitely divisible and has no fixed points.*

If (a) *and* (b) *are satisfied, then Θ is the intensity measure of X, and Θ is translation invariant.*

A subclass of infinitely divisible random closed sets is given by the random sets Z which are *stable* with respect to union. These sets Z have the property that for each $m \in \mathbb{N}$ there are independent random closed sets $Z_1, \dots, Z_m$, distributed as Z, and a constant $\alpha_m > 0$ such that $\alpha_m Z$ equals $Z_1 \cup \cdots \cup Z_m$ in distribution. For stationary random sets Z without fixed points, stability can be characterized by property (b) in the above theorem, where Θ has the additional scaling property

$$\Theta(\mathcal{F}_{tC}) = t^{\beta} \Theta(\mathcal{F}_C)$$

for some $\beta > 0$, all $t > 0$ and all $C \in \mathcal{C}'$.

Stable random closed sets appear as limits of unions of i.i.d. random sets. More precisely, let $Y, Y_1, Y_2, \dots$ be a sequence of i.i.d. random closed sets in $\mathbb{R}^d$, and put $Z_n = Y_1 \cup \dots \cup Y_n$, $n \in \mathbb{N}$. Assume that a suitably normalized sequence $a_n^{-1} Z_n$, $n \in \mathbb{N}$, converges in distribution to some non-trivial random closed set Z (here, Z is trivial if Z equals a.s. the set of its fixed points). If $a_n \to 0$ and $T_Y(\{o\}) = T_Z(\{o\}) = 0$ or if $a_n \to \infty$ and if Y and Z are a.s. bounded by some (joint) half-space, then Z is stable. One can also derive a law of large numbers, namely an a.s. convergence of $a_n^{-1} Z_n$ to a (non-random) limit set using regularly varying capacities (and suitable normalizing factors a_n).

Infinite divisibility or stability for random closed sets in $\mathbb{R}^d$ can also be formulated w.r.t. Minkowski addition. Since the sum of closed sets need not be closed, we concentrate here on random compact sets, and we also present only some limit results. They are in close analogy to the classical limit theorems for real or vector-valued random variables and, in fact, using support functions, can be deduced from Banach space variants of these classical results. Let $Y, Y_1, Y_2, \dots$ be a sequence of i.i.d. random compact sets in $\mathbb{R}^d$ which are integrable, in the sense that $\mathbf{E}\|Y\| < \infty$, $\|Y\| = \max_{x \in Y} \|x\|$. The law of large numbers asserts that $n^{-1}(Y_1 + \dots + Y_n)$ converges a.s. in the Hausdorff metric to the expectation $\mathbf{E}Y$. Here, $\mathbf{E}Y$, also called the Aumann expectation, is defined as

$$\mathbf{E}Y = \{\mathbf{E}\,\xi : \xi : \Omega \to \mathbb{R}^d \text{ measurable and } \xi \in Y \text{ a.s.}\},$$

hence $\mathbf{E}Y$ is the set of all expectations of measurable selections of Y. Since the existence of such an i.i.d. sequence implies that $(\Omega, \mathcal{A}, \mathbf{P})$ does not have atoms, the Aumann expectation is convex, by Lyapounov's theorem, and compact, by our integrability condition, hence it is a convex body. Using the support function h_C of a set $C \in \mathcal{C}'$, we get the representation

$$h_{\mathbf{E}Y} = \mathbf{E}\, h_Y.$$

Now we state the *strong law of large numbers for random compact sets.*

Theorem 1.24 *Let $Y, Y_1, Y_2, \dots$ be a sequence of integrable and i.i.d. random compact sets in $\mathbb{R}^d$. Then,*

$$\frac{1}{n}(Y_1 + \dots + Y_n) \to \mathbf{E}Y \quad \textit{a.s., as } n \to \infty.$$

For a corresponding central limit theorem, we require Y to be square integrable, thus $\mathbf{E}\|Y\|^2 < \infty$, and we use the *covariance function*

$$\Gamma_Y : (u, v) \mapsto \mathbf{E}\,[h_Y(u) h_Y(v)] - \mathbf{E}\, h_Y(u)\, \mathbf{E}\, h_Y(v)$$

of h_Y, the latter viewed as a random element of the Banach space $\mathbf{C}(\mathbb{S}^{d-1})$. The following result is the *central limit theorem for random compact sets.*

Theorem 1.25 *Let $Y, Y_1, Y_2, \ldots$ be a sequence of square integrable and i.i.d. random compact sets in $\mathbb{R}^d$. Then,*

$$\sqrt{n}\rho_H\left(\frac{1}{n}(Y_1 + \cdots + Y_n), \mathbf{E}Y\right) \xrightarrow{d} \max_{x \in \mathbb{S}^{d-1}} \|\zeta(x)\|, \qquad \text{as } n \to \infty,$$

where ζ is a centred Gaussian random function in $\mathbf{C}(\mathbb{S}^{d-1})$ with covariance $\mathbf{E}[\zeta(u)\zeta(v)] = \Gamma_Y(u,v)$, $u, v \in \mathbb{S}^{d-1}$.

Both results, Theorem 1.24 and Theorem 1.25, follow for convex random sets Y from corresponding limit theorems in the Banach space $\mathbf{C}(\mathbb{S}^{d-1})$ (using the linear bijection $K \mapsto h_K$, $K \in \mathcal{K}'$). The extension to non-convex sets Y is based on inequality (1.5).

Hints to the literature The theory of random sets was initiated independently by D.G. Kendall and by G. Matheron. A first account was given in the monograph by Matheron (1975). The vigorous further development can be seen from the books by Molchanov (2005) and Nguyen (2006). We also refer to Molchanov (2005) for details, further results and extensive references concerning limit theorems, stability and infinitely divisible random sets, both with respect to union or Minkowski addition.

1.5 Geometric processes

We now concentrate on simple point processes in $E = \mathcal{F}' = \mathcal{F}'(\mathbb{R}^d)$ which are concentrated either on $\mathcal{C}'$ or on $A(d,q)$. In the first case we speak of a *particle process*, in the second of a *flat process* (or q-flat process). Note the special case of $q = 1$, which yields the case of *line processes.* If we assume stationarity, as we shall do now, the intensity measures of particle processes and flat processes admit basic decompositions.

For $C \in \mathcal{C}'$, let $c(C)$ be the centre of the smallest ball containing C. The mapping $C \mapsto c(C)$ is continuous. Let

$$\mathcal{C}_0 = \{C \in \mathcal{C}' : c(C) = o\}$$

and $\mathcal{K}_0 = \mathcal{K} \cap \mathcal{C}_0$.

Theorem 1.26 *Let X be a stationary particle process in $\mathbb{R}^d$ with intensity measure $\Theta \neq 0$. Then there exist a number $\gamma \in (0,\infty)$ and a probability measure $\mathbf{Q}$ on $\mathcal{C}_0$ such that*

$$\Theta(A) = \gamma \int_{\mathcal{C}_0} \int_{\mathbb{R}^d} \mathbf{1}_A(C + x)\, \lambda_d(\mathrm{d}x)\, \mathbf{Q}(\mathrm{d}C) \qquad (1.26)$$

for $A \in \mathcal{B}(\mathcal{C})$. Here, γ and $\mathbf{Q}$ are uniquely determined.

This decomposition is based on the fact that, according to our general assumption, the intensity measure is locally finite. For the probability measure $\mathbf{Q}$ this has the consequence that

$$\int_{\mathcal{C}_0} V_d(C + rB^d)\,\mathbf{Q}(\mathrm{d}C) < \infty, \quad r > 0. \tag{1.27}$$

We use γ and $\mathbf{Q}$ to define, for any measurable, translation invariant function φ on $\mathcal{C}'$, which is either nonnegative or $\mathbf{Q}$-integrable, a mean value

$$\overline{\varphi}(X) = \gamma \int \varphi(C)\,\mathbf{Q}(\mathrm{d}C). \tag{1.28}$$

This number is called the *φ-density* of X. Campbell's theorem implies that

$$\overline{\varphi}(X) = \frac{1}{\lambda_d(B)}\,\mathbf{E} \sum_{C\in X,\, c(C)\in B} \varphi(C) \tag{1.29}$$

for all $B \in \mathcal{B}(\mathbb{R}^d)$ with $0 < \lambda_d(B) < \infty$. Simple estimates show that also

$$\overline{\varphi}(X) = \lim_{r\to\infty} \frac{1}{\lambda_d(rW)}\,\mathbf{E} \sum_{C\in X,\, C\subset rW} \varphi(C) \tag{1.30}$$

for 'windows' $W \in \mathcal{K}$ with $\lambda_d(W) > 0$, and, under an additional integrability condition,

$$\overline{\varphi}(X) = \lim_{r\to\infty} \frac{1}{\lambda_d(rW)}\,\mathbf{E} \sum_{C\in X,\, C\cap rW\neq\emptyset} \varphi(C). \tag{1.31}$$

Choosing $\varphi = 1$, we see that γ can be interpreted as the mean number of particles per unit volume (of $\mathbb{R}^d$). Therefore, γ is also called the *intensity* of X. The probability measure $\mathbf{Q}$ is called the *grain distribution*. The latter name is motivated by the fact that X can be represented as a (stationary) marked point process $\tilde{X}$ (with mark space $\mathcal{C}_0$), namely through

$$\tilde{X} = \sum_{\substack{(x,C)\in\mathbb{R}^d\times\mathcal{C}_0,\\ C+x\in X}} \delta_{(x,C)}.$$

Then $\mathbf{Q}$ is the mark distribution of $\tilde{X}$.

Important examples of functionals φ are the intrinsic volumes, in the case of processes of polyconvex grains. If X is a stationary process of convex particles, then it follows from (1.27) that the intrinsic volumes V_j, $j = 0, \ldots, d$, are $\mathbf{Q}$-integrable.

Interesting special particle processes arise if the grains satisfy further restrictions, for example, if they are convex or if they are segments. Also random mosaics are subsumed under this notion. A *random mosaic* $X^{(d)}$ is a particle process such that the particles are d-dimensional, tile the space $\mathbb{R}^d$ and do not overlap in interior points. Usually, one also requires that the particles (called *cells* in this case) are convex, hence convex polytopes. For such a random mosaic $X^{(d)}$ and for $k = 0, \ldots, d-1$, the collection $X^{(k)}$ of k-faces is a particle process of k-dimensional polytopes. The facet process $X^{(d-1)}$ determines $X^{(d)}$ uniquely (but this is not true for $X^{(k)}$, $k \in \{0, \ldots, d-2\}$). The union set $Z_{d-1} = \bigcup_{S \in X^{(d-1)}} S$ is a random closed set, which also determines $X^{(d)}$, but does not easily allow to read off parameters like the intensities of $X^{(d)}$ or $X^{(d-1)}$. Alternatively, instead of Z_{d-1}, the random measure $\mathcal{H}_{d-1} \llcorner Z_{d-1}$ can be considered.

For general particle processes X, the union set $Z = \bigcup_{C \in X} C$ carries much less information. For example, any random closed set Z arises as such a union set, and if Z is stationary, X can be chosen to be stationary. But even this can be performed in many different ways. It is perhaps more surprising that a stationary random closed set Z with values in $\mathcal{S}$ (locally polyconvex sets) can be decomposed into convex particles such that the corresponding particle process X is stationary, too.

A particular class of random sets are the *Boolean models* Z. These are union sets of Poisson particle processes X. For Boolean models Z a rich variety of formulae exist. In particular, the capacity functional of Z and the intensity measure Θ of X are connected by

$$T_Z(C) = 1 - \mathrm{e}^{-\Theta(\mathcal{F}_C)}, \quad C \in \mathcal{C}. \tag{1.32}$$

This shows that the underlying Poisson process X is uniquely determined by Z. In particular, Z is stationary (isotropic) if and only if X is stationary (isotropic). For stationary Z, (1.32), (1.26), (1.9) and (1.28) show that

$$T_Z(C) = 1 - \exp\left(-\gamma \int_{\mathcal{C}_0} \lambda_d(K - C)\, \mathbf{Q}(\mathrm{d}K)\right), \quad C \in \mathcal{C}, \tag{1.33}$$

where $K - C = \{x - y : x \in K, y \in C\}$. Some further formulae for stationary Boolean models are given in Section 1.6.

We now come to stationary q-flat processes X. Let $G(d,q)$ be the Grassmannian of q-dimensional linear subspaces of $\mathbb{R}^d$.

Theorem 1.27 *Let X be a stationary q-flat process in $\mathbb{R}^d$ with intensity measure $\Theta \neq 0$. Then there exist a number $\gamma \in (0, \infty)$ and a probability measure $\mathbf{Q}$ on $G(d,q)$ with*

$$\Theta(A) = \gamma \int_{G(d,q)} \int_{L^\perp} \mathbf{1}_A(L + x)\, \lambda_{L^\perp}(\mathrm{d}x)\, \mathbf{Q}(\mathrm{d}L) \tag{1.34}$$

for $A \in \mathcal{B}(A(d,q))$. Here, γ and $\mathbf{Q}$ are uniquely determined.

We call γ the *intensity* and $\mathbf{Q}$ the *directional distribution* of X. If X is isotropic, then $\mathbf{Q}$ is rotation invariant, and hence $\mathbf{Q}$ is equal to ν_q, the rotation invariant probability measure on $G(d,q)$. Whereas $\mathbf{Q}$ controls the directions of

X, the interpretation of γ as the mean number of flats in X per unit volume comes from

$$\gamma = \frac{1}{\kappa_{d-q}} \mathbf{E}\, X\left(\mathcal{F}_{B^d}\right) \tag{1.35}$$

(note that $X(\mathcal{F}_{B^d})$ is the number of q-flats in X meeting the unit ball B^d). A similar formula is

$$\gamma = \binom{d}{q} \frac{\kappa_d}{\kappa_q \kappa_{d-q}^2} \mathbf{E}_\nu \mathbf{E}\, X\left(\mathcal{F}_{\vartheta B^{d-q}}\right), \tag{1.36}$$

where ϑ is a random rotation with distribution ν, independent of X, and B^{d-q} is the unit ball in some subspace $L \in G(d, d-q)$. If X is isotropic, $\mathbf{E}\, X\left(\mathcal{F}_{\vartheta B^{d-q}}\right)$ is independent of ϑ and the choice of L, and the formula holds without the random rotation ϑ on the right side. Note that $\frac{1}{\kappa_{d-q}} \mathbf{E}\, X\left(\mathcal{F}_{B^{d-q}}\right)$ is the intensity $\gamma_{X \cap L}$ of the *intersection process* $X \cap L$. The latter is a.s. an ordinary point process in L. If X is not isotropic, the intensity $\gamma_{X \cap L}$ will depend on the subspace L (and not only on its dimension $d-q$).

Theorem 1.28 *Let X be a stationary q-flat process in $\mathbb{R}^d$ with intensity γ and directional distribution $\mathbf{Q}$, $q \in \{1, \ldots, d-1\}$. For $L \in G(d, d-q)$, let $\gamma_{X \cap L}$ be the intensity of the point process $X \cap L$. Then*

$$\gamma_{X \cap L} = \gamma \int_{G(d,q)} [L, L']\, \mathbf{Q}(\mathrm{d}L').$$

Here, $[L, L']$ denotes the determinant of the orthogonal projection from $L^\perp$ onto L'. In general, the function $L \mapsto \gamma_{X \cap L}$ does not determine γ and $\mathbf{Q}$ uniquely, but uniqueness holds for $q = 1$ or $q = d-1$. In both cases, lines and hyperplanes can be represented by antipodal pairs of points $u, -u \in \mathbb{S}^{d-1}$, hence $\gamma_{X \cap L}$ gives rise to an even function γ_X and $\mathbf{Q}$ to an even probability measure φ on $\mathbb{S}^{d-1}$. Theorem 1.28 then implies that

$$\gamma_X(u) = \gamma \int_{\mathbb{S}^{d-1}} |\langle x, u \rangle|\, \varphi(\mathrm{d}x).$$

The fact that this integral equation has a unique (even) solution φ was mentioned in Section 1.2, in connection with zonoids.

In fact, the function γ_X is the support function of an *associated zonoid* Π_X, where, for a stationary line process X, $h_{\Pi_X}(u)$ gives the intensity of intersection points of lines of X with the hyperplane $L = u^\perp$. A similar interpretation holds for stationary hyperplane processes.

Associated zonoids can be introduced for many geometric processes and random sets. In addition to line and hyperplane processes, this is the case for processes of segments or fibres, processes of plates or surfaces (e.g. boundaries of full-dimensional particles), and for random mosaics. Apart from uniqueness questions, such as the one explained above, which are of a purely analytic nature, often also the geometric properties of associated zonoids are helpful. For example, classical inequalities from convex geometry can be used to obtain extremal properties of geometric processes or random sets.

In order to make this more precise, let $Z = Z_X$ be a stationary Boolean model with convex grains (thus, the grain distribution $\mathbf{Q}$ is concentrated on $\mathcal{K}'$). We consider Z_X as being opaque and assume that Z is nondegenerate, that is, for any point outside Z_X, the range of visible points is bounded a.s. This is, for example, the case if the underlying Poisson process X consists of d-dimensional convex bodies. If $o \notin Z$, we consider the star-shaped set $S_0(Z)$ of all points in $\mathbb{R}^d \setminus Z$ which are visible from o. The conditional expectation

$$\overline{V}_s(Z) = \mathbf{E}\left(\lambda_d(S_0(Z)) \mid o \notin Z\right)$$

is called the *mean visible volume outside* Z. It turns out that $(d!)^{-1}\overline{V}_s(Z)$ equals the volume $V_d(\Pi_X^\circ)$ of the polar body of the associated zonoid Π_X of X. Another geometric parameter for the Poisson process X is the *intersection density* $\gamma_d(X)$ of the particle boundaries. Namely, the points in $\mathbb{R}^d$ arising as intersection points of the boundaries of any d distinct bodies of the process form a stationary point process in $\mathbb{R}^d$, and $\gamma_d(X)$ is defined as the intensity of this process. Using Campbell's theorem and a translative integral formula of Poincaré type, one can show that $\gamma_d(X) = V_d(\Pi_X)$. Applying the Blaschke–Santaló inequality and its inverse for zonoids, together with the corresponding assertions on the equality case, we get the following result. Here we say that the particle process X is *affinely isotropic* if it is the image of an isotropic particle process under an affine transformation of $\mathbb{R}^d$.

Theorem 1.29 *Let $Z = Z_X$ be a nondegenerate stationary Boolean model with convex grains in $\mathbb{R}^d$. Then*

$$4^d \leq \gamma_d(X)\overline{V}_s(Z) \leq d!\kappa_d^2. \tag{1.37}$$

On the right side, equality holds if and only if the process X is affinely isotropic. On the left side, equality holds if and only if the particles of X are almost surely parallelepipeds with edges of d fixed directions.

Inequality (1.37) reflects the intuitively clear fact that a larger visible volume requires more scarcely scattered particles; hence, the intersection density has to be small.

As a second example, we consider a stationary Poisson hyperplane process X with intensity $\gamma > 0$ in $\mathbb{R}^d$. For $k \in \{2, \ldots, d\}$, let γ_k be the kth intersection density of X. This is the intensity of the (stationary) process of $(d-k)$-flats that is obtained by intersecting any k hyperplanes of X which are in general position. Using a suitable associated zonoid and the Aleksandrov–Fenchel inequality from the geometry of convex bodies, one obtains the inequality

$$\gamma_k \leq \frac{\binom{d}{k}\kappa_{d-1}^k}{d^k \kappa_{d-k}\kappa_d^{k-1}}\gamma^k.$$

The equality sign holds if and only if the process X is isotropic.

Hints to the literature Early important contributions are the thesis of R.E. Miles of 1961 and his subsequent publications, and the book by Matheron (1975). The later development is reflected in the books by Stoyan, Kendall and Mecke (1995) and Schneider and Weil (2008). Associated zonoids were introduced by Matheron, under the name of 'Steiner compact', and later employed by several authors. For further results (also for applications to random mosaics), see the last-mentioned book.

1.6 Mean values of geometric functionals

For a stationary random closed set Z in $\mathbb{R}^d$, the volume fraction $\overline{V}_d(Z) = \mathbf{P}\{o \in Z\}$ can be represented by

$$\overline{V}_d(Z) = \frac{\mathbf{E}\,\lambda_d(Z \cap B)}{\lambda_d(B)}, \tag{1.38}$$

with any $B \in \mathcal{B}(\mathbb{R}^d)$ satisfying $\lambda_d(B) > 0$. This mean expected volume is certainly the simplest parameter by which we can measure the average size of Z. A finer quantitative description will require more parameters. However, in order that averages for further functionals (for example, surface area or Euler characteristic) exist, the realizations of the random closed set must be restricted suitably, and an integrability assumption is necessary. For a polyconvex set K, we denote by $N(K)$ the smallest number of convex bodies with union equal to K. The function N is measurable. The unit cube $C^d = [0,1]^d$ used below could be replaced by any other convex body with interior points. We define a class of random sets which are sufficiently general for many purposes, such as modelling real materials, and are mathematically well accessible.

Definition 1.30 *A* standard random set *in $\mathbb{R}^d$ is a stationary random closed set Z with the properties that its realizations are almost surely locally polyconvex and that*

$$\mathbf{E}\,2^{N(Z \cap C^d)} < \infty.$$

For a stationary particle process X and any measurable, translation invariant, nonnegative function φ on $\mathcal{C}'$, we have defined the φ-density of X by means of (1.28). The interpretations (1.29), (1.30), (1.31) show how this density can be obtained by a double averaging, stochastic and spatial. For standard random sets and a more restricted class of functions φ, densities can be defined in a similar way.

Theorem 1.31 *Let Z be a standard random set in $\mathbb{R}^d$, and let φ be a real function on the convex ring $\mathcal{R}$ which is translation invariant, additive, measurable, and is bounded on the set $\{K \in \mathcal{K}' : K \subset C^d\}$. Then, for every convex body $W \in \mathcal{K}'$ with $V_d(W) > 0$, the limit*

$$\overline{\varphi}(Z) = \lim_{r\to\infty} \frac{\mathbf{E}\,\varphi(Z \cap rW)}{V_d(rW)} \tag{1.39}$$

exists and is independent of W.

We call $\overline{\varphi}(Z)$ the *φ-density* of Z. The most important functions φ satisfying the assumptions are the intrinsic volumes V_j, $j = 0, \ldots, d$, additively extended to the convex ring. The density $\overline{V}_j(Z)$ is also known as the *specific* jth *intrinsic volume* of Z. In particular, $\overline{V}_d$ is the *specific volume* (given by (1.38)), $2\overline{V}_{d-1}$ is the *specific surface area*, and $\overline{V}_0$ is the *specific Euler characteristic*. The densities $\overline{V}_d$ and $\overline{V}_{d-1}$ are nonnegative. One can also define densities of functions or measures, by applying Theorem 1.31 argument-wise. For example, for each $u \in \mathbb{S}^{d-1}$, the function φ defined by $\varphi(K) = h^*_K(u)$, $K \in \mathcal{R}$, where h^* is the additive extension of the centred support function, satisfies the assumptions. Hence, its density, denoted by $\overline{h}_Z(u)$, is defined. This yields a function $\overline{h}_Z$ on $\mathbb{S}^{d-1}$, which turns out to be continuous, and a support function if $d = 2$. In a similar way, the surface area measure $S_{d-1}(\cdot, \cdot)$ gives rise to a specific surface area measure $\overline{S}_{d-1}(Z, \cdot)$ on $\mathbb{S}^{d-1}$. If certain degenerate standard random sets are excluded, then the measure $\overline{S}_{d-1}(Z, \cdot)$ satisfies the assumptions of Theorem 1.2 and hence is the surface area measure of a convex body $B(Z)$, called the *Blaschke body* of Z. Thus, Theorem 1.31 allows us to associate also measure valued and body valued parameters with a standard random set.

Returning to the specific intrinsic volumes, as the basic real-valued parameters of a standard random set, we discuss how one can obtain unbiased estimators for them. First, we note that, for standard random sets, the relation (1.38) can be generalized, if the additive extensions of $\Phi_j(K, \cdot)$, defined for $K \in \mathcal{R}$, are employed. If Z is a standard random set and $B \in \mathcal{B}(\mathbb{R}^d)$ is a bounded Borel set with $\lambda_d(B) > 0$, then

$$\overline{V}_j(Z) = \frac{\mathbf{E}\,\Phi_j(Z, B)}{\lambda_d(B)} \tag{1.40}$$

for $j = 0, \ldots, d$. Here, $\Phi_j(Z, B)$ is defined by $\Phi_j(Z \cap W, B)$, where $W \in \mathcal{K}'$ is any convex body containing B in its interior. Since the curvature measures are locally determined, this value does not depend on the choice of W. The proof of relation (1.40) makes use of the local translative formula (1.20) (for $k = 2$). If the standard random set Z is isotropic (its distribution is invariant under rigid motions), then the local principal kinematic formula can be used to show, for any sampling window $W \in \mathcal{K}'$, the expectation formula

$$\mathbf{E}\,\Phi_j(Z \cap W, \cdot) = \sum_{k=j}^{d} c(d, j, k)\,\overline{V}_k(Z)\Phi_{d-k+j}(W, \cdot). \tag{1.41}$$

The coefficients are given by (1.13). The global case of (1.41) reads

$$\mathbf{E}\,V_j(Z \cap W) = \sum_{k=j}^{d} c(d, j, k)\,\overline{V}_k(Z)V_{d-k+j}(W). \tag{1.42}$$

For non-isotropic standard random sets Z, (1.42) still holds if W is a ball. Another possibility to obtain a similar mean value formula in the non-isotropic case is to replace the window W by ϑW, where ϑ is a random rotation independent of Z and with uniform distribution ν, and to take also the expectation over ϑ. This gives

$$\mathbf{E}_\nu \mathbf{E}\, V_j(Z \cap \vartheta W) = \sum_{k=j}^{d} c(d,j,k)\, \overline{V}_k(Z) V_{d-k+j}(W). \tag{1.43}$$

The preceding expectation formulae can be used to obtain unbiased estimators for the specific intrinsic volumes $\overline{V}_j(Z)$ of a standard random set Z. For example, (1.41) yields the unbiased estimator

$$\widehat{V}_j = \frac{\Phi_j(Z \cap W, \operatorname{int} W)}{V_d(W)},$$

for any sampling window $W \in \mathcal{K}'$ with $V_d(W) > 0$. In the case $j = d-1$, for instance, using this estimator requires the evaluation of the surface area of the boundary of Z within the interior of W.

Since the evaluation of curvature measures Φ_j for $j < d-1$ is difficult, it might be desirable to work with $V_j(Z \cap W)/V_d(W)$ as an estimator (for $d = 2$, for example, the determination of the Euler characteristic $V_0(Z \cap W)$, for a given realization of Z, is much easier than the determination of the curvature measure $\Phi_0(Z \cap W, \operatorname{int} W)$). This estimator is asymptotically unbiased, by Theorem 1.31, but not unbiased. Information on the error is obtained from the counterpart to (1.42) for non-isotropic standard random sets, which reads

$$\mathbf{E}\, V_j(Z \cap W) = \sum_{k=j}^{d} \overline{V}^{(j)}_{k,d-k+j}(Z, W).$$

Here $\overline{V}^{(j)}_{k,d-k+j}(Z, W)$ is the density of the mixed functional $V^{(j)}_{k,d-k+j}(\cdot, W)$. From this, it follows that

$$\frac{\mathbf{E}\, V_j(Z \cap rW)}{V_d(rW)} = \overline{V}_j(Z) + \frac{1}{V_d(W)} \sum_{k=j+1}^{d} r^{j-k}\, \overline{V}^{(j)}_{k,d-k+j}(Z, W),$$

which exhibits the bias.

For an isotropic standard random set Z, the formulae (1.42) for $j = 0, \dots, d$ yield a triangular system of linear equations for $\overline{V}_0(Z), \dots, \overline{V}_d(Z)$. The solution is of the form

$$\overline{V}_i(Z) = \mathbf{E}\left(\sum_{j=i}^{d} \beta_{dij}(W) V_j(Z \cap W) \right), \quad i = 0, \dots, d,$$

with certain constants $\beta_{dij}(W)$ (which are easily computed if W is a rectangular parallelepiped or a ball). Therefore,

$$\widehat{V}_i = \sum_{j=i}^{d} \beta_{dij}(W) V_j(Z \cap W)$$

is an unbiased estimator for $\overline{V}_i(Z)$. Alternatively, one can base unbiased estimators for $\overline{V}_i(Z)$ on the determination of the Euler characteristic $V_0(Z \cap W)$ alone, provided one employs more sampling windows in a suitable way. For example, the system of equations (resulting from (1.42))

$$\mathbf{E} V_0(Z \cap r_m W) = \sum_{k=0}^{d} c(d,0,k)\, r_m^k V_k(W) \overline{V}_{d-k}(Z), \quad m = 0, \ldots, d,$$

can be solved for $\overline{V}_0(Z), \ldots, \overline{V}_d(Z)$ if the dilatation factors $r_0, \ldots, r_d$ are chosen such that the matrix with entries $c(d,0,k) r_m^k V_k(W)$ $(k, m = 0, \ldots, d)$ is regular. This yields unbiased estimators for $\overline{V}_i(Z)$ of the form

$$\widehat{V}_i = \sum_{m=0}^{d} \alpha_{dij}(W) V_0(Z \cap r_j W),$$

with certain constants $\alpha_{dij}(W)$.

The preceding expectation formulae referred to the intersection of a standard random set with a sampling or observation window. Similar mean value formulae, the interest in which came originally from stereology, hold for intersections with a fixed flat (affine subspace) E of dimension $q \in \{1, \ldots, d-1\}$. Let Z be a standard random set. Then $Z \cap E$ is a standard random set in E, hence the density $\overline{V}_j(Z \cap E)$ exists for $j \in \{0, \ldots, k\}$. If Z is isotropic, then the relation

$$\overline{V}_j(Z \cap E) = c(d,j,q) \overline{V}_{d-q+j}(Z) \tag{1.44}$$

holds. It shows that an unbiased estimator for the density $\overline{V}_j(Z \cap E)$ is also an unbiased estimator for the density $c(d,j,q)\overline{V}_{d-q+j}(Z)$. Thus, the specific mth intrinsic volume of an isotropic standard random set can be estimated from measurements in the section with a fixed q-flat, if $m \geq d-q$. In the non-isotropic case, a similar procedure is possible if one works with a randomly and independently rotated section flat.

For particle processes, expectation formulae similar to (1.40)–(1.42) can be obtained. We restrict ourselves to stationary processes X of convex particles. For these, the intrinsic volumes have finite densities $\overline{V}_j(X)$, $j = 0, \ldots, d$, defined by (1.39). The following is a further representation. If $B \in \mathcal{B}(\mathbb{R}^d)$ is a Borel set with $\lambda_d(B) > 0$, then

$$\overline{V}_j(X) = \frac{1}{\lambda_d(B)} \mathbf{E} \sum_{K \in X} \Phi_j(K, B). \tag{1.45}$$

If X is isotropic, then for any sampling window $W \in \mathcal{K}'$ with $V_d(W) > 0$ we have

$$\mathbf{E} \sum_{K \in X} \Phi_j(K \cap W, \cdot) = \sum_{k=j}^{d} c(d,j,k) \overline{V}_k(X) \Phi_{d-k+j}(W, \cdot). \tag{1.46}$$

The global version of this is the relation

$$\mathbf{E} \sum_{K \in X} V_j(K \cap W) = \sum_{k=j}^{d} c(d,j,k) V_k(W) \overline{V}_{d-k+j}(X). \tag{1.47}$$

There is also a counterpart to (1.42). Further, if X is isotropic and E is a k-flat, $k \in \{1, \dots, d-1\}$, then $X \cap E$ is a stationary and isotropic particle process in E, and

$$\overline{V}_j(X \cap E) = c(d,j,k) \overline{V}_{d-k+j}(X). \tag{1.48}$$

The consequences as to the construction of unbiased estimators are analogous to those in the case of standard random sets.

The derivation of all the preceding expectation formulae makes essential use of integral geometry. This is also true for the fundamental density relations for stationary Boolean models with convex grains, to which we turn now. A *Boolean model* is a random closed set that is obtained as the union set of a Poisson particle process. Let X be stationary Poisson process of convex particles in $\mathbb{R}^d$, and let $Z = \bigcup_{K \in X} K$. The intensity γ and the grain distribution $\mathbf{Q}$ of X completely determine the intensity measure of X and hence the distribution of the Poisson process X. Therefore, they determine also the distribution of the Boolean model Z and, in particular, its specific intrinsic volumes $\overline{V}_j(Z)$. We sketch how they can be computed. For this, we use (1.39), with some $W \in \mathcal{K}'$ satisfying $V_d(W) > 0$. By the additivity of V_j (which implies an inclusion–exclusion formula), the Campbell formula of Corollary 1.8, the relation (1.23) for Poisson processes, and the decomposition (1.26) of the intensity measure, we obtain, for $r > 0$,

$$\begin{aligned}
&\mathbf{E}\, V_j(Z \cap rW) \\
&= \sum_{k=1}^{\infty} \frac{(-1)^{k-1}}{k!} \mathbf{E} \sum_{(K_1,\dots,K_k) \in X^k_{\neq}} V_j(K_1 \cap \cdots \cap K_k \cap rW) \\
&= \sum_{k=1}^{\infty} \frac{(-1)^{k-1}}{k!} \int_{\mathcal{K}^k} V_j(K_1 \cap \cdots \cap K_k \cap rW)\, \Theta^k(\mathrm{d}(K_1, \dots, K_k)) \\
&= \sum_{k=1}^{\infty} \frac{(-1)^{k-1}}{k!} \gamma^k \int_{\mathcal{K}_0^k} \int_{(\mathbb{R}^d)^k} V_j((K_1 + x_1) \cap \cdots \cap (K_k + x_k) \cap rW) \\
&\quad \times \lambda_d^k(\mathrm{d}(x_1, \dots, x_k))\, \mathbf{Q}^k(\mathrm{d}(K_1, \dots, K_k)).
\end{aligned}$$

The computation of the inner integral over $(\mathbb{R}^d)^k$ is a typical task of translative integral geometry. The global case of (1.20) can be used. We restrict ourselves here to the case where X and Z are isotropic. Then $\mathbf{Q}$ is rotation invariant, and the inner integral can be replaced by

$$\int_{G_d} \cdots \int_{G_d} V_j(g_1 K_1 \cap \cdots \cap g_k K_k \cap rW)\, \mu(\mathrm{d}g_1) \cdots \mu(\mathrm{d}g_k).$$

By iteration of the principal kinematic formula (1.14) one finds that this can be expressed as a sum of products of intrinsic volumes of $K_1, \ldots, K_k$ and W. Then we use (1.28) and obtain an explicit result, which we formulate here only for $d = 3$:

$$\overline{V}_3(Z) = 1 - \mathrm{e}^{-\overline{V}_3(X)}$$

$$\overline{V}_2(Z) = \mathrm{e}^{-\overline{V}_3(X)} \overline{V}_2(X)$$

$$\overline{V}_1(Z) = \mathrm{e}^{-\overline{V}_3(X)} \left(\overline{V}_1(X) - \frac{\pi}{8} \overline{V}_2(X)^2 \right)$$

$$\overline{V}_0(Z) = \mathrm{e}^{-\overline{V}_3(X)} \left(\overline{V}_0(X) - \frac{1}{2} \overline{V}_1(X) \overline{V}_2(X) + \frac{\pi}{48} \overline{V}_2(X)^3 \right).$$

This shows how the densities $\overline{V}_0(Z), \ldots, \overline{V}_3(Z)$ of the Boolean model Z are determined by the densities $\overline{V}_0(X), \ldots, \overline{V}_3(X)$ of the underlying particle process X. But it also shows that, conversely, the densities $\overline{V}_0(X), \ldots, \overline{V}_3(X)$ are determined by the densities $\overline{V}_0(Z), \ldots, \overline{V}_3(Z)$ of the union set. This may seem surprising, but can be explained by the strong independence properties of Poisson processes. In fact, as remarked earlier, the underlying Poisson particle process X is uniquely determined by Z.

The representation of the densities $\overline{V}_j(Z)$ in terms of data of the underlying particle process extends to non-isotropic stationary Boolean models, where the densities $\overline{V}_j(X)$ have to be replaced by densities of mixed functionals. Even an extension to non-stationary Boolean models is possible, where the densities are no longer constants but almost everywhere defined functions. All this holds in arbitrary dimensions d.

For a stationary Boolean model Z in $\mathbb{R}^d$, it is also possible to determine the densities $\overline{V}_j(X)$ of the underlying particle process X from volume densities alone, if parallel sets of Z are taken into account. The *spherical contact distribution function* of a stationary random closed set Z is defined by

$$H(r) = \mathbf{P}\{\rho(o, Z) \leq r \mid o \notin Z\} = \mathbf{P}\{o \in Z + rB^d \mid o \notin Z\}, \quad r > 0.$$

It can be expressed in terms of volume densities, namely

$$H(r) = \frac{\overline{V}_d(Z + rB^d) - \overline{V}_d(Z)}{1 - \overline{V}_d(Z)}.$$

If now Z is a stationary Boolean model with convex grains, then, by (1.33),

$$H(r) = 1 - \exp\left(-\sum_{k=1}^{d} \kappa_k r^k \overline{V}_{d-k}(X)\right), \quad r \geq 0.$$

Simple estimators for $\overline{V}_j(X)$, $j = 0, \ldots, d-1$, can be based on the last two formulae, applying them for different values of r.

Hints to the literature To the introduction of functional densities and the derivation of mean value formulae and estimators for them, many authors have contributed. The beginnings can be seen in work of Matheron, Miles and Davy; the generality increased over the years. After the groundbreaking work of Matheron (1975), the development was reflected in the book by Stoyan, Kendall and Mecke (1995). A detailed treatment and more references are found in the book by Schneider and Weil (2008). For a comprehensive treatment of Boolean models, we refer to Molchanov (1997).

References

Ambartzumian, R.V. (1982). *Combinatorial Integral Geometry. With Applications to Mathematical Stereology.* Wiley, Chichester.

Ambartzumian, R.V. (1990). *Factorization Calculus and Geometric Probability.* Cambridge Univ. Press, Cambridge.

Baddeley, A., Bárány, I., Schneider, R., and Weil, W. (2007). *Stochastic Geometry.* C.I.M.E. Summer School, Martina Franca, Italy, 2004 (edited by W. Weil). *Lecture Notes in Mathematics*, **1892**, Springer, Berlin.

Baddeley, A. and Jensen, E.B.V. (2005). *Stereology for Statisticians.* Chapman & Hall/CRC, Boca Raton.

Czuber, E. (1884). *Geometrische Wahrscheinlichkeiten und Mittelwerte.* Teubner, Leipzig.

Daley, D.J. and Vere-Jones, D. (2005). *An Introduction to the Theory of Point Processes, Vol. 1: Elementary Theory and Methods.* 2nd ed. 2003, 2nd corrected printing, Springer, New York.

Daley, D.J. and Vere-Jones, D. (2008). *An Introduction to the Theory of Point Processes, Vol. 2: General Theory and Structure. 2nd ed.*, Springer, New York.

Deltheil, R. (1926). *Probabilités géométriques.* Gauthiers-Villars, Paris.

Harding, E.F. and Kendall, D.G. (eds.) (1974). *Stochastic Geometry.* Wiley, London.

Illian, J., Penttinen, A., Stoyan, H., and Stoyan, D. (2008). *Statistical Analysis and Modelling of Spatial Point Patterns.* Wiley, Chichester.

Jensen, E.B.V. (1998). *Local Stereology.* World Scientific, Singapore.

Kendall, M.G. and Moran, P.A.P. (1963). *Geometrical Probability.* Griffin, London.

Matheron, G. (1975). *Random Sets and Integral Geometry.* Wiley, New York.
Miles, R.E. and Serra, J. (eds.) (1978). *Geometrical Probability and Biological Structures: Buffon's 200th Anniversary.* Proceedings, Paris 1977. Springer, Berlin.
Molchanov, I.S. (1997). *Statistics of the Boolean Model for Practitioners and Mathematicians.* Wiley, Chichester.
Molchanov, I.S. (2005). *Theory of Random Sets.* Springer, London.
Mosler, K. (2002). *Multivariate Dispersion, Central Regions and Depth. The Lift Zonoid Approach.* Lect. Notes Statist., **165**, Springer, New York.
Nguyen, H.T. (2006). *An Introduction to Random Sets.* Chapman & Hall/CRC, Boca Raton.
Santaló, L.A. (1976). *Integral Geometry and Geometric Probability.* Addison-Wesley, Reading, Mass.
Schneider, R. (1993). *Convex Bodies: the Brunn–Minkowski Theory.* Cambridge University Press, Cambridge.
Schneider, R. and Weil, W. (2008). *Stochastic and Integral Geometry.* Springer, Berlin Heidelberg.
Serra, J.A. (1982). *Image Analysis and Mathematical Morphology.* Academic Press, London.
Stoyan, D., Kendall, W.S., and Mecke, J. (1995). *Stochastic Geometry and its Applications.* 2nd ed., Wiley, Chichester.

PART I

NEW DEVELOPMENTS IN CLASSICAL STOCHASTIC GEOMETRY

2

RANDOM POLYTOPES

Matthias Reitzner

Dedicated to my teacher Christian Buchta on the occasion of his 50th birthday

Abstract

This chapter deals with random polytopes which are the convex hulls of random points. Recent developments concern mainly distributional aspects of functionals of random polytopes: estimates for higher moments, limit theorems, and large deviation inequalities. The geometric tools which led to this progress are the floating body and the visibility regions which are described in detail.

2.1 Introduction

The first occurrence of problems in geometric probabilities are Buffon's needle problem (1733), Sylvester's four point problem (1864), and Bertrand's paradox (1888). Among these Sylvester's problem has attracted particular interest and initiated a large number of contributions. Choose n points independently according to some distribution function in $\mathbb{R}^d$. Denote the convex hull of these points by

$$P_n = \text{conv}[X_1, \dots, X_n].$$

Sylvester asked for the distribution function of the number of vertices of P_4 in the case $d = 2$.

During the last 30 years investigations concerning random polytopes have become of increasing importance due to important applications and connections to other fields. Among the applications are the analysis of the average complexity of algorithms (cf. e.g. the books of Preparata and Shamos (1990, Ch. 4)) and Edelsbrunner (1987, Ch. 8) and optimization (cf. e.g. the book of Borgwardt (1987)), but also problems of biology (cf. e.g. the book of Solomon (1978)) and 'exotic' applications like one in archeology pointed out by Kendall (1977).

Whereas the connections to statistics (extreme points of random samples, convex hull peeling) and convex geometry (approximation of convex sets by random polytopes) are immediate, the connection to functional analysis has only been made in the last 20 years: Milman and Pajor (1989) showed that the expected volume of a (suitably defined) random simplex P_{d+1} is closely connected to the so-called isotropic constant of a convex set which is a fundamental quantity in the local theory of Banach spaces. Recent developments here are work of Klartag and Kozma (2008) and Dafnis, Giannopoulos and Gúedon (2008).

There are only a few results concerning functionals of P_n for fixed n, in particular, results for random polytopes with general distribution. Thus most results in this chapter deal with the asymptotic behaviour of functionals of random polytopes.

One notable exception is the question about the probability that a random polytope P_n contains the origin o. If the points X_i are chosen independently according to a symmetric distribution then for $P_n = \mathrm{conv}[X_1, \ldots, X_n]$ Wendel (1962) proved that

$$\mathbf{P}(o \notin P_n) = 2^{-(n-1)} \sum_{k=0}^{d-1} \binom{n-1}{k}.$$

The essential ingredient is Schläfli's formula which tells us that n linear hyperplanes in general position divide $\mathbb{R}^d$ in precisely

$$2 \sum_{k=0}^{d-1} \binom{n-1}{k}$$

parts. Recently, a generalization of Wendel's result was proved by Wagner and Welzl (2001). If points X_i are chosen independently according to an absolutely continuous distribution then

$$\mathbf{P}(o \notin P_n) \geq 2^{-(n-1)} \sum_{k=0}^{d-1} \binom{n-1}{k}$$

and thus symmetric distributions are extremal in this sense.

Notation We work in d-dimensional Euclidean space $\mathbb{R}^d$, and use the notation introduced in Chapter 1. Let $\mathcal{K}^d$ be the set of d-dimensional compact convex sets with nonempty interior, and for $K \in \mathcal{K}$ denote by ∂K the boundary of K. Let $\mathcal{P}^d \subset \mathcal{K}^d$ be the set of convex polytopes, and $\mathcal{K}^d_{sm} \subset \mathcal{K}^d$ the set of smooth convex sets. Here, a smooth convex set has a twice differentiable boundary with Gaussian curvature $\kappa(x) > 0$ for all $x \in \partial K$.

For a set $A \subset \mathbb{R}^d$, we use $|A|$ for the cardinality of A. The size of a convex set K can be measured using the intrinsic volumes $V_i(K)$, where, for instance, V_d is the volume, $2V_{d-1}$ the surface area and V_1 is a multiple of the mean width. If K is a polytope, information about its boundary structure is given by the numbers of ℓ-dimensional faces $f_\ell(K)$, $\ell = 0, \ldots, d-1$.

An important role in many questions is played by the *affine surface area* $\Omega(K)$ of a convex body K which is given by

$$\Omega(K) = \int_{\partial K} \kappa(x)^{\frac{1}{d+1}} \, \mathcal{H}_{d-1}(\mathrm{d}x),$$

see Leichtweiß (1998), and also Ludwig and Reitzner (1999).

Constants are denoted by $c, c(d), c(\cdot)$. Their values may differ from line to line. We use $\operatorname{lln} x$ as a shorthand for $\ln(\ln x)$.

2.2 Convex hull of uniform random points

In order to study the convex hull of random points, one has to make a reasonable assumption about their probability distribution. The assumption that the points should be uniformly distributed, which appears to be the most natural one, requires the restriction to a bounded set. As one studies the convex hull, one naturally tends to assume that this set should be convex, and thus one is led to work with points which are distributed uniformly in a convex body K.

Fix $K \in \mathcal{K}^d$. We will assume that $V_d(K) = 1$. Choose points $X_1, \dots, X_n$ from K, independently and according to the uniform distribution, and set $P_n = \mathrm{conv}[X_1, \dots, X_n]$. In the following, P_n always denotes a random polytope which is the convex hull of uniformly distributed points. This will change in Section 2.3, where we will be interested in other types of random polytopes.

There are only few results for general convex sets K. Efron (1965) showed, that results concerning $\mathbf{E}V_d(P_n)$ can be used to determine the expected number of vertices $\mathbf{E}f_0(P_n)$:

$$\mathbf{E}f_0(P_n) = n(1 - \mathbf{E}V_d(P_{n-1})). \tag{2.1}$$

This has been generalized by Buchta (2005) to identities between higher moments of $f_0(P_n)$ and $V_d(P_n)$. Cowan (2007) proved an identity between the variances of $V_d(P_{d+1})$, $V_d(P_{d+2})$ and the covariance between $V_d(P_{d+2})$ and $V_d(P'_{d+2})$ where P'_{d+2} is obtained from P_{d+2} by replacing one of the random points by an independent copy.

2.2.1 *Inequalities*

As mentioned above, the question to determine the expected volume $\mathbf{E}V_2(P_3)$ was raised – in a formulation equivalent to the formulation given here – by Sylvester in 1864. After several results concerning specific planar convex bodies the first result of a more general nature, but still for $d = 2$ and $n = 3$, was achieved in 1917 by Blaschke (1917), who proved sharp inequalities for $\mathbf{E}V_2(P_3)$.

To present these inequalities we use the slightly more precise notation P_n^K for the convex hull of n uniformly chosen random points in the convex set K. Denote by B, and $\triangle$ respectively, the centered ball, and the regular simplex respectively, of unit volume in $\mathbb{R}^d$. Blaschke proved that for $K \in \mathcal{K}^2$,

$$\mathbf{E}V_2(P_3^{\triangle}) \geq \mathbf{E}V_2(P_3^K) \geq \mathbf{E}V_2(P_3^B). \tag{2.2}$$

The left hand side of this inequality was generalized to

$$\mathbf{E}V_2(P_n^{\triangle}) \geq \mathbf{E}V_2(P_n^K)$$

for $K \in \mathcal{K}^2$ by Dalla and Larman (1991) and Giannopoulos (1992). Yet to prove this extremal property of the simplex in arbitrary dimensions seems to be difficult and is still an open problem. A positive solution to this problem would immediately imply a solution of the so-called *slicing problem* which is a major problem in the asymptotic theory of Banach spaces, see Milman and Pajor (1989).

The right hand side of inequality (2.2) was generalized to

$$\mathbf{E}V_d(P_n^K) \geq \mathbf{E}V_d(P_n^B)$$

for all $K \in \mathcal{K}^d$, $d \geq 2$, by Groemer (1973, 1974). A further generalization to intrinsic volumes was recently given by Hartzoulaki and Paouris (2003) who showed that for $K \in \mathcal{K}^d$ with $V_d(K) = 1$,

$$\mathbf{E}V_i(P_n^K) \geq \mathbf{E}V_i(P_n^B).$$

From the viewpoint of approximation it would be of interest to prove such an inequality for all convex sets with $V_i(K) = V_i(B)$ instead of $V_d(K) = V_d(B) = 1$, yet this seems to be open.

Note that all these volume inequalities and Efron's identity (2.1) imply corresponding inequalities for $f_0(P_n)$, e.g. it follows from Groemer's result that

$$\mathbf{E}f_0(P_n^K) \leq \mathbf{E}f_0(P_n^B)$$

for all $K \in \mathcal{K}^d$, $d \geq 2$. Similar inequalities for the number of ℓ-dimensional faces are still missing.

We conclude this section with three elementary inequalities. The first inequality says that for all $K \in \mathcal{K}^d$ and $i = 1, \ldots, d$,

$$\mathbf{E}V_i(P_n^K) \leq \mathbf{E}V_i(P_{n+1}^K).$$

This is an immediate consequence of the fact that the intrinsic volumes are monotone with respect to set inclusion. The other two inequalities – yet also innocent looking – are still resisting any kind of proof:

Is it true that

$$\mathbf{E}f_0(P_n^K) \leq \mathbf{E}f_0(P_{n+1}^K) \ ?$$

Is it true that if $K \subset L$

$$\mathbf{E}V_d(P_n^K) \leq \mathbf{E}V_d(P_n^L) \ ?$$

(Clearly, since $V_d(K) < V_d(L)$ we make an exception to the rule that random points are chosen in convex sets of unit volume.)

2.2.2 *Expectations*

Because of the difficulties to derive general and explicit formulae, the investigations focused on the asymptotic behaviour of the expected values as n tends to infinity. It is nearly impossible to state all results dealing with the asymptotic behaviour of the expectation of $V_i(P_n)$ and $f_\ell(P_n)$ as $n \to \infty$. For any convex body K and $\ell \in \{0, \ldots, d-1\}$, the expected number $\mathbf{E}f_\ell(P_n)$ of ℓ-dimensional faces of P_n tends to infinity as n tends to infinity. The shape of the boundary of K determines the order of magnitude of $\mathbf{E}f_\ell(P_n)$. Similarly, for any convex body K and $i \in \{1, \ldots, d\}$, the expected value $\mathbf{E}V_i(P_n)$ tends to $V_i(K)$ as n tends to infinity, and the shape of the boundary of K determines the order of magnitude of $V_i(K) - \mathbf{E}V_i(P_n)$, which measures the quality of the approximation of K by P_n.

The starting point were two famous articles by Rényi and Sulanke (1963, 1964) who obtained in dimension 2 the asymptotic behaviour of the expected volume $\mathbf{E}V_d(P_n)$ if the boundary of K is sufficiently smooth, and if K is a polygon. In a series of papers these formulae were generalized to higher dimensions. Due to work of Wieacker (1978), Schneider and Wieacker (1980), Bárány (1992, 2004), and Reitzner (2004), for $i = 1, \ldots, d$,

$$V_i(K) - \mathbf{E}V_i(P_n) = c_i(K) n^{-\frac{2}{d+1}} + o(n^{-\frac{2}{d+1}}) \tag{2.3}$$

if $K \in \mathcal{K}^d_{sm}$ where in many cases more precise error terms are known. The coefficient $c_i(K)$ is known precisely for $i = d$ where $c_d(K)$ is an explicit constant c_d times the affine surface area $\Omega(K)$.

The corresponding results for polytopes are known only for $i = 1$ and $i = d$. In a long and intricate proof Bárány and Buchta (1993) settled the case $K \in \mathcal{P}^d$.

$$V_d(K) - \mathbf{E}V_d(P_n) = \frac{T(K)}{(d+1)^{d-1}(d-1)!} n^{-1} \ln^{d-1} n + \mathcal{O}(n^{-1} \ln^{d-2} n \operatorname{lln} n),$$

where $T(K)$ is the number of flags of the polytope K. A *flag* is a sequence of i-dimensional faces F_i of K, $i = 0, \ldots, d-1$, such that $F_i \subset F_{i+1}$. For $i = 1$, Buchta (1984) and Schneider (1987) showed that

$$V_1(K) - \mathbf{E}V_1(P_n) = c(K) n^{-\frac{1}{d}} + o\left(n^{-\frac{1}{d}}\right).$$

Surprisingly, the cases $2 \leq i \leq d-1$ are still open. It is conjectured (see Bárány 1989) that

$$V_i(K) - \mathbf{E}V_i(P_n) = c_i(K)\, n^{-\frac{1}{d-i+1}} + \cdots .$$

Bárány (1989) proved that at least the order $n^{-1/(d-i+1)}$ is correct. He deduced this from the behaviour of the *floating body* of a convex set K which we introduce in the next section.

Due to Efron's identity (2.1) the results concerning $\mathbf{E}V_d(P_n)$ can be used to determine the expected number of vertices of P_n. In Reitzner (2005*b*) these results for $\mathbf{E}f_0(P_n)$ were generalized to arbitrary $\ell \in \{0, \ldots, d-1\}$: if $K \in \mathcal{K}^d_{sm}$, then

$$\mathbf{E}f_\ell(P_n) = c(d, \ell)\, \Omega(K)\, n^{\frac{d-1}{d+1}} + o\left(n^{\frac{d-1}{d+1}}\right), \tag{2.4}$$

and if $K \in \mathcal{P}^d$,

$$\mathbf{E}f_\ell(P_n) = c(d, \ell)\, T(K)\, \ln^{d-1} n + o(\ln^{d-1} n). \tag{2.5}$$

Further functionals of interest, which will not be investigated in this contribution, are the Hausdorff distance between K and P_n, see Bingham, Bräker and

Hsing (1998) and Dümbgen and Walther (1996). Also the difference between the diameters of K and P_n has been investigated recently, see Lao (2006), Mayer and Molchanov (2007), Mayer (2008), and Lao and Mayer (2008).

2.2.3 *The floating body*

Let $K \in \mathcal{K}^d$ be a convex set. The intersection of K with a halfspace H is called a *cap* of K. We define the function $v : K \to \mathbb{R}$ which will play an important role throughout:

$$v(z) = \min\{V_d(K \cap H) : H \text{ is a halfspace and } z \in H\}.$$

The *floating body* with parameter t is just the level set

$$K(v \geq t) = \{z \in K : v(z) \geq t\},$$

which is convex. The *wet part* is $K(v \leq t)$, that is, where v is at most t. The name comes from the three-dimensional picture when K is a box containing t units of water. We will use the notation $K(t_1 \leq v < t_2)$ for the set $\{z \in K : t_1 \leq v(z) < t_2\}$.

That this notion is of importance for random polytopes was first observed by Bárány and Larman ((1988)) who proved that P_n is close to $K(v \leq n^{-1})$ in the following sense:

Theorem 2.1 *Let $K \in \mathcal{K}^d$. There are constants $c, c(d), N(d)$ such that*

$$c\, V_d(K(v \leq n^{-1})) \leq V_d(K) - \mathbf{E}(V_d(P_n)) \leq c(d) V_d(K(v \leq n^{-1}))$$

for $n \geq N(d)$.

As an essential tool Bárány and Larman developed and used the so-called *economic cap covering* which we state here in a form which is convenient for our purposes.

Let $K \in \mathcal{K}^d$ with $V_d(K) = 1$. There is a covering of K with sets $\bar{S}_i(t)$ with pairwise disjoint interiors, and caps $K_i(t)$, $i = 1, \ldots, m(t)$, with the following properties: For all $t \in (0, (2d)^{-2d}]$, we have

(P1) $\bar{S}_i(t) \cap K(v \leq t) \subset K_i(t)$,

(P2) $(6d)^{-d} t \leq V_d(\bar{S}_i(t) \cap K(v \leq t)) \leq V_d(K_i(t)) \leq 6^d t$, $i = 1, \ldots, m(t)$,

(P3) $V_d(\bar{S}_i(t) \cap K(v \leq d6^d t)) \leq (3\,6^4\, d^3)^d t$.

Further

$$\frac{1}{6^d t}\, V_d(K(v \leq t)) \leq m(t) \leq \frac{1}{(6d)^{-d} t}\, V_d(K(v \leq t)). \tag{2.6}$$

The upper bound for $m(t)$ follows since the sets $\bar{S}_i(t) \cap K(v \leq t)$ have pairwise disjoint interior, all of them have volume $\geq (6d)^{-d} t$ and are contained in $K(v \leq t)$. The lower bound for $m(t)$ comes from the fact that by (P1) the sets $K_i(t)$ cover $K(v \leq t)$ (forming the economic cap covering of $K(v \leq t)$), and all of them have volume $\leq 6^d t$.

We make this precise in the two important cases $K \in \mathcal{K}^d_{sm}$ and $K \in \mathcal{P}^d$. For $K \in \mathcal{K}^d_{sm}$, it is well known that

$$V_d(K(v \le t)) = \Omega(K) t^{\frac{2}{d+1}} + o\left(t^{\frac{2}{d+1}}\right), \quad \text{as } t \to 0.$$

This even holds for general convex bodies $K \in \mathcal{K}^d$ as was proved by Schütt and Werner (1990). From (2.6) we obtain

$$m(t) \le c(K) t^{-\frac{d-1}{d+1}}$$

for t sufficiently small. If $K \in \mathcal{P}^d$ it was proved by Schütt (1991) and Bárány and Buchta (1993) that

$$V_d(K(v \le t)) = c_d T(K) t \ln^{d-1}(t^{-1}) + o\left(t \ln^{d-1}(t^{-1})\right) \le c(K) t \ln^{d-1}(t^{-1}) \tag{2.7}$$

and according to (2.6) we have

$$m(t) \le c(K) \ln^{d-1}(t^{-1}).$$

It can be proved (see e.g. Bárány and Reitzner (2008)) that there are sets $K_i'(t) \subset \bar{S}_i(t)$ of volume at least $(6d)^{-d}t$ such that if we choose in each $K_i'(t)$ an arbitrary point x_i, we have

$$K(v \ge d6^d t) \subset \operatorname{conv}[x_1, \ldots, x_{m(t)}]. \tag{2.8}$$

Using these facts we show that ∂P_n is sandwiched between two floating bodies with high probability. This precise description of the position of the boundary of P_n will play an important role throughout this chapter.

Denote by B^c the complement of an event B. For $j = 1, \ldots, m(t)$, set

$$S_j(t) = \bar{S}_j(t) \cap K(v \le d6^d t),$$
$$Y_j(t) = \{X_1, \ldots, X_n\} \cap S_j(t),$$

and let $A_1(Y_j(t))$ be the event that $|Y_j(t) \cap K_j'(t)| \ge 1$, i.e. $K_j'(t)$ contains at least one of the random points. It follows from $(6d)^{-d}t \le V_d(K_j'(t))$, that

$$\mathbf{P}(A_1(Y_j(t))^c) = (1 - V_d(K_j'(t))^n \le \mathrm{e}^{-(6d)^{-d}nt}. \tag{2.9}$$

Define $A_1(t) = \bigcap_{j=1}^{m(t)} A_1(Y_j(t))$. The upper bound for $m(t)$ in (2.6) implies that

$$\mathbf{P}(A_1(t)^c) \le \frac{1}{(6d)^{-d}t} V_d(K(v \le t))\, \mathrm{e}^{-(6d)^{-d}nt} \tag{2.10}$$

which is small for nt sufficiently large. Together with (2.8) this shows that with high probability the boundary of P_n is sandwiched between the floating body $K(v \geq d6^d t)$ and K.

$$\mathbf{P}(K(v \geq d6^d t) \subset P_n) \geq \mathbf{P}(A_1(t)). \tag{2.11}$$

Set

$$t = t_n = (\alpha + 1)(6d)^d \frac{\ln n}{n} \tag{2.12}$$

with some $\alpha \geq 1$.

Lemma 2.2 *Choose $X_1, \ldots, X_n$ independently and uniformly in $K \in \mathcal{K}^d$. The random polytope $P_n = \mathrm{conv}[X_1, \ldots, X_n]$ satisfies*

$$K(v \geq d6^d t) \subset P_n \subset K \text{ with probability } \geq 1 - n^{-\alpha}$$

with t given by (2.12).

For polytopes we need a more refined sandwiching where the trivial statement $P_n \subset K$ is replaced by a more precise inclusion. We set

$$s = s_n = \frac{1}{n \ln^{4d^2+d-1} n}, \quad t = t_n = (4d^2 + d - 1)6^d d^d \frac{\ln\ln n}{n}. \tag{2.13}$$

Denote by $A_{2,j}(s)$ the event that $|Y_j(t) \cap K(v \leq s)| = 0$, $j = 1, \ldots, m(t)$, and set $A_2(s) = \bigcap_{j=1}^{m(t)} A_{2,j}(s)$ which is the probability that $K(v \leq s)$ contains no random point. Then

$$\mathbf{P}(A_{2,j}(s)) = (1 - V_d(S_j(t) \cap K(v \leq s)))^n \geq 1 - nV_d(S_j(t) \cap K(v \leq s)) \tag{2.14}$$

and thus

$$\mathbf{P}(A_2(s)) \geq 1 - nV_d(K(v \leq s)). \tag{2.15}$$

Combined with (2.10), (2.11), and (2.7) this shows that ∂P_n is sandwiched between $K(v \geq d6^d t)$ and $K(v \geq s)$ with high probability.

Lemma 2.3 *Choose $X_1, \ldots, X_n$ independently and uniformly in $K \in \mathcal{P}^d$. Then $P_n = \mathrm{conv}[X_1, \ldots, X_n]$ satisfies*

$$K(v \geq d6^d t) \subset P_n \subset K(v \geq s) \text{ with probability } \geq 1 - c(K) \ln^{-4d^2} n$$

with s, t given by (2.13).

The two events A_1 and A_2 defined above ensure that the points $X_1, \ldots, X_n$ do not behave in an irregular way, either far away from or too close to ∂P_n. The following event A_3 ensures in addition that they do not form clusters: Define $A_3(Y_j(t))$ to be the event that $|Y_j(t)| \leq 2nV_d(S_j(t))$, and set

$A_3(t) = \bigcap_{j=1}^{m(t)} A_3(Y_j(t))$. By the inequality $\mathbf{P}(N \geq 2np) \leq \mathrm{e}^{-np}$ for a binomial random variable with mean np, we obtain

$$\mathbf{P}(A_3(Y_j(t))^c) \leq \mathrm{e}^{-nV_d(S_j(t))} \leq \mathrm{e}^{-(6d)^{-d}nt}, \ \mathbf{P}(A_3(t)^c) \leq m(t)\,\mathrm{e}^{-(6d)^{-d}nt}. \tag{2.16}$$

For example, the event $A_1(t) \cap A_3(t)$ implies that the number of vertices of P_n in $S_j(t)$ is bounded by $2nV_d(S_j(t))$.

The following definition is crucial, and was used by Vu (2005) and Bárány and Reitzner (2008) in slightly different form: Set

$$\mathcal{V}_i(s,t) = \{x \in K : [x,z] \subset K(s \leq v \leq d6^d t) \text{ for some } z \in S_i(t)\};$$

this is the set of points that are *visible from* $S_i(t)$ *within* $K(s \leq v \leq d6^d t)$. That this notion is of importance in our investigations follows from the fact that – given the sandwiching – a random point in $S_i(t)$ can influence the shape of P_n only within $\mathcal{V}_i(s,t)$. We are interested in the size of $\mathcal{V}_i(s,t)$.

We bound $V_d(\mathcal{V}_i(s,t))$ using the number of those sets $S_j(t)$ which have nonempty intersection with $\mathcal{V}_i(s,t)$. Set $\mathcal{D}_i(s,t) = \{j : S_j(t) \cap \mathcal{V}_i(s,t) \neq \emptyset\}$, and define

$$D(s,t) = \max_{i=1,\ldots,m(t)} |\mathcal{D}_i(s,t)|. \tag{2.17}$$

Each bound on $D(s,t)$ will immediately bound $V_d(\mathcal{V}_i(s,t))$ by $(3\,6^4\,d^3)^d D(s,t)t$.

Without proof we state the following lemmata:

Lemma 2.4 *For $K \in \mathcal{K}^d_{sm}$ and all $t > 0$, there is a constant $c(K)$ depending only on K such that*

$$D(0,t) \leq c(K),$$

and thus $V_d(\mathcal{V}_i(0,t)) \leq c(K)t$.

Note that the lemma also holds for general $s < t$ since $\mathcal{V}_i(s,t) \subset \mathcal{V}_i(0,t)$. It turns out that in the case of general convex bodies it is no longer true that D is bounded as $s,t \to 0$. In particular, if K is a polytope then D depends on the combinatorial structure of K and on s and t as well.

Lemma 2.5 *For $K \in \mathcal{P}^d$ and all $t > 4s > 0$, there is a constant $c(K)$ depending on K such that*

$$D(s,t) \leq c(K) \ln^{3(d-1)}\left(\frac{t}{s}\right),$$

and thus $V_d(\mathcal{V}_i(s,t)) \leq c(K)t \ln^{3(d-1)}\left(\frac{t}{s}\right)$.

Observe that these estimates are only of interest for $s \in (t^2, t/4)$ since for $s \leq t^2$ the trivial bounds $D(s,t) \leq m(t)$ and $V_d(\mathcal{V}_i(s,t)) \leq V_d(K(v \leq d6^d t))$ turn out to be better. Thus for $s \leq t^2$ we will always replace the bounds of Lemma 2.5 by these trivial bounds.

A proof of Lemma 2.4 is given in Reitzner (2005*a*), the proof of Lemma 2.5 in Bárány and Reitzner (2008); it is difficult and above the scope of this contribution. It would be of high interest to show that Lemma 2.5 holds for all $K \in \mathcal{K}^d$. Yet, this is maybe difficult and seems to be out of reach at the moment.

The bounds in Lemma 2.5 increase for $s \to 0$. Hence in the case of polytopes a careful choice for s is needed to obtain good bounds for the sandwiching procedure and for the visibility region at the same time. Taking s, t as in (2.13) we obtain

$$D(s,t) \le c(K) \operatorname{lln}^{3(d-1)} n.$$

As mentioned above, the event $A_3(t)$ bounds the number of vertices of P_n in $S_j(t)$. To obtain a more general bound for the number of ℓ-dimensional faces assume that also $A_1(t)$ and $A_2(s)$ hold, i.e. $K(v \ge d6^d t) \subset P_n \subset K(v \ge s)$. Set $A = A_1(t) \cap A_2(s) \cap A_3(t)$. Then each face which has nonempty intersection with $S_j(t)$ is contained in $\mathcal{V}_j(s,t)$. Denote the number of faces of P_n having nonempty intersection with a set S by $f_\ell(P_n, S)$. The number of ℓ-dimensional faces of a polytope with k vertices is at most

$$\binom{k}{\ell+1} \le k^{\ell+1}.$$

Hence, the number of ℓ-dimensional faces of P_n meeting S_j is bounded by the $(\ell+1)$-th power of the number of points in $\mathcal{V}_j(s,t)$,

$$f_\ell(P_n, S_j(t)) \le (f_0(P_n, \mathcal{V}_j(s,t)))^{\ell+1} \le (2\,(3\,6^4\,d^3)^d D(s,t) n t)^{\ell+1} \tag{2.18}$$

with probability $\mathbf{P}(A)$. In the case $s = 0$ we obtain the following result.

Lemma 2.6 *Choose $X_1, \dots, X_n$ independently and uniformly in $K \in \mathcal{K}^d$. Then $P_n = \operatorname{conv}[X_1, \dots, X_n]$ satisfies $K(v \ge d6^d t) \subset P_n$ and*

$$f_\ell(P_n, S_j(t)) \le (f_0(P_n, \mathcal{V}_j(0,t)))^{\ell+1} \le (c(d) D(0,t) n t)^{\ell+1}$$

for all j with probability $\ge 1 - n^{-\alpha}$, where t is given by (2.12).

Observe that (2.18) immediately gives a weak form of (2.4) and (2.5).

2.2.4 *Higher moments*

In contrast to the large number of contributions dealing with the expectation of functionals of P_n, only very recently there has been progress on higher moments. For example, the question to determine an asymptotic formula for the variance $\operatorname{Var} V_d(P_n)$ for random points chosen uniformly in a ball seems to be out of reach. In the following we describe and prove some of the more recent results.

The first results for variances go back to work of Groeneboom (1988), Cabo and Groeneboom (1994), and Hsing (1994), who proved central limit theorems in the case that K is either the circle or a polygon. In the course of their proof they determined the asymptotic behaviour of the occurring variances. The precise coefficients in these asymptotic formulae have been determined by Finch and Hueter (2004) and Buchta (2005). Buchta obtained the precise coefficients as a consequence of his new identity for higher moments of $V_d(P_n)$ and $f_0(P_n)$ which generalizes Efron's identity (2.1):

$$\mathbf{E}V_d(P_n)^k = \mathbf{E}\prod_{i=1}^{k}\left(1 - \frac{f_0(P_{n+k})}{n+i}\right). \tag{2.19}$$

This shows that the first k moments of $f_0(P_{n+k})$ determine the first k moments of $V_d(P_n)$.

Even to obtain good estimates for variances turns out to be nontrivial. (Küfer, (1994)) gave an estimate for $\operatorname{Var} V_d(P_n)$ for balls in arbitrary dimensions, and Buchta deduced from (2.19) lower bounds for $\operatorname{Var} V_d(P_n)$ for $d = 2$, and $\operatorname{Var} f_0(P_n)$ for $d \geq 4$.

In the last years several estimates have been obtained from which the order of the variances can be deduced, see Reitzner (2003, 2005*a*, 2005*b*), Vu (2005) and Bárány and Reitzner (2008).

Theorem 2.7 *Choose $X_1, \ldots, X_n$ independently and uniformly in $K \in \mathcal{K}^d_{sm} \cup \mathcal{P}^d$. Then there are constants $\overline{c}(K), \underline{c}(K)$ such that $P_n = \operatorname{conv}[X_1, \ldots, X_n]$ satisfies*

$$\underline{c}(K) \leq \frac{\operatorname{Var} V_d(P_n)}{n^{-1} V_d(K(v \leq n^{-1}))} \leq \overline{c}(K),$$

and

$$\underline{c}(K) \leq \frac{\operatorname{Var} f_\ell(P_n)}{n V_d(K(v \leq n^{-1}))} \leq \overline{c}(K).$$

It is conjectured that these inequalities hold for general convex bodies. That the lower bounds hold in general has been proved in Bárány and Reitzner (2008), but the general upper bounds are missing.

We will not give the proof of Theorem 2.7. Here, we prove the following bound for all moments of $V_d(P_n)$ which gives for $k = 2$ a slightly weaker upper bound than that of Theorem 2.7.

Theorem 2.8 *Choose $X_1, \ldots, X_n$ independently and uniformly in $K \in \mathcal{K}^d$. Then there is a constant $c(p)$ such that $P_n = \operatorname{conv}[X_1, \ldots, X_n]$ satisfies*

$$\mathbf{E}|V_d(P_n) - \mathbf{E}V_d(P_n)|^p \leq c(p) n^{\frac{p}{2}} \left(D(0,t)t\right)^p V_d(K(v \leq d6^d t)),$$

and

$$\mathbf{E}|f_\ell(P_n) - \mathbf{E}f_\ell(P_n)|^p \leq c(p) n^{\frac{p}{2}} \left(D(0,t)^2 nt\right)^{p(\ell+1)} V_d(K(v \leq d6^d t)),$$

where t is chosen as in (2.12).

For example, if $K \in \mathcal{K}^d_{sm}$, then

$$\mathbf{E}|V_d(P_n) - \mathbf{E}V_d(P_n)|^p \leq c(p,K)\, n^{-\frac{p}{2}-\frac{2}{d+1}} \ln^{p+\frac{2}{d+1}} n\,,$$

and

$$\mathbf{E}|f_\ell(P_n) - \mathbf{E}f_\ell(P_n)|^p \leq c(p,K)\, n^{\frac{p}{2}-\frac{2}{d+1}} \ln^{p(\ell+1)+\frac{2}{d+1}} n\,.$$

These bounds follow from Lemma 2.4. Using either integral geometry or slightly more advanced methods from the economic cap covering, the method described here yields for $k = 2$ the precise orders given in Theorem 2.7. The first general estimates for higher moments have been proved by Vu (2005). For example, for $K \in \mathcal{K}^d_{sm}$, he deduced from his large deviation inequality stated in Section 2.2.6 the bounds

$$\mathbf{E}|V_d(P_n) - \mathbf{E}V_d(P_n)|^p \leq c(p,K) n^{-\frac{p}{2}\frac{d+3}{d+1}}$$

which are probably best possible up to the constant.

The appropriate tool for proving upper bounds for the variance is the Efron–Stein jackknife inequality which gives an easy to handle estimate for the variance of $V_d(P_n)$ and $f_\ell(P_n)$. Here, we use a generalization of the Efron-Stein jackknife inequality due to Rhee and Talagrand (1986) which makes use of Burkholder's inequality.

Lemma 2.9 (Rhee and Talagrand) *Let $S = S(Y_1, \dots, Y_n)$ be a real symmetric function of the independent identically distributed random vectors Y_j, $1 \leq j \leq n+2$. Set $S_i = S(Y_1, \dots, Y_{i-1}, Y_{i+1}, \dots, Y_{n+1})$, and $S_{n+1} = S$. Then*

$$\mathbf{E}|S - \mathbf{E}S|^p \leq n^{\frac{p}{2}} c_p \mathbf{E}|S_n - S_{n+1}|^p$$

for $p \geq 1$ with $c_p = (18\sqrt{q}\ p)^p$ where $p^{-1} + q^{-1} = 1$.

Notice that the elementary inequality $|x-y|^p \leq 2^{p-1}(|r-x|^p + |r-y|^p)$ gives the following: for any real symmetric function $R = R(Y_1, \dots, Y_{n+1})$, we have

$$\mathbf{E}|S - \mathbf{E}S|^p \leq 2^p n^{\frac{p}{2}} c_p \mathbf{E}|R - S|^p. \tag{2.20}$$

Proof of Theorem 2.8 Set $Z_n = V_d(P_n)$ or $Z_n = f_\ell(P_n)$. We apply inequality (2.20) to the random variables $S = Z_n$ and $R = Z_{n+1}$. Then

$$\mathbf{E}|Z_n - \mathbf{E}Z_n|^p \leq c(p) n^{\frac{p}{2}} \mathbf{E}|Z_{n+1} - Z_n|^p.$$

To get an upper bound for the moments we need to investigate the difference between Z_n and Z_{n+1}, i.e. the difference between P_n and P_{n+1}. As in (2.12) set $t = (\alpha+1)(6d)^d\, n^{-1} \ln n$, with $\alpha \geq 1$ to be fixed later, and assume that the event A holds, i.e.

$$K(v \geq d6^d t) \subset P_n, \quad f_0(P_n, \mathcal{V}_j(0,t)) \leq c(d) D(0,t) nt\,,$$

which happens with probability $1 - n^{-\alpha}$ according to Lemma 2.6. If the random point X_{n+1} is contained in P_n, then $Z_{n+1} = Z_n$. But, for $X_{n+1} \notin P_n$ (which happens with probability $V_d(K(v \le d6^d t)))$ the point X_{n+1} is contained in some $S_j(t)$ and we obtain the estimate

$$|Z_{n+1} - Z_n| \le \begin{cases} V_d(\operatorname{conv}[X_{n+1}, P_n] \setminus P_n) & \text{for } Z_n = V_d(P_n), \\ f_\ell(\operatorname{conv}[X_{n+1}, P_n], S_j(t)) + f_\ell(P_n, S_j(t)) & \text{for } Z_n = f_\ell(P_n). \end{cases}$$

For $Z_n = V_d(P_n)$, this tells us that $|Z_{n+1} - Z_n| \le V_d(\mathcal{V}_j(0,t))$ and thus

$$\mathbf{E}|Z_n - \mathbf{E}Z_n|^p \le c(p) n^{\frac{p}{2}} \left\{ V_d(K(v \le d6^d t)) \, (c(d) D(0,t)t)^p + n^{-\alpha} \right\},$$

and for $Z_n = f_\ell(P_n)$, we obtain by (2.18)

$$\mathbf{E}|Z_n - \mathbf{E}Z_n|^p \le c(p) n^{\frac{p}{2}} \left\{ V_d(K(v \le d6^d t))(c(d)D(0,t)^2 nt)^{p(\ell+1)} + n^{\ell+1-\alpha} \right\}.$$

Choosing α sufficiently large finishes the proof. □

2.2.5 *Limit theorems*

The first who succeeded in proving a limit theorem for random polytopes was Schneider (1988). He obtained in the planar case a strong law of large numbers for $V_d(P_n)$ if K is smooth. The asymptotic formulae for mean values and the bounds for the variance from the previous sections imply laws of large numbers (LLN). As an example we prove the following Corollary 2.10. Yet in most cases much stronger results can be deduced from the large deviation inequalities in Section 2.2.6.

Corollary 2.10 *Choose a sequence of random points X_i, $i \in \mathbb{N}$, in $K \in \mathcal{K}^d_{sm}$ independently and according to the uniform distribution. Then $P_n = \operatorname{conv}[X_1, \ldots, X_n]$ satisfies*

$$\lim_{n\to\infty} (V_d(K) - V_d(P_n)) n^{\frac{2}{d+1}} = c_d \Omega(K)$$

with probability one.

Proof Chebyshev's inequality together with Theorem 2.7 yields

$$\begin{aligned} p_n &= \mathbf{P}\big(|(V_d(K) - V_d(P_n)) - \mathbf{E}(V_d(K) - V_d(P_n))| n^{\frac{2}{d+1}} \ge \varepsilon\big) \\ &\le \varepsilon^{-2} n^{\frac{4}{d+1}} \operatorname{Var} V_d(P_n) \\ &\le c(K) \varepsilon^{-2} n^{-\frac{d-1}{d+1}}. \end{aligned}$$

Since for $n_k = k^4$ the sum $\sum n_k^{-(d-1)/(d+1)}$ is finite we see that the probabilities p_{n_k} are summable. By the Borel–Cantelli lemma and (2.3) this implies that

$$\lim_{k\to\infty} (V_d(K) - V_d(K_{n_k})) n_k^{\frac{2}{d+1}} = c_d \Omega(K) \tag{2.21}$$

with probability one. Since $V_d(K) - V_d(P_n)$ is decreasing in n,

$$(V_d(K) - V_d(K_{n_k}))n_{k-1}{}^{\frac{2}{d+1}} \le (V_d(K) - V_d(P_n))n^{\frac{2}{d+1}}$$
$$\le (V_d(K) - V_d(K_{n_{k-1}}))n_k{}^{\frac{2}{d+1}}$$

for $n_{k-1} \le n \le n_k$, where by definition $n_{k+1}/n_k \to 1$, and thus the subsequence limit (2.21) suffices to prove Corollary 2.10. □

The main part of this section deals with central limit theorems (CLT) for functionals of P_n. First CLT's have been proved by Groeneboom (1988) and Cabo and Groeneboom (1994) if K is a ball and $d = 2$. Hsing (1994) succeeded in proving a central limit theorem for $V_d(P_n)$ in the case $d = 2$. It seems that the methods cannot be applied to solve the problem in higher dimensions.

In the last three years a new method was developed to prove CLTs for the random variables $V_d(P_n)$ and $f_\ell(P_n)$. For smooth convex sets this was achieved in Reitzner (2005*a*) and in a paper by Vu (2006), and for polytopes in a paper by Bárány and Reitzner (2008).

Theorem 2.11 *Let $K \in \mathcal{K}^d_{sm}$ or $K \in \mathcal{P}^d$. If the random variable $Z_n = Z(P_n)$ equals either $V_d(P_n)$ or $f_\ell(P_n)$ for some $\ell \in \{0, \dots d-1\}$, then there is a constant $c(K)$ and a function $\varepsilon(n)$, tending to zero as $n \to \infty$, such that*

$$\left| \mathbf{P}\left(\frac{Z_n - \mathbf{E}Z_n}{\sqrt{\operatorname{Var} Z_n}} \le x \right) - \Phi(x) \right| \le c(K)\ \varepsilon(n).$$

The error term for smooth convex sets is

$$\varepsilon(n) = n^{-\frac{1}{2} + \frac{1}{d+1} + o(1)},$$

and for polytopes

$$\varepsilon(n) = \ln^{-\frac{d-1}{2} + o(1)} n, \quad \text{as } n \to \infty.$$

Recently, Schreiber and Yukich (2008) succeeded to prove an even stronger result, a multivariate CLT for random polytopes, using the methods explained in Chapter 4.

Proof To give an idea of the method of proof we sketch the proof for $V_d(P_n)$. As a first step we approximate P_n by a random polytope Π_n which is defined in the following way:

Let $X(n)$ be a Poisson point process in $\mathbb{R}^d$ of intensity n. The intersection of K with $X(n)$ consists of random points $X_1, \dots, X_N$. Define

$$\Pi_n = \operatorname{conv}[X(n) \cap K] = \operatorname{conv}[X_1, \dots, X_N].$$

It is not too difficult to show that the distribution of $V_d(P_n)$ is close to the distribution of $V_d(\Pi_n)$ but the question whether also their moments are close turned out to be difficult and was answered only recently by Vu (2005) for smooth convex sets and Bárány and Reitzner (2008) for polytopes.

In the case $K \in \mathcal{K}^d_{sm}$, we set $s = 0$ and t as in (2.12), and for $K \in \mathcal{P}^d$, we set s, t as in (2.13). Following the approach of Section 2.2.3 we define A to be the event that each $K'_j(t)$ contains at least one point of $X(n)$ and that $X(n) \cap K(v \leq s) = \emptyset$. This again implies that the boundary of Π_n is sandwiched between $K(v \geq d6^d t)$ and $K(v \geq s)$. The number of points in $K'_j(t)$, and $K(v \leq s)$ respectively, is Poisson distributed with parameter $nV_d(K'_j(t))$, and $nV_d(K(v \leq s))$ respectively, instead of binomial as in Section 2.2.3. Nevertheless, the methods from there can easily be applied to the present situation and yield precisely the same estimates as in Lemmata 2.2 and 2.3. A careful analysis shows that distribution function and moments of $V_d(\Pi_n)$ and $V_d(\Pi_n)|A$ are close to each other.

We omit the details and just state the following transference lemma:

If a CLT holds for $V_d(\Pi_n)|A$ then a CLT also holds for $V_d(\Pi_n)$ and $V_d(P_n)$.

The basic tool for proving the CLT is a central limit theorem with weakly dependent random variables. This weak dependence is given by the so-called *dependency graph* which is defined as follows: Let ζ_i, $i \in \mathcal{V}$, be a finite collection of random variables. The graph $\mathcal{G} = (\mathcal{V}, \mathcal{E})$ is said to be a dependency graph for ζ_i if for any pair of disjoint sets $\mathcal{W}_1, \mathcal{W}_2 \subset \mathcal{V}$ such that no edge in $\mathcal{E}$ goes between $\mathcal{W}_1$ and $\mathcal{W}_2$, the sets of random variables $\{\zeta_i : i \in \mathcal{W}_1\}$ and $\{\zeta_i : i \in \mathcal{W}_2\}$ are independent. The following central limit theorem with weak dependence is due to Rinott (1994).

Theorem 2.12 (Rinott) *Let ζ_i, $i \in \mathcal{V}$, be random variables having a dependency graph $\mathcal{G} = (\mathcal{V}, \mathcal{E})$. Set $\zeta = \sum_{i \in \mathcal{V}} \zeta_i$ and $\sigma^2(\zeta) = \operatorname{Var} \zeta$. Denote the maximal degree of $\mathcal{G}$ by D and suppose that $|\zeta_i - \mathbf{E}\zeta_i| \leq M$ almost surely. Then*

$$\left| \mathbf{P}\left(\frac{\zeta - \mathbf{E}\zeta}{\sqrt{\operatorname{Var}\zeta}} \leq x \right) - \Phi(x) \right| \leq \frac{1}{\sqrt{2\pi}} \frac{DM}{\sigma(\zeta)} + 16 \frac{|\mathcal{V}|^{\frac{1}{2}} D^{\frac{3}{2}} M^2}{\sigma^2(\zeta)} + 10 \frac{|\mathcal{V}|\, D^2 M^3}{\sigma^3(\zeta)}.$$

When using this theorem one has to define the dependency graph and prove the necessary properties. Also, one needs a lower bound on $\operatorname{Var}\zeta$.

Introduce random variables ζ_j in the following way. Define ζ_j as the missed volume in the set $S_j(t)$,

$$\zeta_j = V_d(S_j(t)) - V_d(S_j(t) \cap \Pi_n),\ j = 1, \ldots, m(t).$$

Given A we have that $K(v \geq d6^d t) \subset \Pi_n$, and thus

$$\zeta = \sum_{j=1}^{m(t)} \zeta_j = V_d(K) - V_d(\Pi_n) = 1 - V_d(\Pi_n).$$

We start to define the dependency graph $\mathcal{G} = (\mathcal{V}, \mathcal{E})$. The vertex set, $\mathcal{V}$, of the dependency graph is just $\{1, \ldots, m(t)\}$. Recall that $\mathcal{V}_i(s, t)$ is the set of points that are visible from $S_i(t)$ within $K(s \le v \le d6^d t)$. Now distinct vertices $i, j \in \mathcal{V}$ form an edge in $\mathcal{G}$ if $\mathcal{V}_i(s,t) \cap \mathcal{V}_j(s,t) \neq \emptyset$. This defines a dependency graph: Since ζ_i is determined by the facets of Π_n with at least one vertex in $S_i(t)$, it is determined by those facets which are contained in $\mathcal{V}_i(s,t)$ and thus determined by $X(n) \cap \mathcal{V}_i(s,t)$. As $X(n)$ is a Poisson point process, $X(n) \cap \mathcal{V}_i(s,t)$ and $X(n) \cap \mathcal{V}_j(s,t)$ and thus ζ_i, ζ_j are independent if there is no edge between i, j, i.e. if $\mathcal{V}_i(s,t) \cap \mathcal{V}_j(s,t) = \emptyset$.

In order to prove the CLT for $V_d(\Pi_n)|A$ we check the conditions of Rinott's theorem. We start with the upper bound on the maximal degree D. If $\mathcal{V}_i(s,t)$ meets $\mathcal{V}_j(s,t)$ then there is a set $S_k(t)$ which meets their intersection, and thus $S_i(t), S_j(t)$ also meet $\mathcal{V}_k(s,t)$. Hence, $\mathcal{V}_i(s,t)$ meets $\mathcal{V}_j(s,t)$, if $\mathcal{V}_i(s,t) \cap S_k(t) \neq \emptyset$ and $\mathcal{V}_k(s,t) \cap S_j(t) \neq \emptyset$. Recall that $\mathcal{D}_i(s,t)$ is the number of sets $S_k(t)$ which meet $\mathcal{V}_i(s,t)$ and that $D(s,t) = \max \mathcal{D}_i(s,t)$. We obtain

$$D \le \max_{i,k} \mathcal{D}_i(s,t)\mathcal{D}_k(s,t) \le D(s,t)^2$$

and use the bounds from Section 2.2.3.

We have to check the other two conditions of Rinott's theorem. Since $\zeta_j \le V_d(S_j(t))$ we see that under condition A,

$$M = \max \zeta_j \le c(d)t.$$

Also, since the variance of $V_d(P_n)$ is close to the variance of $V_d(\Pi_n)|A$ it can be deduced from (2.7) that

$$\mathrm{Var}\,(V_d(\Pi_n)|A) \ge c(K)n^{-1}V_d(K(v \le n^{-1})).$$

These bounds on $|\mathcal{V}| = m(t)$, D, ζ_j and $\mathrm{Var}\,\zeta = \mathrm{Var}\,(V_d(\Pi_n)|A)$ together with Rinott's theorem yield the CLT for $V_d(\Pi_n)|A$ and thus also Theorem 2.11. □

2.2.6 *Large deviation inequalities*

Maybe the best known large deviation inequality which seems to fit into our investigations is Azuma's inequality for martingale differences in connection with Doob's construction of a martingale out of a random variable. Let $Z = Z(Y)$ be a function of m random variables, $Y = (Y_1, \dots, Y_m)$. Denote by

$$d_i(Y_1, \dots, Y_i) = \mathbf{E}(Z|Y_1, \dots, Y_i) - \mathbf{E}(Z|Y_1, \dots, Y_{i-1})$$

the martingale differences. If $|d_i(Y)| \le w$ a.s. then the Hoeffding–Azuma inequality tells us that

$$\mathbf{P}(|Z(Y) - \mathbf{E}Z(Y)| \ge x) \le 2\,\mathrm{e}^{-x^2/(2mw^2)}.$$

Here, we would like to partition K into sets S_i and take $Y_i = \{X_1, \dots, X_n\} \cap S_i$ and $Z(Y) = V_d(P_n)$. Yet, in this case we would have $|d_i| \le 1$ which results only in trivial bounds.

We apply the following version of the Hoeffding–Azuma inequality due to Chalker, Godbole, Hitczenko, Radcliff and Ruehr (1999).

Lemma 2.13 (Chalker, *et al.*) *Let d_i be a martingale difference sequence with $\sum_1^m d_i = Z(Y) - \mathbf{E}Z(Y)$. For all $x, w > 0$ we have*

$$\mathbf{P}(|Z(Y) - \mathbf{E}Z(Y)| > x) \le 2\,\mathrm{e}^{-x^2/(32mw^2)} + \left(1 + \frac{\|d^*\|_\infty}{4\sqrt{mw^2}}\right) \sum_{i=1}^{m} \mathbf{P}(|d_i| > w)$$

where $\|d^*\|_\infty = \sup_i \|d_i\|_\infty$.

Fix t, set $S_{m(t)+1} = K(v \ge d6^d t)$, and observe that the sets $S_i(t)$, $i = 1, \dots, m(t)+1$, form a covering of K with pairwise disjoint interiors. As suggested above set

$$Y_i = \{X_1, \dots, X_n\} \cap S_i(t)$$

for $i = 1, \dots, m(t) + 1$, and $Y = (Y_1, \dots, Y_{m(t)+1})$. We apply the tail inequality of Lemma 2.13 to the function $Z(Y) = Z_n(Y) = f_\ell(P_n)$. To obtain bounds on $|d_i|$ which hold with high probability, we need a coupling argument.

Fix $Y_1, \dots, Y_i$, and define $N_{i,n} = n - \sum_{j=1}^{i-1} |Y_j|$. Assume that Y_i' is an independent copy of Y_i, i.e. choose $N_{i,n}$ points $X_1', \dots, X_{N_{i,n}}'$, independently and uniformly in $\bigcup_{j=i}^{m(t)+1} S_j(t)$, and set

$$Y_i' = \{X_1', \dots, X_{N_{i,n}}'\} \cap S_i(t)\,.$$

Let $|Y_i| = k$, $|Y_i'| = k'$, $0 \le k, k' \le n$ and $l = k - k'$ and assume $l \ge 0$ without loss of generality.

Having thus chosen $Y_1, \dots, Y_i, Y_i'$ we proceed by choosing $N_{i+1,n} = N_{i,n} - k$ points $\tilde{X}_1, \dots, \tilde{X}_{N_{i+1,n}}$, independently and uniformly in $\bigcup_{j=i+1}^{m(t)+1} S_j(t)$. This yields random variables

$$\tilde{Y}_j = \{\tilde{X}_1, \dots, \tilde{X}_{N_{i+1,n}}\} \cap S_j(t),\ j = i+1, \dots, m(t) + 1\,.$$

Distribute $l = k - k'$ further points uniformly in $\bigcup_{j=i+1}^{m(t)+1} S_j(t)$. This yields new random variables $\tilde{Y}_j'$ consisting of the points already in $\tilde{Y}_j$ and the new points falling in $S_j(t)$. Denote by l' the number of points that fall into some $S_j(t)$, $i + 1 \le j \le m(t)$. For $k - k' < 0$, we interchange the roles of $Y_i, \tilde{Y}_{i+1}, \dots$ and $Y_i', \tilde{Y}_{i+1}', \dots$. Then

$$\begin{aligned} d_i = \mathbf{E}\big[&Z_n(Y_1, \dots, Y_i, \tilde{Y}_{i+1}, \dots, \tilde{Y}_{m(t)+1}) \\ &-Z_n(Y_1, \dots, Y_{i-1}, Y_i', \tilde{Y}_{i+1}', \dots, \tilde{Y}_{m(t)+1}') \mid Y_1, \dots, Y_i\big]. \end{aligned}$$

Set again $t = (\alpha + 1)(6d)^d n^{-1} \ln n$ with $\alpha \ge 1$ as in (2.12). Let $A(Y_j)$ be the event that $K_j'(t)$ contains at least one random point and that $|Y_j| \le 2nV_d(S_j(t))$.

Given $A(Y_1), \dots A(Y_i)$, $A(Y_i')$ and $A(\tilde{Y}_{i+1})$, $\dots$,$A(\tilde{Y}_{m(t)})$, l' new points are in some $S_j(t)$, changing the shape of P_n in at most l' sets $S_j(t)$ in addition to the change in $S_i(t)$. Thus we have (see Lemma 2.6)

$$|Z_n(Y_1, \dots, Y_i, \tilde{Y}_{i+1} \dots, \tilde{Y}_{m+1}) - Z_n(Y_1, \dots, Y_i', \tilde{Y}_{i+1}' \dots, \tilde{Y}_{m+1}')| \\ \leq (l' + 1)\,(c(d)nD(0,t)t + l')^{\ell+1}.$$

Let p denote the probability that a point chosen uniformly in $\bigcup_{j=i+1}^{m(t)+1} S_j(t)$ falls into $\bigcup_{j=i+1}^{m(t)} S_j(t)$. Clearly

$$p \leq \frac{V_d(K(v \leq d6^d t))}{V_d(K(v \geq d6^d t))}.$$

Elementary estimates for the binomial distributed variable l' show that

$$\mathbf{E}(l' + 1)\,(c(d)nD(0,t)t + l')^{\ell+1} \leq (2l^2 p + 1)\,(c(d)nD(0,t)t + l)^{\ell+1} \\ \leq c(d)\,(nD(0,t)t)^{\ell+1}$$

since $l \leq 2nV_d(S_i(t)) \leq c(d)nt$ given $A(Y_i), A(Y_i')$, and $t \leq n^{-1} \ln n$.
Thus, given $A(Y_1), \dots, A(Y_i)$, we obtain

$$|\mathbf{E}(Z_n(Y_1, \dots, Y_{i-1}, Y_i, \dots, Y_{m+1}) - Z_n(Y_1, \dots, Y_{i-1}, Y_i', \dots, Y_{m+1}')|\, Y_1, \dots, Y_i)| \\ \leq c(d)\,(nD(0,t)t)^{\ell+1} + n^{\ell+1} \left[\mathbf{P}(A(Y_i')^c | Y_1, \dots, Y_i) + \sum_{i+1}^{m(t)} \mathbf{P}(A(\tilde{Y}_j)^c | Y_1, \dots, Y_i) \right].$$

Given $A(Y_1), \dots, A(Y_i)$ we have that $N_{i,n}, N_{i+1,n} \geq n/2$, and using the results from Section 2.2.3

$$\mathbf{P}(A(Y_i')^c | Y_1, \dots, Y_i) + \sum_{i+1}^{m(t)} \mathbf{P}(A(\tilde{Y}_j)^c | Y_1, \dots, Y_i) \leq 2m(t)\, \mathrm{e}^{-(6d)^{-d} tn/2} \\ \leq t^{\ell+1}$$

by the definition of t for $\alpha \geq 2\ell + 3$.

Combining our results we obtain

$$|d_i| \leq c(d)(nD(0,t)t)^{\ell+1} \tag{2.22}$$

given $A = A(Y_1) \cap \dots \cap A(Y_m)$, which finally leads to

$$\mathbf{P}(|d_i| \geq c(d)(nD(0,t)t)^{\ell+1}) \leq \mathbf{P}(A^c)$$

for $i = 1, \dots, m(t)$. It is immediate that $\mathbf{P}(|d_{m(t)+1}| \neq 0) \leq \mathbf{P}(A^c)$.

Choose $w = w_i = c(d)(nD(0,t)t)^{\ell+1}$, $i = 1, \ldots, m(t)$, and $w_{m(t)+1} = 0$. Combining (2.22) with Lemma 2.13 we get the following:

$$\mathbf{P}(|Z_n(Y) - \mathbf{E}Z_n(Y)| > x) \leq 2\,\mathrm{e}^{-x^2/(m(t)w^2)} + 2\left(1 + \frac{n^{\ell+1}}{4\sqrt{(m(t)+1)w^2}}\right)(m(t)+1)\mathbf{P}(A^c).$$

We make this precise for $K \in \mathcal{K}^d_{sm}$ and $K \in \mathcal{P}^d$ where the order of the variance is known.

Theorem 2.14 *For $\alpha \geq 1$ there is a constant $c(K,\alpha) = \alpha^{-2(\ell+1)}c(K)$ such that*

$$\mathbf{P}\left(\left|\frac{f_\ell(P_n) - \mathbf{E}f_\ell(P_n)}{\sqrt{\operatorname{Var} f_\ell(P_n)}}\right| \geq x\right) \leq 2\,\mathrm{e}^{-c(K,\alpha)x^2 \ln^{-2(\ell+1)} n} + n^{-\alpha+\ell+2}$$

for $K \in \mathcal{K}^d_{sm}$. And for $K \in \mathcal{P}^d$ we have

$$\mathbf{P}\left(\left|\frac{f_\ell(P_n) - \mathbf{E}f_\ell(P_n)}{\sqrt{\operatorname{Var} f_\ell(P_n)}}\right| \geq x\right) \leq 2\,\mathrm{e}^{-c(K,\alpha)x^2 \ln^{-2d(\ell+1)} n} + n^{-\alpha+\ell+1}.$$

For $Z_n(Y) = V_d(P_n)$ the same method with some straightforward modifications gives the following.

Theorem 2.15 *For $\alpha \geq 1$ there is a constant $c(K,\alpha) = \alpha^{-2}c(K)$ such that*

$$\mathbf{P}\left(\left|\frac{V_d(P_n) - \mathbf{E}V_d(P_n)}{\sqrt{\operatorname{Var} V_d(P_n)}}\right| \geq x\right) \leq 2\,\mathrm{e}^{-c(K,\alpha)x^2 \ln^{-2} n} + n^{-\alpha+2}$$

for $K \in \mathcal{K}^d_{sm}$. And for $K \in \mathcal{P}^d$ we have

$$\mathbf{P}\left(\left|\frac{V_d(P_n) - \mathbf{E}V_d(P_n)}{\sqrt{\operatorname{Var} V_d(P_n)}}\right| \geq x\right) \leq 2\,\mathrm{e}^{-c(K,\alpha)x^2 \ln^{-2d} n} + n^{-\alpha+1}.$$

For $K \in \mathcal{P}^d$, we could also set s, t as in (2.13) and use the refined estimates from Lemma 2.5, to get a slightly better exponential term. Yet – since there is a tradeoff between the exponent and the error term – this would yield a larger error term $\mathbf{P}(A^c)$ from Lemma 2.3.

In view of the CLT proved in Section 2.2.5 it should be possible to remove the logarithmic terms in the exponent of the large deviation inequality, and the best possible error term probably should be exponential in a power of n instead of $n^{-\alpha}$.

A sharpening of Theorem 2.15 was proved by Vu (2005) using a different method. His paper was the first proof of a large deviation inequality for random polytopes. Most probably his results are best possible for $K \in \mathcal{K}^d_{sm}$. Set

$$S^2(Y) = \sum_i \mathbf{E}_{Y_i} d_i^2(Y_1, \dots, Y_i), \tag{2.23}$$

note that $\mathbf{E}S^2(Y) = \operatorname{Var} Z(Y)$, and put $W(Y) = \max_i(\sup_{Y_i} |d_i(Y)|)$. The main step is to use a concentration inequality which can be stated in the following form: For $x \le S \operatorname{Var} Z(Y)/(2W)$, we have

$$\mathbf{P}(|Z(Y) - \mathbf{E}Z(Y)| \ge x) \le 2\,\mathrm{e}^{-x^2/(4S\operatorname{Var} Z(Y))} \\ +\mathbf{P}(S^2(Y) \ge S \operatorname{Var} Z(Y) \text{ or } W(Y) \ge W)\,.$$

The advantage in this deviation inequality is to replace the l_∞-estimate (2.22) by the l_2-norm (2.23). In the case that K is smooth and $Z(Y) = V_d(P_n)$, Vu obtains the following precise bound:

$$\mathbf{P}(S^2(Y) \ge S \operatorname{Var} Z(Y) \text{ or } W(Y) \ge W) \le \mathrm{e}^{-cn^{\frac{d-1}{3d+5}}}\,,$$

where $W = c(K)n^{-1-(d-1)/(3d+5)}$ and $S = c'(K)$, with sufficiently large $c(K)$, $c'(K)$.

Theorem 2.16 (Vu) *For $K \in \mathcal{K}^d_{sm}$ and $Z_n = V_d(P_n)$ or $Z_n = f_0(P_n)$, there are constants $c_i(K)$ such that*

$$\mathbf{P}\left(\left|\frac{Z_n - \mathbf{E}Z_n}{\operatorname{Var} Z_n}\right| \ge x\right) \le 2\,\mathrm{e}^{-c_1(K)x^2} + \mathrm{e}^{-c_2(K)n^{\frac{d-1}{3d+5}}}$$

for $x \le c_3(K)n^{-\frac{(d+3)^2}{(d+1)(3d+5)}}$.

2.2.7 *Convex hull peeling*

We introduce the following more refined definition: Set $P_{n,1} = P_n$ and denote by $\mathcal{X}_1$ the set of all vertices of $P_{n,1}$. We remove from our random sample all points in $\mathcal{X}_1$ and take the convex hull of the remaining points:

$$P_{n,2} = \operatorname{conv}[\{X_1, \dots X_n\} \backslash \mathcal{X}_1]\,,$$

and then we proceed in this way: denote by $\mathcal{X}_i$ the vertices of $P_{n,i}$ and set

$$P_{n,i+1} = \operatorname{conv}[\{X_1, \dots X_n\} \backslash (\mathcal{X}_1 \cup \dots \cup \mathcal{X}_i)]$$

until this set is empty. This operation is called *convex hull peeling*. Its importance for multivariate statistics and data depth is described in Chapter 12.

All results above deal with expectations and distributional aspects of $|\mathcal{X}_1|$ as $n \to \infty$, e.g. it follows from the definition of the event $A_3(t)$, see formula (2.16) and Lemma 2.6, that

$$|\mathcal{X}_1| \leq 2nK(v \leq d6^d t) \tag{2.24}$$

with probability $\geq 1 - n^{-\alpha}$, where t is given in (2.12).

Analogous investigations concerning expectations and deviation inequalities for $|\mathcal{X}_i|$, $i \geq 2$, are unknown, for example generalizing (2.24) to $i \geq 2$.

2.3 Gaussian polytopes

Instead of random points uniformly distributed in a convex set, we investigate in this section Gaussian samples $X_1, \ldots, X_n$ in $\mathbb{R}^d$, i.e. independent random points chosen according to the d-dimensional standard normal distribution with density $\varphi_d = (2\pi)^{-d/2} \mathrm{e}^{-\|x\|^2/2}$. The convex hull $P_n = \operatorname{conv}[X_1, \ldots, X_n]$ of these random points is called a *Gaussian polytope*. The recent progress on distributional aspects of the random polytopes described for the uniform model in the sections above led also to new developments on geometric functionals (such as volume, intrinsic volumes, and the number of i-dimensional faces) for Gaussian polytopes.

First results are due to Rényi and Sulanke (1963) who determined the asymptotic behaviour of the expected number of vertices in the plane, and to Raynaud (1970) who investigated the asymptotic behaviour of the mean number of facets in arbitrary dimensions. Both results are only special cases of the formula

$$\mathbf{E} f_\ell(P_n) = \frac{2^d}{\sqrt{d}} \binom{d}{\ell+1} \beta_{\ell,d-1} (\pi \ln n)^{\frac{d-1}{2}} \, (1 + o(1)), \quad \text{as } n \to \infty, \tag{2.25}$$

where $\ell \in \{0, \ldots, d-1\}$ and $d \in \mathbb{N}$. This follows, in arbitrary dimensions, from work of Affentranger and Schneider (1992) and Baryshnikov and Vitale (1994). Here, $\beta_{\ell,d-1}$ is the internal angle of a regular $(d-1)$-simplex at one of its ℓ-dimensional faces. Recently, a more direct proof of (2.25) was given by Hug, Munsonius and Reitzner (2004). The expected values of intrinsic volumes were investigated by Affentranger (1991) who proved that

$$\mathbf{E} V_i(P_n) = \binom{d}{i} \frac{\kappa_d}{\kappa_{d-i}} (2 \ln n)^{\frac{i}{2}} \, (1 + o(1)), \quad \text{as } n \to \infty, \tag{2.26}$$

for $i \in \{1, \ldots, d\}$. Relation (2.26) is related to the result of Geffroy (1961) that the Hausdorff distance between P_n and the d-dimensional ball of radius $\sqrt{2} \ln^{1/2} n$ and centre at the origin converges almost surely to zero.

Using the Efron–Stein jackknife inequality upper bounds for the variance have been given by Hug and Reitzner (2005). A matching lower bound was recently proved by Bárány and Vu (2007). There exist positive constants $\underline{c}(d), \overline{c}(d)$, depending only on the dimension, such that for all $\ell \in \{0, \dots, d-1\}$,

$$\underline{c}(d) \leq \operatorname{Var} f_\ell(P_n)(\ln n)^{-\frac{d-1}{2}} \leq \overline{c}(d) \,. \tag{2.27}$$

A precise asymptotic formula for $\operatorname{Var} f_0(P_n)$ was given by Hueter (1994, 1999). Analogously there exists a positive constant $\overline{c}_d$, depending only on the dimension, such that

$$\operatorname{Var} V_i(P_n) \leq \overline{c}_d (\ln n)^{\frac{i-3}{2}} \tag{2.28}$$

for all $i \in \{1, \dots, d\}$. Yet, it seems that these upper bounds are not best possible for $i \leq d-1$. For $i = d$, a matching lower bound was proved by Bárány and Vu (2007).

In their paper Bárány and Vu proved a CLT for $Z = f_\ell(P_n)$ or $Z = V_d(P_n)$:

$$\left| \mathbf{P}\left(\frac{Z - \mathbf{E}Z}{\sqrt{\operatorname{Var} Z}} \leq x \right) - \Phi(x) \right| \leq c_d \ln^{-\frac{d-1}{4} + o(1)} n, \quad \text{as } n \to \infty.$$

The proof is based again on sandwiching ∂P_n using as a 'floating body', a ball of radius $\sqrt{2} \ln^{1/2} n$. For $T^2 = 2 \ln n - (\operatorname{lln} n + 2 \operatorname{llln} \mathrm{n} + \alpha)$ and $s^2 = 2 \ln n + \beta \operatorname{lln} n$ with α, β sufficiently large, we have

$$B(o, T) \subset P_n \subset B(o, s) \text{ with probability } \geq 1 - \ln^{-c} n$$

with some $c = c(\alpha, \beta, d)$.

2.3.1 *Projections of high-dimensional simplices*

We want to give two interpretations of the results of the last section. The first one uses the fact that any orthogonal projection of a Gaussian sample again is a Gaussian sample. So we make our notation more precise by writing $P_n^{(d)}$ for a Gaussian polytope in $\mathbb{R}^d$. Denote by Π_i the projection to the first i components. Then

$$\varphi(\Pi_i P_n^{(d)}) \overset{d}{=} \varphi(P_n^{(i)}), \tag{2.29}$$

where $\overset{d}{=}$ means equality in distribution and φ is any (measurable) functional on the convex polytopes.

Now let $P_{n+1}^{(n)}$ be a Gaussian simplex in $\mathbb{R}^n$. As a consequence of the results above one obtains for example a law of large numbers for projections of high-dimensional random simplices, see Hug and Reitzner (2005). For $d \geq 1$,

$$V_d(\Pi_d P_{n+1}^{(n)}) \, (\ln n)^{-\frac{d}{2}} \to \kappa_d \,,$$

and

$$f_\ell(\Pi_d P_{n+1}^{(n)}) \, (\ln n)^{-\frac{d-1}{2}} \to \frac{2^d}{\sqrt{d}} \binom{d}{\ell+1} \beta_{\ell, d-1} \pi^{\frac{d-1}{2}} \tag{2.30}$$

in probability, as n tends to infinity.

Another method of generating $n+1$ random points in $\mathbb{R}^d$ goes back to a suggestion of Goodman and Pollack. Let R denote a random rotation of $\mathbb{R}^n$, put $\Pi_d^* = \Pi_d \circ R$, and denote by $T^{(n)}$ the regular simplex in $\mathbb{R}^n$. Then $\Pi_d^*(T^{(n)})$ is a random polytope in $\mathbb{R}^d$ in the *Goodman–Pollack model*. It was proved by Baryshnikov and Vitale (1994) that

$$\varphi\left(\Pi_d^* T^{(n)}\right) \stackrel{d}{=} \varphi\left(P_{n+1}^{(d)}\right), \tag{2.31}$$

for any affine invariant (measurable) functional φ on the convex polytopes. Thus, if f_ℓ denotes the number of ℓ-faces, (2.25) is equivalent to

$$\mathbf{E} f_\ell\left(\Pi_d^* T^{(n)}\right) = \frac{2^d}{\sqrt{d}}\binom{d}{\ell+1}\beta_{\ell,d-1}\left(\pi \ln n\right)^{\frac{d-1}{2}}\left(1+o(1)\right), \quad \text{as } n \to \infty,$$

which is what was actually proved by Affentranger and Schneider (1992). By (2.31), the law of large numbers (2.30) can be rewritten for $\Pi_d^* T^{(n)}$,

$$f_\ell\left(\Pi_d^* T^{(n)}\right)(\ln n)^{-\frac{d-1}{2}} \to \frac{2^d}{\sqrt{d}}\binom{d}{\ell+1}\beta_{\ell,d-1}\,\pi^{\frac{d-1}{2}}$$

in probability, as n tends to infinity.

Recent interest in the 'Goodman–Pollack model' comes from applications in coding theory. It turns out to be of interest to determine the probability that $\Pi_d^* T^{(n)}$ is a k-neighbourly polytope which means that the convex hull of the projection of k arbitrary vertices of $T^{(n)}$ always form a $(k-1)$-dimensional face of $\Pi_d^* T^{(n)}$. Here, k and d should be chosen proportional to n (and not fixed) as $n \to \infty$. First results in this line have been proved by Vershik and Sporyshev (1992), and recent investigations by Donoho and Tanner (2005*a*, 2005*b*) led to new insight in this problem. The main point is to describe precisely the behaviour of $k = k_n$ and $d = d_n$ for which $\Pi_d^* T^{(n)}$ is a k-neighbourly polytope with high probability.

For these applications in coding theory analogous results for random projections of the regular cross-polytope are even more interesting. The expected numbers of ℓ-dimensional faces of a randomly projected crosspolytope have been computed by Böröczky and Henk (1999), for results on k-neighbourly projections see the recent work of Donoho (2006), Candes and Tao (2005), Rudelson and Vershynin (2005), and Candes, Rudelson, Tao, and Vershynin (2005). For an overview of these results relating k-neighbourly projections of polytopes to applications see Donoho and Tanner (2008).

2.4 Random points on convex surfaces

Choose random points $X_1, \ldots, X_n$ according to a density function concentrated on the boundary of a given convex body $K \in \mathcal{K}^d$. Let P_n be the convex hull of $X_1, \ldots, X_n$. If K is a strict convex set, then clearly $f_0(P_n) = n$. Since the random polytope P_n now is forced to be close to the boundary of K, the rate of approximation of $V_(K)$ by $V_i(P_n)$ should be better than $n^{-2/(d+1)}$, the rate obtained for random points chosen in the interior of K.

It turns out that the rate of convergence is of order $n^{-2/(d-1)}$,

$$V_i(K) - \mathbf{E}V_i(P_n) = c_i(K)\, n^{-\frac{2}{d-1}} + o\left(n^{-\frac{2}{d-1}}\right), \quad \text{as } n \to \infty. \tag{2.32}$$

This follows from work of Buchta, Müller and Tichy (1985), Müller (1989), Gruber (1996) and Reitzner (2002). Recently, the case $i = d$ was generalized by Schütt and Werner (2000) to arbitrary convex bodies $K \in \mathcal{K}$ only fulfilling some weak regularity conditions on the boundary of K. Apart from results of Schneider (1988) and Reitzner (2003), no distributional results are known.

As already mentioned $f_0(P_n) = n$, and it seems to be natural to conjecture that $\mathbf{E}f_\ell(P_n) = c_\ell(K)n + o(n)$ as $n \to \infty$ but this seems to be open.

2.5 Convex hull of 0-1-polytopes

It would of interest to choose random points on the boundary of a convex polytope, in particular, from the set of vertices. The only case which has been considered so far is the case of 0-1-polytopes, where the points are chosen independently and uniformly from the vertices of the cube $C^d = [0,1]^d$. Thus each X_j consists of d Bernoulli distributed coordinates. Denote by P_n^d the convex hull $\text{conv}[X_1, \ldots X_n]$.

Interest in this problem came from a question by Fukuda and Ziegler, who asked for the maximum number of facets of a 0-1-polytope. This number is always bounded from below by the expected number of facets of a random 0-1-polytope which led to first investigations by Bárány and Pór (2001). Improved lower bounds were given by Gatzouras, Giannopoulos and Markoulakis (2005, 2007). They proved that

$$\mathbf{E}f_{d-1}(P_n^d) \geq c \ln^{\frac{d}{2}} n$$

for $d^{c_1} < n < (2^d)^{c_2 \ln^{-1} d}$ with some constants c, c_1, c_2. Choosing $n = (2^d)^{c_2 \ln^{-1} d}$ we see that the maximum number of facets of a 0-1-polytope is bounded from below by $c(c_2 d \ln^{-1} d)^{d/2}$ and thus is superexponential in d. Clearly $f_0(P_n^d) \leq \min(n, 2^d)$. To determine the expected values of $f_\ell(P_n^d)$, $\ell = 1, \ldots, d-2$, is probably an extremely difficult task.

Before this Dyer, Füredi and McDiarmid (1992) investigated the volume of P_n^d and proved that

$$\mathbf{E}V_d(P_n^d) \to \begin{cases} 1 & n \geq \left(\frac{2}{\sqrt{e}} + \varepsilon\right)^d \\ 0 & n \leq \left(\frac{2}{\sqrt{e}} - \varepsilon\right)^d \end{cases} \tag{2.33}$$

as $d \to \infty$ for any $\varepsilon > 0$.

Again, at the heart of both results is the construction of some 'floating body'. Denote by ν_d the uniform probability distribution on the vertices of C^d, having mass 2^{-d} at each vertex. In analogy to the definition of the floating body in Section 2.2.3 we use the function

$$q : C^d \to \mathbb{R}, \quad q(z) = \min\{\nu_d(C^d \cap H) : \ H \text{ is a halfspace and } z \in H\}$$

to define

$$C^d(q \geq t) = \{z \in C^d : \ q(z) \geq t\}$$

which is a convex polytope. It turns out that this 'floating body' $C^d(q \geq n^{-1})$ is close to P_n^d with high probability. Here, one has to be careful about the meaning of the word *close*, since P_n by definition always contains some vertices of C^d whereas $C^d(q \geq n^{-1})$ contains no vertex of C^d. It is one of the difficult steps in the above-mentioned papers to make this connection precise.

Recently, Giannopoulos and Hartzoulaki (2002) and Litvak, Pajor, Rudelson and Tomczak-Jaegermann (2005) proved a similar result where $C^d(q \geq n^{-1})$ is replaced by $\frac{1}{8}(C_d \cap RB^d)$ with a suitably defined radius R. The main point in the last paper is a highly interesting connection between the geometry of 0-1-polytopes and the smallest singular value of Bernoulli matrices (i.e. matrices with random 0–1 entries).

2.6 Intersection of halfspaces

Since every polytope is the intersection of finitely many halfspaces, one can generate a random polytope, or better a random polyhedron, by choosing such halfspaces at random. Here, one has to make reasonable assumptions about the distribution of the random halfspaces. Further one has to take care of the possibility that this random polyhedron is unbounded, which in most models occurs with positive probability.

We will describe two models. In the first approach random halfspaces are chosen which do not meet a given convex set $K \in \mathcal{K}^d$. As second model we mention the zero-cell of a hyperplane tessellation of $\mathbb{R}^d$.

2.6.1 *Random polyhedra containing a convex set*

Let $\mathcal{H}$ be the set of hyperplanes in $\mathbb{R}^d$, and $\mathcal{H}_K$ the subset of hyperplanes meeting a given set K. We assume that the origin is an interior point of K. For a hyperplane H we denote by H^- the halfspace bounded by H which contains the origin. The measure μ denotes the suitably normalized Haar measure on $\mathcal{H}$. To generate a random polyhedron $P_{(n)}$ we choose n hyperplanes in $\mathcal{H}_L \setminus \mathcal{H}_K$ according to the measure μ, where $L \subset \mathcal{K}^d$ is a convex set which contains K in its interior, and set

$$P_{(n)} = \bigcap_{i=1,\dots,n} H_i^- \cap L.$$

Suitable choices for L are either a large ball or the set $K + B^d$, the actual choice affects only some normalization constants. With probability one, the random polyhedron will converge to K as the number of halfspaces tends to infinity.

First investigations again go back to Rényi and Sulanke (1968) who treated the planar case. More than 20 years later Kaltenbach (1990) continued this line of research and proved that for sufficiently smooth $K \in \mathcal{K}^d$,

$$\mathbf{E}V_d(P_{(n)}) - V_d(K) = c(K)n^{-\frac{2}{d+1}} + o\left(n^{-\frac{2}{d+1}}\right),$$

and for $K \in \mathcal{P}^d$,

$$\mathbf{E}V_d(P_{(n)}) - V_d(K) = c(K)n^{-\frac{1}{d}} + o\left(n^{-\frac{1}{d}}\right).$$

These two cases are extremal as was shown by Kaltenbach: for $K \in \mathcal{K}^d$ we have

$$\underline{c}(d)n^{-\frac{2}{d+1}} \leq \mathbf{E}V_d(P_{(n)}) - V_d(K) \leq \overline{c}(d)n^{-\frac{1}{d}}.$$

Recently, Böröczky Jr. and Schneider (2008) investigated the mean width of random polyhedra. They proved that if K is a simplicial polytope with r facets then

$$\mathbf{E}V_1(P_{(n)}) - V_1(K) = \frac{2rd}{(d+1)^{d-1}} n^{-1} \ln^{d-1} n + o(n^{-1} \ln^{d-1} n).$$

The case of polytopes is extremal since they could also show that

$$\underline{c}(d)n^{-1} \ln^{d-1} n \leq \mathbf{E}V_1(P_{(n)}) - V_1(K) \leq \overline{c}(d)n^{-\frac{2}{d+1}}, \tag{2.34}$$

where the upper bound equals the asymptotic behaviour if K is a ball and follows from the inequality

$$\frac{\mathbf{E}V_1(P_{(n)})}{V_1(K)} \leq \frac{\mathbf{E}V_1\left(P^B_{(n)}\right)}{V_1(B)}.$$

In the same paper, Böröczky and Schneider also determined the asymptotic behaviour of $\mathbf{E}f_\ell(P_{(n)})$, for $\ell = 0$, $\ell = d-1$, if K is a simplicial polytope with r facets.

The main ingredient in the work of Böröczky and Schneider is a kind of 'outer floating body'. Define

$$w(x) = V_1(\mathrm{conv}[K, x]) - V_1(K), \quad \text{and} \quad K(w \leq t) = \{x \in \mathbb{R}^d : w(x) \leq t\}.$$

In a certain sense it is proved that $P_{(n)}$ is close to $K(w \leq n^{-1})$ as n tends to infinity, which gives the left inequality of (2.34).

Similar results can be obtained for

$$\Pi_{(n)} = \bigcap_{H \in X(n) \cap \mathcal{H}_L} H^-$$

where $X(n)$ is a Poisson hyperplane process with intensity n.

Instead of choosing n random hyperplanes from $\mathcal{H}_K$ one could also choose n random hyperplanes touching K at some boundary points according to some given density function on ∂K. Asymptotic results for the number of vertices of this polyhedron if K is a ball, and for volume, surface area and mean width for sufficiently smooth convex bodies K, have been derived by Buchta (1987) and Böröczky Jr. and Reitzner (2004).

2.6.2 *The zero-cell*

If the set K equals the origin o then the above procedure yields for $P_{(n)}$ the zero-cell of the hyperplane tessellation $H_1, \dots, H_n$. In this case it seems to be more natural to start with a Poisson hyperplane process X and to denote by

$$\Pi_{(n)} = Z_o = \bigcap_{H \in X(n)} H^-$$

the zero-cell Z_o of the hyperplane tessellation. Of interest are the expected intrinsic volumes and the f-vector of Z_o. Explicit formulae for $\mathbf{E} f_0(Z_o)$ and $\mathbf{E} V_d(Z_o)$ can be deduced from work of Schneider (1982) but systematic investigations seem to be missing.

Recent important progress in this area concerns the solution of Kendall's conjecture. For more information we refer to Chapter 5 on tessellations.

References

Affentranger, F. (1991). The convex hull of random points with spherically symmetric distributions. *Rend. Sem. Mat. Univ. Politec. Torino*, **49**, 359–383.

Affentranger, F. and Schneider, R. (1992). Random projections of regular simplices. *Discrete Comput. Geom.*, **7**, 219–226.

Bárány, I. (1989). Intrinsic volumes and f-vectors of random polytopes. *Math. Ann.*, **285**, 671–699.

Bárány, I. (1992). Random polytopes in smooth convex bodies. *Mathematika*, **39**, 81–92.

Bárány, I. (2004). Corrigendum: Random polytopes in smooth convex bodies. *Mathematika*, **51**, 31.

Bárány, I. and Buchta, C. (1993). Random polytopes in a convex polytope, independence of shape, and concentration of vertices. *Math. Ann.*, **297**, 467–497.

Bárány, I. and Larman, D. G. (1988). Convex bodies, economic cap coverings, random polytopes. *Mathematika*, **35**, 274–291.

Bárány, I. and Pór, A. (2001). On 0–1 polytopes with many facets. *Adv. Math.*, **161**, 209–228.

Bárány, I. and Reitzner, M. (2008). Random polytopes. *Manuscript*.

Bárány, I. and Vu, H. V. (2007). Central limit theorems for Gaussian polytopes. *Ann. Probab.*, **35**, 1593–1621.

Baryshnikov, Y. M. and Vitale, R. A. (1994). Regular simplices and Gaussian samples. *Discrete Comput. Geom.*, **11**, 141–147.

Bingham, N. H., Bräker, H., and Hsing, T. (1998). On the Hausdorff distance between a convex set and an interior random convex hull. *Adv. Appl. Prob.*, **30**, 295–316.

Blaschke, W. (1917). Über affine Geometrie XI: Lösung des "Vierpunktproblems" von Sylvester aus der Theorie der geometrischen Wahrscheinlichkeiten. *Ber. Verh. Sächs. Ges. Wiss. Leipzig, Math.-Phys. Kl.*, **69**, 436–453. Reprinted in: Burau, W., *et al.* (eds.): Wilhelm Blaschke. Gesammelte Werke, vol. 3: Konvexgeometrie. pp. 284–301. Essen, Thales 1985.

Borgwardt, K. H. (1987). *The Simplex Method: A Probabilistic Analysis*, Volume 1 of *Algorithms and Combinatorics*. Springer, Berlin.

Böröczky Jr., K. and Henk, M. (1999). Random projections of regular polytopes. *Arch. Math.*, **73**, 465–473.

Böröczky Jr., K. and Reitzner, M. (2004). Approximation of smooth convex bodies by random circumscribed polytopes. *Ann. Appl. Probab.*, **14**, 239–273.

Böröczky Jr., K. and Schneider, R. (2008). Mean width of random polytopes. *Manuscript.*

Buchta, C. (1984). Stochastische Approximation konvexer Polygone. *Z. Wahrsch. Verw. Geb.*, **67**, 283–304.

Buchta, C. (1987). On the number of vertices of random polyhedra with a given number of facets. *SIAM J. Alg. Disc. Meth.*, **8**, 85–92.

Buchta, C. (2005). An identity relating moments of functionals of convex hulls. *Discrete Comput. Geom.*, **33**, 125–142.

Buchta, C., Müller, J., and Tichy, R. F. (1985). Stochastical approximation of convex bodies. *Math. Ann.*, **271**, 225–235.

Cabo, A. J. and Groeneboom, P. (1994). Limit theorems for functionals of convex hulls. *Probab. Theory Relat. Fields*, **100**, 31–55.

Candes, E., Rudelson, M., Tao, T., and Vershynin, R. (2005). Error correction via Linear Programming. *Proceedings of the 46th Annual IEEE Symposium on Foundations of Computer Science (FOCS 2005)*, 295–308.

Candes, E. and Tao, T. (2005). Decoding by linear programming. *IEEE Trans. Inform. Theory*, **51**, 4203–4215.

Chalker, T. K., Godbole, A. P., Hitczenko, P., Radcliff, J., and Ruehr, O. G. (1999). On the size of a random sphere of influence graph. *Adv. Appl. Prob.*, **31**, 596–609.

Cowan, R. (2007). Identities linking volumes of convex hulls. *Adv. Appl. Prob.*, **39**, 630–644.

Dafnis, N., Giannopoulos, A., and Gúedon, O. (2008). On the isotropic constant of random polytopes. *Adv. Geom.*, to appear.

Dalla, L. and Larman, D. G. (1991). Volumes of a random polytope in a convex set. In: Gritzmann, P., Sturmfels, B. (eds.): Applied geometry and discrete mathematics. The Victor Klee Festschrift. *DIMACS Series in Discrete Mathematics and Theoretical Computer Science*, **4**, 175–180.

Donoho, D. L. (2006). High-dimensional centrally symmetric polytopes with neighbourliness proportional to dimension. *Discrete Comput. Geom.*, **35**, 617–652.

Donoho, D. L. and Tanner, J. (2005*a*). Neighborliness of randomly projected simplices in high dimensions. *Proc. Natl. Acad. Sci. USA*, **102**, 9452–9457.

Donoho, D. L. and Tanner, J. (2005*b*). Sparse nonnegative solution of underdetermined linear equations by linear programming. *Proc. Natl. Acad. Sci. USA*, **102**(27), 9446–9451.

Donoho, D. L. and Tanner, J. (2009). Counting faces of randomly projected polytopes when the projection radically lowers dimension. *J. Amer. Math. Soc.*, **22**, 1–53.

Dümbgen, L. and Walther, G. (1996). Rates of convergence for random approximations of convex sets. *Adv. Appl. Prob.*, **28**, 981–986.

Dyer, M. E., Füredi, Z., and McDiarmid, C. (1992). Volumes spanned by random points in the hypercube. *Random Structures Algorithms*, **3**, 91–106.

Edelsbrunner, H. (1987). *Algorithms in Combinatorial Geometry*, Volume 10 of *EATCS Monographs on Theoretical Computer Science*. Springer, Berlin.

Efron, B. (1965). The convex hull of a random set of points. *Biometrika*, **52**, 331–343.

Finch, S. and Hueter, I. (2004). Random convex hulls: a variance revisited. *Adv. Appl. Prob.*, **36**, 981–986.

Gatzouras, D., Giannopoulos, A., and Markoulakis, N. (2005). Lower bound for the maximal number of facets of a 0/1 polytope. *Discrete Comput. Geom.*, **34**, 331–349.

Gatzouras, D., Giannopoulos, A., and Markoulakis, N. (2007). On the maximal number of facets of 0/1 polytopes. in: Milman, V.D., and Schechtman, G. (eds.): *Geometric Aspects of Functional Analysis* (2004–2005). *Lecture Notes in Mathematics*, **1910**, 117–125.

Geffroy, J. (1961). Localisation asymptotique du polyèdre d'appui d'un échantillon Laplacien à k dimensions. *Publ. Inst. Statist. Univ. Paris*, **10**, 213–228.

Giannopoulos, A. (1992). On the mean value of the area of a random polygon in a plane convex body. *Mathematika*, **39**, 279–290.

Giannopoulos, A. and Hartzoulaki, M. (2002). Random spaces generated by vertices of the cube. *Discrete Comput. Geom.*, **28**, 255–273.

Groemer, H. (1973). On some mean values associated with a randomly selected simplex in a convex set. *Pacific J. Math.*, **45**, 525–533.

Groemer, H. (1974). On the mean value of the volume of a random polytope in a convex set. *Arch. Math.*, **25**, 86–90.

Groeneboom, P. (1988). Limit theorems for convex hulls. *Probab. Theory Relat. Fields*, **79**, 327–368.

Gruber, P. M. (1996). Expectation of random polytopes. *Manuscripta Math.*, **91**, 393–419.

Hartzoulaki, M. and Paouris, G. (2003). Quermassintegrals of a random polytope in a convex body. *Arch. Math.*, **80**, 430–438.

Hsing, T. (1994). On the asymptotic distribution of the area outside a random convex hull in a disk. *Ann. Appl. Probab.*, **4**, 478–493.

Hueter, I. (1994). The convex hull of a normal sample. *Adv. Appl. Prob.*, **26**, 855–875.

Hueter, I. (1999). Limit theorems for the convex hull of random points in higher dimensions. *Trans. Amer. Math. Soc.*, **351**, 4337–4363.

Hug, D., Munsonius, G. O., and Reitzner, M. (2004). Asymptotic mean values of Gaussian polytopes. *Beitr. Algebra Geom.*, **45**, 531–548.

Hug, D. and Reitzner, M. (2005). Gaussian polytopes: variances and limit theorems. *Adv. Appl. Prob.*, **37**, 297–320.

Kaltenbach, F. J. (1990). *Asymptotisches Verhalten zufälliger konvexer Polyeder. Dissertation.* Freiburg im Breisgau.

Kendall, D. (1977). Computer techniques and the archival map-reconstruction of Mycenaean Messenia. In: Bintliff, J. (ed.): Mycenaean geography. *Proceedings of the Cambridge Colloquium September 1976*, 83–88.

Klartag, B. and Kozma, G. (2008). On the hyperplane conjecture for random convex sets. *Israel J. Math.*, **170**, 253–268.

Küfer, K.-H. (1994). On the approximation of a ball by random polytopes. *Adv. Appl. Prob.*, **26**, 876–892.

Lao, W. (2006). The limit law of the maximum distance of points in a sphere in $\mathbb{R}^d$. *Technical report 2006/10, University of Karlsruhe.*

Lao, W. and Mayer, M. (2008). *U*-max-statistics. *J. Multivariate Anal.*, **99**, 2039–2052.

Leichtweiß, K. (1998). *Affine geometry of convex bodies.* Johann Ambrosius Barth Verlag, Heidelberg Leipzig.

Litvak, A. E., Pajor, A., Rudelson, M., and Tomczak-Jaegermann, N. (2005). Smallest singular value of random matrices and geometry of random polytopes. *Adv. Math.*, **195**, 491–523.

Ludwig, M. and Reitzner, M. (1999). A characterization of affine surface area. *Adv. Math.*, **147**, 138–172.

Mayer, M. (2008). *Random diameters and other U-Max-Statistics. Dissertation.* Bern.

Mayer, M. and Molchanov, I. (2007). Limit theorems for the diameter of a random sample in the unit ball. *Extremes*, **10**, 129–150.

Milman, V. D. and Pajor, A. (1989). Isotropic position and inertia ellipsoids and zonoids of the unit ball of a normed n-dimensional space. In: *Geometric Aspects of Functional Analysis* (1987–88). *Lecture Notes in Math.*, **1376**, 64–104.

Müller, J. (1989). On the mean width of random polytopes. *Probab. Theory Relat. Fields*, **82**, 33–37.

Preparata, F. P. and Shamos, M. I. (1990). *Computational Geometry: An Introduction*. Texts and Monographs in Computer Science. Springer, New York.

Raynaud, H. (1970). Sur l'enveloppe convexe des nuages de points aléatoires dans $\mathbb{R}^n$ I. *J. Appl. Probab.*, **7**, 35–48.

Reitzner, M. (2002). Random points on the boundary of smooth convex bodies. *Trans. Amer. Math. Soc.*, **354**, 2243–2278.

Reitzner, M. (2003). Random polytopes and the Efron–Stein jackknife inequality. *Ann. Probab.*, **31**, 2136–2166.

Reitzner, M. (2004). Stochastical approximation of smooth convex bodies. *Mathematika*, **51**, 11–29.

Reitzner, M. (2005*a*). Central limit theorems for random polytopes. *Probab. Theory Relat. Fields*, **133**, 483–507.

Reitzner, M. (2005*b*). The combinatorial structure of random polytopes. *Adv. Math.*, **191**, 178–208.

Rényi, A. and Sulanke, R. (1963). Über die konvexe Hülle von n zufällig gewählten Punkten. *Z. Wahrsch. Verw. Geb.*, **2**, 75–84.

Rényi, A. and Sulanke, R. (1964). Über die konvexe Hülle von n zufällig gewählten Punkten II. *Z. Wahrsch. Verw. Geb.*, **3**, 138–147.

Rényi, A. and Sulanke, R. (1968). Zufällige konvexe Polygone in einem Ringgebiet. *Z. Wahrsch. Verw. Geb.*, **9**, 146–157.

Rhee, W. T. and Talagrand, M. (1986). Martingale inequalities and the jackknife estimate of variance. *Statist. Probab. Lett.*, **4**, 5–6.

Rinott, Y. (1994). On normal approximation rates for certain sums of dependent random variables. *J. Comput. Appl. Math.*, **55**, 135–143.

Rudelson, M. and Vershynin, R. (2005). Geometric approach to error correcting codes and reconstruction of signals. *Int. Math. Res. Not.*, **64**, 4019–4041.

Schneider, R. (1982). Random polytopes generated by anisotropic hyperplanes. *Bull. London Math. Soc.*, **14**, 549–553.

Schneider, R. (1987). Approximation of convex bodies by random polytopes. *Aequationes Math.*, **32**, 304–310.

Schneider, R. (1988). Random approximation of convex sets. *J. Microscopy*, **151**, 211–227.

Schneider, R. and Wieacker, J. A. (1980). Random polytopes in a convex body. *Z. Wahrsch. Verw. Geb.*, **52**, 69–73.

Schreiber, T., and Yukich, J. E. (2008). Variance asymptotics and central limit theorems for generalized growth processes with applications to convex hulls and maximal points. *Ann. Probab.*, **36**, 363–396.

Schütt, C. (1991). The convex floating body and polyhedral approximation. *Israel J. Math.*, **73**, 65–77.

Schütt, C. and Werner, E. (1990). The convex floating body. *Math. Scand.*, **66**, 275–290.

Schütt, C. and Werner, E. (2000). Random polytopes with vertices on the boundary of a convex body. *C. R. Acad. Sci. Paris I*, **331**, 1–5.

Solomon, H. (1978). *Geometric Probability*, Volume 28 of *Regional Conference Series in Applied Mathematics*. SIAM, Philadelphia.

Vershik, A. M. and Sporyshev, P. V. (1992). Asymptotic behaviour of the number of faces of random polyhedra and the neighbourliness problem. *Selecta Math. Soviet.*, **11**, 181–201.

Vu, V. H. (2005). Sharp concentration of random polytopes. *Geom. Funct. Anal.*, **15**, 1284–1318.

Vu, V. H. (2006). Central limit theorems for random polytopes in a smooth convex set. *Adv. Math.*, **207**, 221–243.

Wagner, U. and Welzl, E. (2001). A continuous analogue of the upper bound theorem. *Discrete Comput. Geom.*, **26**, 205–219.

Wendel, J. G. (1962). A problem in geometric probability. *Math. Scand.*, **11**, 109–111.

Wieacker, J. A. (1978). *Einige Probleme der polyedrischen Approximation. Diplomarbeit.* Freiburg im Breisgau.

3

MODERN RANDOM MEASURES: PALM THEORY AND RELATED MODELS

Günter Last

Abstract

The theory of Palm measures is developed for the general context of stationary random measures on locally compact second countable groups (not necessarily Abelian). Connections are made with transport kernels, allocation problems and shift coupling.

3.1 Outline

This chapter aims at developing the theory of Palm measures of stationary random measures on locally compact second countable groups. The focus will be on recent developments concerning invariant transport-kernels and related invariance properties of Palm measures, shift-coupling, and mass-stationarity. Stationary partitions and (invariant) matchings will serve as extensive examples. Many recent results will be extended from the Abelian (or $\mathbb{R}^d$) case to general locally compact groups.

Palm probabilities are a very important concept in the theory and application of point processes and random measures, see (Matthes, Kerstan and Mecke, 1978; Kallenberg, 2007; Stoyan, Kendall and Mecke, 1995; Daley and Vere-Jones, 2008; Thorisson, 2000; Kallenberg, 2002). The Palm distribution of a stationary random measure M on an locally compact group G describes the statistical behaviour of M as seen from a *typical* point in the mass of M. Actually it is mathematically more fruitful to consider a general stationary probability measure on an abstract sample space equipped with a flow. This measure is then itself a Palm probability measure, obtained when M is given as the Haar measure on G. This general approach is also supported by many applications that require to consider a random measure together with other (jointly stationary) random measures and fields. Already the basic notions of stochastic geometry (e.g. typical cell, typical face, rose of directions) require the use of Palm probability measure in such a setting. And the *refined Campbell theorem* is a simple, yet powerful tool for handling them.

In a seminal paper Mecke (1967) has introduced and studied Palm measures of stationary random measures on Abelian groups. Although Palm distributions were defined in Tortrat (1969) in case of a group and in Rother and Zähle (1990) in case of a homogeneous space, these more general cases have found little attention in the literature. A very general approach to invariance properties of Palm measures as well as historical comments and further references can be found in Kallenberg (2007).

Recent years have seen some remarkable progress in understanding invariance properties of Palm probability measures and associated transport and coupling questions. Liggett (2002) found a way to remove a head from an i.i.d. sequence of coin tosses so that the rest of the coin tosses are again i.i.d.! This is an explicit *shift-coupling* of an i.i.d. sequence and its Palm version: they are the same up to a shift of the origin. Holroyd and Peres (2005) and Hoffman, Holroyd and Peres (2006) have used a *stable marriage algorithm* for transporting a given fraction of Lebesgue measure on $\mathbb{R}^d$ to an ergodic point process. Again this yields an explicit shift-coupling of a point process with its Palm version. In dimension $d \geq 3$, Chatterjee, Peled, Peres and Romik (2007) found an (invariant) *gravitational allocation rule* transporting Lebesgue measure to a homogeneous Poisson process exhibiting surprisingly nice moment properties. Motivated by Holroyd and Peres (2005) and Hoffman, Holroyd and Peres (2006), Last (2006) has studied general stationary partitions of $\mathbb{R}^d$. Last and Thorisson (2009) have developed a general theory for invariant (weighted) *transport-kernels* on Abelian groups.

Another related line of research can be traced back to the intrinsic characterization of Palm measures via an integral equation in Mecke (1967). Thorisson (2000) called a simple point processes on $\mathbb{R}^d$ *point-stationary* if its distribution is invariant under *bijective point-shifts* against any independent stationary background. He proved that point-stationarity is a characterizing property of Palm versions of stationary point processes. Heveling and Last (2005) and Heveling and Last (2007) showed that the independent stationary background could be removed from the definition of point-stationarity. Bijective point shifts are closely related to invariant graphs and trees on point processes (Holroyd and Peres, 2003; Ferrari, Landim and Thorisson, 2004; Timar, 2004). The extension of point-stationarity to general random measures is not straightforward. Last and Thorisson (2009) call a random measure on an Abelian group *mass-stationary* if, informally speaking, the neutral element is a typical location in the mass. They could prove that mass-stationarity is indeed a characterizing property of Palm measures.

In this chapter we will develop Palm theory for stationary random measures on locally compact second countable groups. In doing so we will extend many recent results from the Abelian (or $\mathbb{R}^d$) case to general groups. After having introduced stationary random measures and their Palm measures we will focus on invariant weighted transport-kernels balancing stationary random measures, invariance and transport properties of Palm measures and mass-stationarity. In the final two sections we will apply parts of the theory to *stationary partitions* and (invariant) *matchings* of point processes.

3.2 Stationary random measures

We consider a topological (multiplicative) group G that is assumed to be a locally compact, second countable Hausdorff space with unit element e and Borel σ-field $\mathcal{G}$. A classical source on such groups is Nachbin (1965), see also Chapter 2 in Kallenberg (2002). A measure η on G is *locally finite* if it is finite on compact sets. There exists a left-invariant Haar measure on G , i.e. a locally finite measure satisfying

$$\int f(hg)\lambda(\mathrm{d}g) = \int f(g)\lambda(\mathrm{d}g), \quad h \in G, \tag{3.1}$$

for all measurable $f : G \to \mathbb{R}_+$, where $\mathbb{R}_+ = [0, \infty)$. This measure is unique up to normalization. The *modular function* is a continuous homomorphism $\Delta : G \to (0, \infty)$ satisfying

$$\int f(gh)\lambda(\mathrm{d}g) = \Delta(h^{-1}) \int f(g)\lambda(\mathrm{d}g), \quad h \in G, \tag{3.2}$$

for all f as above. This modular function has the property

$$\int f(g^{-1})\lambda(\mathrm{d}g) = \int \Delta(g^{-1})f(g)\lambda(\mathrm{d}g). \tag{3.3}$$

The group G is called *unimodular*, if $\Delta(g) = 1$ for all $g \in G$.

We denote by M the set of all locally finite measures on G, and by $\mathcal{M}$ the cylindrical σ-field on M which is generated by the evaluation functionals $\eta \mapsto \eta(B)$, $B \in \mathcal{G}$. The *support* $\operatorname{supp} \eta$ of a measure $\eta \in \mathsf{M}$ is the smallest closed set $F \subset G$ such that $\eta(G \setminus F) = 0$. By $\mathsf{N} \subset \mathsf{M}$ (resp. $\mathsf{N}_s \subset \mathsf{M}$) we denote the measurable set of all (resp. *simple*) *counting measures* on G, i.e. the set of all those $\eta \in \mathsf{M}$ with discrete support and $\eta\{g\} = \eta(\{g\}) \in \mathbb{N}_0$ (resp. $\eta\{g\} \in \{0, 1\}$) for all $g \in G$. We can and will identify N_s with the class of all *locally finite* subsets of G, where a set is called locally finite if its intersection with any compact set is finite.

We will mostly work on a σ-finite measure space $(\Omega, \mathcal{A}, \mathbf{P})$ (see Remark 3.8). Although $\mathbf{P}$ need not be a probability measure, we are still using a probabilistic language. Moreover, we would like to point out already at this early stage, that we will consider several measures on $(\Omega, \mathcal{A})$. A *random measure* on G is a measurable mapping $M : \Omega \to \mathsf{M}$. A random measure is a (simple) *point process* on G if $M(\omega) \in \mathsf{N}$ (resp. $M(\omega) \in \mathsf{N}_s$) for all $\omega \in \Omega$. A random measure M can also be regarded as a *kernel* from Ω to G. Accordingly we write $M(\omega, B)$ instead of $M(\omega)(B)$. If M is a random measure, then the mapping $(\omega, g) \mapsto \mathbf{1}\{g \in \operatorname{supp} M(\omega)\}$ is measurable.

We assume that $(\Omega, \mathcal{A})$ is equipped with a *measurable flow* $\theta_g : \Omega \to \Omega$, $g \in G$. This is a family of measurable mappings such that $(\omega, g) \mapsto \theta_g\omega$ is measurable, θ_e is the identity on Ω and

$$\theta_g \circ \theta_h = \theta_{gh}, \quad g, h \in G, \tag{3.4}$$

where $\circ$ denotes composition. This implies that θ_g is a bijection with inverse $\theta_g^{-1} = \theta_{g^{-1}}$. A random measure M on G is called *invariant* (or *flow-adapted*) if

$$M(\theta_g\omega, gB) = M(\omega, B), \quad \omega \in \Omega,\ g \in G,\ B \in \mathcal{G}, \tag{3.5}$$

where $gB = \{gh : h \in B\}$. This means that

$$\int f(h)M(\theta_g\omega, \mathrm{d}h) = \int f(gh)M(\omega, \mathrm{d}h) \tag{3.6}$$

for all measurable $f : G \to \mathbb{R}_+$. We will often skip the ω in such relations, i.e. we write (3.6) as $\int f(h)M(\theta_g, \mathrm{d}h) = \int f(gh)M(\theta_e, \mathrm{d}h)$ or $\int f(h)M(\theta_g, \mathrm{d}h) = \int f(gh)M(\mathrm{d}h)$. Recall that θ_e is the identity on Ω. Still another way of expressing (3.5) is

$$M \circ \theta_g = gM, \quad g \in G, \tag{3.7}$$

where for $(\eta, g) \in \mathsf{M} \times G$ the measure $g\eta$ is defined by $g\eta(\cdot) = \int \mathbf{1}\{gh \in \cdot\}\eta(\mathrm{d}h)$.

A measure $\mathbf{P}$ on $(\Omega, \mathcal{A})$ is called *stationary* if it is invariant under the flow, i.e.

$$\mathbf{P} \circ \theta_g = \mathbf{P}, \quad g \in G, \tag{3.8}$$

where θ_g is interpreted as a mapping from $\mathcal{A}$ to $\mathcal{A}$ in the usual way:

$$\theta_g A = \{\theta_g\omega : \omega \in A\}, \quad A \in \mathcal{A},\ g \in G.$$

Example 3.1 Consider the measurable space $(\mathsf{M}, \mathcal{M})$ and define for $\eta \in \mathsf{M}$ and $g \in G$ the measure $\theta_g\eta$ by $\theta_g\eta = g\eta$, i.e. $\theta_g\eta(B) = \eta(g^{-1}B)$, $B \in \mathcal{G}$. Then $\{\theta_g : g \in G\}$ is a measurable flow and the identity M on M is an invariant random measure. A stationary probability measure on $(\mathsf{M}, \mathcal{M})$ can be interpreted as the distribution of a *stationary random measure.*

Example 3.2 Let $(E, \mathcal{E})$ be a Polish space and assume that Ω is the space of all measures ω on $G \times E \times G \times E$ such that $\omega(B \times E \times G \times E)$ and $\omega(G \times E \times B \times E)$ are finite for compact $B \subset G$. The σ-field $\mathcal{A}$ is defined as the cylindrical σ-field on Ω. It is stated in Port and Stone (1973) (and can be proved as in Matthes, Kerstan and Mecke 1978) that $(\Omega, \mathcal{A})$ is a Polish space. For $g \in G$ and $\omega \in \Omega$ we let $\theta_g\omega$ denote the measure satisfying

$$\theta_g\omega(B \times C \times B' \times C') = \omega(g^{-1}B \times C \times g^{-1}B' \times C')$$

for all $B, B' \in \mathcal{G}$ and $C, C' \in \mathcal{E}$. The random measures M and N defined by $M(\omega, \cdot) = \omega(\cdot \times E \times G \times E)$ and $N(\omega, \cdot) = \omega(G \times E \times \cdot \times E)$ are invariant. Port and Stone (1973) (see also Harris 1971) call a stationary probability measure on $(\Omega, \mathcal{A})$ concentrated on the set of integer-valued $\omega \in \Omega$ a (translation invariant) *marked motion process.* The idea is that the (marked) points of M move to the points of N in one unit of time.

Example 3.3 Assume that G is an additive Abelian group and consider a flow $\{\tilde{\theta}_g : g \in G\}$ as in Last and Thorisson (2009) (see also Neveu 1977). In our current setting this amounts to define $gh = h + g$ and $\theta_g = \tilde{\theta}_g^{-1}$. It is somewhat unfortunate that in the point process literature it is common to define the shift of a measure $\eta \in \mathsf{M}$ by $g \in G$ by $g^{-1}\eta$ and not (as it would be more natural) by $g\eta$. Here we follow the terminology of Kallenberg (2007).

Remark 3.4 Our setting accommodates stationary marked point processes (see Daley and Vere-Jones 2008, Matthes, Kerstan and Mecke 1978 and Section 3.3) as well as stochastic processes (fields) jointly stationary with a random measure M, see Thorisson (2000). The use of an abstract flow $\{\theta_g : g \in G\}$ acting directly on the underlying sample space is making the notation quite efficient. A more general framework is to work on an abstract measure space and to replace (3.8) by a distributional invariance, see Kallenberg (2007).

We now fix a σ-finite stationary measure $\mathbf{P}$ on $(\Omega, \mathcal{A})$ and an invariant random measure M on G. Let $w : G \to \mathbb{R}_+$ be a measurable function having $\int w(g)\lambda(\mathrm{d}g) = 1$. The measure

$$\mathbf{Q}_M(A) = \iint \mathbf{1}\{\theta_g^{-1}\omega \in A\}\, w(g) M(\omega, \mathrm{d}g)\mathbf{P}(\mathrm{d}\omega), \quad A \in \mathcal{A}, \tag{3.9}$$

is called the *Palm measure* of M (with respect to $\mathbf{P}$). We will see below that this definition does not depend on the choice of the function w. The *intensity* γ_M of M is defined by

$$\gamma_M = \mathbf{Q}_M(\Omega) = \mathbf{E} \int w(g) M(\mathrm{d}g). \tag{3.10}$$

We have $\gamma_M = \mathbf{E}M(B)$ for any $B \in \mathcal{G}$ with $\lambda(B) = 1$. In case $0 < \gamma_M < \infty$ we can define the *Palm probability measure* of M by $\mathbf{Q}_M^0 = \gamma_M^{-1}\mathbf{Q}_M$. These definitions extend the definitions given in in Section 1.3 of Chapter 1. An interpretation of the Palm probability measure will be given in Remark 3.10. In this section we will often work with two (or more) invariant random measures. Therefore we use a lower index to denote the dependence of the Palm (probability) measure on M. A common alternative notation is $\mathbf{P}_M$ (resp. $\mathbf{P}_M^0$).

A more succinct way of writing (3.9) is

$$\mathbf{Q}_M(A) = \mathbf{E} \int \mathbf{1}\{\theta_g^{-1} \in A\}\, w(g) M(\mathrm{d}g), \quad A \in \mathcal{A}, \tag{3.11}$$

where $\mathbf{E}$ denotes integration with respect to $\mathbf{P}$. The following lemma shows that $\mathbf{Q}_M$ is concentrated on $\{e \in \operatorname{supp} M\}$.

Proposition 3.5 *We have* $\mathbf{Q}_M\{e \notin \operatorname{supp} M\} = 0$.

Proof From the definition (3.11) and (3.7) we have

$$\begin{aligned}\mathbf{Q}_M\{e \notin \operatorname{supp} M\} &= \mathbf{E}\int \mathbf{1}\{e \notin \operatorname{supp} M \circ \theta_g^{-1}\}\, w(g)M(\mathrm{d}g)\\ &= \mathbf{E}\int \mathbf{1}\{e \notin \operatorname{supp} g^{-1}M\}w(g)M(\mathrm{d}g)\\ &= \mathbf{E}\int \mathbf{1}\{g \notin \operatorname{supp} M\}w(g)M(\mathrm{d}g) = 0,\end{aligned}$$

proving the assertion. □

Our first result is the *refined Campbell theorem* connecting $\mathbf{P}$ and $\mathbf{Q}_M$. In the canonical case of Example 3.1 the result was derived in Tortrat (1969).

Theorem 3.6 *For any measurable* $f : \Omega \times G \to \mathbb{R}_+$,

$$\mathbf{E}\int f\left(\theta_g^{-1}, g\right) M(\mathrm{d}g) = \mathbf{E}_{\mathbf{Q}_M}\int f(\theta_e, g)\lambda(\mathrm{d}g). \tag{3.12}$$

Proof From definition (3.11) and Fubini's theorem we obtain that

$$\mathbf{E}_{\mathbf{Q}_M}\int f(\theta_e, g)\lambda(\mathrm{d}g) = \mathbf{E}\iint f\left(\theta_h^{-1}, g\right) w(h)\lambda(\mathrm{d}g)M(\mathrm{d}h),$$

Using Fubini's theorem again, as well as (3.3), we get

$$\begin{aligned}\mathbf{E}_{\mathbf{Q}_M}\int f(\theta_e, g)\lambda(\mathrm{d}g) &= \mathbf{E}\iint f\left(\theta_h^{-1}, g^{-1}\right)\Delta(g^{-1})w(h)\lambda(\mathrm{d}g)M(\mathrm{d}h)\\ &= \mathbf{E}\iint f\left(\theta_h^{-1}, g^{-1}h\right)\Delta(g^{-1}h)w(h)\lambda(\mathrm{d}g)M(\mathrm{d}h),\end{aligned}$$

where we have used left-invariance of λ for the second equality. From Fubini's theorem we obtain

$$\begin{aligned}\mathbf{E}_{\mathbf{Q}_M}\int f(\theta_e, g)\lambda(\mathrm{d}g) &= \mathbf{E}\iint f\left(\theta_{gh}^{-1}, h\right)\Delta(h)w(gh)M\circ\theta_g^{-1}(\mathrm{d}h)\lambda(\mathrm{d}g)\\ &= \mathbf{E}\iint f\left(\theta_h^{-1}\circ\theta_g^{-1}, h\right)\Delta(h)w(gh)M\circ\theta_g^{-1}(\mathrm{d}h)\lambda(\mathrm{d}g)\\ &= \mathbf{E}\iint f\left(\theta_h^{-1}, h\right)\Delta(h)w(gh)M(\mathrm{d}h)\lambda(\mathrm{d}g),\end{aligned}$$

where we have used the flow property (3.4) and (3.7) for the second and stationarity of $\mathbf{P}$ for the final equation. The result now follows from

$$\Delta(h)\int w(gh)\lambda(\mathrm{d}g) = \Delta(h)\Delta(h^{-1})\int w(g)\lambda(\mathrm{d}g) = 1. \tag{3.13}$$

□

Remark 3.7 Establish the setting of Example 3.3. Then (3.12) means that

$$\mathbf{E}\int f(\tilde{\theta}_g, g)M(\mathrm{d}g) = \mathbf{E}_{\mathbf{Q}_M}\int f(\tilde{\theta}_e, g)\lambda(\mathrm{d}g), \tag{3.14}$$

cf. also Theorem 1.9 in Chapter 1 of this volume. Equation (3.12) or rather its equivalent version

$$\mathbf{E}\int \mathbf{1}\{(\theta_e, g) \in \cdot\}M(\mathrm{d}g) = \mathbf{E}_{\mathbf{Q}_M}\int \mathbf{1}\{(\theta_g, g) \in \cdot\}\lambda(\mathrm{d}g),$$

is known as *skew factorization* of the *Campbell measure* of M. A general discussion of this technique can be found in Kallenberg (2007).

Let $w' : G \to \mathbb{R}_+$ be another measurable function having $\int w'(g)\lambda(\mathrm{d}g) = 1$ and take $A \in \mathcal{A}$. Then (3.12) implies that

$$\begin{aligned}\mathbf{E}\int \mathbf{1}\left\{\theta_g^{-1} \in A\right\} w'(g)M(\mathrm{d}g) &= \mathbf{E}\int \mathbf{1}\left\{\theta_g^{-1} \in A\right\} w'(g)M(\mathrm{d}g)\\ &= \mathbf{E}_{\mathbf{Q}_M}\int \mathbf{1}\{\theta_e \in A\}w'(g)\lambda(\mathrm{d}g) = \mathbf{Q}_M(A).\end{aligned}$$

Hence the definition (3.9) is indeed independent of the choice of w. We also note that the refined Campbell theorem implies the ordinary Campbell theorem

$$\mathbf{E}\int f(g)M(\mathrm{d}g) = \gamma_M\int f(g)\lambda(\mathrm{d}g), \tag{3.15}$$

for all measurable $f : G \to \mathbb{R}_+$.

Remark 3.8 Working with a (stationary) σ-finite measure $\mathbf{P}$ rather than with probability measure doesn't make the theory more complicated. In fact, some of the fundamental results can even be more easily stated this way. An example is the one-to-one correspondence between $\mathbf{P}$ and the Palm measure $\mathbf{Q}_M$, see e.g. Theorem 3.24. Otherwise, extra technical integrability assumptions are required, see Theorem 11.4 in Kallenberg (2002). Another advantage is that in some applications it is the Palm probability measure that has a probabilistic interpretation. This measure can be well-defined also in case where $\mathbf{P}$ is not a finite measure.

To derive another corollary of the refined Campbell theorem, we take a measurable function $\tilde{w} : \mathsf{M} \times G \to \mathbb{R}_+$ satisfying

$$\int \tilde{w}(\eta, g)\eta(\mathrm{d}g) = 1, \tag{3.16}$$

whenever $\eta \in \mathsf{M}$ is not the null measure. For one example of such a function we refer to Mecke (1967). We then have the *inversion formula*

$$\mathbf{E}\mathbf{1}\{M(G) > 0\}f = \mathbf{E}_{\mathbf{Q}_M}\int \tilde{w}(M \circ \theta_g, g)f \circ \theta_g\lambda(\mathrm{d}g) \tag{3.17}$$

for all measurable $f : \Omega \to \mathbb{R}_+$. This is a direct consequence of the refined Campbell theorem (3.12). A first useful consequence of the inversion formula is the following.

Proposition 3.9 *The Palm measure $\mathbf{Q}_M$ is σ-finite.*

Proof As $\mathbf{P}$ is assumed σ-finite, there is a positive measurable function f on Ω such that $\mathbf{E}f < \infty$. The function $\int \tilde{w}(M \circ \theta_g, g) f(\theta_g) \lambda(\mathrm{d}g)$ is positive on $\{M(G) > 0\}$ and has by (3.17) a finite integral with respect to $\mathbf{Q}_M$. Since $\mathbf{Q}_M\{M(G) = 0\} = 0$ (see Lemma 3.5) we obtain that $\mathbf{Q}_M$ is σ-finite. □

The *invariant* σ-field $\mathcal{I} \subset \mathcal{A}$ is the class of all sets $A \in \mathcal{A}$ satisfying $\theta_g A = A$ for all $g \in G$. Let M be an invariant random measure with finite intensity and define

$$\hat{M} = \mathbf{E}\Big[\int w(g) M(\mathrm{d}g) \Big| \mathcal{I}\Big], \tag{3.18}$$

where the conditional expectation is defined as for probability measures. Since $\hat{M} \circ \theta_g = \hat{M}$, $g \in G$, the refined Campbell theorem (3.12) implies for all $A \in \mathcal{I}$ that $\mathbf{E}\mathbf{1}_A \int w(g) M(\mathrm{d}g) = \mathbf{Q}_M(A)$. Therefore definition (3.18) is independent of the choice of w. If $\mathbf{P}$ is a probability measure and $G = \mathbb{R}^d$, then $\hat{M}$ is called *sample intensity* of M, see Matthes, Kerstan and Mecke (1978) and Kallenberg (2002). Assuming that $\mathbf{P}\{\hat{M} = 0\} = 0$, we define the *modified Palm measure* $\mathbf{Q}_M^*$ (Matthes, Kerstan and Mecke, 1978; Thorisson, 2000; Last, 2006) by

$$\mathbf{Q}_M^*(A) = \mathbf{E}\hat{M}^{-1} \int \mathbf{1}\{\theta_g^{-1} \in A\} w(g) M(\mathrm{d}g). \tag{3.19}$$

Conditioning shows that

$$\mathbf{Q}_M^*(A) = \mathbf{P}(A), \quad A \in \mathcal{I}. \tag{3.20}$$

Comparing (3.19) and (3.9) yields

$$d\mathbf{Q}_M^* = \hat{M}^{-1} d\mathbf{Q}_M. \tag{3.21}$$

The refined Campbell theorem (3.12) takes the form

$$\mathbf{E}\hat{M}^{-1} \int f\left(\theta_g^{-1}, g\right) M(\mathrm{d}g) = \mathbf{E}_{\mathbf{Q}_M^*} \int f(\theta_e, g) \lambda(\mathrm{d}g). \tag{3.22}$$

Remark 3.10 If $\mathbf{P}$ is a probability measure and M is a simple point process with a positive and finite intensity, then the Palm probability measure $\mathbf{Q}_M^0$ can be interpreted as a conditional probability measure given that M has a point in e (Kallenberg, 2007). The modified version describes the underlying stochastic experiment as seen from a randomly chosen point of M (Matthes, Kerstan and Mecke, 1978; Thorisson, 2000). Both measures agree iff $\hat{M}$ is $\mathbf{P}$-a.e. constant and in particular if $\mathbf{P}$ is *ergodic*, i.e. $\mathbf{P}(A) = 0$ or $\mathbf{P}(\Omega \setminus A) = 0$ for all $A \in \mathcal{I}$.

3.3 Stationary marked random measures

Let $(S, \mathcal{S})$ be some measurable space and M' a kernel from Ω to $G \times S$. We call M' *marked random measure* (on G with mark space S) if $M = M'(\cdot \times S)$ is a random measure on G. Daley and Vere-Jones (2008) call M the *ground process* of M'.

A marked random measure M' is *invariant* if $M'(\cdot \times B)$ is invariant for all $B \in \mathcal{S}$. In many applications M' is of the form

$$M' = \int \mathbf{1}\{(g, \delta(g)) \in \cdot\} M(\mathrm{d}g), \tag{3.23}$$

where $\delta : \Omega \times G \to S$ is measurable. If $g \in \operatorname{supp} M$ we think of $\delta(g)$ as *mark* of g. If M is invariant and δ is invariant in the sense that

$$\delta(\omega, g) = \delta\left(\theta_g^{-1}\omega, e\right), \quad \omega \in \Omega,\ g \in G, \tag{3.24}$$

then it is easy to check that M' is invariant.

Let M' be an invariant marked random measure and $\mathbf{P}$ a σ-finite stationary measure on $(\Omega, \mathcal{A})$. The *Palm measure* of M' is the measure $\mathbf{Q}_{M'}$ on $\Omega \times S$ defined by

$$\mathbf{Q}_{M'}(A) = \mathbf{E} \int \mathbf{1}\{(\theta_g^{-1}, z) \in A\} w(g) M'(\mathrm{d}(g, z)), \quad A \in \mathcal{A} \otimes \mathcal{S}. \tag{3.25}$$

Note that $\mathbf{Q}_{M'}(\cdot \times B)$ is the Palm measure of $M'(\cdot \times B)$. The refined Campbell theorem (3.12) takes the form

$$\mathbf{E} \int f\left(\theta_g^{-1}, g, z\right) M'(\mathrm{d}(g, z)) = \iint f(\omega, g, z) \lambda(\mathrm{d}g) \mathbf{Q}_{M'}(\mathrm{d}(\omega, z)) \tag{3.26}$$

for all measurable $f : \Omega \times G \times S \to \mathbb{R}_+$.

Assume that $(\Omega, \mathcal{A})$ is a Borel space. (This is, e.g. the case in Example 3.1.) If $\mathbf{Q}_{M'}(\Omega \times \cdot)$ is a σ-finite measure, then we may *disintegrate* $\mathbf{Q}_{M'}$, to get another form of (3.26). For simplicity we even assume that the intensity γ_M of $M = M'(\cdot \times S)$ is finite. Assuming also $\gamma_M > 0$ we can define the *mark distribution* $\mathbb{W}$ of M' by $\mathbb{W} = \gamma_M^{-1} \mathbf{Q}_{M'}(\Omega \times \cdot)$. There exists a stochastic kernel $(z, A) \mapsto \mathbf{Q}_{M'}^z(A)$ from S to Ω satisfying

$$\mathbf{Q}_{M'}(\mathrm{d}(\omega, z)) = \gamma_M \mathbf{Q}_{M'}^z(\mathrm{d}\omega) \mathbb{W}(\mathrm{d}z).$$

Therefore (3.26) can be written as

$$\mathbf{E} \int f\left(\theta_g^{-1}, g, z\right) M'(\mathrm{d}(g, z)) = \gamma_M \iiint f(\omega, g, z) \lambda(\mathrm{d}g) \mathbf{Q}_{M'}^z(\mathrm{d}\omega) \mathbb{W}(\mathrm{d}z), \tag{3.27}$$

cf. also Theorem 1.12 from Chapter 1 in this volume.

Our next lemma provides an elegant way for handling stationary marked random measures. A kernel κ from $\Omega \times G$ to S is called *invariant* if

$$\kappa(\omega, g, \cdot) = \kappa\left(\theta_g^{-1}\omega, e, \cdot\right), \quad \omega \in \Omega,\, g \in G. \tag{3.28}$$

Lemma 3.11 *Let M' be an invariant marked random measure and assume that $(S, \mathcal{S})$ is a Borel space. Then there this an invariant stochastic kernel κ from $\Omega \times G$ to S such that*

$$M' = \iint \mathbf{1}\{(g, z) \in \cdot\}\kappa(g, \mathrm{d}z)M(\mathrm{d}g) \quad \mathbf{P} - a.e. \tag{3.29}$$

Proof Define a measure C' on $\Omega \times G \times S$ by

$$C'(\cdot) = \mathbf{E}\int \mathbf{1}\{(\theta_e, g, z) \in \cdot\}M'(\mathrm{d}(g, z)).$$

From stationarity of $\mathbf{P}$ and invariance of M' we easily obtain that

$$\int \mathbf{1}\{(\theta_h\omega, hg, z) \in \cdot\}C'(\mathrm{d}(\omega, g, z)) = C', \quad h \in G.$$

Moreover, the measure $C = C'(\cdot \times S)$ is σ-finite. We can now apply Theorem 3.5 in Kallenberg (2007) to obtain an invariant kernel κ satisfying

$$C' = \iint \mathbf{1}\{(\omega, g, z) \in \cdot\}\kappa(\omega, g, \mathrm{d}z)C(\mathrm{d}(\omega, g)). \tag{3.30}$$

(In fact the theorem yields an invariant kernel κ', satisfying this equation. But in our specific situation we have $\kappa'(\omega, g, S) = 1$ for C-a.e. (ω, g), so that κ' can be modified in an obvious way to yield the desired κ.) Equation (3.30) implies that

$$\mathbf{E}\mathbf{1}_A M'(B) = \mathbf{E}\mathbf{1}_A \iint \mathbf{1}\{(g, z) \in B\}\kappa(g, \mathrm{d}z)M(\mathrm{d}g), \quad B \in \mathcal{G} \times \mathcal{S},$$

for all $A \in \mathcal{A}$. Therefore $M'(B) = \iint \mathbf{1}\{(g, z) \in B\}\kappa(g, \mathrm{d}z)M(\mathrm{d}g)$ holds $\mathbf{P}$-a.e. Since $\mathcal{G} \times \mathcal{S}$ is countably generated, (3.29) follows. □

We now assume that (3.29) holds for an invariant kernel κ. Invariance of κ implies that the Palm measure of M' is given by

$$\mathbf{Q}_{M'} = \mathbf{E}_{\mathbf{Q}_M} \int \mathbf{1}\{(\theta_e, z) \in \cdot\}\kappa(e, \mathrm{d}z). \tag{3.31}$$

The refined Campbell theorem (3.27) reads

$$\mathbf{E}\int f\left(\theta_g^{-1}, g, z\right) M'(\mathrm{d}(g, z)) = \mathbf{E}_{\mathbf{Q}_M} \iint f(\omega, g, z)\kappa(e, \mathrm{d}z)\lambda(\mathrm{d}g). \tag{3.32}$$

The mark distribution is given by $W = \gamma_M^{-1}\mathbf{E}_{\mathbf{Q}_M}\kappa(e,\cdot)$. In the special case (3.23) the refined Campbell theorem (3.32) says that

$$\mathbf{E}\int f\left(\theta_g^{-1},g,z\right) M'(\mathrm{d}(g,z)) = \mathbf{E}_{\mathbf{Q}_M}\int f(\omega,g,\delta(e))\lambda(\mathrm{d}g). \tag{3.33}$$

The mark distribution is given by $W = \gamma_M^{-1}\mathbf{Q}_M\{\delta(e)\in\cdot\}$.

3.4 Invariant transport-kernels

We first adapt the terminology from Last and Thorisson (2009) to our present more general setting. A *transport-kernel* (on G) is a Markovian kernel T from $\Omega\times G$ to G. We think of $T(\omega,g,B)$ as proportion of mass transported from location g to the set B, when ω is given. A *weighted transport-kernel* (on G) is a kernel from $\Omega\times G$ to G such that $T(\omega,g,\cdot)$ is locally finite for all $(\omega,g)\in\Omega\times G$. If T is finite then the mass at g is weighted by $T(\omega,g,G)$ before being transported by the normalized T. A weighted transport kernel T is called *invariant* if

$$T(\theta_g\omega,gh,gB) = T(\omega,h,B),\quad g,h\in G,\ \omega\in\Omega,\ B\in\mathcal{G}. \tag{3.34}$$

Quite often we use the short-hand notation $T(g,\cdot) = T(\theta_e,g,\cdot)$. If M is an invariant random measure on G and $N = \int T(\omega,g,\cdot)M(\omega,\mathrm{d}g)$ is locally finite for each $\omega\in\Omega$, then N is again an invariant random measure. Our interpretation is, that T transports M to N in an invariant way.

Example 3.12 Consider a measurable function $t:\Omega\times G\times G\to\mathbb{R}_+$ and assume that t is invariant, i.e.

$$t(\theta_k\omega,kg,kh) = t(\omega,g,h),\quad \omega\in\Omega,\ g,h,k\in G. \tag{3.35}$$

Let M be an invariant random measure on G and define

$$T(g,B) = \int_B t(g,h)M(\mathrm{d}h).$$

Then (3.34) holds. Such functions t occur in the *mass-transport principle*, see Benjamini, Lyons, Peres and Schramm (1999) and Remark 3.17 below. The number $t(\omega,g,h)$ is then interpreted as the mass sent from g to h when the configuration ω is given.

Example 3.13 Consider a measurable mapping $\tau:\Omega\times G\to G$. The transport kernel T defined by $T(g,\cdot)=\delta_{\tau(g)}$ is invariant if and only τ is *covariant* in the sense that

$$\tau(\theta_g\omega,gh) = g\tau(\omega,h),\quad \omega\in\Omega,\ g,h\in G. \tag{3.36}$$

In this case we call τ *allocation rule*. This terminology is taken from Holroyd and Peres (2005).

Example 3.14 Consider the setting of Example 3.2 and let $\mathbf{P}$ be a σ-finite stationary measure on $(\Omega, \mathcal{A})$ concentrated on the set Ω' of all integer-valued $\omega \in \Omega$. Define an invariant transport-kernel T by

$$T(\omega, g, \cdot) = \frac{1}{M(\omega, \{g\})} \sum_{h:\omega(g,h)>0} \omega(g,h)\delta_h, \quad \omega \in \Omega', \, g \in G, \tag{3.37}$$

if $M(\omega, \{g\}) > 0$ and $T(\omega, g, \cdot) = \delta_g$ otherwise, where $\omega(g,h) = \omega(\{g\} \times E \times \{h\} \times E)$. It can be easily checked that T satisfies (3.38) below for $\mathbf{P}$-a.e. $\omega \in \Omega$.

Let M and N be two invariant random measures on G. A weighted transport-kernel T on G is called (M,N)-*balancing* if

$$\int T(\omega, g, \cdot) M(\omega, \mathrm{d}g) = N(\omega, \cdot) \tag{3.38}$$

holds for all $\omega \in \Omega$. In case $M = N$ we also say that T is M-*preserving*. If $\mathbf{Q}$ is a measure on $(\Omega, \mathcal{A})$ such that (3.38) holds for $\mathbf{Q}$-a.e. $\omega \in \Omega$ then we say that T is $\mathbf{Q}$-a.e. (M,N)-balancing.

The next result is a fundamental transport property of Palm measures. It generalizes Theorem 4.2 in Last and Thorisson (2009).

Theorem 3.15 *Let $\mathbf{P}$ be a σ-finite stationary measure on $(\Omega, \mathcal{A})$. Consider two invariant random measures M and N on G and let T and T^* be invariant weighted transport-kernels satisfying*

$$\iint \mathbf{1}\{(g,h) \in \cdot\} T(\omega, g, \mathrm{d}h) M(\omega, \mathrm{d}g) = \iint \mathbf{1}\{(g,h) \in \cdot\} T^*(\omega, h, \mathrm{d}g) N(\omega, \mathrm{d}h) \tag{3.39}$$

for $\mathbf{P}$-a.e. $\omega \in \Omega$. Then we have for any measurable function $f : \Omega \times G \to \mathbb{R}_+$ that

$$\mathbf{E}_{\mathbf{Q}_M} \int f\left(\theta_g^{-1}, g^{-1}\right) \Delta(g^{-1}) T(e, \mathrm{d}g) = \mathbf{E}_{\mathbf{Q}_N} \int f(\theta_e, g) T^*(e, \mathrm{d}g). \tag{3.40}$$

Proof Since $\int w(g^{-1}h)\lambda(\mathrm{d}h) = 1$, $g \in G$, we have from Fubini's theorem that

$$\begin{aligned} I = \mathbf{E}_{\mathbf{Q}_N} \int f(\theta_e, g) T^*(c, \mathrm{d}g) &= \mathbf{E}_{\mathbf{Q}_N} \iint f(\theta_e, g) w(g^{-1}h)\lambda(\mathrm{d}h) T^*(c, \mathrm{d}g) \\ &= \mathbf{E}_{\mathbf{Q}_N} \iint f(\theta_e, g) w(g^{-1}h) T^*(c, \mathrm{d}g)\lambda(\mathrm{d}h). \end{aligned}$$

Next we use the refined Campbell theorem (3.12) and (3.34) (applied to T^*) to get

$$\begin{aligned} I &= \mathbf{E} \iint f\left(\theta_h^{-1}, g\right) w(g^{-1}h) T^*\left(\theta_h^{-1}, e, \mathrm{d}g\right) N(\mathrm{d}h) \\ &= \mathbf{E} \iint f\left(\theta_h^{-1}, h^{-1}g\right) w(g^{-1}hh) T^*(\theta_e, h, \mathrm{d}g) N(\mathrm{d}h) \\ &= \mathbf{E} \iint f\left(\theta_h^{-1}, h^{-1}g\right) w(g^{-1}hh) T(\theta_e, g, \mathrm{d}h) M(\mathrm{d}g), \end{aligned}$$

where the last equality is due to assumption (3.39). We now make the above steps in the reversed direction. Again by (3.34),

$$I = \mathbf{E} \iint f\left(\theta_{gh}^{-1}, h^{-1}\right) w(hgh) T\left(\theta_g^{-1}, e, \mathrm{d}h\right) M(\mathrm{d}g).$$

Since $\theta_{gh}^{-1} = \theta_h^{-1} \circ \theta_g^{-1}$ we can use the refined Campbell theorem, to obtain

$$I = \mathbf{E}_{\mathbf{Q}_M} \iint f\left(\theta_h^{-1}, h^{-1}\right) w(hgh) T(\theta_e, e, \mathrm{d}h) \lambda(\mathrm{d}g).$$

The result now follows from

$$\int w(hgh)\lambda(\mathrm{d}g) = \int w(gh)\lambda(\mathrm{d}g) = \Delta(h^{-1}), \quad h \in G. \tag{3.41}$$

□

Theorem 3.15 will play a key role in establishing an invariance property of Palm measures, see Theorem 3.19 below. One interesting special case is the following generalization of the exchange formula (Neveu, 1977) from Abelian to locally compact groups, see also Theorem 1.10 in Chapter 1 of this volume. Another special case is Theorem 3.27 below.

Corollary 3.16 *Let* $\mathbf{P}$ *be a* σ*-finite stationary measure on* $(\Omega, \mathcal{A})$ *and* M, N *invariant random measure on* G*. Then we have for any measurable function* $f : \Omega \times G \to \mathbb{R}_+$ *that*

$$\mathbf{E}_{\mathbf{Q}_M} \int f\left(\theta_g^{-1}, g^{-1}\right) \Delta(g^{-1}) N(\mathrm{d}g) = \mathbf{E}_{\mathbf{Q}_N} \int f(\theta_e, g) M(\mathrm{d}g). \tag{3.42}$$

Proof Apply Theorem 3.15 with $T(g, \cdot) = M$ and $T^*(g, \cdot) = N$. □

Remark 3.17 Let $B \in \mathcal{G}$ have positive and finite Haar measure. Using the definition (3.11) of Palm measures (applied to $w = \lambda(B)^{-1}\mathbf{1}_B$) as well as invariance of M and N, we can rewrite the exchange formula (3.42) as

$$\begin{aligned} \mathbf{E} \iint & \mathbf{1}\{h \in B\} f\left(\theta_g^{-1}, g^{-1}h\right) \Delta(g^{-1}) \Delta(h) N(\mathrm{d}g) M(\mathrm{d}h) \\ &= \mathbf{E} \iint \mathbf{1}\{h \in B\} f\left(\theta_h^{-1}, h^{-1}g\right) M(\mathrm{d}g) N(\mathrm{d}h). \end{aligned} \tag{3.43}$$

The function $t(\omega, g, h) = f\left(\theta_g^{-1}\omega, g^{-1}h\right)$ is invariant in the sense of (3.35). Equation (3.43) implies

$$\mathbf{E}\iint \mathbf{1}\{g \in B\} t(h,g)\Delta(g^{-1})\Delta(h)N(\mathrm{d}h)M(\mathrm{d}g)$$
$$= \mathbf{E}\iint \mathbf{1}\{g \in B\} t(g,h)N(\mathrm{d}g)M(\mathrm{d}h) \qquad (3.44)$$

for all invariant t. In case $M = N$ this gives a version of the *mass-transport principle* (Benjamini, Lyons, Peres and Schramm, 1999) for stationary random measures on groups. It is shown in Last (2008) that Neveu's exchange formula (3.42) can be generalized to jointly stationary random measures on a homogeneous space. In fact, Benjamini, Lyons, Peres and Schramm (1999), Aldous and Lyons (2007), and Gentner and Last (2009) show that the mass-transport principle can be extended beyond this setting.

Corollary 3.18 *Let the assumptions of Theorem 3.15 be satisfied. Assume also that M and N have finite intensities and that $\mathbf{P}\{\hat{M} = 0\} = \mathbf{P}\{\hat{N} = 0\} = 0$. Then we have for any measurable function $f : \Omega \times G \to \mathbb{R}_+$ that*

$$\hat{M}\mathbf{E}_{\mathbf{Q}_M^*}\left[\int f\left(\theta_g^{-1}, g^{-1}\right)\Delta(g^{-1})T(c,\mathrm{d}g)\Big|\mathcal{I}\right] = \hat{N}\mathbf{E}_{\mathbf{Q}_N^*}\left[\int f(\theta_e, g)T^*(e,\mathrm{d}g)\Big|\mathcal{I}\right],$$

$\mathbf{P}$*-a.e. for any choice of the conditional expectations.*

Proof Define the variables X and X' by $X = \int f\left(\theta_g^{-1}, g^{-1}\right)\Delta(g^{-1})T(e,\mathrm{d}g)$ and $X' = \int f(\theta_e, g)T^*(c,\mathrm{d}g)$. Due to (3.20) we have $\mathbf{Q}_M^* = \mathbf{Q}_N^*$ on $\mathcal{I}$. Hence we have to show that

$$\mathbf{E}_{\mathbf{Q}_M^*}\mathbf{1}_A\hat{M}X = \mathbf{E}_{\mathbf{Q}_N^*}\mathbf{1}_A\hat{N}X', \quad A \in \mathcal{I}.$$

By (3.21) this amounts to $\mathbf{E}_{\mathbf{Q}_M}\mathbf{1}_A X = \mathbf{E}_{\mathbf{Q}_N}\mathbf{1}_A X'$, i.e. to a consequence of (3.40). □

3.5 Invariance properties of Palm measures

In this section we fix a stationary σ-finite measure $\mathbf{P}$ on $(\Omega, \mathcal{A})$. In the special case of an Abelian group the following fundamental invariance property of Palm measures has recently been established in Last and Thorisson (2009).

Theorem 3.19 *Consider two invariant random measures M and N on G and an invariant weighted transport-kernel T. Then T is $\mathbf{P}$-a.e. (M,N)-balancing iff*

$$\mathbf{E}_{\mathbf{Q}_M}\int f\left(\theta_g^{-1}\right)\Delta(g^{-1})T(e,\mathrm{d}g) = \mathbf{E}_{\mathbf{Q}_N} f \qquad (3.45)$$

holds for all measurable $f : \Omega \to \mathbb{R}_+$.

Proof Assume first that T is $\mathbf{P}$-a.e. (M,N)-balancing. Lemma 3.20 below shows that there exists an invariant transport-kernel T^* satisfying (3.39) for $\mathbf{P}$-a.e. $\omega \in \Omega$. Applying (3.40) to a function not depending on the second argument, yields (3.45).

Let us now assume that (3.45) holds. Take a measurable function $f : \Omega \times G \to \mathbb{R}_+$. By the refined Campbell theorem and (3.45),

$$\mathbf{E}\int f\left(\theta_g^{-1}, g\right) N(\mathrm{d}g) = \mathbf{E}_{\mathbf{Q}_M} \iint f\left(\theta_h^{-1}, g\right) \Delta(h^{-1}) T(e, \mathrm{d}h)\lambda(\mathrm{d}g).$$

By Fubini's theorem and (3.2) this equals

$$\mathbf{E}_{\mathbf{Q}_M} \iint f\left(\theta_h^{-1}, gh\right)\Delta(h)T(e,\mathrm{d}h)\lambda(\mathrm{d}g)$$
$$= \mathbf{E}\iint f\left(\theta_h^{-1}\circ\theta_g^{-1}, gh\right) T\left(\theta_g^{-1}, e, \mathrm{d}h\right) M(\mathrm{d}g),$$

where the equality is again due to the refined Campbell theorem, this time applied to M. By (3.34),

$$\mathbf{E}\int f\left(\theta_g^{-1}, g\right) N(\mathrm{d}g) = \mathbf{E}\iint f\left(\theta_h^{-1}, h\right) T(g,\mathrm{d}h)M(\mathrm{d}g).$$

A straightforward substitution yields

$$\mathbf{E}\int \tilde{f}(\theta_e, g)N(\mathrm{d}g) = \mathbf{E}\iint \tilde{f}(\theta_e, g)T(h,\mathrm{d}g)M(\mathrm{d}h),$$

for all measurable functions $\tilde{f} : \Omega \times G \to \mathbb{R}_+$. From this we obtain by a standard procedure that $N(B) = \int T(\theta_e, g, B)M(\mathrm{d}g)$ $\mathbf{P}$-a.e. for all $B \in \mathcal{G}$. Since $\mathcal{G}$ is countably generated, this concludes the proof of the theorem. □

The above proof has used the following lemma:

Lemma 3.20 *Assume that T is a $\mathbf{P}$-a.e. (M,N)-balancing invariant weighted transport-kernel. Then there exists an invariant transport-kernel T^* on G such that* (3.39) *holds for $\mathbf{P}$-a.e. $\omega \in \Omega$.*

Proof Consider the following measure W on $\Omega \times G \times G$:

$$W = \iiint \mathbf{1}\{(\omega, g, h) \in \cdot\}T(\omega, g, \mathrm{d}h)M(\omega, \mathrm{d}g)\mathbf{P}(\mathrm{d}\omega).$$

Stationarity of $\mathbf{P}$, (3.5), and (3.34) easily imply that

$$\int \mathbf{1}\{(\theta_g\omega, kg, kh) \in \cdot\}W(\mathrm{d}(\omega, g, h)) = W, \quad k \in G.$$

Moreover, as T is a $\mathbf{P}$-a.e. (M, N)-balancing, we have

$$W' = \int \mathbf{1}\{(\omega, h) \in \cdot\} W(\mathrm{d}(\omega, g, h)) = \iint \mathbf{1}\{(\omega, h) \in \cdot\} N(\omega, \mathrm{d}h)\mathbf{P}(\mathrm{d}\omega). \quad (3.46)$$

This is a σ-finite measure on $\Omega \times G$. As in the proof of Lemma 3.11 we can now apply Theorem 3.5 in Kallenberg (2007) to obtain an invariant transport-kernel T^* satisfying

$$W = \iint \mathbf{1}\{(\omega, g, h) \in \cdot\} T^*(\omega, h, \mathrm{d}g) W'(\mathrm{d}(\omega, h)).$$

Recalling the definition of W and the second equation in (3.46) we get that

$$\iiint \mathbf{1}\{(\omega, g, h) \in \cdot\} T(\omega, g, \mathrm{d}h) M(\omega, \mathrm{d}g)\mathbf{P}(\mathrm{d}\omega)$$
$$= \iiint \mathbf{1}\{(\omega, g, h) \in \cdot\} T^*(\omega, h, \mathrm{d}g) N(\omega, \mathrm{d}h)\mathbf{P}(\mathrm{d}\omega)$$

and hence the assertion of the lemma. □

Example 3.21 Consider the setting of Example 3.2 and let $\mathbf{P}$ be a σ-finite stationary measure on $(\Omega, \mathcal{A})$ concentrated on the set Ω' of all integer-valued $\omega \in \Omega$. Applying Theorem 3.19 with T given by (3.37) yields

$$\int \frac{1}{M(\omega, \{e\})} \sum_{h:\omega(e,h)>0} f(\theta_h \omega)\omega(e, h)\Delta(g^{-1})\mathbf{Q}_M(\mathrm{d}\omega) = \mathbf{E}_{\mathbf{Q}_N}[f] \quad (3.47)$$

for any measurable $f : \Omega \to \mathbb{R}_+$. Specializing to the case of a function f depending only on $N(\omega)$ (and to an Abelian subgroup of $\mathbb{R}^d$) yields Theorem 6.5 in Port and Stone (1973). In the special case $G = \mathbb{R}$ (and under further restrictions on the support of $\mathbf{P}$), (3.47) is Theorem (6.5) in Harris (1971).

In fact, Port and Stone (1973) use a specific form of the Palm measure $\mathbf{Q}_M$. To explain this, we assume that $\mathbf{P}\{M' \in \cdot\}$ is σ-finite, where M' is the marked random measure defined by $M'(\omega) = \omega(\cdot \times G \times E)$. Then there is a Markov kernel K from M_E to Ω satisfying

$$\mathbf{P} = \int K(\eta, \cdot)\mathbf{P}\{M' \in \mathrm{d}\eta\}.$$

Here M_E denotes the space of all measures η on $G \times E$ such that $\eta(\cdot \times E) \in \mathsf{M}$. (Similarly as in Example 3.2 this space can be equipped with a σ-field and a flow.) By Theorem 3.5 in Kallenberg (2007) we can assume that K is invariant. Of course, if $\mathbf{P}$ is a probability measure, then $K(M', A)$ is a version of the conditional probability of $A \in \mathcal{A}$ given M'. Using invariance of K, it is straightforward to check that

$$\mathbf{Q}_M = \int K(\eta, \cdot)\mathbf{Q}_M\{M' \in \mathrm{d}\eta\}. \quad (3.48)$$

Let M and N be random measures on G. We call an allocation rule τ (see Example 3.13) **P**-a.e. *(M,N)-balancing*, if

$$\int \mathbf{1}\{\tau(g) \in \cdot\} M(\mathrm{d}g) = N \tag{3.49}$$

holds **P**-a.e., i.e. if the transport T defined by $T(g,\cdot) = \delta_{\tau(g)}$ is **P**-a.e. (M,N)-balancing. For an allocation rule τ it is convenient to introduce the measurable mapping $\theta_\tau : \Omega \to \Omega$ by

$$\theta_\tau(\omega) = \theta_{\tau(\omega,e)}\omega. \tag{3.50}$$

Similarly we define $\theta_\tau^{-1} : \Omega \to \Omega$ by $\theta_\tau^{-1}(\omega) = \theta_{\tau(\omega,e)}^{-1}\omega$.

Corollary 3.22 *Consider two invariant random measures M and N and let τ be an allocation rule. Then τ is **P**-a.e. (M,N)-balancing iff*

$$\mathbf{E}_{\mathbf{Q}_M} f\left(\theta_\tau^{-1}\right) \Delta(\tau(e)^{-1}) = \mathbf{E}_{\mathbf{Q}_N} f \tag{3.51}$$

holds for all measurable $f : \Omega \to \mathbb{R}_+$.

In the Abelian case Corollary 3.22 can be found in Last and Thorisson (2009). If in addition $M = N$, then one implication is (essentially) a consequence of Satz 4.3 in Mecke (1975). The special case $M = \lambda$ was treated in Geman and Horowitz (1975).

3.6 Existence of balancing weighted transport-kernels

We fix a stationary σ-finite measure **P** on $(\Omega, \mathcal{A})$ and consider two invariant random measures M, N on G. Our aim is to establish a necessary and sufficient condition for the existence of (M,N)-balancing invariant weighted transport-kernels T satisfying

$$\int \Delta(g^{-1}) T(e, \mathrm{d}g) = 1. \tag{3.52}$$

Theorem 3.23 *Assume that M and N have positive and finite intensities. Then there exists a **P**-a.e. (M,N)-balancing invariant weighted transport-kernel satisfying* (3.52) *iff*

$$\mathbf{E}[M(B)|\mathcal{I}] = \mathbf{E}[N(B)|\mathcal{I}] \quad \mathbf{P}-a.e. \tag{3.53}$$

for some $B \in \mathcal{G}$ satisfying $0 < \lambda(B) < \infty$.

Proof Let $B \in \mathcal{G}$ satisfy $0 < \lambda(B) < \infty$. For any $A \in \mathcal{I}$ we have from the refined Campbell theorem (3.12) that

$$\lambda(B)\mathbf{Q}_M(A) = \mathbf{E}[\mathbf{1}_A M(B)], \quad \lambda(B)\mathbf{Q}_N(A) = \mathbf{E}[\mathbf{1}_A N(B)]. \tag{3.54}$$

Assume now that T is a **P**-a.e. (M,N)-balancing invariant weighted transport-kernel satisfying (3.52). Then Theorem 3.19 implies for all $A \in \mathcal{I}$

the equality $\mathbf{Q}_M(A) = \mathbf{Q}_N(A)$. Thus (3.54) implies $\mathbf{E}\mathbf{1}_A M(B) = \mathbf{E}\mathbf{1}_A N(B)$ and hence (3.53).

Let us now assume that (3.53) holds for some $B \in \mathcal{G}$ satisfying $0 < \lambda(B) < \infty$. Since $\mathbf{E}M(\cdot)$ and $\mathbf{E}N(\cdot)$ are multiples of λ, M and N have the same intensities. We assume without loss of generality that these intensities are equal to 1. From (3.54) and conditioning we obtain that $\mathbf{Q}_M = \mathbf{Q}_N$ on $\mathcal{I}$. Using the group-coupling result in Thorisson (1996) as in Last and Thorisson (2009), we obtain a stochastic kernel $\tilde{T}$ from Ω to G satisfying

$$\mathbf{Q}_N(A) = \mathbf{E}_{\mathbf{Q}_M} \int \mathbf{1}\left\{\theta_g^{-1} \in A\right\} \tilde{T}(\theta_e, \mathrm{d}g), \quad A \in \mathcal{A}. \tag{3.55}$$

Let T' be the kernel from Ω to G defined by

$$T'(\omega, \cdot) = \int \mathbf{1}\{g \in \cdot\}\Delta(g)\tilde{T}(\omega, \mathrm{d}g).$$

Since Δ is continuous, $T'(\omega, B)$ is finite for all $\omega \in \Omega$ and all compact $B \subset G$. The invariant weighted transport-kernel T defined by $T(\omega, g, B) = T'\left(\theta_g^{-1}\omega, g^{-1}B\right)$ satisfies (3.52) by definition. And because of (3.55) it does also satisfy (3.45). Theorem 3.19 implies that T is $\mathbf{P}$-a.e. (M, N)-balancing. □

3.7 Mecke's characterization of Palm measures

Let M be an invariant random measure on G. In contrast to the previous sections we do not fix a stationary measure on $(\Omega, \mathcal{A})$. Instead we consider here a measure $\mathbf{Q}$ on $(\Omega, \mathcal{A})$ as a candidate for a Palm measure $\mathbf{Q}_M$ of M w.r.t. some stationary measure $\mathbf{P}$ on $(\Omega, \mathcal{A})$. In case G is an Abelian group (and within a canonical framework) the following fundamental characterization theorem was proved in Mecke (1967). In a canonical framework (and for finite intensities) the present extension has been established in Rother and Zähle (1990), even in the more general case of a random measure on a homogeneous space.

Theorem 3.24 *The measure $\mathbf{Q}$ is a Palm measure of M with respect to some σ-finite stationary measure iff $\mathbf{Q}$ is σ-finite, $\mathbf{Q}\{M(G) = 0\} = 0$, and*

$$\mathbf{E}_{\mathbf{Q}} \int f\left(\theta_g^{-1}, g^{-1}\right) \Delta(g^{-1}) M(\mathrm{d}g) = \mathbf{E}_{\mathbf{Q}} \int f(\theta_e, g) M(\mathrm{d}g) \tag{3.56}$$

holds for all measurable $f : \Omega \times G \to \mathbb{R}_+$.

Proof If $\mathbf{Q}$ is a Palm measure of M, then $\mathbf{Q}$ is σ-finite by Proposition 3.9. Equation $\mathbf{Q}\{M(G) = 0\} = 0$ holds by Proposition 3.5, while the Mecke equation (3.56) is a special case of (3.40).

Let us now conversely assume the stated conditions. Using the function $\tilde{w}$ occurring in (3.17), we define a measure $\mathbf{P}$ on Ω by

$$\mathbf{P}(A) = \mathbf{E_Q} \int \mathbf{1}\{\theta_g \in A\} \tilde{w}(gM, g)\lambda(\mathrm{d}g), \quad A \in \mathcal{A}. \tag{3.57}$$

Take a measurable $f : \Omega \times G \to \mathbb{R}_+$. By (3.16),

$$\begin{aligned}\mathbf{E_Q} \int f(\theta_g, g)\lambda(\mathrm{d}g) &= \mathbf{E_Q} \iint f(\theta_g, g)\tilde{w}(gM, h)gM(\mathrm{d}h)\lambda(\mathrm{d}g) \\ &= \mathbf{E_Q} \iint f(\theta_g, g)\tilde{w}(M \circ \theta_g, gh)\lambda(\mathrm{d}g)M(\mathrm{d}h).\end{aligned}$$

Using assumption (3.56), we get

$$\begin{aligned}&\mathbf{E_Q} \int f(\theta_g, g)\lambda(\mathrm{d}g) \\ &= \mathbf{E_Q} \iint f(\theta_g \circ \theta_{h^{-1}}, g)\tilde{w}(M \circ \theta_g \circ \theta_{h^{-1}}, gh^{-1})\Delta(h^{-1})\lambda(\mathrm{d}g)M(\mathrm{d}h) \\ &= \mathbf{E_Q} \iint f(\theta_{g^{-1}h^{-1}}, g^{-1})\tilde{w}(M \circ \theta_{g^{-1}h^{-1}}, g^{-1}h^{-1})\Delta(h^{-1})\Delta(g^{-1})\lambda(\mathrm{d}g)M(\mathrm{d}h) \\ &= \mathbf{E_Q} \iint f(\theta_{g^{-1}}, g^{-1}h)\tilde{w}(M \circ \theta_{g^{-1}}, g^{-1})\Delta(g^{-1})\lambda(\mathrm{d}g)M(\mathrm{d}h) \\ &= \mathbf{E_Q} \iint f(\theta_g, gh)\tilde{w}(M \circ \theta_g, g)\lambda(\mathrm{d}g)M(\mathrm{d}h),\end{aligned}$$

where we have used properties of the Haar measure λ for the second, third, and fourth equality. This implies

$$\mathbf{E_Q} \int f(\theta_g, g)\lambda(\mathrm{d}g) = \mathbf{E_Q} \iint f(\theta_g, h)\tilde{w}(M \circ \theta_g, g)M \circ \theta_g(\mathrm{d}h)\lambda(\mathrm{d}g),$$

so that by definition (3.57)

$$\mathbf{E_Q} \int f(\theta_g, g)\lambda(\mathrm{d}g) = \mathbf{E} \int f(\theta_e, h)M(\mathrm{d}h). \tag{3.58}$$

Since $\mathbf{Q} \otimes \lambda$ is σ-finite, there is a measurable function $\tilde{f} : \Omega \times G \to (0, \infty)$ having $\mathbf{E_Q} \int \tilde{f}(\theta_e, g)\lambda(\mathrm{d}g) < \infty$. Defining a positive measurable function $f : \Omega \times G \to \mathbb{R}_+$ by $f(\omega, h) = \tilde{f}(\theta_{h^{-1}}\omega, h)$, we obtain

$$\mathbf{E_Q} \int f(\theta_g, g)\lambda(\mathrm{d}g) = \mathbf{E_Q} \int \tilde{f}(\theta_e, g)\lambda(\mathrm{d}g) < \infty.$$

Hence (3.58) implies that the $\mathbf{P}$-a.e. positive measurable function $\int f(\theta_e, g)$ $M(\mathrm{d}g)$ has a finite integral with respect to $\mathbf{P}$. Therefore $\mathbf{P}$ is σ-finite.

Next we show that $\mathbf{P}$ is stationary. By invariance of λ we have for $A \in \mathcal{A}$ and $h \in G$

$$\begin{aligned}\mathbf{P}(A) &= \mathbf{E}_{\mathbf{Q}} \int \mathbf{1}\{\theta_{h^{-1}g} \in A\}\tilde{w}(M \circ \theta_{h^{-1}g}, h^{-1}g)\lambda(\mathrm{d}g) \\ &= \mathbf{E}_{\mathbf{Q}} \int \mathbf{1}\{\theta_g \in \theta_h A\}\tilde{w}(M \circ \theta_{h^{-1}} \circ \theta_g, h^{-1}g)\lambda(\mathrm{d}g).\end{aligned}$$

Applying (3.58), yields

$$\mathbf{P}(A) = \mathbf{E} \int \mathbf{1}\{\theta_e \in \theta_h A\}\tilde{w}(M \circ \theta_{h^{-1}}, h^{-1}g)M(\mathrm{d}g) = \mathbf{P}(\theta_h A),$$

since $\int \tilde{w}(M \circ \theta_{h^{-1}}, h^{-1}g)M(\mathrm{d}g) = 1$ $\mathbf{P}$-a.e.

It remains to show that $\mathbf{Q}$ is the Palm measure $\mathbf{Q}_M$ of M with respect to $\mathbf{P}$. By (3.11) and (3.58),

$$\mathbf{Q}_M(A) = \mathbf{E} \int \mathbf{1}\left\{\theta_h^{-1} \in A\right\} w(h)M(\mathrm{d}h) = \mathbf{E}_{\mathbf{Q}} \int \mathbf{1}\left\{\theta_g^{-1} \circ \theta_g \in A\right\} w(g)\lambda(\mathrm{d}g).$$

Hence $\mathbf{Q}_M(A) = \mathbf{Q}(A)$, as desired. □

3.8 Mass-stationarity

We fix an invariant random measure M on G and a σ-finite measure $\mathbf{Q}$ on $(\Omega, \mathcal{A})$. Our aim is to establish *mass-stationarity* as another characterizing property of Palm measures of M. This generalizes one of the main results in Last and Thorisson (2009) from Abelian to arbitrary groups satisfying the general assumptions.

Let $C \in \mathcal{G}$ be relatively compact and define a Markovian transport kernel T_C by

$$T_C(g, B) = M(gC)^{-1}M(B \cap gC), \quad g \in G,\ B \in \mathcal{G}, \tag{3.59}$$

if $M(gC) > 0$, and by letting $T_C(g, \cdot)$ equal some fixed probability measure, otherwise. In the former case $T_C(g, \cdot)$ is just governing a G-valued stochastic experiment that picks a point uniformly in the mass of M in gC. Since M is invariant, it is immediate that T_C is invariant too. If $0 < \lambda(C) < \infty$ we also define the uniform distribution λ_C on C by $\lambda_C(B) = \lambda(B \cap C)/\lambda(C)$. The interior (resp. boundary) of a set $C \subset G$ is denoted by $\operatorname{int} C$ (resp. ∂C).

A σ-finite measure $\mathbf{Q}$ on $(\Omega, \mathcal{A})$ is called *mass-stationary for M* if $\mathbf{Q}\{M(G) = 0\} = 0$ and

$$\begin{aligned}\mathbf{E}_{\mathbf{Q}} \iint \mathbf{1}\{(\theta_g^{-1}, hg) \in A\}T_C(h^{-1}, \mathrm{d}g)\lambda_C(\mathrm{d}h) \\ = \mathbf{Q} \otimes \lambda_C(A), \quad A \in \mathcal{A} \otimes \mathcal{G},\end{aligned} \tag{3.60}$$

holds for all relatively compact sets $C \in \mathcal{G}$ with $\lambda(C) > 0$ and $\lambda(\partial C) = 0$.

Remark 3.25 Assume that $\mathbf{Q}$ is a probability measure. Let C be as assumed in (3.60). Extend the space $(\Omega, \mathcal{A}, \mathbf{Q})$, so as to carry random elements U, V in G such that θ_e and U are independent, U has distribution λ_C, and the conditional distribution of V given (θ_e, U) is uniform in the mass of M on $U^{-1}C$. (The mappings θ_g, $g \in G$, are extended, so that they still take values in the original space Ω.) Then (3.60) can be written as

$$\left(\theta_V^{-1}, UV\right) \stackrel{d}{=} (\theta_e, U). \tag{3.61}$$

Mass-stationarity of $\mathbf{Q}$ requires that this holds for all such pairs (U, V).

Theorem 3.26 *There exists a σ-finite stationary measure $\mathbf{P}$ on $(\Omega, \mathcal{A})$ such that $\mathbf{Q} = \mathbf{Q}_M$ iff $\mathbf{Q}$ is mass-stationary for M.*

Proof Assume first that $\mathbf{Q} = \mathbf{Q}_M$ is the Palm measure of M with respect to a σ-finite stationary measure $\mathbf{P}$. Let C be as in (3.60) and D a measurable subset of C. Define a weighted transport-kernel T by

$$T(g, B) = \iint \mathbf{1}\{k \in B, h^{-1}k \in D\}\mathbf{1}\{h^{-1}g \in C\}T_C(h, \mathrm{d}k)\lambda(\mathrm{d}h).$$

By invariance of T_C and left-invariance of λ,

$$\begin{aligned} T(\theta_g, g, gB) &= \iint \mathbf{1}\{k \in B, h^{-1}gk \in D\}\mathbf{1}\{h^{-1}g \in C\}T_C(g^{-1}h, \mathrm{d}k)\lambda(\mathrm{d}h) \\ &= T(\theta_e, e, B). \end{aligned}$$

Hence T is invariant. Let $g \in G$. From the properties of the set C and continuity of group multiplication we have for λ-a.e. h such that $h^{-1}g \in C$ the relationship $h^{-1}g \in \operatorname{int} C$. If $h^{-1}g \in \operatorname{int} C$ and $g \in \operatorname{supp} M$, then $g \in \operatorname{int} hC$ and $M(hC) > 0$. Using this and the definition of T_C, we obtain

$$\begin{aligned} \int T(g, B)M(\mathrm{d}g) &= \iint \mathbf{1}\{k \in B, h^{-1}k \in D\}\lambda(\mathrm{d}h)M(\mathrm{d}k) \\ &= \iint \mathbf{1}\{k \in B, h^{-1} \in D\}\lambda(\mathrm{d}h)M(\mathrm{d}k) = M(B)\lambda^*(D), \end{aligned}$$

where $\lambda^*(D) = \int \mathbf{1}\{h^{-1} \in D\}\lambda(\mathrm{d}h)$. Therefore we obtain from Theorem 3.19 that

$$\mathbf{E}_\mathbf{Q} \int f\left(\theta_g^{-1}\right) \Delta(g^{-1})T(e, \mathrm{d}g) = \lambda^*(D)\mathbf{E}_\mathbf{Q} f \tag{3.62}$$

for all measurable $f : \Omega \to \mathbb{R}_+$. The above left-hand side equals

$$\begin{aligned} &\mathbf{E}_\mathbf{Q} \iint f\left(\theta_g^{-1}\right) \Delta(g^{-1})\mathbf{1}\{g \in hD, h^{-1} \in C\}T(e, \mathrm{d}g)\lambda(\mathrm{d}h) \\ &\quad = \mathbf{E}_\mathbf{Q} \iint f\left(\theta_g^{-1}\right) \Delta(g^{-1})\mathbf{1}\{hg \in D, h \in C\}T(e, \mathrm{d}g)\lambda^*(\mathrm{d}h). \end{aligned}$$

Hence we obtain from (3.62) and a monotone class argument

$$\mathbf{E}_{\mathbf{Q}} \iint f\left(\theta_g^{-1}, hg\right) \Delta(g^{-1}) \mathbf{1}\{h \in C\} T(e, \mathrm{d}g) \lambda^*(\mathrm{d}h)$$
$$= \mathbf{E}_{\mathbf{Q}} \int f(\theta_e, h) \mathbf{1}\{h \in C\} \lambda^*(\mathrm{d}h)$$

for all measurable $f : \Omega \times G \to \mathbb{R}_+$. By (3.3) this means

$$\mathbf{E}_{\mathbf{Q}} \iint f\left(\theta_g^{-1}, hg\right) \Delta(g^{-1}) \Delta(h^{-1}) \mathbf{1}\{h \in C\} T(e, \mathrm{d}g) \lambda(\mathrm{d}h)$$
$$= \mathbf{E}_{\mathbf{Q}} \int f(\theta_e, h) \Delta(h^{-1}) \mathbf{1}\{h \in C\} \lambda(\mathrm{d}h).$$

Applying this with $f(\omega, k)$ replaced by $f(\omega, k)\Delta(k)$ gives

$$\mathbf{E}_{\mathbf{Q}} \iint f\left(\theta_g^{-1}, hg\right) \mathbf{1}\{h \in C\} T(e, \mathrm{d}g) \lambda(\mathrm{d}h) = \mathbf{E}_{\mathbf{Q}} \int f(\theta_e, h) \mathbf{1}\{h \in C\} \lambda(\mathrm{d}h).$$

This is equivalent to (3.60).

Now we assume, conversely, that $\mathbf{Q}$ is mass-stationary. Our strategy is to derive the Mecke equation (3.56) and to apply Theorem 3.24. Let $f : \Omega \to \mathbb{R}_+$ be measurable, C be as in (3.60) and D a measurable subset of C. By (3.60)

$$\mathbf{E}_{\mathbf{Q}} \iint f\left(\theta_g^{-1}\right) M(h^{-1}C)^{-1} \mathbf{1}\{hg \in D, h \in C\} M(\mathrm{d}g) \lambda(\mathrm{d}h) = \lambda(D) \mathbf{E}_{\mathbf{Q}} f,$$

where $1/0 = 0$. Therefore,

$$\mathbf{E}_{\mathbf{Q}} \iint f\left(\theta_g^{-1}\right) M(gh^{-1}C)^{-1} \mathbf{1}\{h \in D, hg^{-1} \in C\} \Delta(g^{-1}) M(\mathrm{d}g) \lambda(\mathrm{d}h)$$
$$= \lambda(D) \mathbf{E}_{\mathbf{Q}} f.$$

Since D can be any measurable subset of C, we get

$$\mathbf{E}_{\mathbf{Q}} \int f\left(\theta_g^{-1}\right) M(gh^{-1}C)^{-1} \mathbf{1}\{hg^{-1} \in C\} \Delta(g^{-1}) M(\mathrm{d}g) = \mathbf{E}_{\mathbf{Q}} f$$

for λ-a.e. $h \in C$. In particular we also have for λ-a.e. $h \in C$ that

$$\mathbf{E}_{\mathbf{Q}} \int f\left(\theta_g^{-1}\right) \tilde{f}(M \circ \theta_g^{-1}) M(gh^{-1}C)^{-1} \mathbf{1}\{hg^{-1} \in C\} \Delta(g^{-1}) M(\mathrm{d}g)$$
$$= \mathbf{E}_{\mathbf{Q}} f \tilde{f}(M),$$

for any measurable $\tilde{f} : \mathsf{M} \to \mathbb{R}_+$. Since $\mathcal{M}$ is countably generated, we may choose the corresponding λ-null set independent of $\tilde{f}$. In particular we may take for any

h outside a λ-null set, $\tilde{f}(\eta) = \eta(h^{-1}C)$. Then $\tilde{f}\left(M \circ \theta_g^{-1}\right) = M \circ \theta_g^{-1}(h^{-1}C) = M(gh^{-1}C)$, so that

$$\mathbf{E_Q} \int f\left(\theta_g^{-1}\right) \mathbf{1}\{hg^{-1} \in C\}\Delta(g^{-1})M(\mathrm{d}g) = \mathbf{E_Q} f M(h^{-1}C)$$

for λ-a.e. $h \in C$. From here we can proceed as in Last and Thorisson (2009), to get the full Mecke equation (3.56). $\square$

3.9 Stationary partitions

The topic of this section are (extended) stationary partitions as introduced (in case $G = \mathbb{R}^d$) and studied in Last (2006). The motivation of Last (2006) was to connect stationary tessellations of classical stochastic geometry (Stoyan, Kendall and Mecke, (1995); 1) with recent work in Holroyd and Peres (2005) and Hoffman, Holroyd and Peres (2006) on allocation rules transporting Lebesgue measure to a simple point process. Our terminology here is closer to stochastic geometry.

Let N be an invariant simple point process on G. As mentioned earlier, we identify N with its support. A *stationary partition* (based on N) is a pair (Z, τ) consisting of a *measurable* set $Z : \Omega \to \mathcal{G}$ and an allocation rule τ such that $\tau(g) \in N$ whenever $g \in Z$. We also assume that $\{Z = \emptyset\} = \{N = \emptyset\}$. Measurability of Z just means that $(\omega, g) \mapsto \mathbf{1}\{g \in Z(\omega)\}$ is measurable, while covariance of Z means that

$$Z(\theta_g \omega) = gZ(\omega), \quad \omega \in \Omega,\ g \in G. \tag{3.63}$$

For convenience we also assume that $\tau(g) = g$, $g \in G$, whenever $N = \emptyset$. Define

$$C(\omega, g) = \{h \in Z(\omega) : \tau(\omega, h) = g\}, \quad \omega \in \Omega,\ g \in G. \tag{3.64}$$

Note that $C(g) = \emptyset$ whenever $g \notin N \neq \emptyset$. The system $\{C(g) : g \in N\}$ forms a partition of Z into measurable sets, provided that $N \neq \emptyset$. Equations (3.63) and (3.36) imply the following covariance property:

$$C(\theta_h \omega, hg) = hC(\omega, g), \quad \omega \in \Omega,\ g, h \in G. \tag{3.65}$$

Although we do not make any topological or geometrical assumptions, we refer to $C(g)$ as *cell* with (generalized) *centre* $g \in N$. We do not assume that $g \in C(g)$ and some of the cells might be empty.

We now fix a σ-finite stationary measure $\mathbf{P}$ on $(\Omega, \mathcal{A})$. The following theorem generalizes Theorem 7.1 in Last (2006) from the case $G = \mathbb{R}^d$ to general groups. The former case is also touched by Lemma 16 in Hoffman, Holroyd and Peres (2006).

Theorem 3.27 *Let (Z, τ) be a stationary partition. Then we have for any measurable $f, \tilde{f} : \Omega \to \mathbb{R}_+$,*

$$\mathbf{E}\mathbf{1}\{e \in Z\}\Delta(\tau^*(e))\tilde{f}f(\theta_{\tau^*}) = \mathbf{E}_{\mathbf{Q}_N} f \int_{C(e)} \tilde{f}\left(\theta_g^{-1}\right)\lambda(\mathrm{d}g), \tag{3.66}$$

where $\tau^*(g) = \tau(g)^{-1}$, $g \in G$, and θ_{τ^*} is defined by (3.50).

Proof Consider the random measure $M = \lambda(Z \cap \cdot)$. By covariance (3.63) of Z and invariance of λ we have $M(\theta_g, gB) = \int \mathbf{1}\{h \in gB \cap gZ\}\lambda(\mathrm{d}h) = M(B)$ for all $g \in G$ and $B \in \mathcal{S}$. Hence M is invariant. An equally simple calculation shows that the Palm measure of M is given by

$$\mathbf{Q}_M(A) = \mathbf{P}(A \cap \{e \in Z\}), \quad A \in \mathcal{A}. \tag{3.67}$$

Define transport-kernels T and T^* by $T(g, \cdot) = \delta_{\tau(g)}$ and $T^*(g, \cdot) = \lambda(C(g) \cap \cdot)$. Since $\tau(g) \in N$ whenever $g \in Z$, it is straightforward to check that (3.39) holds, even for all $\omega \in \Omega$. By (3.36), T is invariant. Invariance of T^* follows from (3.65) and invariance of λ:

$$T^*(\theta_g, gh, gB) = \lambda(C(\theta_g, gh) \cap gB) = \lambda(gC(h) \cap gB) = T^*(h, B).$$

Theorem 3.15 implies that (3.40) holds. Applying this formula to the measurable function $(\omega, g) \mapsto f(\omega)\tilde{f}\left(\theta_g^{-1}\omega\right)$ and taking into account (3.67) as well as the definitions of T and T^*, yields the assertion (3.66). □

The special choice $\tilde{f} \equiv 1$ in (3.66) yields the following relationship between the measure $\mathbf{E}\mathbf{1}\{e \in Z\}\Delta(\tau^*(e))\mathbf{1}\{\theta_{\tau^*} \in \cdot\}$ and a *volume-weighted* version of the Palm measure $\mathbf{Q}_N$. In case of $G = \mathbb{R}^d$ we refer to Last (2006, Sec. 4).

Proposition 3.28 *Let (Z, τ) be a stationary partition. Then we have for any measurable $f : \Omega \to \mathbb{R}_+$,*

$$\mathbf{E}\mathbf{1}\{e \in Z\}\Delta(\tau^*(e))f(\theta_{\tau^*}) = \mathbf{E}_{\mathbf{Q}_N}\lambda(C(e))f. \tag{3.68}$$

Remark 3.29 The special case $f \equiv 1$ of (3.68) gives

$$\mathbf{E}_{\mathbf{Q}_N}\lambda(C(e)) = \mathbf{E}\mathbf{1}\{e \in Z\}\Delta(\tau^*(e)). \tag{3.69}$$

If N has a positive and finite intensity γ_N and G is unimodular, this yields the intuitively obvious formula

$$\mathbf{E}_{\mathbf{Q}_N^0}\lambda(C(e)) = \mathbf{P}\{e \in Z\}\gamma_N^{-1}. \tag{3.70}$$

Define

$$V(g) = \{h \in G : \tau(h) = \tau(g)\}$$

as the cell containing $g \in G$.

Corollary 3.30 *Let (Z, τ) be a stationary partition. Then we have for any $\beta \geq 0$ that*

$$\mathbf{E1}\{e \in Z\}\Delta(\tau^*(e))\lambda(V(e))^\beta = \mathbf{E}_{\mathbf{Q}_N}\lambda(C(e))^{\beta+1}. \tag{3.71}$$

Proof For $h \in N$ we get from (3.65) that $C(\theta_{h^{-1}}, e) = h^{-1}C(h)$. Assuming $e \in Z$, we can apply this fact to $h = \tau(e) \in N$, to obtain

$$C(\theta_{\tau^*}, e) = \tau^*(e)V(e). \tag{3.72}$$

We now apply (3.68) to $f = \lambda(C(e))^\beta$. Using (3.72) together with invariance of λ, yields (3.71). □

A stationary partition (Z, τ) is called *proper*, if

$$\mathbf{Q}_N\{\lambda(C(e)) = 0\} = \mathbf{Q}_N\{\lambda(C(e)) = \infty\} = 0. \tag{3.73}$$

In the unimodular case the second equation is implied by (3.69). The following two results can be proved as in Last (2006, Sec. 5). The details are left to the reader.

Proposition 3.31 *Let (Z, τ) be a stationary and proper partition. Then we have for any measurable $f : \Omega \to \mathbb{R}_+$,*

$$\mathbf{E1}\{e \in Z\}\Delta(\tau^*(e))\lambda(V(e))^{-1}f(\theta_{\tau^*}) = \mathbf{E}_{\mathbf{Q}_N}f. \tag{3.74}$$

Corollary 3.32 *Let (Z, τ) be a stationary and proper partition. Then (3.71) holds for all $\beta \in \mathbb{R}$. If the intensity γ_N of N is finite, then we have in particular,*

$$\gamma_N = \mathbf{E1}\{e \in Z\}\Delta(\tau^*(e))\lambda(V(e))^{-1}. \tag{3.75}$$

From now on we assume that N has a finite intensity and $\mathbf{P}\{\hat{N} = 0\} = 0$. Just for simplicity we also assume that $\mathbf{P}$ (and hence also $\mathbf{Q}_N^*$) is a probability measure. We first note the following consequence of the proof of Theorem 3.33 and Corollary 3.18:

Corollary 3.33 *Let (Z, τ) be a stationary partition. We have for any measurable $f, \tilde{f} : \Omega \to \mathbb{R}_+$,*

$$\mathbf{E}[\mathbf{1}\{e \in Z\}\tilde{f}f(\theta_{\tau^*})\Delta(\tau^*(e))|\mathcal{I}] = \hat{N}\mathbf{E}_{\mathbf{Q}_N^*}\left[f\int_{C(e)}\tilde{f}\left(\theta_g^{-1}\right)\lambda(\mathrm{d}g)\Big|\mathcal{I}\right], \tag{3.76}$$

$\mathbf{P}$*-a.e. for any choice of the conditional expectations. In particular,*

$$\hat{N}^{-1}\mathbf{E}[\mathbf{1}\{e \in Z\}f(\theta_{\tau^*}\Delta(\tau^*(e))|\mathcal{I}] = \mathbf{E}_{\mathbf{Q}_N^*}[f\lambda(C(e))|\mathcal{I}]. \tag{3.77}$$

Let $\alpha > 0$. Essentially following Hoffman, Holroyd and Peres (2006) (dealing with the case $G = \mathbb{R}^d$), we call a stationary partition (Z, τ) (based on N) α-*balanced*, if

$$\mathbf{P}\{\lambda(C(g)) = \alpha \hat{N}^{-1} \text{ for all } g \in N\} = 1. \tag{3.78}$$

The significance of α-balanced stationary partitions is due to the following theorem. The result extends Theorem 13 in Holroyd and Peres (2005) and Theorem 9.1 in Last (2006) (both dealing with $\alpha = 1$) from $\mathbb{R}^d$ to general groups.

Theorem 3.34 *Let $\alpha > 0$. A stationary partition (Z, τ) is α-balanced iff*

$$\mathbf{Q}_N^* = \alpha^{-1}\mathbf{E}\mathbf{1}\{e \in Z\}\Delta(\tau^*(e))\mathbf{1}\{\theta_{\tau^*} \in \cdot\}. \tag{3.79}$$

Proof If (Z, τ) is a α-balanced stationary partition, then (3.79) follows from (3.77).

Assume now that (3.79) holds. Since $\mathbf{Q}_N^*$ has the invariant density $\hat{N}$ with respect to $\mathbf{Q}_N$, (3.79) implies that

$$\mathbf{Q}_{\alpha N} = \mathbf{E}\mathbf{1}\{e \in Z\}\hat{N}\mathbf{1}\{\theta_{\tau^*} \in \cdot\}\Delta(\tau^*(e)), \tag{3.80}$$

where we have used that $\mathbf{Q}_{\alpha N} = \alpha \mathbf{Q}_N$. Using the invariant weighted transport-kernels $T(g, \cdot) = \hat{N}\mathbf{1}\{g \in Z\}\delta_{\tau(g)}$, this reads

$$\mathbf{Q}_{\alpha N} = \mathbf{E}\int \mathbf{1}\left\{\theta_g^{-1} \in \cdot\right\}\Delta(g^{-1})T(e, \mathrm{d}g).$$

Since $\mathbf{P}$ is the Palm measure of λ, we get from Theorem 3.19 that T is $\mathbf{P}$-a.e. $(\lambda, \alpha N)$-balancing. Therefore we have $\mathbf{P}$-a.e. that

$$\int \mathbf{1}\{g \in Z, \tau(g) \in \cdot\}\lambda(\mathrm{d}g) = \alpha \hat{N}^{-1} N(\cdot).$$

This is just saying that (Z, τ) is α-balanced. □

Remark 3.35 If a α-balanced stationary partition (Z, τ) is given, then (3.79) provides an explicit method for constructing the modified Palm probability measure $\mathbf{Q}_N^*$ by a *shift-coupling* with the stationary measure $\mathbf{P}$. In case G is a unimodular group, (3.79) simplifies to

$$\mathbf{Q}_N^* = \alpha^{-1}\mathbf{E}\mathbf{1}\{e \in Z\}\mathbf{1}\{\theta_{\tau^*} \in \cdot\}. \tag{3.81}$$

Since $\alpha = \mathbf{P}\{e \in Z\}$, this means that $\mathbf{Q}_N^* = \mathbf{P}\{\theta_{\tau^*} \in \cdot | e \in Z\}$.

The actual construction of α-balanced partitions is an interesting topic in its own right. Triggered by Liggett (2002), the case $G = \mathbb{R}^d$ was discussed in Holroyd and Peres (2005) and Hoffman, Holroyd and Peres (2006). Among many other things it was shown there that α-balanced partitions do actually exist for any $\alpha \le 1$. The occurrence of the sample intensity $\hat{N}$ in (3.78) is explained by the spatial ergodic theorem, see Proposition 9.1 in Last (2006), at least in case $G = \mathbb{R}^d$. Holroyd and Peres (2005) have also results on discrete groups in case $\alpha = 1$. It might be conjectured that α-balanced partitions exist for all $\alpha \le 1$, provided that the Haar measure λ is diffuse.

3.10 Matchings and point stationarity

We consider an invariant simple point process N on G. A *point-allocation* for N is an allocation rule τ having $\tau(g) \in N$ whenever $g \in N$. (Recall that we identify N with its support.) We also assume that

$$\tau(g) = g, \quad g \notin N. \tag{3.82}$$

A point-allocation for N is called *bijective* if $g \mapsto \tau(g)$ is a bijection on N whenever $N(G) > 0$. In the Abelian (or $\mathbb{R}^d$) case the following special case of Corollary 3.22 is discussed in (Thorisson, (2000); Heveling and Last, (2005); Heveling and Last, (2007)). Recall the notation introduced at (3.50).

Corollary 3.36 *Let* $\mathbf{P}$ *be a* σ*-finite stationary measure* $\mathbf{P}$ *on* $(\Omega, \mathcal{A})$ *and* τ *be a point-allocation for* N*. Then* τ *is* $\mathbf{P}$*-a.e. bijective iff*

$$\mathbf{E}_{\mathbf{Q}_N} f\left(\theta_\tau^{-1}\right) \Delta(\tau(e)^{-1}) = \mathbf{E}_{\mathbf{Q}_N} f \tag{3.83}$$

holds for all measurable $f : \Omega \to \mathbb{R}_+$.

A N*-matching* is a point-allocation τ such that $\tau(\tau(g)) = g$ for all $g \in N$. (We don't require that $\tau(g) \neq g$ for $g \in N$.) In the canonical case $\Omega = \mathsf{N}_s$ with N being the identity on Ω (and with the flow given as in Example 3.1) we just say that τ is a matching. Our next result is generalizing the point process case of Theorem 1.1 in Heveling and Last (2007) (dealing with an Abelian group). We assume that there exist a measurable and injective function $I : \Omega \to [0, 1]$. As any Borel space has this property, this is no serious restriction of generality.

Theorem 3.37 *A measure* $\mathbf{Q}$ *on* $(\Omega, \mathcal{A})$ *is a Palm measure of* N *with respect to some* σ*-finite stationary measure iff* $\mathbf{Q}$ *is* σ*-finite,* $\mathbf{Q}\{e \notin N\} = 0$*, and*

$$\mathbf{E}_{\mathbf{Q}} f\left(\theta_\tau^{-1}\right) \Delta(\tau(e)^{-1}) = \mathbf{E}_{\mathbf{Q}} f \tag{3.84}$$

holds for all N*-matchings* τ *and all measurable* $f : \Omega \to \mathbb{R}_+$.

Our proof of Theorem 3.37 requires the following generalization of a result in Heveling and Last (2007), that is of interest in its own right.

Proposition 3.38 *There exist* N*-matchings* τ_k, $k \in \mathbb{N}$*, such that for* $e \in N$

$$\{e\} \cup \{g \in N : \theta_g \neq \theta_e\} \subset \{\tau_k(e) : k \in \mathbb{N}\}. \tag{3.85}$$

The proof of Proposition 3.38 is based on several lemmas. A *subprocess* of N is an invariant simple point process S such that $S \subset N$.

Lemma 3.39 *Let* S *be a subprocess of* N *and* τ *a matching. Then* $\tau'(\omega, g) = \tau(S(\omega), g)$ *is a* N*-matching.*

Proof Covariance of τ' is a direct consequence of the invariance of N and covariance of τ. We have to show that $\tau'(g) \in N$ and $\tau'(\tau'(g))$ if $g \in N$. If $g \notin S$ then (3.82) (for τ') implies $\tau(g) = g \in N$ and in particular $\tau'(\tau'(g)) = g$. If $g \in S$, then $\tau'(g) = \tau(S, g) \in S \subset N$. Moreover, since τ is a matching we have $\tau'(\tau'(g)) = \tau(S, \tau(S, g)) = g$.

To check the asserted matching property we take $g \in N$. If $g \notin S$ then (3.82) (for τ') implies $\tau(g) = g$ and in particular $\tau(\tau(g)) = g$. □

Lemma 3.40 *There exists a countable family of subprocesses $\{S_n : n \in \mathbb{N}\}$ of N such that, for any $B \in \mathcal{G}$ with compact closure, $\omega \in \Omega$ and $g, h \in N(\omega)$, there exists $n \in \mathbb{N}$ with*

$$I\left(\theta_k^{-1}\omega\right) \in \{I\left(\theta_g^{-1}\omega\right), I\left(\theta_h^{-1}\omega\right)\}, \quad k \in S_n(\omega) \cap B. \tag{3.86}$$

Proof Let $\{(q_n, r_n, s_n) : n \in \mathbb{N}\}$ be dense in $[0,1]^3$. Define the simple point processes S_n, $n \in \mathbb{N}$, by

$$S_n(\omega) = \{k \in N(\omega) : \min\{|I\left(\theta_k^{-1}\omega\right) - q_n|, |I\left(\theta_k^{-1}\omega\right) - r_n|\} \le s_n\}.$$

The measurability of S_n follows from the measurability of I. Invariance of S_n follows from invariance of N and the flow property (3.4).

Fix $B \in \mathcal{G}$ with compact closure, $\omega \in \Omega$ and $g, h \in N(\omega)$. From the local finiteness of $N(\omega)$ we deduce that the set

$$\{I\left(\theta_k^{-1}\omega\right) : k \in N(\omega) \cap B\}$$

is a finite subset of $[0,1]$. Hence, there exists $\varepsilon > 0$ such that

$$\{|I\left(\theta_k^{-1}\omega\right) - I\left(\theta_g^{-1}\omega\right)|, |I\left(\theta_k^{-1}\omega\right) - I\left(\theta_h^{-1}\omega\right)| : k \in N(\omega) \cap B\} \cap (0, \varepsilon) = \emptyset.$$

Moreover, there exists $n \in \mathbb{N}$ such that $0 < s_n < \varepsilon/2$ and both $|I\left(\theta_g^{-1}\omega\right) - q_n| < s_n$ and $|I\left(\theta_h^{-1}\omega\right) - r_n| < s_n$ hold, finishing the proof of the proposition. □

Let $\eta \in \mathsf{N}_s$ and take a Borel set $B \in \mathcal{G}$. Then call $h \in \eta$ a *B-neighbour* of $g \in \eta$ if $h \in gB$. Clearly, h is a B-neighbour of g if and only if g is a B^*-neighbour of h, where the inverted set B^* is defined by $B^* = \{g^{-1} : g \in B\}$. We say that h is the *unique B-neighbour* of g if $gB \cap \eta = \{h\}$. A point h can be the unique B-neighbour of g, and g a non-unique B^*-neighbour of h. We then define an allocation rule $\pi_B : \mathsf{N}_s \times G \to G$ by

$$\pi_B(\eta, g) = \begin{cases} h, & \text{if } g \text{ and } h \text{ are mutual unique } B \cup B^*\text{-neighbours}, \\ g, & \text{otherwise.} \end{cases} \tag{3.87}$$

Lemma 3.41 *There exists a countable family of N-matchings (σ_n) such that*

$$\begin{aligned} \{e\} \cup \{g \in N(\omega) : I\left(\theta_g^{-1}\omega\right) \ne I(\omega) \quad \textit{and} \quad I\left(\theta_g^{-1}\omega\right) \ne I(\theta_g\omega)\} \\ \subset \{\sigma_n(\omega, e) : n \in \mathbb{N}\} \end{aligned} \tag{3.88}$$

for all $\omega \in \Omega$ such that $e \in N(\omega)$.

Proof Let $\{B_m\}$ be a countable base of the topology of G. We consider the matchings $\tau_m = \pi_{B_m}, m \in \mathbb{N}$, defined in (3.87) and the family of subprocesses $\{S_n : n \in \mathbb{N}\}$ from Lemma 3.40.

Let $\omega \in \Omega$ such that $e \in N(\omega)$. Let $g \in N(\omega)$ such that $I\left(\theta_g^{-1}\omega\right) \neq I(\omega)$ and $I\left(\theta_g^{-1}\omega\right) \neq I(\theta_g\omega)$. There exists an open set U with compact closure in G such that $e, g, g^{-1}, g^2 \in U$. By Proposition 3.40 there exists $n \in \mathbb{N}$ such that, for all $k \in S_n(\omega) \cap U$, we have $I\left(\theta_k^{-1}\omega\right) \in \{I(\omega), I\left(\theta_g^{-1}\omega\right)\}$. From $I\left(\theta_g^{-1}\omega\right) \notin \{I(\omega), I(\theta_g\omega)\}$ and injectivity of I we obtain $\theta_g^{-1}\omega \neq \omega$ and $\theta_g^{-1}\omega \neq \theta_g\omega$. This gives $\theta_{g^{-2}}\omega \neq \theta_{g^{-1}}\omega$ and $\theta_{g^{-2}} \neq \omega$. We deduce that $I(\theta_{g^2}^{-1}\omega) \notin \{I(\omega), I\left(\theta_g^{-1}\omega\right)\}$, so $g^2 \notin S_n(\omega)$. In the same (even simpler) way we get $g^{-1} \notin S_n(\omega)$

Since $\{B_m\}$ is a base of the topology of G there exists $m \in \mathbb{N}$ such that $B_m \cap N(\omega) = \{g\}$, $gB_m^* \cap N(\omega) = \{e\}$, $B_m^* \cap N(\omega) \subset \{g^{-1}\}$ and $gB_m \cap N(\omega) \subset \{g^2\}$. We conclude that $\tau_m(S_n(\omega), e) = g$.

Re-enumeration of the N-matchings $(\omega, h) \mapsto \tau_m(S_n(\omega), h)$ (see Lemma 3.39) yields a family $\{\sigma_n : n \in \mathbb{N}\}$ that satisfies (3.88). □

Lemma 3.42 *There exists a countable family of N-matchings (σ_n) such that*

$$\begin{aligned}\{g \in N(\omega) : I\left(\theta_g^{-1}\omega\right) \neq I(\omega) \text{ and } I\left(\theta_g^{-1}\omega\right) = I(\theta_g\omega)\} \\ \subset \{\sigma_n(\omega, e) : n \in \mathbb{N}\}\end{aligned} \tag{3.89}$$

for all $\omega \in \Omega$ with $e \in N(\omega)$.

Proof For $n \in \mathbb{N}$, define a covariant mapping $\sigma_n : \Omega \times G \to G$ by

$$\sigma_n(e) = \begin{cases} g, & \text{if } N \cap B_n = \{g\}, N \cap B_n^* = \{g^{-1}\}, N \cap gB_n^* = \{e\}, \\ & N \cap gB_n = \{g^2\} \text{ and } I(\theta_g) = I\left(\theta_g^{-1}\right) > I, \\ g, & \text{if } N \cap B_n^* = \{g\}, N \cap B_n = \{g^{-1}\}, N \cap gB_n = \{e\}, \\ & N \cap gB_n^* = \{g^2\} \text{ and } I(\theta_g) = I\left(\theta_g^{-1}\right) < I, \\ e, & \text{otherwise,} \end{cases} \tag{3.90}$$

and $\sigma_n(\omega, g) = g\sigma_n\left(\theta_g^{-1}\omega, e\right)$. Take $\omega \in \Omega$ with $e \in N(\omega)$ and $g \in N(\omega)$ such that $I\left(\theta_g^{-1}\omega\right) = I(\theta_g\omega)$. Then $g^{-1}, g^2 \in N(\omega)$. Moreover, there exists $n \in \mathbb{N}$ such that g is the unique B_n-neighbour of e in $N(\omega)$, g^{-1} the unique B_n^*-neighbour of e, e the unique B_n^*-neighbour of g in $N(\omega)$ and g^2 the unique B_n-neighbour of g. Hence, $\sigma_n(\omega, e) = g$ and the inclusion (3.89) holds.

It remains to show that σ_n is for each $n \in \mathbb{N}$ a N-matching. It is clearly sufficient to assume $e \in N$ and to prove that $\sigma_n(g) = e$ if $\sigma_n(e) = g \neq e$. Let

$g \in N$. By definition of σ_n and invariance of N we have for $h \in G$ that

$$\sigma_n(g) = \begin{cases} gh, & \text{if } g^{-1}N \cap B_n = \{h\}, g^{-1}N \cap B_n^* = \{h^{-1}\}, g^{-1}N \cap hB_n^* = \{e\}, \\ & g^{-1}N \cap gB_n = \{h^2\} \text{ and } I(\theta_{hg}) = I\left(\theta_{hg}^{-1}\right) > I(\theta_g), \\ gh, & \text{if } g^{-1}N \cap B_n^* = \{h\}, g^{-1}N \cap B_n = \{h^{-1}\}, g^{-1}N \cap gB_n = \{e\}, \\ & g^{-1}N \cap gB_n^* = \{h^2\} \text{ and } I(\theta_{hg}) = I\left(\theta_{hg}^{-1}\right) < I(\theta_g), \\ g, & \text{otherwise.} \end{cases}$$

Putting $h = g^{-1}$ and using $\sigma_n(e) = g$ (i.e. the conditions on the right-hand side of (3.90)) we see that indeed $\sigma_n(g) = e$. □

Combining Lemma 3.41 with Lemma 3.42 yields Proposition 3.38.

Proof of Theorem 3.37: One implication follows from Corollary 3.36. So we assume that $\mathbf{Q}$ is σ-finite, $\mathbf{Q}\{e \notin N\} = 0$, and that (3.83) holds for all τ and f as stated there. We will check the Mecke equation (3.56).

Without loss of generality we assume that $e \in N$ everywhere on Ω. Let τ be a N-matching. Since $\tau(e) \in N$ we get from the matching property and covariance (writing $g = \tau(e)$) that $\tau\left(\theta_g^{-1}, e\right) = g^{-1}\tau(\theta_e, g) = g^{-1}$, i.e.

$$\tau\left(\theta_\tau^{-1}, e\right) = \tau(\theta_e, e)^{-1}. \tag{3.91}$$

Therefore we obtain for all measurable $f : \Omega \times G \to \mathbb{R}_+$ from (3.83) that

$$\mathbf{E_Q}\mathbf{1}\{\theta_\tau \neq \theta_e\}f(\theta_e, \tau(e)) = \mathbf{E_Q}\mathbf{1}\left\{\theta_\tau^{-1} \neq \theta_e\right\} f(\theta_e, \tau(e)^{-1})\Delta(\tau(e)^{-1}). \tag{3.92}$$

Applying Proposition 3.38 yields

$$\mathbf{E_Q} \int \mathbf{1}\{\theta_g \neq \theta_e\}f(\theta_e, g)N(\mathrm{d}g) = \sum_{k=1}^{\infty} \mathbf{E_Q} f_k(\theta_e, \tau_k(e))f(\theta_e, \tau_k(e)), \tag{3.93}$$

where

$$f_k(g) = \mathbf{1}\{\theta_g \neq \theta_e\}\mathbf{1}\{\tau_m(e) \neq g \text{ for } 1 \leq m \leq k-1\}.$$

We claim that

$$f_k\left(\theta_{\tau_k}^{-1}, \tau_k(e)^{-1}\right) = f_k(\theta_e, \tau_k(e)). \tag{3.94}$$

To see this, we take $\omega \in \Omega$ and set $g = \tau_k(\omega, e)$. We have to show that

$$f_k\left(\theta_g^{-1}\omega, g^{-1}\right) = f_k(\omega, g). \tag{3.95}$$

For $k \leq 2$ and $m \leq k-1$ we have from covariance that $\tau_m\left(\theta_g^{-1}\omega, e\right) \neq g^{-1}$ is equivalent to $g^{-1}\tau_m(\omega, g) \neq g^{-1}$, i.e. to $\tau_m(\omega, g) \neq e$. Since τ_m is matching the latter is also equivalent to $\tau_m(\omega, e) \neq g$. Hence (3.95) follows. Using this fact, we obtain from (3.93) and (3.92)

$$\begin{aligned}\mathbf{E_Q} \int \mathbf{1}\{\theta_g \neq \theta_e\} f(\theta_e, g) N(\mathrm{d}g) \\ &= \sum_{k=1}^{\infty} \mathbf{E_Q} f_k(\theta_e, \tau_k(e)) f\left(\theta_{\tau_k}^{-1}, \tau_k(e)^{-1}\right) \Delta(\tau_k(e)^{-1}) \\ &= \mathbf{E_Q} \int \mathbf{1}\{\theta_g \neq \theta_e\} f\left(\theta_g^{-1}, g^{-1}\right) \Delta(g^{-1}) N(\mathrm{d}g).\end{aligned}$$

In order to show (3.56) it remains to verify

$$\begin{aligned}\mathbf{E_Q} \int \mathbf{1}\{\theta_g = \theta_e\} f(\theta_e, g) N(\mathrm{d}g) \\ &= \mathbf{E_Q} \int \mathbf{1}\{\theta_g = \theta_e\} f\left(\theta_g^{-1}, g^{-1}\right) \Delta(g^{-1}) N(\mathrm{d}g). \qquad (3.96)\end{aligned}$$

For any fixed $\omega \in \Omega$ the closed set $H_\omega = \{g \in G : \theta_g \omega = \omega\}$ is a subgroup of G. Indeed, if $g, h \in H_\omega$ then $\theta_{gh}\omega = \theta_g \circ \theta_h \omega = \omega$, while $g \in H_\omega$ is equivalent to $g^{-1} \in H_\omega$. Since $e \in N(\omega)$ we have, moreover, for all $g \in H_\omega$ that $g \in gN(\omega) = N(\theta_g \omega) = N(\omega)$. Hence H_ω is a subset of $N(\omega)$ and in particular unimodular. On the other hand we clearly have that $h^{-1}gh \in H_\omega$ for all $h \in G$ and $g \in H_\omega$, so that H_ω is a normal subgroup of G. Therefore we obtain from Proposition II.21 in Nachbin (1965) that $\Delta(g) = 1$ for all $g \in H_\omega$. Hence (3.96) is equivalent to

$$\mathbf{E_Q} \int \mathbf{1}\{\theta_g = \theta_e\} f(\theta_e, g) N(\mathrm{d}g) = \mathbf{E_Q} \int \mathbf{1}\{\theta_g = \theta_e\} f\left(\theta_e, g^{-1}\right) N(\mathrm{d}g). \quad (3.97)$$

This follows from $\{g \in N : \theta_g = \theta_e\} = \{g^{-1} : \theta_g = \theta_e, g \in N\}$. □

One might wonder whether the N-matchings in Theorem 3.37 may be chosen *N-measurable*, i.e. measurable with respect to $\sigma(N) \otimes \mathcal{G}$. In the canonical case $\Omega = \mathsf{N}_s$ this is trivially true. Otherwise, applying Lemma 3.41 to the canonical case and composing with N, yields the following result without any assumptions on $(\Omega, \mathcal{A})$.

Proposition 3.43 *There exist N-matchings τ_k, $k \in \mathbb{N}$, such that for $e \in N$*

$$\{e\} \cup \{g \in N : N \circ \theta_g \neq N\} \subset \{\tau_k(e) : k \in \mathbb{N}\}. \qquad (3.98)$$

It is easy to see (Heveling and Last, 2007) that the (random) set $\{g \in N : N \circ \theta_g^{-1} = N\}$ cannot be exhausted by matchings as in Proposition 3.43. In case $G = \mathbb{R}^d$, however, the set can be exhausted by bijective point-allocations (Heveling and Last, 2005). Proposition 3.43 together with a suitable modification of the arguments of the proof of Theorem 3.37 then leads to the following Theorem 4.1 in Heveling and Last (2005). We skip further details.

Theorem 3.44 *Let N be an invariant simple point process on $\mathbb{R}^d$ and $\mathbf{Q}$ be a measure on $(\Omega, \mathcal{A})$. Then $\mathbf{Q}$ is a Palm measure of N with respect to some σ-finite stationary measure iff $\mathbf{Q}$ is σ-finite, $\mathbf{Q}\{0 \notin N\} = 0$, and*

$$\mathbf{E}_{\mathbf{Q}} f\left(\theta_\tau^{-1}\right) = \mathbf{E}_{\mathbf{Q}} f \tag{3.99}$$

holds for all N-measurable bijective point-allocations τ and all measurable $f : \Omega \to \mathbb{R}_+$.

Thorisson (2000) refers to (3.99) as *point-stationarity* of $\mathbf{Q}$ (with respect to N). We conjecture that Theorem 3.44 holds for general groups G, i.e. that the validity of (3.84) for all N-measurable bijective point-allocations τ is characterizing Palm measures.

The matchings considered in this section are examples of what Holroyd, Pemantle, Peres and Schramm (2008) call *one-colour matching scheme. Mutual nearest neighbour matching* (Ferrari, Landim and Thorisson, 2004; Holroyd and Peres, 2003; Daley and Last, 2005) is another example of such matchings. The *two-colour matching* schemes studied in Holroyd, Pemantle, Peres and Schramm (2008) are interesting examples of allocation rules balancing different jointly stationary (simple) point processes.

Acknowledgements

I wish to thank Hermann Thorisson for allowing me to benefit freely from our joint work (Last and Thorisson 2009). Thanks are also due to Daniel Gentner for his careful and critical reading of a preliminary version of this work. Finally I would like to thank a referee for pointing out the relevance of Harris (1971) and Port and Stone (1973) for the present work.

References

Aldous, D. and Lyons, R. (2007). Processes on unimodular random networks. *Electron. J. Probab.* **12**, 1454–1508.

Benjamini, I., Lyons, R., Peres, Y. and Schramm, O. (1999). Group-invariant percolation on graphs *Geom. Funct. Anal.* **9**, 29–66.

Chatterjee, S., Peled, R., Peres, Y. and Romik, R. (2007). Gravitational allocation to Poisson points. To appear in *Ann. of Math.*

Daley, D.J. and Last, G. (2005). Descending chains, the lilypond model, and mutual nearest neighbour matching. *Adv. Appl. Probab.* **37**, 604–628.

Daley, D.J. and Vere-Jones, D. (2008). *An Introduction to the Theory of Point Processes.* 2nd ed., Springer, New York.

Ferrari, P.A., Landim, C. and Thorisson, H. (2004). Poisson trees, succession lines and coalescing random walks. *Ann. Inst. H. Poincaré Probab. Statist.* **40** no. 2, 141–152.

Geman, D. and Horowitz, J. (1975). Random shifts which preserve measure. *Proc. Amer. Math. Soc.* **49**, 143–150.

Gentner, D. and Last, G. (2009). Palm pairs and the general mass-transport principle. Submitted for publication.

Harris, T.E. (1971). Random measures and motions of point processes. *Z. Wahrsch. verw. Gebiete* **18**, 85–115.

Heveling, M. and Last, G. (2005). Characterization of Palm measures via bijective point-shifts. *Ann. Probab.* **33**, 1698–1715.

Heveling, M. and Last, G. (2007). Point shift characterization of Palm measures on Abelian groups. *Electron. J. Probab.* **12**, 122–137.

Hoffman, C. Holroyd, A.E. and Peres, Y. (2006). A stable marriage of Poisson and Lebesgue. *Ann. Probab.* **34**, 1241–1272.

Holroyd, A.E. and Peres, Y. (2003). Trees and matchings from point processes. *Electron. Comm. Probab.* **8**, 17–27.

Holroyd, A.E. and Peres, Y. (2005). Extra heads and invariant allocations. *Ann. Probab.* **33**, 31–52.

Holroyd, A.E., Pemantle, R., Peres, Y. and Schramm, O. (2009). Poisson matching. *Ann. Inst. H. Poincaré Probab. Statist.*, **45**, 266–287.

Kallenberg, O. (1983). *Random Measures.* Akademie-Verlag Berlin and Academic Press, London.

Kallenberg, O. (2002). *Foundations of Modern Probability.* Second Edition, Springer, New York.

Kallenberg, O. (2007). Invariant measures and disintegrations with applications to Palm and related kernels. *Probab. Theory Related Fields* **139**, 285–310.

Last, G. (2006). Stationary partitions and Palm probabilities. *Adv. Appl. Probab.* **37**, 603–620.

Last, G. (2009). Stationary random measures on homogeneous spaces. To appear in *J. Theor. Prob.*

Last, G. and Thorisson, H. (2009). Invariant transports of stationary random measures and mass-stationarity. *Ann. Probab.* **37**, 790–813.

Liggett, T.M. (2002). Tagged particle distributions or how to choose a head at random. In *In and Out of Equlibrium* (V. Sidoravicious, ed.) 133–162, Birkhäuser, Boston.

Matthes, K., Kerstan, J. and Mecke, J. (1978). *Infinitely Divisible Point Processes.* Wiley, Chichester.

Mecke, J. (1967). Stationäre zufällige Maße auf lokalkompakten Abelschen Gruppen. *Z. Wahrsch. verw. Gebiete* **9**, 36–58.

Mecke, J. (1975). Invarianzeigenschaften allgemeiner Palmscher Maße. *Math. Nachr.* **65**, 335–344.

Nachbin, L. (1965). *The Haar Integral.* D. van Nostrand Company, Inc., Princeton.

Neveu, J. (1977). Processus ponctuels. École d'Eté de Probabilités de Saint-Flour VI. *Lecture Notes in Mathematics* **598**, 249–445, Springer, Berlin.

Port, S.C. and Stone, C.J. (1973). Infinite particle systems. *Trans. Amer. Math. Soc.* **178**, 307–340.

Rother, W. and Zähle, M. (1990). Palm distributions in homogeneous spaces. *Math. Nachr.* **149**, 255–263.

Schneider, R. and Weil, W. (2008). *Stochastic and Integral Geometry.* Springer, Berlin.

Stoyan, D., Kendall, W.S. and Mecke, J. (1995). *Stochastic Geometry and Its Applications.* Second Edition, Wiley, Chichester.

Thorisson, H. (1996). Transforming random elements and shifting random fields. *Ann. Probab.* **24**, 2057–2064.

Thorisson, H. (2000). *Coupling, Stationarity, and Regeneration.* Springer, New York.

Timar, A. (2004). Tree and Grid Factors of General Point Processes. *Electron. Comm. Probab.* **9**, 53–59.

Tortrat, A. (1969). Sur les mesures aléatoires dans les groupes non abéliens. *Ann. Inst. H. Poincaré Probab. Statist.* **5**, 31–47.

4

LIMIT THEOREMS IN STOCHASTIC GEOMETRY

Tomasz Schreiber

Abstract

This chapter gives a survey of limit techniques used in stochastic geometry, emphasizing the interplay of classical techniques with geometric and algorithmic ideas.

4.1 Introduction

Large scale random geometric structures tend to exhibit extremely complicated behaviour, which only seldom admits effective and explicit finite size descriptions. In analogy to other fields of probability theory, a natural and elegant way to overcome this difficulty is provided by the remarkable asymptotic phenomena arising while letting the system size grow to infinity and going under the common name of limit theorems. Understandably, a strong interest in limit theorems has always existed in stochastic geometry and it yielded a considerable wealth of theorems due to numerous authors. The resulting developments were only to some extent parallel to those made in other areas of probability theory, due to the particular geometric nature of the considered problems, often eluding effective analytic description and requiring special methods reflecting their geometric ingredients. In response to this need a class of new techniques of asymptotic study have emerged encompassing and taking advantage of the geometric structure of the studied systems and often also intimately related to their algorithmic construction. This combination of classical probabilistic techniques with geometric as well as algorithmic ideas is in our opinion one of the remarkable features of modern stochastic-geometric asymptotic theory. The present article is an attempt to provide a (far from exhaustive) survey of these techniques, with emphasis put on their geometric and algorithmic features. It should be stressed at this point that these methods are not meant to replace but to complement the classical results, as seen in the structure of our presentation below, and that the possibly smaller-than-due fraction of the material devoted to classical limit theory at the expense of new developments comes from the fact that an incomparably richer literature is currently available on the former than on the latter.

To provide a unified formal framework for our survey below we let $\xi(\cdot,\cdot)$ denote a generic real-valued *geometric functional* defined on pairs $(x, \mathcal{X})$ for $\mathcal{X}$ ranging through locally finite point configurations in $\mathbb{R}^d$ and with $x \in \mathcal{X}$. We shall abuse the notation writing $\xi(x, \mathcal{X}) = \xi(x, \mathcal{X} \cup \{x\})$ for $x \notin \mathcal{X}$. Often we shall assume that ξ is *translation invariant*, that is to say

$$\xi(x+u, \mathcal{X}+u) = \xi(x, \mathcal{X}), \ u \in \mathbb{R}^d.$$

To a geometric functional ξ and a locally finite point configuration $\mathcal{X}$ we associate the corresponding *empirical measure*

$$\mu^\xi[\mathcal{X}] = \sum_{x \in \mathcal{X}} \xi(x, \mathcal{X})\delta_x, \tag{4.1}$$

with δ_x standing for the unit Dirac mass at x. The point configuration $\mathcal{X}$ will usually be a realization of a point process $\mathcal{P}$, in which case we shall often abbreviate $\mu^\xi[\mathcal{P}]$ to μ^ξ if no ambiguities arise. For $\lambda > 0$ we define $Q_\lambda = [-\lambda^{1/d}/2, \lambda^{1/d}/2]^d$ to be the cube of volume λ centred at 0. Once the input process $\mathcal{P}$ (often Poissonian) is fixed, we shall write

$$\mu^\xi_\lambda = [\mu^\xi[\mathcal{P} \cap Q_\lambda]] \circ (\lambda^{-1/d} *)^{-1}, \tag{4.2}$$

where $(\lambda^{-1/d}*) : Q_\lambda \to Q_1$ is the operation of multiplication by $\lambda^{-1/d}$ whereas $\circ$ stands for composition. In other words, μ^ξ_λ is a random measure on Q_1 arising by $\lambda^{-1/d}$–re-scaling the domain of μ^ξ,

$$\mu^\xi_\lambda = \sum_{x \in \mathcal{P} \cap Q_\lambda} \xi(x, \mathcal{P} \cap Q_\lambda)\delta_{\lambda^{-1/d}x}.$$

We shall also consider centred empirical measures $\bar{\mu}^\xi_\lambda = \mu^\xi_\lambda - \mathbf{E}\mu^\xi_\lambda$. We shall write $H^\xi[\cdot]$ and H^ξ_λ for the total mass of $\mu^\xi[\cdot]$ and μ^ξ_λ respectively, so that in particular

$$H^\xi[\mathcal{X}] = \sum_{x \in \mathcal{X}} \xi(x, \mathcal{X})$$

and

$$H^\xi_\lambda = H^\xi[\mathcal{P} \cap Q_\lambda].$$

It is the large λ asymptotic behaviour of such measures and their respective total masses, corresponding to cumulative values of ξ over large samples, which is the main object of our interest in this section.

4.2 Theory of stabilization

An elegant and important tool of geometric limit theory is the concept of *stabilization*, introduced in its present form by Penrose and Yukich (2001, 2002, 2003) and prior to that already signalled in the work of Kesten and Lee (1996) and Lee (1997). Roughly speaking, a geometric functional stabilizes if its behaviour at a given point is locally determined in that it depends on the environment only within a certain finite but possibly random distance. As we shall see below, there exists a quite complete collection of limit theorems for stabilizing geometric functionals on Poisson input, including laws of large numbers, central limit theorems as well as moderate and large deviation principles.

4.2.1 *Point-value stabilization*

Among various versions of the stabilization concept the one discussed in this paragraph is, according to the author's experience and opinion, right in the middle of the range of stabilization variants ordered by increasing proving difficulty – it is usually easier to establish than the add-one cost stabilization and more difficult than the localization as defined below. Thus, it appears, it can be regarded as the most standard one and will be referred to as *the stabilization* in the sequel, whereas its variants, often also termed *stabilization* in the literature, will be always qualified with additional adjectives to avoid terminological confusion. As weak as this subjective criterion may be, we feel excused in the presence of numerous meanings the notion of stabilization has been given in the literature. A geometric functional ξ is said to be *stabilizing* on input point process $\mathcal{P}$ iff for each $x \in \mathbb{R}^d$ there exists an a.s. finite random variable $R[x] = R^{\xi}[x;\mathcal{P}]$ such that a.s.

$$\xi(x,\mathcal{P}) = \xi\left(x, [\mathcal{P} \cap B_{R[x]}(x)] \cup \mathcal{A}\right) \text{ for all finite } \mathcal{A} \subset \mathbb{R}^d \setminus B_{R[x]}(x), \qquad (4.3)$$

where $B_r(x)$ is the ball of radius r around x in $\mathbb{R}^d$. In other words, there exists a finite random *radius of stabilization* $R[x]$ such that the value of $\xi(x,\mathcal{P})$ is fully determined by the configuration of $\mathcal{P}$ inside $B_{R[x]}(x)$, see Baryshnikov and Yukich (2005, Sec. 2.1). Technically, it is often the case that geometric functionals are initially defined only on finite rather than locally finite samples and it is stabilization which guarantees their extension to locally finite configurations through the a.s. valid formula $\xi(x,\mathcal{X}) = \lim_{r\to\infty} \xi(x,\mathcal{X} \cap B_r(x))$. The notion of stabilization is naturally associated with Poisson input $\mathcal{P}$, in which case a stabilizing functional ξ enjoys the property that its value at x does not depend on the configuration of $\mathcal{P}$ outside $B_{R[x]}(x)$. This is not the case if the input is non-Poisson because, even though given the configuration of $\mathcal{P}$ inside $B_{R[x]}(x)$ the value of ξ at x is not affected by the outside configuration, the point configuration inside $B_{R[x]}(x)$ may itself depend on the configuration outside. For this reason we shall mainly consider Poisson input in this subsection, with exceptions though when presenting de-Poissonization techniques allowing us to transfer certain Poisson-input results also for binomial input; more general Gibbsian input can be dealt with using the concept of stabilization as well, see Schreiber and Yukich (2008*a*). The speed of decay of tail probabilities for the stabilization radius is a natural criterion characterizing the strength of stabilization. In particular, if there exist constants $M, C > 0$ such that for all $x \in \mathbb{R}^d$ and $r > 0$

$$\mathbf{P}\{R[x] > r\} \leq M\mathrm{e}^{-Cr},$$

we say that the geometric functional ξ is *exponentially stabilizing* with rate C. Likewise, if we have for all $x \in \mathbb{R}^d$ and $r > 0$

$$\mathbf{P}\{R[x] > r\} \leq cr^{-q}$$

for some $c, q > 0$, we say that ξ *stabilizes polynomially with exponent* q. Often we will consider stabilization over classes of input processes: we shall say that a geometric functional ξ *stabilizes exponentially on Poisson input with rate C* (*stabilizes polynomially on Poisson input with exponent q*) if for each $0 < a < b < \infty$ the tails $\mathbf{P}\{R[x] > r\}$ of stabilization radii admit uniform exponential bounds with rate C (polynomial bounds with exponent q) on the following class of processes

- all non-homogeneous Poisson point processes with absolutely continuous intensity measures on $\mathbb{R}^d$ with density ranging between a and b,
- all restrictions of the above processes to positive size cubes and balls in $\mathbb{R}^d$ and finite intersections thereof.

4.2.2 *Add-one cost stabilization*

Another important variant of the stabilization concept is the *add-one cost stabilization*, present in two standard *strong* and *weak* versions. To define it, we first need the notion of add-one cost, see Penrose and Yukich (2002, Sec. 3). The add-one cost of a point x to a configuration $\mathcal{X} \not\ni x$ with respect to a geometric functional ξ is given by

$$\Delta(x;\mathcal{X}) = \sum_{y \in \mathcal{X} \cup \{x\}} \xi(y, \mathcal{X} \cup \{x\}) - \sum_{y \in \mathcal{X}} \xi(y, \mathcal{X}) = H^{\xi}[\mathcal{X} \cup \{x\}] - H^{\xi}[\mathcal{X}]. \quad (4.4)$$

In other words, it is the difference of cumulative value of ξ caused by introducing x into the configuration $\mathcal{X}$. Note that the add-one cost is itself a geometric functional. The functional ξ is *strongly stabilizing for add-one cost* on input process $\mathcal{P}$ iff its add-one cost stabilizes in the standard sense on $\mathcal{P}$, see Penrose and Yukich (2002, Sec. 3). This version of stabilization is also referred to as *external stabilization* because, roughly speaking, it is corresponds to a situation where not only is the value of ξ at x unaffected by the configuration beyond the stabilization distance from x (*external* configuration) but also the values of ξ for these *external* points remain themselves unaffected by the presence of x. If the stabilization of the add-one cost functional is exponential or polynomial, we respectively speak of strong exponential or strong polynomial stabilization. A considerably weaker notion is also considered: the functional ξ is said to be *weakly stabilizing for add-one cost* on $\mathcal{P}$ iff for all $x \in \mathbb{R}^d$ the limit

$$\Delta(x;\mathcal{P}) = \lim_{r \to \infty} \Delta(x; \mathcal{P} \cap B_r(x))$$

exists almost surely, see Penrose and Yukich (2002).

Clearly, the weak add-one cost stabilization is implied by strong add-one cost stabilization, as follows directly from definitions. The relationship between the standard point-value stabilization and its weak and strong add-one cost variants are a more delicate issue though and no general results are available at the moment. Nevertheless, on Poisson input it is easily checked that if ξ is either exponentially stabilizing or polynomially stabilizing with sufficiently high exponent (exceeding the dimensionality d) then it is also weakly stabilizing for add-one cost. Indeed, the value of $\Delta(x; \mathcal{P} \cap B_r(x))$ is easily seen to stay constant for r larger than the maximum of the stabilization radius at x and the distance from x to the furthest point y in $\mathcal{P}$ whose stabilization ball $B_{R[y]}(y)$ contains x. By exponential stabilization of ξ and standard properties of the Poisson process this distance exhibits exponentially decaying tail probabilities. Likewise, it exhibits polynomially decaying tail probabilities if ξ stabilizes polynomially with large enough exponent. Note that this argument yields much more than just the existence of the limit $\Delta(x; \mathcal{P})$, in fact it respectively ensures the exponential and polynomial localization of $\Delta(\cdot;\cdot)$ as defined in the next paragraph.

4.2.3 *Localization*

One further variant of stabilization, going under the name of *localization*, arises as a weaker version of the standard notion, see Schreiber and Yukich (2008*b*, Sec. 3.1). To define it, we consider finite volume approximations $\xi^{[r]}$, $r > 0$, of ξ, given by

$$\xi^{[r]}(x, \mathcal{X}) = \xi(x, \mathcal{X} \cap B_r(x))$$

and we say that a random variable $\hat{R}[x] = \hat{R}^{\xi}(x, \mathcal{P})$ is a radius of localization for ξ on input process $\mathcal{P}$ iff a.s.

$$\xi(x, \mathcal{P}) = \xi^{[\hat{R}[x]]}(x, \mathcal{P}) \text{ and } \forall_{s > \hat{R}[x]}\, \xi^{[s]}(x, \mathcal{P}) = \xi^{[\hat{R}[x]]}(x, \mathcal{P}). \tag{4.5}$$

Should the tail probabilities for the radius of localization exhibit exponential or polynomial decay, we speak of exponential or polynomial localization respectively. As easily follows by definition, localization as defined above is implied by standard stabilization. Often in applications though, the finite volume approximations $\xi^{[r]}$ may be replaced by some other model-specific local functionals, in which case this relationship becomes a more delicate question.

4.2.4 *Laws of large numbers*

When complemented with appropriate moment bounds imposed on a geometric functional, stabilization and its variants usually imply laws of large numbers providing a first order description of the asymptotic behaviour of empirical measures μ_λ^ξ. We survey results of this kind in this paragraph and give a generic outline of their proofs, referring the reader to original papers Penrose and Yukich (2001, 2002, 2003) and Baryshnikov and Yukich (2005) for full details.

We say that a geometric functional ξ satisfies the bounded p-th moment condition on input $\mathcal{P}$ iff

$$\sup_{x \in \mathbb{R}^d,\ \mathcal{A} \in \mathrm{Fin}(\mathbb{R}^d)} \mathbf{E}|\xi(x, \mathcal{P} \cup \mathcal{A})|^p < \infty, \tag{4.6}$$

with $\mathrm{Fin}(\mathbb{R}^d)$ standing for the family of finite subsets of $\mathbb{R}^d$. Consider an almost everywhere continuous function $\kappa : Q_1 \to \mathbb{R}_+$, taking uniformly bounded values and, for $\lambda > 0$, let $\mathcal{P}_{\lambda,\kappa}$ be the Poisson point process on Q_λ with intensity

$\kappa(\lambda^{-1/d}x)\mathrm{d}x$, which equivalently arises by taking the Poisson point process of intensity $\lambda\kappa$ on Q_1 and blowing it up by a factor $\lambda^{1/d}$. The measure μ_λ^ξ is the usual re-scaled empirical measure on input $\mathcal{P}_{\lambda,\kappa}$, as given in (4.2). Also, for $x \in Q_1$ denote by $\mathcal{P}_{\kappa(x)}$ the homogeneous Poisson point process on $\mathbb{R}^d$ with intensity $\kappa(x)$. The following law of large numbers arose in a series of developments by Penrose and Yukich (2003), Baryshnikov and Yukich (2005) and Penrose (2007*b*).

Theorem 4.1 *Assume that ξ is translation invariant, stabilizing on Poisson input and satisfies the bounded moment condition (4.6) for some $p > 1$. Then for each bounded measurable function $f : Q_1 \to \mathbb{R}$*

$$\lim_{\lambda\to\infty} \lambda^{-1}\mathbf{E}\left\langle f, \mu_\lambda^\xi \right\rangle = \int f(x)[\mathbf{E}\xi(o, \mathcal{P}_{\kappa(x)})]\kappa(x)\mathrm{d}x. \tag{4.7}$$

Moreover, if for some $q = 1, 2$ the bounded moment condition (4.6) is satisfied for $p > q$ then

$$\lim_{\lambda\to\infty} \lambda^{-1}\left\langle f, \mu_\lambda^\xi \right\rangle = \int f(x)[\mathbf{E}\xi(o, \mathcal{P}_{\kappa(x)})]\kappa(x)\mathrm{d}x \tag{4.8}$$

in L^q.

Various other laws of large numbers for stabilizing functionals are available in the literature. We discuss some of them briefly below mentioning certain further generalizations available in the literature.

- The results in the papers quoted above are in fact much more general than our Theorem 4.1 and they include treatment of non-homogeneous functionals, general marked point processes as well as more general form of empirical measures, see Penrose (2007*b*). These extensions fall beyond the scope of our present survey due to space limitations.
- A strong (almost sure) law of large numbers has also been obtained for stabilizing functionals, see e.g. Penrose and Yukich (2002, Th. 3.2) and Penrose (2007*b*, Th. 2.2). It requires certain extra bounded moment conditions on the add-one-cost under which it states that (4.8) holds almost surely, see Penrose (2007*b*). With some further assumptions (strong add-one cost stabilization and further bounded moment and polynomial growth conditions) the so-called *complete convergence* can be obtained (for homogeneous input, translation invariant ξ and constant test functions, see Penrose and Yukich (2002)), where a sequence of r.v. X_n is declared to completely converge to some X iff

$$\sum_{n=1}^{\infty} \mathbf{P}\{|X_n - X| > \epsilon\} < \infty \tag{4.9}$$

for all $\epsilon > 0$, in particular the complete convergence is stronger than the a.s. one.

- In some cases the right hand side of (4.7) can be made fully explicit. This is the case for many instances of the nearest neighbour graph problems, see Wade (2007) and Example 4.9 in Section 4.3 below. These and related issues are discussed in more detail in Chapter 7 of this volume. Likewise, explicit expressions for the right hand side of (4.7) as well as for limit variances have also been obtained in some examples involving the so-called spacing statistics, see Baryshnikov, Penrose and Yukich (2009).

The idea underlying the proofs of the expectation convergence (4.7) is to write

$$\lambda^{-1}\mathbf{E}\left\langle f, \mu_\lambda^\xi \right\rangle = \lambda^{-1}\mathbf{E}\int_{Q_\lambda} f(y)\xi(y, \mathcal{P}_{\lambda,\kappa})\kappa(\lambda^{-1/d}y)dy,$$

where standard Palm theory (see Chapters 1 and 3 of this volume) is used stating that the Poisson point process $\mathcal{P}_{\lambda,\kappa}$ conditioned on containing a point at y coincides in law with $\mathcal{P}_{\lambda,\kappa} \cup \{y\}$. Now, changing the variable $x = \lambda^{-1/d}y$ and moving the expectation under the integral sign (valid by bounded moment conditions) we see that the above equals

$$\int_{Q_1} f(x)[\mathbf{E}\xi(\lambda^{1/d}x, \mathcal{P}_{\lambda,\kappa})]\kappa(x)\mathrm{d}x.$$

The final step is to show that $\mathbf{E}\xi(\lambda^{1/d}x, \mathcal{P}_{\lambda,\kappa})$ is well enough approximated by $\mathbf{E}\xi(o, \mathcal{P}_{\kappa(x)})$ and this is where the essential use of stabilization takes place. Roughly speaking, this amounts to the intuitive observation that *from the viewpoint of an observer located at* $\lambda^{1/d}x$ the local behaviour of the process $\mathcal{P}_{\lambda,\kappa}$ closely resembles that of $\mathcal{P}_{\kappa(x)}$. From the formal viewpoint this usually requires some extra work based on stabilization and possibly also involving some regularity of f (usually its continuity). Some additional effort is also required to get rid of the boundary effects because the stabilization yields the above approximation only for points at a distance $\gg \lambda^{-1/d}$ from the boundary. We refer the readers to the papers quoted above for full details, which vary depending on the particular set of assumptions imposed. It should be emphasized at this point that the local homogeneous Poisson approximation of the behaviour of the input process, which is easily obtained here under Poisson input, is also available for binomial input, yet upon a considerable additional effort, see below for more detail. This approximation technique for the input process carries over for its stabilizing functionals – this idea often goes under the name of *objective method*, see Aldous and Steele (2003), and is one of the cornerstones of limit theory for stabilizing geometric functionals as making it possible to explicitly relate the macroscopic (large-scale) and microscopic (local) behaviour of the random empirical measures μ_λ^ξ generated by stabilizing functionals.

Further, to prove the L^2-convergence one controls the variance of the considered empirical integral, as made possible by the decay of covariances implied by stabilization, see Penrose and Yukich (2003). To establish the L^1 convergence truncation methods can be used combined with the L^2-convergence statements for truncated variables, whence one recovers the L^1-convergence using appropriate integrability properties of ξ, see ibidem. Finally, the Azuma inequality for martingale differences has been used by Penrose and Yukich (2002) to establish the complete convergence. This is an example of a combination of martingale and measure concentration techniques which do in fact yield much more than just a law of large numbers. Likewise, large deviation results can be applied to get the complete convergence, see Schreiber and Yukich (2005). These methods will be discussed in further paragraphs below.

4.2.5 *Convergence of variance*

In the previous LLN paragraph stabilization and the objective method were used to provide an integral representation for the limit of suitably scaled expectations of random measures μ_λ^ξ in terms of the local geometry of the process. Here we present analogous results for the variance asymptotics. To this end, with notation as in the LLN paragraph above, for a translation invariant geometric functional ξ put

$$\begin{aligned} V^\xi(\tau) &= \mathbf{E}\xi^2(o, \mathcal{P}_\tau) \\ &\quad + \tau \int_{\mathbb{R}^d} \left[\mathbf{E}\xi(o, \mathcal{P}_\tau \cup \{y\})\xi(y, \mathcal{P}_\tau \cup \{o\}) - [\mathbf{E}\xi(o, \mathcal{P}_\tau)]^2\right] dy, \qquad (4.10) \end{aligned}$$

whenever the integral converges. The following result characterizes the variance asymptotics for integrals of bounded measurable test functions against the random measures μ_λ^ξ, see Penrose (2007*a*, Th. 2.1) and Baryshnikov and Yukich (2005, Th. 2.1).

Theorem 4.2 *Suppose that ξ is translation invariant and assume one of the following*

- *either ξ satisfies the bounded moment condition* (4.6) *for some $p > 2$ and is polynomially stabilizing on Poisson input with some exponent $q > dp/(p-2)$, see Penrose* (2007*a*)*;*
- *or ξ satisfies the bounded moment condition for $p = 4$ and is polynomially stabilizing on Poisson input with some exponent $q > d$, see Baryshnikov and Yukich* (2005).

Then the integral in (4.10) *converges for each $\tau > 0$ and $V^\xi(\cdot)$ is well defined. Moreover, for each bounded measurable $f : Q_1 \to \mathbb{R}$ we have*

$$\lim_{\lambda\to\infty} \lambda^{-1}\mathrm{Var}\left(\left\langle f, \mu_\lambda^\xi \right\rangle\right) = \left[\sigma_\kappa^\xi[f]\right]^2 = \int_{Q_1} V^\xi(\kappa(x)) f^2(x)\kappa(x)\mathrm{d}x. \qquad (4.11)$$

While referring the reader to Penrose (2007*a*) and Baryshnikov and Yukich (2005) for (far from trivial) details of the proof, we present here an intuitive idea of why (4.11) should hold. To this end, we write

$$\lambda^{-1}\mathrm{Var}\left(\left\langle f,\mu_\lambda^\xi\right\rangle\right) = \lambda^{-1}\int_{Q_\lambda} \mathbf{E}f^2(\lambda^{-1/d}u)\xi^2(u;\mathcal{P}_{\lambda,\kappa})\kappa(\lambda^{-1/d}u)\mathrm{d}u$$

$$+\lambda^{-1}\int_{Q_\lambda\times Q_\lambda} f(\lambda^{-1/d}u)f(\lambda^{-1/d}v)\sigma^\xi_{\lambda,\kappa}[u,v]\kappa(\lambda^{-1/d}u)\kappa(\lambda^{-1/d}v)\mathrm{d}u\mathrm{d}v,$$

where

$$\sigma^\xi_{\lambda,\kappa}[u,v] = \mathbf{E}\xi(u;\mathcal{P}_{\lambda,\kappa}\cup\{v\})\xi(v;\mathcal{P}_{\lambda,\kappa}\cup\{u\}) - \mathbf{E}\xi(u;\mathcal{P}_{\lambda,\kappa})\mathbf{E}\xi(v;\mathcal{P}_{\lambda,\kappa}).$$

Now, substituting $x=\lambda^{-1/d}u$ and $y=v-u$ we come to

$$\lambda^{-1}\mathrm{Var}\left(\left\langle f,\mu_\lambda^\xi\right\rangle\right) = \int_{Q_1} f^2(x)\mathbf{E}\xi^2(\lambda^{1/d}x,\mathcal{P}_{\lambda,\kappa})\kappa(x)\mathrm{d}x + \int_{Q_1}\int_{Q_\lambda-\lambda^{1/d}x} f(x)$$

$$\times f(x+\lambda^{-1/d}y)\sigma^\xi_{\lambda,\kappa}[\lambda^{1/d}x,\lambda^{1/d}x+y]\kappa(x)\kappa(x+\lambda^{-1/d}y)\mathrm{d}y\mathrm{d}x.$$

At this point we resort again to the afore-mentioned objective method combined with translation invariance of ξ to approximate for λ large enough

- the local geometry of $\mathcal{P}_{\lambda,\kappa}$ around $\lambda^{1/d}x$ by the homogeneous process $\mathcal{P}_{\kappa(x)}$ around o,
- the behaviour of the functional $\xi(\lambda^{1/d}x,\mathcal{P}_{\lambda,\kappa})$ by that of $\xi(o,\mathcal{P}_{\kappa(x)})$,
- the covariance kernel $\sigma^\xi_{\lambda,\kappa}[\lambda^{1/d}x,\lambda^{1/d}x+y]$ by

$$\sigma^\xi_{\kappa(x)}[o,y] = \mathbf{E}\xi(o;\mathcal{P}_{\kappa(x)}\cup\{y\})\xi(y;\mathcal{P}_{\kappa(x)}\cup\{o\}) - [\mathbf{E}\xi(o;\mathcal{P}_{\kappa(x)})]^2.$$

Making these steps formal requires considerable effort and multiple use of stabilization (to carry over the input-level local approximations to the level of its functional ξ and to get rid of boundary effects) as well as moment bounds (to ensure suitable limit behaviour of the integrals). Note that the convergence of covariance integrals is again determined by stabilization as implying rapid enough decay of covariances. This leads us to

$$\lim_{\lambda\to\infty}\lambda^{-1}\mathrm{Var}\left(\left\langle f,\mu_\lambda^\xi\right\rangle\right) = \int_{Q_1} f^2(x)[\mathbf{E}\xi^2(o,\mathcal{P}_{\kappa(x)})]\kappa(x)\mathrm{d}x$$

$$+\int_{Q_1} f^2(x)\int_{\mathbb{R}^d}\sigma^\xi_{\kappa(x)}[o,y]dy\kappa^2(x)\mathrm{d}x.$$

Comparing this with the definition (4.10) yields the required relation (4.11).

It should be emphasized at this point that considerable progress has been and is currently being made in asymptotic variance characterization also for geometric functionals which are *not translation invariant*, see Baryshnikov and Yukich (2005), Penrose (2007*a*) and Baryshnikov, Penrose and Yukich (2009). These important developments fall beyond the scope of this article though, due to space limitations.

4.2.6 *Central limit theorem*

With the variance asymptotics established, the next natural question is to get Gaussian limits for the empirical measures μ_λ^ξ. The central limit problem for stabilizing functionals has been thoroughly studied in the literature, see e.g. Penrose and Yukich (2001, 2002, 2005), Baryshnikov and Yukich (2005), Penrose (2007*a*) and references therein. With notation introduced in previous paragraphs we have, see Penrose (2007*a*, Th. 2.2) and Baryshnikov and Yukich (2005, Th. 2.1),

Theorem 4.3 *Suppose that ξ is translation invariant and assume one of the following*

- *either ξ stabilizes exponentially on Poisson input and satisfies the bounded moment condition* (4.6) *for some exponent $p > 2$;*
- *or ξ satisfies the bounded moment condition* (4.6) *for some exponent $p > 3$ and is polynomially stabilizing on Poisson input with some exponent $q > d(150 + 6/p)$.*

Then, for bounded measurable f, the random variables $\lambda^{-1/2}\langle f, \bar{\mu}_\lambda^\xi\rangle$ converge in law to the normal $\mathcal{N}(0, [\sigma_\kappa^\xi[f]]^2)$. Moreover, under the first assumption we have in addition, for each fixed $q \in (2, \min(p, 3))$,

$$\sup_{t\in\mathbb{R}} \left| \mathbf{P}\left\{ \frac{\left\langle f, \bar{\mu}_\lambda^\xi \right\rangle}{\sqrt{\mathrm{Var}\left(\left\langle f, \mu_\lambda^\xi \right\rangle\right)}} \le t \right\} - \Phi(t) \right|$$

$$= O\left((\log \lambda)^{qd} \lambda \left[\mathrm{Var}\left(\left\langle f, \mu_\lambda^\xi \right\rangle\right)\right]^{-q/2} \right), \tag{4.12}$$

where $\Phi(\cdot)$ is a distribution of the standard normal.

The bound (4.12) on rates of normal approximation, although improving on many previous results, is not yet optimal and there is an ongoing research on making it tighter, see e.g. Barbour and Xia (2006, Sec. 3.2) where, by application of Stein's method, the power of $\log \lambda$ on the right hand side of (4.12) has been reduced to $(q-1)d$ under appropriate assumptions.

There are several possible ways of obtaining asymptotic Gaussianity for μ_λ^ξ, including

- Standard martingale difference techniques, see e.g. Penrose (2003, Ch. 2) for more details.

- Cumulant methods, see Baryshnikov and Yukich (2005). The idea is to show that cumulant measures of μ_λ^ξ of order higher than 2 decay to 0 as $\lambda \to \infty$ under the CLT scaling, which ensures the asymptotic normality by standard argument. To establish the decay of cumulant measures one decomposes them into linear combinations of the so-called cluster measures, which are directly controllable in terms of stabilization, see ibidem.
- Stein–Chen method or, more precisely its consequences for sums of random variables with sparse dependency graphs, see Baldi and Rinott (1989), Penrose (2003, Sec 2.3) and Chen and Shao (2004). This method, originally first used in geometric probability by Avram and Bertsimas (1993) relying upon Baldi and Rinott (1989), provides the best up-to-date control on CLT convergence rates for stabilizing functionals, see Penrose and Yukich (2005), Barbour and Xia (2006) and Penrose (2007*a*). To be more specific, consider a finite collection $(\eta_i)_{i\in I}$ of random variables and say that a graph with vertex set I is a dependency graph for (η_i) if for any two disjoint subsets $I_1, I_2 \subseteq I$ the collection $(\eta_i)_{i\in I_1}$ is independent from $(\eta_i)_{i\in I_2}$ then there are no edges in the graph connecting vertices from I_1 and I_2. Assume now that the collection $(\eta_i)_{i\in I}$ admits a dependency graph with maximum vertex degree D and that

    - $\mathbf{E}\eta_i = 0,\ i \in I$,
    - $\mathbf{E}W^2 = 1$ for $W = \sum_{i\in I} \eta_i$,
    - For some $2 < p \leq 3$ the L^p-norm $||\eta_i||_p$ is bounded above by $\theta > 0$ for all $i \in I$,

  Then we have for all $t \in \mathbb{R}$, see Chen and Shao (2004, Th. 2.7),

$$|\mathbf{P}\{W \leq t\} - \Phi(t)| \leq 75 D^{5(p-1)} |I| \theta^p. \tag{4.13}$$

  The argument in Penrose and Yukich (2005) uses this statement for η_i's taken to be, roughly speaking, cumulative sums of a geometric functional ξ over logarithmic-size sub-cubes of Q_λ as $\lambda \to \infty$, subject to some extra corrections due to the fact that stabilization radii of ξ may admit very large although extremely improbable values. This difficulty is dealt with by coupling the original system with one in which no dependencies occur beyond a logarithmic order cut-off (in fact the cut-off agrees with the cube size there) and then applying (4.13) for the coupled system and using stabilization to estimate the variational distance between the original and coupled system. The final bound for the original process is then obtained by combining the above estimates upon optimizing the precise basic sub-cube size and the related stabilization cut-off. We refer the reader to Penrose and Yukich (2005) for full details of this non-trivial argument. See also Barbour and Xia (2006) and Penrose and Wade (2008) for further developments.

We note that the results in Baryshnikov and Yukich (2005) and Penrose (2007*a*) are stated in larger generality there, including non-translation invariant functionals and more general form of empirical measures. We refer the reader to these papers for more details, keeping our considerations restricted to translation invariant functionals here for presentational convenience.

4.2.7 *Binomial input and de-Poissonization*

In this paragraph we shall discuss a general technique of making the above results for stabilizing functionals work also on binomial input. This technique goes under the name of *de-Poissonization*, see Penrose and Yukich (2001), Baryshnikov and Yukich (2005), Penrose (2007*a*) and references therein. To proceed with its presentation we shall use some extra notation. Let $\kappa(\cdot)$ be an almost everywhere continuous probability density on Q_1 and let $\mathcal{X}^{(n)} = \{X_1^{(n)}, \ldots, X_n^{(n)}\}$ be an i.i.d. sample of size n in Q_n drawn from density $n^{-1}\kappa(n^{-1/d}x)\mathrm{d}x$. We introduce a special notation for the empirical measure generated by ξ on this binomial input:

$$\nu_n^\xi = \sum_{i=1}^n \xi\left(X_i^{(n)}, \mathcal{X}^{(n)}\right)\delta_{n^{-1/d}X_i^{(n)}}.$$

As usual, we write $\bar{\nu}_n^\xi = \nu_n^\xi - \mathbf{E}\nu_n^\xi$ for its centred version. We have the following limit theorem for ν_n^ξ

Theorem 4.4 *Assume that ξ is translation invariant, stabilizing on Poisson input and satisfies the bounded moment condition (4.6) for some $p > 1$. Then for each bounded measurable function $f : Q_1 \to \mathbb{R}$*

$$\lim_{n\to\infty} n^{-1}\mathbf{E}\left\langle f, \nu_n^\xi\right\rangle = \int f(x)[\mathbf{E}\xi(o, \mathcal{P}_{\kappa(x)})]\kappa(x)\mathrm{d}x. \tag{4.14}$$

Moreover, if ξ satisfies any of the two alternative conditions in Theorem 4.3 *then the limit*

$$\left[\tau_\kappa^\xi[f]\right]^2 = \lim_{n\to\infty} \mathrm{Var}\left(\langle f, \nu_n^\xi\rangle\right) \tag{4.15}$$

exists and the random variables $n^{-1/2}\langle f, \bar{\nu}_n^\xi\rangle$ converge in law to the normal $\mathcal{N}(0, [\tau_\kappa^\xi[f]]^2)$.

Note that the convergence of expectations can be strengthened to a law of large numbers fully analogous to that in Theorem 4.1 above, see Penrose (2007*b*) for full details. Moreover, the limit variance $\tau_\kappa^\xi[f]$ in (4.15) above admits an explicit representation, to avoid technicalities we provide it here in a slightly restricted set-up where we impose on ξ the additional strong add-one-cost stabilization requirement on Poisson input. This assumption ensures the existence of the add-one-cost $\Delta^\xi[\tau] = \Delta^\xi(o, \mathcal{P}_\tau)$, $\tau > 0$. We have then, see Penrose and Yukich (2001), Baryshnikov and Yukich (2005), Penrose (2007*a*) and the references therein,

$$\left[\tau_\kappa^\xi[f]\right]^2 = \left[\sigma_\kappa^\xi[f]\right]^2 - \left(\int_{Q_1} f(x)[\mathbf{E}\Delta^\xi[\kappa(x)]]\kappa(x)\mathrm{d}x\right)^2. \tag{4.16}$$

Moreover, it can be shown that whenever the add-one cost $\Delta^{\xi}[\tau]$ is of non-degenerate distribution, the limit variance $[\tau_{\kappa}^{\xi}[f]]^2$ is non-zero as soon as neither κ nor f are identically zero, see Penrose and Yukich (2001, Th. 2.1). The proof of Theorem 4.4 is quite complicated in its full details, see ibidem, yet the main idea is rather intuitive. Upon appropriate coupling using the properties of Poisson processes, the binomial process $\mathcal{X}^{(n)}$ is represented by taking the Poisson process $\mathcal{P}_{n,\kappa}$ of the same intensity and then removing or adding Poiss$(n) - n$ randomly chosen extra points. Denoting the process of these removed or added extra points by Γ_n one proceeds as follows.

- Since the points of Γ_n are rather scarcely scattered in Q_n, as their number is of order $o(n)$, each such locally *feels as if it were* in homogeneous Poisson environment of intensity given by the local value of κ. Thus, using appropriate coupling techniques and the translation invariance, the contribution the removal/addition of any such point $x \in \Gamma_n$ brings to the empirical measure μ_{λ}^{ξ} is well approximated by a copy of $-/+\Delta(o, \mathcal{P}_{\kappa(x)})\delta_x$. This observation is again an instance of the objective method.
- Using again that the points of Γ_n are usually at large distances from each other, one argues that the afore-mentioned copies of the add-one cost variables are only very weakly dependent and one establishes a sort of a law of large numbers for the correction to $\langle f, \mu_n^{\xi} \rangle$ due to removing or adding points of Γ_n. In other words, one shows that

$$\langle f, \mu_n^{\xi} - \nu_n^{\xi} \rangle \approx (\text{Poiss(n)} - n) \left[\int_{Q_1} f(x) [\mathbf{E}\Delta^{\xi}[\kappa(x)]] \kappa(x) \mathrm{d}x \right]$$

  with possible lower order fluctuations, where Poiss(n) stands for the Poisson random variable of mean n.
- Arguing that the correction Poiss(n) $- n$ of the number of points is strongly decorrelated with $\langle f, \nu_n^{\xi} \rangle$ one gets the asymptotic relation

$$\langle f, \mu_n^{\xi} \rangle \approx \gamma_n \tau_n^{\xi}[f]$$

  with $\gamma_n \sim$ Poiss$(n) - n$ asymptotically independent of $\mathcal{X}^{(n)}$.
- Finally, one uses that $\mathrm{Var}(\gamma_n) = n$ to obtain the desired results.

4.2.8 *Moderate and large deviations*

In addition to the knowledge of deviations of the empirical measures μ_{λ}^{ξ} on the CLT scale $\lambda^{1/2}$, a lot is also known on larger scale deviations. As is often the case in the theory of weakly dependent random variables, also here two natural and essentially different regimes emerge:

- deviations under the volume-order scaling λ, usually going under the name of large deviations,

- deviations under the scaling of order strictly between $\lambda^{1/2}$ and λ, referred to as the *moderate deviations.*

The large deviations for the empirical measures μ_λ^ξ are characterized by the following result, see Schreiber and Yukich (2005, Th. 1.1, 1.2) and the applications discussed in Section 2 there. It should be emphasized that this is a measure-level and not just scalar-level statement. In order to avoid introducing extra notation specific only to this paragraph, instead of formulating general assumptions on the functional ξ as ibidem, here we simply list prominent examples for which the theorem below holds.

Theorem 4.5 *Assume that ξ is a non-negative stabilizing functional belonging to one of the following classes*

- *acceptance indicators for random sequential packing, see Example* 4.7 *in Section* 4.3 *below,*
- *acceptance indicators for spatial birth and growth models, see Example* 4.8 *in Section* 4.3 *below,*
- *nearest neighbour graph empirical functionals and edge length functionals as discussed in Example* 4.9 *in Section* 4.3 *below,*
- *sublinear edge length functionals on discretized Delaunay graphs and discretized sphere of influence graphs, see Examples* 4.10 *and* 4.11 *in Section* 4.3 *below.*

Moreover, suppose the density profile κ is constant. Then there exists a lower semicontinuous convex function $I^\xi : \mathbb{R} \to \mathbb{R}_+ \cup \{+\infty\}$ enjoying the additional properties that $I^\xi(t) > 0$ for $t \neq 0$ and that $\lim_{t\to\infty} I^\xi(t)/t = +\infty$ and such that the family $(\bar{\mu}_\lambda^\xi)$ of centred empirical measures generated by ξ satisfies on the space of Borel measures on Q_1 endowed with the usual weak topology the full large deviation principle with speed λ and with good rate function J given by $J^\xi(\gamma) = \int_{Q_1} I^\xi(d\gamma/dl)dl$ if γ is absolutely continuous with respect to the Lebesgue measure l and $J(\gamma) = +\infty$ otherwise.

Recall that, as in Dembo and Zeitouni (1998, Sec. 1.2), we say that a family of random elements $(\mathbb{Y}_\lambda)_{\lambda>0}$ taking values in a general topological space $\mathcal{Y}$ satisfies on $\mathcal{Y}$ a full large deviation principle with speed $s(\lambda)$ and with a good rate function $I : \mathcal{Y} \to \mathbb{R}_+ \cup \{+\infty\}$ iff the level sets $\{I \leq M\}$, $M > 0$ are compact, for each open $\mathcal{O} \subseteq \mathcal{Y}$ we have $\lim_{\lambda\to\infty} \frac{1}{s(\lambda)} \log \mathbf{P}\{\mathcal{Y} \in \mathcal{O}\} \geq -\inf_{y\in\mathcal{O}} I(y)$ and for each closed $\mathcal{F} \subseteq \mathcal{Y}$ $\lim_{\lambda\to\infty} \frac{1}{s(\lambda)} \log \mathbf{P}\{\mathcal{Y} \in \mathcal{F}\} \leq -\inf_{y\in\mathcal{F}} I(y)$. From the formal viewpoint a moderate deviation principle is just a large deviation principle with speed between the square root CLT scaling and the volume order LDP scaling.

The following result characterizes the moderate deviations for random measures $\bar{\mu}_\lambda^\xi$, see Baryshnikov, Eichelsbacher, Schreiber and Yukich (2008).

Theorem 4.6 *Assume that the functional ξ belongs to one of the following classes of exponentially stabilizing functionals:*

- *acceptance indicator for random sequential packing, see Example* 4.7 *in Section* 4.3 *below,*
- *acceptance indicator for spatial birth and growth model with initial particle sizes admitting uniform deterministic bounds, see Example* 4.8 *in Section* 4.3 *below,*
- *nearest neighbour graph empirical functionals as discussed in Example* 4.9 *in Section* 4.3 *below,*
- *volume and vacancy functionals for germ-grain models with deterministically bounded grains, see Baryshnikov, Eichelsbacher, Schreiber and Yukich* (2008) *and Chapter* 1.

Suppose that numbers $(\alpha_\lambda)_{\lambda>0}$ *are chosen so that* $\alpha_\lambda \to +\infty$ *and* $\alpha_\lambda \lambda^{-1/2} \to 0$ *as* $\lambda \to \infty$. *Then the family of re-scaled empirical measures* $(\alpha_\lambda^{-1}\lambda^{-1/2}\mu_\lambda^\xi)_{\lambda>0}$ *satisfies on the space of signed Borel measures on* Q_1 *endowed with the weak topology the full moderate deviation principle with speed* α_λ^2 *and with good rate function*

$$K_\kappa^\xi(\gamma) = \frac{1}{2}\int_{Q_1}\left(\frac{d\nu}{V^\xi(\kappa(x))\kappa(x)\mathrm{d}x}\right)^2 V^\xi(\kappa(x))\kappa(x)\mathrm{d}x$$

if ν *is absolutely continuous with respect to* $V^\xi(\kappa(x))\kappa(x)\mathrm{d}x$ *and* $K_\kappa^\xi(\gamma) = +\infty$ *otherwise.*

The proof of Theorem 4.5 strongly relies upon the large deviation result of Seppäläinen and Yukich (2001), see Theorem 4.16 below, stated for nearly additive geometric functionals to which the stabilizing functionals are reduced via the so-called finite range corrections. We refer the reader to Schreiber and Yukich (2005) for further details. The moderate deviation principle in Theorem 4.6 is established by Baryshnikov, Eichelsbacher, Schreiber and Yukich (2008) using local exponential modifications of the input Poisson process, whose cumulants are controlled by methods inspired by Fernández–Ferrari–Garcia graphical constructions, see Section 4.7 below. This provides local control of the log-Laplace transform, as required to get the moderate deviation principle via the Gärtner–Ellis theorem and can be lifted to the measure level via projective limit techniques, see Theorem 2.3.6 and Section 4.6 of Dembo and Zeitouni (1998).

4.3 Some applications of stabilization theory

We supplement the preceding subsection on stabilizing functionals by providing a (far from exhaustive) list of examples. It should be emphasized at this point that even though here we limit ourselves to continuum applications as better illustrating the use of stabilization in stochastic geometry, the stabilization techniques have been also successfully applied to lattice-based problems, see e.g. Penrose (2005) and the references therein. Note also that in addition to the examples below stabilization does also yield limit theory for Boolean models (Baryshnikov and Yukich, 2005), convex hulls (Schreiber and Yukich, 2008*b*), maximal points

(Baryshnikov and Yukich, 2006; Schreiber and Yukich, 2008*b*), generalized spacings in high dimensions (Baryshnikov, Penrose and Yukich, 2009) and many other models, which illustrates the broad and rich scope of stabilization techniques.

Example 4.7 (Random sequential packing) Consider a finite point configuration $\mathcal{X}$ and to each $x \in \mathcal{X}$ attach a unit ball centred at x. Moreover, to all points in $\mathcal{X}$ attach i.i.d. uniform time marks taking values in some finite time interval $[0, T]$, $T > 0$. This establishes a *chronological* order on the points of $\mathcal{X}$. Declare the first point in this ordering *accepted* and proceed recursively, each time accepting the consecutive point if the ball it carries does not overlap the previously accepted (packed) balls and rejecting it otherwise. The functional $\xi(x, \mathcal{X})$ is defined to be 1 is the ball centred at x has been accepted and 0 otherwise. This defines the prototypical random sequential packing/adsorption (RSA) process, which originates in its one-dimensional version from the classical Rényi car parking model, see Rényi (1958), and which is used in statistical physics along with its many variants and modifications as a model for adsorption and deposition, see Evans (1993), Senger, Voegel and Schaaf (2000) and Torquato (2002) for a survey. There are many mathematical results for the RSA model, see Penrose (2001*a*, 2001*b*), Penrose and Yukich (2002), Baryshnikov and Yukich (2003) and Schreiber, Penrose and Yukich (2007). The random packing model satisfies the assumptions and hence also the statements of all Theorems 4.1, 4.2, 4.3, 4.4, 4.5 and 4.6 above, see ibidem.

Example 4.8 (Spatial birth and growth models) Consider the following generalization of the basic RSA model: the balls attached to subsequent points, further interpreted as *particles*, are allowed to have their initial radii random i.i.d. rather than fixed. Moreover, at the moment of its birth each particle begins to *grow* radially with constant speed v until it hits another particle or reaches a certain maximal admissible size – in both these cases it stops growing in the offending directions, possibly continuing its growth in the remaining admissible directions though. In analogy to the basic RSA, a particle is accepted if it does not overlap any previously accepted one and is discarded otherwise. The functional of interest is again given by $\xi(x, \mathcal{X}) = 1$ if the particle centred at x has been accepted and 0 otherwise. This model, going also under the name of the Johnson–Mehl growth process in the particular case where the initial radii are 0, has attracted a lot of interest in the literature, see Stoyan, Kendall and Mecke (2005), Chiu and Quine (1997), Penrose and Yukich (2002), Baryshnikov and Yukich (2005). The spatial birth and growth model satisfies the assumptions and hence also the statements of Theorems 4.1, 4.2, 4.3, 4.4 and 4.5. If in addition the initial radii are deterministically bounded away from 0, Theorem 4.6 is applicable as well. It should also be mentioned at this point that generalized variants of growth processes can be constructed, fitting neatly into the stabilization/localization set-up modified by admitting an unbounded time direction and using localizing space-time cylinders rather than balls. In this way one may deduce limit theory for the number of points in the convex hull of a Poisson sample, albeit with a non-standard variance, see Schreiber and Yukich (2008*b*).

Example 4.9 (Nearest neighbour graphs) Fix a positive integer k. By the k-nearest neighbour graph $NG^{\rightarrow}(\mathcal{X}) = NG^{\rightarrow}(\mathcal{X}; k)$ on a locally finite point configuration $\mathcal{X}$ we mean the directed graph where an edge is present from x to y in $\mathcal{X}$ whenever y is among the k nearest neighbours of x. We also consider its undirected version $NG^{\leftrightarrow}(\mathcal{X})$ arising by forgetting the direction of edges and collapsing possible double edges into single ones. A broad family of functionals, further referred to as *edge length functionals*, arise as

$$\xi(x, \mathcal{X}) = \sum_{e \in \mathrm{Edges}(x; NG^*(\mathcal{X}))} \phi(|e|),$$

with $\mathrm{Edges}(x; NG^*(\mathcal{X}))$ standing for the collection of edges outgoing from x in the graph $NG^{\rightarrow}(x; \mathcal{X})$ or $NG^{\leftrightarrow}(x; \mathcal{X})$ and where ϕ is a non-negative real function with at most polynomial growth. Another collection of natural functionals are the *empirical functionals of nearest neighbour graphs* which either admit the above representation with 0–1 – valued ϕ indicating whether the edge length exceeds a certain threshold, or are defined as an indicator function of some event involving the degree of the graph at x and possibly also the edge length, such as 'the total length of edges incident to x exceeds a certain multiplicity of the graph degree of x', etc. Such functionals of the nearest neighbour graphs have been thoroughly investigated in the literature, see Avram and Bertsimas (1993), Penrose and Yukich (2001), Penrose and Yukich (2003), Baryshnikov and Yukich (2005), Wade (2007) and the geometry of nearest neighbour like random graphs will be covered in much more detail in Chapter 7, whence we only briefly conclude our discussion here by noting that the aforementioned functionals satisfy the assumptions and hence also the statements of Theorems 4.1, 4.2, 4.3, 4.4 and 4.5. In addition, for the empirical functionals of nearest neighbour graphs Theorem 4.6 is applicable as well. The reader is referred ibidem and to Schreiber and Yukich (2005) and Baryshnikov, Eichelsbacher, Schreiber and Yukich (2008) for proofs.

Example 4.10 (Voronoi tessellations and Delaunay graphs) By the Delaunay graph of a planar input point process we mean the dual graph of the Voronoi tessellation it induces, where two cell centres are connected with an edge whenever the corresponding Voronoi cells are adjacent. We refer the reader to Chapter 5 for full details of these notions. The edge length functionals and empirical functionals of Delaunay graphs can be considered in full analogy with those for nearest neighbour graphs above and the set of literature references for limit theorems in this context is precisely the same as in Example 4.9 above. Again, these functionals satisfy the assumptions and hence also the statements of Theorems 4.1, 4.2, 4.3, 4.4. Moreover, appropriate discrete modifications thereof satisfy the LDP Theorem 4.5 whereas the MDP Theorem 4.6 holds for empirical functionals, see ibidem for details.

Example 4.11 (Sphere of influence graphs) By the *sphere of influence* of a given point $x \in \mathcal{X}$ we understand the largest ball centred at x which does not contain any other points of $\mathcal{X}$ in its interior. The sphere of influence graph on $\mathcal{X}$ is built by connecting with an undirected edge each pair of points in $\mathcal{X}$ whose spheres of influence overlap. These graphs have been treated ibidem and the above discussion of Delaunay graphs applies also verbatim to sphere of influence graphs.

4.4 Nearly additive Euclidean functionals

Another interesting class of geometric functionals, for which also a rich supply of limit theorems is available, are the so-called Euclidean functionals often enjoying variants of approximate subadditivity and superadditivity properties. Our discussion of this important family of functionals provided in this subsection will be rather brief and by no means exhaustive because a detailed survey of this subject is available in the monograph Yukich (1998), which collects the contributions of many authors working in the subject. Let us just mention Rhee (1992, 1993, 1994), Redmond (1993), Redmond and Yukich (1994, 1996), Steele (1990, 1993, 1997), Talagrand (1995, 1996*a*, 1996*b*) Yukich (1995, 1996) noting that this list is clearly not exhaustive and recommending Yukich (1998, 1999) for further reference. We follow Yukich (1998) in our presentation below, with a somewhat modified notation to keep it compatible with the remaining parts of the present survey.

4.4.1 *Euclidean functionals*

We say that a geometric functional ξ on $\mathbb{R}^d$ is a *Euclidean functional of order* p, $p > 0$, if ξ is translation invariant and homogeneous of order p, that is to say $\xi(\alpha x, \alpha \mathcal{X}) = \alpha^p \xi(x, \mathcal{X})$ for all $x, \mathcal{X}$ and $\alpha > 0$. A Euclidean functional ξ of order p is said to enjoy *geometric subadditivity* if there exists a constant C such that for each rectangle $\mathcal{R} \subseteq \mathbb{R}^d$ partitioned into two sub-rectangles $\mathcal{R}_1$ and $\mathcal{R}_2$ and for arbitrary locally finite point configuration $\mathcal{X}$ we have (cf Definition 3.1 ibidem)

$$H^{\xi}[\mathcal{X} \cap \mathcal{R}] \leq H^{\xi}[\mathcal{X} \cap \mathcal{R}_1] + H^{\xi}[\mathcal{X} \cap \mathcal{R}_2] + C(\mathrm{diam}\mathcal{R})^p. \tag{4.17}$$

Moreover, we say that a Euclidean functional is *superadditive* (cf ibidem) iff, with $\mathcal{R}, \mathcal{R}_1$ and $\mathcal{R}_2$ as above,

$$H^{\xi}[\mathcal{X} \cap \mathcal{R}] \geq H^{\xi}[\mathcal{X} \cap \mathcal{R}_1] + H^{\xi}[\mathcal{X} \cap \mathcal{R}_2]. \tag{4.18}$$

Whereas the geometric subadditivity (4.17) is satisfied by many functionals, as illustrated by a number of examples below, oftentimes the superadditivity property holds for an appropriate modification of the original functional rather than for the functional itself. Quite often the Euclidean functionals arise in graph optimization problems, see ibidem and examples in Section 4.5 below, in which case the considered $\xi(x, \mathcal{X})$ is usually expressed in terms of (a power of) the lengths of edges of some optimized graph on vertex set $\mathcal{X}$ incident with x. In this case one considers the so-called *boundary modification* of the functional in a (usually rectangular) region $\mathcal{R}$, which is defined using the same rules for graph optimization on $\mathcal{X}$, but with the standard Euclidean distance replaced by the *wired* metric $d_{\mathcal{R}}(x, y) = \min(|x-y|, \mathrm{dist}(x, \partial\mathcal{R}) + \mathrm{dist}(y, \partial\mathcal{R}))$. This corresponds to collapsing the boundary to one point or, alternatively, making the movement on the boundary 'free of cost'. Such modification admits a rather intuitive interpretation for the examples discussed below, for which we shall argue that their boundary modifications do satisfy the superadditivity property even though the original functionals in general do not. For a Euclidean functional ξ we will use the standard notation $\xi^{\partial\mathcal{R}}$ to denote the boundary version of ξ in a rectangle $\mathcal{R}$. It is of crucial importance to the theory discussed below that many natural Euclidean functionals are very well approximated by their boundary modifications. In this context, we say that a Euclidean functional ξ of order p is *pointwise close* to its boundary modification $\xi^{\partial[\cdot]}$ iff for all finite subsets $\mathcal{X} \subseteq Q_1$

$$\left| H^{\xi}[\mathcal{X}] - H^{\xi^{\partial Q_1}}[\mathcal{X}] \right| = o\big((\mathrm{card}\mathcal{X})^{(d-p)/d}\big), \tag{4.19}$$

where, recall, $Q_1 = [-1/2, 1/2]^d$. This notion has its natural probabilistic counterpart – we say that ξ is close in mean to its boundary modification $\xi^{\partial[\cdot]}$ on a (random) input $\mathcal{P} \subseteq Q_1$ iff

$$\left| H^{\xi}[\mathcal{P}] - H^{\xi^{\partial Q_1}}[\mathcal{P}] \right| = o\big((\mathbf{E}\mathrm{card}\mathcal{P})^{(d-p)/d}\big). \tag{4.20}$$

One further useful notion is *smoothness.* We say that a Euclidean functional of order p is *smooth of order* p iff for all $\mathcal{X}, \mathcal{Y} \subseteq Q_1$ we have

$$\left| H^{\xi}[\mathcal{X} \cap \mathcal{Y}] - H^{\xi}[\mathcal{X}] \right| = \mathcal{O}\big((\mathrm{card}\mathcal{Y})^{(d-p)/d}\big). \tag{4.21}$$

4.4.2 *Laws of large numbers*

As already signalled above, the standard situation is that a Euclidean functional itself satisfies the geometric subadditivity (4.17) whereas its boundary modification is superadditive (4.18). The combination of these properties with smoothness and closeness to the boundary functional is also quite common. This is already enough to guarantee the following law of large numbers on homogeneous input, see Yukich (1998, Th. 4.1).

Theorem 4.12 *Assume that ξ is a Euclidean functional of order p on $\mathbb{R}^d$ with $1 \le p < d$. If $\xi^{\partial[\cdot]}$ is a smooth and superadditive Euclidean functional of order p then*

$$\lim_{n\to\infty} H^{\xi^{\partial Q_1}}[\mathcal{X}^{(n)}]/n^{(d-p)/p} = \alpha[\xi^{\partial[\cdot]}] \quad c.c., \tag{4.22}$$

where c.c. stands for the complete convergence as in (4.9) whereas $\mathcal{X}^{(n)}$ *is the homogeneous binomial point process of cardinality* n *in* Q_1 *and* $\alpha[\xi^{\partial[\cdot]}]$ *is a constant. Moreover, under the same assumptions,*

$$\lim_{\lambda\to\infty} H^{\xi^{\partial Q_1}}[\mathcal{P}_\lambda]/\lambda^{(d-p)/p} = \alpha[\xi^{\partial[\cdot]}] \ a.s., \tag{4.23}$$

where $\mathcal{P}_\lambda = \mathcal{P}_{\lambda,1}$ *is the homogeneous Poisson point process of intensity* λ *in* Q_1*. If in addition* ξ *is pointwise close to its boundary modification* $\xi^{\partial[\cdot]}$*, the relations* (4.22) *and* (4.23) *hold with* $\xi^{\partial[\cdot]}$ *replaced there by* ξ*.*

The proof of this theorem makes standard use of superadditivity, with smoothness used to deal neatly with boundary effects, moreover it is enough to establish convergence of expectations whence the required complete convergence follows by suitable measure concentration results discussed in the sequel of this survey, see ibidem for further details and Theorem 4.15 below. Note that even though formally only the binomial version of the result is stated there, the Poisson version is just a straightforward modification thereof. This result can be extended in several natural directions. The first one is related to the requirement that the order of homogeneity fall below the dimensionality d. The behaviour of functionals with critical ($= d$) or supercritical ($> d$) homogeneity order is a deep issue falling beyond the scope of the present survey, we only mention here that the convergence of means can be obtained with $d \geq 2$ under the assumptions of Theorem 4.12 combined with additional assumptions of closeness in mean on Poisson input between ξ and its boundary modification as well as of smoothness in mean for ξ, see Yukich (1998, Th. 4.3, 4.5). In general, one should not hope for any kind of strong law of large numbers in such case because the order of fluctuations is not smaller than that of the mean there. We refer the reader to Steele (1997) for further details which are beyond our scope here.

Another natural direction in which Theorem 4.12 can be extended is to estimate the rates of convergence of means. This can indeed be done, and the right tool here is the geometric subadditivity (4.17) not used so far. We have, see Yukich (1998, Th. 5.1, 5.2).

Theorem 4.13 *Suppose that* ξ *and its boundary approximation* $\xi^{\partial[\cdot]}$ *are Euclidean functionals of order* p, $1 \leq p < d$ *on* $\mathbb{R}^d$, $d \geq 2$*. Further, assume that* ξ *is subadditive* (4.17) *whereas* $\xi^{\partial[\cdot]}$ *is superadditive* (4.18) *and that* ξ *and* $\xi^{\partial[\cdot]}$ *are close in mean* (4.20)*. Then*

$$\left|\mathbf{E}H^{\xi}[\mathcal{P}_\lambda] - \alpha[\xi]\lambda^{(d-p)/d}\right| = \mathcal{O}(\lambda^{(d-p-1)/d} \wedge 1). \tag{4.24}$$

If, moreover, ξ *satisfies the* add-one bound

$$\left|\mathbf{E}H^{\xi}[\mathcal{X}^{(n+1)}] - \mathbf{E}H^{\xi}[\mathcal{X}^{(n)}]\right| = \mathcal{O}(n^{-p/d}),$$

then (4.24) *holds also for the Poisson input* $\mathcal{P}_\lambda$ *replaced by the uniform binomial point process* $\mathcal{X}^{(n)}$ *on* Q_1*.*

We refer the reader to Yukich (1998) for details and discussion of this result and its consequences.

Finally, the third important direction of extension for Theorem 4.12 is to consider non-homogeneous input. This is done in the following result, see Yukich (1998, Th. 7.1).

Theorem 4.14 *Let π be a probability measure on Q_1 and κ the density of its absolutely continuous part. Let ξ and $\xi^{\partial[\cdot]}$ be Euclidean functionals of order p, $p < d$ on $\mathbb{R}^d$, $d \geq 2$, respectively subadditive and superadditive and close in mean on uniform binomial samples* (4.20). *Write $\mathcal{X}_\pi^{(n)}$ for the cardinality n binomial point process in Q_1 with individual point distribution π and let $\mathcal{P}_{\lambda\pi}$ be the Poisson point process with intensity measure $\lambda\pi$ on Q_1. Then*

$$\lim_{n\to\infty} H^{\xi}\left[\mathcal{X}_\pi^{(n)}\right]/n^{(d-p)/d} = \alpha[\xi]\int_{Q_1}[\kappa(x)]^{(d-p)/d}\mathrm{d}x \ \ c.c.$$

and

$$\lim_{\lambda\to\infty} H^{\xi}[\mathcal{P}_{\lambda\pi}]/\lambda^{(d-p)/d} = \alpha[\xi]\int_{Q_1}[\kappa(x)]^{(d-p)/d}\mathrm{d}x \ \ a.s.$$

The proof makes use of the homogeneous input Theorem 4.12 combined with a blocking device provided by Steele (1988) allowing for local sub-cube-wise homogeneous approximations of the underlying distribution π, this is then put together with the measure concentration Theorem 4.15 to conclude the complete convergence from the convergence of means, see Yukich (1998) for further details. Note that the asymptotic formulae for expectations are strongly reminiscent of the limit expression in (4.7) in Theorem 4.1, upon taking into account the homogeneity property of ξ.

4.4.3 *Concentration inequalities and large deviations*

Over the past two decades considerable progress has been made in understanding the so-called measure concentration phenomenon and powerful tools have been constructed allowing one to establish very general concentration inequalities even for extremely complex probabilistic systems. These methods include martingale techniques, isoperimetric inequalities, log-Sobolev inequalities, transportation inequalities and many other ones falling far beyond the scope of this survey. Among many authors who contributed to these developments one can never overemphasize the role of Michel Talagrand, whose works were crucial to the numerous breakthroughs achieved in this area of probability theory. We refer the reader to the monograph of Ledoux (2001) for further reference in this rapidly developing subject. Not unexpectedly, these methods proved also very useful in the theory of stabilizing functionals, see Rhee (1993), Talagrand (1995), Yukich (1998, Ch. 6) and references therein. Here we provide only one result in this vein, due to Rhee (1993) and relying on isoperimetric inequalities, see in Yukich (1998, Th. 6.3).

Theorem 4.15 *Assume that ξ is a Euclidean functional of order $0 < p < d$ on $\mathbb{R}^d$, $d \geq 2$, enjoying in addition the smoothness property* (4.21) *of order p. Further, let $\mathcal{X}_\pi^{(n)}$ be as in Theorem* 4.14 *above. Then there exists a constant C, depending on ξ and d but not on π and such that for all n and $t > 0$ we have*

$$\mathbf{P}\left\{\left|H^\xi[\mathcal{X}^{(n)}] - \mathbf{E}H^\xi[\mathcal{X}^{(n)}]\right| \geq t\right\} \leq C \exp\left(\frac{-t^{2d/(d-p)}}{Cn}\right).$$

As shown below, the assumptions of this general result are satisfied for many of our examples. The proof of this theorem uses the general isoperimetric results for product measures, with smoothness (4.21) used to control the variation of the considered functional in terms of Hamming distance between samples. We refer the reader ibidem for further details as well as for a survey of many other important concentration inequalities available for stabilizing functionals.

Another interesting question is whether a general Varadhan-style large deviation result can be provided for a broad class of nearly additive. The positive answer to this question has been given by Seppäläinen and Yukich (2001, Th. 2.1). As in case of Theorem 4.5 above, also here we avoid introducing any extra notions and simply list the principal example in our statement below.

Theorem 4.16 *Assume that ξ is a Euclidean functional belonging to one of the following classes*

- *Travelling salesman functionals, see Example* 4.17 *in Section* 4.5 *below,*
- *Minimum spanning tree functionals, see Example* 4.18 *in Section* 4.5 *below,*
- *Minimal matching functionals, see Example* 4.19 *in Section* 4.5 *below.*

Then there exists a good and convex rate function $I^\xi : \mathbb{R} \to \mathbb{R}_+ \cup \{+\infty\}$ enjoying the additional properties that $I^\xi(t) > 0$ for $t \neq 0$ and that $\lim_{t\to\infty} I^\xi(t)/t = +\infty$ and such that the family $\bar{H}^\xi[\mathcal{P}_\lambda] = H^\xi[\mathcal{P}_\lambda] - \mathbf{E}H^\xi[\mathcal{P}_\lambda]$ of centred cumulative functionals satisfies on $\mathbb{R}$ the full large deviation principle with speed λ and with rate function I^ξ.

The proof of this theorem relies on the near-additivity property of the considered functionals and uses suitable exponential approximations by sums of i.i.d. random variables arising as cumulative functionals over sub-cubes of Q_1, see Seppäläinen and Yukich (2001) for details.

4.5 Examples of nearly additive Euclidean functionals

Below, we discuss some examples of Euclidean functionals, constituting a representative but modest illustration of the class of models to which the discussed theory applies.

Example 4.17 (Travelling salesman problem (TSP)) Given a finite point configuration $\mathcal{X}$ we are looking for a closed polygonal tour visiting each point of $\mathcal{X}$ exactly once and minimizing the sum of (pth powers of) lengths of edges constituting the tour. The functional $\xi(x, \mathcal{X})$, $x \in \mathcal{X}$, is defined here as the sum of (pth powers of) lengths of edges in the optimal tour incident to x. This is a classical and fundamental problem in geometric optimization and it has a long history, see Frieze and Yukich (2002), Gutin and Punnen (2002), and Applegate, Bixby, Chvátal and Cook (2006). One reason for its importance is that this is an NP-complete problem, see ibidem. The TSP functional is known to be a Euclidean functional of order p with p as above, and its probabilistic theory has been widely discussed by Yukich (1998). Some new developments for a variant of the TSP (multiple depot vehicle routing problem) have been recently reported by Baltz (2007). Here we only mention that the travelling salesman problem does satisfy all assumptions and hence also the statements of Theorems 4.12, 4.14, 4.15 and 4.16, see ibidem and the references therein. Strong and general concentration results applying in particular for the TSP have been obtained by Talagrand (1995), see also the references therein.

Example 4.18 (Minimum spanning tree (MST)) By the minimum spanning tree (MST) of a point configuration $\mathcal{X}$ we understand the tree spanning all points in $\mathcal{X}$ and minimizing the sum of (pth powers of) lengths of its edges. In sharp contrast to the TSP, this problem is easily solved by a classical greedy algorithm due to Kruskal (1956). The MST functional $\xi(x, \mathcal{X})$ is defined as the sum of (pth powers of) lengths of edges in the minimum spanning tree incident to x. Again, this is an important and classical problem in geometric optimization, see Wu and Chao (2004) and the references therein. The probabilistic theory of the MST has been treated in detail by Yukich (1998). The MST functional is easily seen to be Euclidean and does satisfy all assumptions and hence also the statements of Theorems 4.12, 4.14, 4.15 and 4.16, see ibidem and the references therein. In contrast to the TSP, for the MST functional the central limit theorem is known – this remarkable result has been achieved by Kesten and Lee (1996) and Lee (1997).

Example 4.19 (Minimal matching) For a given point configuration $\mathcal{X}$ of even cardinality by minimal matching we mean splitting the whole configuration into pairs of points so that the sum of (pth powers of) distances between the points constituting each pair are minimal. The value of this minimum is referred to as the minimal matching functional and is easily checked to be Euclidean. The minimal matching problem can be solved in polynomial time, see Tarjan (1983) and the references therein for these classical algorithms. Again, all assumptions and thus also statements of Theorems 4.12, 4.14, 4.15 and 4.16 are satisfied for the minimal matching functional, see Yukich (1998) for further details.

4.6 Limit theory for germ-grain models

The stochastic-geometric limit theory is particularly well developed in the context of germ-grain models with the important results due to Heinrich, Molchanov and other authors, see Molchanov and Stoyan (1994) and Molchanov (1995), Molchanov (1997) for a statistically oriented exposition, Heinrich (1992) for general mixing properties, Heinrich and Molchanov (1999) for a central limit theorem for stationary β-mixing germ processes, Heinrich (2005) for very detailed moderate deviation analysis of Poisson grain models, as well as the references therein. Our presentation is rather brief here since the germ-grain models and related important statistical problems are discussed in other places in this volume, see Chapters 1 and 9 of this volume. Recall (cf. Chapter 1 and the discussion on Boolean models there) that by a germ-grain model with germ point process $\mathcal{P}$ we understand the union $\Xi = \bigcup_{x\in\mathcal{P}} x + \Xi_x$, where Ξ_x are independent random closed sets, here assumed in addition to be compact. Unless otherwise stated, we shall assume that $\mathcal{P}$ is stationary and Ξ_x's are i.i.d. which makes Ξ translation invariant. Note that, as everywhere throughout this survey, also here we restrict our considerations to germ-grain models in Euclidean spaces $\mathbb{R}^d$. The usual functional of our interest is given by

$$\xi(x, \mathcal{P}) = |\Xi \cap C(x, \mathcal{P})|,$$

where $C(x, \mathcal{P})$ is the Voronoi cell of x in $\mathcal{P}$. This functional is easily shown to be stabilizing and thus can also be studied by means of stabilization theory, see Baryshnikov and Yukich (2005) and Section 4.2, but we focus on different methods in the present subsection. Note that this definition of ξ means that the total-mass functional H^{ξ}_{λ} does in fact coincide with the so-called *empirical volume fraction*

$$\hat{p}_\lambda = \frac{|Q_\lambda \cap \Xi|}{\lambda}$$

constituting a natural unbiased estimator of the occupancy probability $p_\Xi = \mathbf{P}\{o \in \Xi\}$. We shall stick to the $\hat{p}_\lambda$ notation in our discussion below in the particular context of germ-grain models. A number of other functionals of Boolean models are also of considerable interest, both from the theoretical and applied viewpoint, some of them are briefly mentioned below.

4.6.1 *Results for stationary Poisson germ-grain process*

For the special case of stationary Poisson germ-grain processes, also known as Boolean models, extremely precise asymptotic results have been obtained by Heinrich (2005). Assume $\mathcal{P} = \mathcal{P}_\tau$ to be a homogeneous Poisson point process in $\mathbb{R}^d$ with intensity $\tau > 0$ and let Ξ_x's be i.i.d. copies of a random compact set Ξ_0 referred to as the *typical grain.* The following assumption will be imposed throughout this paragraph

$$M(a) = \mathbf{E}\exp(a|\Xi_0|) < \infty \text{ for some } a > 0. \tag{4.25}$$

We have then the following free energy analyticity result and Cramér full asymptotic series development for the deviation probabilities of $\hat{p}_\lambda$ from its mean p_Ξ, see Heinrich (2005, Th. 1,2). In the statement of this result we will use notation for the variance of $\hat{p}_\lambda$,

$$\sigma_\lambda = \lambda^{-1}\mathrm{Var}\hat{p}_\lambda$$

for its re-scaled distribution function

$$F_\lambda(x) = \mathbf{P}\left\{\sqrt{\lambda}(\hat{p}_\lambda - p_\Xi) \leq \sigma_\lambda x\right\}$$

as well as for its re-scaled log-Laplace transform

$$L_\lambda(z) = \lambda^{-1}\log \mathbf{E}\exp[\lambda z \hat{p}_\lambda], \; z \in \mathbb{C},$$

whenever existing.

Theorem 4.20 *Assume* (4.25) *and the non-degeneracy condition* $\mathbf{E}|\Xi_0| > 0$ *for the typical grain. Then the following statements hold.*

- *The limit* $L(z) = \lim_{\lambda\to\infty} L_\lambda(z)$ *exists and is analytic in a complex disk around* 0.
- *The variance* σ_λ *converges to a non-zero limit.*
- *For all* $0 \leq x \leq C(a)\sqrt{\lambda}\sigma_\lambda$ *with* $C(a) > 0$ *depending on* a *in* (4.25) *we have the asymptotic relations*

$$\frac{1 - F_\lambda(x)}{1 - \Phi(x)} = \exp\left(\frac{x^3}{\sqrt{\lambda}\sigma_\lambda}\sum_{k=0}^{\infty}\mu_k^{(\lambda)}\left(\frac{x}{\sqrt{\lambda}\sigma_\lambda}\right)^k\right)\left(1 + O\left(\frac{1+x}{\sqrt{\lambda}}\right)\right)$$

  and

$$\frac{F_\lambda(-x)}{\Phi(-x)} = \exp\left(\frac{-x^3}{\sqrt{\lambda}\sigma_\lambda}\sum_{k=0}^{\infty}\mu_k^{(\lambda)}\left(\frac{-x}{\sqrt{\lambda}\sigma_\lambda}\right)^k\right)\left(1 + O\left(\frac{1+x}{\sqrt{\lambda}}\right)\right)$$

  with $\Phi(\cdot)$ *standing for the distribution function of the standard normal and where* $\mu_k^{(\lambda)}$ *can be explicitly expressed in terms of correlation functions of the field* Ξ *and converge as* $\lambda \to \infty$.
- *There exists a constant depending on the parameters of the field* Ξ *such that*

$$\sup_{x\in\mathbb{R}} |F_\lambda(x) - \Phi(x)| \leq \frac{c}{\sqrt{\lambda}}.$$

Note that this remarkable result subsumes a central limit theorem with very precise information on convergence rates (including Berry-Esséen-type bounds) as well as a moderate deviation principle and even a large deviation principle, although restricted to some non-empty interval. Further results including a concentration inequality (Heinrich, 2005, Th. 3) and volume order large deviations (Theorem 4 ibidem) are also available. See also Heinrich (2007) for a discussion of the particular one-dimensional case.

The proof techniques applied to establish the above results are very deep and go far beyond the scope of the present survey. Roughly speaking, the main difficulty lies in obtaining appropriate bounds on total variation of cumulant measures of all orders for the empirical measure μ_λ^ξ (see in Theorem 1 ibidem) whereupon general limit theory can be used, see Saulis and Statulavičius (1991). The crucial bounds on cumulants are themselves obtained via rather involved recursive representation of higher order correlation densities of the field Ξ, somewhat reminiscent of the cluster expansion techniques employed in statistical mechanics, see Malyshev and Minlos (1999). In the particular case where the grains are bounded, the volume functional of the field Ξ admits an m-dependent representation and general theory of m-dependent fields can be used to establish similar results, see Götze, Heinrich and Hipp (1995).

4.6.2 *Results for general germ-grain models*

A very general central limit theorem for stationary random empirical measures generated by stationary germ-grain models on regularly growing domains with β-mixing germ process has been established by Heinrich and Molchanov (1999). The family of measures and functionals considered there is far richer than just empirical volume fractions and includes surface and curvature measures and, more generally, vectors of Minkowski measures of arbitrary orders, see ibidem for details. The asymptotic normality is established for properly re-scaled empirical measures generated by the field Ξ under certain bounded moment conditions on the volume of the typical grain Ξ_0 and appropriate polynomial decay assumptions imposed on the β-mixing coefficients of the germ point process, see Theorem 6.2 ibidem. Moreover, integral formulae for variance asymptotics are given there. Due to the high generality of this result we do not quote it here in detail but refer the reader to the original paper of Heinrich and Molchanov (1999). It should be noted that a general central limit theorem for Poisson grain processes (Boolean models) follows from these results and it only requires the existence of the second moment of typical grain volume, i.e. $\mathbf{E}|\Xi_0|^2 < \infty$, rather than the existence of its exponential moment as in Theorem 4.20. The proof of Heinrich and Molchanov (1999, Th. 6.2) relies on general results for β-mixing fields.

4.7 Clan-of-ancestors graphical construction

In this final section we present an interesting stochastic-geometric tool invented by Fernández, Ferrari and Garcia (1998, 2001, 2002) which has, in our opinion, a remarkable potential for applications in stochastic geometry of dependent

(Gibbs) point processes. This potential is due to several attractive features of the *FFG* graphical representation, which are

- its fully geometric nature, allowing for geometric proofs of limit theorems in settings previously dominated by classical more analytic techniques,
- fully algorithmisable construction of the representation, which greatly facilitates stochastic simulation and often yields perfect simulation schemes,
- broad scope and flexibility combined with elegance and simplicity of the construction.

In sharp contrast to previous topics of this survey, where we discussed theories already well established and were able to provide a rich supply of references, here we focus on a presentation of this new technique whose applications so far have mainly concentrated in the area of geometric models of statistical mechanics, and we advocate its use in stochastic geometry. This approach affects our presentation below, where we show how the FFG construction works and what conclusions it yields for a very particular class of Gibbs point processes – the so-called hard-object processes – rather than providing a detailed discussion of the scope of its applicability which can be found in Fernández, Ferrari and Garcia (1998, 2001, 2002) and which is also a subject of our current research in progress. In particular, the FFG construction can be used to provide limit theorems for stabilizing functionals on Gibbsian input, as shown by Schreiber and Yukich (2008*a*).

To proceed with a brief presentation of the FFG graphical construction, assume a fixed compact *grain* $K \ni 0$ is given and, in a given domain D, we consider the Gibbs point process $\mathcal{G}_D^K$ arising by letting the points of a homogeneous Poisson point process $\mathcal{P}_\tau \cap D$ of some fixed intensity $\tau > 0$ interact by hard-core exclusions upon attaching a copy of the grain K to each point of the process. In other words, $\mathcal{G}_D^K$ coincides in law with $\mathcal{P}_\tau \cap D$ conditioned on the event that for no two points x, y of the process do we have $x + K \cap y + K \neq \emptyset$ with $+$ standing here for the usual Minkowski addition, see Chapter 1. Such processes have been widely considered in statistical mechanics and are also known under the name of a *hard sphere gas*. To see how such processes can be obtained via FFG representation, consider first a stationary homogeneous *free birth and death process* $\left(\rho_t^D\right)_{t\in\mathbb{R}}$ in D with the following dynamics:

- A new point $x \in D$ is born in ρ_t during the time interval $[t - \mathrm{d}t, t]$ with probability $\tau \mathrm{d}x\mathrm{d}t$,
- An existing point $x \in \rho_t^D$ dies during the time interval $[t - \mathrm{d}t, t]$ with probability $\mathrm{d}t$, that is to say the lifetimes of points of the process are independent standard exponential.

Clearly, the unique stationary and reversible measure for this process is just the law of the Poisson point process $\mathcal{P}_\tau \cap D$.

Consider now the following *trimming* procedure performed on ρ_t^D, based on the ideas developed in Fernández, Ferrari and Garcia (1998, 2001, 2002). Choose a birth site for a point $x \in D$ and accept it if its grain does not overlap the grains carried by other points alive at the time t, otherwise reject it. Clearly, this can only be done if the acceptance/rejection statuses of all points in $\rho_{t-}^D \cap x + (K + (-K))$ are determined, otherwise we proceed recursively backwards in time to determine the statuses of points in $x + (K + (-K))$.

Before discussing any further properties of this procedure, we have to ensure first that it does terminate. To this end, note that each point x with the property of having its *dependency region* $x + (K + (-K))$ devoid of points from ρ_{t-}^D at its birth time t has its acceptance status determined – it is accepted. More generally, the acceptance status of a point x at its birth time t only depends on statuses of points in $\rho_{t-}^D \cap x + (K + (-K))$, that is to say points born before and falling into the dependency region $x + (K + (-K))$. We call these points *causal ancestors* of x and, in general, for a subset $A \subseteq D$ by $\mathrm{An}_t[A]$ we denote the set of all points in $\rho_t^D \cap A$, their causal ancestors, the causal ancestors of their ancestors and so forth throughout all past generations. The set $\mathrm{An}_t[A]$ is referred to as the *causal ancestor cone* or *causal ancestor clan* of A with respect to the birth and death process $(\rho_t^D)_{t \in \mathbb{R}}$, whence the name *clan-of-ancestors construction* often used in the context of the FFG representation.

It is now clear that in order for our recursive status determination procedure to terminate for all points of ρ_t^D in A, it is enough to have the causal ancestor cone $\mathrm{An}_t[A]$ finite. This is easily checked to be a.s. the case for each $A \subseteq D$ – indeed, since D is bounded, a.s. there exists some $s < t$ such that $\rho_s^D = \emptyset$ and thus no ancestor clan of a point alive at the time t can go past s backwards in time.

Having defined the trimming procedure above, we recursively remove from ρ_t^D the points rejected at their birth, and we write $(\gamma_t^D)_{t \in \mathbb{R}}$ for the resulting process. Clearly, γ_t^D is stationary because so was ρ_t^D and the acceptance/rejection procedure is time-invariant as well. Moreover, the process γ_t^D is easily seen to evolve according to the following dynamics:

- Add a new point x with intensity $\tau \mathrm{d}x \mathrm{d}t$ provided the grain at x does not overlap any of the already accepted grains,
- Remove an existing point with intensity $\mathrm{d}t$.

These are the standard Monte-Carlo dynamics for $\mathcal{G}_D^K$ and the law of $\mathcal{G}_D^K$ is its unique invariant distribution. Consequently, in full analogy with Fernández, Ferrari and Garcia (1998, 2001, 2002) the point process γ_t^D coincides in law with $\mathcal{G}_D^K$ for all $t \in \mathcal{R}$.

So far, no particular advantages of the above construction are evident. However, it is known, see ibidem where a proof based on subcritical branching process domination is given, that if the reference intensity τ is chosen small enough, then not only are all causal ancestor cones a.s. finite but in addition there is a $C > 0$

such that for all $t, R \in \mathbb{R}_+$ and $A \subset D$ we have the crucial bound

$$\mathbf{P}\{\mathrm{diamAn}_t[A] \geq R + \mathrm{diam}(A)\} \leq \mathrm{vol}(A)\exp(-CR). \tag{4.26}$$

Moreover, most importantly the constant C in (4.26) *can be chosen not to depend on* D! A typical use of this fact to establish limit theorems for functionals of $\mathcal{G}_D^K$ goes as follows

- We identify the point process $\mathcal{G}_D^K$ with γ_0^D.
- Then we note that the behaviour of the point process γ_0^D on two disjoint regions $A, B \subseteq D$ is conditionally independent on the event $\mathcal{E}[A, B; R]$ that $\mathrm{An}_0[A] \subseteq A + B_R$ and $\mathrm{An}_0[B] \subseteq B + B_R$ for some R smaller than the halved distance between A and B – indeed, on this event the respective restrictions of $\mathcal{G}_D^K$ to A and B arise from disjoint and hence independent portions of the free birth-and-death process ρ_t^D.
- If the distance between A and B is large enough, the above event $\mathcal{E}[A, B; R]$ fails to happen only with probability exponentially decaying with R. This yields a version of α- and β-mixing and can be readily used to produce central limit theorems.

To state a central limit theorem for functionals of $\mathcal{G}_D^K$, write $\mathcal{G}_\lambda$ for $\mathcal{G}_{Q_\lambda}^K$ and let ξ be a *bounded range geometric functional*, that is to say a stabilizing geometric functional with deterministic stabilization radius. We have then, see Fernández, Ferrari and Garcia (1998, Th. R5).

Theorem 4.21 *With the notation introduced above, assume in addition that* ξ *satisfies the bounded moment condition* (4.6) *on Gibbsian input* $\mathcal{G}^K$ *for some* $p > 2$. *Then*

$$\frac{H^\xi[\mathcal{G}_\lambda] - \mathbf{E}H^\xi[\mathcal{G}_\lambda]}{\mathrm{Var}(H^\xi[\mathcal{G}_\lambda])}$$

converges in law to standard centred Gaussian $\mathcal{N}(0, 1)$ *as* $\lambda \to \infty$.

Note at this point that the notation $\mathcal{G}^K$ used above refers to the infinite-volume (thermodynamic) limit for $\mathcal{G}_D^K$, see ibidem for further details. As has already been indicated above, the scope to which the central limit theory obtained via the FFG construction applies is much broader than what has been presented above and we refer the reader to Fernández, Ferrari and Garcia (1998, 2001, 2002) for further applications, also taking the opportunity to announce our further results in Schreiber and Yukich (2008*a*) in the context of stabilizing functionals.

4.8 Further limit theory

As mentioned in the introduction above, this is not an exhaustive survey of limit theory present in stochastic geometry – in fact many other limit theorems can also be found in other sections of this volume. In particular, we did not discuss material already thoroughly described in existing monographs. In this context, we refer the reader to Matheron (1975) and Molchanov (1993) for deep and rich limit theory for unions of random closed sets, including such topics as union-infinitely divisible random closed sets and union-stable random closed sets. We also recommend Molchanov (2005) monograph on random sets, which contains a rich collection of limit theorems in its broad scope. Another book of interest is Hall (1988) book on coverage processes. A further example of a rich and important research area not discussed in this survey are limit theorems for random tessellations, see e.g. Heinrich (1994), Heinrich and Muche (2008) and the references therein. Random tessellations are considered in more detail in Chapter 5 of this volume.

Acknowledgements

I wish to express my gratitude to J.E. Yukich for his numerous comments and suggestions which were very helpful in improving this survey. I am also grateful to M.D. Penrose for important bibliographical suggestions. Special thanks are due as well to anonymous referees for their helpful and constructive comments.

I gratefully acknowledge the support from the Polish Minister of Science and Higher Education grant N N201 385234 (2008–2010).

References

Aldous, D. and Steele, J.M. (2003). The objective method: probabilistic combinatorial optimization and local weak convergence. In *Discrete and combinatorial probability* (ed. H. Kesten), pp. 1–72. Springer-Verlag.

Applegate, D.L., Bixby, R., Chvátal, V., and Cook, W.J. (2006). *The traveling salesman problem. A computational study.* Princeton Series in Applied Mathematics, Princeton University Press.

Avram, F. and Bertsimas, D. (1993). On central limit theorems in geometrical probability. *Ann. Appl. Probab.*, **3**, 1033–1046.

Baldi, P. and Rinott, Y. (1989). Asymptotic normality of some graph related statistics. *J. Appl. Probab.*, **26**, 171–175.

Baltz (2007). Probabilistic analysis for a multiple depot vehicle routing problem. *Random Structures and Algorithms*, **30**, 206–225.

Barbour, A.D. and Xia, A. (2006). Normal approximation for random sums. *Adv. Appl. Probab.*, **38**, 693–728.

Baryshnikov, Y., Eichelsbacher, P., Schreiber, T., and Yukich, J.E. (2008). Moderate deviations for some point measures in geometric probability. *Annals Inst. H. Poincaré*, **44**, 422–446.

Baryshnikov, Y., Penrose, M.D., and Yukich, J.E. (2009). Gaussian limits for generalized spacings. *Ann. Appl. Probab.*, **19**, 158–185.

Baryshnikov, Y. and Yukich, J.E. (2003). Gaussian fields and random packing. *J. Stat. Phys.*, **111**, 443–463.

Baryshnikov, Y. and Yukich, J.E. (2005). Gaussian limits for random measures in geometric probability. *Ann. Appl. Probab.*, **15**, 213–253.

Baryshnikov, Y. and Yukich, J.E. (2006). Gaussian limits and maximal points. preprint.

Chen, L.H.Y. and Shao, Q.-M. (2004). Normal approximation under local dependence. *Ann. Probab.*, **32**, 1985–2028.

Chiu, S.N. and Quine, M.P. (1997). Central limit theory for the number of seeds in a growth model in $\mathbb{R}^d$ with inhomogeneous poisson arrivals. *Ann. Appl. Probab.*, **7**, 802–814.

Dembo, A. and Zeitouni, O. (1998). *Large deviations techniques and applications*, Second Edition. Springer.

Evans, J.W. (1993). Random and cooperative adsorption. *Rev. Modern Phys.*, **65**, 1281–1329.

Fernández, R., Ferrari, P., and Garcia, N. (1998). Measures on contour, polymer or animal models. a probabilistic approach. *Markov Processes and Related Fields*, **4**, 479–497.

Fernández, R., Ferrari, P., and Garcia, N. (2001). Loss network representation of ising contours. *Ann. Probab.*, **29**, 902–937.

Fernández, R., Ferrari, P., and Garcia, N. (2002). Perfect simulation for interacting point processes, loss networks and ising models. *Stoch. Proc. Appl.*, **102**, 63–88.

Frieze, A. and Yukich, J.E. (2002). Probabilistic analysis of the traveling salesman problem. In *Traveling Salesman Problem and its Variations*, pp. 257–307. Kluwer Academic Publishers.

Götze, F., Heinrich, L., and Hipp, C. (1995). m-dependent random fields with analytic cumulant generating function. *Scandinavian Journal of Statistics*, **22**, 183–195.

Gutin, G. and Punnen, A. P. (eds) (2002). *The traveling salesman problem and its variations.* Combinatorial Optimization **12**, Kluwer Academic Publishers.

Hall, P. (1988). *An introduction to the theory of coverage processes.* Wiley & Sons, New York.

Heinrich, L. (1992). On existence and mixing properties of germ-grain models. *Statistics*, **23**, 271–286.

Heinrich, L. (1994). Normal approximation for some mean-value estimates of absolutely regular tessellations. *Math. Methods Statist.*, **3**, 1–24.

Heinrich, L. (2005). Large deviations of the empirical volume fraction for stationary poisson grain models. *Ann. Appl. Probab.*, **15**, 392–420.

Heinrich, L. (2007). An almost-markov type mixing condition and large deviations for boolean models on the line. *Acta Applicandae Math.*, **96**, 247–262.

Heinrich, L. and Molchanov, I. (1999). Central limit theorem for a class of random measures associated with germ-grain models. *Adv. Appl. Probab.*, **31**, 283–314.

Heinrich, L. and Muche, L. (2008). Second-order properties of the point process of nodes in a stationary voronoi tessellation. *Math. Nachr.*, **281**, 350–375.

Kesten, H. and Lee, S. (1996). The central limit theorem for weighted minimal spanning trees on random points. *Ann. Appl. Probab.*, **6**, 495–527.

Kruskal, J.B. (1956). On the shortest spanning subtree of a graph and the traveling salesman problem. *Proc. Amer. Math. Soc.*, **7**, 48–50.

Ledoux, M. (2001). *The Concentration of Measure Phenomenon.* Mathematical Surveys and Monographs, **89**, American Mathematical Society, Providence, RI.

Lee, S. (1997). The central limit theorem for euclidean minimum spanning trees. *Ann. Probab.*, **7**, 996–1020.

Malyshev, V.A. and Minlos, R.A. (1999). *Gibbs Random Fields: Cluster Expansions.* Springer.

Matheron, G. (1975). *Random Sets and Integral Geometry.* Wiley, New York.

Molchanov, I.S. (1993). *Limit Theorems for Unions of Random Closed Sets. Lecture Notes in Mathematics*, **1561**, Springer-Verlag.

Molchanov, I.S. (1995). Statistics of the boolean model: from the estimation of means to the estimation of distributions. *Adv. Appl. Probab.*, **27**, 63–86.

Molchanov, I.S. (1997). *Statistics of the Boolean Model for Practicioners and Mathematicians.* Wiley, Chichester.

Molchanov, I.S. (2005). *Theory of Random Sets.* Springer-Verlag, London.

Molchanov, I.S. and Stoyan, D. (1994). Asymptotic properties of estimators for parameters of the boolean model. *Adv. Appl. Probab.*, **26**, 301–323.

Penrose, M.D. (2001*a*). Limit theorems for monolayer ballistic deposition in the continuum. *J. Stat. Phys.*, **105**, 561–583.

Penrose, M.D. (2001*b*). Random parking, sequential adsorption and the jamming limit. *Comm. Math. Phys.*, **218**, 153–176.

Penrose, M.D. (2003). *Random geometric graphs.* Oxford University Press, Oxford, New York.

Penrose, M.D. (2007*a*). Gaussian limits for random geometric measures. *Electronic Journal of Probability*, **12**, 989–1035.

Penrose, M.D. (2007*b*). Laws of large numbers in stochastic geometry with statistical applications. *Bernoulli*, **13**, 1124–1150.

Penrose, M.D. and Wade, A. (2008). Multivariate normal approximation in geometric probability. *Journal of Statistical Theory and Practice*, **2**, 293–326.

Penrose, M.D. and Yukich, J.E. (2001). Central limit theorems for some graphs in computational geometry. *Ann. Appl. Probab.*, **11**, 1005–1041.

Penrose, M.D. and Yukich, J.E. (2002). Limit theory for random sequential packing and deposition. *Ann. Appl. Probab.*, **12**, 272–301.

Penrose, M.D. and Yukich, J.E. (2003). Weak laws of large numbers in geometric probability. *Ann. Appl. Probab.*, **13**, 277–303.

Penrose, M.D. and Yukich, J.E. (2005). Normal approximation in geometric probability. In *Proceedings of the Workshop 'Stein's Method and Applications', Lecture Notes Series,* **5**, Singapore, pp. 37–58. Institute for Mathematical Sciences: World Scientific Press.

Redmond, C. (1993). *Boundary rooted graphs and Euclidean matching algorithms.* Ph.D. thesis, Department of Mathematics, Lehigh University, Bethlehem, PA.

Redmond, C. and Yukich, J.E. (1994). Limit theorems and rates of convergence for subadditive euclidean functionals. *Ann. Appl. Probab.*, **4**, 1057–1073.

Redmond, C. and Yukich, J.E. (1996). Asymptotics for euclidean functionals with power-weighted edges. *Stoch. Proc. Appl.*, **61**, 289–304.

Rényi, A. (1958). On a one dimensional space filling problem. *MTA Mat. Kut. Int. Közl.*, **3**, 109–127. (in Russian).

Rhee, W. (1992). On the traveling salesperson problem in many dimensions. *Random Structures and Algorithms*, **3**, 227–233.

Rhee, W. (1993). A matching problem and subadditive euclidean functionals. *Ann. Appl. Probab.*, **3**, 794–801.

Rhee, W. (1994). Boundary effects in the traveling salesperson problem. *Oper. Research Letters*, **16**, 19–25.

Saulis, L. and Statulavičius, V. (1991). *Limit theorems for large deviations.* Kluwer Academic, Dodrecht.

Schreiber, T., Penrose, M.D., and Yukich, J.E. (2007). Gaussian limits for random sequential packing at saturation. *Comm. Math. Phys.*, **272**, 167–183.

Schreiber, T. and Yukich, J.E. (2005). Large deviations for functionals of spatial point processes with applications to random packing and spatial graphs. *Stoch. Proc. Appl.*, **115**, 1332–1356.

Schreiber, T. and Yukich, J.E. (2008*a*). Decay of spatial dependencies and limit theorems for stabilizing functionals of gibbs point processes with localized potentials. preprint.

Schreiber, T. and Yukich, J.E. (2008*b*). Variance asymptotics and central limit theorems for generalized growth processes with applications to convex hulls and maximal points. *Ann. Probab.*, **36**, 363–396.

Senger, B., Voegel, J.-C., and Schaaf, P. (2000). Irreversible adsorption of colloidal particles on solid substrates. *Colloids and Surfaces A*, **165**, 255–285.

Seppäläinen, T. and Yukich, J.E. (2001). Large deviation principles for euclidean functionals and other nearly additive processes. *Probability Theory Related Fields*, **120**, 309–345.

Steele, J.M. (1988). Growth rates of euclidean minimal spanning trees with power weighted edges. *Ann. Probab.*, **16**, 1767–1787.

Steele, J.M. (1990). Probabilistic and worst case analyses of classical problems of combinatorial optimization in euclidean space. *Math. Oper. Research*, **15**, 749–770.

Steele, J.M. (1993). Probability and problems in euclidean combinatorial optimization. *Statistical Science*, **8**, 48–56.

Steele, J.M. (1997). *Probability Theory and Combinatorial Optimization.* SIAM.

Stoyan, D., Kendall, W., and Mecke, J. (2005). *Stochastic Geometry and its Applications.* John Wiley and Sons. Second Ed.

Talagrand, M. (1995). Concentration of measure and isoperimetric inequalities in product spaces. *Publications Mathématiques de l'I.H.E.S.*, **81**, 73–205.

Talagrand, M. (1996*a*). New concentration inequalities in product spaces. *Invent. Math.*, **126**, 505–563.

Talagrand, M. (1996*b*). A new look at independence. *Ann. Probab.*, **24**, 1–34.

Tarjan, R.E. (1983). *Data Structures and Network Algorithms.* SIAM.

Torquato, S. (2002). *Random Heterogeneous Materials.* Springer Interdisciplinary Applied Mathematics, Springer-Verlag, New York.

Wade, A. (2007). Explicit laws of large numbers for random nearest-neighbour-type graphs. *Adv. Appl. Probab.*, **39**, 326–342.

Wu, B.Y. and Chao, K.-M. (2004). *Spanning Trees and Optimization Problems.* Chapman & Hall/CRC.

Yukich, J.E. (1995). Asymptotics for the stochastic tsp with power-weighted edges. *Probab. Theory Related Fields*, **102**, 203–220.

Yukich, J.E. (1996). Ergodic theorems for some classical optimization problems. *Ann. Appl. Probab.*, **6**, 1006–1023.

Yukich, J.E. (1998). *Probability Theory of Classical Euclidean Optimization Problems.* Lecture Notes in Mathematics, **1675**, Springer.

Yukich, J.E. (1999). Limit theorems for random euclidean graphs. *Modelos Estocásticos*, **19**, 121–180. ed. by Sociedad Matemática Mexicana, 1999.

5

TESSELLATIONS

Pierre Calka

Abstract

Random tessellations and cellular structures occur in many domains of application, such as astrophysics, ecology, telecommunications, biochemistry and naturally cellular biology (see Stoyan, Kendall and Mecke 1987 or Okabe, Boots, Sugihara and Chiu 2000 for complete surveys). The theoretical study of these objects was initiated in the second half of the twentieth century by D. G. Kendall, J. L. Meijering, E. N. Gilbert and R. E. Miles, notably. Two isotropic and stationary models have emerged as the most basic and useful: the Poisson hyperplane tessellation and the Poisson–Voronoi tessellation. Since then, a large majority of questions raised about random tessellations have concerned statistics of the population of cells ('how many cells are triangles in the plane?', 'how many cells have a volume greater than one?') or properties of a specific cell (typically the one containing the origin). Two types of results are presented below: exact distributional calculations and asymptotic estimations.

In the first part, we describe the two basic constructions of random tessellations (i.e. by throwing random hyperplanes or by constructing Voronoi cells around random nuclei) and we introduce the fundamental notion of typical cell of a stationary tessellation. The second part is devoted to the presentation of exact distributional results on basic geometrical characteristics (number of hyperfaces, typical k-face, etc.). The following part concerns asymptotic properties of the cells. It concentrates in particular on the well-known D. G. Kendall conjecture which states that large planar cells in a Poisson line tessellation are close to the circular shape. In the last part, we present some recent models of iterated tessellations which appear naturally in applied fields (study of crack structures, telecommunications).

Intentionally, this chapter does not contain an exhaustive presentation of all the models of random tessellations existing in the literature (in particular, dynamical constructions such as Johnson–Mehl tessellations will be omitted). The aim of the text below is to provide a selective view of recent selected methods and results on a few specific models.

5.1 Definitions and basic properties of random tessellations

5.1.1 *Introduction*

Let $\mathcal{T} = \{C_i\}_{i\geq 1}$ be a locally finite collection of closed sets of $\mathbb{R}^d$, $d \geq 1$. The family $\mathcal{T}$ is said to be a *tessellation* of $\mathbb{R}^d$ if C_i and C_j have disjoint interiors for $i \neq j$ and $\bigcup_{i\geq 1} C_i = \mathbb{R}^d$. The sets C_i, $i \geq 1$, are called the *cells* of the tessellation $\mathcal{T}$. In this chapter, we will consider the particular case of a *convex tessellation* where each cell is a convex polyhedron.

We endow the set $\mathbb{T}$ of all convex tessellations of $\mathbb{R}^d$ with the σ-algebra generated by sets of the form

$$\{\mathcal{T} = \{C_i\}_{i\geq 1} : [\cup_{i\geq 1}\partial C_i] \cap K \neq \emptyset\}$$

where K is any compact subset of $\mathbb{R}^d$.

A *random convex tessellation* is then defined as a random variable with values in $\mathbb{T}$, see Stoyan, Kendall and Mecke (1987). In the following, we will focus on two fundamental examples: the hyperplane tessellation and the Voronoi tessellation.

Let X be a point process in $\mathbb{R}^d$ which does not contain the origin almost surely. For every $x \in X \setminus \{0\}$, we denote by H_x the affine hyperplane orthogonal to x and containing x, i.e.

$$H_x = \{y \in \mathbb{R}^d : \langle y - x, x\rangle = 0\}, \tag{5.1}$$

where $\langle \cdot, \cdot \rangle$ denotes the usual scalar product in $\mathbb{R}^d$. The *hyperplane tessellation* induced by X is the convex tessellation constituted by the closure of each convex component of $\mathbb{R}^d \setminus \cup_{x\in X} H_x$.

Let X be a point process in $\mathbb{R}^d$. For every $x \in X$, we define the cell $C(x)$ associated with x as

$$C(x) = \{y \in \mathbb{R}^d : \|y - x\| \leq \|y - x'\| \text{ for all } x' \in X\}.$$

The *Voronoi tessellation* induced by X is the tessellation $\{C(x) : x \in X\}$. The points in X are called the *nuclei* of the tessellation.

In particular, if X is a stationary point process in $\mathbb{R}^d$, the associated Voronoi tessellation is stationary (invariant under any translation). If the particular case where X is a homogeneous Poisson point process, it is stationary and isotropic (invariant under any rotation): we speak of a *Poisson–Voronoi tessellation*, see Okabe, Boots, Sugihara and Chiu (2000) and Møller (1994).

Let us consider the measure Θ_0 on $\mathbb{R}^d$ such that its density with respect to the Lebesgue measure is $\|\cdot\|^{-(d-1)}$. If X is a Poisson point process with intensity measure Θ_0 (up to a multiplicative constant), the associated hyperplane tessellation is isotropic and stationary. It is called a (stationary) *Poisson hyperplane tessellation*, see Gilbert (1962) and Miles (1964*a*, 1964*b*, 1969, 1971).

There are obviously many other types of tessellations which are of great interest and could not be discussed here, for instance Johnson–Mehl tessellations (Møller, 1992) or Laguerre tessellations (Lautensack, 2007; Lautensack and Zuyev, 2008). Though the only Voronoi tessellation that will receive attention in the rest of the chapter is the (homogeneous) Poisson–Voronoi tessellation, it should be noted that Voronoi tessellations generated by other types of more complicated point processes have also been investigated and have led to significant results, see e.g. Heinrich (1998), Błaszczyszyn and Schott (2003) and Heinrich and Muche (2008).

5.1.2 *Zero-cell, ergodic means and typical cell*

One of the fundamental questions raised in the study of random tessellations is to find a way to isolate a particular cell which will be a good descriptor of the collection of all cells, i.e. to define a uniform sample among all the cells.

5.1.2.1 *Zero-cell* A first idea is to fix a point in $\mathbb{R}^d$ and consider the cell containing that point.

If there is almost surely a unique cell containing the origin 0, it is called the *zero-cell* of the tessellation and will be denoted by C_0. It is correctly defined for a Poisson–Voronoi tessellation or for a hyperplane tessellation where the associated point process has an intensity measure which does not charge the origin. In particular, the *zero-cell* of the stationary Poisson hyperplane tessellation is called the *Crofton cell.*

Let us remark that C_0 is not a 'mean cell', in the sense that it does not have the mean characteristics of the whole population of cells. It is in particular bigger than the typical cell defined below, see the ergodic convergence (5.2) applied to $f = \lambda_d$.

5.1.2.2 *Ergodic means* It is therefore intuitive to consider a finite set of cells included in a non-empty compact set W and to calculate the mean over these cells of a real-valued, bounded, measurable and invariant-under-translations function f defined on the family $\mathcal{K}$ of convex and compact sets in $\mathbb{R}^d$. If the tessellation is stationary and ergodic (as in both Poisson–Voronoi and Poisson hyperplane cases), Wiener's ergodic theorem (Wiener, 1939) and a proper treatment of 'edge regions' ensure that this mean converges when the size of W goes to infinity, see Cowan (1978, 1980).

Let N_R be the number of cells of a Poisson–Voronoi or stationary Poisson hyperplane tessellation included in RW for every $R > 0$. Then for any bounded, measurable and invariant-under-translations function $f : \mathcal{K} \to \mathbb{R}$,

$$\lim_{R\to\infty} \frac{1}{N_R} \sum_{i=1}^{N_R} f(C_i) = \frac{1}{\mathbf{E}(\lambda_d(C_0)^{-1})} \mathbf{E}\left(\frac{f(C_0)}{\lambda_d(C_0)}\right), \tag{5.2}$$

where λ_d is the d-dimensional Lebesgue measure. The *typical cell* $\mathscr{C}$ is then defined as a random variable which takes values in the set $\mathcal{K}$ and has a density with respect to the distribution of C_0 equal to $(1/\lambda_d)$ up to a multiplicative constant.

The use of this kind of convergence as an approximation of the typical cell requires the existence of central limit theorems: in the two-dimensional Poisson–Voronoi case, it was proved by Avram and Bertsimas (1993) when f is the perimeter of a polygon, through a stabilization-type method. Afterwards, Paroux (1998) obtained with the method of moments a similar result for the Poisson line tessellation in the plane and for several functionals among which the perimeter and the number of vertices. More recently, Heinrich, Schmidt and Schmidt (2005) used Hoeffding's decomposition of U-statistics to derive multivariate central limit theorems for a d-dimensional Poisson hyperplane tessellation ($d \geq 2$) and for the number and volume of k-faces ($0 \leq k \leq (d-1)$) of the tessellation.

5.1.2.3 *Typical cell and Palm measure* Defining the typical cell through ergodic means may not be the most convenient way to study its specific properties. Stationarity allows us to deduce an equivalent definition through the use of *Palm measures*, see Mecke (1967), Neveu (1977) and Chapter 1 of this volume.

Indeed, let us suppose that we can assign to any cell C in the tessellation a unique centroid $z(C)$ such that $z(C+x) = z(C) + x$ and the point process Y of all these centroids is stationary (of intensity γ). For instance, in the Poisson–Voronoi case, $z(C)$ can be the nucleus associated with C. In the general case of a stationary tessellation, we can take the centre of mass, or the centre of the largest ball included in the cell or equivalently the lowest point of the cell (with respect to one of the coordinates).

The typical cell $\mathscr{C}$ is then equivalently defined as a random variable such that for every bounded, measurable and invariant-under-translations function $f : \mathcal{K} \to \mathbb{R}$ and every $B \in \mathcal{B}$ such that $0 < \lambda_d(B) < \infty$

$$\mathbf{E}(f(\mathscr{C})) = \frac{1}{\gamma \lambda_d(B)} \mathbf{E} \left\{ \sum_{\{C : z(C) \in B\}} f(C - z(C)) \right\}. \tag{5.3}$$

That definition (Mecke, 1967, 1975; Møller, 1986) does not depend on the chosen centroid process and the method is equivalent to the ergodic procedure, since $\mathbf{E}(f(\mathscr{C}))$ is precisely the limit of the ergodic means, i.e.

$$\mathbf{E}(f(\mathscr{C})) = \frac{1}{\mathbf{E}(\lambda_d(C_0)^{-1})} \mathbf{E} \left(\frac{f(C_0)}{\lambda_d(C_0)} \right). \tag{5.4}$$

This last equality from Møller (1989) can also be seen as a generalization of a formula which can be found in (Gilbert, 1962, p. 964) and would here correspond to the choice $f = {\lambda_d}^2$.

In simple words, $\mathscr{C}$ is the cell containing the origin 0 when the point process of centroids is conditioned on containing 0. It is also the cell 'seen from a typical centroid'.

In the same way, it is possible to define the *typical k-face* $\mathscr{C}_k$, $0 \leq k \leq (d-1)$ by associating with any k-face of the tessellation a precise centroid and by using the underlying Palm probability measure $\widetilde{P}_k^0$, see e.g. Møller (1989, 1994).

Finally, a different approach consists in defining a typical point on a k-face by considering the stationary random measure $M_k = \sum_{\{F \ k\text{-face}\}} \mathcal{H}_k(F \cap \cdot)$ and its associated Palm probability measure P_k^0 on the set of locally finite subsets of $\mathbb{R}^d$:

$$\int f(\omega) \mathrm{d}P_k^0(\omega) = \frac{1}{\gamma_k \lambda_d(B)} \mathbf{E} \left\{ \int_B f(\{z(C) - x, C \text{ cell}\}) \mathrm{d}M_k(x) \right\}, \tag{5.5}$$

where f is a measurable bounded function defined on locally finite subsets of $\mathbb{R}^d$ and γ_k is defined as the multiplicative constant such that the deterministic measure $\mathbf{E}(M_k)$ which is invariant under translations is equal to $\gamma_k \lambda_d$, see e.g. Baumstark and Last (2007). We say that P_k^0 is the distribution of the point process of the centroids of the cells 'seen from a typical point on a k-face'. We denote by $C_{0,k}$ the k-face containing the origin under P_k^0.

The two Palm procedures (seeing the tessellation from the centroid of a k-face or from a typical point of a k-face) are related in the same way as $\mathscr{C}$ and C_0 in equation (5.4). Indeed, Neveu's exchange formula (Neveu, 1977) provides the relation

$$\mathbf{E}(f(\mathscr{C}_k)) = \frac{1}{\mathbf{E}(\mathcal{H}_k(C_{0,k})^{-1})}\mathbf{E}\left(\frac{f(C_{0,k})}{\mathcal{H}_k(C_{0,k})}\right) \tag{5.6}$$

for any bounded, measurable and invariant-under-translations $f : \mathcal{K} \to \mathbb{R}$, see Baumstark and Last (2007). As a matter of fact, (5.6) is generally applied to functions f which are invariant-under-rotations as well.

5.1.2.4 *Realizations of the typical cell* The typical cell is not a particular cell isolated from a realization of the tessellation. Nevertheless, it can be explicitly constructed in both Poisson–Voronoi and Poisson hyperplane cases.

The key result in both cases is the well-known and very useful Slivnyak's formula for Poisson processes (see Chapter 1). If X is a Poisson point process in $\mathbb{R}^d$ of intensity measure μ, then for every $n \geq 1$

$$\mathbf{E}\left\{\sum_{\{x_1,\ldots,x_n\}\subset X} f(x_1,\ldots,x_n,X)\right\}$$
$$= \frac{1}{n!}\int \mathbf{E}\left\{f(x_1,\ldots,x_n,X\cup\{x_1,\ldots,x_n\})\right\} \mathrm{d}\mu(x_1)\ldots\mathrm{d}\mu(x_n), \tag{5.7}$$

where f is a bounded, measurable and invariant-under-permutations function defined on the product $(\mathbb{R}^d)^n \times \mathcal{N}$, $\mathcal{N}$ being the set of locally finite subsets of $\mathbb{R}^d$, see Møller (1994).

If the set of centroids is a homogeneous Poisson point process, a basic use of Slivnyak's formula implies that the associated Palm measure is the distribution of the same process with an extra point at the origin. In particular, the typical cell $\mathscr{C}$ of a Poisson–Voronoi tessellation generated by a homogeneous Poisson point process X is equal in distribution to the zero-cell of a Voronoi tessellation constructed with the new set of nuclei $X \cup \{0\}$. In the Poisson hyperplane tessellation case, a specific choice of a centroid process is required. The result below is a particular example of a possible realization of the typical cell. It is based on Calka (2001).

If we take for $z(C)$ the center of the largest ball included in the cell C, Slivnyak's formula allows us to obtain a generalization in any dimension of the construction given by Miles (1973) in dimension two.

- Let R and $(U_0, \ldots, U_d)$ be independent random variables with values in $\mathbb{R}_+$ and $(\mathbb{S}^{d-1})^{(d+1)}$ respectively such that R is exponentially distributed with mean $1/\omega_d$ and $(U_0, \ldots, U_d)$ has a density with respect to the uniform measure on $(\mathbb{S}^{d-1})^{(d+1)}$ which is proportional to the volume of the simplex constructed with these $(d+1)$ vectors (multiplied by the indicator function that this simplex contains the origin).
- Let Y be a point process such that conditionally on $\{R = r\}$, $r > 0$, Y is a Poisson point process of intensity measure $\mathbf{1}_{\mathbb{R}^d \setminus B_r(0)}(x)\|x\|^{-(d-1)}dx$.
- Let $\mathscr{C}_1$ be the polyhedron containing the origin obtained as the intersection of the $(d+1)$ half-spaces bounded by the hyperplanes H_{RU_i}, $0 \leq i \leq d$.
- Let $\mathscr{C}_2$ be the zero-cell of the hyperplane tessellation associated with Y.

Then the typical cell of the stationary Poisson hyperplane tessellation is distributed as the intersection $\mathscr{C}_1 \cap \mathscr{C}_2$.

The following section provides some precise results on the distribution of several geometrical characteristics of the typical cell and the typical k-faces.

5.2 Exact distributional results

5.2.1 *Number of hyperfaces and distribution of the cell conditioned on the number of hyperfaces*

In this section, we consider the zero-cell C_0 of a Poisson hyperplane process such that the intensity measure of the underlying point process is (in spherical coordinates)

$$\mathrm{d}\Theta(t,u) = \gamma \mathbf{1}_{\mathbb{R}_+}(t) t^{\alpha-1} \mathrm{d}t \mathrm{d}\sigma_d(u),$$

where σ_d is the uniform measure on $\mathbb{S}^{d-1}$, $\alpha \geq 1$ and $\gamma > 0$.

In particular, if $\alpha = 1$, we obtain the Crofton cell and if $\alpha = d$ and $\gamma = 2^d$, C_0 is distributed as the typical cell of a Poisson–Voronoi tessellation of unit intensity.

Let us denote by N_{d-1} the number of hyperfaces of C_0. In the Poisson–Voronoi case, it can be identified as the number of neighbours of the typical nucleus 0. We explain here how to calculate the probability of having n hyperfaces as a multiple integral of order n. The formula can be made fully explicit in dimension two. The lines below are based on Calka (2003*a*, 2003*b*).

For n points $x_1, \ldots, x_n \in \mathbb{R}^d \setminus \{0\}$, we define $D(x_1, \ldots, x_n)$ as the connected component containing the origin of $\mathbb{R}^d \setminus \cup_{i=1}^n H_{x_i}$, see (5.1) for the definition of H_{x_i}.

In any dimension, Slivnyak's formula (5.7) yields that for any $n \geq (d+1)$,

$$\mathbf{P}\{N_{d-1} = n\} = \frac{1}{n!} \int \mathrm{e}^{-\gamma\Phi(x_1,\ldots,x_n)} \mathbf{1}_{A_n}(x_1, \ldots, x_n) \mathrm{d}\Theta(x_1) \ldots \mathrm{d}\Theta(x_n), \quad (5.8)$$

where $\gamma\Phi(x_1, \ldots, x_n) = \Theta(\{x \in \mathbb{R}^d : H_x \cap D(x_1, \ldots, x_n) \neq \emptyset\})$ and A_n is the subset of n-tuples $(x_1, \ldots, x_n)$ such that $D(x_1, \ldots, x_n)$ is a convex polyhedron with n hyperfaces and containing the origin 0.

Let us remark that this formula was heuristically obtained in dimension two and in the Poisson–Voronoi case by Miles and Maillardet (1982). The functional Φ as well as the indicator function $\mathbf{1}_{\mathbf{A_n}}$ need to be made more explicit in function of $x_1, \ldots, x_n$. For any convex subset K of $\mathbb{R}^d$ which contains the origin, we denote by $h(K,u) = \sup\{\langle x, u\rangle : x \in K\}$, $u \in \mathbb{S}^{d-1}$, its support function. Then for any real $\alpha \geq 1$,

$$\Phi(x_1, \ldots, x_n) = \frac{1}{\alpha} \int h(D(x_1, \ldots, x_n), u)^{\alpha} \mathrm{d}\sigma_d(u).$$

In the two-dimensional case, if the points $x_1 = (r_1, \theta_1), \ldots, x_n = (r_n, \theta_n)$ are sorted by angular coordinate such that $0 \leq \theta_1 < \cdots < \theta_n < 2\pi$, we have for every $u_\theta = (\cos(\theta), \sin(\theta))$ with $\theta_i \leq \theta < \theta_{i+1}$ $(1 \leq i < n)$

$$h(D(x_1, \ldots, x_n), u_\theta) = \frac{1}{\sin(\theta_{i+1} - \theta_i)} (\sin(\theta_{i+1} - \theta) r_i + \sin(\theta - \theta_i) r_{i+1})$$

and the integration of h^α can be carried out for $\alpha = 1$ and $\alpha = 2$. If $\gamma = 1$, we then obtain in the first case the perimeter of the set $D(x_1, \ldots, x_n)$ and in the second case the area of the fundamental domain of $D(x_1, \ldots, x_n)$ (often called Voronoi flower), i.e. the union of the n discs centred at $y_i/2$ and of radius $\|y_i\|/2$, where the y_i, $1 \leq i \leq n$, are the vertices of $D(x_1, \ldots, x_n)$. Lastly, we can express the indicator function with the use of polar coordinates: $(x_1, \ldots, x_n) \in C_n$ if and only if for every $1 \leq i \leq n$,

$$r_{i-1} \sin(\theta_{i+1} - \theta_i) + r_{i+1} \sin(\theta_i - \theta_{i-1}) > r_i \sin(\theta_{i+1} - \theta_{i-1})$$

with the conventions $x_0 = (r_0, \theta_0) = x_n$ and $x_{n+1} = x_1$.

Consequently, we obtain an explicit formula which provides a way to do numerical calculations or to look for asymptotic estimates and limit shapes for many-sided cells, see e.g. Hilhorst (2005, 2006) and Hilhorst and Calka (2008). Nevertheless it is of little help for estimating the moments, for instance, it is very hard to verify the well-known equality $\mathbf{E}(N_1) = 6$ in the planar Poisson–Voronoi case. In this connection, it should be noted that a calculation of the second moment of the number of vertices of the typical Poisson–Voronoi cell in any dimension has been recently provided by Heinrich and Muche (2008) via the use of second-order properties of the point process of nodes of the tessellation, see also Heinrich, Körner, Mehlhorn and Muche (1998).

Going back to (5.8), let us add that the functional of $(x_1, \ldots, x_n)$ inside the integral is up to a multiplicative constant the density of the distribution of the respective positions of the n hyperplanes which surround the zero-cell conditioned on having n hyperfaces. In particular, it provides an easy way to verify that conditionally on $\{N_{d-1} = n\}$, $n \geq (d+1)$, $\Theta(\{x \in \mathbb{R}^d : H_x \cap C_0 \neq \emptyset\})$ is Gamma-distributed with parameters n and 1, see Zuyev (1992), Cowan, Quine and Zuyev (2003) and also the work by S. Zuyev on 'stopping sets' techniques (Zuyev, 1999).

5.2.2 *Typical k-face of a section of a Poisson–Voronoi tessellation*

This section is a quick survey of a number of papers (Møller, 1989; Muche and Stoyan, 1992; Mecke and Muche, 1995; Muche, 1996; Schlather, 2000; Muche, 2005; Baumstark and Last, 2007), which concern exclusively the d-dimensional Poisson–Voronoi tessellation or its sections with deterministic affine subspaces and propose mainly explicit formulae for the distribution of the typical k-face or typical edge.

Let $\mathcal{T}$ be a Voronoi tessellation generated by a set of nuclei distributed as a homogeneous Poisson point process of intensity 1. Miles (1970), Mecke and Muche (1995) and Muche (1996) provided a precise description of the Palm measure P_0^0 as defined in (5.5), i.e. the distribution of the point process of the nuclei as seen from a typical vertex. Baumstark and Last (2007) prove a generalization of their result, i.e. a full characterization of the Palm measure P_k^0 $(0 \leq k \leq d)$ which is explained below.

- The random set P_k^0 contains no point (from the homogeneous Poisson point process) apart from the origin 0 in a random ball $B_{R_k}(0)$ such that R_k is Gamma-distributed variable with parameters $(d-k+k/d)$ and κ_d.
- Conditioned on $\{R_k = r\}$, $r > 0$, $P_k^0 \cap \{x \in \mathbb{R}^d : \|x\| > r\}$ is distributed as a homogeneous Poisson point process of intensity 1.
- The intersection $P_k^0 \cap (R_k \mathbb{S}^{d-1})$ contains exactly $(d+1-k)$ points from the homogeneous Poisson point process which are independent with $(R_k, \{x \in P_k^0 : \|x\| > r\})$ and distributed as follows: we denote by Z_k and R'_k the centre and the radius of the unique $(d-1-k)$-dimensional sphere containing these $(d+1-k)$ nuclei. The $(d+1-k)$ neighbours of the origin are then distributed as $\sqrt{R_k^2 - R_k'^2}U + R'_k U_i$, $0 \leq i \leq (d-k)$ (up to a special orthogonal transformation), where
  - $U \in \mathbb{S}^{d-1}$ and $(U_0, \ldots, U_{d-k}) \in (\mathbb{S}^{d-1-k})^{(d-k+1)}$;
  - the $(d-k+1)$-tuple $(U_0, \ldots, U_{d-k})$ has a density (with respect to the uniform measure on $(\mathbb{S}^{d-1-k})^{(d-k+1)}$) which is proportional to the $(d-k)$-dimensional Hausdorff measure raised to the power $(k+1)$ of the simplex spanned by the $(d-k+1)$ vectors;
  - conditioned on $(U_0, \ldots, U_{d-k})$, the direction U is uniformly distributed on $\mathbb{S}^{d-1} \cap \{U_0, \ldots, U_d\}^{\perp}$,
  - the quantity $R_k'^2/R_k^2$ is independent with the vector of directions $(U, U_0, \ldots, U_{d-k})$ and is Beta-distributed with parameters $d(d-k)/2$ and $k/2$.

The main tools for proving this decomposition are Slivnyak's formula (5.7) and the Blaschke–Petkantschin change of variables formula, see Schneider and Weil (2000, Satz 7.2.1). It can be thought as a generalization of the previously known distributional results about the Poisson–Delaunay typical cell which would correspond here to the case $k = 0$, see e.g. Miles (1970) and Møller (1994). Let us remark finally that this results have been recently extended to Laguerre tessellations by Lautensack (2007).

This description combined with the relation expressed by (5.6) between the Palm measure P_k^0 and the typical k-face $\mathscr{C}_k$ implies some precise distributional results on $\mathscr{C}_k$. If the centroid of a k-face is chosen as the equidistant point from the $(d-k+1)$ neighbours of the k-face and in the affine subspace generated by the neighbours, it is possible to calculate explicitly the joint distribution of the vector constituted with $\mathcal{H}_k(\mathscr{C}_k)$, the distance ρ_k from the centroid 0 of the typical k-face to the $(d-k+1)$ neighbours of the k-face and the $(d-k+1)$ unit-vectors which determine the directions from the centroid to the neighbours. In particular, these directions are shown to be independent with $(\mathcal{H}_k(\mathcal{C}_k), \rho_k)$ and have the same distribution as $(U_0, \ldots, U_{d-k})$ in the construction above.

If $k = 1$, the calculation can be simplified and provides the same formulae as in Muche (2005), which will be given afterwards. In that paper, L. Muche investigates more generally the typical edge of a section of a Poisson–Voronoi tessellation of intensity 1 by a deterministic s-dimensional affine subspace where $1 \leq s \leq d$ and $d \geq 2$. This work unifies and extends previous efforts due to Brakke (1985), Møller (1989), Muche and Stoyan (1992) and Schlather (2000).

Exploiting the fact that the typical edge is equal in distribution to a randomly chosen edge emanating from the typical vertex, L. Muche makes explicit the joint distribution of the vector constituted with the length L of the typical edge and the two adjacent angles $B_1 = \angle(v_1, v_2, x)$ and $B_2 = \angle(v_2, v_1, x)$ where v_1 and v_2 are the vertices of the edge and x is one of the s neighbours of the edge. The density of (L, B_1, B_2) is up to a multiplicative constant equal to

$$f(l, \beta_1, \beta_2) = \frac{l^{d(s+2)-s-1}(\sin(\beta_1)\sin(\beta_2))^d}{\sin^{ds+2d-s}(\beta_1+\beta_2)} f_B(\beta_1) f_B(\beta_2) \exp(-\nu(l, \beta_1, \beta_2)),$$

where

- $\nu(l, \beta_1, \beta_2)$ is the d-dimensional Lebesgue measure of the union of two balls such that their centres v_1 and v_2 are at distance l and the angles $\angle(v_2, v_1, x)$ and $\angle(v_1, v_2, x)$ are equal to β_1 and β_2, x being any point at the intersection of the boundaries of the balls;
- the function f_B is defined by

$$f_B(\beta) = \frac{(d-1)s\Gamma\left(\frac{d+1}{2}\right)\Gamma\left(\frac{ds+d-s}{2}\right)}{\Gamma\left(\frac{d}{2}\right)\Gamma\left(\frac{ds+d-s+1}{2}\right)} \sin^{ds-s-1}(\beta) \sum_{i=0}^{\lfloor \frac{d-1}{2} \rfloor} b_i(\beta),$$

for every $0 \leq \beta \leq \pi$ with

$$b_0 = \begin{cases} \frac{1}{\pi}[(\pi-\beta)\cos(\beta)+\sin(\beta)] & \text{if } d \text{ even} \\ \\ \cos^2(\beta/2) & \text{otherwise,} \end{cases}$$

$$b_i = \begin{cases} -\frac{\Gamma(i)}{4\sqrt{\pi}\Gamma(i+\frac{3}{2})}\sin^{2i+1}(\beta) & \text{if } d \text{ even} \\ \\ -\frac{\Gamma(i-\frac{1}{2})}{4\sqrt{\pi}\Gamma(i+1)}\sin^{2i}(\beta) & \text{otherwise,} \end{cases} \qquad 1 \leq i \leq \lfloor (d-1)/2 \rfloor.$$

Let us remark that the explicit calculation of all the moments of the variables can be deduced from the formulae above. Additional calculations were added in Muche (2005) about the relative positions of the s neighbours. It concerns in particular the distance from any neighbour to the affine subspace generated by the typical edge and the conditional distribution of the length of the edge conditionally on the fact that the projection of any neighbour on the spanned affine subspace is inside the edge or not.

5.2.3 *The circumscribed radius*

This section provides a different kind of information on the zero-cell C_0 defined in section 5.2.1. Indeed, we are now interested in putting in an optimal way the boundary of C_0 between two spheres centred at 0. The main result is that in any dimension, the joint distribution of the radii of the two spheres can be expressed in terms of covering probabilities of the unit-sphere by random caps and in dimension two, it can be explicitly calculated. The following ideas are basic generalizations of the results by Calka (2002).

We introduce the quantity $R_m = \sup\{r > 0 : B_r(0) \subset C_0\}$, i.e. the radius of the largest disk centred at 0 and contained in the cell. Clearly, we have

$$\mathbf{P}\{R_m \geq r\} = \exp\left(-\frac{\gamma\omega_d}{\alpha}r^\alpha\right), \quad r > 0,$$

and the distribution of the hyperplane process conditioned on $\{R_m \geq r\}$, $r > 0$, is a new hyperplane process of intensity measure $\mathbf{1}_{\mathbb{R}^d \setminus B_r(0)} d\Theta$.

In order to have a more precise idea of the shape of C_0, it is relevant to consider the radius of the circumscribed ball centred at the origin, i.e. $R_M = \inf\{r > 0 : B_r(0) \supset C_0\}$. It can be shown that the distribution of R_M is related to the covering probability of $\mathbb{S}^{d-1}$ by a Poissonian number of independent random circular caps such that their centres on $\mathbb{S}^{d-1}$ are uniformly distributed and their angular radii (divided by π) have the distribution given below:

$$\mathrm{d}\nu(\theta) = \alpha\pi \sin(\pi\theta) \cos^\alpha(\pi\theta) \mathbf{1}_{[0,1/2]}(\theta) \mathrm{d}\theta.$$

Indeed, having $R_M \geq r$ for a fixed $r > 0$ means that there is a non-empty portion of the sphere $r\mathbb{S}^{d-1}$ which is not covered by the intersection with the hyperplanes of the Poisson hyperplane process. To be more precise, let us denote by $P(\nu, n)$, for every $n \geq 0$, the covering probability of $\mathbb{S}^{d-1}$ by n independent random circular caps which are isotropic and of angular radius distributed as $\nu(\cdot/\pi)$. We obtain

$$\mathbf{P}\{R_M \geq r\} = \exp\left(-\frac{\gamma\omega_d}{\alpha}r^\alpha\right) \sum_{n=0}^{\infty} \frac{\left(\frac{\gamma\omega_d}{\alpha}r^\alpha\right)^n}{n!}(1 - P(\nu, n)).$$

This relation is easily generalized if C_0 is conditioned on $\{R_m = r\}$.

We now concentrate on the two-dimensional case. If $d = 2$, the covering probabilities $P(\nu, n)$ can be calculated (Stevens, 1939; Siegel and Holst, 1982), which provides us an expression for $\mathbf{P}\{R_M \geq r\}$. In particular, when $r \to \infty$, we use a basic ordering relation between the covering probabilities (conjectured by Siegel (1978) and proved by Calka (2002)) in order to 'replace' the distribution ν by its mean $I_\alpha = \frac{1}{\pi}\int_0^{\pi/2} \cos^\alpha(u)du$. We obtain that there exist two constants $0 < C_1 < C_2 < \infty$ such that for r large enough,

$$C_1 r^\alpha \exp\left(-\frac{2\pi\gamma I_\alpha}{\alpha} r^\alpha\right) \leq \mathbf{P}\{R_M \geq r\} \leq C_2 r^\alpha \exp\left(-\frac{2\pi\gamma I_\alpha}{\alpha} r^\alpha\right).$$

The calculation of the joint distribution of the couple (R_M, R_m) leads us to obtain an asymptotic estimation of $\mathbf{P}\{R_M \geq r + r^\delta | R_m = r\}$ for $1 - \frac{2}{3}\alpha < \delta < 1$, see Section 5.3.2 for a generalization of that result in any dimension. In fact, this probability is proved to decrease exponentially fast to zero, which indicates that the zero-cell converges to a circle when its inradius goes to infinity. It is a particular case of D. G. Kendall's conjecture which will be the centre of our next section.

5.3 Asymptotic results

5.3.1 *D. G. Kendall's conjecture*

A very-well known conjecture due to D. G. Kendall states that cells of large area in an isotropic Poisson line tessellation are close to the circular shape, see e.g. the foreword of Stoyan, Kendall and Mecke (1987). The conjecture can be rephrased in a modern setting as follows: the conditional distribution of the two-dimensional Crofton cell converges weakly when its area goes to infinity to the degenerate law concentrated at the circular shape.

Works due notably to Miles (1995) or Goldman (1998) were first advancements on the subject (see also Goldman 1996; Goldman and Calka 2003 for an interpretation of the conjecture in terms of the first eigenvalue of the Dirichlet–Laplacian on the cell). Kovalenko (1997, 1998, 1999) proved the conjecture in the case of a two-dimensional isotropic and stationary Poisson line tessellation. Since then, Hug, Schneider and Reitzner (2004*a*, 2004*b*) have obtained far more precise results which generalize D. G. Kendall's conjecture in four different ways: more general Poisson hyperplane tessellations, more general functionals to measure the largeness of a cell, explicit estimates for deviations from asymptotic shapes and identification of the cases where limit shapes do not exist. Their proofs mix precise arguments from geometry of convex bodies (in particular, isoperimetric inequalities and existence of associated extremal bodies) combined with probabilistic estimations which make good use of the Poissonian distribution. This section is devoted to a basic presentation of the main results and the underlying methods contained in papers by Hug, Reitzner and Schneider (2004*a*, 2004*b*) and by Hug and Schneider (2004, 2007*a*).

5.3.1.1 *Context and useful functionals* We consider the zero-cell C_0 of a hyperplane tessellation in $\mathbb{R}^d$, $d \geq 2$, such that its intensity measure denoted by Θ is defined in spherical coordinates by the equality

$$\mathrm{d}\Theta(t,u) = \gamma t^{\alpha-1}\mathrm{d}t\varphi(\mathrm{d}u),$$

where $\gamma > 0$, $\alpha \geq 1$ and φ is a probability measure on $\mathcal{S}^{d-1}$ such that its support is not contained in a half-sphere. As previously mentioned, this general model interpolates the cases of the Crofton cell of an isotropic Poisson hyperplane tessellation and of the typical Poisson–Voronoi cell.

Here we follow the scheme developed in the papers cited above. Three types of functionals are used to study size and shape of the zero-cell C_0. They are defined on the set denoted by $\mathcal{K}_0$ of all convex bodies K such that K contains 0 and is the intersection of its supporting halfspaces which have an outer unit normal vector in the support of φ.

- The *parameter functional* Φ (already used in Section 5.2.1) is defined by the equality
$$\Phi(K) = \frac{1}{\alpha}\int h(K,u)^{\alpha}\varphi(\mathrm{d}u), \quad K \in \mathcal{K}_0.$$
It is a continuous function related to the intensity measure in such a way that the probability $\mathbf{P}\{K \subset C_0\}$ is equal to $\exp(-\gamma\Phi(K))$.
- The *size functional* denoted by Σ is a function aimed at 'measuring' the size of C_0. The only properties that Σ has to satisfy are that it has to be continuous, increasing and homogeneous of some degree $k > 0$ (i.e. $\Sigma(rK) = r^k\Sigma(K)$ for every $K \in \mathcal{K}$ and $r > 0$). Volume, surface area or inradius (see Section 5.3.2) are basic examples of such a function.
- The *deviation functional* $\mathcal{V}$ is related to the two previous functionals Φ and Σ and will measure the distance between C_0 and the potential limit shape. It is defined in the following way: since Φ is homogeneous of degree α, the two previous functionals Φ and Σ satisfy an isoperimetric inequality of the form
$$\Phi(K) \geq \tau\Sigma(K)^{\alpha/k}, \quad K \in \mathcal{K}_0, \tag{5.9}$$
where τ is some positive constant which can be chosen such that there exist convex bodies $K \in \mathcal{K}_0$, called *extremal bodies*, for which the equality holds. The functional $\mathcal{V}$ is then introduced as a continuous, non-negative and homogeneous of degree zero function such that $\mathcal{V}(K) = 0$ implies that K is extremal. For instance, $\mathcal{V}$ can be defined by the equality
$$\mathcal{V}(K) = \frac{\Phi(K)}{\tau\Sigma(K)^{\alpha/k}} - 1, \quad K \in \mathcal{K}. \tag{5.10}$$

Let us remark that the isoperimetric inequality (5.9) can be strengthened as follows: there exists a non-negative continuous function f on $\mathbb{R}_+$ with a unique zero at zero such that if $\mathcal{V}(K) \geq \varepsilon > 0$ for $K \in \mathcal{K}_0$, K must satisfy the inequality $\Phi(K) \geq \tau(1 + f(\varepsilon))\Sigma(K)^{\alpha/k}$.

5.3.1.2 *Estimates of conditional probabilities* If $\varepsilon > 0$ and $a > 0$ are fixed, the aim is now to evaluate conditional probabilities

$$\mathbf{P}\{\mathcal{V}(C_0) \geq \varepsilon | \Sigma(C_0) \geq a\} = \frac{\mathbf{P}\{\mathcal{V}(C_0) \geq \varepsilon; \Sigma(C_0) \geq a\}}{\mathbf{P}\{\Sigma(C_0) \geq a\}} \tag{5.11}$$

for a fixed $\varepsilon > 0$ and a sufficiently large.

A first remark is that it is easier to give a lower bound of the denominator in (5.11). Indeed, for an extremal body K, the probability of having $K \subset C_0$ is equal to $\exp(-\tau\gamma\Sigma(K)^{\alpha/k})$. Consequently, it suffices to compare C_0 with an extremal body included in it of size a to obtain that $\mathbf{P}\{\Sigma(C_0) \geq a\} \geq \exp(-\tau\gamma a^{\alpha/k})$ for every $a > 0$.

The estimation of the numerator in (5.11) is a much more delicate matter. Following the structure of the proofs from Hug, Reitzner and Schneider (2004*a*) and Hug and Schneider (2004), we describe below some of the key arguments in order to do it.

- The range of $\Sigma(C_0)$ in the event has first to be limited to an interval $[a, a(1+h)]$ for some h 'not too large' (afterwards, the range is extended by a covering argument).
- An additional condition is added in the event in order to guarantee that the diameter of C_0 is bounded and that C_0 is included in a deterministic ball B (afterwards, the sum over all possible intervals for the diameter is taken).
- The cell C_0 is defined as the intersection of all half-spaces coming from the initial hyperplane process. Hyperplanes which have a non-empty intersection with the boundary of C_0 must also intersect the deterministic ball B introduced above. Consequently, the event we are interested in can be rewritten in terms of the set $\mathcal{G}_B$ of all hyperplanes which hit B. Fortunately the distribution of $\mathcal{G}_B$ is known: its cardinality is shown to be Poisson distributed, of mean $\gamma\Phi(B)$ and all hyperplanes in $\mathcal{G}_B$ are i.i.d. and distributed as $(\gamma\Phi(B))^{-1}\mathbf{1}_{\{H_x \cap B \neq \emptyset\}}\mathrm{d}\Theta$.
- By an argument of convex geometry, for a fixed $\alpha > 0$, the polyhedron C_0 can be replaced by the convex hull $\widetilde{C}_0$ of a finite number of its vertices in such a way that the ratio $\Phi(\widetilde{C}_0)/\Phi(C_0)$ is more than $1 - \alpha$.
- The numerator in (5.11) is finally overestimated by the probability that $\widetilde{C}_0$ is not hit by the majority of the hyperplanes involved.

5.3.1.3 *Results and examples* Various extensions of the estimations presented above imply the following general result: there exists a constant c_0 depending only on the dimension d such that for every $\varepsilon > 0$ and $0 < a < b \leq \infty$ with $a^{\alpha/k}\gamma \geq \sigma_0$, we have

$$\mathbf{P}\{\mathcal{V}(C_0) \geq \varepsilon | \Sigma(C_0) \in [a, b)\} \leq c \exp(-c_0 f(\varepsilon) a^{\alpha/k}\gamma), \tag{5.12}$$

where c is a constant depending only on the measure Θ and the choices of Σ, f, ε and σ_0.

Moreover, the question of the existence of a limit shape has also been investigated: the shape of a convex set K is defined as the equivalence class of K under the action of a subgroup of similarities. The zero-cell C_0 is said to have a limit shape if the distribution of the shape of C_0 conditioned on $\{\Sigma(C_0) \geq a\}$ converges weakly to a Dirac measure when $a \to \infty$. In particular, if there exists a subgroup of similarities which preserves $\mathcal{K}_0$ and the function defined in (5.10) and such that all extremal bodies are in the same equivalence class, then the zero-cell C_0 admits a limit shape. It should be noticed that this limit shape depends not only on the distribution Θ but also on the chosen size functional.

The general result (5.12) can be applied in the particular case where C_0 is the typical Poisson–Voronoi cell ($\gamma = 2^d \omega_d$, $\alpha = d$ and φ is the uniform probability measure on $\mathbb{S}^{d-1}$). As previously seen in Section 5.2.1, Φ is the Lebesgue measure of the union of all the balls $B_{\|x\|/2}(x/2)$ where x is any point of C_0. If the size functional is the k-th intrinsic volume, the limit shape is a ball and a convenient choice of deviation functional is $\mathcal{V}(K) = (R_M - R_m)/(R_M + R_m)$ where the radii R_m and R_M are the same as in Section 5.2.3. In particular, if $\Sigma(C_0) = \lambda_d(C_0)$, the estimation (5.12) becomes

$$\mathbf{P}\left\{ \frac{R_M - R_m}{R_M + R_m} \geq \varepsilon \,\middle|\, \lambda_d(C_0) \in [a, b) \right\} \leq c \exp(-c_0 \varepsilon^{(d+1)/2} a \gamma). \tag{5.13}$$

In the same way, if $\Sigma(C_0) = R_m$, we have

$$\mathbf{P}\left\{ \frac{R_M - R_m}{R_M + R_m} \geq \varepsilon \,\middle|\, R_m \in [a, b) \right\} \leq c \exp(-c_0 \varepsilon^{(d+1)/2} a^d \gamma). \tag{5.14}$$

When C_0 is the Crofton cell ($\gamma = \omega_d$, $\alpha = 1$ and φ is the uniform probability measure on $\mathbb{S}^{d-1}$), D. G. Kendall's conjecture (i.e. convergence to the limit shape of a ball) is solved in any dimension with the particular choice of the d-dimensional Lebesgue measure as the size functional and a deviation function $\mathcal{V}$ defined in the following way: the quantity $\mathcal{V}(K)$ is the infimum of $(s/r - 1)$ over all couples $(s, r) \in (\mathbb{R}_+^*)^2$ such that there exists a translate of K which is between $B_r(0)$ and $B_s(0)$ for the inclusion. Interestingly, the limit shape is not a ball but a segment if the size is measured by the diameter.

These results for the Crofton cell can be extended to the typical cell of a stationary Poisson hyperplane tessellation through the use of a new realization of the typical cell based on the choice of the lowest points of the cells as centroids (Hug and Schneider, 2007*b*), see Section 5.1.2.4. Other extensions of this work concern the determination of a logarithmic equivalent for the distribution tail of $\Sigma(C_0)$ (Goldman, 1998; Hug and Schneider, 2007*a*) and large typical cells in Poisson–Delaunay tessellations (Hug and Schneider, 2005).

5.3.2 *Cells with a large inradius*

We go back to the model introduced in Section 5.2.1, i.e. we concentrate on the particular case where the hyperplane process is isotropic, of intensity measure $d\Theta = t^{\alpha-1}dtd\sigma_d(u)$ where $1 \leq \alpha \leq d$. We suppose now that the inball centred at the origin is large. The preceding results show that the cell is close to a ball but some specifications can be added. Indeed, the boundary of the cell can be proved to be inside an annulus around the origin with a decreasing thickness when the inradius goes to infinity. Moreover, limit theorems can be deduced from this fact for both the number N_{d-1} of hyperfaces and the volume V_d between the boundary of the cell and the inball. We give below a description of the methods involved and of the main asymptotic results. The results and methods developed below are almost direct generalizations of a joint work with Schreiber (Calka and Schreiber, 2005; Calka and Schreiber, 2006).

In any dimension, an asymptotic estimation of the probability $\mathbf{P}\{R_M \geq r+s|R_m = r\}$ $(s > 0)$ when r goes to infinity, is made possible by a method introduced by Calka and Schreiber (2006). This procedure allows us to estimate the quantities N_{d-1} and V_d as well and can be described as follows.

- *Step 1.* After a homothetic transformation on the zero-cell C_0 conditioned on $\{R_m = r\}$, we obtain the zero-cell associated with a deterministic hyperplane at distance one from 0 and a hyperplane process outside $B_1(0)$ of intensity measure $r^\alpha \mathbf{1}_{\mathbb{R}\setminus B_1(0)}d\Theta$. The number N_{d-1} is preserved whereas V_d is multiplied by r^{-d}.
- *Step 2.* Let us apply the inversion defined by $I(x) = x/\|x\|^2$ for every $x \in \mathbb{R}^d \setminus \{0\}$. It transforms the zero-cell into a germ-grain model inside the unit-ball. More precisely, the image of the hyperplane process outside $B_1(0)$ is a process of spheres centered at $y/2$ and of radius $\|y\|/2$ where y is an element of a Poisson point process Ψ inside $B_1(0)$ of intensity measure $r^\alpha \mathbf{1}_{B_1(0)}t^{-(\alpha+1)}dtd\sigma_d(u)$. The number N_{d-1} can be seen as the number of extreme points of the convex hull of Ψ and the volume V_d as $r^{-d}\mu(B_1(0) \setminus \cup_{y\in\Psi\cup\{y_0\}} B_{\|y\|/2}(y/2))$ where y_0 is a deterministic point on $\mathbb{S}^{d-1}$ and $\mu = \mathbf{1}_{B_1(0)}t^{-(d+1)}dtd\sigma_d(u)$.
- *Step 3.* We consider a Poisson point process Ψ in $B_1(0)$ of intensity measure λdx $(\lambda > 0)$ or in a more general context $\lambda f(t)dtd\sigma_d(u)$ with $\lim_{t\to 1} f(t) = 1$. Then
  - the convex hull of this process contains the ball $B_{(1-Kt^{-\delta})}(0)$ (for a fixed constant $K > 0$ and with $0 < \delta < 2/(d+1)$) with a probability going to 1 exponentially fast, see Calka and Schreiber (2006);
  - the number of vertices of the convex hull follows a law of large numbers (Rényi and Sulanke, 1963; Reitzner, 2003), a central limit theorem (Reitzner, 2005) and a large deviation-type result (Calka and Schreiber, 2006; Vu, 2005);

- the volume or the μ-measure of the set between the unit-sphere and the union $\cup_{y\in\Psi} B_{\|y\|/2}(y/2)$ satisfies a law of large numbers, a central limit theorem and a moderate deviation principle (Schreiber, 2002; Schreiber, 2003).

We obtain from the three steps above the following results:

1. There exists a constant $c > 0$ such that for every $\frac{d+1-2\alpha}{d+1} < \delta < 1$, we have when r goes to infinity
$$\mathbf{P}\{R_M \geq r + r^{\delta} | R_m = r\} = \mathcal{O}(\exp(-cr^{\beta})) \tag{5.15}$$
where $\beta = \frac{1}{2}[(2\alpha - (d+1)) + \delta(d+1)]$.
In other words, the boundary of the zero-cell conditioned on $\{R_m = r\}$ is typically included in an annulus of thickness $r^{(d+1-2\alpha)/(d+1)}$.
2. There exists a constant $a > 0$ (known explicitly) depending only on d and α such that
$$N_{d-1}(ar^{\frac{\alpha(d-1)}{d+1}})^{-1} \longrightarrow 1 \quad \text{in } L^1 \text{ and a.s. when } r \to \infty.$$
3. The number N_{d-1} satisfies a central limit theorem when $r \to \infty$ as well as a moderate-deviation result: for every $\varepsilon > 0$,
$$\liminf_{r\to\infty} \frac{1}{\log(r)} \log\left(-\log\left(\mathbf{P}\left\{\left|\frac{N_{d-1}}{\mathbf{E}N_{d-1}} - 1\right| \geq \varepsilon\right\}\right)\right) \geq \frac{\alpha(d-1)}{3d+5}.$$
4. The same type of limit theorems holds for the quantity V_d which grows as $r^{\frac{\alpha(d-1)}{d+1}}$ (up to an explicit multiplicative constant).

It emerges that in the context of a large inradius, supplementary information on the growth of the number of hyperfaces and of the volume outside the inball can be obtained to specify the convergence of the random polyhedron to the ball-shape. Besides, we may notice that the asymptotic result (5.15) could not be deduced from the previous estimation (5.14) since the constant c depends on ε in the latter inequality.

The last part is independent of the rest of the chapter: it concerns different types of tessellation called iterated tessellations, which are natural models in several concrete situations and have been recently investigated for application purposes.

5.4 Iterated tessellations

5.4.1 *Tessellations stable with respect to iteration*

Real tessellations may present 'hierarchical' structures, which occur in some crack structures, as the 'craquelé' on pottery surfaces. In order to provide a good approximation, W. Nagel and V. Weiss have investigated the iteration of tessellations. They aim at determining the existence of tessellations which are stable (in distribution) with respect to iteration (STIT) and at characterizing these tessellations.

An explicit model called the crack STIT tessellation is given via an algorithmic construction and it is proved that such a model is indeed STIT and conversely, that any STIT tessellation is a crack STIT tessellation. This section is devoted to a formal description of the work due to Nagel and Weiss (2003, 2004, 2005).

5.4.1.1 *Construction of a crack STIT tessellation in a window* Let φ be a probability measure on $\mathbb{S}^{d-1}$ such that its support contains a basis of $\mathbb{R}^d$. As we previously did, we consider the associated measure $\mathrm{d}\Theta = \mathrm{d}t\phi(\mathrm{d}u)$ on the set of hyperplanes of $\mathbb{R}^d$ which is supposed to be invariant with respect to the translations of $\mathbb{R}^d$. For a bounded Borel set $C \subset \mathbb{R}^d$, we denote by $[C]$ the set of all hyperplanes that hit C and by Θ_C the probability measure on $[C]$ defined by the relation $\Theta_C = \frac{1}{\Theta([C])}\Theta(\cdot \cap [C])$ (if $0 < \Theta([C]) < \infty$).

Let $W \subset \mathbb{R}^d$ be a d-dimensional compact and convex domain such that $0 < \Theta([W]) < \infty$. The *crack STIT tessellation* $\mathcal{T}(a, W)$ is constructed in W and on a time interval $[0, a]$, $a > 0$, as follows: an i.i.d sequence (τ_i, γ_i), $i \geq 1$, is given where τ_i is a random time which is exponentially distributed with parameter $\Theta([W])$ and γ_i is a random hyperplane with distribution Θ_W.

- If $\tau_1 > a$, the algorithm does not begin and the tessellation is W itself.
- If $\tau_1 \leq a$, the algorithm starts with a first cutting of W at time τ_1 into two parts W_+ and W_-. W_+ and W_- are then treated in the same way, separately and independently. Let us describe the evolution of W_+:
  - if $\tau_1 + \tau_2 > a$, W_+ is conserved as it is and will be a part of the final tessellation;
  - if $\tau_1 + \tau_2 \leq a$, W_+ is divided at time $\tau_1 + \tau_2$ by γ_2 if γ_2 intersects W_+. If W_+ has not been divided by γ_2, the next potential division of W_+ occurs at time $\tau_1 + \tau_2 + \tau_3$ (if that time is less than a). If W_+ has been divided by γ_2, the algorithm goes on with the two subsections $W_{+,+}$ and $W_{+,-}$.

This construction can be carried over equivalently by means of a formal description based on random binary trees.

5.4.1.2 *Extension to a crack STIT tessellation of* $\mathbb{R}^d$ *and stability with respect to iteration* The capacity functional of the tessellation $\mathcal{T}(a, W)$,

$$T_{\mathcal{T}(a,W)} : C \longmapsto \mathbf{P}\{\mathcal{T}(a, W) \cap C = \emptyset\}, \quad C \in \mathcal{K} \cap \mathscr{C},$$

can be calculated if C is connected and recursively if C has a finite number of connected components. Moreover, this computation does not depend on the window W which contains C and is invariant with respect to translations of $\mathbb{R}^d$. Consequently, there exists a random stationary tessellation $\mathcal{T}(a)$ of the whole space $\mathbb{R}^d$ such that its intersection with a compact convex window W is equal in distribution to $\mathcal{T}(a, W)$, see Schneider and Weil (2000, Satz 2.3.1). In particular, $\mathcal{T}(a)$ satisfies the scaling property, i.e. for $a > 0$, $\mathcal{T}(a)$ coincides in distribution with $\frac{1}{a}\mathcal{T}(1)$.

A fundamental property of the tessellation $\mathcal{T}(a)$ is that it is stable with respect to iteration. More precisely, let us define the operation of iteration: an initial stationary tessellation $\mathcal{T}^{(0)}$ of the whole space $\mathbb{R}^d$ and a sequence of i.i.d. tessellations $\mathcal{Y} = \{\mathcal{T}^{(i)} : i \geq 1\}$ are given. We denote by $C_1^{(0)}, C_2^{(0)}, \ldots$ the cells of $\mathcal{T}^{(0)}$. The iterated tessellation $I(\mathcal{T}^{(0)}, \mathcal{Y})$ is then obtained by replacing in $\mathcal{T}^{(0)}$ the interior of each cell $C_i^{(0)}$ by $C_i^{(0)} \cap \mathcal{T}^{(i)}$, $i \geq 1$.

In order to preserve the same surface intensity of the tessellation, a rescaling is needed. Consequently, if $\{\mathcal{Y}_m : m \geq 1\}$ is a sequence of i.i.d. sequences of tessellations (distributed as $\mathcal{T}^{(0)}$), we define $I_2(\mathcal{T}^{(0)}) = I(2\mathcal{T}^{(0)}, 2\mathcal{Y}_1)$ and recursively, for every $m \geq 3$,

$$I_m(\mathcal{T}^{(0)}) = I\left(\frac{m}{m-1} I_{m-1}(\mathcal{T}^{(0)}), m\mathcal{Y}_{m-1}\right).$$

In other words, at step m, the tessellation $\dfrac{m}{m-1} I_{m-1}(\mathcal{T}^{(0)})$ is 'iterated' with the sequence of tessellations $\mathcal{Y} = m\mathcal{Y}_{m-1}$.

A tessellation is said to be *stable with respect to iteration* (STIT) if $I_m(\mathcal{T}^{(0)})$ and $\mathcal{T}^{(0)}$ are equal in law for every $m \geq 2$.

Going back to the crack STIT tessellation $\mathcal{T}(a)$, we can observe that for any $a, b > 0$, the process of iterating $\mathcal{T}(a)$ with an independent sequence $\mathcal{T}(b)$ of i.i.d. tessellations distributed as $\mathcal{T}(b)$ is equivalent to constructing the crack STIT tessellation over the time interval $[0, a + b]$, i.e. $I(\mathcal{T}(a), \mathcal{Y}(b))$ coincides in distribution with $\mathcal{T}(a + b)$. This property comes from the Markov property of $(\mathcal{T}(t, W))_{t>0}$ for a fixed window W and combined with the scaling property of $\mathcal{T}(a)$, it implies that $\mathcal{T}(a)$, $a > 0$, is stable with respect to iteration.

It can be proved conversely that any tessellation which has the STIT property is a crack STIT tessellation. Indeed, a modified version of Korolyuk's theorem on processes of facets (Daley and Vere-Jones, 1988, Ch. 3) can be used to show that for any stationary tessellation $\mathcal{T}$, the sequence $I_m(\mathcal{T})$, $m \geq 1$, converges weakly to a crack STIT tessellation with the same surface intensity and directional distribution as $\mathcal{T}$.

Numerous properties of STIT tessellations have been derived by Nagel and Weiss (2004). Let us cite in particular the preservation of the STIT property for every section of a STIT tessellation, as well as the equality in distribution of the interior of the typical cell with the typical cell of a homogeneous Poisson hyperplane tessellation with the same surface intensity and directional distribution. Mean values in dimension two and three have also been calculated.

5.4.2 *Iterated tessellations in telecommunications*

We end this last section with a small introduction on the use of iterated tessellations in telecommunications. Models and results cited below are due notably to Maier, Schmidt and Mayer (2004, 2003) and Heinrich, Schmidt and Schmidt (2006).

Classical tessellations have been of great use in that specific domain of application for several years (see e.g. Baccelli, Gloaguen and Zuyev 2000*a*; Baccelli, Tchoumatchenko and Zuyev 2000*b*; Baccelli and Błaszczyszyn 2001; Błaszczyszyn and Schott 2003). In order to make the models more realistic and take into account the fact that a network may contain two levels of roads, Maier and Schmidt (2003) introduced a stationary iterated tessellation in the following way.

- Let $\mathcal{T}^{(0)} = \{C_i^{(0)}\}_{i\geq 1}$ be a random stationary tessellation called the *initial tessellation*.
- Let $\{\mathcal{T}^{(n)}\}_{n\geq 1}$ be a sequence of random stationary tessellations which is independent with $\mathcal{T}^{(0)}$ and such that the $\mathcal{T}^{(n)}$, $n \geq 1$, are i.i.d. or at least exchangeable. The $\mathcal{T}^{(n)} = \{C_i^{(n)}\}_{i\geq 1}$ are called the *component tessellations*.

Then the tessellation $\mathcal{T}$ constituted with all the intersections $C_i^{(n)} \cap C_n^{(0)}$, $n \geq 1$, $i \geq 1$, with a non-empty interior, is the associated *stationary iterated tessellation*. Basic examples which are concretely used are obtained when the initial and component tessellations are distributed as Poisson–Voronoi tessellations or stationary Poisson hyperplane tessellations (Gloaguen, Fleischer, Schmidt and Schmidt, 2006).

Let us denote by $\mathscr{C}^{(0)}$ and $\lambda^{(0)}$ (resp. $\mathscr{C}$ and λ) the typical cell and the intensity of $\mathcal{T}^{(0)}$ (resp. of $\mathcal{T}$). The use of Neveu's exchange formula (Neveu, 1977) provides a precise link between $\mathscr{C}^{(0)}$ and $\mathscr{C}$, i.e. for every bounded and measurable function $f : \mathcal{K} \to \mathbb{R}$,

$$\mathbf{E}(f(\mathscr{C})) = \frac{\lambda^{(0)}}{\lambda}\mathbf{E}\sum_{i\geq 1} f(C_i^{(1)} \cap \mathscr{C}^{(0)})\mathbf{1}_{\{\mathrm{Int}(C_i^{(1)})\cap \mathrm{Int}(\mathscr{C}^{(0)})\neq\emptyset\}}.$$

Quantities of interest for such an iterated tessellation are measurements of inner structure of the initial cells, such as the number or the k-dimensional Hausdorff measure of the k-faces of the component tessellation inside an initial cell. Heinrich, Schmidt and Schmidt (2006) obtained a more general law of large numbers and a multivariate central limit theorem which can be applied to the quantities above. Indeed, for a fixed $m \geq 1$, they consider a sequence of i.i.d. vectors $J_i = (J_{i,1}, \ldots, J_{i,m})$, $i \geq 1$, whose coordinates are stationary random measures. In particular, J_i is the 'descriptor' of the inner structure of the i-th cell of the initial tessellation and it is supposed to have a finite intensity vector $(\lambda_1, \ldots, \lambda_m)$. For a fixed window W which is a convex set of $\mathbb{R}^d$ with a non-empty interior, we denote by $Z_{k,\rho}$ the quantity $\sum_{i\geq 0} J_{i,k}(C_i^{(0)} \cap \rho W)$ where $1 \leq k \leq m$

and $\rho > 0$. If the initial tessellation is ergodic, under some integrability conditions upon the typical cell of $\mathcal{T}^{(0)}$ and J_i, we have for every $1 \leq k \leq m$

$$\frac{1}{\lambda_d(\rho W)} Z_{k,\rho} \longrightarrow \lambda_k, \quad \text{as } \rho \to \infty.$$

The proof of this result is based on classical methods related to Wiener's ergodic theorem associated with a precise treatment of the contribution of the cells hitting the boundary of the window.

Moreover, the authors use a refinement of the Berry–Esseen inequality to prove under certain conditions of integrability related to J_i and the typical cell $\mathscr{C}^{(0)}$ that the vector

$$\frac{1}{\sqrt{\lambda_d(\rho W)}}(Z_{1,\rho} - \lambda_1 \lambda_d(\rho W), \ldots, Z_{m,\rho} - \lambda_m \lambda_d(\rho W))$$

converges to a mean-zero normal distribution with an explicit covariance matrix.

These convergence results are used to estimate the quantities λ_k and decide which model of iterated tessellation fits the best in concrete situations, see Gloaguen, Fleischer, Schmidt and Schmidt (2006).

References

Avram, F. and Bertsimas, D. (1993). On central limit theorems in geometrical probability. *Ann. Appl. Probab.*, **3**, 1033–1046.

Baccelli, F. and Błaszczyszyn, B. (2001). On a coverage process ranging from the Boolean model to the Poisson–Voronoi tessellation with applications to wireless communications. *Adv. in Appl. Probab.*, **33**, 293–323.

Baccelli, F., Gloaguen, C., and Zuyev, S. (2000*a*). Superposition of planar Voronoi tessellations. *Comm. Statist. Stochastic Models*, **16**, 69–98.

Baccelli, F., Tchoumatchenko, K., and Zuyev, S. (2000*b*). Markov paths on the Poisson–Delaunay graph with applications to routing in mobile networks. *Adv. in Appl. Probab.*, **32**, 1–18.

Baumstark, V. and Last, G. (2007). Some distributional results for Poisson–Voronoi tessellations. *Adv. in Appl. Probab.*, **39**, 16–40.

Błaszczyszyn, B. and Schott, R. (2003). Approximate decomposition of some modulated-Poisson Voronoi tessellations. *Adv. in Appl. Probab.*, **35**, 847–862.

Brakke, K. A. (1985). Statistics of three dimensional random Voronoi tessellations. *Dept. of Math Sciences. Susquehanna University Selinsgrove, Pennsylvania*, 1–30.

Calka, P. (2001). Mosaïques poissoniennes de l'espace euclidian. Une extension d'un résultat de R. E. Miles. *C. R. Acad. Sci. Paris Sér. I Math.*, **332**, 557–562.

Calka, P. (2002). The distributions of the smallest disks containing the Poisson–Voronoi typical cell and the Crofton cell in the plane. *Adv. in Appl. Probab.*, **34**, 702–717.

Calka, P. (2003*a*). An explicit expression for the distribution of the number of sides of the typical Poisson–Voronoi cell. *Adv. in Appl. Probab.*, **35**, 863–870.

Calka, P. (2003*b*). Precise formulae for the distributions of the principal geometric characteristics of the typical cells of a two-dimensional Poisson–Voronoi tessellation and a Poisson line process. *Adv. in Appl. Probab.*, **35**, 551–562.

Calka, P. and Schreiber, T. (2005). Limit theorems for the typical Poisson–Voronoi cell and the Crofton cell with a large inradius. *Ann. Probab.*, **33**, 1625–1642.

Calka, P. and Schreiber, T. (2006). Large deviation probabilities for the number of vertices of random polytopes in the ball. *Adv. in Appl. Probab.*, **38**, 47–58.

Cowan, R. (1978). The use of the ergodic theorems in random geometry. *Adv. Appl. Probab.* (suppl.), 47–57. *Spatial patterns and processes (Proc. Conf., Canberra, 1977).*

Cowan, R. (1980). Properties of ergodic random mosaic processes. *Math. Nachr.*, **97**, 89–102.

Cowan, R., Quine, M., and Zuyev, S. (2003). Decomposition of gamma-distributed domains constructed from Poisson point processes. *Adv. in Appl. Probab.*, **35**, 56–69. In honor of Joseph Mecke.

Daley, D. J. and Vere-Jones, D. (1988). *An Introduction to the Theory of Point Processes.* Springer Series in Statistics. Springer-Verlag, New York.

Gilbert, E. N. (1962). Random subdivisions of space into crystals. *Ann. Math. Statist.*, **33**, 958–972.

Gloaguen, C., Fleischer, F., Schmidt, H., and Schmidt, V. (2006). Fitting of stochastic telecommunication network models via distance measures and Monte Carlo tests. *Telecommunications Systems*, **31**, 353–377.

Goldman, A. (1996). Le spectre de certaines mosaïques poissoniennes du plan et l'enveloppe convexe du pont brownien. *Probab. Theory Related Fields*, **105**, 57–83.

Goldman, A. (1998). Sur une conjecture de D. G. Kendall concernant la cellule de Crofton du plan et sur sa contrepartie brownienne. *Ann. Probab.*, **26**, 1727–1750.

Goldman, A. and Calka, P. (2003). On the spectral function of the Poisson–Voronoi cells. *Ann. Inst. H. Poincaré Probab. Statist.*, **39**, 1057–1082.

Heinrich, L. (1998). Contact and chord length distribution of a stationary Voronoĭ tessellation. *Adv. in Appl. Probab.*, **30**, 603–618.

Heinrich, L., Körner, R., Mehlhorn, N., and Muche, L. (1998). Numerical and analytical computation of some second-order characteristics of spatial Poisson–Voronoi tessellations. *Statistics*, **31**, 235–259.

Heinrich, L. and Muche, L. (2008). Second-order properties of the point process of nodes in a stationary Voronoi tessellation. *Math. Nachr.*, **281**, 350–375.

Heinrich, L., Schmidt, H., and Schmidt, V. (2005). Limit theorems for stationary tessellations with random inner cell structures. *Adv. in Appl. Probab.*, **37**, 25–47.

Heinrich, L., Schmidt, H., and Schmidt, V. (2006). Central limit theorems for Poisson hyperplane tessellations. *Ann. Appl. Probab.*, **16**, 919–950.

Hilhorst, H. J. (2005). Asymptotic statistics of the n-sided planar Poisson–Voronoi cell. I. Exact results. *J. Stat. Mech. Theory Exp.* (9), P09005, 45 pp. (electronic).

Hilhorst, H. J. (2006). Planar Voronoi cells: the violation of Aboav's law explained. *J. Phys. A*, **39**, 7227–7243.

Hilhorst, H. J. and Calka, P. (2008). Random line tessellations of the plane: statistical properties of many-sided cells. *J. Stat. Phys.*, **132**, 627–647.

Hug, D., Reitzner, M., and Schneider, R. (2004*a*). Large Poisson–Voronoi cells and Crofton cells. *Adv. in Appl. Probab.*, **36**, 667–690.

Hug, D., Reitzner, M., and Schneider, R. (2004*b*). The limit shape of the zero cell in a stationary Poisson hyperplane tessellation. *Ann. Probab.*, **32**, 1140–1167.

Hug, D. and Schneider, R. (2004). Large cells in Poisson–Delaunay tessellations. *Discrete Comput. Geom.*, **31**, 503–514.

Hug, D. and Schneider, R. (2005). Large typical cells in Poisson–Delaunay mosaics. *Rev. Roumaine Math. Pures Appl.*, **50**, 657–670.

Hug, D. and Schneider, R. (2007*a*). Asymptotic shapes of large cells in random tessellations. *Geom. Funct. Anal.*, **17**, 156–191.

Hug, D. and Schneider, R. (2007*b*). Typical cells in Poisson hyperplane tessellations. *Discrete Comput. Geom.*, **38**, 305–319.

Kovalenko, I. N. (1997). A proof of a conjecture of David Kendall on the shape of random polygons of large area. *Kibernet. Sistem. Anal.* (4), 3–10, 187.

Kovalenko, I. N. (1998). On certain random polygons of large areas. *J. Appl. Math. Stochastic Anal.*, **11**, 369–376.

Kovalenko, I. N. (1999). A simplified proof of a conjecture of D. G. Kendall concerning shapes of random polygons. *J. Appl. Math. Stochastic Anal.*, **12**, 301–310.

Lautensack, C. (2007). *Random Laguerre Tessellations.* Ph. D. thesis, Karlsruhe University.

Lautensack, C. and Zuyev, S. (2008). Random Laguerre tessellations. *Adv. in Appl. Probab.*, **40**, 630–650.

Maier, R., Mayer, J., and Schmidt, V. (2004). Distributional properties of the typical cell of stationary iterated tessellations. *Math. Methods Oper. Res.*, **59**, 287–302.

Maier, R. and Schmidt, V. (2003). Stationary iterated tessellations. *Adv. in Appl. Probab.*, **35**, 337–353.

Mecke, J. (1967). Stationäre zufällige Masse auf lokalkompakten Abelschen Gruppen. *Z. Wahrscheinlichkeitstheorie und Verw. Gebiete*, **9**, 36–58.

Mecke, J. (1975). Invarianzeigenschaften allgemeiner Palmscher Maße. *Math. Nachr.*, **65**, 335–344.

Mecke, J. and Muche, L. (1995). The Poisson Voronoi tessellation. I. A basic identity. *Math. Nachr.*, **176**, 199–208.

Miles, R. E. (1964*a*). Random polygons determined by random lines in a plane. *Proc. Nat. Acad. Sci. U.S.A.*, **52**, 901–907.

Miles, R. E. (1964*b*). Random polygons determined by random lines in a plane. II. *Proc. Nat. Acad. Sci. U.S.A.*, **52**, 1157–1160.

Miles, R. E. (1969). Poisson flats in Euclidean spaces. I. A finite number of random uniform flats. *Adv. in Appl. Probab.*, **1**, 211–237.

Miles, R. E. (1970). A synopsis of 'Poisson flats in Euclidean spaces'. *Izv. Akad. Nauk Armjan. SSR Ser. Mat.*, **5**, 263–285.

Miles, R. E. (1971). Poisson flats in Euclidean spaces. II. Homogeneous Poisson flats and the complementary theorem. *Adv. in Appl. Probab.*, **3**, 1–43.

Miles, R. E. (1973). The various aggregates of random polygons determined by random lines in a plane. *Advances in Math.*, **10**, 256–290.

Miles, R. E. (1995). A heuristic proof of a long-standing conjecture of D. G. Kendall concerning the shapes of certain large random polygons. *Adv. in Appl. Probab.*, **27**, 397–417.

Miles, R. E. and Maillardet, R. J. (1982). The basic structures of Voronoï and generalized Voronoï polygons. *J. Appl. Probab.* (Special Vol. 19A), 97–111. Essays in statistical science.

Møller, J. (1986). *Random Tessellations in* $\mathbf{R}^d$, Volume 9 of *Memoirs*. Aarhus University Institute of Mathematics Department of Theoretical Statistics, Aarhus.

Møller, J. (1989). Random tessellations in $\mathbf{R}^d$. *Adv. in Appl. Probab.*, **21**, 37–73.

Møller, J. (1992). Random Johnson–Mehl tessellations. *Adv. in Appl. Probab.*, **24**, 814–844.

Møller, J. (1994). *Lectures on Random Voronoï Tessellations*, Volume 87 of *Lecture Notes in Statistics*. Springer-Verlag, New York.

Muche, L. (1996). The Poisson–Voronoï tessellation. II. Edge length distribution functions. *Math. Nachr.*, **178**, 271–283.

Muche, L. (2005). The Poisson–Voronoi tessellation: relationships for edges. *Adv. in Appl. Probab.*, **37**, 279–296.

Muche, L. and Stoyan, D. (1992). Contact and chord length distributions of the Poisson Voronoï tessellation. *J. Appl. Probab.*, **29**, 467–471.

Nagel, W. and Weiss, V. (2003). Limits of sequences of stationary planar tessellations. *Adv. in Appl. Probab.*, **35**, 123–138. In honor of Joseph Mecke.

Nagel, W. and Weiss, V. (2004). Crack STIT tessellations – existence and uniqueness of tessellations that are stable with respect to iterations. *Izv. Nats. Akad. Nauk Armenii Mat.*, **39**, 84–114.

Nagel, W. and Weiss, V. (2005). Crack STIT tessellations: characterization of stationary random tessellations stable with respect to iteration. *Adv. in Appl. Probab.*, **37**, 859–883.

Neveu, J. (1977). Processus ponctuels. In *École d'Été de Probabilités de Saint-Flour, VI–1976*, pp. 249–445. Lecture Notes in Math., Vol. 598. Springer-Verlag, Berlin.

Okabe, A., Boots, B., Sugihara, K., and Chiu, S. N. (2000). *Spatial Tessellations: Concepts and Applications of Voronoi Diagrams* (Second edn). Wiley Series in Probability and Statistics. John Wiley & Sons Ltd., Chichester. With a foreword by D. G. Kendall.

Paroux, K. (1998). Quelques théorèmes centraux limites pour les processus Poissoniens de droites dans le plan. *Adv. in Appl. Probab.*, **30**, 640–656.

Reitzner, M. (2003). Random polytopes and the Efron–Stein jackknife inequality. *Ann. Probab.*, **31**, 2136–2166.

Reitzner, M. (2005). Central limit theorems for random polytopes. *Probab. Theory Related Fields*, **133**, 483–507.

Rényi, A. and Sulanke, R. (1963). Über die konvexe Hülle von n zufällig gewählten Punkten. *Z. Wahrscheinlichkeitstheorie und Verw. Gebiete*, **2**, 75–84 (1963).

Schlather, M. (2000). A formula for the edge length distribution function of the Poisson Voronoi tessellation. *Math. Nachr.*, **214**, 113–119.

Schneider, R. and Weil, W. (2000). *Stochastische Geometrie.* Teubner Skripten zur Mathematischen Stochastik. [Teubner Texts on Mathematical Stochastics]. B. G. Teubner, Stuttgart.

Schreiber, T. (2002). Variance asymptotics and central limit theorems for volumes of unions of random closed sets. *Adv. in Appl. Probab.*, **34**, 520–539.

Schreiber, T. (2003). Asymptotic geometry of high-density smooth-grained Boolean models in bounded domains. *Adv. in Appl. Probab.*, **35**, 913–936.

Siegel, A. F. (1978). Random space filling and moments of coverage in geometrical probability. *J. Appl. Probab.*, **15**, 340–355.

Siegel, A. F. and Holst, L. (1982). Covering the circle with random arcs of random sizes. *J. Appl. Probab.*, **19**, 373–381.

Stevens, W. L. (1939). Solution to a geometrical problem in probability. *Ann. Eugenics*, **9**, 315–320.

Stoyan, D., Kendall, W. S., and Mecke, J. (1987). *Stochastic Geometry and its Applications.* Wiley Series in Probability and Mathematical Statistics: Applied Probability and Statistics. John Wiley & Sons Ltd., Chichester. With a foreword by D. G. Kendall.

Vu, V. H. (2005). Sharp concentration of random polytopes. *Geom. Funct. Anal.*, **15**, 1284–1318.

Wiener, N. (1939). The ergodic theorem. *Duke Math. J.*, **5**, 1–18.

Zuyev, S. (1999). Stopping sets: gamma-type results and hitting properties. *Adv. in Appl. Probab.*, **31**, 355–366.

Zuyev, S. A. (1992). Estimates for distributions of the Voronoĭ polygon's geometric characteristics. *Random Structures Algorithms*, **3**, 149–162.

PART II

STOCHASTIC GEOMETRY AND MODERN PROBABILITY

6

PERCOLATION AND RANDOM GRAPHS

Remco van der Hofstad

Abstract

In this chapter, we define percolation and random graph models, and survey the features of these models.

6.1 Introduction and notation

In this section, we discuss random networks. In Section 6.2, we study *percolation*, which is obtained by independently removing vertices or edges from a graph. Percolation is a model of a porous medium, and is a paradigm model of statistical physics. Think of the bonds in an infinite graph that are kept as indicating whether water can flow through this part of the medium. Then, the interesting question is whether water can *percolate*, or, alternatively, whether there is an infinite connected component of bonds that are kept? As it turns out, the answer to this question depends sensitively on the fraction of bonds that are kept. When we keep most bonds, then the kept or occupied bonds form most of the original graph. In particular, an infinite connected component may exist, and if this happens, we say that the system *percolates*. On the other hand, when most bonds are removed or vacant, then the connected components tend to be small and insignificant. Thus, percolation admits a *phase transition*. Despite the simplicity of the model, the results obtained up to date are far from complete, and many aspects of percolation, particularly of its critical behaviour, are only poorly understood. In Section 6.2 we shall discuss the basics of percolation, and highlight some important open questions. The key challenge in percolation is to uncover the relation between the percolation critical behaviour and the properties of the underlying graph from which we obtain percolation by removing edges.

In Section 6.3, we discuss *random graphs*. While in percolation, the random network considered naturally lives on an *infinite graph*, in random graph theory one considers random *finite* graphs. Thus, all random graphs are obtained by removing edges from the complete graph, or by adding edges to an empty graph. An important example of a random graph is obtained by *independently* removing bonds from a finite graph, which makes it clear that there is a strong link to percolation. However, also other mechanisms are possible to generate a random graph. We shall discuss some of the basics of random graph theory, focussing on the phase transition of the largest connected component and the distances in random graphs. The random graph models studied here are inspired by applications, and we shall highlight real-world networks that these random graphs aim to model to some extent.

The fields that this contribution covers, percolation and random graph theory, have attracted tremendous attention in the past decades, and enormous progress has been made. It is impossible to cover all material appearing in the literature, and we believe that one should not aim to do so. Thus, we have strived to cover the main results which have been proved, as well as recent results in which we expect that more progress shall be made in the (near?) future, and we list open problems which we find of interest. We have tried to give particular attention to results that are of interest to the stochastic geometry community, thus giving detailed accounts of the recent progress on two-dimensional percolation and percolation on tessellations, as well as on continuum percolation and random geometric graphs and its relations to telecommunications. For a more detailed discussion of telecommunication applications in stochastic geometry, we refer to Chapter 16 and the references therein. For a specific class of spatial random networks, the so-called *random directed and on-line networks,* we refer the reader to Chapter 7. We hope that this contribution gives an idea of the breadth and depth of percolation and random graphs, as well as on the challenges for the future.

We now start by introducing some notation. Let $\mathbb{G} = (\mathbb{V}, \mathbb{E})$ be a graph, where $\mathbb{V}$ is the vertex set and $\mathbb{E} \subseteq \mathbb{V} \times \mathbb{V}$ is the edge set. For percolation, the number of vertices, denoted by $|\mathbb{V}|$, is naturally infinite, while for random graphs, $|\mathbb{V}|$ is naturally finite. A random network is obtained by a certain rule that determines which subset of the edges $\mathbb{E}$ is occupied, the remaining edges being vacant. Let $v \in \mathbb{V}$, and denote by $C(v)$ the set of vertices which can be reached from v by occupied edges. More precisely, for $u \in \mathbb{V}$, we say that $u \longleftrightarrow v$ when there exists a path of occupied edges that connects u and v, and we write

$$C(v) = \{u \in \mathbb{V} : u \longleftrightarrow v\}. \tag{6.1}$$

The central question in the study of random networks involves the cluster size distributions, i.e. for percolation whether there exists an infinite connected component, and for random graphs what the distribution is of the largest connected component.

6.2 Percolation

In this section, we discuss percolation. For more details, see the books by Grimmett (1999), Hughes (1996), Kesten (1982) and Bollobás and Riordan (2006*b*). For an expository account of recent progress with a long list of open problems, we refer to Kesten (2002). There is a nice account of the history of percolation in Hughes (1996, Section 1.1.1), the foundation of percolation as a mathematical discipline being generally ascribed to Broadbent and Hammersley (1957).

Introduction of the model In this section, $\mathbb{G} = (\mathbb{V}, \mathbb{E})$ shall denote an infinite graph. We shall assume that $\mathbb{G} = (\mathbb{V}, \mathbb{E})$ is *transitive*, i.e., the neighbourhoods of all points are the same. More precisely, transitivity means that for every $x, y \in \mathbb{V}$, there exists a bijection $\phi \colon \mathbb{V} \to \mathbb{V}$ for which $\phi(x) = y$ and $\{\phi(u), \phi(v)\} \in \mathbb{E}$ precisely when $\{u, v\} \in \mathbb{E}$. Such a bijection $\phi \colon \mathbb{V} \to \mathbb{V}$ is called an *automorphism.* In particular, transitivity of a graph $\mathbb{G}$ implies that each vertex has the same degree (which could possibly be infinite). We shall denote the degree of $\mathbb{G}$ by r. We sometimes assume the weaker condition of *quasi-transitivity,* which means that there is a *finite* set of vertices such that for each vertex v, there is a graph automorphism of the graph which maps v to one of the vertices in the finite set. We note that if a graph is quasi-transitive and each vertex has a bounded degree, then the degrees are uniformly bounded.

Bond percolation is obtained by independently removing each of the bonds or edges with a fixed probability $1 - p$. Thus, each edge is *occupied* with probability p, and *vacant* otherwise, and the edge statuses are independent. We shall restrict to the setting where the probability that an edge is occupied is fixed. In the literature, also the setting is studied where $\mathbb{E} = \mathbb{V} \times \mathbb{V}$ and, for $b \in \mathbb{E}$, the probability that b is occupied depends on b in a translation invariant way. For simplicity, we refrain from this generality. The resulting probability measure is denoted by $\mathbf{P}_p$, and $\mathbf{E}_p$ denotes expectation w.r.t. $\mathbf{P}_p$.

We define the *percolation function* $p \mapsto \theta(p)$ by

$$\theta(p) = \mathbf{P}_p\{|C(v)| = \infty\}, \tag{6.2}$$

where $v \in \mathbb{V}$ is an arbitrary vertex. By transitivity, the above probability does not depend on the choice of v. We shall therefore often write $C = C(o)$ where $o \in \mathbb{V}$ is an appropriately chosen origin. When $\mathbb{G}$ is quasi-transitive, then $\theta^v(p) = \mathbf{P}_p\{|C(v)| = \infty\}$ possibly depends on $v \in \mathbb{V}$, but these is a finite collection of vertices V_0 such that for every v there exists a $v_0 \in V_0$ such that $\theta^v(p) = \theta^{v_0}(p)$.

When $\theta(p) = 0$, then the probability that o is inside an infinite connected component is 0, so that there will be no infinite connected component a.s. When $\theta(p) > 0$, on the other hand, then, by ergodicity, the proportion of vertices in infinite connected components equals $\theta(p) > 0$, and we say that the system *percolates*. For quasi-transitive graphs, if $\theta^v(p) = 0$ for *some* v, then, in fact, $\theta^v(p) = 0$ for *all* v.

We define the *critical value* by

$$p_c = p_c(\mathbb{G}) = \inf\{p : \theta(p) > 0\}. \tag{6.3}$$

For quasi-transitive graphs, it might appear that this definition depends on the choice of v for which $\theta(p) = \theta^v(p)$. However, since $\theta^v(p) = 0$ for some v implies that $\theta^v(p) = 0$ for all v, in fact, $p_c(\mathbb{G})$ is independent of the choice of v. The above critical value is sometimes written as $p_c(\mathbb{G}) = p_{\scriptscriptstyle H}(\mathbb{G})$ in honour of Hammersley, who defined it in Hammersley (1961).

An important question is whether the critical value is non-trivial, i.e. whether $p_c \in (0, 1)$. We shall study this question in detail below. When $\theta(p) > 0$, then a natural question is *how many* infinite connected components there can be. Denote this number of infinite connected components by N. We shall now first prove that, for any infinite graph, $N \in \{0, 1, \infty\}$ a.s., and that N is constant a.s. Indeed, by ergodicity arguments and the fact that N is a *translation invariant* random variable, $N = k$ almost surely for some $k \in \{0, 1, 2, \ldots\} \cup \{\infty\}$. Moreover, since N is a.s. constant, it is not changed when we change the status of a *finite* number of edges. Indeed, for $B \subseteq \mathbb{E}$, let $N_B(0)$ denote the number of infinite components when all edges in B are declared vacant, and $N_B(1)$ the number of infinite components when all edges in B are declared occupied. Then $N_B(0) = N_B(1) = k$ a.s. When $k < \infty$, $N_B(0) = N_B(1) = k$ only when B intersects *at most one* infinite connected component, and we conclude that the number of infinite connected components which intersect B is at most 1 a.s. When $B \uparrow \mathbb{V}$, this number increases to N, and we arrive at the claim that if $k < \infty$, then $k \leq 1$. This completes the proof that $N = k$ a.s., for some $k \in \{0, 1, \infty\}$.

Instead of considering bond percolation, one can also study *site percolation*, for which we independently and with fixed probability $1 - p$ remove the vertices in $\mathbb{V}$, and we are only allowed to make use of edges for which *both* endpoints are kept. In the literature (see e.g. Grimmett 1999), the main focus has been on bond percolation, despite the fact that, as we shall show now, site percolation is more general. Indeed, we shall show that for each bond percolation model indicated by $\mathbb{G}$, there exists a site percolation model, denoted by $\mathbb{G}_s$, which is equivalent to bond percolation on $\mathbb{G}$. Indeed, we take $\mathbb{V}_s$ to contain the edges in $\mathbb{E}$, and say that $\{a_s, b_s\} \in \mathbb{E}_s$ precisely when, in $\mathbb{G}$, the edges to which a_s and b_s are identified share a common endpoint. In this setting, bond percolation on $\mathbb{G}$ becomes site percolation on $\mathbb{G}_s$, and the connected component of $v \in \mathbb{V}$ is infinite precisely when there exists an edge $b \in \mathbb{E}$ such that the connected component of $b \in \mathbb{V}_s$ is infinite. The reverse is not true, i.e., not every site percolation model is equivalent to a bond percolation model. In this paper, we shall restrict to bond percolation, having in mind that almost all arguments can be straightforwardly adapted to the site percolation setting, possibly except for certain *duality* arguments which play an important role in two dimensions, and are described in more detail in Section 6.2.4. Interestingly, already in Hammersley (1961) it was shown that the critical value for bond percolation is never larger than the one for site percolation.

It is useful to note that obviously percolation is *monotone* in the parameter p. To make this notion precise, we say that an event E is *increasing*, when for every $\omega \in E$ and $\eta \geq \omega$, where $\eta \geq \omega$ when $\eta(e) \geq \omega(e)$ for every edge $e \in \mathbb{E}$, we have that also $\eta \in E$. For example, the event $\{|C(v)| = \infty\}$ is increasing. Then, we can couple percolation for all probabilities p simultaneously as follows. We let $\{U_e\}_{e \in \mathbb{E}}$ be i.i.d. uniform random variables, and note that percolation with percolation probability p is obtained by declaring an edge e occupied precisely when $U_e \leq p$ (see Hammersley 1963). This implies that when $p_1 \leq p_2$ and E an arbitrary increasing event, we have that

$$\mathbf{P}_{p_1}(E) \leq \mathbf{P}_{p_2}(E). \tag{6.4}$$

In particular, we obtain that $\theta(p_1) \leq \theta(p_2)$, i.e., $p \mapsto \theta(p)$ is non-decreasing. We also say that a random variable X is increasing when $\{X \geq x\}$ is increasing for all $x \in \mathbb{R}$.

We mention two inequalities that play a profound role in percolation theory, namely the FKG and BK inequalities. The FKG inequality in the context of percolation is called the *Harris inequality* and was first proved in Harris (1960). The more general FKG inequality, which, for example, also applies to the Ising model, was derived in Fortuin, Kasteleyn and Ginibre (1971). The Harris inequality states that for two increasing events E and F,

$$\mathbf{P}_p(E \cap F) \geq \mathbf{P}_p(E)\mathbf{P}_p(F), \tag{6.5}$$

the FKG inequality gives the same conclusion under weaker assumptions on the measure involved. In words, for increasing events E and F, the occurrence of E makes the simultaneous occurrence of F more likely. For example, the FKG inequality yields that, for every $x, y, u, v, \in \mathbb{V}$, we have

$$\mathbf{P}_p\{x \longleftrightarrow y, u \longleftrightarrow v\} \geq \mathbf{P}_p\{x \longleftrightarrow y\}\mathbf{P}_p\{u \longleftrightarrow v\}. \tag{6.6}$$

The intuition for the FKG inequality is that if the increasing event E holds, then this makes it more likely for edges to be occupied, and, therefore, it becomes more likely that the increasing event F also holds. Thus, $\mathbf{P}_p(F|E) \geq \mathbf{P}_p(F)$, which is equivalent to (6.5). See Häggström (2007) for a Markov chain proof of the FKG-inequality.

The BK-inequality gives, in a certain sense, an opposite inequality. We shall only state it in the case of increasing events, for which it was proved in van den Berg and Kesten (1985). The most general version is proved in Reimer (2000). For $K \subseteq \mathbb{E}$ and $\omega \in \{0,1\}^{\mathbb{E}}$, we write $\omega_K(e) = \omega(e)$ for $e \in K$, and $\omega_K(e) = 0$ otherwise. Let E and F again be increasing events, and write

$$E \circ F = \{\omega : \exists K \subseteq \mathbb{E} \text{ such that } \omega_K \in E, \omega_{K^c} \in F\}. \tag{6.7}$$

Then, the van den Berg-Kesten (BK) inequality states that

$$\mathbf{P}_p(E \circ F) \leq \mathbf{P}_p(E)\mathbf{P}_p(F). \tag{6.8}$$

For example, the event $\{x \longleftrightarrow y\} \circ \{u \longleftrightarrow v\}$ is the event that there are edge-disjoint occupied paths from x to y and from u to v, and (6.8) implies that

$$\mathbf{P}_p\big\{\{x \longleftrightarrow y\} \circ \{u \longleftrightarrow v\}\big\} \leq \mathbf{P}_p\{x \longleftrightarrow y\}\mathbf{P}_p\{u \longleftrightarrow v\}. \tag{6.9}$$

Intuitively, this can be understood by noting that, if $x \longleftrightarrow y$ and $u \longleftrightarrow v$ must occur *disjointly*, then we can first fix an occupied path connecting x and y in a certain arbitrary manner, and remove the occupied edges used in this path. Then $\{x \longleftrightarrow y\} \circ \{u \longleftrightarrow v\}$ occurs when in the configuration with the edges removed, we still have that $u \longleftrightarrow v$. Since we have removed the edges in the occupied path from x to y, this event now has smaller probability than $\mathbf{P}_p\{u \longleftrightarrow v\}$.

Many objects we shall study are increasing or decreasing. For example, $|C(v)|$ is obviously increasing. The number of infinite connected components N is an example of a random variable that is neither increasing nor decreasing, and we shall see the complications of this fact later on.

We next discuss an important tool to study probabilities which goes under the name of Russo's formula (Russo, 1981). Let E be an increasing event. Then we say that the bond (u, v) is *pivotal* for the event E when E occurs when the status of (u, v) in the (possibly modified) configuration where (u, v) is turned occupied, while E does not occur in the (possibly modified) configuration where (u, v) is turned vacant. Thus, the bond (u, v) is essential for the occurrence of the event E. The set of pivotal bonds for an event is *random*, as it depends on which other bonds are occupied and vacant in the configuration. Russo's Formula states that for every increasing event E which depends on a *finite number of bonds*,

$$\frac{\mathrm{d}}{\mathrm{d}p}\mathbf{P}_p(E) = \sum_{(u,v)\in\mathbb{E}} \mathbf{P}_p\{(u, v) \text{ is pivotal for } E\}. \tag{6.10}$$

Russo's formula allows us to study how the probability of an event changes as p varies. The fact that (6.10) is only valid for events that only depend on a finite number of bonds is a nuisance, and there are many settings in which Russo's formula can be extended to events depending on infinitely many bonds by an appropriate cutting procedure.

We shall be interested in several key functions that describe the connections in bond percolation. The *susceptibility* $p \mapsto \chi(p)$ is the expected cluster size

$$\chi(p) = \mathbf{E}_p|C(v)|. \tag{6.11}$$

Clearly, we have that $\chi(p) = \infty$ for $p > p_c$, since then, with probability $\theta(p) > 0$, we have that $|C(v)| = \infty$. Further, $p \mapsto \chi(p)$ is clearly increasing. Define the critical value $p_T = p_T(\mathbb{G})$ by

$$p_T = \sup\{p : \chi(p) < \infty\}. \tag{6.12}$$

The subscript T in $p_T(\mathbb{G})$ is in honour of H.N.V. Temperley.

A natural question is whether $p_T(\mathbb{G}) = p_c(\mathbb{G})$, i.e., is $\chi(p) < \infty$ for every $p < p_c$? The latter indeed turns out to be true by the celebrated results independently proved by Menshikov (1986) and Aizenman and Barsky (1987), as we shall discuss in more detail below.

For $p \in [0, 1]$, let

$$\chi^{\mathrm{f}}(p) = \mathbf{E}_p[|C(v)|\mathbf{1}_{\{|C(v)|<\infty\}}] \tag{6.13}$$

Also, we shall often work with the related two-point function

$$\tau_p(x, y) = P_p\{x \longleftrightarrow y\}.$$

denote the mean finite cluster size. Clearly, for $p < p_C = p_H$, we have that $\chi^{\mathrm{f}}(p) = \chi(p)$, but for $p > p_C = p_H$, this may not be true. We define the *finite two point function* $\tau_p^{\mathrm{f}}\colon \mathbb{V}\times\mathbb{V}\to[0,1]$ by

$$\tau_p^{\mathrm{f}}(x,y)=\mathbf{P}_p\{x\longleftrightarrow y,|C(x)|<\infty\}. \tag{6.14}$$

On $\mathbb{Z}^d$, when the model is translation invariant, we have that $\tau_p^{\mathrm{f}}(x,y)=\tau_p^{\mathrm{f}}(y-x,0)\equiv\tau_p^{\mathrm{f}}(y-x)$. Also for transitive $\mathbb{G}$, the two-point function is characterized by $\tau_p^{\mathrm{f}}(x)=\tau_p^{\mathrm{f}}(o,x)$, where o is an appropriately chosen origin in the model. In terms of $\tau_p^{\mathrm{f}}(x)$, we can identify $\chi^{\mathrm{f}}(p)$ as

$$\chi^{\mathrm{f}}(p)=\sum_{x\in\mathbb{V}}\tau_p^{\mathrm{f}}(x). \tag{6.15}$$

We shall also be interested in the *mean number of clusters per vertex* $\kappa(p)$, which is defined as

$$\kappa(p)=\mathbf{E}_p[1/|C(v)|]. \tag{6.16}$$

The significance of $\kappa(p)$ is that it measures the average number of connected components in large volumes. Indeed, let $B(n)=\{x\in\mathbb{V}: d(o,x)\leq n\}$ denote a ball of radius n around o, where $d(x,y)$ denotes an appropriately chosen distance function on $\mathbb{G}$. Then, with C_n the number of different connected components obtained when making every edges not entirely in $B(n)$ vacant, and when $\partial B(n)=o(B(n))$, $C_n/|B(n)|\to\kappa(p)$ a.s.

An important measure of the *spatial extent* of clusters is the *correlation length* $\xi(p)$ defined by

$$\xi(p)=\sqrt{\frac{1}{\chi^{\mathrm{f}}(p)}\sum_{x\in\mathbb{V}}|x|^2\tau_p^{\mathrm{f}}(x)}, \tag{6.17}$$

where we write $|x|=d(o,x)$ for the distance of x to an appropriately chosen origin $o\in\mathbb{V}$. For many graphs $\mathbb{G}$, there are several ways of defining the correlation length, many of them being equivalent in the sense that they are bounded above and below by finite and positive constants times $\xi(p)$ defined in (6.17). The correlation length measures the *dependence* between finite clusters at a given distance. If $d(x,y)\gg\xi(p)$ and, for $p>p_c$, x and y are in different finite clusters, then we can think of $C(x)$ and $C(y)$ as being close to independent, while if $d(x,y)\ll\xi(p)$, then $C(x)$ and $C(y)$ are quite dependent.

6.2.1 *Critical behaviour*

The behaviour of percolation models is most interesting and richest for p values which are close to the critical value. Clearly, the precise value of $p_c(\mathbb{G})$ depends sensitively on the nature of the underlying graph $\mathbb{G}$. By drawing an analogy to physical systems, physicists predict that the behaviour of percolative systems close to criticality is rather insensitive to the precise details of the model, and it is only characterized by the macroscopic behaviour. Thus, percolation is expected to behave in a *universal* manner. For example, it is predicted that the critical nature of finite-range percolation systems on $\mathbb{Z}^d$, under suitable symmetry conditions, is similar in the sense that the nature of the critical behaviour are similar. While this prediction is far from rigorous, it does offer us a way of summarizing percolation models by only looking at their simplest examples. One of the key challenges of percolation theory is to make rigorous sense of this universality paradigm.

We shall now make the notion of universality more tangible, by discussing *critical exponents*. The critical nature of many physical systems is believed to be characterized by the validity of *power laws*, the exponent of which is a robust or universal measure of the underlying critical behaviour. We start by giving an example of a critical exponent. It is predicted that

$$\theta(p) \sim (p - p_c)^\beta \quad \text{as} \quad p \downarrow p_c, \tag{6.18}$$

for some $\beta > 0$. The value of β is expected to be different for different $\mathbb{G}$, but (6.18) remains valid. The symbol $\sim$ in (6.18) can have several meanings, which we now elaborate on. We say that the critical exponent β exists in the *logarithmic form* if

$$\lim_{p \downarrow p_c} \frac{\log \theta(p)}{\log (p - p_c)} = \beta, \tag{6.19}$$

while β exists in the *bounded-ratios form* if there exist $0 < c_1 < c_2 < \infty$ such that, uniformly for $p \geq p_c$,

$$c_1(p - p_c)^\beta \leq \theta(p) \leq c_2(p - p_c)^\beta. \tag{6.20}$$

Finally, we say that β exists in the *asymptotic form* if, as $p \downarrow p_c$, there exists a $c > 0$ such that

$$\theta(p) = c(p - p_c)^\beta(1 + o(1)). \tag{6.21}$$

The existence of a critical exponent is *a priori* unclear, and needs a mathematical proof. Unfortunately, in general such a proof is missing, and we can only give proofs of the existence in special cases, which we shall discuss below in quite some detail. Indeed, the existence of the critical exponent $\beta > 0$ is stronger than continuity of $p \mapsto \theta(p)$, which is unknown in general, and is arguably the holy grail of percolation theory. Indeed, $p \mapsto \theta(p)$ is clearly continuous on $[0, p_c)$, and it is also continuous (and even infinitely differentiable) on $(p_c, 1]$ by the results of van den Berg and Keane (1984) (for infinite differentiability of $p \mapsto \theta(p)$ for $p \in (p_c, 1]$, see Russo 1978). Thus, continuity of $p \mapsto \theta(p)$ is equivalent to the statement that $\theta(p_c(\mathbb{G})) = 0$.

We now introduce several more critical exponents. The critical exponent γ for the expected cluster size is given by

$$\chi^{\mathrm{f}}(p) \sim |p - p_c|^{-\gamma}, \quad p \to p_c. \tag{6.22}$$

More precisely, we can think of (6.22) as defining the critical exponents $\gamma, \gamma' > 0$ defined by

$$\chi(p) \sim (p - p_c)^{-\gamma}, \quad p \uparrow p_c, \qquad \chi^{\mathrm{f}}(p) \sim (p - p_c)^{-\gamma'}, \quad p \downarrow p_c, \tag{6.23}$$

with the predicted equality $\gamma = \gamma'$. For (6.22) and (6.23), we are implicitly assuming that $p_T(\mathbb{G}) = p_c(\mathbb{G})$, this equality shall be discussed in more detail below.

Further, ν, ν' are defined by

$$\xi(p) \sim (p - p_c)^{-\nu}, \quad p \uparrow p_c, \qquad \xi(p) \sim (p - p_c)^{-\nu'}, \quad p \downarrow p_c, \tag{6.24}$$

again with the prediction that $\nu = \nu'$. The exponent $-1 \leq \alpha < 0$ is defined by the blow up of the third derivative of $p \mapsto \kappa(p)$ at $p = p_c$, i.e.

$$\kappa'''(p) \sim |p - p_c|^{-1-\alpha}, \tag{6.25}$$

while the *gap exponent* $\Delta > 0$ is defined by, for $k \geq 1$,

$$\frac{\mathbf{E}_p\left[|C|^{k+1}\mathbf{1}_{\{|C|<\infty\}}\right]}{\mathbf{E}_p\left[|C|^{k+1}\mathbf{1}_{\{|C|<\infty\}}\right]} \sim |p - p_c|^{-\Delta}. \tag{6.26}$$

Also α and Δ can be defined, similarly to (6.23), as an exponent for $p \uparrow p_c$ and one for $p \downarrow p_c$, the values being equal, and we shall always use the convention that α and Δ denote the $p \uparrow p_c$ versions, while α' and Δ' denote the $p \downarrow p_c$ versions.

As mentioned before, it is highly unclear that these critical exponents are well-defined, and that the value of $\Delta > 0$ does not depend on k. However, there are good physical reasons why these exponents are defined as they are. The exponents $\beta, \gamma, \nu, \alpha, \Delta$ can be thought of as *approach exponents* which measure the blow-up of various aspects of the cluster size as p approaches the critical value $p = p_c(\mathbb{G})$.

We finally define three critical exponents at criticality. The exponent $\delta \geq 1$ measures the power-law exponent of the critical cluster tail, i.e.

$$\mathbf{P}_{p_c}\{|C| \geq n\} \sim n^{-1/\delta}, \quad n \to \infty, \tag{6.27}$$

the assumption that $\delta \geq 1$ following from the prediction that $\chi(p_c) = \infty$. Further, we define the exponent $\rho > 0$ by

$$\mathbf{P}_{p_c}\{o \longleftrightarrow \partial B(n)\} \sim n^{-1/\rho}, \quad n \to \infty. \tag{6.28}$$

Finally, η is defined by

$$\mathbf{E}_{p_c}\left[|C \cap B(n)|\right] = \sum_{x:|x| \leq n} \tau_{p_c}(x) \sim n^{2-\eta}, \quad n \to \infty, \tag{6.29}$$

where we recall that $|x| = d(o, x)$ is the distance of x to o in $\mathbb{G}$.

The above definitions give rise to eight critical values that each describe a certain aspect of the (near-)critical behaviour of the percolation system. Several relations between these critical exponents are predicted by (non-rigorous) physics argument, and go under the name of *scaling relations.* These scaling relations assert that

$$\gamma + 2\beta = \beta(\delta + 1) = 2 - \alpha = \Delta + \beta, \qquad \gamma = \nu(2 - \eta). \tag{6.30}$$

The validity of the scaling relations in (6.30) is widely accepted, but few proofs exist, in particular since the existence of the critical exponents is in general unknown. We can intuitively understand the scaling relations in (6.30) by assuming the existence of certain *scaling functions*, which describe certain percolation quantities close to criticality. An example is to assume that there exist scaling functions f_+ and f_-, as well as some exponents $\sigma, \tau > 0$, such that

$$\mathbf{P}_p\{|C| = n\} \sim \begin{cases} n^{-\sigma} f_-(n/\xi(p)^\tau) & \text{for} \quad p \le p_c \\ n^{-\sigma} f_+(n/\xi(p)^\tau) & \text{for} \quad p \ge p_c, \end{cases} \tag{6.31}$$

for some sufficiently smooth and integrable functions f_+ and f_-. When working out the consequences of (6.31), we can see that it must imply the first three scaling relations in (6.30), and when also assuming that a scaling function exists for $x \mapsto \tau_p^{\mathrm{f}}(x)$, the relation $\gamma = \nu(2-\eta)$ follows. The existence of scaling functions is, as far as we know, unknown except for percolation on a tree. Nice discussions of scaling functions can be found in Grimmett (1999, Sections 9.1-9.2) or Hughes (1996, Section 4.2). The most complete reference to results on critical exponents until 1996 is Hughes (1996, Chapter 4).

Percolation is a paradigm model in statistical physics. As discussed before, a central notion in this field is *universality.* An example of universality in the setting of percolation is the prediction that for any fixed d, any finite-range percolation model on $\mathbb{Z}^d$ has the *same* critical behaviour. In particular, it has the same critical exponents and these critical exponents only depend on d. While universality is quite plausible when describing real physical systems from the viewpoint of statistical physics, and while universality is a very useful notion since it allows us to study only the simplest finite-range model available, there are very few examples where universality can be rigorously proved. We shall discuss a few universality results below.

So far, we have discussed percolation in full generality. We shall now treat examples of percolation models. In Section 6.2.2, we shall describe percolation on regular trees, in Section 6.2.3 we discuss percolation on $\mathbb{Z}^d$ for general d, in Section 6.2.4 we specialize to the two-dimensional setting, in Section 6.2.5, we study the high-dimensional case for which $d > 6$ and in Section 6.2.6, we study oriented or directed percolation. Finally, in Section 6.2.7, we study the case of percolation on non-amenable graphs and we close this section in Section 6.2.8 by discussing continuum percolation and its applications.

6.2.2 *Percolation on the regular tree*

In this section, we study percolation on the regular tree. Let $\mathbb{T}_r$ denote the r-regular tree of degree r. The advantage of trees is that they do not contain cycles, which makes explicit computations possible. In order to compute the critical exponents ν and η, we first identify

$$|x| = \sqrt{h(x)}, \tag{6.32}$$

where $h(x)$ is the *height* of the vertex x in the tree, i.e. the length of the shortest path linking o and x, so that (6.32) is the *Euclidean distance* in the tree. We shall first prove that the critical exponents for percolation on a regular tree exist and identify their values in the following theorem:

Theorem 6.1 (Critical behaviour on the r-regular tree) *On the r-regular tree $\mathbb{T}_r$, $p_c = p_T = 1/(r-1)$, and $\beta = \gamma = \gamma' = 1$, $\delta = \Delta = \Delta' = \rho = 2$, $\nu = \nu' = 1/2$ and $\alpha = \alpha' = -1$ in the asymptotic sense.*

Proof We shall make substantial use of the fact that percolation on a tree can be described in terms of *branching processes.* Recall that o is the root of the tree. For $x \neq o$, we write $C_{\rm BP}(x)$ for the forward cluster of x, i.e. the vertices $y \in \mathbb{T}_r$ which are connected to x and for which the unique path from x to y only moves away from the root o. Then, clearly,

$$|C(o)| = 1 + \sum_{e \sim o} I_{(o,e)} |C_{\rm BP}(e)|, \tag{6.33}$$

where the sum is over all neighbours e of o, $I_{(o,e)}$ is the indicator that the edge (o, e) is occupied, and $C_{\rm BP}(e)$ is the forward cluster of e. The random vector $\{I_{(o,e)}\}_{e\sim o}$ forms a collection of r independent Bernoulli random variables with success probability p, and $\{|C_{\rm BP}(e)|\}_{e\sim o}$ is an i.i.d. sequence independent of $\{I_{(o,e)}\}_{e\sim o}$. Equation (6.33) allows us to deduce all information concerning $|C(o)|$ from the information of $|C_{\rm BP}(e)|$. Also, for each e, $|C_{\rm BP}(x)|$ satisfies the formula

$$|C_{\rm BP}(x)| = 1 + \sum_{e\sim x: h(e)>h(x)} I_{(x,e)} |C_{\rm BP}(e)|, \tag{6.34}$$

where $h(x)$ is the height of x in $\mathbb{T}_r$, and $\{|C_{\rm BP}(e)|\}_{e\sim x:h(e)>h(x)}$ is a set of $r-1$ independent copies of $|C_{\rm BP}(x)|$. Thus, $|C_{\rm BP}(x)|$ is the *total progeny* of a branching process. As a result, since the expected total progeny of a branching process with mean offspring $\mu < 1$ equals $1/(1-\mu)$,

$$\chi_{\rm BP}(p) = \mathbf{E}|C_{\rm BP}(x)| = \frac{1}{1-(r-1)p}. \tag{6.35}$$

From (6.33), we obtain that, for $p < 1/(r-1)$,

$$\chi(p) = 1 + rp\chi_{\rm BP}(p) = 1 + \frac{rp}{1-(r-1)p} = \frac{1+p}{1-(r-1)p}, \tag{6.36}$$

while, for $p > 1/(r-1)$, $\chi(p) = \infty$. In particular, $p_T = 1/(r-1)$, and $\gamma = 1$ in the asymptotic sense. The computation of $\chi(p)$ can also be performed without the use of (6.34), by noting that, for $p \in [0,1]$,

$$\tau_p(x) = p^{h(x)}, \tag{6.37}$$

and the fact that, for $n \geq 1$, there are $r(r-1)^{n-1}$ vertices in $\mathbb{T}_r$ at height n, so that, for $p < 1/(r-1)$,

$$\chi(p) = 1 + \sum_{n=1}^{\infty} r(r-1)^{n-1} p^n = \frac{rp + 1 - (r-1)p}{1-(r-1)p}. \tag{6.38}$$

However, for related results for percolation on a tree, the connection to branching processes in (6.34) is vital. We defer the proof that $\gamma' = 1$ to later.

Let $\theta_{\mathrm{BP}}(p) = \mathbf{P}_p\{|C_{\mathrm{BP}}(x)| = \infty\}$. Then $\theta_{\mathrm{BP}}(p)$ is the survival probability of a branching process with a binomial offspring distribution with parameters $r-1$ and p. Thus, $\theta_{\mathrm{BP}}(p)$ satisfies the equation

$$\theta_{\mathrm{BP}}(p) = 1 - \big(1 - p + p(1-\theta_{\mathrm{BP}}(p))\big)^{r-1} = 1 - \big(1 - p\theta_{\mathrm{BP}}(p)\big)^{r-1}. \tag{6.39}$$

To compute $\theta_{\mathrm{BP}}(p)$, it is more convenient to work with the extinction probability $\zeta_{\mathrm{BP}}(p) = 1 - \theta_{\mathrm{BP}}(p)$, which is the probability that the branching process dies out. The extinction probability $\zeta_{\mathrm{BP}}(p)$ satisfies

$$\zeta_{\mathrm{BP}}(p) = \big(1 - p + p\zeta_{\mathrm{BP}}(p)\big)^{r-1}. \tag{6.40}$$

This equation can be solved explicitly when $r = 2$, when the unique solution is $\theta_{\mathrm{BP}}(p) = 0$ for $p \in [0,1)$ and $\theta_{\mathrm{BP}}(1) = 1$, so that $p_c = 1$. When $r = 3$, we obtain that

$$p^2\zeta_{\mathrm{BP}}(p)^2 + (2p(1-p) - 1)\zeta_{\mathrm{BP}}(p) + (1-p)^2 = 0, \tag{6.41}$$

so that

$$\zeta_{\mathrm{BP}}(p) = \frac{1 - 2p(1-p) \pm |2p-1|}{2p^2}. \tag{6.42}$$

Since $\zeta_{\mathrm{BP}}(0) = 1, \zeta_{\mathrm{BP}}(1) = 0$, we must have that

$$\zeta_{\mathrm{BP}}(p) = \frac{1 - 2p(1-p) - |2p-1|}{2p^2}, \tag{6.43}$$

so that $\zeta_{\mathrm{BP}}(p) = 1$ for $p \in [0, 1/2]$, while, for $p \in [1/2, 1]$,

$$\zeta_{\mathrm{BP}}(p) = \frac{1 - 2p(1-p) + (1-2p)}{2p^2} = \frac{2 - 4p + 2p^2}{2p^2} = \left(\frac{1-p}{p}\right)^2. \tag{6.44}$$

As a result, we have the explicit form $\theta_{\mathrm{BP}}(p) = 0$ for $p \in [0, 1/2]$ and

$$\theta_{\mathrm{BP}}(p) = 1 - \left(\frac{1-p}{p}\right)^2 = \frac{2p-1}{p^2}, \tag{6.45}$$

for $p \in [1/2, 1]$, so that $p_c = 1/2$. In particular, $p \mapsto \theta_{\mathrm{BP}}(p)$ is continuous, and, for $p \downarrow p_c$,

$$\theta_{\mathrm{BP}}(p) = 8(p - p_c)(1 + o(1)). \tag{6.46}$$

It is not hard to see that (6.46) together with (6.33) implies that

$$\theta(p) = 12(p - p_c)(1 + o(1)). \tag{6.47}$$

Thus, for $r = 3$, the percolation function is continuous, and $\beta = 1$ in the asymptotic sense. It is not hard to extend the asymptotic analysis in (6.46)–(6.47) to $r \geq 4$, for which $p_c(\mathbb{T}_r) = p_T(\mathbb{T}_r) = 1/(r-1)$, but we omit the details.

In order to study $\chi^{\mathrm{f}}_{\mathrm{BP}}(p) = \mathbf{E}_p[|C_{\mathrm{BP}}(x)|\mathbf{1}_{\{|C_{\mathrm{BP}}(x)|<\infty\}}]$ for $p > p_c = 1/(r-1)$, we make use of the fact that

$$\chi^{\mathrm{f}}_{\mathrm{BP}}(p) = (1 - \theta_{\mathrm{BP}}(p))\mathbf{E}_p\big[|C_{\mathrm{BP}}(x)|\big||C_{\mathrm{BP}}(x)| < \infty\big], \tag{6.48}$$

and the conditional law of percolation on the tree given that $|C_{\mathrm{BP}}(x)| < \infty$ is percolation on a tree with p replaced by the *dual* percolation probability p_d given by

$$p_d = p(1 - \theta_{\mathrm{BP}}(p)). \tag{6.49}$$

The crucial fact is that $p_d < p_c(\mathbb{T}_r)$, which follows from the equality $1 - \theta_{\mathrm{BP}}(p) = \zeta_{\mathrm{BP}}(p)$, (6.40) and the fact that

$$\begin{aligned}(r-1)p\zeta_{\mathrm{BP}}(p) &= (r-1)p\big(1 - p + p\zeta_{\mathrm{BP}}(p)\big)^{r-1} \\ &< (r-1)p\big(1 - p + p\zeta_{\mathrm{BP}}(p)\big)^{r-2} \\ &= \frac{\mathrm{d}}{\mathrm{d}s}(1 - p + ps)^{r-1}\Big|_{s=\zeta_{\mathrm{BP}}(p)},\end{aligned} \tag{6.50}$$

which, since $\zeta_{\mathrm{BP}}(p)$ is the smallest solution of $(1 - p + ps)^{r-1} = s$, implies that the derivative of $(1 - p + ps)^{r-1}$ at $s = \zeta_{\mathrm{BP}}(p)$ is strictly bounded above by 1 for $p > p_c(\mathbb{T}_r)$. Thus, by conditioning a supercritical cluster in percolation on a tree to die out, we obtain a subcritical cluster at an appropriate subcritical p_d which is related to the original percolation parameter. This fact is sometimes called the *discrete duality principle*.

We conclude that

$$\begin{aligned}\chi^{\mathrm{f}}_{\mathrm{BP}}(p) &= (1 - \theta_{\mathrm{BP}}(p))\mathbf{E}_p\big[|C|\big||C| < \infty\big] \\ &= (1 - \theta_{\mathrm{BP}}(p))\frac{1}{1 - (r-1)p(1 - \theta_{\mathrm{BP}}(p))}.\end{aligned} \tag{6.51}$$

Using that $\beta = 1$ in the asymptotic sense then gives that

$$\chi^{\mathrm{f}}_{\mathrm{BP}}(p) = \frac{C_{\gamma'} + o(1)}{p - p_c}. \tag{6.52}$$

By (6.33), this can easily be transferred to $\chi^{\mathrm{f}}(p)$, so that $\gamma' = 1$ in the asymptotic sense.

To compute ν, we note that, by (6.32) and (6.37), we have that, for $p < 1/(r-1)$,

$$\xi(p)^2 = \frac{1}{\chi^{\mathrm{f}}_{\mathrm{BP}}(p)} \sum_{n=1}^{\infty} nr(r-1)^{n-1}p^n = \frac{1}{\chi^{\mathrm{f}}_{\mathrm{BP}}(p)} \frac{r}{[1-(r-1)p]^2}, \tag{6.53}$$

so that $\nu = 1/2$ in the asymptotic sense by (6.52). We refrain from proving that $\nu' = \nu$, and only remark that this follows again by using the duality used in (6.49).

To compute η, we note that

$$\mathbf{E}_{p_c}\big[|C_{\mathrm{BP}}(x) \cap B(n)|\big] = \sum_{m=1}^{n^2} (r-1)^m p_c^m = n^2, \tag{6.54}$$

so that $\eta = 0$. We can compute δ by using the *random walk hitting time theorem*, see Grimmett (1999, Prop. 10.22) and van der Hofstad and Keane (2007), where a very simple proof is given applying to general branching processes. This result yields that

$$\mathbf{P}_p\{|C_{\mathrm{BP}}(x)| = k\} = \frac{1}{k}\mathbf{P}\{X_1 + \ldots + X_k = k-1\}, \tag{6.55}$$

where $\{X_i\}_{i=1}^{\infty}$ is an i.i.d. sequence of binomial random variables with parameter $r-1$ and success probability p. Thus,

$$\mathbf{P}_p\{|C_{\mathrm{BP}}(x)| = k\} = \frac{1}{k}\binom{k(r-1)}{k-1} p^{k-1}(1-p)^{k(r-1)-(k-1)}. \tag{6.56}$$

To prove that $\delta = 2$, we note that for $p = p_c = 1/(r-1)$, by a local limit theorem, we obtain

$$\mathbf{P}_{p_c}\{|C_{\mathrm{BP}}(x)| = k\} = (C_\delta + o(1))\frac{1}{\sqrt{k^3}}, \tag{6.57}$$

so that, by summing over $k \geq n$, we obtain $\delta = 2$ in an asymptotic sense. We can compute ρ by noting that

$$\theta_n = \mathbf{P}_{p_c}\{\exists y \in C_{\mathrm{BP}}(x) \text{ such that } h(y) = n + h(x)\} \tag{6.58}$$

satisfies the recursion relation

$$1 - \theta_n = (1 - p_c\theta_{n-1})^{r-1}. \tag{6.59}$$

It is not hard to see that (6.59) together with $p_c(\mathbb{T}_r) = 1/(r-1)$ implies that $\theta_n = (C_\rho + o(1))/n$. Thus, $\rho = 1/2$ in the asymptotic sense, since $|x| = \sqrt{h(x)}$ and $\mathbf{P}_{p_c}(o \longleftrightarrow \partial B(n)) = \theta_{\sqrt{n}} \sim (C_\rho + o(1))/\sqrt{n}$. □

The computation of the key objects for percolation on a tree is feasible due to the close relationship to branching processes, a topic which has attracted substantial interest in the probability community. See Athreya and Ney (1972), Harris (1963) and Jagers (1975) for detailed discussions about branching processes. As it turns out, the computations on a tree also have direct consequences for percolation on general graphs, with and without loops. We shall now discuss some of these consequences, the first being that $p_c > 0$ on any graph with bounded degree:

Theorem 6.2 (Percolation threshold is strictly positive) *Let $\mathbb{G} = (\mathbb{V}, \mathbb{E})$ be a graph for which the degree of every vertex is bounded by r. Then, for every $x \in \mathbb{V}$, and $p < 1/(r-1)$,*

$$\mathbf{E}_p|C(x)| < \infty. \tag{6.60}$$

In particular, for transitive graphs $\mathbb{G}$ with degree equal to r, $p_c(\mathbb{G}) \geq p_T(\mathbb{G}) \geq 1/(r-1)$.

Proof Let $\omega = (\omega_1, \ldots, \omega_n)$ be a nearest-neighbour path in $\mathbb{G}$. We call ω *self-avoiding* when $\omega_i \neq \omega_j$ for all $0 \leq i < j \leq n$. We let $c_n(x, y)$ denote the number of n-step nearest-neighbour self-avoiding walks starting at x with endpoint y. Then, we note that if $x \longleftrightarrow y$, then there must be a self-avoiding walk path consisting of occupied bonds. As a result,

$$\tau_p(x, y) \leq \sum_{n=0}^{\infty} p^n c_n(x, y). \tag{6.61}$$

Therefore,

$$\mathbf{E}_p[|C(x)|] = \sum_{y \in \mathbb{V}} \tau_p(x, y) \leq \sum_{y \in \mathbb{V}} \sum_{n=0}^{\infty} p^n c_n(x, y) = \sum_{n=0}^{\infty} p^n c_n(x), \tag{6.62}$$

where $c_n(x)$ denotes the number of n-step self-avoiding walk paths starting at x. Since the degree of $\mathbb{G}$ is bounded by r, we have that, uniformly in $x \in \mathbb{V}$,

$$c_n(x) \leq r(r-1)^{n-1}. \tag{6.63}$$

Thus, we arrive at

$$\mathbf{E}_p[|C(x)|] \leq \sum_{n=0}^{\infty} p^n r(r-1)^{n-1} = \frac{r}{r-1} \frac{1}{1-p(r-1)}, \tag{6.64}$$

so that $\mathbf{E}_p[|C(x)|] < \infty$ for $p < 1/(r-1)$. This completes the proof. □

For supercritical percolation on a tree, it is not hard to see that the number of infinite clusters N equals infinity a.s. To see this, for $p > p_c(\mathbb{T}_r)$, fix a root o of the tree $\mathbb{T}_r$. For any vertex v unequal to o, let u be the unique vertex in $\mathbb{T}_r$ that is closer to the root. Then, the probability that the bond (u, v) is vacant is strictly positive. If $p > p_c = 1/(r-1)$, then with probability $\theta_{\rm BP}(p) > 0$, the vertex v will lie in an infinite component. Thus, with strictly positive probability, there will be at least two bonds (o, e_1) and (o, e_2) that are vacant, and of which e_1 and e_2 lie in an infinite cluster. Thus, $\mathbf{P}_p\{N \geq 2\} > 0$, so that, since $N \in \{0, 1, \infty\}$ a.s., we must have that $N = \infty$ a.s.

We conclude that percolation on a tree is very well understood, that all of its critical exponents exist in an asymptotic sense and can be explicitly identified. Moreover, for $p \leq p_c(\mathbb{T}_r) = 1/(r-1)$, there is no infinite connected component, while, for $p \in (p_c, 1)$, there are infinitely many of them. For $p = 1$, there is a unique infinite cluster. We now proceed to study percolation on graphs with loops, starting with the paradigm example of $\mathbb{G} = \mathbb{Z}^d$.

6.2.3 *Percolation on $\mathbb{Z}^d$*

In this section, we study percolation on $\mathbb{Z}^d$. We start by proving that the phase transition is non-trivial:

Theorem 6.3 (Phase transition on $\mathbb{Z}^d$ is non-trivial) *For nearest-neighbour percolation on $\mathbb{Z}^d$ with $d \geq 2$, $p_c(\mathbb{Z}^d) \in [\frac{1}{2d-1}, 1)$.*

Proof The lower bound on $p_c(\mathbb{Z}^d)$ is immediate from Theorem 6.2, so that we are left to prove that $p_c(\mathbb{Z}^d) < 1$. For this, we must prove that, on $\mathbb{Z}^d$, $\theta(p) > 0$ for p sufficiently close to 1. We first show that it suffices to prove this for $d = 2$. For this, we fix $p \in (0, 1)$ and note that $|C| = |C(o)|$ for percolation on $\mathbb{Z}^d$ with parameter p is stochastically larger than $|C| = |C(o)|$ for percolation on $\mathbb{Z}^2$ with parameter p, since by only using the nearest-neighbour edges in $\mathbb{Z}^2 \times \{\vec{0}_{d-2}\}$, where $\vec{0}_{d-2} \in \mathbb{Z}^{d-2}$ denotes the origin in $\mathbb{Z}^{d-2}$, we obtain a cluster which is not larger than the one using all nearest-neighbour edges in $\mathbb{Z}^d$. Thus, it suffices to prove that $\theta(p) > 0$ for p sufficiently close to 1 for $d = 2$.

We shall make use of *duality*, a notion which is crucial for $d = 2$ (see also Section 6.2.4 below). For a set of vertices A, let the *boundary edges* of A be $\partial_e A = \{\{x, y\} : x \in A, y \notin A\}$. Clearly, for $C(o)$ the cluster of the origin, we have that all edges in $\partial_e C(o)$ are vacant. Define the *dual lattice* $\mathbb{L}^*$ to consist of the vertices $\mathbb{V}_d = \{x + (1/2, 1/2) : x \in \mathbb{Z}^2\}$ and an edge exists between $x + (1/2, 1/2)$ and $y + (1/2, 1/2)$ if and only if x and y are nearest-neighbours in $\mathbb{Z}^2$. We note that each edge in $\mathbb{L}^*$ intersects *precisely* one nearest-neighbour edge in $\mathbb{Z}^d$. We perform percolation on $\mathbb{L}^*$ by identifying the status of the edge $e \in \mathbb{L}^*$ by the status of the unique edge it intersects in $\mathbb{Z}^d$. Then, the set of vacant edges $\partial_e C$ is identified with a set of vacant edges in $\mathbb{L}^*$. We next study the structure of the set of edges.

We call a path $\omega = (\omega_0, \ldots, \omega_n)$ a *self-avoiding polygon of length n* on $\mathbb{L}^*$ when $\omega_i \in \mathbb{V}_d$, $\omega_0 = \omega_n$ and when $\omega_i \neq \omega_j$ for every $0 \leq i < j \leq n$ for which $i, j \neq 0, n$. A self-avoiding polygon separates $\mathbb{L}^*$ into the outside and the inside of the polygon, and we say that a self-avoiding polygon surrounds a point $x \in \mathbb{Z}^2$ when the point lies on the inside of the polygon. For $\mathbb{Z}^2$, we observe that if $|C(o)| < \infty$, then there must be a self-avoiding polygon $\omega = (\omega_1, \ldots, \omega_n)$ of length $n \geq 4$ on $\mathbb{L}^*$ which surrounds the origin $o = (0,0)$ of which each edge is vacant in $\mathbb{L}^*$. This is worked out in detail in Kesten (1982, page 386). Thus,

$$1 - \theta(p) \leq \sum_{n=4}^{\infty} m_n (1-p)^n, \tag{6.65}$$

where m_n is the number of self-avoiding loops of length n on $\mathbb{L}$ surrounding the origin $o = (0,0)$. Clearly, we have that $m_n \leq n3^{n-1}$, since there are at most n possible positions where the self-avoiding loop crosses the half-line $\{(x,0) : x \geq 0\}$, and the number of self-avoiding loops starting from any vertex is bounded by 3^{n-1}, so that

$$1 - \theta(p) \leq \sum_{n=4}^{\infty} n3^{n-1}(1-p)^n < 1, \tag{6.66}$$

when $p < 1$ is sufficiently close to 1. Thus, for such p, we have that $\theta(p) > 0$, proving that $p > p_c(\mathbb{Z}^2)$. □

Having established that the phase transition is non-trivial, the natural question is what the critical value is. Below, we shall give some results on critical values, particularly in 2 dimensions (see Section 6.2.4). An excellent reference to both numerical values as well as rigorous bounds is Hughes (1996, Chapter 3). For example, see Hughes (1996, Table 3.3) for some exact values of critical values, and Hughes (1996, Table 3.6) for numerical values of $p_c(\mathbb{Z}^d)$ on the nearest-neighbour lattice, showing that the inequality $p_c(\mathbb{Z}^d) \geq 1/(2d-1)$ is only a few percent off in dimensions $d \geq 5$. We now move on to the problem of the *uniqueness of the phase transition*, i.e., whether $p_c(\mathbb{Z}^d) = p_T(\mathbb{Z}^d)$:

Theorem 6.4 (Phase transition on $\mathbb{Z}^d$ is unique) *For nearest-neighbour percolation on $\mathbb{Z}^d$ with $d \geq 2$, $p_c(\mathbb{Z}^d) = p_T(\mathbb{Z}^d)$. In particular, $\chi(p) < \infty$ for all $p < p_c(\mathbb{Z}^d)$.*

The proof of Theorem 6.4 was given independently in Menshikov (1986) and Aizenman and Barsky (1987). Menshikov investigates $\mathbf{P}_p(A_n)$, where A_n is the event that there exists an open path from o to $\partial S(n)$ and $S(n) = \{x \in \mathbb{Z}^d : |x_1| + \cdots + |x_d| \leq n\}$. Since $p < p_c$, we have that $\mathbf{P}_p(A_n) \downarrow 0$ as $n \to \infty$. In order to prove that $\chi(p) < \infty$, it suffices to prove that $\mathbf{P}_p(A_n)$ converges to zero sufficiently fast, since

$$\chi(p) = \sum_{n=0}^{\infty} \mathbf{P}_p\{|C(o)| > n\} \leq \sum_{n=0}^{\infty} \mathbf{P}_p(A_{\lceil n^{1/d} \rceil}). \tag{6.67}$$

Menshikov shows that $\mathbf{P}_p(A_n)$ is close to exponentially small in n for rather general graphs, Grimmett (1999, Theorem 5.4) improves this to

$$\mathbf{P}_p(A_n) \leq \mathrm{e}^{-n\psi(p)} \tag{6.68}$$

for some $\psi(p)$ which is strictly positive for $p < p_c$. The proof of (6.68) makes clever use of Russo's formula, together with an appropriate renewal type argument to bound the expected number of pivotal bonds for A_n.

The proof in Aizenman and Barsky (1987) makes use of *differential inequalities*, a powerful technique which can also be used to prove bounds on critical exponents. In more detail, Aizenman and Barsky (1987) prove that for any p for which $\chi^{\mathrm{f}}(p) = \infty$, we have that either $\theta(p) > 0$ or $\theta(p) = 0$ and $\theta(p') \geq \frac{1}{p'}(p' - p)$ for all $p' > p$. This shows that $\chi(p) < \infty$ for $p < p_c$. Indeed, suppose that $\chi(p) = \infty$ for some $p < p_c$. Then, since $p < p_c$, we have that

$$\chi^{\mathrm{f}}(p) = \mathbf{E}_p[|C(o)|\mathbf{1}_{\{|C(o)|<\infty\}}] = \mathbf{E}_p|C(o)| = \infty, \tag{6.69}$$

where we made use of the fact that since $p < p_c$, we have $\theta(p) = 0$. Then, by the above result and again since $\theta(p) = 0$, $\theta(p') \geq \frac{1}{p'}(p' - p) > 0$ for all $p' > p$, so that $p \geq p_c = p_c(\mathbb{Z}^d)$. This is in contradiction to $p < p_c$. We shall not present further details of the proof of Theorem 6.4, as this is beyond the scope of this paper. We do mention that both the proofs in Aizenman and Barsky (1987) and in Menshikov (1986) are valid more generally. The proof in Aizenman and Barsky (1987) holds for general partially-oriented models, including also finite-range models as well as oriented percolation models. The proof in Menshikov (1986) does not rely on a lattice structure, but rather on the fact that the boundary of a ball of size n grows not faster than e^{cn^a} for some $a < 1$. These extensions will be relevant below.

We have seen that, in general, the number of infinite clusters N is a.s. constant, and that $N \in \{0, 1, \infty\}$ a.s. When $p < p_c$, then clearly $N = 0$ a.s. This leaves the question what the value of N is in the supercritical regime. On the tree, we have seen that $N = \infty$ a.s., and the main question is whether this can also occur on finite-dimensional lattices. This question was beautifully answered by Burton and Keane (1989) using a geometric argument. A nice survey of the implications of the Burton–Keane proof can be found in Meester (1994).

Theorem 6.5 (Uniqueness of infinite cluster on $\mathbb{Z}^d$) *For nearest-neighbour percolation on $\mathbb{Z}^d$ with $d \geq 2$, there is at most one infinite cluster. More precisely, for p such that $\theta(p) > 0$,*

$$\mathbf{P}_p(\textit{there is exactly one infinite cluster}) = 1. \tag{6.70}$$

A first proof of this result appeared in Aizenman, Kesten and Newman (1987).

Proof As discussed on page 176, we only need to rule out the possibility that N, the number of infinite clusters, is equal to $+\infty$ a.s. Indeed, when $\theta(p) = 0$, then $N = 0$ a.s., while if $\theta(p) > 0$, then the proportion of vertices in infinite clusters will be strictly positive, so that $N \geq 1$ a.s. Since N is a.s. constant and $N \in \{0, 1, \infty\}$ a.s., we conclude that $N = 1$ a.s. when $N < \infty$.

We will argue by contradiction. We call $x \in \mathbb{V}$ a *trifurcation* when it satisfies

(1) $|C(x)| = \infty$;
(2) there are *precisely* three occupied edges incident to x;
(3) the removal of the three edges incident to x splits the infinite cluster $C(x)$ into precisely three disjoint infinite clusters, and no finite clusters.

We write I_x for the indicator that x is a trifurcation. By translation invariance, $\mathbf{P}_p(I_x) = \mathbf{P}_p(I_0)$ for every $x \in \mathbb{V}$, and any reference point 0. We shall first argue that when $N = \infty$ a.s., then $\mathbf{P}_p(I_0) > 0$. Let M_B denote the number of infinite clusters touching B, and $M_B(0)$ the number of infinite clusters touching B when all edges with both vertices in B are declared vacant. Since $M_B(0) \geq M_B$, and $\mathbf{P}_p(M_{B(n)} \geq 3) \to \mathbf{P}_p(N \geq 3) = 1$, as $n \to \infty$ by our assumption, we can take n so large that $\mathbf{P}_p(M_{B(n)}(0) \geq 3) \geq 1/2$. Let ω be such that $\{M_{B(n)}(0) \geq 3\}$ occurs, and fix $x, y, z \in \partial B(n)$ such that x, y, z lie in three *distinct* infinite clusters in the configuration $\tilde{\omega}$ for which all edges with both vertices in B are declared vacant. We make the choice of x, y, z unique in an arbitrary way. Denote by $K_{x,y,z}$ the event that $M_{B(n)}(0) \geq 3$ *and* the three points are x, y, z.

Then, there exist three self-avoiding paths that only cross in the origin, and have only x, y, z, respectively, as elements in $\partial B(n)$. Let $J_{x,y,z}$ be the event that all the edges along the three paths are occupied, while all other edges are vacant. Since $B(n)$ is finite, and letting $R(n)$ denote the number of edges in $B(n)$, we obtain that

$$\mathbf{P}_p(J_{x,y,z}) \geq (\min\{p, 1-p\})^{R(n)} = \varepsilon > 0. \tag{6.71}$$

Finally, since, for all fixed x, y, z, the events $J_{x,y,z}$ and $K_{x,y,z}$ are independent, we obtain that

$$\begin{aligned}
\mathbf{P}_p(I_0) &\geq \sum_{x,y,z \in \partial B(n)} \mathbf{P}_p(J_{x,y,z} \cap K_{x,y,z}) \\
&= \sum_{x,y,z \in \partial B(n)} \mathbf{P}_p(J_{x,y,z}) \mathbf{P}_p(K_{x,y,z}) \\
&\geq \varepsilon \sum_{x,y,z \in \partial B(n)} \mathbf{P}_p(K_{x,y,z}) = \varepsilon \mathbf{P}_p(M_{B(n)}(0) \geq 3) \geq \varepsilon/2 > 0,
\end{aligned} \tag{6.72}$$

i.e. the probability that x is a trifurcation is strictly positive. By translation invariance, we therefore obtain that

$$\mathbf{E}_p\Big[\sum_{x \in B(n)} I_x\Big] \geq |B(n)|\varepsilon/2. \tag{6.73}$$

We now derive a contradiction by noting that the number of trifurcations in $B(n)$ can be no more than $\partial B(n+1)$. Let τ denote the number of trifurcations and $u_1, u_2, \ldots, u_\tau$ denote the trifurcations in $B(n)$. Next, we informally argue that we can identify to each trifurcation u_k a point $y_k \in \partial B(n+1)$ such that $y_i \neq y_j$ for each $1 \leq i < j \leq \tau$. This can be argued by noting that each trifurcation $u_i \in B(n)$ in a cluster K splits the intersection $K \cap \partial B(n+1)$ into three disjoint parts. This can be formalized into a family of so-called *compatible families of distinct 3-partitions* of $K \cap \partial B(n+1)$, and their number is restricted by $|K \cap \partial B(n+1)| - 2$, see Grimmett (1999, Lemma 8.5). As a result, the number of trifurcations in $B(n)$ of a cluster K is bounded by $|K \cap \partial B(n+1)| - 2 \leq |K \cap \partial B(n+1)|$. Summing out over the distinct clusters yields

$$\sum_{x \in B(n)} I_x \leq |\partial B(n+1)|. \tag{6.74}$$

By (6.73), we obtain

$$|B(n)|\varepsilon/2 \leq |\partial B(n+1)|, \tag{6.75}$$

for some $\varepsilon > 0$, which causes a contradiction when $n \to \infty$ whenever $|\partial B(n+1)|/|B(n)| \to 0$. As we see, this argument applies more generally than merely on $\mathbb{Z}^d$. □

Theorems 6.3, 6.4 and 6.5 answer many of the important questions for percolation on $\mathbb{Z}^d$, namely, the phase transition is non-trivial, it is unique in the sense that $p_c(\mathbb{Z}^d) = p_T(\mathbb{Z}^d)$ and the infinite cluster, when it exists, is unique. However, the critical behaviour is less well understood. One of the fundamental questions in percolation is whether there is no infinite component at criticality, i.e., is $\theta(p_c) = 0$, so that $p \mapsto \theta(p)$ is continuous? Remarkably, this question is in general wide open, even though in $d = 2$ and in high dimensions, it is affirmatively answered. We shall discuss these special cases in more detail below. In general, the behaviour of percolation systems at criticality is still ill understood, and there are no general arguments for the existence of critical exponents. Proofs so far concentrate on proving the existence by computing their values in special cases.

Interestingly, several relations can be drawn between the critical exponents on the tree and the ones on $\mathbb{Z}^d$. This is the content of the following theorem:

Theorem 6.6 (Inequalities for critical exponents on $\mathbb{Z}^d$) *For nearest-neighbour bond percolation on $\mathbb{Z}^d$, $\beta \leq 1, \gamma \geq 1$ and $\beta \geq 2/\delta$, when, for each of the inequalities, we assume that the critical exponent in question exists.*

We note that on $\mathbb{T}_r$, the bounds in Theorem 6.6 hold with equality, since $\beta = \gamma = 1, \delta = 2$. In general, the equalities are thus changed to inequalities. We shall prove that $\gamma \geq 1$ below, making use of Russo's formula as well as the BK-inequality. See Hughes (1996, Sections 4.3 and 4.4) for a more complete account of inequalities for critical exponents.

Proof that $\gamma \geq 1$. We use Russo's formula (6.10) to obtain

$$\frac{\mathrm{d}}{\mathrm{d}p}\tau_p(x) = \sum_{(u,v)\in\mathbb{E}} \mathbf{P}_p\{(u,v) \text{ is pivotal for } o \longleftrightarrow x\}. \tag{6.76}$$

Summing over x yields

$$\frac{\mathrm{d}}{\mathrm{d}p}\chi(p) = \sum_{x\in\mathbb{V}} \sum_{(u,v)\in\mathbb{E}} \mathbf{P}_p\{(u,v) \text{ is pivotal for } o \longleftrightarrow x\}. \tag{6.77}$$

We next note that if (u,v) is pivotal for $o \longleftrightarrow x$, then there exist two disjoint paths of occupied bonds connecting 0 and u, and v and x, respectively. Thus, $\{o \longleftrightarrow u\} \circ \{v \longleftrightarrow x\}$ occurs. By the BK inequality (6.8), we obtain

$$\frac{\mathrm{d}}{\mathrm{d}p}\chi(p) \leq \sum_{x\in\mathbb{V}} \sum_{(u,v)\in\mathbb{E}} \tau_p(u)\tau_p(x-v) = r\chi(p)^2, \tag{6.78}$$

where r is the degree of the graph $\mathbb{G}$. To make (6.78) precise, one has to overcome a few technicalities, since Russo's formula in (6.10) can only be applied to events E depending on a *finite* number of bonds, which is not the case for $\{o \longleftrightarrow x\}$. For details, see e.g., Heydenreich, van der Hofstad and Sakai (2008).

To complete the argument, note that we can rewrite (6.78) as

$$\frac{\mathrm{d}}{\mathrm{d}p}\chi(p)^{-1} \geq -r. \tag{6.79}$$

Integrating (6.79) over $[p, p_c]$, and noting that $\chi(p_c)^{-1} = 0$ by Theorem 6.4, we obtain that

$$\chi(p_c)^{-1} - \chi(p)^{-1} = -\chi(p)^{-1} \geq -r(p_c - p), \tag{6.80}$$

so that

$$\chi(p) \geq \frac{1}{r(p_c - p)}. \tag{6.81}$$

This proves that $\gamma \geq 1$ if γ exists. □

As remarked before, there are no general arguments that guarantee the existence of critical exponents. As we shall explain in more detail below, there are results that guarantee their existence in special cases, and that the critical exponents satisfy the scaling relations.

For percolation on $\mathbb{Z}^d$, apart from the scaling relations in (6.30), theoretical physics arguments predict that for low enough dimensions, two further scaling relations going under the name *hyperscaling relations*, which involve the dimension d, hold:

$$d\rho = \delta + 1, \qquad d\nu = 2 - \alpha. \tag{6.82}$$

More precisely, it is predicted that the scaling relations in (6.30) hold for *all* d, while the hyperscaling relations in (6.82) only hold for d below the so-called *upper critical dimension*. Above the upper critical dimension, the critical exponents are equal to the ones on the tree as appearing in Theorem 6.1. There has been some work in the direction of establishing the scaling and hyperscaling relations. In Borgs, Chayes, Kesten and Spencer (1999), it is shown that *if* the critical exponent ρ exists in the logarithmic sense, and if the probability at criticality $p = p_c$ that there is an easy-way crossing in the cube $[0, n] \times [0, 3n]^{d-1}$ is uniformly bounded by $1 - \varepsilon$ for some $\varepsilon > 0$, then also δ and η exist in the logarithmic sense, and satisfy $d\rho = \delta + 1$ and $2 - \eta = d(\delta - 1)/(\delta + 1)$. The latter equation is, when we assume the scaling relations in (6.30), equivalent to $d\nu = 2 - \alpha$.

In Tasaki (1987*a*, 1987*b*) and Chayes and Chayes (1987) it is shown that when the critical exponents take on the values on a tree, then $d \geq 6$, suggesting that the upper critical dimension is at least 6.

After this discussion on percolation on $\mathbb{Z}^d$ in general dimension, we now move to two special cases where much more is rigorously known, namely, two dimensions and high dimensions.

6.2.4 *Percolation in two dimensions*

In this section, we study percolation in the plane. We start with the fact that $p_c(\mathbb{Z}^2) = 1/2$ for bond percolation on the square lattice:

Theorem 6.7 (Harris–Kesten theorem: $p_c(\mathbb{Z}^2) = 1/2$) *For nearest-neighbour bond percolation on $\mathbb{Z}^2$, $\theta(1/2) = 0$ and $\theta(p) > 0$ for $p > 1/2$.*

The fact that $\theta(1/2) = 0$, which implies that $p_c(\mathbb{Z}^2) \geq 1/2$, is sometimes called *Harris' theorem* and was proved in Harris (1960). The proof that $\theta(p) > 0$ for $p > 1/2$ is sometimes called *Kesten's theorem* Kesten (1980), and is considerably more involved. We shall prove Harris' theorem below, making use of an important tool in two-dimensional percolation going under the name of *Russo–Seymour–Welsh* or RSW theory (Russo, 1981; Seymour and Welsh, 1978). We state the version of the RSW theorem from Russo (1981). We first introduce some notation. For $n \geq 2$ and k such that kn is an integer we let $R_{n,k}(p)$ be the probability that there is an occupied path from left to right crossing the rectangle $[0, kn] \times [0, n]$.

Theorem 6.8 (RSW theorem) *For any $p \in (0, 1)$ and $n, k \geq 1$ with n even, and for percolation on the square lattice $\mathbb{Z}^2$, the following bounds hold:*

$$R_{n,2}(p) \geq R_{n,1}(p)\Big[1 - \sqrt{1 - R_{n,1}(p)}\Big]^6,$$
$$R_{n,3}(p) \geq R_{n,1}(p)^3\Big[1 - \sqrt{1 - R_{n,1}(p)}\Big]^{12}. \tag{6.83}$$

We start by proving the RSW theorem:

Proof of the RSW Theorem 6.8. The main tool will be the Harris inequality in (6.5). We start by deducing the second inequality from the first. We note that if we take the two rectangles $[0,2n]\times[0,n]$ and $[n,3n]\times[0,n]$, then their intersection is $[n,2n]\times[0,n]$. Further, if there are left-to-right crossings in $[0,2n]\times[0,n]$ and in $[n,3n]\times[0,n]$, and a top-to-bottom crossing in $[n,2n]\times[0,n]$, then there is also a left-to-right crossing in $[0,3n]\times[0,n]$. Let $\mathrm{LR}_{n,1}, \mathrm{LR}_{n,2}, \mathrm{TB}_{n,3}$ be the events that the three respective crossings exist. Then,

$$R_{n,3}(p) \geq \mathbf{P}_p(\mathrm{LR}_{n,1} \cap \mathrm{LR}_{n,2} \cap \mathrm{TB}_{n,3}). \tag{6.84}$$

Now, $\mathrm{LR}_{n,1}, \mathrm{LR}_{n,2}$ and $\mathrm{TB}_{n,3}$ are all increasing events with

$$\mathbf{P}_p(\mathrm{LR}_{n,1}) = \mathbf{P}_p(\mathrm{LR}_{n,2}) = R_{n,2}(p), \quad \text{and} \quad \mathbf{P}_p(\mathrm{TB}_{n,3}) = R_{n,1}(p). \tag{6.85}$$

Thus, we arrive at

$$R_{n,3}(p) \geq R_{n,1}(p) R_{n,2}(p)^2, \tag{6.86}$$

showing that the first inequality in (6.83) implies the second.

To see the first inequality in (6.83), in exactly the same way as above, we can show that

$$R_{n,2}(p) \geq R_{n,1}(p) R_{n,3/2}(p)^2. \tag{6.87}$$

We shall frequently make use of a clever consequence of Harris' inequality, sometimes called the *square root trick*. This trick states that, for any two increasing events A_1 and A_2 with equal probability $\mathbf{P}_p(A_1) = \mathbf{P}_p(A_2)$, we have

$$\mathbf{P}_p(A_1) \geq 1 - \sqrt{1 - \mathbf{P}_p(A_1 \cup A_2)}. \tag{6.88}$$

To see (6.88), we note that

$$\begin{aligned} [1 - \mathbf{P}_p(A_1)]^2 &= \mathbf{P}_p(A_1^c)^2 = \mathbf{P}_p(A_1^c)\,\mathbf{P}_p(A_2^c) \\ &\leq \mathbf{P}_p(A_1^c \cap A_2^c) = 1 - \mathbf{P}_p(A_1 \cup A_2), \end{aligned} \tag{6.89}$$

so that (6.88) follows by taking the square root and rearranging terms. We shall now apply (6.88). For this, let H_u be the event that there exists a left-to-right crossing in $[0,n]\times[0,n]$ starting in the line $a_u = \{0\}\times[n/2,n]$ and H_l the event that there exists a left-to-right crossing in $[0,n]\times[0,n]$ starting in the line $a_l = \{0\}\times[0,n/2]$. Then, by (6.88), and the fact that there is a left-to-right crossing in $[0,n]\times[0,n]$ precisely when $H_u \cup H_l$ holds, we have

$$\mathbf{P}_p(H_u) \geq 1 - \sqrt{1 - R_{n,1}(p)}. \tag{6.90}$$

Now we introduce some more notation. Let $s = (s_1, \ldots, s_m)$ be a path connecting the left and right sides of $[0,n]\times[0,n]$, starting on the left side and taking values in $[0,n]\times[0,n]$. We let S_l denote the set of such paths for which $s^a \in a_l$. Let E_s be the event that s is the *lowest* occupied path connecting the left and right sides of $[0,n]\times[0,n]$. For such an s, and with $a = a_u \cup a_l = \{0\}\times[0,n]$, we let s^a be the last intersection with a, and we write $s^r = (s^a, \ldots, s_m)$ for the part of the path s

after its last visit to a, and $s^{r'}$ for the reflection of s^r in the line $\{n\} \times [0, n]$. Let F_s be the event that there exists a path in $[n/2, 3n/2] \times [0, n]$ connecting the top of $[n/2, 3n/2] \times [0, n]$ to the path s^r, and which always remains above $s^r \cup s^{r'}$. It is not hard to see that, for every s, we have that $E_s \cap F_s \cap H_u \subset \mathrm{LR}_{n,3/2}$, where we write $\mathrm{LR}_{n,3/2}$ for the event that there is a left-to-right crossing in $[0, 3n/2] \times [0, n]$. Thus, with

$$G = \bigcup_{s \in S_l} E_s \cap F_s, \tag{6.91}$$

we obtain that $G \cap H_u \subset \mathrm{LR}_{n,3/2}$, which implies that

$$R_{n,3/2}(p) \geq \mathbf{P}_p(G \cap H_u). \tag{6.92}$$

Applying the Harris' inequality once more and noting that both G and H_u are increasing, we obtain that

$$R_{n,3/2}(p) \geq \mathbf{P}_p(G)\mathbf{P}_p(H_u). \tag{6.93}$$

We have already derived a lower bound on $\mathbf{P}_p(H_u)$ in (6.90), so we are left to lower bound $\mathbf{P}_p(G)$. Since G is a disjoint union, we have that

$$\mathbf{P}_p(G) = \sum_{s \in S_l} \mathbf{P}_p(E_s \cap F_s). \tag{6.94}$$

Now, for a fixed s, F_s only depends on edges in $[0, 3n/2] \times [0, n]$ that are above $s^r \cup s^{r'}$, while E_s only depends on edges in $[0, n] \times [0, n]$ that are below $s^r \cup s^{r'}$, so that E_s and F_s are independent. Thus,

$$\mathbf{P}_p(G) \geq \sum_{s \in S_l} \mathbf{P}_p(F_s)\mathbf{P}_p(E_s). \tag{6.95}$$

For fixed s, denote by F_s' the event that there exists a path in $[n/2, 3n/2] \times [0, n]$ connecting the top of $[n/2, 3n/2] \times [0, n]$ to the path $s^{r'}$, and which always remains above $s^r \cup s^{r'}$. Then, clearly, $\mathbf{P}_p(F_s) = \mathbf{P}_p(F_s')$ and both F_s and F_s' are increasing, so that by the square root trick (6.88), we have

$$\mathbf{P}_p(F_s) \geq 1 - \sqrt{1 - \mathbf{P}_p(F_s \cup F_s')} \geq 1 - \sqrt{1 - R_{n,1}(p)}, \tag{6.96}$$

since $\mathbf{P}_p(F_s \cup F_s') \leq R_{n,1}(p)$. Thus, using that the union over $s \in S_l$ of E_s equals H_u,

$$\mathbf{P}_p(G) \geq \left[1 - \sqrt{1 - R_{n,1}(p)}\right] \sum_{s \in S_l} \mathbf{P}_p(E_s) = \left[1 - \sqrt{1 - R_{n,1}(p)}\right] \mathbf{P}_p(H_u). \tag{6.97}$$

Combining (6.93) with (6.97) yields,

$$R_{n,3/2}(p) \geq \left[1 - \sqrt{1 - R_{n,1}(p)}\right] \mathbf{P}_p(H_u)^2 \geq \left[1 - \sqrt{1 - R_{n,1}(p)}\right]^3, \tag{6.98}$$

which, combined with (6.87), yields the first claim in (6.83). □

We continue by discussing *duality*, a notion which has been extremely important in two-dimensional percolation. We shall assume that we are working on a two-dimensional *planar lattice* $\mathbb{L}$, i.e., a graph that can be embedded into $\mathbb{R}^2$ in such a way that different edges can only meet at the vertices of the lattice. We shall assume that the graph is translation invariant, and that the embedding of the lattice divides $\mathbb{R}^2$ into an infinite tiling of identical and bounded faces. Then, the vertices of the dual lattice $\mathbb{L}^*$ are the *faces* of the embedding of $\mathbb{L}$, and we connect two vertices in $\mathbb{L}^*$ when their corresponding faces share a boundary, which is a bond in $\mathbb{L}$. Thus, to each bond in $\mathbb{L}^*$ we can identify a *unique* bond in $\mathbb{L}$. Thus, to a bond percolation configuration on $\mathbb{L}$, we have identified a bond percolation configuration on $\mathbb{L}^*$ by identifying the occupation status of a bond on $\mathbb{L}^*$ to the one of the bond on $\mathbb{L}$ to which it is identified. It is sometimes convenient to identify the vertices of the dual lattice with the *centres* of the faces, and in this representation, we have that the nearest-neighbour square lattice on $\mathbb{Z}^2$ is dual to the nearest-neighbour square lattice on $\mathbb{Z}^2 + (1/2, 1/2)$, as discussed in Section 6.2.3. As a result, percolation on the nearest-neighbour square lattice $\mathbb{Z}^2$ is self-dual.

We now investigate the event LR_n that there is a left-right crossing of occupied bonds in the rectangle $[0, n+1] \times [0, n]$. Also, denote by TB_n^* the event that there is a top to bottom crossing of *vacant* bonds in the dual lattice on $[1/2, n+1/2] \times [-1/2, n+1/2]$. Clearly, one of the two must happen, so that

$$\mathbf{P}_p(\mathrm{LR}_n) + \mathbf{P}_p(\mathrm{TB}_n^*) = 1. \tag{6.99}$$

By construction, we have that $\mathbf{P}_p(\mathrm{TB}_n^*) = \mathbf{P}_{1-p}(\mathrm{LR}_n)$, so that we obtain, for all $p \in [0,1]$ and $n \geq 0$,

$$\mathbf{P}_p(\mathrm{LR}_n) + \mathbf{P}_{1-p}(\mathrm{LR}_n) = 1. \tag{6.100}$$

When we pick $p = 1/2$, we thus obtain that $\mathbf{P}_{1/2}(\mathrm{LR}_n) = 1/2$ for every $n \geq 1$. Thus, $R_{n,1}(1/2) \geq \mathbf{P}_{1/2}(\mathrm{LR}_n) = 1/2$ for every $n \geq 1$, so that $R_{n,2}(1/2) \geq [1 - \sqrt{1/2}]^6/2 > 0$ uniformly in $n \geq 1$. We shall see that this is sufficient to show that $\theta(1/2) = 0$. Followed by the proof of $\theta(1/2) = 0$, we shall discuss some further consequences of duality.

Proof that $\theta(1/2) = 0$ for bond percolation on the two-dimensional square lattice Fix $n \geq 1$, and let G_n be the event that there exists an occupied path from $\partial[-n, n] \times [-n, n]$ to $\partial[-3n, 3n] \times [-3n, 3n]$, where we write $\partial[a, b] \times [c, d]$ for the boundary of the rectangle $[a, b] \times [c, d]$. Then,

$$\begin{aligned}\mathbf{P}_{1/2}(o \longleftrightarrow \partial B(3^k)) &\leq \mathbf{P}_{1/2}\left(\cap_{l=1}^k G_{3^l}\right)\\ &= \prod_{l=1}^k \mathbf{P}_{1/2}(G_{3^l}) = \prod_{l=1}^k \left[1 - \mathbf{P}_{1/2}\left(G_{3^l}^c\right)\right],\end{aligned} \tag{6.101}$$

since the events $\{G_{3^l}\}_{l=1}^\infty$ depend on the occupation status of disjoint sets of bonds, and are therefore independent. By duality, the event G_n^c occurs *precisely* when there is a dual path of vacant edges in $[-3n, 3n]\times[-3n, 3n]\setminus[-n, n]\times[-n, n]$ surrounding the square $[-n, n] \times [-n, n]$. Denote this event by $O(n)$. Then, $\mathbf{P}_{1/2}(G_n^c) = \mathbf{P}_{1/2}(O(n))$. The event $O(n)$ is a subset of the event that there exists left-to-right crossings in the rectangles $[-3n, 3n] \times [n, 3n]$ and $[-3n, -3n] \times [-n, -3n]$, and top-to-bottom crossings in the rectangles $[n, 3n] \times [-3n, 3n]$ and $[-n, -3n] \times [-3n, 3n]$. The probability of each of these events is equal to $R_{n,3}(1/2)$. By Harris' inequality, the crossings are positively correlated:

$$\mathbf{P}_{1/2}(O(n)) \geq R_{n,3}(1/2)^4. \tag{6.102}$$

By Theorem 6.8 and since $R_{n,1}(1/2) \geq 1/2$, we obtain that $R_{n,3}(1/2) \geq a$ for some explicit $a > 0$, and uniformly in $n \geq 1$, so that

$$\mathbf{P}_{1/2}\{o \longleftrightarrow \partial B(3^k)\} \leq \prod_{l=1}^{k}(1-a) = (1-a)^k. \tag{6.103}$$

As a result, we even obtain that $\mathbf{P}_{1/2}\{o \longleftrightarrow \partial B(n)\} \leq (1-a)^{\lfloor \log n/ \log 3 \rfloor} = O(n^{-a/\log 3})$, so that, if $\rho > 0$ exists, we obtain that $\rho \leq \log 3/a$. In particular, we have that $\theta(1/2) \leq \mathbf{P}_{1/2}\{o \longleftrightarrow \partial B(n)\}$ for each $n \geq 1$, which tends to 0 when $n \to \infty$. □

Duality can be used in several more convenient ways. For example, we can define an alternative *correlation length* $\tilde{\xi}(p)$ by the limit

$$\tilde{\xi}(p)^{-1} = -\lim_{n\to\infty} \frac{\log \tau_p^{\mathrm{f}}(0, ne_1)}{n}. \tag{6.104}$$

It can be expected that $\tilde{\xi}(p)$ is of the same order of magnitude as $\xi(p)$ in (6.17) when p is close to critical, as they both describe the maximal distance between vertices for which there is dependence between their clusters. Then, Chayes, Chayes, Grimmett, Kesten and Schonmann (1989) use duality to show that $\tilde{\xi}(p) = \tilde{\xi}(1-p)/2$ for $p > 1/2$. Thus, in particular, if ν exists, then so does ν' and it takes the same value.

Using arguments such as duality and RSW theory, the critical values of several other lattices in two dimensions have been established. An example is $p_c^{\mathrm{site}}(T) = 1/2$, where T is the triangular lattice, which will play an important role later on. Also site percolation on the triangular lattice turns out to be self-dual, which explains why its critical value is 1/2.

Recently, Bollobás and Riordan adapted the RSW ideas in such a way that they are more generally applicable. For example, in Bollobás and Riordan (2006*c*), they used this new methodology to give a simpler proof of the Harris–Kesten theorem. The adapted RSW methods have also been crucial in order to prove that certain critical values for site percolation on certain tessellations equal 1/2. For example, take a Poisson point process. For each vertex in

$\mathbb{R}^2$, attach it to the closest point(s) in the Poisson point process. This divides $\mathbb{R}^2$ in cells, and we draw an edge between two cells when they share a line of their respective boundaries. Then, when we colour the cells independently green with probability p and yellow with probability $1-p$, Bollobás and Riordan (2006*a*) shows that the critical value of the occurrence of an infinite green connected component again is $1/2$. In Bollobás and Riordan (2008), this result is extended to other two-dimensional tessellations, such as the Johnson–Mehl tessellation and two-dimensional slices of three-dimensional Voronoi tessellations.

In the remainder of this section, we shall work with site percolation on the triangular lattice, for which in the past decade tremendous and remarkable progress has been made. We start by giving the exact values of the critical exponents, the values of which have been predicted early on in the physics community (see Nienhuis 1984) for general two-dimensional lattices:

Theorem 6.9 (Critical exponents on the triangular lattice) *For site percolation on the two-dimensional triangular lattice, the critical exponents $\beta, \gamma, \nu, \eta, \delta$ and ρ exist in the logarithmic sense, and take on the values*

$$\beta = \frac{5}{36}, \quad \gamma = \gamma' = \frac{43}{18}, \quad \nu = \nu' = \frac{4}{3}, \quad \eta = \frac{5}{24}, \quad \delta = \frac{91}{5} \quad \rho = \frac{48}{5}. \tag{6.105}$$

Theorem 6.9 is one of the major breakthroughs in modern probability theory, particularly since its proof has shed light not only on the existence and scaling limit of two-dimensional percolation, but rather of the critical behaviour of a wide class of two-dimensional statistical physical models. So far, this technology has not only been used for percolation, but also for loop-erased random walk and uniform spanning trees (Schramm, 2000), and a proof for the Ising model has been announced by Smirnov. The proof of Theorem 6.9 is a consequence of the connection between critical percolation and so-called *stochastic Loewner evolution* (SLE), a topic which we will discuss in some detail below. The history is that (Schramm, 2000) first identified a class of continuous models, so-called SLE, which are conformally invariant models in the plane of which the properties depend on its parameter $\kappa > 0$. Schramm continued by noting that *if* the scaling limit of two-dimensional percolation would be conformally invariant, then it must be equal to SLE with parameter $\kappa = 6$. Smirnov (2001) proved that indeed the scaling limit of critical percolation on the triangular lattice is conformally invariant. This is the celebrated result by Smirnov (2001), which we shall discuss in more detail below. Schramm already noted that SLE with parameter $\kappa = 6$ has similar critical exponents as in (6.105), when defined in an appropriate way. Smirnov and Werner (2001) realized that the values listed in (6.105) follow from the two statements

$$\mathbf{P}_{1/2}\left(A_R^1\right) = R^{-5/48+o(1)} \tag{6.106}$$

and

$$\mathbf{P}_{1/2}\left(A_R^2\right) = R^{-5/4+o(1)}, \tag{6.107}$$

where A_R^1 is the event that the origin is connected to the boundary of a ball of radius R, while A_R^2 is the probability that there exist two neighbours of the origin of which one has a green connection and another has a yellow connection to the boundary of the ball of width R. The first of these identities simply states that ρ exists and takes the value 48/5. The fact that these two statements imply the existence and values of the critical exponents listed in (6.105) is non-trivial and due to Kesten (1987). The asymptotics in (6.106) was shown in Lawler, Schramm and Werner (2002), the one in (6.107) in Smirnov and Werner (2001). The equalities $\gamma = \gamma'$ and $\nu = \nu'$ follow from the self-duality of site percolation on the triangular lattice. Theorem 6.9 identifies almost all critical exponents, an exception being α, which has always been somewhat mysterious. While Δ does not appear in Theorem 6.9, we believe that its derivation should be easier than that of α.

We now discuss the recent work on the scaling limit of critical percolation on the triangular lattice in more detail. In order to do so, we must start with the notion of *conformal invariance*, a notion which is crucial in the study of two-dimensional critical systems. We work on $\mathbb{C}$, and we let $D \subset \mathbb{C}$ be a simply connected domain. We say that a map $f\colon \mathbb{C} \to \mathbb{C}$ is *conformally invariant* when it preserves angles. In order to explain when a map f is angle-preserving we introduce some notation. Let $t \mapsto \gamma_1(t)$ and $t \mapsto \gamma_2(t)$ be two crossing curves with $\gamma_1(t), \gamma_2(t) \in \mathbb{C}$ for all t. Suppose that γ_1 and γ_2 are sufficiently smooth, then the curves $\gamma_1(t)$ and $\gamma_2(t)$ cross each other at a certain angle. f is called a *conformal map* or *preserves angles* when the curves $f(\gamma_1(t))$ and $f(\gamma_2(t))$ cross at the same angle as γ_1 and γ_2 for every pair of crossing curves γ_1 and γ_2 in D. Important examples of conformal maps are *Möbius transformations* given by

$$f(z) = \frac{az+b}{cz+d}, \tag{6.108}$$

where $a, b, c, d \in \mathbb{C}$ with $ad - bc \neq 0$. Let $D \subset \mathbb{C}$ be a (sufficiently smooth) domain, with four points P_1, P_2, P_3 and P_4 on the boundary which are such that P_i is in between P_{i-1} and P_{i+1} (where, by convention, $P_5 \equiv P_1$). We call $D_4 = (D; P_1, P_2, P_3, P_4)$ a 4-marked domain.

We shall investigate general lattices $\mathbb{L}$ in two dimensions, and we shall rescale the lattice with a small factor δ that shall later tend to zero (note that this δ has nothing to do with the critical exponent δ in (6.27)). For example, for the triangular lattice, we can think of δ as being the width of the edges in the lattice. Then, Langlands, Pouliot and Saint-Aubin (1994) studied crossing probabilities of the form that the boundary of D between P_1 and P_2 has an occupied path to the part of the boundary between P_3 and P_4. Denote this event by $C_\delta(D_4)$. When we work on a rectangle with $P_1 = (0,0), P_2 = (0,n), P_3 = (n,m), P_4 = (m,0)$, then this is nothing but the statement that the rectangle has a left-to-right crossing. In Langlands, Pouliot and Saint-Aubin (1994), the hypothesis was made that

$$P(D_4) = \lim_{\delta \downarrow 0} \mathbf{P}_{p_c}(C_\delta(D_4)) \tag{6.109}$$

exists and lies in $(0,1)$ when the points P_1, P_2, P_3, P_4 are different. These assumptions are already highly non-trivial, but the main assumption in Langlands, Pouliot and Saint-Aubin (1994) is that the limit $P(D_4)$ is *conformally invariant.* This is what is often meant with the assumption that the scaling limit of percolation is conformally invariant. Let us now explain what this assumption means in more detail. The limit $P(D_4)$ is conformally invariant when, for D_4' being the image under a conformal map of D_4, we have that $P(D_4) = P(D_4')$. This means that if we would consider the intersection of D_4' with the discretized lattice of width δ, and we compute the limit

$$P(D_4') = \lim_{\delta \downarrow 0} \mathbf{P}_{p_c}(C_\delta(D_4')), \tag{6.110}$$

then in fact this limit exists and equals $P(D_4') = P(D_4)$.

The above 'hypothesis' now goes under the name of *conformal invariance* of percolation, and is in fact what the celebrated paper Smirnov (2001) has proved on the triangular lattice. Since there are many conformal maps, the conformal invariance hypothesis is actually quite strong. In fact, Cardy (1992) used it to make a prediction of the exact limit of crossing probabilities in various domains, using mathematically non-rigorous arguments from conformal field theory. In order to explain Cardy's conjecture, we note that, since the limit of crossing probabilities is invariant under conformal maps, and for any 4-marked domain D_4 there exists a conformal map that maps D to the circle $B(1)$ and P_i to z_i on the boundary of the circle, the limit of crossing probabilities is determined by the crossing probabilities on the circle. For such special 4-marked domains, we define the *cross-ratio* η (which again has nothing to do with the critical exponent η defined in (6.29)) by

$$\eta = \frac{(z_4 - z_3)(z_2 - z_1)}{(z_4 - z_2)(z_3 - z_1)}. \tag{6.111}$$

It turns out that $\eta \in (0,1)$ and that two 4-marked domains on the circle can be mapped to one another by a conformal map *if and only if* they have the same cross-ratio η. Thus, 4-marked domains on the circle can be characterized by their cross-ratios. In turn, there is a unique conformal map mapping any 4-marked domain to a 4-marked domain on the circle, so that we can define the cross-ratio of a general 4-marked domain to be the cross-ratio of the image under the unique conformal map to the circle. Thus, we see that two 4-marked domains are conformally equivalent precisely when their cross-ratios are equal, and we can reformulate the hypothesis of Langlands, Pouliot and Saint-Aubin (1994) to say that the limiting crossing probabilities are a function of their cross-ratio, i.e., there exists a function $f\colon (0,1) \mapsto (0,1)$ such that $P(D_4) = f(\eta(D_4))$ when $\eta(D_4)$ is the cross-ratio of the 4-marked domain D_4. Based on this assumption, Cardy (1992) shows that in fact

$$P(D_4) = \frac{3\Gamma(2/3)}{\Gamma(1/3)^2}\eta^{1/3}\ {}_2F_1(1/3, 2/3; 4/3; \eta), \tag{6.112}$$

where ${}_2F_1$ is a hypergeometric function. Carleson noted that Cardy's conjecture takes a particularly appealing form on an equilateral triangle, i.e., we take $P_1 = (1,0), P_2 = (1/2, \sqrt{3}/2), P_3 = (0,0)$ and $P_4 = (x,0)$ where $x \in (0,1)$, and D is the interior of the equilateral triangle spanned by P_1, P_2, P_3. Then, (6.112) is equivalent to the statement that, for all $x \in (0,1)$,

$$P(D_4) = x. \tag{6.113}$$

Note that any 4-marked domain can be conformally mapped to the equilateral triangle, the only degree of freedom being the value of x in $P_x = (x,0)$. In his seminal paper, Smirnov (2001) showed (6.113). We shall not go into the proof in Smirnov (2001), as this would take up many more pages than were allotted to us. Instead, we give some more references to the literature. The first idea of a possible scaling limit of critical percolation is in the seminal paper of Schramm, which introduce the limiting stochastic process, which goes under the name of *Stochastic Loewner Evolution*, or sometimes *Schramm Loewner Evolution* (SLE). Now, SLE has developed into the main tool for studying two-dimensional critical systems, and it is likely that much more progress shall be made in this direction in the coming years. We refer to Bollobás and Riordan (2006*b*, Chapter 7) for an expository discussion of Smirnov's proof, as well as its consequences. Reviews on SLE and its consequences can be found in Lawler (2004, 2005), Kager and Nienhuis (2004), Werner (2004, 2005), and we refer there for more detailed discussions. We close this discussion by noting that, while the physics community has always predicted that the *same* critical behaviour should be valid for a wide range of two-dimensional critical percolation models, the proof on the triangular lattice is, to date, still basically the only proof of conformal invariance for two-dimensional percolation models. In particular, the corresponding result for two-dimensional bond percolation on the square lattice is unknown. One reason for this is that the proof in Smirnov (2001) makes essential use of the three-fold rotational symmetry of the triangular lattice, and it is, up to date, unclear how these symmetries can be replaced by the different sets of symmetries on the square lattice.

SLE has also proved useful to understand the so-called *near-critical phase* of percolation, where $p = 1/2 + \theta\delta^{4/3}$, a nice survey of these results, as well as of the proof of convergence of the percolation exploration process which explores the boundary of clusters can be found in Camia (2008).

The two-dimensional percolation problem is, after percolation on the tree, the best understood percolation problem, and the results described above give a rather complete overview of the depth and wealth of two-dimensional percolation theory. Yet, several results are not yet known and are worth considering: (a) proof of existence of *scaling functions* (see e.g. Hughes 1996, (4.295)); (b) a closer investigation of near-critical percolation (see Camia 2008 for an overview); (c) improvement of our understanding of *universality* in two-dimensional percolation, for example, by proving conformal invariance of two-dimensional bond percolation on the square lattice.

6.2.5 *Percolation in high dimensions*

In this section, we study percolation in high-dimensions. We consider $\mathbb{G} = (\mathbb{V}, \mathbb{E})$ with $\mathbb{V} = \mathbb{Z}^d$ and with edge set $\mathbb{E}$ either the nearest-neighbour bonds in sufficiently high dimension, or the *spread-out* bonds

$$\mathbb{E} = \{\{x, y\} : 0 < \|x - y\|_\infty \leq L\} \tag{6.114}$$

for some L sufficiently large. The main result in high dimensions is the following:

Theorem 6.10 (Mean-field critical exponents for high-d percolation) *For percolation on $\mathbb{Z}^d$, for either d sufficiently large in the nearest-neighbour model, or $d > 6$ and L sufficiently large in the spread-out model, $\beta = \gamma = 1$, $\nu = 1/2$ and $\delta = \Delta = 2$ in the bounded-ratio sense, while $\eta = 0$ in the asymptotic sense.*

It is believed, by invoking the paradigm of *universality*, that the critical exponents for any finite-range system which has sufficient symmetry are equal. Thus, Theorem 6.10 suggests that also $\beta = \gamma = 1$, $\nu = 1/2$ and $\delta = \Delta = 2$ for the nearest-neighbour model with $d > 6$. However, since the paradigm of universality is not mathematically rigorous, we cannot conclude this. Note that Theorem 6.10 *does* support the prediction of universality, since, in particular, the values of the critical exponents do not depend on the precise values of L, when L is sufficiently large. Also, for d sufficiently large, the critical values agree for *all* values of L.

The reason for the fact that L or d needs to be big in Theorem 6.10 is that the proof of Theorem 6.10 makes use of a perturbation expansion called the *lace expansion*. We shall now first discuss the history of the proof, before discussing some details.

In Aizenman and Newman (1984), it was proved that $\gamma = 1$ when the so-called *triangle condition*, a condition on the percolation model, holds. The triangle condition states that

$$\nabla(p_c) = \sum_{x,y \in \mathbb{V}} \tau_{p_c}(0, x)\tau_{p_c}(x, y)\tau_{p_c}(y, 0) < \infty, \tag{6.115}$$

where we recall that $\tau_p(x, y) = \mathbf{P}_p\{x \longleftrightarrow y\}$ is the two-point function. In Barsky and Aizenman (1991), it was shown that, under the same condition, $\beta = 1$ and $\delta = 2$. Needless to say, without the actual *verification* of the triangle condition, this would not prove anything. The triangle condition was proved to hold in the setting in Theorem 6.10 in Hara and Slade (1990) by the use of the lace expansion, a method which has since proved to be extremely powerful in order to characterise mean-field behaviour of various models in high dimensions. Later, the results for ν, Δ and η were proved in Hara (1990), Nguyen (1987), Hara (2005) and Hara, van der Hofstad and Slade (2003), again using lace expansion arguments. Several related results on high-dimensional percolation, in particular suggesting that the scaling limit of large critical clusters is a process called *Integrated Brownian excursion*, can be found in Hara and Slade (2000*a*, 2000*b*), where also the fact that $\delta = 2$ in the asymptotic sense was proved.

We now discuss the methodology in high dimensions. We start with the proof that $\gamma = 1$ if the triangle condition holds. Recall the argument that shows that $\gamma \geq 1$ below Theorem 6.6, and (6.77) in particular. The BK-inequality gives an upper bound on (6.77), and, in order to prove that $\gamma = 1$, a matching lower bound needs to be obtained. For this, we can use the independence of the occupation status of the bonds to explicitly write

$$\mathbf{P}_p\{(u,v) \text{ is pivotal for } o \longleftrightarrow x\} = \mathbf{E}_p[\mathbf{1}_{\{o \longleftrightarrow u\}} \tau^{\tilde{C}^{(u,v)}(o)}(v,x)], \qquad (6.116)$$

where, for a set of sites A, the restricted two-point function $\tau^A(v,x)$ is the probability that v is connected to x using only bonds with both endpoints outside A, and $\tilde{C}^{(u,v)}(o)$ consists of those sites which are connected to 0 without using the bond (u,v). Clearly, $\tau^{\tilde{C}^{(u,v)}(o)}(v,x) \leq \tau(v,x)$, and this reproves the upper bound previously obtained using the BK-inequality. We note that

$$\tau(v,x) - \tau^A(v,x) = \mathbf{P}_p\{v \overset{A}{\longleftrightarrow} x\}, \qquad (6.117)$$

where we write that $v \overset{A}{\longleftrightarrow} x$ when every path of occupied bonds from v to x contains a bond containing a vertex in A. Thus,

$$\begin{aligned}\frac{\mathrm{d}}{\mathrm{d}p}\chi(p) &= \sum_{x\in\mathbb{V}}\sum_{(u,v)\in\mathbb{E}} \mathbf{E}_p[\mathbf{1}_{\{o\longleftrightarrow u\}}]\tau_p(v,x) \\ &\quad - \sum_{x\in\mathbb{V}}\sum_{(u,v)\in\mathbb{E}} \mathbf{E}_p\left[\mathbf{1}_{\{o\longleftrightarrow u\}}\mathbf{P}_p\left\{v \overset{\tilde{C}^{(u,v)}(o)}{\longleftrightarrow} x\right\}\right] \\ &= r\chi(p)^2 - \sum_{x\in\mathbb{V}}\sum_{(u,v)\in\mathbb{E}} \mathbf{E}_p\left[\mathbf{1}_{\{o\longleftrightarrow u\}}\mathbf{P}_p\left\{v \overset{\tilde{C}^{(u,v)}(o)}{\longleftrightarrow} x\right\}\right]. \end{aligned} \qquad (6.118)$$

Now, for any $A \subseteq \mathbb{Z}^d$,

$$\mathbf{P}_p\left(v \overset{A}{\longleftrightarrow} x\right) \leq \sum_{a\in A} \mathbf{P}_p\{\{v \longleftrightarrow a\} \circ \{a \longleftrightarrow x\}\}, \qquad (6.119)$$

using the BK-inequality, which leads to

$$\frac{\mathrm{d}}{\mathrm{d}p}\chi(p) \geq r\chi(p)^2 - \chi(p) \sum_{(u,v)\in\mathbb{E}} \mathbf{E}_p[\mathbf{1}_{\{o\longleftrightarrow u,a\}}]\mathbf{P}_p\{v \longleftrightarrow a\}. \qquad (6.120)$$

Applying the BK inequality yields that

$$\begin{aligned}\mathbf{P}_p\{o \longleftrightarrow u,a\} &\leq \sum_z \mathbf{P}_p\{\{o \longleftrightarrow z\} \circ \{z \longleftrightarrow u\} \circ \{z \longleftrightarrow a\}\} \\ &\leq \sum_z \tau_p(z)\tau_p(u-z)\tau_p(a-z),\end{aligned} \qquad (6.121)$$

so that

$$\sum_{(u,v)\in\mathbb{E}} \mathbf{P}_p\{o \longleftrightarrow u, a\}\mathbf{P}_p\{v \longleftrightarrow a\} \le \chi(p)[\nabla(p) - 1], \tag{6.122}$$

implying that

$$\frac{\mathrm{d}}{\mathrm{d}p}\chi(p) \ge \chi(p)^2[2 - \nabla(p_c)]. \tag{6.123}$$

If we know that $\nabla(p_c) < 2$, then we can integrate the above equation in a similar fashion as around (6.79) to obtain that $\gamma = 1$. When we only have the *finiteness* of the triangle, then some more work is necessary to make the same conclusion (see Aizenman and Newman 1984 for details).

In the lace expansion, the above argument is adapted to deal with $\tau_p(x)$ directly, using rewrites as in (6.116)–(6.117) repeatedly, instead of using the upper bound in (6.119). We refer to Hara and Slade (1990) or the monograph Slade (2006) for detailed derivations of the lace expansion.

The lace expansion can also be used to prove asymptotics of the critical value in high dimensions, either when $L \to \infty$ for $d > 6$ fixed or for the nearest-neighbour model and $d \to \infty$. In van der Hofstad and Sakai (2005), the asymptotics of the critical point for percolation, as well as for self-avoiding walk, the contact process and oriented percolation, were investigated for $d > 6$ and $L \to \infty$. It was shown that

$$p_c(L, \mathbb{Z}^d) = \frac{1 + c_d L^d + o(L^{-d})}{(2L+1)^d - 1}, \tag{6.124}$$

for some explicit constant $c_d > 0$, and where $p_c(L, \mathbb{Z}^d)$ is the critical value of spread-out percolation with edge set $\mathbb{E}$ in (6.114). The best asymptotics of $p_c(\mathbb{Z}^d)$ when $d \to \infty$ are in Hara and Slade (1993, 1995), van der Hofstad and Slade (2005, 2006), where it is shown that, when $d \to \infty$, $p_c(\mathbb{Z}^d)$ has an asymptotic expansion in terms of inverse powers of $(2d)$ with rational coefficients, i.e. for each n, we can write

$$p_c(\mathbb{Z}^d) = \sum_{i=1}^{n} a_i (2d)^{-i} + \mathcal{O}((2d)^{n+1}), \tag{6.125}$$

where the a_i are rational coefficients with $a_1 = a_2 = 1, a_3 = 7/2$. We refer to the references in van der Hofstad and Slade (2005, 2006) for the literature on asymptotics of percolation critical points.

We close this section by discussing finite-size scaling in high-dimensional percolation. In Aizenman (1997), it was assumed that a version of $\eta = 0$ holds (more precisely, that $\tau_{p_c}(x)$ is bounded above and below by positive and finite constants times $|x|^{-(d-2)}$) in order to show that, at criticality, the largest intersection of a cluster with a cube of width $2r + 1$ grows like r^4 times logarithmic corrections. This corresponds to the *bulk* boundary condition. The condition used was verified in Hara (2005), Hara, van der Hofstad and Slade (2003) in the setting of Theorem 6.10.

Aizenman proceeds to conjecture that at criticality, with *periodic* boundary conditions, the largest critical cluster grows like $r^{2d/3}$, i.e., like $V^{2/3}$, where $V = (2r+1)^d$ is the volume of the cube. This was proved in Heydenreich and van der Hofstad (2007) making crucial use of the combined results in Borgs, Chayes, van der Hofstad, Slade and Spencer (2005*a*, 2005*b*). Such behaviour is dubbed *random graph asymptotics*, as $V^{2/3}$ growth at criticality is best known for the Erdős–Rényi random graph discussed in Section 6.3 below. Related results in this direction can be found in van der Hofstad and Luczak (2006) and Nachmias (2007), using alternative methods. An interesting question is what proper definitions of the critical value or window are in the general context of percolation on finite graphs, and what the proper conditions on the graph are such that percolation on it has random graph asymptotics close to criticality. While the combined examples in Borgs, Chayes, van der Hofstad, Slade and Spencer (2005*a*, 2005*b*), Heydenreich and van der Hofstad (2007), Nachmias (2007) provide some initial ideas, the general picture is not yet clear.

Despite the fact that detailed results are available in high dimensions, several results are not yet known and are worth considering: (a) proof of existence and mean-field values of ν', γ', ρ (particularly the supercritical regime in high-dimensions is still ill understood); (b) proof of existence of *scaling functions* (see e.g. Hughes (1996, (4.295)); (c) improvement of our understanding of *universality* in high-dimensional percolation, by, for example, showing that bond percolation on the nearest-neighbour lattice has the critical exponents in Theorem 6.10 for any $d > 6$.

6.2.6 *Oriented percolation*

In this section, we study so-called *oriented* or *directed* percolation. In this case, $\mathbb{G} = (\mathbb{V}, \mathbb{E})$ is given by $\mathbb{V} = \mathbb{Z}^d \times \mathbb{Z}_+$, and $\mathbb{E} = \{((x,n),(y,n+1)) : |x-y| = 1\}$, and $\mathbb{G}$ is considered as a *directed* graph, i.e., we remove all bonds independently with probability p and $(x,m) \longrightarrow (y,n)$ is only possible when $m \geq n$. Thus, we can only traverse edges in the direction of increasing last coordinate, and this last coordinate has the convenient interpretation of *time*. We define the *forward cluster* $C(x,n)$ of $(x,n) \in \mathbb{Z}^d \times \mathbb{Z}_+$ to be

$$C(x,n) = \{(z,l) : (x,n) \longrightarrow (z,l)\}, \tag{6.126}$$

so that, in particular, $C(x,n) \subset \mathbb{Z}^d \times \{n, n+1, \ldots\}$.

In some cases, we shall, similarly to the setting in Section 6.2.5, also deal with the *spread-out model*, in which $\mathbb{E} = \{((x,n),(y,n+1)) : 0 < \|x-y\|_\infty \leq L\}$ for some $L \geq 1$. While one might expect that percolation on oriented lattices is quite similar to percolation on unoriented lattices, this turns out not to be the case:

Theorem 6.11 (Continuity of oriented percolation) *For oriented percolation on $\mathbb{Z}^d \times \mathbb{Z}_+$, for $d \geq 1$, there is no infinite cluster at $p = p_c(\mathbb{Z}^d \times \mathbb{Z}_+)$, i.e. $\theta(p_c(\mathbb{Z}^d \times \mathbb{Z}_+)) = 0$.*

Theorem 6.11 was first proved in Bezuidenhout and Grimmett (1990) for directed percolation, which is a slight variation of the model defined here. The results were extended to the oriented percolation setting described above in Grimmett and Hiemer (2002).

The proof of Theorem 6.11 makes use of a block renormalization which was also used in Barsky, Grimmett and Newman (1991*a*, 1991*b*) to prove that percolation does not occur in half-spaces. The proof in Bezuidenhout and Grimmett (1990) also applies to the *contact process*, a continuous-time adaptation of oriented percolation. The deep relation between the contact process and oriented percolation has proved to be quite useful, and results in one model can typically also be proved for the other.

In Durrett (1980), the one-dimensional contact process and oriented percolation models were studied, focussing on the growth of the vertices in the cluster of the origin $(0,0) \in \mathbb{Z} \times \mathbb{Z}_+$ at time n. These results basically show that when the cluster of the origin is infinite, then the part of it at time n grows linearly in n with a specific growth constant. In Sakai (2002), the hyperscaling inequalities for oriented percolation and the contact process were proved, indicating that mean-field critical exponents can only occur for $d > 4$, thus suggesting that the upper critical dimension of oriented percolation equals $d_c = 4$. Indeed, as proved thereafter, the orientation of the percolation problem implies that mean-field behaviour already occurs for $d > 4$:

Theorem 6.12 (Mean-field critical exponents for oriented percolation) *For oriented percolation on $\mathbb{Z}^d \times \mathbb{Z}_+$, for either d sufficiently large in the nearest-neighbour model, or $d>4$ and L sufficiently large in the spread-out model, $\beta = \gamma = 1$ and $\delta = \Delta = 2$ in the bounded-ratio sense, while $\eta = 0$ in the Fourier-asymptotic sense.*

The proof of Theorem 6.12 is given in Nguyen and Yang (1993, 1995), van der Hofstad and Slade (2003) and follows a similar strategy as the proof of Theorem 6.10 by employing the results in Aizenman and Barsky (1987), Aizenman and Newman (1984) assuming the triangle condition, and using the lace expansion as in Hara and Slade (1990). In van der Hofstad, den Hollander and Slade (2007), it is proved that in the spread-out setting, for $d > 4$, the probability that there is an occupied path at criticality connecting $(0,0)$ to $\{(x,n) : x \in \mathbb{Z}^d\}$ is asymptotic to $1/(Bn)(1+o(1))$. This can be seen as a version of the statement that the critical exponent ρ exists and takes the mean-field value $\rho = 1/2$. In Sakai (2002), hyperscaling inequalities are shown that imply that critical exponents cannot take their mean-field values when $d < 4$. The main results in van der Hofstad and Slade (2003) make a connection between clusters at criticality for the spread-out oriented percolation model above 4 dimensions, and a measure-valued process called *super-Brownian motion*, a model which can be seen as the scaling limit of critical branching random walk. See Dawson (1993), Dynkin (1994), Etheridge (2000), Le Gall (1999), Perkins (2002) for expositions on super-processes.

It would be of interest to prove that *scaling functions* exist for high-dimensional oriented percolation, and to prove further results concerning the critical exponents. For example, we do not know ν, or that γ' exists and $\gamma' = \gamma = 1$ for oriented percolation above four dimensions.

6.2.7 *Percolation on non-amenable graphs*

We start by defining what an *amenable* graph is. For a finite set of vertices V, we denote its *edge boundary* by

$$\partial_E V = \{(u, v) : u \in V, v \notin V\}. \tag{6.127}$$

The notion of amenability is all about whether the size of $\partial_E V$ is of equal order as that of V, or is much smaller. To formalize this, we denote the *Cheeger constant* of a graph $\mathbb{G}$ by

$$h(\mathbb{G}) = \inf_{V \subset \mathbb{V}: |V| < \infty} \frac{|\partial_E V|}{|V|}. \tag{6.128}$$

A graph is called amenable when $h(\mathbb{G}) = 0$, and is it called non-amenable otherwise. Key examples of amenable graphs are finite-range translation invariant graphs $\mathbb{G}$ with vertex set $\mathbb{Z}^d$, the simplest example of a non-amenable graph is the regular tree $\mathbb{T}_r$ with $r \geq 3$. For the regular tree $\mathbb{T}_r$ with $r \geq 3$, it is not hard to see that $h(\mathbb{T}_r) = r - 2$. Benjamini and Schramm (1996) contains certain preliminary results of percolation on non-amenable graphs, and many open questions, some of which have been settled in the mean time. For example, Benjamini and Schramm (1996, Theorem 1) prove that $p_c(\mathbb{G}) \leq 1/(h(\mathbb{G}) + 1)$, so that $p_c(\mathbb{G}) < 1$ for every non-amenable graph.

A related definition of non-amenability can be given in terms of the *spectral radius* of a graph. Let $p_n(u, v)$ be the probability that simple random walk on $\mathbb{G}$ starting at $u \in \mathbb{V}$ is at time n at $v \in \mathbb{V}$. The spectral radius of $\mathbb{G}$ is defined as

$$\rho(\mathbb{G}) = \lim_{n \to \infty} p_{2n}(u, u)^{1/(2n)}. \tag{6.129}$$

By Kesten's theorem (Kesten 1959*a*, 1959*b*), see also Dodziuk (1984), when $\mathbb{G}$ has bounded degree, $\rho(\mathbb{G}) < 1$ precisely when $h(\mathbb{G}) > 0$. This exemplifies the fact that there is a close relationship between graph theoretic properties, and the behaviour of stochastic processes on the graph. A similar relation between the existence of invariant site percolation and amenability of Cayley graphs is proved in Benjamini, Lyons, Peres and Schramm (1999*b*, Theorem 1.1).

As we have seen for percolation on $\mathbb{Z}^d$ in Theorem 6.5, in the super-critical regime, the infinite cluster is unique. It turns out (see e.g. the discussion following Kesten (2002, Theorem 4)) that the uniqueness of the infinite cluster is valid for all amenable graphs. As the proof of Theorem 6.5 shows, there is a close relation between the ratio of the size of the boundary and its volume and the uniqueness of the infinite cluster, which helps to explain the uniqueness for all amenable graphs. On the other hand, for trees, the number of infinite components equals $N = \infty$ a.s. in the supercritical phase, which can be attributed to the fact that if we remove one edge, then a tree falls apart into two infinite graphs which will each have at least one infinite component a.s., so that, in total there will be infinitely many infinite clusters. Thus, this phenomenon is more related to there not being any cycles rather than the boundary being large.

In order to investigate the number of infinite clusters, we define the *uniqueness critical value* by

$$p_u = p_u(\mathbb{G}) = \inf\{p : \mathbf{P}_p - \text{a.s. there is a unique infinite cluster}\}. \tag{6.130}$$

For the regular tree with $r \geq 3$, $p_u = 1$, while for $\mathbb{Z}^d$, $p_u = p_c$. Below, we shall give examples where $p_c < p_u < 1$.

While the existence of an infinite cluster is clearly an increasing event, the uniqueness of the infinite cluster is not. Therefore, it is a priori not obvious that for all $p > p_u$, the infinite cluster will be unique. This is the main content of the following theorem. In its statement, we will write $N(p)$ for the number of infinite clusters in the coupling of the percolation models for all $p \in [0, 1]$ described above (6.4).

Theorem 6.13 (Uniqueness transition) *For percolation on a connected, quasi-transitive, infinite graph of bounded degree, a.s.,*

$$N(p) = \begin{cases} 0 & \text{for } p \in [0, p_c); \\ \infty & \text{for } p \in (p_c, p_u); \\ 1 & \text{for } p \in (p_u, 1]. \end{cases} \tag{6.131}$$

The proof of this theorem can be found in Schonmann (1999), and related results appeared in Häggström and Peres (1999), Häggström, Peres and Schonmann (1999). Note that, in general, not much is known for the critical cases $p = p_c$ and $p = p_u$.

Benjamini and Schramm (1996, Theorem 4) give a criterion in terms of the spectral radius which implies that $p_c < p_u$. Indeed, it shows that if $\rho(\mathbb{G})pr < 1$, where r is the maximal degree of $\mathbb{G}$, then there are a.s. infinitely many infinite clusters. Thus, if $\rho(\mathbb{G})p_c r < 1$, then $p_c < p_u$. Interesting examples arise by looking at Cartesian products of graphs. Let $\mathbb{G}_1 = (\mathbb{V}_1, \mathbb{E}_1)$ and $\mathbb{G}_2 = (\mathbb{V}_1, \mathbb{E}_1)$ be two graphs, and let $\mathbb{G} = \mathbb{G}_1 \times \mathbb{G}_2$ have vertex set $\mathbb{V} = \mathbb{V}_1 \times \mathbb{V}_2$ and edge set

$$\begin{aligned} \mathbb{E} = \Big\{ &\{(u_1, v_1), (u_2, v_2)\} : (u_1, v_1) \in \mathbb{E}_1 \\ &\text{and } (u_2 = v_2) \text{ or } (u_2, v_2) \in \mathbb{E}_2 \text{ and } u_1 = v_1 \Big\}. \end{aligned} \tag{6.132}$$

Then, clearly, when the maximal degrees of $\mathbb{G}_1$ and $\mathbb{G}_2$, respectively, are denoted by r_1 and r_2 respectively, the maximal degree of $\mathbb{G} = \mathbb{G}_1 \times \mathbb{G}_2$ is $r_1 + r_2$. Also, $p_c(\mathbb{G}_1 \times \mathbb{G}_2) \leq p_c(\mathbb{G}_2)$. Benjamini and Schramm (1996, Corollary 1) give many examples of graphs with $p_c < p_u$ by looking at $\mathbb{G} = \mathbb{G}_1 \times \mathbb{T}_r$, where $\mathbb{G}_1$ is quasi transitive and $\mathbb{T}_r$ is the r-regular tree of degree $r \geq 3$. Indeed, Benjamini and Schramm (1996, Theorem 4) prove that $\rho(\mathbb{G})p_c r < 1$ when r is sufficiently large, since $p_c(\mathbb{G}_1 \times \mathbb{T}_r) \leq p_c(\mathbb{T}_r) = (r-1)^{-1}$ and $\rho(\mathbb{G}) \to 0$ as $r \to \infty$. A simple example of a graph where $p_c < p_u < 1$ is $\mathbb{Z} \times \mathbb{T}_r$ as proved in Grimmett and Newman (1990).

We continue by studying the nature of the phase transition on non-amenable graphs. The first result concerns the continuity of the phase transition for *Cayley graphs of non-amenable groups.* We start by defining what a Cayley graph is. Let Γ be a group, and let $S = \{g_1, \ldots, g_n\} \cup \{g_1^{-1}, \ldots, g_n^{-1}\}$ be a finite set of generators. The Cayley graph $\mathbb{G} = \mathbb{G}(\Gamma)$ has vertex set $\mathbb{V} = \Gamma$, and edge set $\mathbb{E} = \{\{g, h\} : g^{-1}h \in S\}$.

Theorem 6.14 (Continuity on non-amenable Cayley graphs) *For percolation on a Cayley graph of a finitely generated non-amenable group, there is no infinite cluster at $p = p_c(\mathbb{G})$, i.e., $\theta(p_c(\mathbb{G})) = 0$.*

This result was proved in Benjamini, Lyons, Peres and Schramm (1999*a*, 1999*b*) and generalized earlier work by Wu (1993) on $\mathbb{Z} \times \mathbb{T}_r$ with $r \geq 7$. The proof makes use of the *mass-transport technique* to a clever choice of the mass-transport function.

We complete this section by describing some results on critical exponents. For this, we need to introduce the notions of *planar* and *unimodular* graphs. A graph is called *planar* when it can be embedded into $\mathbb{R}^2$ with vertices being represented by points in $\mathbb{R}^2$ and edges by lines between the respective vertices such that the edges only intersect at their end-points. For $x \in \mathbb{V}$, let the *stabilizer of x $S(x)$* be the set of automorphisms of $\mathbb{G}$ that keep x fixed, i.e., $S(x) = \{\gamma : \gamma(x) = x\}$. The graph $\mathbb{G}$ is called *unimodular* if $|\{\gamma(y) : \gamma \in S(x)\}| = |\{\gamma(x) : \gamma \in S(y)\}|$ for every $x, y \in \mathbb{V}$. Unimodularity turns out to be an extremely useful notion, particularly since it turns out to be imply the so-called *mass-transport principle* (MTP), see also Chapter 3, in particular Example 3.12, for a more general discussion. Indeed, let $\mathbb{G}$ be a transitive unimodular graph. We say that $f\colon \mathbb{V} \times \mathbb{V} \to [0, \infty)$ is diagonally invariant under the automorphisms of $\mathbb{G}$ when $f(x, y) = f(\gamma(x), \gamma(y))$ for every automorphism $\gamma\colon \mathbb{V} \to \mathbb{V}$. Then, the MTP states that, for any diagonally invariant function $f\colon \mathbb{V} \times \mathbb{V} \to [0, \infty)$, we have

$$\sum_{y \in \mathbb{V}} f(x, y) = \sum_{y \in \mathbb{V}} f(y, x). \tag{6.133}$$

For a proof, see e.g. Benjamini, Lyons, Peres and Schramm (1999*a*). In most cases, the MTP in (6.133) is used in the following way. We take $\phi\colon \mathbb{V} \times \mathbb{V} \times 2^{\mathbb{G}} \to [0, \infty)$ such that $\phi(x, y, \omega) = \phi(\gamma(x), \gamma(y), \gamma \circ \omega)$, for any configuration ω and where $\gamma \circ \omega$ is the configuration $(\gamma \circ \omega)(x) = \omega(\gamma(x))$. We interpreted $\phi(x, y, \omega)$ as the mass which x sends to y in the configuration κ. Then, we take $f(x, y) = \mathbf{E}_p[\phi(x, y, \omega)]$, and the MTP implies that the mass sent out by x is equal to the amount of mass x receives, which explains the name mass-transport principle. For non-unimodular graphs, an adaptation of (6.133) holds, where the left hand side is multiplied by $w(x)$ and the right hand side by $w(y)$, where $w(x) = |S_x o|/|S_o x|$, but this relation is not as powerful. All amenable graphs are unimodular, an example of a non-unimodular graph is the so-called *grandmother graph*, which is obtained by adding a connection between any vertex of the tree to its grandmother (i.e., the unique vertex which is two steps closer to the root than the vertex itself). Timár (2006) gives a wealth of related examples of non-unimodular graphs.

Finally, the *number of ends* of a graph $\mathbb{G}$ is

$$\mathcal{E}(\mathbb{G}) = \sup_{S\subset\mathbb{V}:|S|<\infty} \{\text{number of infinite connected components of } \mathbb{G}\setminus S\}. \quad (6.134)$$

Then, Schonmann(2001, 2002) proves that percolation has mean-field critical exponents in the following cases:

Theorem 6.15 (Mean-field critical exponent on non-amenable graphs) *For percolation on a locally finite, connected, transitive, non-amenable graph $\mathbb{G}$, $\beta = \gamma = 1, \delta = \Delta = 2$ in the bounded-ratio sense, in the following cases:*

1. *graphs $\mathbb{G}$ for which $h(\mathbb{G}) > (\sqrt{2r^2-1}-1)/2$, where r is the degree of the graph;*
2. *graphs $\mathbb{G}$ which are planar and have one end;*
3. *graphs $\mathbb{G}$ which are unimodular and have finitely many ends.*

Since percolation in high-dimensions is known to have mean-field critical exponents, which are the critical exponents on the tree, one would expect that, in great generality, percolation on non-amenable graphs do so too. Theorem 6.15 is a step into the direction of proving this belief, but a general result to this extent is still missing. It would be of interest to investigate this form of universality in more detail, as well as the existence of scaling functions for general non-amenable graphs.

An application of percolation on non-amenable graphs to image analysis can be found in Kendall and Wilson (2003).

6.2.8 *Continuum percolation*

Continuum percolation is a close brother of percolation, where instead of working on a lattice, we work in the continuum. While there are many possible models, we shall restrict to the simplest version, the so-called *Boolean model.* For details on the model, see the monograph by Meester and Roy (1996), or Penrose (2003, Section 9.6), and the references therein. In the Boolean model, we start with a Poisson point process PPP of a given intensity $\lambda > 0$, and each point $x \in$ PPP is assigned a radius. The radii of the different vertices are independent random variables, an important special case is when all radii are fixed. For $x \in$ PPP, let R_x be its corresponding radius. We create an *occupied region* by looking at all vertices contained in the union over x of the balls of radius R_x, i.e. the occupied region is given by

$$O = \bigcup_{x \in \mathrm{PPP}} B(x, R_x), \tag{6.135}$$

where $B(x,r)$ is the ball of radius r centred at $x \in \mathbb{R}^d$. We denote by $C(x)$ the connected part of O that contains x, we let $\theta(\lambda)$ be the probability that $C(x)$ is unbounded, and we define

$$\chi(\lambda) = \mathbf{E}_\lambda[|C(o)|], \tag{6.136}$$

where, for a region $C \subset \mathbb{R}^d$, we write $|C|$ for the Lebesgue measure of C. The function $\theta(\lambda)$ plays a similar role in continuum percolation as the percolation function in (discrete) percolation, while $\chi(\lambda)$ plays a similar role as the expected cluster size. With the above definitions at hand, continuum percolation as a model is quite similar to discrete percolation described above, and most of the results discussed above for percolation on $\mathbb{Z}^d$ carry over to continuum percolation on $\mathbb{R}^d$. In fact, many proofs make crucial use of the discrete result, by an appropriate discretization procedure. However, by varying the random radii, certain phenomena arise that are not present in bond percolation, such as the possibility that cluster have, a.s., finitely many Poisson points, yet the expected number of Poisson points is infinite.

Continuum percolation is an important model from the point of applications, as it can be seen as a simple model for a communication network where transmitters have a finite transmitting distance. When the points in the Poisson point process PPP mark the locations of sensors in an ad hoc network, and the radii correspond to their transmission distance, then the fact that $|C(x)| = \infty$ corresponds to the fact that the sensors can jointly transmit over an unbounded domain. In this light, continuum percolation is becoming an important tool in the investigation of various telecommunication networks. For examples of the application of continuum percolation ideas to communications, see e.g. Dousse, Franceschetti, Macris, Meester and Thiran (2006), Baccelli and Błaszczyszyn (2001) or the recent book by Franceschetti and Meester (2008). In Grossglauser and Thiran (2006), you can find a discussion on the relation between percolation problems and the engineering of wireless telecommunication models, touching upon navigability of networks, and the relation between connectivity and capacity. We also refer to Chapter 16 for a discussion on telecommunication aspects of random networks.

Random geometric graphs are obtained by taking only a bounded domain and performing a similar strategy as described above on it. More precisely, random geometric graphs are characterized by two parameters and can be constructed as follows. We consider the square $[0,1]^2$ and we put n points in it, uniformly at random. We can think of these n points as being a population of adhoc or peer-to-peer network users, who wish to communicate data to each other. Then we connect pairs of points within distance r for some appropriate r (possibly depending on n). Questions of interest are how large the radius should be as a function of the intensity in order for all points to be connected, and what the minimal degree of the vertices in this graph is. The connectivity of the graph is essential for each of the network users to be able to transmit data to each other. Clearly, this model is only a mere caricature of reality, and a better understanding of the real-world properties of networks is necessary. However, this caricature model does already shed light on some of the basic problems in geometric wireless networks. We refer to Penrose (2003) for detailed results on this model, as well as on the literature.

6.3 Random graphs

6.3.1 *Motivation*

6.3.1.1 *Real-world networks* In the past decade, many examples have been found of real-world networks that are *small worlds*, *scale-free* and *highly clustered*. We shall start by discussing these notions one by one.

The small-world phenomenon states that distances in a network are relatively small. This is related to the well-known 'Six degrees of separation' paradigm in social networks, which conjectures that any pair of individuals in the world can be connected by a chain of persons knowing each other on a first name basis, the chain consisting of at most six intermediary individuals. This paradigm has attracted considerable attention, see e.g. Newman, Watts and Barabási (2006) for a historical account, including the original papers. The notion of real networks being small worlds is inherently a bit imprecise, as we do not define what 'relatively' small is. For the time being one can have in mind that distances in many real networks are at most 10–20, later, we shall give a more precise definition what it means for a process of random graphs to be a small-world process.

The scale-free phenomenon states that the degree sequences in a network satisfies a power law. In more detail, the degree sequence of a network of size n, which we denote by $\{p_k^{(n)}\}_{k=0}^{\infty}$, is given by

$$p_k^{(n)} = \frac{1}{n}\sum_{i=1}^{n} \mathbf{1}_{\{d_i=k\}}, \tag{6.137}$$

where d_i is the degree of vertex i in the network, i.e., the number of neighbours of vertex i. Then, a real network is scale free when $p_k^{(n)}$ is approximately proportional to a power law with a certain exponent $\tau \in [1, \infty)$, i.e. for k sufficiently large, we have that there is a constant $C > 0$ such that

$$p_k^{(n)} \approx Ck^{-\tau}. \tag{6.138}$$

Naturally, also the notion of a real network being scale free is inherently somewhat vague. For example, how precise must the approximation in (6.138) be? Equation (6.138) can only be valid up to a certain point, since, for a simple graph of size n (i.e., a graph without self-loops and multiple edges), the maximal degree is equal to $n-1$, so that the left hand side of (6.138) equals 0 for $k \geq n$, while the right hand side remains positive for *all* k. In practice, (6.138) is verified by taking the logarithm on both sides and noticing that $\log p_k^{(n)}$ is close to linear in $\log k$:

$$\log p_k^{(n)} \approx \log C - \tau \log k, \tag{6.139}$$

i.e. a loglog plot of the degree sequence should be close to linear, and the slope is given by $-\tau$, where τ denotes the power-law exponent.

A network is highly clustered when two typical neighbours of an arbitrary vertex are more likely to be connected to each other as well than an arbitrary pair of vertices, i.e. many wedges are in fact closed to become triangles. When we draw two different vertices uniformly at random from a graph of size n, the probability that there is an edge between the drawn vertices is equal to

$$\frac{2E}{n(n-1)}, \tag{6.140}$$

where E is the number of edges. Note that $2E/n = \bar{d}$, where

$$\bar{d} = \frac{1}{n}\sum_{i=1}^{n} d_i \tag{6.141}$$

is the average degree of all vertices in the network. Thus, we see that $\frac{2E}{n(n-1)} = \bar{d}/(n-1)$. In most real-world networks, the average degree is much smaller than the size of the network, so that the probability that two uniformly drawn vertices share an edge is rather small. The *clustering coefficient* $C_{\mathbb{G}}$ of a graph $\mathbb{G} = (\mathbb{V}, \mathbb{E})$ is defined by

$$C_{\mathbb{G}} = \frac{\sum_{i,j,k \in \mathbb{V}} \mathbf{1}_{\{ij,ik,jk \in \mathbb{E}\}}}{\sum_{i,j,k \in \mathbb{V}} \mathbf{1}_{\{ij,ik \in \mathbb{E}\}}}, \tag{6.142}$$

i.e. the proportion of wedges that forms a triangle. A network is highly clustered when $C_{\mathbb{G}}$ is much larger than $\frac{2E}{n(n-1)}$. Again, this notion is inherently imprecise, as we do not define *how much* larger $C_{\mathbb{G}}$ needs to be. The reason why many networks are highly clustered is that there often is a certain group structure. For example, in a collaboration network, if a mathematician has published with another mathematician, then they are likely to be in the same community. Thus, when a mathematician has published with two other mathematicians, then they are likely to all be in the same community, which increases the likelihood that the two mathematicians have also published together. Similar effects play a role in the World Wide Web and in social networks.

As explained above, all these notions are merely empirical, and we shall give a proper mathematical definition before we introduce the mathematical random graph models that we shall consider in this section, and which are aimed at describing real networks. The aim of this section is not to define these notions precisely for empirical networks, but instead to define these notions precisely for random graph models for them. See Albert and Barabási (2002), Dorogovtsev and Mendes (2002), Newman (2003) for reviews on complex networks, and Barabási (2002) for a more expository account. We do notice that these real-world complex networks are not at all like classical random graphs (see Alon and Spencer 2000; Bollobás 2001; Janson, Łuczak and Rucinski 2000 and the references therein), particularly since the classical models do not have power-law degrees. As a result, the empirical findings of real-world networks have ignited enormous research on adaptations of the classical random graph that do have power-law degree sequences. We shall survey some of these results, and we shall start by defining the notions of small-world, scale-free and highly-clustered random graphs in a precise mathematical way.

6.3.1.2 *Small-world, scale-free and highly-clustered random graph processes* As described in the above motivation, many real-world complex networks are large. Many of them, such as the World Wide Web and collaboration networks, grow in size as time proceeds. Therefore, it is reasonable to consider graphs of growing size, and to define the notions of scale-free, small-world and highly-clustered random graphs as a limiting statement when the size of the random graphs tends to infinity. This naturally leads us to study graph sequences. We shall denote a sequence of random graphs by $\{\mathbb{G}_n\}_{n=1}^{\infty}$, where n denotes the size of the graph $\mathbb{G}_n$, i.e., the number of vertices in $\mathbb{G}_n$.

Denote the proportion of vertices with degree k by $P_k^{(n)}$, i.e.

$$P_k^{(n)} = \frac{1}{n}\sum_{i=1}^{n} \mathbf{1}_{\left\{D_i^{(n)}=k\right\}}, \tag{6.143}$$

where $D_i^{(n)}$ denotes the degree of vertex $i \in \{1,\ldots,n\}$ in the graph $\mathbb{G}_n$, and recall that the degree sequence of $\mathbb{G}_n$ is given by $\left\{P_k^{(n)}\right\}_{k=0}^{\infty}$. We use capital letters in our notation to indicate that we are dealing with *random variables*, due to the fact that $\mathbb{G}_n$ is a random graph. This explains why there are capitals in (6.143), but not in (6.137). Now we are ready to define what it means for a random graph process $\{\mathbb{G}_n\}_{n=1}^{\infty}$ to be scale free.

We call a random graph process $\{\mathbb{G}_n\}_{n=1}^{\infty}$ *sparse* when

$$\lim_{n\to\infty} P_k^{(n)} = p_k, \tag{6.144}$$

for some *deterministic* limiting probability distribution $\{p_k\}_{k=0}^{\infty}$. Since the limit p_k in (6.144) is deterministic, the convergence in (6.144) can equivalently be taken as convergence in probability or in distribution. Also, since $\{p_k\}_{k=0}^{\infty}$ sums up to one, for large n, most of the vertices have a bounded degree, which explains the phrase sparse random graphs.

We further call a random graph process $\{\mathbb{G}_n\}_{n=1}^{\infty}$ *scale free with exponent* τ when it is sparse and when

$$\lim_{k\to\infty} \frac{\log p_k}{\log(1/k)} = \tau \tag{6.145}$$

exists. Thus, for a scale-free random graph process its degree sequence converges to a limiting probability distribution as in (6.144), and the limiting distribution has asymptotic power-law tails described in (6.145). This gives a precise mathematical meaning to a random graph process being scale free. In some cases, the definition in (6.145) is a bit too restrictive, particularly when the probability mass function $k \mapsto p_k$ is not very smooth. Instead, we can also replace it by

$$\lim_{k\to\infty} \frac{\log[1 - F(k)]}{\log(1/k)} = \tau - 1, \tag{6.146}$$

where $F(x) = \sum_{y \le x} p_y$ denotes the distribution function corresponding to the probability mass function $\{p_k\}_{k=0}^{\infty}$. In particular models, we shall use the version that is most appropriate in the setting under consideration.

We say that a graph process $\{\mathbb{G}_n\}_{n=1}^{\infty}$ is *highly clustered* when

$$\lim_{n\to\infty} C_{\mathbb{G}_n} = C_{\mathbb{G}_\infty} > 0. \tag{6.147}$$

We finally define what it means for a graph process $\{\mathbb{G}_n\}_{n=1}^{\infty}$ to be a small world. Intuitively, a small world should have distances that are much smaller than those in a lattice or torus. When we consider the nearest-neighbour torus $\mathbb{T}_{r,d}$, and we draw two vertices uniformly at random, then their distance will be of the order r. Denote the size of the torus by $n = (2r+1)^d$, then the typical distance between two uniformly chosen vertices is of the order $n^{1/d}$, so that it grows as a positive power of n. We shall be dealing with random graph processes $\{\mathbb{G}_n\}_{n=1}^{\infty}$ for which $\mathbb{G}_n$ is not necessarily connected. Let H_n denote the distance between two uniformly chosen *connected* vertices, i.e., we pick a pair of vertices uniformly at random from all pairs of connected vertices, and we let H_n denote the graph distance between these two vertices. We shall call H_n the *typical distance of* $\mathbb{G}_n$. Then, we say that a random graph process $\{\mathbb{G}_n\}_{n=1}^{\infty}$ is a *small world* when there exists a constant K such that

$$\lim_{n\to\infty} \mathbf{P}\{H_n \le K \log n\} = 1. \tag{6.148}$$

Note that, for a graph with a *bounded* degree $d_{\max}$, the typical distance is *at least* $(1-\varepsilon)\log n/\log d_{\max}$, with high probability, so that a random graph process with bounded degree is a small world precisely when the order of the typical distance is optimal.

For a graph $\mathbb{G}_n$, let $\mathrm{diam}(\mathbb{G}_n)$ denote the *diameter* of $\mathbb{G}_n$, i.e. the maximal graph distance between any pair of connected vertices. Then, we could also have chosen to replace H_n in (6.148) by $\mathrm{diam}(\mathbb{G}_n)$. However, the diameter of a graph is a rather sensitive object which can easily be changed by making small changes to a graph in such a way that the scale-free nature and the typical distance H_n do not change. For example, by adding a sequence of m vertices in a line, which are not connected to any other vertex, the diameter of the graph becomes at least m, whereas, if m is much smaller than n, H_n is not changed very much. This explain why we have a preference to work with the typical distance H_n rather than with the diameter $\mathrm{diam}(\mathbb{G}_n)$.

In some models, we shall see that typical distances can be even much smaller than $\log n$, and this is sometimes called an *ultra-small world.* More precisely, we say that a random graph process $\{\mathbb{G}_n\}_{n=1}^{\infty}$ is an *ultra-small world* when there exists a constant K such that

$$\lim_{n\to\infty} \mathbf{P}\{H_n \le K \log\log n\} = 1. \tag{6.149}$$

There are many models for which (6.149) is satisfied, but for which at the same time $\mathrm{diam}(\mathbb{G}_n)/\log n$ converges in probability to a positive limit. This once more explains our preference to work with the typical graph distance H_n.

We have given precise mathematical definitions for the notions of random graphs being scale free, highly clustered and small worlds. This has not been done in the literature so far, and our definitions are based upon a summary of the relevant results proved for random graph models. We believe it to be a sound step forward to make the connection between the theory of random graphs and the empirical findings on real-life networks.

The remainder of this section is organised as follows. In Section 6.3.2, we study three models for random graphs without geometry that can have rather general degree sequences, namely, inhomogeneous random graphs, the configuration model and preferential attachment models. We discuss results concerning the phase transitions and distances in such models. In Section 6.3.5, we shall discuss random graphs with geometry. The results in Section 6.3.5 are closely related to percolation questions discussed in Section 6.2. The main distinction between the random networks discussed in this section and the percolation networks discussed in Section 6.2 is that the random graphs discussed here shall be *finite*, while the networks in Section 6.2 are all *infinite*. This raises interesting new questions, such as how the phase transition can be defined (a cluster can never be infinite), to what extent the phase transition is *unique*, and what the distance between two uniformly chosen vertices is.

6.3.2 *Models without geometry*

Extensive discussions of scale-free random graphs are given in Chung and Lu (2006*a*), Durrett (2007), monographs on classical random graphs are Bollobás

(2001), Janson, Łuczak and Rucinski (2000). We now discuss three particular examples of random graphs with power-law degree sequences, namely, the inhomogeneous random graph, the configuration model, and preferential attachment models.

6.3.2.1 *Inhomogeneous random graphs* The simplest imaginable random graph is the so-called *Erdős–Rényi random graph*, which consists of n vertices and each of the $n(n-1)/2$ edges is present or occupied with probability p, independently of the occupation status of the other edges. Denote the resulting graph by $G(n,p)$. This model was introduced by Gilbert (1959), while Erdős and Rényi (1959) introduced a model where a fixed number of edges is chosen uniformly at random and without replacement. The two models are quite comparable, and most asymptotic results in one of the two models can easily be transferred to asymptotic results in the other. A model with a fixed number of edges being chosen *with* replacement, so that possibly multiple edges between vertices arise, can be found in Austin, Fagen, Penney and Riordan (1959). The name Erdős–Rényi random graph is given to this class of models due to the fact that the first rigorous results were derived in the seminal paper Erdős and Rényi (1960), which can be seen as having founded the field of random graphs, and which has inspired research questions for decades to follow (see also the books Bollobás 2001, Janson, Łuczak and Rucinski 2000). The above model with independent edges can be viewed as *percolation on the complete graph*, the main difference to the theory in Section 6.2 being that the graph is *finite.*

One of the charming features of the Erdős–Rényi random graph is the fact that its vertices are completely exchangeable. For example, every vertex $v \in [n]$, where we write $[n] = \{1,\ldots,n\}$ for the vertex set, has a degree that is distributed as a binomial random variable with parameters $n-1$ and p. Thus, when $np \to \infty$, the average degree tends to infinity, while if $np = \lambda$, for some λ, the average degree remains uniformly bounded. When $p = \lambda/n$, the average degree of each vertex is roughly equal to λ, and the degree is a vertex converges in distribution to a Poisson random variable with parameter λ. It can be seen that also the proportion of vertices with degree k, as defined in (6.143), converges in probability to the Poisson probability mass function $p_k = \mathrm{e}^{-\lambda}\lambda^k/k!$. Thus, the Erdős–Rényi random graph process is *sparse.*

Note that the tails of a Poisson distribution are quite thin, even subexponentially thin. As a result, the Erdős–Rényi random graph process is *not* scale free. This problem can be overcome by stepping away from the assumption that the edge probabilities are *equal*, instead taking them unequal. This is the celebrated *inhomogeneous random graph* (IRG), about which the seminal paper Bollobás, Janson and Riordan (2007) proves substantial results in full generality. See also the references in Bollobás, Janson and Riordan (2007) for several examples which have been studied in the literature, and which they generalize. We shall not go into the precise definition of the model in Bollobás, Janson and Riordan (2007), but rather look at some simpler examples which already allow for general degree sequences.

To give the general setting, we let $G(n, \mathbf{p})$ denote a general inhomogeneous random graph, where $\mathbf{p} = \{p_{ij}\}_{1 \leq i < j \leq n}$ is such that p_{ij} is the probability that the edge $ij = (i, j)$ is occupied, and where different edges are independent. The Erdős–Rényi random graph is retrieved when taking $p_{ij} = p$ for all ij. We now generalize the definition in such a way that power-law degree sequences can be obtained. We assign *weights* $\{w_i\}_{i=1}^n$ to the vertices, w_i being the weight of vertex i. Let $l_n = \sum_{i=1}^n w_i$ be the total weight. Then, we can take

$$p_{ij} = 1 - \mathrm{e}^{-w_i w_j / l_n}. \tag{6.150}$$

In this way, we retrieve the *Poisson random graph* Norros and Reittu (2006). Alternatively, for $p_{ij} = \max\{w_i w_j / l_n, 1\}$ we retrieve the *random graph with given expected degree* as studied in detail by Chung and Lu (2002*a*, 2002*b*, 2003, 2006*a*, 2006*b*); Chung, Lu and Vu (2004).

Note that, in this model, if we assume that $\max_{i=1}^n w_i^2 < n$, and if we allow for a single self-loop at vertex i with probability w_i^2 / l_n, then the expected degree of vertex i is precisely equal to w_i, which explains the name of this model. A final example occurs when we take $p_{ij} = w_i w_j / (l_n + w_i w_j)$, which is called the *generalized random graph* (Britton, Deijfen and Martin-Löf, 2006). In Janson (2008), conditions are given as to when two inhomogeneous random graph processes $\{G(n, \mathbf{p})\}_{n=1}^\infty$ and $\{G(n, \mathbf{p}')\}_{n=1}^\infty$ are equivalent, i.e. when events have asymptotically the same probability for $G(n, \mathbf{p})$ and $G(n, \mathbf{p}')$ as $n \to \infty$.

Since the expected degree of vertex i is close to w_i, in order to obtain a specified degree sequence in the graph, we need to pick the weights $\{w_i\}_{i=1}^n$ appropriately. In the sequel, we shall take

$$w_i = w_i^{(n)} = [1 - F]^{-1}(i/n), \quad i \in [n], \tag{6.151}$$

where $[1 - F]^{-1}$ is the generalized inverse function of the survival function $x \mapsto 1 - F(x)$ given by

$$(1 - F)^{-1}(u) = \inf\{s : [1 - F](s) \leq u\}, \quad u \in (0, 1). \tag{6.152}$$

We call $G(n, \mathbf{p})$, with $\mathbf{p}$ as in (6.150) and $\{w_i\}_{i=1}^n$ as in (6.151) the *rank-1 inhomogeneous random graph with deterministic weights according to F*. In this case, the degrees of $G(n, \mathbf{p})$ can be seen to have a mixed Poisson distribution with mixing distribution F (Britton, Deijfen and Martin-Löf, 2006), i.e. the asymptotic probability that a uniform vertex has degree k is

$$p_k = \mathbf{E}\left[\frac{W^k}{k!}\mathrm{e}^{-W}\right], \tag{6.153}$$

where W has distribution function F (see, for example, Britton, Deijfen and Martin-Löf (2006, Theorem 3.1) or van der Hofstad (2008, Chapter 6)). Thus, the rank-1 inhomogeneous random graph is sparse. It is not hard to see, using concentration techniques on a Poisson random variable, that the mixed-Poisson distribution in (6.153) satisfies (6.145) when W is a continuous random variable with density f_W satisfying

$$\lim_{w\to\infty} \frac{\log f_W(w)}{\log(1/w)} = \tau. \tag{6.154}$$

The condition in (6.154) is sufficient, but not necessary. However, it does illustrate that there is a close relation between the tails of a mixed Poisson random variable and that of its mixing distribution.

Instead of taking the weights deterministic as in (6.151), we can also take $w_i = W_i$, where $\{W_i\}_{i=1}^n$ is an i.i.d. sequence of random variables with distribution function F. We call $G(n, \mathbf{p})$ with $\mathbf{p}$ as in (6.150) and $\{W_i\}_{i=1}^n$ an i.i.d. sequence of random variables with distribution F, the *rank-1 inhomogeneous random graph with random weights according to F*. The advantage of random degrees is that the vertices in the resulting graph are exchangeable, the disadvantage is that the edges are no longer independent, their dependence being moderated by their random weights.

IRGs give a flexible class of models for random graphs with flexible degree sequences of mixed Poisson form.

6.3.2.2 *The configuration model* The class of IRGs is a flexible class of random graphs, allowing for rather general degree sequences. One slight disadvantage of the model is the fact that one cannot specify the degree sequence more precisely. For example, one cannot have fixed degrees (so-called random regular graphs), or force the degrees to be at least 1 (which might be appropriate when using the model for certain real networks). An alternative model, allowing for more general degree sequences, is the configuration model (CM). We shall first introduce the model.

Fix an integer n. Consider a sequence of degrees $d_1, d_2, \ldots, d_n$. We will construct an undirected graph with n vertices where vertex j has degree d_j. Obviously, we will assume that $l_n = \sum_{j=1}^n d_j$ is even, since in any graph, the total degree is twice the number of edges.

To construct the graph, we have n separate vertices and incident to vertex j, we take d_j stubs. All stubs need to be connected to another stub to build the graph. The stubs are numbered in an arbitrary order from 1 to l_n. We start by connecting at random the first stub with one of the $l_n - 1$ remaining stubs. Once paired, two stubs form a single edge of the graph. We continue the procedure of randomly choosing and pairing the stubs without replacement, until all stubs are connected. Unfortunately, vertices having self-loops may occur. However, self-loops are scarce when $n \to \infty$. See Janson (2006) for the best results on the simplicity of the CM. In general, it turns out that the vector of the number of cycles of all lengths (i.e. the vector consisting of the number of self-loops, the number of multiple edges, the number of triangles, the number of squares, etc.) converges in distribution to an independent sequence of Poisson random variables, when $\nu_n = \frac{1}{l_n}\sum_{j=1}^{n} d_j(d_j - 1)$ converges to a certain $\nu < \infty$, and $\max_i d_i = o(\sqrt{n})$. See also Bollobás (2001, Section 2.4). The CM with fixed degrees has a long history. It was introduced in Bender and Canfield (1978) to study uniform random graphs with a given degree sequence (see also Bollobás 2001, Section 2.4). One specific example is to take the degrees all equal, in which case we speak of a *random regular graph.*

Similarly to the rank-1 case of IRG described above, there are two cases which have been studied for the CM, depending on whether the degrees are deterministic or random. In the case of deterministic degrees, a possible choice to obtain a degree sequence with distribution function F is to take F the distribution function of an integer random variable, and to take the number of vertices with degree k to be equal to $\lceil nF(k)\rceil - \lceil nF(k-1)\rceil$. We shall call this the *configuration model with deterministic degrees according to F*. For random degrees, we can take the degrees $\{D_i\}_{i=1}^{n}$ to be i.i.d. random variables with distribution function F, which we shall call the *configuration model with random degrees according to F*. In the latter case, unless D_1 is *even* with probability 1, the sum $L_n = \sum_{i=1}^{n} D_i$ will be *odd* with probability exponentially close to 1/2. To avoid this problem, we can increase the degree of vertex n by 1, so that the degree of vertex n equals $D_n + \mathbf{1}_{\{D_n \text{ is odd}\}}$. This hardly makes any difference to the properties of the graph under consideration, and we will ignore this effect in the sequel.

In Britton, Deijfen and Martin-Löf (2006, Theorem 2.1), it is shown that for the CM with random degrees according to F, where F is a distribution function of a random variable with finite $(1+\varepsilon)$th moment, if we erase all self-loops and contract all multiple edges to a single edge, the resulting graph is simple and sparse (recall (6.144)), where $p_k = F(k) - F(k-1)$. Thus, multiple edges and self-loops only form a small proportion of the total number of edges. As a result, in order to obtain a scale-free random graph process, we need to take $p_k = F(k) - F(k-1)$ to satisfy (6.145). In Britton, Deijfen and Martin-Löf (2006), also other constructions to create a CM with a prescribed degree sequence are analysed.

6.3.2.3 *Preferential attachment models* While inhomogeneous random graphs and the configuration model can have power-law degree sequences when the edge probabilities or degrees are chosen appropriately, they do not explain *why* many real-world networks are scale free. A possible explanation was given by Barabási and Albert (1999) by a phenomenon called *preferential attachment.* Preferential attachment models the growth of the network in such a way that new vertices are more likely to add their edges to already present vertices with a high degree. For example, in a social network, a newcomer is more likely to get acquainted to a person who is socially active, and, therefore, is likely to already possess a high degree. Thus, large degrees are likely to become even larger, which explains why this model is sometimes called the *Rich-get-Richer model.*

Interestingly, in certain cases, preferential attachment models (PAMs) have power-law degree sequences, and, therefore, preferential attachment offers a convincing explanation why many real-world networks have power-law degree sequences. As a result, many papers appeared that study such models. (See e.g. Aiello, Chung and Lu 2002; Bollobás, Borgs, Chayes and Riordan 2003, 2003*a*, 2003*b*, 2004*a*, 2004*b*; Bollobás, Riordan, Spencer and Tusnády 2001; Cooper and Frieze 2003 and the references therein.) The literature primarily focusses on three main questions. The first is to prove that such random graphs are indeed scale free (Aiello, Chung and Lu 2002; Bollobás, Borgs, Chayes and Riordan 2003; Bollobás and Riordan 2003*a*, 2003*b*; Bollobás, Riordan, Spencer and Tusnády 2001; Cooper and Frieze 2003). The second is to show that the resulting models are small worlds by investigating the distances in them. See, for example, Bollobás and Riordan (2004*b*) for a result on the diameter. In non-rigorous work, it is often suggested that many of the scale-free models, such as the CM or IRGs, the models in Bollobás, Janson and Riordan (2007) and the PAMs, have similar properties for their distances. Distances in the CM have been shown to depend on the number of finite moments of the degree distribution. The natural question is therefore whether the same applies to PAMs. A third key question for PAMs is their vulnerability, for example to deliberate attack (Bollobás and Riordan, 2003*b*) or to the spread of a disease (Berger, Bollobás, Borgs, Chayes and Riordan, 2003).

We now formulate the preferential attachment model, which is an extension of the Barabási–Albert model, formulated rigorously in Bollobás, Riordan, Spencer and Tusnády (2001), by constructing a graph process $\{G_m(n)\}_{n=1}^{\infty}$. We start by formulating the model for $m = 1$, for which we start with $G_1(1)$ consisting of a single vertex with a single self-loop. We denote the vertices of the graph $G_m(n)$ by $v_1^{(m)}, v_2^{(m)}, \ldots$. We denote the degree of vertex $v_i^{(m)}$ by $D_i^{(n)}$, where a self-loop increases the degree by 2.

Then, for $m = 1$, and conditionally on $G_1(n)$, the growth rule to obtain $G_1(n+1)$ is as follows. We add a single vertex $n+1$ having a single edge. The other end of this edge is equal to $i = n+1$ with probability proportional to $1+\delta$, and to $i \in [n]$ with probability proportional to $D_i^{(n)} + \delta$, where $\delta \geq -1$ is a parameter of the model. Thus,

$$\mathbf{P}\left(v_{n+1}^{(1)} \to v_i^{(1)} \middle| G_1(n)\right) = \begin{cases} \dfrac{1+\delta}{n(2+\delta)+(1+\delta)}, & \text{for } i = n+1, \\ \dfrac{D_i^{(n)}+\delta}{n(2+\delta)+(1+\delta)}, & \text{for } i = 1, 2, \ldots, n. \end{cases} \tag{6.155}$$

The model with integer $m > 1$, is defined in terms of the model for $m = 1$ as follows. We take $\delta \geq -m$, and then start with $G_1(mn)$, with $\delta' = \delta/m \geq -1$ by identifying the vertices $v_1^{(1)}, v_2^{(1)}, \ldots, v_m^{(1)}$ in $G_1(mn)$ to be vertex $v_1^{(m)}$ in $G_m(n)$, and for $1 < j \leq n$, the vertices $v_{(j-1)m+1}^{(1)}, \ldots, v_{jm}^{(1)}$ in $G_1(mn)$ to be vertex

$v_j^{(m)}$ in $G_m(n)$; in particular the degree $D_j(n)$ of vertex $v_j^{(m)}$ in $G_m(n)$ is equal to the sum of the degrees of the vertices $v_{(j-1)m+1}^{(1)}, \ldots, v_{jm}^{(1)}$ in $G_1(mn)$. This defines the model for integer $m \geq 1$. Observe that the range of δ is $[-m, \infty)$.

The resulting graph $G_m(n)$ has precisely mn edges and n vertices at time n, but is not necessarily connected. For $\delta = 0$, we obtain the original model studied in Bollobás, Riordan, Spencer and Tusnády (2001), and further studied in Bollobás and Riordan (2003*b*, 2004*a*, 2004*b*). The extension to $\delta \neq 0$ is crucial in our setting, as we shall explain in more detail below.

There are several related ways in which we can define the model. For example, we can disallow self-loops when $m = 1$ by setting the probability that $v_{n+1}^{(1)}$ connects to $v_{n+1}^{(1)}$ to be 0 in (6.155), and construct the model for $m > 1$ as in the paragraph above. Alternatively, we can let the m edges incident to vertex n to be attached *independently* of each other (in particular, in this case, vertex $v_{n+1}^{(m)}$ cannot connect to itself, so that the graph is connected). For many of the results, this precise choice is irrelevant, and we shall stick to the model in (6.155). The last two versions have the nice feature that they lead to *connected* random graphs.

We continue to discuss the degree sequence of the above preferential attachment model. Recall (6.143) for the definition of the degree sequence. Much of the available literature on PAMs centers around the proof that the asymptotic degree sequence obeys a power law, where the exponent τ depends in a sensitive way on the parameters of the model. Thus, the PAM is scale free. For the PAM considered here, the power-law exponent equals $\tau = 3 + \delta/m$, so that it can take any value $\tau \in (2, \infty)$ by adjusting the parameter $\delta > -m$. It is here that we rely on the choice of the model in (6.155). A form of bias in growing networks towards vertices with higher degree is, from a practical point of view, quite likely to be present in various real networks, but it is unclear why the PA scheme should be *affine* as in (6.155). However, only affine PA schemes give rise to power-law degree sequences. See Oliveira and Spencer (2005), Rudas, Tóth and Valkó (2007) for examples of PAMs with (possibly) non-linear PA-mechanisms and their degree sequences. We now explain how the affine PA-mechanism in (6.155) gives rise to power-law degree sequences and highlight the proof. We start by introducing some notation.

For $m \geq 1$ and $\delta > -m$, we define $\{p_k\}_{k=0}^{\infty}$ to be the probability distribution given by $p_k = 0$ for $k = 0, \ldots, m-1$ and, for $k \geq m$,

$$p_k = (2 + \delta/m)\frac{\Gamma(k+\delta)\Gamma(m+2+\delta+\delta/m)}{\Gamma(m+\delta)\Gamma(k+3+\delta+\delta/m)}. \tag{6.156}$$

Then the main result on the scale-free nature of preferential attachment models is the following:

Theorem 6.16 (Degree sequence in the PAM) *Fix $\delta > -m$ and $m \geq 1$. Then, there exists a constant $C > 0$ such that, as $n \to \infty$,*

$$\mathbf{P}\left\{\max_k |P_k^{(n)} - p_k| \geq C\sqrt{\frac{\log n}{n}}\right\} = o(1). \tag{6.157}$$

Furthermore, there exists a constant $C = C(m, \delta) > 0$ *such that, as* $k \to \infty$,

$$p_k = Ck^{-\tau}\left(1 + \mathcal{O}\left(\frac{1}{k}\right)\right), \tag{6.158}$$

where

$$\tau = 3 + \frac{\delta}{m} > 2. \tag{6.159}$$

In particular, Theorem 6.16 implies that the PA-random graph process is scale free. Theorem 6.16 appears in many forms in various settings. The statement which is closest to Theorem 6.16 is Deijfen, van den Esker, van der Hofstad and Hooghiemstra (2007, Theorem 1.3), where also the setting where each vertex enters the graph process with a *random* number of edges is considered. The first proof of a result as in Theorem 6.16 appeared in Bollobás, Riordan, Spencer and Tusnády (2001), they show a slightly weaker version of Theorem 6.16 when $\delta = 0$.

Virtually all proofs of asymptotic power laws in preferential attachment models consist of two steps: one step where it is proved that the degree sequence is concentrated around its mean, and one where the mean degree sequence is identified. We shall now give an intuitive explanation of Theorem 6.16.

Let $N_k^{(n)} = nP_k^{(n)}$ be the number of vertices with degree k in $G_m(n)$ (recall (6.143)). We are interested in the limiting distribution of $P_k^{(n)}$ as $n \to \infty$. This distribution arises as the solution of a certain recurrence relation, of which we will now give a short heuristic derivation. First note that, obviously,

$$\mathbf{E}\left[N_k^{(n+1)} | G_m(n)\right] = N_k^{(n)} + \mathbf{E}\left[N_k^{(n+1)} - N_k^{(n)} | G_m(n)\right]. \tag{6.160}$$

Asymptotically, for n large, it is quite unlikely that a vertex will be hit by more than one of the m edges added upon the addition of vertex n. Let us hence ignore this possibility for the moment. The difference $N_k^{(n+1)} - N_k^{(n)}$ between the number of vertices with degree k at time $n + 1$ and time n respectively, is then obtained as follows:

(a) Vertices with degree k in $G_m(n)$ that are hit by one of the m edges emanating from vertex n are subtracted from $N_k^{(n)}$. The conditional probability that a fixed edge is attached to a vertex with degree k is $(k + \delta)N_k^{(n)}/(n(2m + \delta))$, so that (ignoring multiple attachments to a single vertex) the mean number of vertices to which this happens is approximately $m(k + \delta)N_k^{(n)}/(n(2m + \delta))$. We note that we have replaced the numerator, which is $n(2m + \delta) + (e - 1)(2 + \delta/m) + 1 + \delta$ in the attachment of the eth edge emanating from vertex n, by its approximate value $n(2m + \delta)$ for large n.

(b) Vertices with degree $k - 1$ in $G_m(n)$ that are hit by one of the m edges emanating from vertex n are added to $N_k^{(n)}$. By reasoning as above, it follows that the mean number of such vertices is approximately $m(k - 1 + \delta)N_k^{(n)}/(n(2m + \delta))$.

(c) The new vertex n should be added if it has degree k. When we ignore the case that vertex n attaches edges to itself, this happens precisely when $k = m$.

Combining this gives

$$\mathbf{E}\left[N_k^{(n+1)} - N_k^{(n)} | G_m(n)\right] \approx \frac{m(k-1+\delta)}{n(2m+\delta)} N_{k-1}^{(n)} - \frac{m(k+\delta)}{n(2m+\delta)} N_k^{(n)} + \mathbf{1}_{\{k=m\}}. \tag{6.161}$$

Substituting (6.161) into (6.160) and taking expectations, we arrive at

$$\mathbf{E}\left[N_k^{(n+1)}\right] \approx \mathbf{E}\left[N_k^{(n)}\right] + \frac{m(k-1+\delta)}{n(2m+\delta)} \mathbf{E}\left[N_{k-1}^{(n)}\right] - \frac{m(k+\delta)}{n(2m+\delta)} \mathbf{E}\left[N_k^{(n)}\right] + \mathbf{1}_{\{k=m\}}. \tag{6.162}$$

Now assume that $P_k^{(n)}$ converges to some limit p_k as $n \to \infty$, so that hence $N_k^{(n)} \sim np_k$. Then, for $n \to \infty$, and observing that $\mathbf{E}[N_k^{(n+1)}] - \mathbf{E}[N_k^{(n)}] \to p_k$ implies that $\frac{1}{n}\mathbf{E}[N_k^{(n)}] \to p_k$, for all k, yields the recursion

$$p_k = \frac{m(k-1+\delta)}{2m+\delta} p_{k-1} - \frac{m(k+\delta)}{2m+\delta} p_k + \mathbf{1}_{\{k=m\}}, \tag{6.163}$$

or, equivalently,

$$p_k = \frac{k-1+\delta}{k+2+\delta+\delta/m} p_{k-1} + \frac{2+\delta/m}{k+2+\delta+\delta/m} \mathbf{1}_{\{k=m\}}. \tag{6.164}$$

By iteration, it can be seen that this recursion is solved by $p_k = 0$ when $k < m$ and

$$p_k = \frac{2+\delta/m}{k+2+\delta+\delta/m} \prod_{j=m}^{k} \frac{k-j+\delta}{k-j+2+\delta+\delta/m}, \quad k \geq m. \tag{6.165}$$

By rewriting the products in terms of Gamma-functions, we see that (6.165) is equal to (6.156). It is not hard to see that, when $k \to \infty$, $p_k \sim Ck^{-\tau}$ with $\tau = 3 + \delta/m$. This explains the occurrence of power-law degree sequences in affine PAMs. The above argument can be made rigorous in order to show that $\max_k |\mathbf{E}[N_k^{(n)}] - np_k|$ remains uniformly bounded (see e.g. Deijfen, van den Esker, van der Hofstad and Hooghiemstra 2007).

In order to prove concentration of $N_k^{(n)}$, all proofs in the literature make use of a clever martingale argument from Bollobás, Riordan, Spencer and Tusnády (2001). Define the Doob martingale M_t by

$$M_t = \mathbf{E}\left[N_k^{(n)} | G_m(t)\right]. \tag{6.166}$$

Then, $M_0 = \mathbf{E}[N_k^{(n)}]$ while $M_n = N_k^{(n)}$, so that $N_k^{(n)} - \mathbf{E}[N_k^{(n)}] = M_n - M_0$. The key ingredient is the observation that, for all $t \in [n]$, $|M_t - M_{t-1}| \leq 2m$ a.s., since the only vertices that are affected by the information of $G_m(t)$ instead of $G_m(t-1)$ are the vertices affected by the attachment of the edges incident to vertex t. Together, the concentration and the asymptotic mean give that $G_m(n)$ has an asymptotic degree sequence $\{p_k\}_{k\geq m}$, where p_k is close to a power-law for k large.

6.3.2.4 *A prediction of universality* In non-rigorous work, it is often suggested the various scale-free random graph models, such as the CM or various models with conditional independence of the edges as in Bollobás, Janson and Riordan (2007), behave similarly. For scale-free random graph processes, this informal statement can be made precise by conjecturing that the phase transition or distances have the same behaviour in graphs with the same power-law degree exponent. We shall discuss some of the results in this direction below.

6.3.3 *Phase transition in models without geometry*

In this section, we study the *phase transition* in random graphs. We first introduce some notation. For the CM with deterministic or random degrees according to F, we define

$$\nu = \frac{\mathbf{E}[D(D-1)]}{\mathbf{E}[D]}, \tag{6.167}$$

where D has distribution function F. For the rank-1 inhomogeneous random graph with deterministic or random weights according to F, we define

$$\nu = \frac{\mathbf{E}[W^2]}{\mathbf{E}[W]}, \tag{6.168}$$

where W has distribution function F. For the PAM, we let $\nu = m$. Below, we say that a sequence of events $\{E_n\}_{n=1}^{\infty}$ occurs *with high probability* (**whp**) when $\lim_{n\to\infty} \mathbf{P}(E_n) = 1$. It turns out that $\nu = 1$ plays the role of a critical value for all these random graphs:

Theorem 6.17 (Phase transition in random graphs)

(a–b) *For the configuration model with deterministic or random degrees according to F, and for the rank-1 inhomogeneous random graph with deterministic or random weights according to F, the largest connected component has,* **whp**, *size $o(n)$ when $\nu \leq 1$, and size $\zeta n(1 + o(1))$ for some $\zeta > 0$ when $\nu > 1$, where n is the size of the graph.*

(c) *For the PAM of size n,* **whp**, *the largest connected component has size $o(n)$ when $\nu = m = 1$, while the probability that the PAM is connected converges to 1 for $n \to \infty$ when $\nu = m > 1$.*

We write Theorem 6.17(a-b) to indicate that the result holds both for the IRG (which is model (a)) and for the CM (which is model (b)). This notation shall be used frequently below.

The result for IRG is a special case of Bollobás, Janson and Riordan (2007, Theorem 3.1). Earlier versions for the random graph with given expected degrees appeared in Chung and Lu (2002*b*, 2006*b*) (see also the monograph Chung and Lu 2006*a*). For the CM, the first result in the generality of Theorem 6.17 appeared in Molloy and Reed (1995, 1998) under stronger conditions than mentioned here. For the sharpest result, see Janson and Luczak (2007). The connectivity of PAMs was investigated for $\delta = 0$ in Bollobás and Riordan (2004*b*), it was extended to all $\delta > -m$ in van der Hofstad (2008, Chapter 11).

In Deijfen, van den Esker, van der Hofstad and Hooghiemstra (2007), also the setting where the number of edges with which a vertex enters the random graph is *random*. Indeed, denote the number of edges of vertex t by W_t, then in Deijfen, van den Esker, van der Hofstad and Hooghiemstra (2007) it is assumed that $\{W_t\}_{t=1}^{\infty}$ is an i.i.d. sequence. In general, such models are also scale-free with power-law exponent $\tau = \min\{\tau_{\mathrm{P}}, \tau_{\mathrm{W}}\}$, where $\tau_{\mathrm{P}} = 3 + \delta/\mu$ and $\mu = \mathbf{E}[W_t]$ is the PA exponent, while τ_{W} is the power-law exponent of the weight distribution, i.e.

$$\lim_{k\to\infty} \frac{\log \mathbf{P}\{W_1 \geq k\}}{\log(1/k)} = \tau_{\mathrm{W}} - 1. \tag{6.169}$$

Thus, one can summarize this by the fact that the effect with the least corresponding power-law exponent determines the power-law exponent of the graph. It would be of interest to study the phase transition for such more general models, and to verify under what condition a giant component exists.

6.3.4 *Distances in models without geometry*

In this section, we summarize the results on distances in power-law random graphs. We combine the results in the three models discussed in Section 6.3.2 by the value of their respective power-law exponent. We define H_n to be the typical distance in the graph of size n, i.e. the number of edges in the shortest path between two uniformly chosen connected vertices. Note that even in a fixed graph, H_n is a *random variable*, as it depends on the uniform vertices which are chosen. We shall also discuss results on the *diameter* of the graph, which is the maximum of the shortest path distances between any pair of connected vertices. Both give information about distances in graphs, the typical distance being a more robust and informative feature of the graph than the diameter.

6.3.4.1 *Distances in random graphs with finite variance degrees* The main results on distances in power-law random graphs with power-law exponent $\tau > 3$ are summarized in the following theorem:

Theorem 6.18 (Distances in graphs with finite variance degrees)

(a–b) For the configuration model and the rank-1 inhomogeneous random graph of size n, $H_n/\log n$ converges in probability to $1/\log \nu$, where ν is given by (6.167) for the CM, and by (6.168) for the rank-1 IRG, when F in the definition of the models satisfies that there exist $c > 0$ and $\tau > 3$ such that

$$1 - F(x) \leq cx^{-(\tau-1)}. \tag{6.170}$$

(c) For the affine PAM of size n with $\delta > 0$, so that $\tau = 3 + \delta/m > 3$, **whp**, *$H_n/\log n$ is bounded above and below by positive and finite constants.*

The result for the rank-1 IRG can be found in van den Esker, van der Hofstad and Hooghiemstra (2006*a*), where it is also shown that the fluctuations of H_n around $\log_\nu n$ remain bounded in probability, both in the case of i.i.d. degrees as well as for deterministic weights under a mild further condition on the distribution function. The first result in this direction was proved in Chung and Lu (2002*a*, 2003) for the expected degree random graph, in the case of rather general deterministic weights. A special case of the IRG with finite variance degrees is the Erdős–Rényi random graph with edge probability $p = \lambda/n$, for which $\nu = \lambda$.

The result for the CM can be found in van der Hofstad, Hooghiemstra and Van Mieghem (2005) in the case of i.i.d. degrees, where again also the fluctuations are determined. The results for deterministic degrees in the CM is conjectured in van den Esker, van der Hofstad and Hooghiemstra (2006*a*), but is not proved anywhere. We expect that the methodology in van der Hofstad, Hooghiemstra and Van Mieghem (2005) can be simply adapted to this case. The result for the affine PAM was proved in van der Hofstad, Hooghiemstra and Dommers (2009). Unfortunately the proof of convergence in probability is missing in this case. It would be of interest to identify the constant to which $H_n/\log n$ converges in this setting.

6.3.4.2 *Distances in random graphs with finite mean and infinite variance degrees* When $\tau \in (2, 3)$, the variance of the degrees becomes infinite, which is equivalent to the statement that, with $D_i^{(n)}$ denoting the degree of vertex i in the graph of size n, $\sum_{i=1}^{n}(D_i^{(n)})^2$ grows much faster than n. The following theorem shows that, in such cases, the distances are much smaller than $\log n$:

Theorem 6.19 (Distances in graphs with $\tau \in (2, 3)$)

(a–b) For the configuration model and the rank-1 inhomogeneous random graph, $H_n/\log\log n$ converges in probability to $2/|\log(\tau - 2)|$, when F in the definition of the models satisfies that there exist $c > 0$ and $\tau \in (2, 3)$ such that

$$1 - F(x) = cx^{-(\tau-1)}(1 + o(1)). \tag{6.171}$$

(c) *For the affine PAM with $\delta < 0$ and $m \geq 2$, so that $\tau = 3 + \delta/m \in (2,3)$,* **whp** *$H_n/\log\log n$ is bounded above and below by finite and positive constants.*

Theorem 6.19 shows that all three models are ultra-small worlds when the power-law exponent τ satisfies $\tau \in (2,3)$.

The result for the rank-1 IRG is proved in Chung and Lu (2002*a*, 2003) for the expected degree random graph, in the case of certain deterministic weights. The result for the CM can be found in van der Hofstad, Hooghiemstra and Znamenski (2007) in the case of i.i.d. degrees, where again also the fluctuations are determined and are proved to be bounded. The restrictions on F are somewhat weaker than (6.171), as they also allow $x \mapsto x^{\tau-1}[1 - F(x)]$ to be slowly varying under certain conditions on the regularly varying function. The results in Fernholz and Ramachandran (2007) mentioned earlier apply in this case as well, and show that, when the proportion of vertices with degrees 1 and 2 is positive, the diameter divided by $\log n$ converges in probability to a positive constant. In van der Hofstad, Hooghiemstra and Znamenski (2009), it is shown that the diameter in the CM is bounded above by a constant times $\log\log n$ when there are no vertices of degree 1 and 2.

The result for the affine PAM was proved in van der Hofstad, Hooghiemstra and Dommers (2009). Again, it would be of interest to identify the constant to which $H_n/\log\log n$ converges in this setting.

For the affine PA-model, we refer to Bollobás and Riordan (2004*b*) for a proof of the fact that the diameter of the PA-model $\text{diam}(G_m(n))$, for $\delta = 0$, satisfies that $\text{diam}(G_m(n)) \times \log\log n/\log n$ converges in probability to 1. This result is much sharper than the ones in Theorem 6.18(c) and Theorem 6.19(c), and it would be of interest to investigate whether the methodology used there can be adapted to the case where $\delta \neq 0$.

6.3.4.3 *Distances in random graphs with infinite mean degrees* Only in the CM and the IRG, it is possible that the power-law exponent of the degrees or the weights of the vertices τ satisfies $\tau \in (1,2)$. In general, this is not very realistic, as it means that either there are extremely many multiple edges (in the CM) or the power-law exponent in the graph does not match the value of τ (in the IRG). Distances in infinite mean random graphs have been studied in van den Esker, van der Hofstad, Hooghiemstra and Znamenski (2006*b*), Norros and Reittu (2006) and show that distances remain uniformly bounded by three. The intuition behind this is clear: all vertices are connected to vertices with extremely high degree, and these vertices form a complete graph, so that typical distances are at most 3. We refrain from a further discussion of random graphs with infinite mean degrees.

6.3.4.4 *Conclusion on phase transition and distances* The main tool in order to study the phase transition and distances in the CM and IRGs is a comparison of the neighbourhood of a vertex to a two-stage (multi-type) branching process. In order to prove distance results, one then has to further investigate the growth of the number of vertices at a given distance using limit laws for branching processes. When (6.170) holds, the number of vertices at a given distance k grows proportionally to ν^k, which suggests that distances are of the order $\log_\nu n$, as stated in Theorem 6.18.

Specifically, for the CM, with deterministic or i.i.d. degrees, the two-stage branching process, which we denote by $\{Z_k\}_{k\geq 0}$, starts from $Z_0 = 1$, has offspring distribution $\{f_n\}_{n=1}^\infty$, where $f_n = F(n) - F(n-1)$ are the jump sizes of the distribution F, in the first generation, and offspring distribution

$$g_n = \frac{(n+1)f_{n+1}}{\mathbf{E}[D]}, \qquad n \geq 0, \tag{6.172}$$

in the second and further generations. It is not hard to verify that the parameter ν in (6.167) is the expectation of the size-biased distribution $\{g_n\}_{n=0}^\infty$. For the rank-1 IRG with deterministic or random weights, the branching process $\{Z_k\}$ has a mixed-Poisson distribution with *random* parameter W in the first generation and a mixed Poisson distribution with random parameter W_e, which has the size-biased distribution of W, in the second and further generations. Thus, when W has a continuous density $w \mapsto f(w)$, the density of W_e is equal to $f_e(w) = wf(w)/\mathbf{E}[W]$. It can be seen that these two mixed Poisson distribution are again related through (6.172), and that again the parameter ν in (6.168) equals the expectation of the size-biased distribution used as offspring distribution in the second and further generations.

The condition $\nu > 1$ assures that the branching process $\{Z_k\}$ is supercritical, so that it can grow to a large size with positive probability (recall Theorem 6.17). Intuitively, all vertices for which the connected component are large (say larger than n^ε for some $\varepsilon > 0$) are connected and thus form a single giant component. The constant ζ in Theorem 6.17 is the survival probability of the two stage (multi-type) branching process $\{Z_k\}_{k\geq 0}$. Now, for a branching process $\{Z_k\}_{k\geq 0}$ for which the offspring distribution has finite $(1+\varepsilon)^{\text{th}}$ moment, we have that $Z_k\nu^{-k}$ converges a.s. to a limiting random variable W that is not identically 0 (by the Kesten–Stigum theorem). The core of the proof is to use that H_n is the graph distance between two uniformly chosen connected vertices V_1 and V_2. Then, the neighbourhood shells $Z_k^{(1,n)}$ and $Z_k^{(2,n)}$ consist of those vertices that are at graph distance *precisely* equal to k from V_1 and V_2, respectively. H_n is equal to k_n+1, where k_n is the first time that any of the vertices in $Z_{\lfloor k_n/2\rfloor}^{(1,n)}$ connects to a vertex in $Z_{\lceil k_n/2\rceil}^{(2,n)}$. The proof is then completed by coupling these neighbourhood shells to two independent two-stage branching processes as described above.

When (6.171) holds, then, by results of Davies (1978), the growth is *super-exponential*, i.e. $(\tau-2)^k \log(Z_k+1)$ converges a.s. to a limiting random variable Y, where $Y > 0$ precisely when the branching process survives. Thus, conditionally on $Y = y > 0$, the number of individuals in generation k grows like

$e^{(\tau-2)^{-k}y(1+o(1))}$ suggesting that distances are of order $\log\log n/|\log(\tau-2)|$. The additional factor 2 in Theorem 6.19 is due to the fact that in order for two vertices to meet, *each* of their neighbourhoods needs to have size at least n^{ε} for some $\varepsilon > 0$, each of which can be expected to occur in a generation k with $k \approx \log\log n/|\log(\tau-2)|$.

6.3.5 *Models with geometry*

In this section, we study some models of finite or bounded random graphs with geometry. We note that two of such models have been considered already in Section 6.2 as they are close relatives of the (infinite) percolation models. Indeed, in Section 6.2.5, the critical nature of percolation on high-dimensional tori was discussed, while in Section 6.2.8, random geometric graphs, which can be viewed as the restriction of continuum percolation to a bounded domain, have been described. Note that models for random directed and on-line networks are discussed in Chapter 7.

We now start by studying a spatial model for small worlds:

6.3.5.1 *Small-world networks* The models described in Section 6.3.2 have flexible degree sequences and small distances, but they tend not to be very highly clustered. Also, these models do not incorporate geometry at all. An alternative approach of explaining the small-world phenomenon is to start with a finite torus, and to add random long range connections to them, independently for each pair of vertices. This gives rise to a graph which is a small perturbation of the original lattice, but has occasional long range connections that are crucial in order to shrink distances. From a practical point of view, we can think of the original graph as being the local description of acquaintances in a social network, while the shortcuts describe the occasional acquaintances in the population living far apart. The main idea is that, even though the shortcuts only form a tiny part of the connections in the graph, they are crucial in order to make it a small world.

There are various ways of adding long-range connections (for example by rewiring the existing edges), and we shall focus on the models in Barbour and Reinert (2001, 2004, 2006), for which the strongest mathematical results have been obtained. Small-world models were first introduced and analysed non-rigorously in Moore and Newman (2000), Newman, Moore and Watts (2000), Newman and Watts (1999), and a non-rigorous mean-field analysis of distances in small-world models was performed in Newman, Moore and Watts (2000). See Barbour and Reinert (2001) for a discussion of the differences between the exact and mean-field analyses.

The simplest version of the model studied in Barbour and Reinert (2001) is obtained by taking the circle of circumference n, and adding a Poisson number of shortcuts with parameter $n\rho/2$, where the starting and endpoints of the shortcuts are chosen uniformly at random independently of each other. This model is called the *continuous circle model* in Barbour and Reinert (2001). Distance is measured as usual along the circle, and the shortcuts have, by convention, length zero. Thus, one can think of this model as the circle where the points along the random shortcut are identified, thus creating a puncture in the circle. Multiple shortcuts then lead to multiple puncturing of the circle, and the distance is then the usual distance along the punctured graph. Denote by H_n the distance between two uniformly chosen points along the punctured circle. Then, Barbour and Reinert (2001, Theorem 3.9) states that as $n \to \infty$, $H_n(2\rho)/\log \rho n$ converges in probability to 1 when $n\rho \to \infty$, and that $\rho(H_n - \log \rho n/2)$ converges in distribution to a random variable T satisfying

$$\mathbf{P}\{T > t\} = \int_0^\infty \frac{\mathrm{e}^{-y} dy}{1 + \mathrm{e}^{2t}}. \tag{6.173}$$

The random variable T can also be described by

$$\mathbf{P}\{T > t\} = \mathbf{E}\left[\mathrm{e}^{-\mathrm{e}^{2t} W^{(1)} W^{(2)}}\right], \tag{6.174}$$

where $W^{(1)}$, $W^{(2)}$ are two independent exponential random variables with parameter 1. Alternatively, it can be see that $T = (G_1 + G_2 - G_3)/2$, where G_1, G_2, G_3 are three independent Gumbel distributions (see Barbour and Reinert 2006, page 1242).

Interestingly, the method of proof of Barbour and Reinert (2001, Theorem 3.9) is quite close to the method of proof for Theorem 6.18. Indeed, again the parts of the graph that can be reached in distance at most t are analyzed. Let P_1 and P_2 be two uniform points along the circle, so that D_n has the same distribution as the distance between P_1 and P_2. Denote by $R^{(1)}(t)$ and $R^{(2)}(t)$ the parts of the graph that can be reached within distance t. Then, $D_n = 2T_n$, where T_n is the first time that $R^{(1)}(t)$ and $R^{(2)}(t)$ have a non-zero intersection. The proof then consists of showing that, up to time T_n, the processes $R^{(1)}(t)$ and $R^{(2)}(t)$ are close to certain continuous-time branching processes, primarily due to the fact that the probability that there are two intervals that are overlapping in quite small. Then, $W^{(1)}$ and $W^{(2)}$ can be viewed as appropriate martingale limits of these branching processes. In Barbour and Reinert (2001, Theorem 4.2), also an extension to higher dimensions is given.

The proof was extended in Barbour and Reinert (2006) to deal with discrete tori where the shortcuts also contribute one to the graphs distance, so that distances are the usual distances on discrete graphs. For this, it was necessary that the average number of shortcuts per vertex $\rho \downarrow 0$, a restriction that does not appear in Barbour and Reinert (2001). It would be of interest to extend the results to the case of *fixed* ρ as well in the discrete setting.

A related model was considered in Turova and Vallier (2006). Indeed, Turova and Vallier (2006) study a mixture between subcritical percolation on a finite cube and the Erdős–Rényi random graph. Using the methodology in Bollobás, Janson and Riordan (2007), it is shown that the phase transition is similar to the one described in Theorem 6.17. It would be of interest to verify whether the distance results in Bollobás, Janson and Riordan (2007) can also be used to prove that the distances grow like $\log_\nu n$, where n is the size of the graph, and $\nu > 1$ an appropriate constant.

6.3.5.2 *A scale-free percolation network* In this section, we discuss the results in Yukich (2006) on an infinite scale-free percolation model. Note that, for a transitive graph with fixed degree r and percolation with a fixed percolation parameter p, the degree of each vertex has a binomial distribution with parameters r and p. Since r is fixed, this does not allow for a power-law degree sequence. As a result, it is impossible to have a scale-free random graph when dealing with *independent* percolation, so that we shall abandon the assumption of independence of the different edges, while keeping the assumption of translation invariance.

The model considered in Yukich (2006) is on $\mathbb{Z}^d$, and, thus, the definition of a scale-free graph process does not apply so literally. We adapt the definition slightly by saying that an infinite random graph is scale-free when, instead,

$$p_k = \mathbf{P}\{D_o = k\}, \tag{6.175}$$

where D_x is the degree of vertex $x \in \mathbb{Z}^d$ and $o \in \mathbb{Z}^d$ is the origin, satisfies (6.145). This is a reasonable definition, since if we let $B_r = [-r, r]^d \cap \mathbb{Z}^d$ be a cube of width r around the origin, and denote $n = (2r+1)^d$, then, for each $k \geq 0$,

$$P_k^{(n)} = \frac{1}{n} \sum_{x \in B_r} \mathbf{1}_{\{D_x = k\}}, \tag{6.176}$$

which, assuming translation invariance and ergodicity, converges to p_k.

We next describe the model in Yukich (2006). We start by taking an i.i.d. sequence $\{U_x\}_{x\in\mathbb{Z}^d}$ of uniform random variables on $[0, 1]$. Fix $\delta \in (0, 1]$ and $q \in (1/d, \infty)$. The edge $\{x, y\} \in \mathbb{Z}^d \times \mathbb{Z}^d$ appears in the random graph precisely when

$$|x - y| \leq \delta \min\{U_x^{-q}, U_y^{-q}\}. \tag{6.177}$$

We can think of the ball of radius δU_x^{-q} as being the region of influence of x, and two vertices are connected precisely when each of them lies into the region of influence of the other. This motivates the choice in (6.177). The parameter δ can be interpreted as the probability that nearest-neighbours are connected, and in the sequel we shall restrict ourselves to $\delta = 1$, in which case the infinite connected component equals $\mathbb{Z}^d$. We denote the resulting (infinite) random graph by $\mathbb{G}_q$.

We next discuss the properties of this model, starting with its scale-free nature. In Yukich (2006, Theorem 1.1), it is shown that, with $\tau = qd/(qd-1) \in (1, \infty)$, the limit

$$\lim_{k\to\infty} k^{\tau-1} \mathbf{P}\{D_o \geq k\} \tag{6.178}$$

exists, so that the model is scale free with degree power-law exponent τ (recall (6.146)). The intuitive explanation of (6.178) is as follows. Suppose we condition on the value of $U_o = u$. Then, the conditional distribution of D_o given that $U_o = u$ is equal to

$$D_o = \sum_{x\in\mathbb{Z}^d} \mathbf{1}_{\{|x|\leq \min\{U_o^{-q}, U_x^{-q}\}\}} = \sum_{x:|x|\leq u^{-q}} \mathbf{1}_{\{|x|\leq U_x^{-q}\}}. \tag{6.179}$$

Note that the random variables $\left\{\mathbf{1}_{\{|x|\leq U_x^{-q}\}}\right\}_{x\in\mathbb{Z}^d}$ are independent Bernoulli random variables with probability of success equal to

$$\mathbf{P}\{\mathbf{1}_{\{|x|\leq U_x^{-q}\}} = 1\} = \mathbf{P}\{U \leq |x|^{-1/q}\} = |x|^{-1/q}. \tag{6.180}$$

In order for $D_o \geq k$ to occur, for k large, we must have that $U_o = u$ is quite small, and, in this case, a central limit theorem should hold for D_o, with mean equal to

$$\mathbf{E}[D_o|U_o = u] = \sum_{x:|x|\leq u^{-q}} |x|^{-1/q} = cu^{-(qd-1)}(1 + o(1)), \tag{6.181}$$

for some explicit constant $c = c(q, d)$. Furthermore, the conditional variance of D_o given that $U_o = u$ is bounded above by its conditional expectation, so that the conditional distribution of D_o given that $U_o = u$ is highly concentrated. We omit the details, and merely note that this can be made precise by using standard large deviations result. Assuming sufficient concentration, we obtain that the probability that $D_o \geq k$ is asymptotically equal to the probability that $U \leq u_k$, where u_k is determined by the equation that

$$\mathbf{E}[D_o|U_o = u_k] = cu_k^{-(qd-1)}(1 + o(1)) = k, \tag{6.182}$$

so that $u_k = (k/c)^{-1/(qd-1)}$. This suggests that

$$\mathbf{P}\{D_o \geq k\} = \mathbf{P}\{U \leq u_k\}(1 + o(1)) = (ck)^{-1/(qd-1)}(1 + o(1)), \tag{6.183}$$

which explains (6.178).

We next turn to distances in this scale-free percolation model. For $x, y \in \mathbb{Z}^d$, we denote by $d_{\mathbb{G}_q}(x, y)$ the graph distance (or chemical distance) between the vertices x and y, i.e., the minimal number of edges in $\mathbb{G}_q$ connecting x and y. The main result in Yukich (2006) is the following theorem:

Theorem 6.20 (Ultra-small distances for scale-free percolation) *For all $d \geq 1$ and all $q \in (1/d, \infty)$,* **whhp** *as $|x| \to \infty$,*

$$d_{\mathbb{G}_q}(o, x) \leq 8 + 4\log\log|x|. \tag{6.184}$$

The result in Theorem 6.20 shows that distances in the scale-free percolation model are *much* smaller than those in normal percolation models. It would be of interest to investigate whether the limit $d_{\mathbb{G}_q}(o, x)/\log\log|x|$ exists, and, if so, what this limit is.

While Theorem 6.20 resembles the results in Theorem 6.19, there are a few essential differences. First of all, $\mathbb{G}_q$ is an *infinite* graph, whereas the models considered in Theorem 6.19 are all *finite*. It would be of interest to extend Theorem 6.20 to the setting on finite tori, where the Euclidean norm $|x - y|$ in (6.177) is replaced by the Euclidean norm on the torus, and the typical distance H_n is considered. This result is not immediate from the proof of Theorem 6.20. Secondly, in Theorems 6.18 and 6.19, it is apparent that the behaviour for $\tau > 3$ is rather different compared to the behaviour for $\tau \in (2, 3)$. This feature is missing in Theorem 6.20. It would be of interest to find a geometric random graph model where the difference in behaviour between $\tau > 3$ and $\tau \in (2, 3)$ also appears.

The result in Theorem 6.20 can be compared to similar results for *long-range percolation*, where edges are present independently, and the probability that the edge $\{x, y\}$ is present equals $|x - y|^{-s+o(1)}$ for some $s > 0$. In this case, detailed results exist for the limiting behaviour of $d(o, x)$ depending on the value of s. For example, in Benjamini, Kesten, Peres and Schramm (2004), it is shown that the diameter of this infinite percolation model is equal to $\lceil d/(d - s)\rceil$ a.s. See also Biskup (2004) and the references therein.

6.3.5.3 *Spatial preferential attachment models* In the past years, several spatial preferential attachment models have been considered. We shall now discuss three of such models.

In Flaxman, Frieze and Vera (2006, 2007), a class of geometric preferential attachment models that combines aspects of random geometric graphs and preferential attachment graphs is introduced and studied. Let $G_t = (V_t, E_t)$ denote the graph at time t. Let S be the sphere S in $\mathbb{R}^3$ with area equal to 1. Then, we let V_t be a subset of S of size t.

The process $\{G_t\}_{t=0}^{\infty}$ evolves as follows. At time $t = 0$, G_0 is the empty graph. At time $t+1$, given G_t, we obtain G_{t+1} as follows. Let x_{t+1} be chosen uniformly at random from S, and denote $V_{t+1} = V_t \cup \{x_{t+1}\}$. We assign m edges to the vertex x_{t+1}, which we shall connect independently of each other to vertices in $V_t(x_{t+1}) \equiv V_t \cap B_r(x_{t+1})$, where $B_r(u) = \{x \in S : \|x - u\| \leq r\}$ denotes the spherical cap of radius r around u. Let

$$D_t(x_{t+1}) = \sum_{v \in V_t(x_t)} D_v^{(t)}, \tag{6.185}$$

where $D_v^{(t)}$ denotes the degree of vertex $v \in V_t$ in G_t. The m edges are connected to vertices $(y_1, \ldots, y_m)$ conditionally independently given (G_t, x_{t+1}), so that, for all $v \in V_t(x_{t+1})$,

$$\mathbf{P}_p\{y_i = v \mid G_t\} = \frac{D_v^{(t)}}{\max(D_t(x_{t+1}), \alpha m A_r t)}, \tag{6.186}$$

while

$$\mathbf{P}_p\{y_i = | G_t\} = 1 - \frac{D_t(x_{t+1})}{\max(D_t(x_{t+1}), \alpha m A_r t)}, \tag{6.187}$$

where A_r is the area of $B_r(u)$, $\alpha \geq 0$ is a parameter, and r is a radius which shall be chosen appropriately. Similarly to the situation of geometric random graphs, the parameter r shall depend on the size of the graph, i.e. we shall be interested in the properties of G_n when $r = r_n$ is chosen appropriately. The main result in Flaxman, Frieze and Vera (2006) is the study of the degree sequence of the arising model. Take $r_n = n^{\beta-1/2} \log n$, where $\beta \in (0, 1/2)$ is a constant. Finally, let $\alpha > 2$. Then, there exists a probability distribution $\{p_k\}_{k=m}^{\infty}$ such that, **whp**,

$$P_k^{(n)} = p_k(1 + o(1)), \tag{6.188}$$

where $\{p_k\}_{k=m}^{\infty}$ satisfies (6.145) with $\tau = 1 + \alpha \in (3, \infty)$. The precise result is in Flaxman, Frieze and Vera (2006, Theorem 1(a)) and is quite a bit sharper, as detailed concentration results are proved as well. Further results involve the proof of connectivity of G_n and an upper bound on the diameter when $r \geq n^{-1/2} \log n$, $m \geq K \log n$ for some large enough K and $\alpha \geq 0$ of order $\mathcal{O}(\log(n/r))$. In Flaxman, Frieze and Vera (2007), these results were generalized to the setting where, instead of a unit ball, a smoother version is used.

In Aiello, Bonato, Cooper, Janssen and Pralat (2007), a spatial preferential attachment model with local influence regions is studied, as a model for the Web graph. The model is *directed*, but it can be easily adapted to an undirected setting. The idea behind the model in Aiello, Bonato, Cooper, Janssen and Pralat (2007) is that for normal preferential attachment models, new vertices should be aware of the degrees of the already present vertices. In reality, it is quite hard to observe the degrees of vertices, and, therefore, in Aiello, Bonato, Cooper, Janssen and Pralat (2007), vertices instead have a region of influence in some metric space, for example the torus $[0, 1]^m$ for some dimension m, for which the metric equals

$$d(x, y) = \min\{\|x - y + u\|_\infty : u \in \{0, 1, -1\}^m\}. \tag{6.189}$$

When the new vertex arrives, it is uniformly located somewhere in the unit cube, and it connects to each of the older vertices in which region of influence they land independently and with fixed probability p. These regions of influence evolve as time proceeds, in such a way that the volume of the influence region of the vertex i at time t is equal to

$$R(i, t) = \frac{A_1 D_i^{(t)} + A_2}{t + A_3}, \tag{6.190}$$

where now $D_i^{(t)}$ is the in-degree of vertex i at time t, and A_1, A_2, A_3 are parameters which are chosen such that $pA_1 \leq 1$. One of the main results of the paper is that this model is a scale-free graph process. Indeed, denote

$$p_k = \frac{p^k}{1 + kpA_1 + pA_2} \prod_{j=0}^{k-1} \frac{jA_1 + A_2}{1 + A_2 + pA_2}, \tag{6.191}$$

then Aiello, Bonato, Cooper, Janssen and Pralat (2007, Theorem 1.1) show that **whp**, for $k \leq (n^{1/8}/\log n)^{4pA_1/(2pA_1+1)}$, the degree sequence of the graph of size n satisfies (recall (6.143))

$$P_k^{(n)} = p_k(1 + o(1)), \tag{6.192}$$

and $\{p_k\}_{k=0}^{\infty}$ satisfies (6.145) with $\tau = 1 + 1/(pA_1) \in [2, \infty)$. Further results involve the study of maximal in-degrees and the total number of edges.

For a relation between preferential attachment graphs with so-called fertility and aging, and a geometric competition-induced growth model for networks, we refer to Berger, Borgs, Chayes, D'Souza and Kleinberg (2004, 2005) and the references therein.

Acknowledgements

The work of RvdH was supported in part by Netherlands Organisation for Scientific Research (NWO). I am grateful to Gordon Slade for introducing me to the field of percolation, which has proved to be an extremely beautiful and rich world indeed! It has always been a delight to attack percolation problems with Gordon, and I have enjoyed this dearly. I warmly thank Gerard Hooghiemstra for jointly taking the first steps (and then many more) in the field of random graphs. Let's take plenty more steps together!

References

Aiello, W., Bonato, A, Cooper, C., Janssen, J., and Pralat, P. (2007). A spatial web graph model with local influence regions. In: *Algorithms and Models for the Web-Graph*, Lecture Notes in Computer Science, Volume 4863, Springer, Berlin, pp. 96–107.

Aiello, W., Chung, F., and Lu, L. (2002). Random evolution in massive graphs. In *Handbook of massive data sets*, Volume **4** of *Massive Comput.*, pp. 97–122. Kluwer Acad. Publ., Dordrecht.

Aizenman, M. (1997). On the number of incipient spanning clusters. *Nucl. Phys. B*, **485**, 551–582.

Aizenman, M. and Barsky, D.J. (1987). Sharpness of the phase transition in percolation models. *Commun. Math. Phys.*, **108**, 489–526.

Aizenman, M., Kesten, H., and Newman, C.M. (1987). Uniqueness of the infinite cluster and continuity of connectivity functions for short and long range percolation. *Commun. Math. Phys.*, **111**, 505–531.

Aizenman, M. and Newman, C.M. (1984). Tree graph inequalities and critical behaviour in percolation models. *J. Stat. Phys.*, **36**, 107–143.

Albert, R. and Barabási, A.-L. (2002). Statistical mechanics of complex networks. *Rev. Modern Phys.*, **74**(1), 47–97.

Alon, N. and Spencer, J. (2000). *The Probabilistic Method*, second edn. Wiley-Interscience Series in Discrete Mathematics and Optimization. John Wiley & Sons, New York.

Athreya, K. and Ney, P. (1972). *Branching Processes.* Springer-Verlag, New York. Die Grundlehren der mathematischen Wissenschaften, Band 196.

Austin, T. L., Fagen, R. E., Penney, W. F., and Riordan, J. (1959). The number of components in random linear graphs. *Ann. Math. Statist*, **30**, 747–754.

Baccelli, F. and Błaszczyszyn, B. (2001). On a coverage process ranging from the Boolean model to the Poisson-Voronoi tessellation with applications to wireless communications. *Adv. in Appl. Probab.*, **33**(2), 293–323.

Barabási, A.-L. (2002). *Linked: The New Science of Networks.* Perseus Publishing, Cambridge, Massachusetts.

Barabási, A.-L. and Albert, R. (1999). Emergence of scaling in random networks. *Science*, **286**(5439), 509–512.

Barbour, A. D. and Reinert, G. (2001). Small worlds. *Random Structures Algorithms*, **19**(1), 54–74.

Barbour, A. D. and Reinert, Gesine (2004). Correction: 'Small worlds' [Random Structures Algorithms **19** (2001), no. 1, 54–74; mr1848027]. *Random Structures Algorithms*, **25**(1), 115.

Barbour, A. D. and Reinert, G. (2006). Discrete small world networks. *Electron. J. Probab.*, **11**, no. 47, 1234–1283 (electronic).

Barsky, D.J. and Aizenman, M. (1991). Percolation critical exponents under the triangle condition. *Ann. Probab.*, **19**, 1520–1536.

Barsky, D.J., Grimmett, G.R., and Newman, C.M. (1991*a*). Dynamic renormalization and continuity of the percolation transition in orthants. In *Spatial stochastic processes*, Volume **19** of *Progr. Probab.*, pp. 37–55. Birkhäuser Boston, Boston, MA.

Barsky, D.J., Grimmett, G.R., and Newman, C.M. (1991*b*). Percolation in half-spaces: equality of critical densities and continuity of the percolation probability. *Probab. Theory Related Fields*, **90**(1), 111–148.

Bender, E.A. and Canfield, E.R. (1978). The asymptotic number of labelled graphs with a given degree sequences. *J. Combinat. Theory (A)*, **24**, 296–307.

Benjamini, I., Kesten, H., Peres, Y., and Schramm, O. (2004). Geometry of the uniform spanning forest: transitions in dimensions $4, 8, 12, \ldots$. *Ann. of Math. (2)*, **160**(2), 465–491.

Benjamini, I., Lyons, R., Peres, Y., and Schramm, O. (1999*a*). Critical percolation on any nonamenable group has no infinite clusters. *Ann. Probab.*, **27**(3), 1347–1356.

Benjamini, I., Lyons, R., Peres, Y., and Schramm, O. (1999*b*). Group-invariant percolation on graphs. *Geom. Funct. Anal.*, **9**(1), 29–66.

Benjamini, I. and Schramm, O. (1996). Percolation beyond $\mathbf{Z}^d$, many questions and a few answers. *Electron. Comm. Probab.*, **1**, no. 8, 71–82 (electronic).

van den Berg, J. and Keane, M. (1984). On the continuity of the percolation probability function. In *Conference in Modern Analysis and Probability (New Haven, Conn., 1982)*, Volume **26** of *Contemp. Math.*, pp. 61–65. Amer. Math. Soc., Providence, RI.

Berg, J. van den and Kesten, H. (1985). Inequalities with applications to percolation and reliability. *J. Appl. Prob.*, **22**, 556–569.

Berger, N., Bollobás, B., Borgs, C., Chayes, J., and Riordan, O. (2003). Degree distribution of the FKP network model. In *Automata, Languages and Programming*, Volume **2719** of *Lecture Notes in Comput. Sci.*, pp. 725–738. Springer, Berlin.

Berger, N., Borgs, C., Chayes, J. T., D'Souza, R. M., and Kleinberg, R. D. ((2004)). Competition-induced preferential attachment. In *Automata, Languages and Programming*, Volume **3142** of *Lecture Notes in Comput. Sci.*, pp. 208–221. Springer, Berlin.

Berger, N., Borgs, C., Chayes, J. T., D'Souza, R. M., and Kleinberg, R. D. (2005). Degree distribution of competition-induced preferential attachment graphs. *Combin. Probab. Comput.*, **14**(5-6), 697–721.

Bezuidenhout, C. and Grimmett, G. (1990). The critical contact process dies out. *Ann. Probab.*, **18**, 1462–1482.

Biskup, M. (2004). On the scaling of the chemical distance in long-range percolation models. *Ann. Probab.*, **32**(4), 2938–2977.

Bollobás, B. (2001). *Random Graphs*, second edn, Volume **73** of *Cambridge Studies in Advanced Mathematics*. Cambridge University Press, Cambridge.

Bollobás, B., Borgs, C., Chayes, J., and Riordan, O. (2003). Directed scale-free graphs. In *Proceedings of the Fourteenth Annual ACM-SIAM Symposium on Discrete Algorithms (Baltimore, MD, 2003)*, New York, pp. 132–139. ACM.

Bollobás, B., Janson, S., and Riordan, O. (2007). The phase transition in inhomogeneous random graphs. *Random Structures Algorithms*, **31**(1), 3–122.

Bollobás, B. and Riordan, O. (2003*a*). Mathematical results on scale-free random graphs. In *Handbook of Graphs and Networks*, pp. 1–34. Wiley-VCH, Weinheim.

Bollobás, B. and Riordan, O. (2003*b*). Robustness and vulnerability of scale-free random graphs. *Internet Math.*, **1**(1), 1–35.

Bollobás, B. and Riordan, O. (2004*a*). Coupling scale-free and classical random graphs. *Internet Math.*, **1**(2), 215–225.

Bollobás, B. and Riordan, O. (2004*b*). The diameter of a scale-free random graph. *Combinatorica*, **24**(1), 5–34.

Bollobás, B. and Riordan, O. (2006*a*). The critical probability for random Voronoi percolation in the plane is 1/2. *Probab. Theory Related Fields*, **136**(3), 417–468.

Bollobás, B. and Riordan, O. (2006*b*). *Percolation.* Cambridge University Press, New York.

Bollobás, B. and Riordan, O. (2006*c*). A short proof of the Harris–Kesten theorem. *Bull. London Math. Soc.*, **38**(3), 470–484.

Bollobás, B. and Riordan, O. (2008). Percolation on random Johnson-Mehl tessellations and related models. *Probab. Theory Related Fields*, **140**(3-4), 319–343.

Bollobás, B., Riordan, O., Spencer, J., and Tusnády, G. (2001). The degree sequence of a scale-free random graph process. *Random Structures Algorithms*, **18**(3), 279–290.

Borgs, C., Chayes, J., Hofstad, R. van der, Slade, G., and Spencer, J. (2005*a*). Random subgraphs of finite graphs. I. The scaling window under the triangle condition. *Random Structures Algorithms*, **27**(2), 137–184.

Borgs, C., Chayes, J., Hofstad, R. van der, Slade, G., and Spencer, J. (2005*b*). Random subgraphs of finite graphs. II. The lace expansion and the triangle condition. *Ann. Probab.*, **33**(5), 1886–1944.

Borgs, C., Chayes, J. T., Kesten, H., and Spencer, J. (1999). Uniform boundedness of critical crossing probabilities implies hyperscaling. *Random Structures Algorithms*, **15**(3-4), 368–413. Statistical physics methods in discrete probability, combinatorics, and theoretical computer science (Princeton, NJ, 1997).

Britton, T., Deijfen, M., and Martin-Löf, A. (2006). Generating simple random graphs with prescribed degree distribution. *J. Stat. Phys.*, **124**(6), 1377–1397.

Broadbent, S. R. and Hammersley, J. M. (1957). Percolation processes. I. Crystals and mazes. *Proc. Cambridge Philos. Soc.*, **53**, 629–641.

Burton, R. M. and Keane, M. (1989). Density and uniqueness in percolation. *Comm. Math. Phys.*, **121**(3), 501–505.

Camia, F. (Preprint (2008)). Scaling limits of two-dimensional percolation: an overview.

Cardy, John L. (1992). Critical percolation in finite geometries. *J. Phys. A*, **25**(4), L201–L206.

Chayes, J. T. and Chayes, L. (1987). On the upper critical dimension of Bernoulli percolation. *Comm. Math. Phys.*, **113**(1), 27–48.

Chayes, J. T., Chayes, L., Grimmett, G. R., Kesten, H., and Schonmann, R. H. (1989). The correlation length for the high-density phase of Bernoulli percolation. *Ann. Probab.*, **17**(4), 1277–1302.

Chung, F. and Lu, L. (2002*a*). The average distances in random graphs with given expected degrees. *Proc. Natl. Acad. Sci. USA*, **99**(25), 15879–15882 (electronic).

Chung, F. and Lu, L. (2002*b*). Connected components in random graphs with given expected degree sequences. *Ann. Comb.*, **6**(2), 125–145.

Chung, F. and Lu, L. (2003). The average distance in a random graph with given expected degrees. *Internet Math.*, **1**(1), 91–113.

Chung, F. and Lu, L. (2006*a*). *Complex Graphs and Networks*, Volume **107** of *CBMS Regional Conference Series in Mathematics*. Published for the Conference Board of the Mathematical Sciences, Washington, DC.

Chung, F. and Lu, L. (2006*b*). The volume of the giant component of a random graph with given expected degrees. *SIAM J. Discrete Math.*, **20**, 395–411.

Chung, F., Lu, L., and Vu, V. (2004). The spectra of random graphs with given expected degrees. *Internet Math.*, **1**(3), 257–275.

Cooper, C. and Frieze, A. (2003). A general model of web graphs. *Random Structures Algorithms*, **22**(3), 311–335.

Davies, P. L. (1978). The simple branching process: a note on convergence when the mean is infinite. *J. Appl. Probab.*, **15**(3), 466–480.

Dawson, D.A. (1993). Measure-valued Markov processes. In *Ecole d'Eté de Probabilités de Saint-Flour 1991. Lecture Notes in Mathematics #1541*, Berlin. Springer.

Deijfen, M., Esker, H. van den, Hofstad, R. van der, and Hooghiemstra, G. (Preprint (2007)). A preferential attachment model with random initial degrees.

Dodziuk, J. (1984). Difference equations, isoperimetric inequality and transience of certain random walks. *Trans. Amer. Math. Soc.*, **284**(2), 787–794.

Dorogovtsev, S.N. and Mendes, J.F.F. (2002). Evolution of networks. *Advances in Physics*, **51**, 1079–1187.

Dousse, O., Franceschetti, M., Macris, N., Meester, R., and Thiran, P. (2006). Percolation in the signal to interference ratio graph. *J. Appl. Probab.*, **43**(2), 552–562.

Durrett, R. (1980). On the growth of one-dimensional contact processes. *Ann. Probab.*, **8**(5), 890–907.

Durrett, R. (2007). *Random Graph Dynamics*. Cambridge Series in Statistical and Probabilistic Mathematics. Cambridge University Press, Cambridge.

Dynkin, E. B. (1994). *An Introduction to Branching Measure-Valued Processes*, Volume **6** of *CRM Monograph Series*. American Mathematical Society, Providence, RI.

Erdős, P. and Rényi, A. (1959). On random graphs. I. *Publ. Math. Debrecen*, **6**, 290–297.

Erdős, P. and Rényi, A. (1960). On the evolution of random graphs. *Magyar Tud. Akad. Mat. Kutató Int. Közl.*, **5**, 17–61.

Esker, H. van den, Hofstad, R. van der, and Hooghiemstra, G. (Preprint (2006*a*). Universality for the distance in finite variance random graphs.

Esker, H. van den, Hofstad, R. van der, Hooghiemstra, G., and Znamenski, D. (2006*b*). Distances in random graphs with infinite mean degrees. *Extremes*, **8**, 111–140.

Etheridge, A. M. (2000). *An Introduction to Superprocesses*, Volume **20** of *University Lecture Series*. American Mathematical Society, Providence, RI.

Fernholz, D. and Ramachandran, V. (2007). The diameter of sparse random graphs. *Random Structures Algorithms*, **31**(4), 482–516.

Flaxman, A., Frieze, A., and Vera, J. (2006). A geometric preferential attachment model of networks. *Internet Math.*, **3**(2), 187–205.

Flaxman, A., Frieze, A., and Vera, J. (2007). A geometric preferential attachment model of networks II. In *Proceedings of WAW 2007*.

Fortuin, C. M., Kasteleyn, P. W., and Ginibre, J. (1971). Correlation inequalities on some partially ordered sets. *Comm. Math. Phys.*, **22**, 89–103.

Franceschetti, M. and Meester, R. (2008). *Random Networks for Communication: From Statistical Physics to Information Systems*. Cambridge University Press.

Gilbert, E. N. (1959). Random graphs. *Ann. Math. Statist.*, **30**, 1141–1144.

Grimmett, G. (1999). *Percolation*, 2nd edn. Springer, Berlin.

Grimmett, G. and Hiemer, P. (2002). Directed percolation and random walk. In *In and out of equilibrium (Mambucaba, 2000)*, Volume **51** of *Progr. Probab.*, pp. 273–297. Birkhäuser Boston, Boston, MA.

Grimmett, G. R. and Newman, C. M. (1990). Percolation in $\infty + 1$ dimensions. In *Disorder in Physical Systems*, Oxford Sci. Publ., pp. 167–190. Oxford Univ. Press, New York.

Grossglauser, M. and Thiran, P. (2006). Networks out of control: Models and methods for random networks. `http://icawww1.epfl.ch/class-nooc/nooc2006.pdf`

Häggström, O. (2007) Problem solving is often a matter of cooking up an appropriate Markov chain. *Scandinavian Journal of Statistics*, **34**, 768–780.

Häggström, O. and Peres, Y. (1999). Monotonicity of uniqueness for percolation on Cayley graphs: all infinite clusters are born simultaneously. *Probab. Theory Related Fields*, **113**(2), 273–285.

Häggström, O., Peres, Y., and Schonmann, R. H. (1999). Percolation on transitive graphs as a coalescent process: relentless merging followed by simultaneous uniqueness. In *Perplexing problems in probability*, Volume **44** of *Progr. Probab.*, pp. 69–90. Birkhäuser Boston, Boston, MA.

Hammersley, J. M. (1961). Comparison of atom and bond percolation processes. *J. Mathematical Phys.*, **2**, 728–733.

Hammersley, J. M. (1963). A Monte Carlo solution for percolation in a cubic lattice. In *Methods in Computational Physics, Vol. I*, pp. 281–298. Academic Press.

Hara, T. (1990). Mean field critical behaviour for correlation length for percolation in high dimensions. *Probab. Th. Rel. Fields*, **86**, 337–385.

Hara, T. (2005). Decay of correlations in nearest-neighbour self-avoiding walk, percolation, lattice trees and animals. Preprint. Available on `http://arxiv.org/abs/math-ph/0504021`.

Hara, T., Hofstad, R. van der, and Slade, G. (2003). Critical two-point functions and the lace expansion for spread-out high-dimensional percolation and related models. *Ann. Probab.*, **31**(1), 349–408.

Hara, T. and Slade, G. (1990). Mean-field critical behaviour for percolation in high dimensions. *Commun. Math. Phys.*, **128**, 333–391.

Hara, T. and Slade, G. (1993). Unpublished appendix to Hara and Slade (1995). Available on `http://www.ma.utexas.edu/mp_arc/index-93.html`.

Hara, T. and Slade, G. (1995). The self-avoiding-walk and percolation critical points in high dimensions. *Combinatorics, Probability and Computing*, **4**, 197–215.

Hara, T. and Slade, G. (2000*a*). The scaling limit of the incipient infinite cluster in high-dimensional percolation. I. Critical exponents. *J. Statist. Phys.*, **99**(5-6), 1075–1168.

Hara, T. and Slade, G. (2000*b*). The scaling limit of the incipient infinite cluster in high-dimensional percolation. II. Integrated super-Brownian excursion. *J. Math. Phys.*, **41**(3), 1244–1293.

Harris, T. (1963). *The theory of branching processes.* Die Grundlehren der Mathematischen Wissenschaften, Bd. 119. Springer-Verlag, Berlin.

Harris, T. E. (1960). A lower bound for the critical probability in a certain percolation process. *Proc. Cambridge Philos. Soc.*, **56**, 13–20.

Heydenreich, M. and Hofstad, R. van der (2007). Random graph asymptotics on high-dimensional tori. *Comm. Math. Phys.*, **270**(2), 335–358.

Heydenreich, M., Hofstad, R. van der, and Sakai, A. (Preprint (2008)). Mean-field behaviour for long- and finite range Ising model, percolation and self-avoiding walk.

Hofstad, R. van der (2008). Random graphs and complex networks. In preparation, see `http://www.win.tue.nl/~rhofstad/NotesRGCN.pdf`.

Hofstad, R. van der, Hollander, F. den, and Slade, G. (2007). The survival probability for critical spread-out oriented percolation above 4+1 dimensions. I. Induction. *Probab. Theory Related Fields*, **138**(3-4), 363–389.

Hofstad, R. van der, Hooghiemstra, G., and Dommers, S. (Preprint (2009)). Diameters in preferential attachment graphs.

Hofstad, R. van der, Hooghiemstra, G., and Znamenski, D. (2009). A phase transition for the diameter of the configuration model. *Internet Mathematics*, **4**(1): 113–128.

Hofstad, R. van der, Hooghiemstra, G., and Van Mieghem, P. (2005). Distances in random graphs with finite variance degrees. *Random Structures Algorithms*, **27**(1), 76–123.

Hofstad, R. van der, Hooghiemstra, G., and Znamenski, D. (2007). Distances in random graphs with finite mean and infinite variance degrees. *Electron. J. Probab.*, **12**(25), 703–766 (electronic).

Hofstad, R. van der and Keane, M. (Preprint (2007)). An elementary proof of the hitting-time theorem.

Hofstad, R. van der and Luczak, M. (Preprint (2006)). Random subgraphs of the 2D Hamming graph: The supercritical phase.

Hofstad, R. van der and Sakai, A. (2005). Critical points for spread-out self-avoiding walk, percolation and the contact process above the upper critical dimensions. *Probab. Theory Related Fields*, **132**(3), 438–470.

Hofstad, R. van der and Slade, G. (2003). Convergence of critical oriented percolation to super-Brownian motion above $4+1$ dimensions. *Ann. Inst. H. Poincaré Probab. Statist.*, **39**(3), 413–485.

Hofstad, R. van der and Slade, G. (2005). Asymptotic expansions in n^{-1} for percolation critical values on the n-cube and $\mathbb{Z}^n$. *Random Structures Algorithms*, **27**(3), 331–357.

Hofstad, R. van der and Slade, G. (2006). Expansion in n^{-1} for percolation critical values on the n-cube and $\mathbb{Z}^n$: the first three terms. *Combin. Probab. Comput.*, **15**(5), 695–713.

Hughes, B. D. (1996). *Random Walks and Random Environments. Vol. 2. Random Environments.* Oxford Science Publications. The Clarendon Press, Oxford University Press, New York.

Jagers, P. (1975). *Branching Processes with Biological Applications.* Wiley-Interscience, London. Wiley Series in Probability and Mathematical Statistics – Applied Probability and Statistics.

Janson, S. (Preprint (2006)). The probability that a random multigraph is simple.

Janson, S. (Preprint (2008)). Asymptotic equivalence and contiguity of some random graphs.

Janson, S. and Luczak, M. (Preprint (2007)). A new approach to the giant component problem.

Janson, S., Łuczak, T., and Rucinski, A. (2000). *Random Graphs.* Wiley-Interscience Series in Discrete Mathematics and Optimization. Wiley-Interscience, New York.

Kager, W. and Nienhuis, B. (2004). A guide to stochastic Löwner evolution and its applications. *J. Statist. Phys.*, **115**(5-6), 1149–1229.

Kendall, W. S. and Wilson, R. G. (2003). Ising models and multiresolution quad-trees. *Adv. in Appl. Probab.*, **35**(1), 96–122. In honor of Joseph Mecke.

Kesten, H. (1959*a*). Full Banach mean values on countable groups. *Math. Scand.*, **7**, 146–156.

Kesten, H. (1959*b*). Symmetric random walks on groups. *Trans. Amer. Math. Soc.*, **92**, 336–354.

Kesten, H. (1980). The critical probability of bond percolation on the square lattice equals $\frac{1}{2}$. *Comm. Math. Phys.*, **74**(1), 41–59.

Kesten, H. (1982). *Percolation Theory for Mathematicians*, Volume **2** of *Progress in Probability and Statistics*. Birkhäuser Boston, Mass.

Kesten, H. (1987). Scaling relations for 2D-percolation. *Comm. Math. Phys.*, **109**(1), 109–156.

Kesten, H. (2002). Some highlights of percolation. In *Proceedings of the International Congress of Mathematicians, Vol. I (Beijing, 2002)*, Beijing, pp. 345–362. Higher Ed. Press.

Langlands, R., Pouliot, P., and Saint-Aubin, Y. (1994). Conformal invariance in two-dimensional percolation. *Bull. Amer. Math. Soc. (N.S.)*, **30**(1), 1–61.

Lawler, G. F. (2004). An introduction to the stochastic Loewner evolution. In *Random Walks and Geometry*, pp. 261–293. Walter de Gruyter GmbH & Co. KG, Berlin.

Lawler, G. F. (2005). *Conformally Invariant Processes in the Plane*, Volume **114** of *Mathematical Surveys and Monographs*. American Mathematical Society, Providence, RI.

Lawler, G. F., Schramm, O., and Werner, W. (2002). One-arm exponent for critical 2D percolation. Volume **7**, pp. no. 2, 13 pp. (electronic).

Le Gall, J.-F. (1999). *Spatial Branching Processes, Random Snakes, and Partial Differential Equations*. Birkhäuser, Basel.

Meester, R. (1994). Uniqueness in percolation theory. *Statist. Neerlandica*, **48**(3), 237–252.

Meester, R. and Roy, R. (1996). *Continuum Percolation*, Volume **119** of *Cambridge Tracts in Mathematics*. Cambridge University Press, Cambridge.

Menshikov, M.V. (1986). Coincidence of critical points in percolation problems. *Soviet Mathematics, Doklady*, **33**, 856–859.

Molloy, M. and Reed, B. (1995). A critical point for random graphs with a given degree sequence. In *Proceedings of the Sixth International Seminar on Random Graphs and Probabilistic Methods in Combinatorics and Computer Science, 'Random Graphs '93' (Poznań, 1993)*, Volume **6**, pp. 161–179.

Molloy, M. and Reed, B. (1998). The size of the giant component of a random graph with a given degree sequence. *Combin. Probab. Comput.*, **7**(3), 295–305.

Moore, C. and Newman, M.E.J. (2000). Epidemics and percolation in small-world networks. *Phys. Rev. E*, **61**, 5678–5682.

Nachmias, A. (Preprint (2007)). Mean-field conditions for percolation in finite graphs.

Newman, M. E. J. (2003). The structure and function of complex networks. *SIAM Rev.*, **45**(2), 167–256 (electronic).

Newman, M.E.J., Moore, C., and Watts, D.J. (2000). Mean-field solution of the small-world network model. *Phys. Rev. Lett.*, **84**, 3201–3204.

Newman, M.E.J. and Watts, D.J. (1999). Scaling and percolation in the small-world network model. *Phys. Rev. E*, **60**, 7332–7344.

Newman, M. E. J., Watts, D. J., and Barabási, A.-L. (2006). *The Structure and Dynamics of Networks*. Princeton Studies in Complexity. Princeton University Press.

Nguyen, B. G. (1987). Gap exponents for percolation processes with triangle condition. *J. Statist. Phys.*, **49**(1-2), 235–243.

Nguyen, B.G. and Yang, W-S. (1993). Triangle condition for oriented percolation in high dimensions. *Ann. Probab.*, **21**, 1809–1844.

Nguyen, B.G. and Yang, W-S. (1995). Gaussian limit for critical oriented percolation in high dimensions. *J. Stat. Phys.*, **78**, 841–876.

Nienhuis, B. (1984). Critical behaviour of two-dimensional spin models and charge asymmetry in the Coulomb gas. *J. Statist. Phys.*, **34**(5-6), 731–761.

Norros, I. and Reittu, H. (2006). On a conditionally Poissonian graph process. *Adv. in Appl. Probab.*, **38**(1), 59–75.

Oliveira, R. and Spencer, J. (2005). Connectivity transitions in networks with super-linear preferential attachment. *Internet Math.*, **2**(2), 121–163.

Penrose, M.D. (2003). *Random Geometric Graphs*, Volume **5** of *Oxford Studies in Probability*. Oxford University Press, Oxford.

Perkins, E. (2002). Dawson-Watanabe superprocesses and measure-valued diffusions. In *Lectures on Probability Theory and Statistics (Saint-Flour, 1999)*, Volume **1781** of *Lecture Notes in Math.*, pp. 125–324. Springer, Berlin.

Reimer, D. (2000). Proof of the van den Berg–Kesten conjecture. *Combin. Probab. Comput.*, **9**(1), 27–32.

Rudas, A., Tóth, B., and Valkó, B. (2007). Random trees and general branching processes. *Random Structures Algorithms*, **31**(2), 186–202.

Russo, L. (1978). A note on percolation. *Z. Wahrscheinlichkeitstheorie und Verw. Gebiete*, **43**(1), 39–48.

Russo, L. (1981). On the critical percolation probabilities. *Z. Wahrsch. Verw. Gebiete*, **56**(2), 229–237.

Sakai, A. (2002). Hyperscaling inequalities for the contact process and oriented percolation. *J. Statist. Phys.*, **106**(1-2), 201–211.

Schonmann, R. H. (1999). Stability of infinite clusters in supercritical percolation. *Probab. Theory Related Fields*, **113**(2), 287–300.

Schonmann, R. H. (2001). Multiplicity of phase transitions and mean-field criticality on highly non-amenable graphs. *Comm. Math. Phys.*, **219**, 271–322.

Schonmann, R. H. (2002). Mean-field criticality for percolation on planar non-amenable graphs. *Comm. Math. Phys.*, **225**(3), 453–463.

Schramm, O. (2000). Scaling limits of loop-erased random walks and uniform spanning trees. *Israel J. Math.*, **118**, 221–288.

Seymour, P. D. and Welsh, D. J. A. (1978). Percolation probabilities on the square lattice. *Ann. Discrete Math.*, **3**, 227–245. Advances in Graph Theory (Cambridge Combinatorial Conf., Trinity College, Cambridge, 1977).

Slade, G. (2006). *The Lace Expansion and its Applications*, Volume **1879** of *Lecture Notes in Mathematics*. Springer-Verlag, Berlin. Lectures from the 34th Summer School on Probability Theory held in Saint-Flour, July 6–24, 2004, Edited and with a foreword by Jean Picard.

Smirnov, S. (2001). Critical percolation in the plane: conformal invariance, Cardy's formula, scaling limits. *C. R. Acad. Sci. Paris Sér. I Math.*, **333**(3), 239–244.

Smirnov, S. and Werner, W. (2001). Critical exponents for two-dimensional percolation. *Math. Res. Lett.*, **8**(5-6), 729–744.

Tasaki, H. (1987*a*). Geometric critical exponent inequalities for general random cluster models. *J. Statist. Phys.*, **49**(3-4), 841–847.

Tasaki, H. (1987*b*). Hyperscaling inequalities for percolation. *Comm. Math. Phys.*, **113**(1), 49–65.

Timár, Á. (2006). Percolation on nonunimodular transitive graphs. *Ann. Probab.*, **34**(6), 2344–2364.

Turova, T.S. and Vallier, T. (Preprint (2006)). Merging percolation on $\mathbf{Z}^d$ and classical random graphs: Phase transition.

Werner, W. (2004). Random planar curves and Schramm–Loewner evolutions. In *Lectures on probability theory and statistics*, Volume **1840** of *Lecture Notes in Math.*, pp. 107–195. Springer, Berlin.

Werner, W. (2005). SLE, conformal restriction, loops. In *European Congress of Mathematics*, pp. 515–528. Eur. Math. Soc., Zürich.

Wu, C. C. (1993). Critical behaviour or percolation and Markov fields on branching planes. *J. Appl. Probab.*, **30**(3), 538–547.

Yukich, J. E. (2006). Ultra-small scale-free geometric networks. *J. Appl. Probab.*, **43**(3), 665–677.

7

RANDOM DIRECTED AND ON-LINE NETWORKS

Mathew D. Penrose and Andrew R. Wade

Abstract

Various random spatial graphs defined on partially ordered point sets have been treated recently in the literature: these are discussed in the framework of the minimal directed spanning forest. Global distributional limits are discussed, and examples given both of Gaussian and non-Gaussian limiting cases.

7.1 Introduction and definitions

We survey recent developments in the probability theory of a class of random spatial graphs defined on partially ordered point sets in $\mathbb{R}^d$, drawing together particular examples from the literature in a unified setting.

The model that we will concentrate on is the minimal directed spanning forest, introduced in a particular case by Bhatt and Roy (2004). Into this framework fit models such as the on-line nearest-neighbour graph (Berger, Bollobás, Borgs, Chayes and Riordan, 2003) and the radial spanning tree (Baccelli and Bordenave, 2007).

In this section we give some formal definitions, and collect some examples from the literature. Motivation for studying these models originates with various applied problems, including communication, transport, and drainage networks; we discuss this in Section 7.2. We now define the model.

For $d \in \mathbb{N}$, let $\mathcal{X}$ be a finite subset of $\mathbb{R}^d$, endowed with a partial order '$\preccurlyeq$' and a symmetric function $w : \mathcal{X} \times \mathcal{X} \to [0, \infty)$, a weight function on edges. $x \in \mathcal{X}$ is a *minimal* element if there is no $y \in \mathcal{X} \setminus \{x\}$ for which $y \preccurlyeq x$. Let $m(\mathcal{X})$ denote the set of minimal elements of $\mathcal{X}$; $m(\mathcal{X})$ is nonempty if $\mathcal{X}$ is finite.

For $x \in \mathcal{X} \setminus m(\mathcal{X})$, we say that $y \in \mathcal{X} \setminus \{x\}$ is a *directed nearest neighbour* of x if $y \preccurlyeq x$ and

$$w(x, y) = \min_{z \in \mathcal{X} \setminus \{x\} : z \preccurlyeq x} w(x, z).$$

For $x \in \mathcal{X} \setminus m(\mathcal{X})$, let $n(x, \mathcal{X})$ (or n_x if $\mathcal{X}$ is clear from the context) denote a directed nearest neighbour of x, chosen arbitrarily if x has more than one.

A *minimal directed spanning forest* (MDSF) on $(\mathcal{X}, \preccurlyeq, w)$, or simply 'on $\mathcal{X}$' if there is no ambiguity over the partial order and weight function, is a (directed) graph with vertex set $\mathcal{X}$ and edge set

$$\{(x, n_x) : x \in \mathcal{X} \setminus m(\mathcal{X})\};$$

that is, each non-minimal vertex is joined by an edge to a directed nearest neighbour.

Ignoring the directedness of edges, an MDSF is a forest having the same number of connected components as the number of minimal vertices $m(\mathcal{X})$. If $m(\mathcal{X})$ consists of a single element, we use the term *minimal directed spanning tree* (MDST) instead of MDSF.

An MDSF can also be viewed as the solution to a problem in Euclidean combinatorial optimization (cf Steele 1997; Yukich 1998): construct a minimal-length directed graph on $\mathcal{X}$ such that from each vertex there is a directed path to a minimal vertex and subject to the constraint that all directed edges must respect the partial order. See Bhatt and Roy (2004), Penrose and Wade (2004) for a fuller discussion.

We now list some examples from the literature.

Example 1 *Bhatt and Roy's 'south-west' MDST.* The terminology 'MDST' was introduced in Bhatt and Roy (2004) for the particular example we now describe. Let '$\preccurlyeq^*$' denote the coordinatewise partial order on $\mathbb{R}^d$: i.e., $(x_1, \ldots, x_d) \preccurlyeq^* (y_1, \ldots, y_d)$ if and only if $x_i \leq y_i$ for each i. For $x, y \in \mathbb{R}^d$, let $D(x, y) = \|x - y\|$; here and subsequently $\| \cdot \|$, or $\| \cdot \|_{(d)}$ when we want to emphasize the dimension d, denotes the Euclidean norm on $\mathbb{R}^d$. For $\mathcal{X} \subset \mathbb{R}^d$, set $\mathcal{X}^0 = \mathcal{X} \cup \{0\}$, where 0 is the origin in $\mathbb{R}^d$. Let $\mathcal{X}$ be a finite point set in $(0, 1)^d$. Bhatt and Roy's model is the MDST on $(\mathcal{X}^0, \preccurlyeq^*, D)$, specifically where $d = 2$. Note that here $m(\mathcal{X}^0) = \{0\}$.

Example 2 *The 'south' MDST.* Let '$\preccurlyeq_*$' be a binary relation on $\mathbb{R}^d$ such that

$$(x_1, \ldots, x_{d-1}, x_d) \preccurlyeq_* (y_1, \ldots, y_{d-1}, y_d) \iff x_d \leq y_d.$$

If $\mathcal{X} \in \mathbb{R}^d$ is a finite point set with distinct d-th coordinates, '$\preccurlyeq_*$' is a partial order (in fact, a total order resembling lexicographic order) on $\mathcal{X}$: we call this the 'one-coordinate' partial order. The MDST on $(\mathcal{X}, \preccurlyeq_*, D)$ is then similar to Bhatt and Roy's model except that each non-minimal vertex is joined to its nearest neighbour to the 'south' rather than 'south-west'.

Example 3 *The on-line nearest-neighbour graph (ONG).* The ONG (at least in $d = 2$) is a simple version of a model of sequential network growth studied in Berger, Bollobás, Borgs, Chayes and Riordan (2003), based on the so-called FKP model of preferential attachment due to Fabrikant, Koutsoupias and Papadimitriou (2002). Let '$\preccurlyeq_{\mathrm{ONG}}$' be a binary relation such that for $x, y \in \mathbb{R}^d$ and $s, t \in [0, \infty)$

$$(x, s) \preccurlyeq_{\mathrm{ONG}} (y, t) \iff s \leq t.$$

Let $\mathcal{X}^*$ be a finite subset of $\mathbb{R}^d \times [0, \infty)$ such that all the elements of $\mathcal{X}^*$ have distinct $[0, \infty)$-coordinate, and let $\mathcal{X}$ denote the projection of $\mathcal{X}^*$ onto $\mathbb{R}^d$. Then '$\preccurlyeq_{\mathrm{ONG}}$' is a partial order (in fact, a total order) on $\mathcal{X}^*$, which induces a total order on $\mathcal{X}$. Let D_{ONG} denote the weight function on $\mathcal{X}^*$ induced by the Euclidean weight function on $\mathcal{X}$, i.e.

$$D_{\mathrm{ONG}}((x, s), (y, t)) = D(x, y) = \|x - y\|.$$

The ONG may be defined as the MDST on $(\mathcal{X}^*, \preccurlyeq_{\text{ONG}}, D_{\text{ONG}})$. We view the ONG as a graph in $\mathbb{R}^d$, however, in which case it is convenient to use the notation $m(\mathcal{X})$ and n_x for the $\mathbb{R}^d$-projections of the corresponding $m(\mathcal{X}^*)$ and $n_{(x,s)}$.

A more natural alternative description of the ONG is given by viewing the last coordinate as 'time' and the $\mathbb{R}^d$-coordinate as 'space'. Then the ONG can be viewed as a graph on the set $\mathcal{X}$ ordered according to the time order induced by '$\preccurlyeq_{\text{ONG}}$' on $\mathcal{X}^*$: the graph is constructed by joining each point of $\mathcal{X}$ after the first to its nearest predecessor in $\mathcal{X}$, when the points are viewed as arriving in time order. In this description, the first point is the point with smallest time-coordinate, i.e. the minimal vertex in the MDST description.

Example 4 *The radial spanning tree (RST).* This was introduced in Baccelli and Bordenave (2007). Let '$\preccurlyeq_{\text{RST}}$' be a binary relation on $\mathbb{R}^d$ defined by

$$x \preccurlyeq_{\text{RST}} y \iff \|x\| \leq \|y\|.$$

For finite $\mathcal{X} \subset \mathbb{R}^d$ such that $\|x\|$ is distinct for each $x \in \mathcal{X}$, '$\preccurlyeq_{\text{RST}}$' is a partial order (again, a total order) on $\mathcal{X}$; the RST is the MDST on $(\mathcal{X}^0, \preccurlyeq_{\text{RST}}, D)$.

Example 5 *The directed linear forest (DLF) and directed linear tree (DLT).* The DLF is a variant of the ONG with one spatial dimension, introduced in Penrose and Wade (2006), motivated by the study of boundary effects in Bhatt and Roy's south-west MDSF, see Penrose and Wade (2006, Sec. 3). For convenience, we describe the sequential version of the construction as in the ONG (Example 3 above). Let $(X_1, X_2, \ldots)$ be a sequence of points in $[0, 1]$. The DLF joins each point after the first in the sequence to its nearest neighbour amongst its predecessors *to the left*. That is, X_i, $i \geq 2$, is joined by an edge to the closest X_j such that $1 \leq j < i$ and $X_j \leq X_i$, arbitrarily breaking ties. The DLT is defined similarly but in terms of the sequence $(X_0, X_1, \ldots)$ with $X_0 = 0$, i.e. with the original sequence of points $X_1, X_2, \ldots$ preceded by an initial point (the root) placed at the left endpoint of the unit interval.

Example 6 *The small world navigation tree.* This graph is discussed in Bordenave (2008). Assume the values of $\{\|x\|, x \in \mathcal{X}\}$ are distinct. The partial order used is $\preccurlyeq_{\text{RST}}$. Let $f : [0, \infty) \to [0, 1]$ be a non-increasing function, assumed to satisfy a polynomial decay condition at infinity. Form a random connection model (see e.g. Franceschetti and Meester 2007) on $\mathcal{X}^0$, with connection function f, subject to the condition that each point x of $\mathcal{X}$ is connected by an edge to at least one point y with $\|y\| < \|x\|$. Then let each included edge (x, y), with $\|y\| < \|x\|$, be assigned the weight $w(x, y) = \|y\|$ (so in this case the weight function is not symmetric). Equivalently, if working with the complete graph on $\mathcal{X}^0$, the weight of edge (x, y) with $\|y\| < \|x\|$ is either $\|y\|$ or $+\infty$ (so here the weight function is also not finite). Given vertex $x \in \mathcal{X}$, the joint distribution of the variables $(\mathbb{1}_{\{w(x,y)=\|y\|\}}, y \in \mathcal{X}^0, \|y\| < \|x\|)$ is that of a collection of independent Bernoulli random variables with parameters $f(\|y - x\|)$, subject to the constraint that at least one of them takes the value 1 (and these collections are independent for different choices of x). The resulting MDST is called the small world navigation graph on $\mathcal{X}^0$. Note that the edge from x in this graph is to the vertex closest to the origin (out of those connected to x by edges in the random connection model), rather than to the one closest to x as in most of the other models considered here, giving this example a different flavour.

In the planar case ($d = 2$), Examples 1 and 2 above are special cases of a family of MDSFs on partial orders indexed by cones in $\mathbb{R}^2$; this direction has received some attention in Wade (2007).

We will be interested in the case where $\mathcal{X}$ is a *random* point set in $(0,1)^d$. In particular, let $X_1, X_2, \ldots$ be independent random vectors on $(0,1)^d$ with a common density function f, and for $n \in \mathbb{N}$ set $\mathcal{X}_n = \{X_1, \ldots, X_n\}$, a binomial point process. We will often consider the special case of uniform points on the unit cube, where $f \equiv 1$ on $(0,1)^d$ and $f \equiv 0$ elsewhere; in this case we write U_i for X_i and set $\mathcal{U}_n = \{U_1, \ldots, U_n\}$. Also, let $\mathcal{P}_n$ denote a homogeneous Poisson point process of intensity $n > 0$ on $(0,1)^d$; often we will take $\mathcal{P}_n = \mathcal{U}_{N(n)}$ where $N(n)$ is Poisson with mean n and independent of $U_1, U_2, \ldots$.

Most of our results will be concerned with the case $\mathcal{X} = \mathcal{U}_n$ or $\mathcal{X} = \mathcal{P}_n$ (sometimes augmented by 0). In particular, we consider the total edge length of the graph in $\mathbb{R}^d$, or, more generally, the total power-weighted edge length defined by

$$\mathcal{L}^{d,\alpha}(\mathcal{X}) = \sum_{x \in \mathcal{X} \setminus m(\mathcal{X})} \|x - n(x, \mathcal{X})\|_{(d)}^{\alpha}.$$

We are interested in large-sample asymptotics, i.e. the limiting behaviour of $\mathcal{L}^{d,\alpha}$ and related geometric quantities as $n \to \infty$. Graph-theoretic properties of these models are mentioned only in passing.

The MDSF models on $\mathcal{U}_n$ or $\mathcal{P}_n$ mentioned above have several common features that distinguish their distributional limit theory. The most striking of these features is inhomogeneity of edge-lengths: edges from vertices low down in the partial order are often considerably longer, due to what we will generally describe as *boundary effects*, which may be spatial (as in the 'south-west' and 'south' MDSTs) or temporal (as in the ONG and DLT). Concretely, in the 'south' MDST for instance, the second-lowest point in the partial order '$\preccurlyeq_*$' will always be joined to the lowest point, and the length of this edge is $\mathcal{O}(1)$; similarly for the edge from the second point in the ONG. Typical 'nearest-neighbour-type' edge lengths in such point sets are $\mathcal{O}(n^{-1/d})$.

Thus the *local* distributional behaviour from points near to the boundaries is typically non-Gaussian. For example, for the 'south-west' MDST, the structure near the boundary has a recursive nature that leads to the appearance of Dickman-type distributions as limits for various boundary-dominated quantities (see Section 7.3).

The impact of these atypically long edges is amplified by increasing the weight exponent α: as α increases, so does the relative importance of long edges. Thus a common theme for the *global* distributional limits (see Section 7.4) is a *phase transition*: behaviour is Gaussian for small α, where the boundary-effects are 'washed-out', and non-Gaussian for large α.

In this article we give an overview of results of this flavour, where both Gaussian and non-Gaussian limits can appear. Proving theorems in this context often requires treating the two possible contributions separately, via rather different techniques, and then combining them at the end.

The outline of the rest of the article is as follows. In Section 7.2 we discuss the motivation for MDSF models, and also mention related models that have appeared in the literature. The main results on the MDSF are presented in Sections 7.3 and 7.4. Section 7.3 deals specifically with Example 1 above and the connection with record values and Dickman-type distributions. Section 7.4 gives results on the asymptotics for $\mathcal{L}^{d,\alpha}$ for several MDSF models; these results typically involve the use of *stabilization* techniques (see e.g. Penrose and Yukich 2001, 2003; Penrose 2007 and also Chapter 4). Section 7.5 gives a brief overview of results on the RST (Example 4 above) obtained in Baccelli and Bordenave (2007). Finally, in Section 7.6, we discuss avenues for future research and give some open problems.

7.2 Motivation and related models

We briefly describe some of the physical motivation behind the MDSF, and mention some related constructions that have appeared in the literature.

The primary motivation for the MDSF is as a model for real-world spatial networks: in particular, networks associated with communication and drainage (Bhatt and Roy, 2004; Penrose and Wade, 2004). For general background on the physical and mathematical modelling of drainage networks, see for instance Gangopadhyay, Roy and Sarkar (2004), Rodriguez-Iturbe and Rinaldo (1997) and references therein; for communications networks, see for instance Franceschetti and Meester (2007). The on-line nearest-neighbour graph is particularly motivated by *evolution* of communications networks; the literature on the modelling of network evolution is vast (see for instance Caldarelli 2007; Dorogovstev and Mendes 2003; Franceschetti and Meester 2007).

There are similarities between the structure of the MDSF and other structures in computational geometry. For instance, nearest-neighbours in certain restricted senses have been studied in the computational geometry context; see for instance Smith (1988).

When $d = 1$ there is a connection to the classical theory of (Dirichlet) spacings (Darling, 1953; Pyke, 1965). For the on-line graphs (the ONG and DLT) this connection extends to certain sequential interval division, fragmentation, and stick-breaking processes; see Penrose and Wade (2006, 2008) for more details.

Lattice models related to the MDSF have been studied. For instance Gangopadhyay, Roy and Sarkar (2004), motivated by drainage network modelling, consider a percolation-style model on $\mathbb{Z}^d$ analogous to the 'south' MDSF; they demonstrate, amongst other things, a tree/forest dichotomy. Another model with some similarities to certain versions of the MDSF is that of so-called Poisson trees (Ferrari, Landim and Thorisson, 2004; Ferrari, Fontes, and Wu, 2005). More distant connections include those to increasing subsequences (Aldous and Diaconis, 1995; Steele, 1995; Wüthrich, 2002).

7.3 Bhatt and Roy's MDST and Dickman-type distributions

In this section we describe results, starting with those of Bhatt and Roy, on the 'south-west' MDST (Example 1 in Section 7.1 above): specifically, lengths of rooted edges, the longest edge length, and the relations to maximal points, order statistics, and the Poisson–Dirichlet and Dickman distributions (named after the work of Dickman (1930) on the asymptotic distribution of large prime factors). The *total length* we discuss along with results for other MDST variants in Section 7.4.

Throughout this section, $\mathcal{X}$ will be a point set in $(0,1)^d$, and we consider the MDST on $(\mathcal{X}^0, \preccurlyeq^*, D)$; we will mostly (as do Bhatt and Roy) concentrate on $d = 2$ and will take $\mathcal{X} = \mathcal{U}_n$ or $\mathcal{X} = \mathcal{P}_n$.

We consider the limiting behaviour, as $n \to \infty$, of the total length of some subsets of the edges in the MDST. In particular, we deal with the edges *incident to the origin*, and the *longest edge*. Limiting distributions for these quantities are given in terms of certain Dickman-type distributions, which emerge from the Poisson–Dirichlet distribution: their appearance here arises from the structure of the MDST near the lower and left boundaries of the unit square.

7.3.1 *Record values and rooted vertices in the MDST*

Here we consider the minimal elements, under '$\preccurlyeq^*$', of random points in $(0,1)^d$. There is a natural and classically established link with the theory of record values when $d = 2$: in the context of the MDST, this connection was noted early on by Bhatt and Roy (2004). Note that the $\preccurlyeq^*$-minimal elements in the point set $\mathcal{X} \subset (0,1)^d$ are exactly those vertices that are connected to the origin in the MDST on $(\mathcal{X}^0, \preccurlyeq^*, D)$. This fact will help to explain the appearance of the Dickman distribution later on.

For $d \in \mathbb{N}$, let $M_d^*(\mathcal{X})$ denote the number of minimal elements under '$\preccurlyeq^*$' of a finite point set $\mathcal{X} \subset \mathbb{R}^d$. When $d = 2$, $M_2^*(\mathcal{X})$ is related to the *record values* in a sample of size n. Given a sequence of real numbers $(x_1, x_2, \ldots, x_n)$, we say that x_i is a *lower record value* if $x_i \leq x_j$ for $j = 1, \ldots, i-1$ (thus x_1 is always a lower record). See e.g. Nevzorov (1987) for a survey on record values.

Let $N_R(x_1, x_2, \ldots, x_n)$ denote the number of lower records in the sequence of real numbers $(x_1, \ldots, x_n)$. The following well-known result gives the connection between M_2^* and N_R; see e.g. Bhatt and Roy (2004, Lemma 2.1).

Lemma 7.1 *Let $\mathcal{Y}_n$ be a point set in $\mathbb{R}^2$ enumerated as (x_i, y_i), $i = 1, \ldots, n$, such that $y_1 \leq y_2 \leq \cdots \leq y_n$. Then*

$$M_2^*(\mathcal{Y}_n) = N_R(x_1, x_2, \ldots, x_n).$$

When $(X_1, \ldots, X_n)$ is an i.i.d. sequence, many properties of the number of records are known, and so by Lemma 7.1 the corresponding properties of $M_2^*(\mathcal{X}_n)$ follow immediately. For example, the following result of Rényi (1976) applies; see also Theorems 1.1 and 2.1 of Bhatt and Roy (2004). Denote by $\mathcal{N}(0, \sigma^2)$ the normal distribution with mean 0 and variance $\sigma^2 \geq 0$; this includes the degenerate case $\mathcal{N}(0,0) \equiv 0$. Here and subsequently '$\xrightarrow{d}$' denotes convergence in distribution.

Lemma 7.2 *Let $U_1, U_2, \ldots$ be independent uniform random variables on* $(0,1)$. *The following results hold, with both $Y_n = N_R(U_1, \ldots, U_n)$ and $Y_n = M_2^*(\mathcal{U}_n)$:*

$$(\log n)^{-1} Y_n \xrightarrow{\text{a.s.}} 1, \quad \text{as } n \to \infty;$$

$$(\log n)^{-1/2}(Y_n - \log n) \xrightarrow{d} \mathcal{N}(0,1), \quad \text{as } n \to \infty;$$

$$\limsup_{n\to\infty} \left[\frac{Y_n - \log n}{(2\log n)^{1/2} (\log\log\log n)^{1/2}} \right] = 1, \text{ a.s.};$$

$$\liminf_{n\to\infty} \left[\frac{Y_n - \log n}{(2\log n)^{1/2} (\log\log\log n)^{1/2}} \right] = -1, \text{ a.s.}$$

We will revisit record values later where *sums* of records will be related to Dickman-type distributions. It is worth noting that $M_d^*(\mathcal{X}_n)$ is of independent interest (see e.g. Barndorff-Nielsen and Sobel 1966) as a so-called distribution-free statistic: the distribution of $M_d^*(\mathcal{X}_n)$ does not depend on the underlying common (continuous) distribution of the points of $\mathcal{X}_n$, provided that the underlying d-dimensional density f is a pure product, i.e. the coordinates of each point are independent.

Also of interest is the asymptotic behaviour of the expectation $\mathbf{E}[M_d^*(\mathcal{U}_n)]$ for general d. The next result is the case $r = 1$ of (3.39) in Barndorff-Nielsen and Sobel (1966); see also the discussion and references in Bai, Devroye, Hwang and Tsai (2005). The asymptotic result in (7.1) is also true (but not very informative) when $d = 1$, for in that case $M_1^*(\mathcal{U}_n) = 1$ almost surely.

Lemma 7.3 *Let $d \in \{2, 3, \ldots\}$. For $n \in \mathbb{N}$, with $i_1 = n$, as $n \to \infty$,*

$$\mathbf{E}\left[M_d^*(\mathcal{U}_n)\right] = \sum_{i_2=1}^{n} \sum_{i_3=1}^{i_2} \cdots \sum_{i_d=1}^{i_{d-1}} \frac{1}{i_2 i_3 \cdots i_d}$$
$$= \frac{(\log n)^{d-1}}{(d-1)!} + \mathcal{O}\left((\log n)^{d-2}\right). \qquad (7.1)$$

7.3.2 *Dickman limits*

For $\mathcal{X}$ a finite subset of $(0,1)^d$, and $\alpha > 0$, let

$$\mathcal{L}_0^{d,\alpha}(\mathcal{X}) = \sum_{x \in m(\mathcal{X})} \|x\|_{(d)}^{\alpha}, \qquad (7.2)$$

the total power-weighted length of edges incident to 0 in the MDST on $\mathcal{X}^0$. Theorem 1.2 of Bhatt and Roy (2004) showed that, as $n \to \infty$, $\mathcal{L}_0^{2,1}(\mathcal{U}_n) \xrightarrow{d} Y$, for some random variable Y with $\mathbf{E}[Y] = 2$ and $\mathrm{Var}[Y] = 1$. This was extended, via a different proof, to all $\alpha > 0$ in Penrose and Wade (2004), where Y was characterized in terms of a *Dickman distribution.* Given $\theta > 0$, we say that a nonnegative integrable random variable X has a *generalized Dickman* distribution with shape parameter θ, and write $X \sim \mathrm{GD}(\theta)$, if it satisfies the distributional fixed-point identity

$$X \stackrel{d}{=} U^{1/\theta}(1 + X), \tag{7.3}$$

where U is uniform on $(0, 1)$, and is independent of the X on the right. (Here and subsequently '$\stackrel{d}{=}$' denotes equality in distribution.) The following is contained in Theorem 1 of Penrose and Wade (2004).

Theorem 7.4 *Let $\alpha > 0$. Let $Z(\theta) \sim \mathrm{GD}(\theta)$. Then as $n \to \infty$,*

$$\mathcal{L}_0^{2,\alpha}(\mathcal{P}_n) \xrightarrow{d} Z(2/\alpha); \quad \textit{and} \quad \mathcal{L}_0^{2,\alpha}(\mathcal{U}_n) \xrightarrow{d} Z(2/\alpha).$$

The natural connection of the $\mathrm{GD}(\theta)$ distribution to the length of rooted edges in the MDST is via *sums of uniform records* (see the discussion in Section 7.1 above); although the Dickman connection is not mentioned explicitly, see Arnold and Villaseñor (1998, Sec. 6) (note the typo γ for $1/\gamma$ after (6.9) there) and also compare Darling (1998). This can be seen as a marginal version of the fact that the MDST is well-approximated by the DLT near to the boundaries; see Section 7.4.2 and Penrose and Wade (2006) for more detail.

In Penrose and Wade (2004), we said that Bhatt and Roy gave only the first two moments of their limit Y; in the final published version of Bhatt and Roy (2004) however, they do give all the moments of X where Y can be characterized as a sum of two independent copies of X, see Bhatt and Roy (2004, Sec. 5).

Bai, Lee and Penrose (2006) show that this two dimensional case is rather special: for $d \geq 3$ the corresponding limits for the length of the rooted edges in the MDST (under $\preccurlyeq^*$) are normally distributed. On the other hand, when $d = 1$, we simply have the first Dirichlet spacing.

A second kind of result concerns the length of the *longest* edge in the MDST on $\mathcal{X}^0$. Bhatt and Roy (2004) considered the *maximum length of edges joined to the origin.* For $\mathcal{X} \subset (0, 1)^d$, set

$$\mathcal{M}_0^d(\mathcal{X}) = \max_{x \in m(\mathcal{X})} \|x\|_{(d)},$$

the length of the longest edge joined to 0 in the MDST $(\mathcal{X}^0, \preccurlyeq^*, D)$. Bhatt and Roy (2004) proved the following result:

Theorem 7.5 *As $n \to \infty$,*

$$\mathcal{M}_0^2(\mathcal{U}_n) \xrightarrow{d} \max\{U_1, U_2\} \stackrel{d}{=} U_1^{1/2}; \quad \mathcal{M}_0^2(\mathcal{P}_n) \xrightarrow{d} U_1^{1/2}, \tag{7.4}$$

where U_1, U_2 are independent uniform random variables on $(0, 1)$.

The global maximum of *all* Euclidean edge lengths in the MDST on $\mathcal{X}^0$ was studied in Penrose and Wade (2004). Denote this maximal edge length by

$$\mathcal{M}^d(\mathcal{X}) = \max_{x \in \mathcal{X}} \|x - n(x, \mathcal{X}^0)\|_{(d)}.$$

The distributional limit for $\mathcal{M}^2(\mathcal{U}_n)$ is expressed in terms of the *max-Dickman* distribution, which can be characterized as the distribution of a nonnegative integrable random variable M satisfying the fixed-point equation

$$M \stackrel{d}{=} \max\{1 - U, UM\}, \tag{7.5}$$

where U is uniform on $(0, 1)$ and independent of the M on the right.

Theorem 7.6 *Suppose that M and M' are independent max-Dickman random variables. As $n \to \infty$,*

$$\mathcal{M}^2(\mathcal{P}_n) \xrightarrow{d} \max\{M, M'\}; \quad \text{and} \quad \mathcal{M}^2(\mathcal{U}_n) \xrightarrow{d} \max\{M, M'\}.$$

Much is known about Dickman-type distributions, which have previously arisen in such fields as probabilistic number theory (e.g. Billingsley 1999; Dickman 1930; Donnelly and Grimmett 1993), population genetics (e.g. Watterson 1976), and the theory of randomized algorithms (e.g Chen and Hwang 2003; Hwang and Tsai 2002). Properties of the GD(θ) distributions are scattered throughout the literature; see for example Penrose and Wade (2004), Arratia (1998, Sec. 11), and references therein.

The GD(1) and max-Dickman distributions are more closely related than might at first be apparent. In probabilistic terms, they can both be expressed in terms of a Poisson point process on $(0, 1)$ with mean measure μ given by $\mathrm{d}\mu = (1/x)\mathrm{d}x$; the Poisson–Dirichlet process arises here (see e.g. Arratia 1998; Billingsley 1999; Griffiths 1988; Kingman 1993; Kingman 2006; Penrose and Wade 2004; Watterson 1976). In more analytical terms, both the GD(1) and the max-Dickman probability density functions are defined in terms of the *Dickman function*, see e.g. Hensley (1986), Tenenbaum (1995).

The following result (see Proposition 3 of Penrose and Wade 2004) collects a few of the more important facts about the GD(θ) distributions.

Proposition 7.7 *Suppose that $\theta > 0$ and $X \sim$ GD(θ). Then*

(a) X has Laplace transform

$$\mathbf{E}\left[\mathrm{e}^{-tX}\right] = \exp\left(\theta \int_0^t \frac{\mathrm{e}^{-s} - 1}{s} \mathrm{d}s\right) = \exp\left(\theta \int_0^1 \frac{\mathrm{e}^{-tu} - 1}{u} \mathrm{d}u\right), \quad (t \in \mathbb{R}).$$

(b) X has infinitely divisible distribution with Lévy–Khinchin measure $(\theta/x)\mathbb{1}\{x \in (0, 1)\}$.

(c) If $Y \sim \mathrm{GD}(\theta')$, for $\theta' > 0$, is independent of X, then $X+Y \sim \mathrm{GD}(\theta+\theta')$.
(d) X has moments $m_k = \mathbf{E}[X^k]$ which satisfy $m_0 = 1$ and, for $k \in \mathbb{N}$,

$$m_k = \frac{\theta}{k} \sum_{j=0}^{k-1} \binom{k}{j} m_j.$$

In particular, X has expected value θ and variance $\theta/2$.

Information about the pdf of $\mathrm{GD}(\theta)$ is given for instance in Watterson (1976) and Penrose and Wade (2004, Sec. 3.3). A curiosity is the fact that GD(1) has constant pdf on the interval $(0, 1)$: if $X = UY$ with U, Y independent and U uniform on $(0, 1)$, then for $x > 0$

$$\mathbf{P}\{X \le x\} = \mathbf{E}[\mathbf{P}(\{U \le (x/Y)\} \mid Y)] = x\mathbf{E}[(1/Y)\mathbb{1}\{Y \ge x\}];$$

in the particular case $Y \stackrel{d}{=} 1 + X$, $Y \ge 1$ a.s., and this gives $\mathbf{P}\{X \le x\} = x\mathbf{E}[1/(1+X)]$ for $x \in (0, 1)$. In other words, by conditioning on Y we see that the distribution of X is a mixture of uniforms on intervals which all contain $(0, 1)$, so that the distribution of X restricted to $(0, 1)$ is itself uniform.

The max-Dickman distribution has appeared in many contexts, and a picture of part of its density function is on the front cover of the second edition of Billingsley's book (Billingsley, 1999). In particular, we note that M can be characterized as the first component of the Poisson–Dirichlet distribution with parameter 1, and $\mathbf{E}[M] \approx 0.6243299$ is Dickman's constant, see Dickman (1930, p. 9), Watterson (1976), and compare Golomb (1964).

The intuition behind Theorem 7.4 goes as follows. If there exists a minimal point of $\mathcal{U}_n$ near to the origin, then there is no minimal point lying to the north-east of that point. Hence, the minimal points are likely to all lie near to either the x-axis or the y-axis, and the contributions from these two axes are nearly independent. Near the x-axis, the x-coordinates of successive minimal points (taken in order of increasing y-coordinate) form a sequence of products of uniforms $U_1, U_1U_2, U_1U_2U_3, \ldots$ and summing these gives a Dickman distribution. Similarly for the y-axis. This intuition is formalized in the following result, which is essentially implicit in Penrose and Wade (2004).

Let $\mathcal{P}^x$ denote a Poisson point process on the line segment $\{x \in (0, 1), y = 0\}$ with one-dimensional intensity $1/x$, and let $\mathcal{P}^y$ be an independent Poisson point process on $\{x = 0, y \in (0, 1)\}$ with one-dimensional intensity $1/y$.

Theorem 7.8 *Let $d = 2$. As $n \to \infty$*

$$m(\mathcal{U}_n) \xrightarrow{d} \mathcal{P}^x \cup \mathcal{P}^y.$$

Here the convergence means convergence of joint distributions of the counting functions for disjoint Borel sets not containing 0.

7.3.3 *Dickman process*

Following his 1930 paper, Dickman (1930) has acquired several distributions and a constant; here is a process.

Let $\mathcal{P}$ be a Poisson process on $(0,\infty)^2$ with intensity measure $(1/x)\mathrm{d}x\mathrm{d}y$. For $s \geq 0$, $t \geq 0$, let $R(s,t)$ denote the region $[0,s] \times [0,t]$, and set

$$D(s,t) = \sum_{(x,y)\in\mathcal{P}\cap R(s,t)} x. \tag{7.6}$$

(Since $(1/x)\mathrm{d}x\mathrm{d}y$ blows up at $x = 0$, one should interpret (7.6) as the sum of the x-coordinates of the points of $\mathcal{P} \cap R(s,t)$ listed in order of decreasing x-coordinate.)

Then $(D(s,t))_{s\geq 0, t\geq 0}$ is a two-parameter stochastic process; denote its centred version by $\tilde{D}(s,t) = D(s,t) - \mathbf{E}[D(s,t)]$.

Let $(X(t))_{t\geq 0}$ be an $\mathbb{R}$-valued stochastic process. If $(Y(t))_{t\geq 0}$ is another $\mathbb{R}$-valued stochastic process, we write $(X(t))_{t\geq 0} \stackrel{d}{=} (Y(t))_{t\geq 0}$ for equality of all finite-dimensional distributions. We will use some standard terminology for stochastic processes, see e.g. Sato (1999, Ch. 1). Recall that the process $(X(t))_{t\geq 0}$ is *stochastically continuous* if, for any $t \geq 0$ and $\varepsilon > 0$,

$$\lim_{h\to 0} \mathbf{P}\{|X(t+h) - X(t)| > \varepsilon\} = 0.$$

7.3.3.1 *The process $(D(s,1))_{s\geq 0}$*

Theorem 7.9 *The process $(D(s,1))_{s\geq 0}$ has the following properties:*

(i) It is self-similar with exponent 1 *(cf Sato 1999, Ch. 3), i.e. for any $a > 0$*

$$(D(as,1))_{s\geq 0} \stackrel{d}{=} (aD(s,1))_{s\geq 0}. \tag{7.7}$$

(ii) It has independent increments.

(iii) It is stochastically continuous, and as $s \downarrow 0$, $D(s,1) \xrightarrow{\text{a.s.}} 0$, and $D(0,1) = 0$ almost surely.

(iv) For $0 \leq q \leq r < \infty$, the increment $W(q,r) = D(r,1) - D(q,1)$ has Laplace transform

$$\mathbf{E}[\mathrm{e}^{-tW(q,r)}] = \int_q^r \left(\frac{\mathrm{e}^{-tu} - 1}{u}\right) \mathrm{d}u, \quad (t \in \mathbb{R}), \tag{7.8}$$

expectation $\mathbf{E}[W(q,r)] = r - q$, *and variance* $\mathrm{Var}\,[W(q,r)] = (r^2 - q^2)/2$.

In particular, part (iv) demonstrates that the increments are not stationary. Moreover, (7.7) together with Proposition 7.7(a) and (7.8) implies that for any $a > 0$, $a^{-1}D(a,1) \sim \mathrm{GD}(1)$. For some general results on self-similar processes with independent increments, see e.g. Embrechts and Maejima (2002), especially Chapter 5.

Proof of Theorem 7.9 We first prove part (ii). By (7.6), for $0 \le q \le r < \infty$, the increment $W(q,r)$ is given by

$$W(q,r) = \sum_{(x,y)\in\mathcal{P}\cap R(r,1)\backslash R(q,1)} x. \tag{7.9}$$

Suppose $0 \le q_1 \le r_1 < q_2 \le r_2 < \cdots < q_k \le r_k < \infty$. Consider $\{W(q_i,r_i) : i = 1,\ldots,k\}$. The regions $R(r_j,1)\backslash R(q_j,1)$ and $R(r_i,1)\backslash R(q_i,1)$, for $i \ne j$, are disjoint. Hence, from (7.9), the $W(q_i,r_i)$ are determined by Poisson points in disjoint regions, and thus independent. This proves part (ii) of the theorem. Part (iv) is an application of Campbell's theorem: see e.g. Kingman (1993, p. 28).

Fix $a > 0$. By the mapping theorem (see Kingman 1993, p. 18) $\mathcal{P}$ under the map $(x,y) \mapsto (ax,y)$ is also a Poisson process on $(0,\infty)^2$ of intensity $(1/x)\mathrm{d}x\mathrm{d}y$, denoted $\mathcal{P}'$, while $R(s,1)$ maps to $R(as,1)$ for all s. Hence,

$$aD(s,1) = \sum_{(x,y)\in\mathcal{P}\cap R(s,1)} ax = \sum_{(x',y)\in\mathcal{P}'\cap R(as,1)} x'. \tag{7.10}$$

On the other hand, $D(as,1) = \sum_{(x,y)\in\mathcal{P}\cap R(as,1)} x$, and part (i) of the theorem follows from comparing this with (7.10).

It remains to prove part (iii). Suppose $s > 0$ and $h > 0$. We have, for $\varepsilon > 0$,

$$\begin{aligned}\mathbf{P}\{|D(s+h,1) - D(s,1)| > \varepsilon\} &= \mathbf{P}\{W(s,s+h) > \varepsilon\} \\ &\le \varepsilon^{-1}\mathbf{E}[W(s,s+h)] = \varepsilon^{-1}h,\end{aligned}$$

by Markov's inequality and part (iv) of the theorem. Thus the process $(D(s,1))_{s\ge 0}$ is stochastically continuous. Moreover, by parts (i) and (ii) of the theorem, and Proposition 7.7(d), for $m \in \mathbb{N}$,

$$\begin{aligned}\mathbf{P}\{D(m^{-1},1) > \varepsilon\} &= \mathbf{P}\{D(1,1) > m\varepsilon\} \\ &\le \frac{\mathbf{E}[D(1,1)^2]}{m^2\varepsilon^2} = \frac{3}{2m^2\varepsilon^2}.\end{aligned}$$

Hence the Borel–Cantelli lemma implies that $D(m^{-1},1)$ converges almost surely to 0 as $m \to \infty$. $D(s,1)$ is non-decreasing, and so $D(s,1) \downarrow 0$ almost surely as $s \downarrow 0$. Since $(D(s,1))_{s\ge 0}$ is monotone, we have $D(0,1) = 0$ a.s. Thus the proof of the theorem is complete. □

7.3.3.2 *The process* $(D(1,t))_{t\ge 0}$ We recall (see e.g. Sato 1999) that a real-valued stochastic process $(X(t))_{t\ge 0}$ is called a *Lévy process* if $X(0) = 0$ a.s., it is stochastically continuous, it has independent and stationary increments, and its sample paths are right-continuous having left limits almost surely.

Theorem 7.10 $(D(1,t))_{t\ge 0}$ *is a Lévy process, with Lévy–Khinchin measure* $(1/x)\mathbb{1}\{x \in (0,1)\}$. *Moreover, for any* $t > 0$,

$$D(1,t) \sim \mathrm{GD}(t), \tag{7.11}$$

i.e. satisfies (7.3) *with* $\theta = t$.

Note that since $(D(1,t))_{t\geq 0}$ is non-decreasing almost surely, Theorem 7.10 implies that it is a subordinator, see e.g. Sato (1999, p. 137). Theorem 7.10 shows that $(D(1,t))_{t\geq 0}$ satisfies standard theorems for Lévy processes, see e.g. Sato (1999).

For $0 \leq q \leq r < \infty$, let $H(q,r)$ denote the increment

$$H(q,r) = D(1,r) - D(1,q).$$

Then, by (7.6), $H(q,r)$ is given by

$$H(q,r) = \sum_{(x,y)\in\mathcal{P}\cap R(1,r)\setminus R(1,q)} x. \tag{7.12}$$

Proposition 7.11 *For* $0 \leq q < \infty$ *and* $0 \leq h < \infty$, *the increment* $H(q,q+h)$ *satisfies* (7.3) *with* $\theta = h$:

$$H(q,q+h) \sim \mathrm{GD}(h). \tag{7.13}$$

Proof Let $\lambda : \mathbb{R}^2 \to \mathbb{R}$ denote the projection $(x,y) \mapsto x$. List the points of $\mathcal{P} \cap R(1,q+h)\setminus R(1,q)$ in order of decreasing x-coordinate as $Z_1, Z_2, \ldots$. In coordinates, write $Z_i = (V_i, W_i)$ for $i = 1, 2, \ldots$. Then,

$$H(q,q+h) = \sum_i V_i. \tag{7.14}$$

Let $\mathcal{P}'$ be the image of the Poisson process $\mathcal{P} \cap R(1,q+h)\setminus R(1,q)$ under the map λ. Then $\lambda(Z_i) = V_i$ for $i = 1, 2, \ldots$. By the mapping theorem (Kingman, 1993), $\mathcal{P}'$ is a Poisson process of intensity $(h/x)\mathrm{d}x$ on $(0,1)$. So we have, by (7.14), that $H(q,q+h)$ is given by the sum of the V_i, which are the points of a Poisson process on $(0,1)$ with intensity $(h/x)\mathrm{d}x$, taken in decreasing order. Hence Penrose and Wade (2004, Prop. 2) implies that $H(q,q+h)$ has the generalized Dickman distribution with parameter h. □

Proof of Theorem 7.10 As in the argument in the proof of Theorem 7.9(ii), disjoint increments are determined by Poisson arrivals in disjoint regions, and the independence of the increments follows.

Next, suppose that $t \geq 0$ and $h \geq 0$. We have, for $\varepsilon > 0$,

$$\begin{aligned}\mathbf{P}\{|D(1,t+h) - D(1,t)| \geq \varepsilon\} &= \mathbf{P}\{H(t,t+h) \geq \varepsilon\}\\ &\leq \varepsilon^{-1}\mathbf{E}[H(t,t+h)] = \varepsilon^{-1}h,\end{aligned}$$

by Markov's inequality and Proposition 7.11. Thus the process $(D(1,t))_{t\geq 0}$ is stochastically continuous.

The statement (7.11) follows from Proposition 7.11 and the fact that $D(1,0) = 0$ a.s.. Thus we have from Proposition 7.11 and the argument in the first paragraph of this proof that the stochastic process $(D(1,t))_{t\geq 0}$ has independent and stationary increments, each with the generalized Dickman distribution.

The Lévy–Khinchin measure of the process is given by the Lévy–Khinchin measure of the infinitely divisible distribution $D(1,1)$, which leads to the stated result by Propositions 7.11 and 7.7. This completes the proof of the theorem.

□

7.4 Total length, stabilization, and phase transitions

In this section we describe results on the *total* power-weighted edge-length of MDSFs on random points. In an applied setting, where the MDSF represents some spatial network, the total length may be viewed as a measure of the network 'throughput' (the rate of transmission of information), see Bhatt and Roy (2004).

In this section we begin with the one-dimensional directed linear tree and then move on to the two-dimensional 'south-west' MDSF, since the boundary effects in the latter are characterized in terms of the former. A similar relationship exists between the on-line nearest-neighbour graph in d dimensions and the 'south' MDSF in $d+1$ dimensions, which we treat next.

An important technique of many of the proofs of the results described in this section is *stabilization* and general limit results of Penrose and Yukich, cf. Chapter 4.

7.4.1 *The directed linear tree*

For this section, we use the notation $\mathcal{U}_n = (U_1, U_2, \ldots, U_n)$ for a sequence of $n \in \mathbb{N}$ independent uniform random variables on $(0,1)$. On such a sequence we can construct the directed linear forest as described in Example 5 of Section 7.1.

Write $\mathcal{U}_n^0$ for the augmented sequence $(0, U_1, U_2, \ldots, U_n)$. Then the directed linear forest on $\mathcal{U}_n^0$ is a tree, and for $\alpha > 0$ we denote its total power-weighted length by

$$\mathcal{D}^{\alpha}\left(\mathcal{U}_n^0\right) = \sum_{i=1}^{n} (U_i - \max\{U_j : 0 \le j < i, U_j \le U_i\})^{\alpha},$$

with the convention $U_0 = 0$ here. Write $\tilde{\mathcal{D}}^{\alpha}\left(\mathcal{U}_n^0\right) = \mathcal{D}^{\alpha}\left(\mathcal{U}_n^0\right) - \mathbf{E}\left[\mathcal{D}^{\alpha}\left(\mathcal{U}_n^0\right)\right]$.

Analysis of $\mathcal{D}^{\alpha}\left(\mathcal{U}_n^0\right)$ starts with the observation that it possesses a very useful recursive structure. This is based on simple scaling and a key independence property: loosely speaking, given U_1, the future evolution of the DLT looks like two independent scaled versions of the DLT on the sub-intervals $[0, U_1]$ and $[U_1, 1]$. Formally we have

$$\mathcal{D}^{\alpha}\left(\mathcal{U}_n^0\right) \stackrel{d}{=} U_1^{\alpha} + U_1^{\alpha}\mathcal{D}^{\alpha}\left(\hat{\mathcal{U}}_M^0\right) + (1-U_1)^{\alpha}\mathcal{D}^{\alpha}\left(\tilde{\mathcal{U}}_{n-1-M}^0\right), \tag{7.15}$$

where given U_1, M is binomial $(n-1, U_1)$, representing the number of points of $(U_2, \ldots, U_n)$ that fall in $[0, U_1]$, and $\hat{\mathcal{U}}_m^0$ and $\tilde{\mathcal{U}}_m^0$ are independent copies of $\mathcal{U}_m^0$.

The analysis of (7.15) is carried out in Penrose and Wade (2006). In order to describe the limit theory for $\mathcal{D}^{\alpha}\left(\mathcal{U}_n^0\right)$, we introduce some more notation. The random variables Y_α, $\alpha \geq 1$ have distributions with mean zero and finite second moment given uniquely (see Rösler 1992, Th. 3) by the fixed-point equations

$$Y_1 \stackrel{d}{=} UY_1^{\{1\}} + (1-U)Y_1^{\{2\}} + U \log U + (1-U)\log(1-U) + U; \tag{7.16}$$

and for $\alpha > 1$

$$Y_\alpha \stackrel{d}{=} U^\alpha Y_\alpha^{\{1\}} + (1-U)^\alpha Y_\alpha^{\{2\}} + \frac{1}{\alpha - 1}(\alpha U^\alpha + (1-U)^\alpha - 1), \tag{7.17}$$

where in each of (7.16) and (7.17) U is uniform on $(0,1)$, $Y_\alpha^{\{1\}}$ and $Y_\alpha^{\{2\}}$ have the distribution of Y_α, and the variables $U, Y_\alpha^{\{1\}}, Y_\alpha^{\{2\}}$ are independent.

Such fixed-point equations are reminiscent of those appearing in the analysis of algorithms such as Quicksort, and they have received much attention recently (see for instance Aldous and Bandyopadhyay 2005; Neininger and Rüschendorf 2004; Rösler 1992; Rösler and Rüschendorf 2001 and references therein). In particular, the fixed-point equation associated with the limiting normalized number of comparisons in randomized Quicksort is similar to (7.16) but with the term $U \log U + (1-U)\log(1-U) + U$ on the right-hand side replaced by $2U \log U + 2(1-U)\log(1-U) + 1$, see e.g. Neininger and Rüschendorf (2004, pp. 400–401). The latter fixed-point equation also appears in connection with the normalized internal path length of a random binary search tree; see e.g. Neininger and Rüschendorf (2004, Cor. 5.5). The authors have not seen (7.16) itself in any other context in the literature.

The following is contained in Theorem 3.1 of Penrose and Wade (2006).

Theorem 7.12 *(i) As $n \to \infty$, $\tilde{\mathcal{D}}^1(\mathcal{U}_n^0) \xrightarrow{d} Y_1$, where Y_1 has the distribution characterized by* (7.16).

(ii) Suppose $\alpha > 1$. As $n \to \infty$, $\tilde{\mathcal{D}}^\alpha(\mathcal{U}_n^0) \xrightarrow{d} Y_\alpha$, where Y_α has the distribution characterized by (7.17).

7.4.2 *Bhatt and Roy's 'south-west' MDST*

For $\mathcal{X}$ a finite subset of $(0,1)^d$, and $\alpha > 0$, let

$$\mathcal{L}^{d,\alpha}(\mathcal{X}^0) = \sum_{x \in \mathcal{X}} \|x - n(x, \mathcal{X}^0)\|^\alpha, \tag{7.18}$$

the total power-weighted edge length of the MDST on $(\mathcal{X}^0, \preccurlyeq^*, D)$. Denote the centred version $\tilde{\mathcal{L}}^{d,\alpha}(\mathcal{X}^0) = \mathcal{L}^{d,\alpha}(\mathcal{X}^0) - \mathbf{E}[\mathcal{L}^{d,\alpha}(\mathcal{X}^0)]$.

When $d = 1$, $\mathcal{L}^{1,\alpha}(\mathcal{U}_n^0)$ corresponds to a sum of powers of uniform spacings; such sums were considered in Darling (1953). In particular, Darling (1953, p. 245) essentially gives a central limit theorem for $\mathcal{L}^{1,\alpha}(\mathcal{U}_n^0)$ (although the model there is not quite the MDST, as the last spacing is also included in the sum).

The *total* edge length (or power-weighted edge length) of Bhatt and Roy's MDST for $d = 2$ was first studied in Penrose and Wade (2006). Here the dominance of boundary effects apparent in the results in Section 7.3.2 persists only up to a point: there is a phase transition. In the limit law there appear both normal and non-normal components. The non-normal part has distribution characterized by a fixed-point equation.

The following result is contained in Penrose and Wade (2006, Th. 2.1), which also gives a similar result where the point set is not augmented by 0.

Theorem 7.13 *Let $\alpha > 0$. Let $Z \sim \mathcal{N}(0,1)$ and $Y_\alpha^{\{1\}}$, $Y_\alpha^{\{2\}}$ be independent random variables, $Y_\alpha^{\{1\}}$ and $Y_\alpha^{\{2\}}$ having the distribution of Y_α characterized by (7.16) for $\alpha = 1$ and (7.17) for $\alpha > 1$. Then there exist constants $0 < s_\alpha \le t_\alpha < \infty$ such that as $n \to \infty$*

$$\tilde{\mathcal{L}}^{2,\alpha}\left(\mathcal{U}_n^0\right) \xrightarrow{d} s_\alpha Z, \quad \tilde{\mathcal{L}}^{2,\alpha}\left(\mathcal{P}_n^0\right) \xrightarrow{d} t_\alpha Z \quad \alpha \in (0,1);$$

$$\tilde{\mathcal{L}}^{2,1}\left(\mathcal{U}_n^0\right) \xrightarrow{d} Y_1^{\{1\}} + Y_1^{\{2\}} + s_1 Z, \quad \tilde{\mathcal{L}}^{2,1}\left(\mathcal{P}_n^0\right) \xrightarrow{d} Y_1^{\{1\}} + Y_1^{\{2\}} + t_1 Z;$$

$$\tilde{\mathcal{L}}^{2,\alpha}\left(\mathcal{U}_n^0\right) \xrightarrow{d} Y_\alpha^{\{1\}} + Y_\alpha^{\{2\}}, \quad \tilde{\mathcal{L}}^{2,\alpha}\left(\mathcal{P}_n^0\right) \xrightarrow{d} Y_\alpha^{\{1\}} + Y_\alpha^{\{2\}} \quad \alpha > 1.$$

The random variables Y_α, $\alpha \ge 1$, emerge as limits of the directed linear tree (Example 5 in Section 7.1, see Section 7.4.1) on uniform random points. The distributional fixed-point equations arise from the recursive nature of the DLT. The intuition behind this result is that close to the lower boundary segment of $(0,1)^2$, the MDST is close in some appropriate sense to the one-dimensional DLT defined on the horizontal coordinates of $\mathcal{U}_n$ close to the boundary taken in order of increasing vertical coordinate; similarly near to the left boundary of $(0,1)^2$.

The authors believe that higher-dimensional analogues of Theorem 7.13 hold. However, for $d \ge 3$ it seems more natural to consider the partial order '$\preccurlyeq_*$' rather than '$\preccurlyeq^*$'; see Section 7.4.4 below. The boundary effects in a $d \ge 3$ version of Theorem 7.13 would be characterized in terms of an on-line version of the $(d-1)$-dimensional $\preccurlyeq^*$-MDST.

7.4.3 *The on-line nearest-neighbour graph*

Let $d \in \mathbb{N}$. With a slight change in notation, we write $\mathcal{U}_n = (U_1, \ldots, U_n)$ for a sequence of n independent uniform random vectors on $(0,1)^d$. Fix $\alpha > 0$. Then for $n \ge 2$

$$\mathcal{O}^{d,\alpha}(\mathcal{U}_n) = \sum_{i=2}^{n} \min_{j<i} \|U_i - U_j\|^\alpha$$

is the *total power-weighted edge-length* of the ONG on $\mathcal{U}_n$. (We do not stress the MDST formulation of the ONG at this point.) Similarly, for $N(n)$ Poisson with mean n and independent of $U_1, U_2, \ldots$, take $\mathcal{P}_n = \mathcal{U}_{N(n)}$, so that the points of $\mathcal{P}_n$ constitute a Poisson process of intensity n on $(0,1)^d$.

Also, we define the centred versions $\tilde{\mathcal{O}}^{d,\alpha}(\mathcal{U}_n) = \mathcal{O}^{d,\alpha}(\mathcal{U}_n) - \mathbf{E}[\mathcal{O}^{d,\alpha}(\mathcal{U}_n)]$ and $\tilde{\mathcal{O}}^{d,\alpha}(\mathcal{P}_n) = \mathcal{O}^{d,\alpha}(\mathcal{P}_n) - \mathbf{E}[\mathcal{O}^{d,\alpha}(\mathcal{P}_n)]$. For $d \in \mathbb{N}$ let v_d denote the volume of the unit-radius Euclidean d-ball, i.e.

$$v_d = \pi^{d/2} \left[\Gamma\left(1 + (d/2)\right)\right]^{-1},$$

see e.g. Huang (1987, Eq. (6.50)).

The limit theory for $\mathcal{O}^{d,\alpha}(\mathcal{U}_n), \mathcal{O}^{d,\alpha}(\mathcal{P}_n)$ has recently been developed in Penrose (2005), Penrose and Wade (2008), Wade (2007, 2009). First-order behaviour (laws of large numbers and expectation asymptotics) was studied in Penrose and Wade (2008) and Wade (2007).

The distributional limit theory was studied in Penrose (2005), and also in Penrose and Wade (2008) primarily in the case $d = 1$. When $d = 1$, there is a recursive structure, analogous to that of the DLT shown by (7.15), see Penrose and Wade (2008) for more detail.

Theorem 7.14 (Penrose (2005)) *Suppose $d \in \mathbb{N}$ and $\alpha \in (0, d/4)$. Then there exists $\sigma^2_{d,\alpha} \in [0, \infty)$ such that*

$$\lim_{n\to\infty} n^{(2\alpha-d)/d}\mathrm{Var}\,[\tilde{\mathcal{O}}^{d,\alpha}(\mathcal{P}_n)] = \sigma^2_{d,\alpha}, \tag{7.19}$$

and as $n \to \infty$

$$n^{(2\alpha-d)/(2d)}\tilde{\mathcal{O}}^{d,\alpha}(\mathcal{P}_n) \xrightarrow{d} \mathcal{N}(0, \sigma^2_{d,\alpha}). \tag{7.20}$$

It is conjectured that Theorem 7.14 is in fact true for $\alpha \in (0, d/2)$. The critical case $\alpha = d/2$ also remains open; see Section 7.6.2 below. On the other hand, the following result for $\alpha > d/2$ was obtained in Wade (2009).

Theorem 7.15 *Suppose $d \in \mathbb{N}$ and $\alpha > d/2$. Then there exists a mean-zero random variable $Q(d, \alpha)$ (which is non-Gaussian for $\alpha > d$) such that as $n \to \infty$*

$$\tilde{\mathcal{O}}^{d,\alpha}(\mathcal{P}_n) \xrightarrow{d} Q(d, \alpha). \tag{7.21}$$

When $\alpha = 1$, Theorem 7.14 gives a CLT for the total length of the random ONG in dimensions $d \geq 5$. On the other hand, when $d = 1$ (with $\alpha = 1$) there is no CLT because $Q(1, 1)$ is not Gaussian (see below).

The fact that for $\alpha > d$ the random variables $Q(d, \alpha)$ in Theorem 7.15 are not normal follows since convergence also holds without any centring, see Penrose and Wade (2008, Th. 2.1(ii)). In the special case $d = 1$, (7.21) was given for $\alpha > 1/2$ in Penrose and Wade (2008, Th. 2.2). In the $d = 1$ case, more information can be obtained about the distribution of $Q(1, \alpha)$ using a 'divide-and-conquer' technique, see Penrose and Wade (2008), in particular Theorem 2.2, where the distribution of $Q(1, \alpha)$, $\alpha > 1/2$ is given. Indeed, $Q(1, \alpha)$, $\alpha > 1/2$, is given by the (unique) solution to a distributional fixed-point equation, and in particular is not Gaussian, see Penrose and Wade (2008) for details.

7.4.4 *The 'south' MDST*

The limit theory for the total power-weighted edge-length $\mathcal{L}^{d,\alpha}(\mathcal{P}_n)$ of the MDST on Poisson points under '$\preccurlyeq_*$' is studied in Penrose and Wade (2009).

When $d = 2$, it is not surprising that the limit theory is analogous to that for the 'south-west' MDST described in Section 7.4.2 above; but now the boundary effects are described by the one-dimensional ONG rather than the DLF. We also give results for the 'south' model with $d > 2$; here again boundary effects are manifest as certain ONG limits.

Theorem 7.16 presents convergence in distribution results for $\mathcal{L}^{d,\alpha}(\mathcal{P}_n)$; the distributional limits contain Gaussian random variables and also random variables defined as distributional limits of certain on-line nearest-neighbour graphs. In general we do not give an explicit description of these distributions. However, in the case of $d = 2$, the limits in question can be characterized as solutions to distributional fixed-point equations.

Theorem 7.16 *Suppose $d \in \{2, 3, 4, \ldots\}$ and $\alpha > 0$. Then there exists a constant $s_\alpha^2 < \infty$ such that, for normal random variables $W_\alpha \sim \mathcal{N}(0, s_\alpha^2)$, as $n \to \infty$:*

$$n^{(\alpha/d)-(1/2)}\tilde{\mathcal{L}}^{d,\alpha}(\mathcal{P}_n) \xrightarrow{d} W_\alpha \quad (0 < \alpha < d/2);$$
$$\tilde{\mathcal{L}}^{d,d/2}(\mathcal{P}_n) \xrightarrow{d} W_{d/2} + Q(d-1, d/2);$$
$$\tilde{\mathcal{L}}^{d,\alpha}(\mathcal{P}_n) \xrightarrow{d} Q(d-1, \alpha) \quad (\alpha > d/2).$$

Here all the random variables in the limits are independent, and the $Q(d-1, \alpha)$ are mean-zero random variables.

The normal random variables W_α arise from the edges away from the boundary. The variables $Q(d-1, \alpha)$ arise from the edges very close to the boundary, where the MDST is asymptotically close to a $(d-1)$-dimensional on-line nearest-neighbour graph; they are the distributions appearing in Theorem 7.15.

Theorem 7.16 indicates a phase transition in the character of the limit law as α increases. The normal contribution (from the points away from the boundary) dominates for $\alpha \in (0, d/2)$, while the boundary contribution dominates for $\alpha > d/2$. In the critical case $\alpha = d/2$ (such as the natural case $d = 2$ and $\alpha = 1$) neither effect dominates and both terms contribute significantly to the asymptotic behaviour.

7.5 The radial spanning tree

Baccelli and Bordenave (2007) study the radial spanning tree (Example 4 in Section 7.1) on a homogeneous Poisson point process on the plane, with a point inserted at the origin. Thus the underlying vertex set in any finite ball is a.s. finite, and the model described in Section 7.1 extends naturally to the whole plane. In the RST, each Poisson point x is joined by an edge to its nearest-neighbour amongst those points that are at distance less than $\|x\|$ from the origin.

The main results of Baccelli and Bordenave (2007) present some global and local properties of the RST, with a somewhat different emphasis from the results mentioned in Section 7.4.

Theorem 1 of Baccelli and Bordenave (2007) states that (a.s.) semi-infinite paths in the RST have an asymptotic direction, that every possible direction is achieved at least once, and that there is a dense set of directions that are achieved more than once. Theorem 2 states that (a.s.) the subtree of the RST containing only nodes at graph-distance k or less from 0 is asymptotically circular in some sense, and the number of nodes is asymptotically of order k^2.

The local behaviour of the RST at long distances from 0 is related to a secondary model, the so-called 'directed spanning forest' (DSF). Again defined on Poisson points on the plane, each point in the DSF is joined by an edge to its nearest neighbour with strictly less x-coordinate (for instance). This model is another MDST-variant on an infinite point set. The motivation for studying the DSF is that it resembles the RST at distances far from the origin, in a sense made precise by some of the results of Baccelli and Bordenave (2007).

Theorem 3 of Baccelli and Bordenave (2007) states that for a functional that is stabilizing (locally determined on the DSF), the distribution of the functional evaluated at a point $(0, x)$ in the RST converges to the distribution of the functional evaluated at the origin in the DSF as x tends to infinity. It is in this local sense that the RST is shown to be asymptotically similar to the DSF at large distances from the origin. Theorem 4 gives an a.s. limit result for graph-distance averaged sums of functionals over edges along long finite paths in the RST. The limit is in terms of a probability measure that is some stationary measure on the infinite edge process in the DSF.

The question of distributional limits for sums of stabilizing functionals (such as the total power-weighted edge length of the RST) over increasing finite subgraphs of the RST is left open: see Section 7.6.3 below.

In Bordenave (2008), results of a similar spirit are presented for a general class of 'navigations' on Poisson points. Specific examples treated include the small-world navigation tree (our Example 6 in Section 7.1) which shares some features with the MDSF.

7.6 Future directions and open problems

7.6.1 *More general point sets*

Most of the results on the MDSF that we have given here are stated in the case where the underlying distribution of the points is uniform on $(0, 1)^d$. (It is mentioned in Wade (2007) that the laws of large numbers for total length considered there carry over to more general underlying densities.)

It is of interest to extend the distributional limit theory for the various kinds of MDSF to more general point sets. Two possible directions are: (i) consider uniform points on more general regions; (ii) consider non-uniform densities of points. Both approaches may be interesting in view of the importance of boundary effects in the MDSF. In the case of the 'south-west' MDSF, relevant to point (i) for the rooted edges may be results on the number of maximal points in general domains, see e.g. Schreiber and Yukich (2008) and the references therein.

In terms of total length limit theory, central limit theorems for functionals on non-uniform point sets can be handled by stabilization methods (e.g. Baryshnikov and Yukich 2005; Penrose 2007; Penrose and Yukich 2001), which should be able to deal with behaviour away from the boundary, but the study of the boundary effects for general distributions is still open. We make some more related remarks in Section 7.6.4 below.

7.6.2 *CLT for the length of the planar ONG*

Consider the set-up of Section 7.4.3. The critical case $\alpha = d/2$ seems to be particularly delicate. This includes perhaps the most natural example, when $d = 2$ and $\alpha = 1$. We conjecture:

Conjecture 1 *Let $d \in \mathbb{N}$. There exists a constant $\sigma^2_{d,d/2} \in (0, \infty)$ such that*

$$(\log n)^{-1/2}\tilde{\mathcal{O}}^{d,d/2}(\mathcal{U}_n) \xrightarrow{d} \mathcal{N}\left(0, \sigma^2_{d,d/2}\right), \quad \text{as } n \to \infty.$$

7.6.3 *CLT for total length of the RST*

Let $C_0 \subset \mathbb{R}^d$ be bounded and convex with the origin in its interior, let $\mathcal{P}_\lambda$ be a homogeneous Poisson process of intensity λ in C_0, and let T_λ denote the total length of the RST on $\mathcal{P}_\lambda$. It is of interest to seek a central limit theorem for T_λ. Here the approach based on stabilization (see Chapter 4) seems to be relevant, although existing results do not appear to be directly applicable.

It may be possible to usefully approximate T_λ by something to which existing general results can be applied. For non-zero $x \in \mathbb{R}^d$, define the half-space

$$H(x) = \{y \in \mathbb{R}^d : y \cdot x \leq x \cdot x\},$$

and for finite $\mathcal{X} \subset \mathbb{R}^d$ with $x \in \mathcal{X}$ define $\xi(x, \mathcal{X})$ to be the distance from x to its nearest neighbour in $(\mathcal{X} \cup \{0\}) \cap H(x)$. Then by a direct application of Penrose (2007, Th. 2.3), one can obtain a central limit theorem for the sum

$$\sum_{x \in \mathcal{P}_\lambda} \lambda^{1/d}\xi(x, \mathcal{P}_\lambda),$$

scaled and centred (the scaling factor is $\lambda^{-1/2}$). For large λ, this sum ought to be a close approximation to $\lambda^{1/d}T_\lambda$.

Either by making use of this approximation, or by adapting the proof of Penrose (2007, Th. 2.2) directly to T_λ, it seems likely that one can derive a central limit theorem for T_λ; likewise for power weighted edges.

7.6.4 *Convergence of measures*

Given a triple $(\mathcal{X}, \preccurlyeq, w)$ as in Section 7.1, define the point measure $\mu_{\mathcal{X}}$ on $\mathbb{R}^d$ by

$$\mu_{\mathcal{X}} = \sum_{x \in \mathcal{X} \setminus m(\mathcal{X})} w(x, n_x)\delta_x$$

where δ_x is the Dirac measure. The total mass of $\mu_{\mathcal{X}}$ is the total weight of the MDSF on $\mathcal{X}$, but $\mu_{\mathcal{X}}$ also keeps track of the locations of the contributions to this total weight.

In the spirit of Baryshnikov and Yukich (2005) and Penrose (2007), for each case where we study the limit theory of the distribution (for a large random point set $\mathcal{X}$) of the total weight of the MDSF, it may be of interest also to consider the limiting behaviour of the associated measure $\mu_{\mathcal{X}}$, acting on continuous test functions.

Consider in particular the 'south-west' MDST on points in $(0,1)^2$, with edge-weight given by Euclidean distance, and take $\mathcal{X} = \mathcal{P}_n^0$. As discussed in Section 7.4.2, and in more detail in Penrose and Wade (2006), the centred total length is known to converge in distribution to the sum of a normally distributed term (arising from the interior points) and two terms each distributed as the solution to a fixed-point equation (arising from points near the lower and left boundaries of $(0,1)^2$). Given $f \in C([0,1]^2)$, write $\bar{\mu}_n(f)$ for the integral of f with respect to $\mu_{\mathcal{P}_n^0}$, minus its expected value.

If f is zero in a neighbourhood of the boundary of $[0,1]^2$, then there should not be any difficulty in using results from Baryshnikov and Yukich (2005) or Penrose (2007) to show that $\bar{\mu}_n(f)$ is asymptotically centred normal, with variance given by a constant times the integral of $f(x)^2$ over the unit square.

Of greater interest is the case where f is non-zero on the boundary. In this case, one would expect, in the limit, a contribution to $\bar{\mu}_n(f)$ from near the lower and left boundaries of $[0,1]^2$. To understand this, one needs to understand the limiting behaviour of the analogously defined point measure on $[0,1]$ associated with the DLT. The empirical cdf of this measure is given by $Y_n = (Y_n(s); 0 \leq s \leq 1)$ with $Y_n(s)$ denoting the total length of those edges in the DLT on a sequence consisting of 0 followed by n uniform random points in $(0,1)$, whose edges start at points located in the interval $[0,s]$. Then Y_n takes values in the Skorohod space $D[0,1]$ of real-valued functions on the unit interval that are right continuous with left limits, and satisfies a recursion in $D[0,1]$ analogous to (7.15):

$$Y_n \stackrel{d}{=} U\mathbb{1}_{[U,1]} + UT_{0U}\left(Y'_M\right) + (1-U)T_{U1}\left(Y''_{n-1-M}\right)$$

where the distribution of M (given U) is binomial with parameters $n-1$ and U, where Y_n' and Y_n'' are independent copies of Y_n, and where (for $0 \leq a < b \leq 1$) the transformation T_{ab} on $D[0,1]$ is given by

$$T_{ab}(Y)(t) = \begin{cases} Y\left(\frac{t-a}{b-a}\right) & a \leq t < b \\ 0 & t < a \\ Y(1) & t \geq b. \end{cases}$$

To understand the limiting behaviour of the measure associated with the south-west MDST in two dimensions, it appears to be necessary to understand the limiting behaviour of the sequence of processes Y_n under this recursion, which may require some infinite-dimensional extension of the theory of distributional fixed points considered in Neininger and Rüschendorf (2004).

Similar questions may be asked about the limiting behaviour of the measure induced by the 'south' MDST in two dimensions, for which one ends up with a similar but more complicated recursion of processes associated with the empirical cdf of the measure associated with the one-dimensional ONG. For both types of MDST one might also consider the associated measure for power-weighted edges with exponent greater than 1, in which case the boundary effects should dominate.

7.6.5 *Degree distributions*

Much of the work described here has been concerned with quantities obtained by combining lengths of edges for the graphs under consideration. Also of interest are certain purely graph-theoretic aspects of these graphs. One of these is the *degree distribution*, much discussed in the random networks literature, especially with regard to the *scale free* property whereby the proportion of nodes of degree greater than k exhibits power-law decay as k becomes large, see e.g. Caldarelli (2007) and Dorogovstev and Mendes (2003).

For most or all of the graphs considered here, it seems likely that the expected proportion of vertices with degree greater than k converges to a limit, ρ_k, as the number of vertices (n) grows large, and that the ρ_k does *not* typically exhibit power-law decay. Unlike in the case of the ordinary nearest-neighbour graph where there is a bound on the maximum degree (see e.g. Lemma 8.4 of Yukich 1998), it seems that ρ_k is strictly positive for all k.

In the case of the ONG, these assertions have been verified in Berger, Bollobás, Borgs, Chayes and Riordan (2003), where it is shown that ρ_k exhibits exponential decay with a lower bound that is also exponentially decaying with a possibly different rate (but non-zero for all k). We conjecture that the correct exponent is unity, i.e. that ρ_k decays as e^{-k} (or at least that $-\log \rho_k \sim k$ as $k \to \infty$). More detailed information about the degree distribution for all of the graphs under consideration here would be of interest.

7.6.6 *Diameter and radius*

Another graph-theoretic quantity worthy of study is the *diameter* of the graph, which is the maximum of the inter-vertex graph-distances. Since our trees are rooted, it is perhaps more natural to study the *radius*, defined to be the maximum, over all vertices, of their graph distances from the root. For the ONG on n random points in the unit cube, it is proved in Berger, Bollobás, Borgs, Chayes and Riordan (2003) that with high probability the radius is bounded by $3 \log n$.

On the other hand, for the MDST ('south' or 'south-west') or radial spanning tree on n random points in d dimensions, we expect the radius to be linear in $n^{1/d}$. In the case of the radial spanning tree, a law of large numbers to this effect follows from Baccelli and Bordenave (2007, Th. 2). It might be of interest to seek an associated central limit theorem. In the 'south-west' MDST, the path from an inserted point at an arbitrary location in $(0,1)^d$ to the root should consist of a 'random walk' of $\mathcal{O}(n^{1/d})$ steps getting to near the boundary of the cube, followed by a further $\mathcal{O}(\log n)$ steps to get to the root. In the case of the 'south' MDST there is some dependence between the successive random walk steps; in either type of MDST, there is dependence between paths from different starting points.

7.6.7 *Path convergence*

Path convergence to Brownian motion, and convergence to Brownian web, have been conjectured for the radial spanning tree in Baccelli and Bordenave (2007). One could seek similar properties for the MDST. In some cases (such as the 'south' MDST), this may be harder than for Poisson trees (Ferrari, Fontes, and Wu 2005) due to the 'excluded volume' issue; in the 'south-west' MDST, this issue does not arise so proving convergence of paths to Brownian motion should be simplest in this case.

7.6.8 *Tessellations*

Consider the following tessellation problem related to the ONG. We are given two initial 'seeding' points and add subsequent points randomly into the box $(0,1)^d$ one by one. We join each new point to its nearest neighbour amongst those points already present. Thus we obtain two trees, one for each of the two seeding points. If we consider the 'territory' claimed by each tree, i.e. the Voronoi cells of its points, we partition the box into two domains. What can be said about the large-sample asymptotic theory of these domains? Or their interface (when $d = 2$, say)? The simulation in Fig. 7.1 shows an instance of the evolution of these two domains (one hatched in the figure) over time. It seems that the general shape rapidly stabilizes; can a result of this sort be proved?

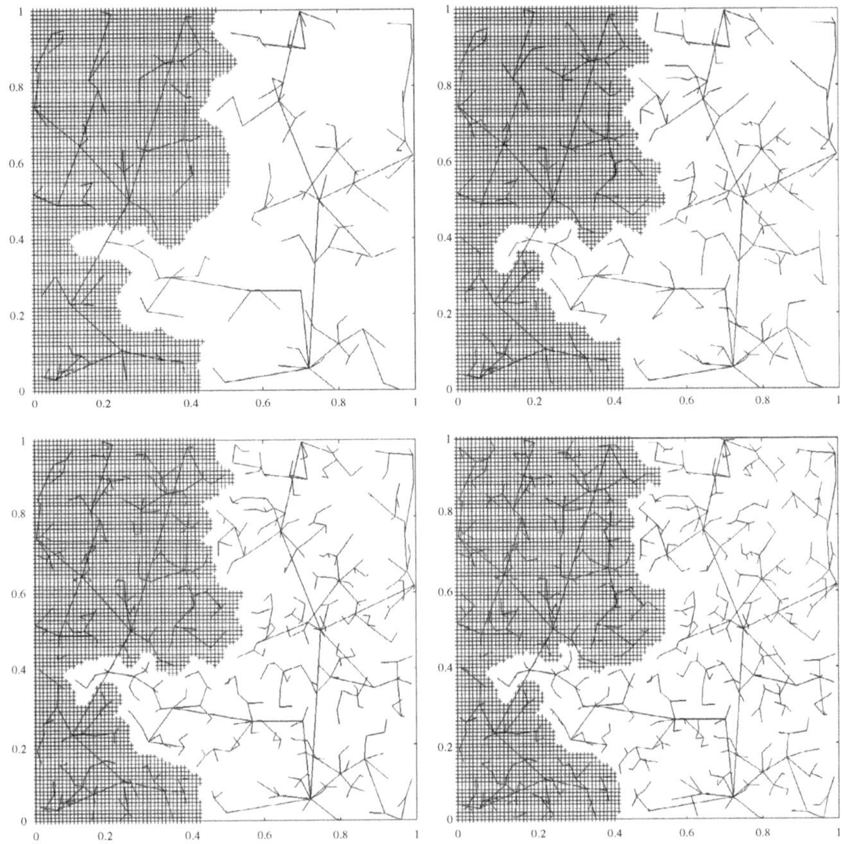

FIG. 7.1. Simulated evolution of an ONG-tessellation on uniform random points in $(0,1)^2$ with two seeding vertices

References

Aldous, D.J. and Bandyopadhyay, A. (2005). A survey of max-type recursive distributional equations. *Ann. Appl. Probab.*, **15**, 1047–1110.

Aldous, D. and Diaconis, P. (1995). Hammersley's interacting particle process and longest increasing subsequences. *Prob. Theory Relat. Fields*, **103**, 199–213.

Arnold, B.C. and Villaseñor, J.A. (1998). The asymptotic distribution of sums of records. *Extremes*, **1**, 351–363.

Arratia, R. (1998). On the central role of scale invariant Poisson processes on $(0, \infty)$. In *Microsurveys in Discrete Probability (Princeton, NJ, 1997)*, Volume 41 of *DIMACS Series in Discrete Mathematics and Theoretical Computer Science*, pp. 21–41. AMS, Providence, RI.

Baccelli, F. and Bordenave, C. (2007). The radial spanning tree of a Poisson point process. *Ann. Appl. Probab.*, **17**, 305–359.

Bai, Z.-D., Devroye, L., Hwang, H.-K., and Tsai, T.-H. (2005). Maxima in hypercubes. *Random Structures Algorithms*, **27**, 290–309.

Bai, Z.-D., Lee, S., and Penrose, M.D. (2006). Rooted edges in a minimal directed spanning tree. *Adv. Appl. Probab.*, **38**, 1–30.

Barndorff-Nielsen, O. and Sobel, M. (1966). On the distribution of the number of admissible points in a vector random sample. *Theory Probab. Appl.*, **11**, 249–269.

Baryshnikov, Yu. and Yukich, J.E. (2005). Gaussian limits for random measures in geometric probability. *Ann. Appl. Probab.*, **15**, 213–253.

Berger, N., Bollobás, B., Borgs, C., Chayes, J., and Riordan, O. (2003). Degree distribution of the FKP model. In *Automata, Languages and Programming* (ed. J. Baeten, J. Lenstra, J. Parrow, and G. Woeginger), Volume 2719 of *Lecture Notes in Computer Science*, pp. 725–738. Springer, Heidelberg.

Bhatt, A.G. and Roy, R. (2004). On a random directed spanning tree. *Adv. Appl. Probab.*, **36**, 19–42.

Billingsley, P. (1999). *Convergence of Probability Measures*, second edn. Wiley, New York.

Bordenave, C. (2008). Navigation on a Poisson point process. *Ann. Appl. Probab.*, **18**, 708–746.

Caldarelli, G. (2007). *Scale-Free Networks*. OUP, Oxford.

Chen, W.-M. and Hwang, H.-K. (2003). Analysis in distribution of two randomized algorithms for finding the maximum in a broadcast communication model. *J. Algorithms*, **46**, 140–177.

Darling, D.A. (1953). On a class of problems related to the random division of an interval. *Ann. Math. Statist.*, **24**, 239–253.

Darling, D.A. (1998). Accumulating success: 10436. *Amer. Math. Monthly*, **105**, 561–562.

Dickman, K. (1930). On the frequency of numbers containing prime factors of a certain relative magnitude. *Ark. Math. Astr. Fys.*, **22**, 1–14.

Donnelly, P. and Grimmett, G. (1993). On the asymptotic distribution of large prime factors. *J. London Math. Soc.*, **47**, 395–404.

Dorogovstev, S.N. and Mendes, J.F.F. (2003). *Evolution of Networks*. OUP, Oxford.

Embrechts, P. and Maejima, M. (2002). *Selfsimilar Processes*. Princeton Series in Applied Mathematics. Princeton University Press, Princeton, NJ.

Fabrikant, A., Koutsoupias, E., and Papadimitriou, C.H. (2002). Heuristically optimized trade-offs: a new paradigm for power laws in the internet. In *Automata, Languages and Programming*, Volume 2380 of *Lecture Notes in Computer Science*, pp. 110–122. Springer, Berlin.

Ferrari, P.A., Fontes, L.R.G., and Wu, X.-Y. (2005). Two-dimensional Poisson trees converge to the Brownian web. *Ann. I. H. Poincaré – PR*, **41**, 851–858.

Ferrari, P.A., Landim, C., and Thorisson, H. (2004). Poisson trees, succession lines and coalescing random walks. *Ann. I. H. Poincaré – PR*, **40**, 141–152.

Franceschetti, M. and Meester, R. (2007). *Random Networks for Communication.* CUP, Cambridge.

Gangopadhyay, S., Roy, R., and Sarkar, A. (2004). Random oriented trees: a model of drainage networks. *Ann. Appl. Probab.*, **14**, 1241–1266.

Golomb, S.W. (1964). Random permutations. *Bull. Amer. Math. Soc.*, **70**, 747.

Griffiths, R.C. (1988). On the distribution of points in a Poisson Dirichlet process. *J. Appl. Probab.*, **25**, 336–345.

Hensley, D. (1986). The convolution powers of the Dickman function. *J. London Math. Soc.*, **33**, 395–406.

Huang, K. (1987). *Statistical Mechanics*, second edn. Wiley, New York.

Hwang, H.-K. and Tsai, T.-H. (2002). Quickselect and Dickman function. *Combin. Probab. Comput.*, **11**, 353–371.

Kingman, J.F.C. (1993). *Poisson Processes.* Oxford Studies in Probability. OUP, Oxford.

Kingman, J.F.C. (2006). Poisson processes revisited. *Probab. Math. Statist.*, **26**, 77–95.

Neininger, R. and Rüschendorf, L. (2004). A general limit theorem for recursive algorithms and combinatorial structures. *Ann. Appl. Probab.*, **14**, 378–418.

Nevzorov, V.B. (1987). Records. *Theory Probab. Appl.*, **32**, 201–228.

Penrose, M.D. (2005). Multivariate spatial central limit theorems with applications to percolation and spatial graphs. *Ann. Probab.*, **33**, 1945–1991.

Penrose, M.D. (2007). Gaussian limits of random geometric measures. *Electron. J. Probab.*, **12**, 989–1035.

Penrose, M.D. and Wade, A.R. (2004). Random minimal directed spanning trees and Dickman-type distributions. *Adv. Appl. Probab.*, **36**, 691–714.

Penrose, M.D. and Wade, A.R. (2006). On the total length of the random minimal directed spanning tree. *Adv. Appl. Probab.*, **38**, 336–372. Extended version available from `arXiv:math.PR/0409201`.

Penrose, M.D. and Wade, A.R. (2008). Limit theory for the random on-line nearest-neighbour graph. *Random Structures Algorithms*, **32**, 125–156.

Penrose, M.D. and Wade, A.R. (2009). Limit theorems for random spatial drainage networks. Available from `arXiv:0901.3297`.

Penrose, M.D. and Yukich, J.E. (2001). Central limit theorems for some graphs in computational geometry. *Ann. Appl. Probab.*, **11**, 1005–1041.

Penrose, M.D. and Yukich, J.E. (2003). Weak laws of large numbers in geometric probability. *Ann. Appl. Probab.*, **13**, 277–303.

Pyke, R. (1965). Spacings. *J. Roy. Statist. Soc. Ser. B*, **27**, 395–436.

Rényi, A. (1976). *On the Extreme Elements of Observations*, Volume 3 of *Selected Papers of Alfred Rényi*, pp. 50–65. Akadémiai Kiadó.

Rodriguez-Iturbe, I. and Rinaldo, A. (1997). *Fractal River Basins: Chance and Self-Organization.* CUP, Cambridge.

Rösler, U. (1992). A fixed point theorem for distributions. *Stochastic Process. Appl.*, **42**, 195–214.

Rösler, U. and Rüschendorf, L. (2001). The contraction method for recursive algorithms. *Algorithmica*, **29**, 3–33.

Sato, K.-I. (1999). *Lévy Processes and Infinitely Divisible Distributions.* CUP, Cambridge.

Schreiber, T. and Yukich, J.E. (2008). Variance asymptotics and central limit theorems for generalized growth processes with applications to convex hulls and maximal points. *Ann. Probab.*, **36**, 363–396.

Smith, W.D. (1988). *Studies in Computational Geometry Motivated by Mesh Generation.* Ph.D. thesis, Princeton University.

Steele, J.M. (1995). Variations on the monotone subsequence theme of Erdős and Szekeres. In *Discrete Probability and Algorithms (Minneapolis, MN, 1993)*, Volume 72 of *IMA Vol. Math. Appl.*, pp. 111–132. Springer, New York.

Steele, J.M. (1997). *Probability Theory and Combinatorial Optimization.* Society for Industrial and Applied Mathematics, Philadelphia.

Tenenbaum, G. (1995). *Introduction to Analytic and Probabilistic Number Theory.* CUP, Cambridge.

Wade, A.R. (2007). Explicit laws of large numbers for random nearest-neighbour-type graphs. *Adv. Appl. Probab.*, **39**, 326–342.

Wade, A.R. (2009). Asymptotic theory for the multidimensional random on-line nearest-neighbour graph. *Stochastic Process. Appl.*, **119**, 1889–1911.

Watterson, G.A. (1976). The stationary distribution of the infinitely-many alleles diffusion model. *J. Appl. Probab.*, **13**, 639–651.

Wüthrich, M.V. (2002). Asymptotic behaviour of semi-infinite geodesics for maximal increasing subsequences in the plane. In *In and Out of Equilibrium* (ed. V. Sidoravicius), pp. 205–226. Birkhäuser, Boston.

Yukich, J.E. (1998). *Probability Theory of Classical Euclidean Optimization Problems*, Volume 1675 of *Lecture Notes in Mathematics.* Springer, Berlin.

8

RANDOM FRACTALS

Peter Mörters

Abstract

This chapter expounds the theory of random fractals, using tree representation as a unifying principle. Applications to the fine structure of Brownian motion are discussed.

8.1 Introduction

The term *fractal* usually refers to sets which, in some sense, have a self-similar structure. Already in the 1970s B. Mandelbrot made a compelling case for the importance of this concept in mathematical modelling. Indeed, some form of self-similarity is common in random sets, in particular those arising from stochastic processes. Therefore studying fractal aspects is an important feature of modern stochastic geometry.

Early progress in fractal geometry often referred to sets with obvious self-similarity, like the fixed points of iterated function systems. These are toy examples, tailor-made to study self-similarity in its tidiest form. An overview of the achievements in this period can be obtained from Falconer (2003).

Starting with the work of S.J. Taylor in the 1960s researchers were also looking at sets where self-similarity is more hidden. Such sets often arise in the context of stochastic processes. A beautiful survey of the state of the art in the mid 1980s, written by the protagonist in this area, is Taylor (1986). In the last ten years interest in this area has increased considerably, powerful techniques have been developed, and very substantial progress has been made. Typical examples of the fractals studied today are level sets of stochastic processes, the double points of random curves, or the boundary of excursions of random fields. The self-similar nature of these examples is typically less tidy and exploiting it means entering deep into the geometry of the sets.

Very roughly speaking, a set is *self-similar* if it can be decomposed into parts which look like scaled copies of the original set. This definition becomes particularly powerful when 'look like' is interpreted in a statistical sense, i.e. if it can be decomposed into parts which have (up to scaling) the same distribution as the whole set. This idea is naturally linked to trees: Starting from the root we identify the parts in the decomposition as the children of the root. Each part is itself a scaled copy of the whole picture and hence has a decomposition of the same kind as its parent, proceeding like this each point of the fractal has a natural address in the tree.

A crucial tool to bring the self-similarity of a random set to light is therefore its representation in terms of a tree, or sometimes a process on a tree. This technique has been exploited with great success in the last 10 years and continues to be a vital tool. My main aim in this chapter is to show that the first step in many deep geometric problems for random sets is to find the self-similarity of the problem and capture it in form of a tree picture. This picture determines the key direction of the argument, although the formalised proof often does not make the tree structure explicit.

Questions in geometry are very often related to the *size* of sets. Other than in classical geometry, random sets can often already be distinguished by the crudest measure of size, which is dimension. The most powerful concept of dimension, but by far not the only one, is *Hausdorff dimension*, introduced almost a century ago by Hausdorff (1918). This concept extends the classical notion of dimension to arbitrary metric spaces, allowing non-integer dimensions for sufficiently irregular sets. The notion is based on a family of measures $\mathcal{H}_s$, $s \geq 0$, the s-Hausdorff measures, which for integer values $s = 1, 2, 3$ coincide with the classical measures of length, area, and volume. The Hausdorff dimension of a set A is the critical value $s > 0$ where the function $s \mapsto \mathcal{H}_s(A)$ jumps from infinity to zero. We do not give a precise definition of Hausdorff measure and dimension here, but refer the reader instead to the excellent book by Falconer (2003).

The first section of this chapter is devoted to representing self-similarity in terms of trees, and we initially confine ourselves to simple examples. We show how to obtain the Hausdorff dimension of a set from a suitable tree representation and apply this to finding the Hausdorff dimension of the zero set of a linear Brownian motion.

In the second section we move to more sophisticated examples and present two more recent results on the fine structure of planar Brownian motion, which make great use of tree representations. On the one hand we look at the problem of the favourite sites, solved by Dembo, Peres, Rosen, and Zeitouni (2001), and on the other hand we study the multifractal spectrum of the intersection of two paths, a result of Klenke and Mörters (2005). In both cases, rather than giving details of the proof, we emphasize the underlying tree structure. We complete the section with a discussion of an open problem initialised by work of Bass, Burdzy, and Khoshnevisan (1994).

The second example introduces the notion of *probability exponents*, the general use of which we discuss in the third section. This particular aspect of random fractals has gained momentum through the discovery of an explicit formula for the intersection and disconnection exponents by Lawler, Schramm, and Werner (2001) and the subsequent award of the Fields medal to Werner in 2006. Here we discuss work of Lawler (1996a) on the Hausdorff dimension of the Brownian frontier and some closely related results.

8.2 Representing fractals by trees

There is no generally accepted definition of a statistically self-similar set, and we do not attempt to give one. Instead, we define a class of statistically self-similar sets, the *Galton–Watson fractals*, which comprises a number of interesting examples. We prove a formula for the Hausdorff dimension of Galton–Watson fractals, which gives us the opportunity to explore the relationship between branching processes and self-similarity and introduce basic ideas about probability on trees. The forthcoming book by Lyons and Peres (2009) gives a comprehensive account of this subject, on which much of this section is based.

8.2.1 *Fractals and trees*

We start with a general approach to capture the self-similar nature of fractals by means of *trees* with weights, so called *capacities*, associated to the edges, and investigate how the Hausdorff dimension of the fractal can be derived from the tree and the capacities.

A *tree* $T = (V, E)$ consists of a finite or countable set V of *vertices* and a set $E \subset V \times V$ of *edges*. For every $v \in V$ the set of *parents* $\{w \in V : (w, v) \in E\}$ consists of exactly one element, denoted by $\bar{v}$, except for exactly one distinguished element, called the *root* $\rho \in V$, which has no parent. For every $v \in V$ there is a unique self-avoiding path from the root to v, called the *ancestral line*, and the number of edges in this path is the *generation* $|v|$ of the vertex $v \in V$. For every vertex $v \in V$ we assume that the set of *children* $\{w \in V : (v, w) \in E\}$ is finite; see Fig. 8.1 for an illustration.

The *offspring* of a vertex v is the collection of vertices having v on their ancestral line. These vertices naturally form a subtree $T(v)$ of T. The *siblings* of $v \in V$ are the vertices $u \neq v$ with $\bar{u} = \bar{v}$. A sequence $(v_0, v_1, \ldots)$ of vertices such that $v_0 = \rho$ and $\bar{v_i} = v_{i-1}$ for all $i \geq 1$ is called a *ray* in the tree. The set of rays in T is denoted by ∂T. Finally, a set $\Pi \subset E$ is called a *cutset* if every ray includes an edge from the set Π.

We now describe a way to represent sets by marked trees. Let $T = (V, E)$ be an infinite tree and associate to each vertex $v \in V$ a nonempty, compact set $I_v \subset \mathbb{R}^d$ such that $I_v = \mathrm{cl}\big(\mathrm{int}\, I_v\big)$ and

- if v is a child of u, then $I_v \subset I_u$;
- if u and v are siblings, then $\mathrm{int}\, I_u \cap \mathrm{int}\, I_v = \emptyset$;
- for all rays $\xi = (v_0, v_1, \ldots)$ we have $\lim_{n \to \infty} \mathrm{diam}(I_{v_n}) = 0$.

Then the set

$$I(T) = \bigcup_{\xi \in \partial T} \bigcap_{v \in \xi} I_v$$

is represented by the tree T and the marks $\{I_v : v \in V\}$. Observe that, except for a possible boundary effect, there is a one-to-one relationship between the points of $I(T)$ and the rays of the tree, which can be interpreted as addresses.

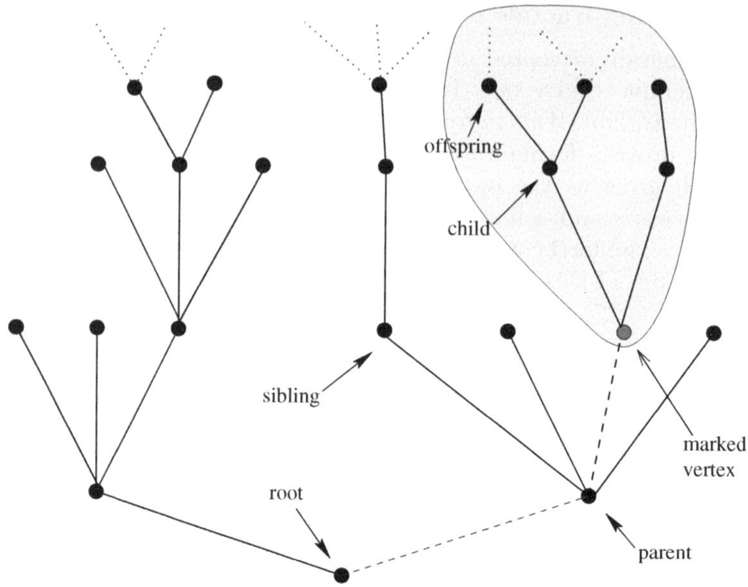

FIG. 8.1. A tree, with a vertex in the second generation marked; its ancestral line is dashed and the tree of its offspring shaded. One of its three siblings, and one of its two children are pointed out, as well as its parent and the root.

It is easy to see that for every compact subset of $\mathbb{R}^d$ there are many representations, but the idea of the method is to pick one which captures the structure of the set and leads to a simple tree.

We now give a formula for the Hausdorff dimension of sets in terms of the parameters of the tree representation. To this end we have to discuss the notion of flows on trees. Fix a mapping $C\colon E \to [0,\infty]$ representing the *capacities* of the edges. A mapping $\theta\colon E \to [0,c]$ such that

- we have $\sum_{\bar{w}=\rho} \theta\big((\rho,w)\big) = c$,
- for every vertex $v \neq \rho$ we have $\theta\big((\bar{v},v)\big) = \sum_{\bar{w}=v} \theta\big((v,w)\big)$,
- for every $e \in E$ we have $\theta(e) \leq C(e)$,

is called a *flow of strength* $c > 0$ through the tree with capacities C.

Theorem 8.1 *Suppose that a set $A \subset \mathbb{R}^d$ is represented by a tree T and sets $\{I_v\colon v \in V\}$. Assume additionally that*

$$\inf_{v\neq\rho} \frac{\operatorname{diam}(I_v)}{\operatorname{diam}(I_{\bar{v}})} > 0 \quad \text{and} \quad \inf_{v} \frac{\operatorname{vol}(\operatorname{int} I_v)}{\operatorname{diam}(I_v)^d} > 0, \tag{8.1}$$

and, for every $s \geq 0$*, define capacities* $C_s(e) = \operatorname{diam}(I_v)^s$ *if* $e = (\bar{v}, v)$*. Then*

$$\dim A = \inf\left\{ s \colon \inf_{\Pi \text{ cutset}} \sum_{e \in \Pi} C_s(e) = 0 \right\}$$
$$= \sup\left\{ s \colon \text{ there is a flow with capacities } C_s \right\}.$$

Theorem 8.1 is not hard to prove. The first equality is little more than the definition of Hausdorff dimension, the second is the famous *max-flow min-cut theorem* from graph theory, which, when applied to trees, states that the maximal strength of a flow with capacities C equals the minimal sum of capacities over the edges in a cutset, see Ford and Fulkerson (1962).

Example 8.2 The ternary Cantor set can be canonically represented by a binary tree such that I_v is an interval of length $3^{-|v|}$. Assigning capacities $C_s = 3^{-sn}$ to edges with end-vertex in the nth generation, it is easy to see that a necessary and sufficient condition for a flow to exists is $3^s \leq 2$. Hence we obtain that the dimension of the Cantor set is $\log 2/\log 3$.

8.2.2 *Galton–Watson fractals*

We now look at random sets given in terms of representations with randomly chosen tree and marks. For this purpose let $X = (N, A_1, \ldots, A_N)$ be a random variable consisting of a nonnegative integer N and weights $0 < A_i \leq 1$. We construct a *(weighted) Galton–Watson tree* by sampling, successively for each vertex, an independent copy of X and assigning N children carrying weights $A_1, \ldots, A_N$. We will be concerned with tree representations with the property that the diameter of the set associated with a vertex v is the product of the weights along the ancestral line of v.

We now recall some well-known facts about Galton–Watson trees. The first question is when a Galton–Watson tree can be infinite and hence suitable for representing a set. Excluding the trivial case $P\{N = 1\} = 1$, we get that

$$p = \mathbf{P}\{ \text{tree infinite} \} > 0 \quad \text{if and only if} \quad EN > 1.$$

A slightly less known important fact is the following zero-one-law for Galton–Watson trees. Let A be a set of trees or, equivalently, a property of trees. We say that A is *inherited* if

- every finite tree is in A, and
- if the tree $T \in A$ and $v \in V$ is a vertex of the tree, then $T(v) \in A$.

Then every inherited property A has $\mathbf{P}\{T \in A\} \in \{1 - p, 1\}$ or, equivalently,

$$\mathbf{P}\{T \in A \,|\, \text{tree infinite}\} \in \{0, 1\}.$$

Suppose now that (random) sets $\{I_v : v \in V\}$ are assigned to the vertices of the Galton–Watson tree in the way of a tree representation such that additionally

$$\inf_v \frac{\mathrm{vol}(\mathrm{int}\, I_v)}{\mathrm{diam}(I_v)^d} > 0,$$

and the normalized diameters correspond to the weights in the sense that

$$\frac{\mathrm{diam}(I_v)}{\mathrm{diam}(I_\rho)} = \prod_{i=1}^{n} A(v_i),$$

where $(\rho, v_1, \ldots, v_n)$ are the vertices on the ancestral line of the vertex $v = v_n$ and $A(v_1), \ldots, A(v_n)$ are the associated weights. Then the set $I(T)$ represented by this tree is a *Galton–Watson fractal.*

By Theorem 8.1, to find the Hausdorff dimension of the Galton–Watson fractals, we first need to study the existence of flows on Galton–Watson trees with edge capacities

$$C_s\big((\bar{v}, v)\big) = \prod_{i=1}^{n} A(v_i)^s.$$

The answer to this question is given by the following theorem of Falconer (1986). Note that the excluded case is trivial.

Theorem 8.3 (Falconer's theorem) *Suppose that a weighted Galton–Watson tree is given by the generating variable $X = (N, A_1, \ldots, A_N)$, let $s > 0$ and assume that $\sum_{i=1}^N A_i^s \neq 1$ with positive probability. Let*

$$\gamma = E\left[\sum_{i=1}^{N} A_i^s\right].$$

(a) If $\gamma \leq 1$ then almost surely no *flow is possible.*

(b) If $\gamma > 1$ then flow is possible almost surely given that the tree is infinite.

Note that in the special case when $A_i = 1$ almost surely, we recover the criterion for trees being finite. We now give a proof of Theorem 8.3, which is due to Falconer (part (a)) and Lyons and Peres (part (b)). The second part of the proof uses the idea of *percolation*, which is another important technique in fractal geometry.

Proof of (a) If $(v_0, \ldots, v_n)$ are the vertices on the ancestral line of $w = v_n$ and let $v = v_j$ for some $j \leq n$, we equip the tree $T(v)$ with capacities $C_s^v((\bar{w}, w)) = \prod_{i=j+1}^n A(v_i)^s$, and let $\theta(v)$ be the maximal strength of a flow in this subtree. Abbreviating $\theta = \theta(\rho)$ we have

$$\theta = \sum_{\bar{v}=\rho} \big(A(v)^s \wedge (A(v)^s \theta(v))\big) = \sum_{\bar{v}=\rho} A(v)^s \big(1 \wedge \theta(v)\big). \tag{8.2}$$

Now suppose that $\gamma \le 1$ and suppose $X = (N, A_1, \dots, A_N)$ describes the children of the root and their weights. Using independence, and the fact that θ and $\theta(v)$ have the same distribution for every edge v,

$$\begin{aligned}\mathbf{E}\theta &= \sum_{n=1}^{\infty} \mathbf{E}\left[\theta \mathbb{1}_{\{N=n\}}\right] = \sum_{n=1}^{\infty}\sum_{v=1}^{n} \mathbf{E}\left[A(v)^s(1\wedge\theta(v))\mathbb{1}_{\{N=n\}}\right] \\ &= \sum_{n=1}^{\infty}\sum_{v=1}^{n} \mathbf{E}\left[A(v)^s\mathbb{1}_{\{N=n\}}\right]\mathbf{E}\left[1\wedge\theta(v)\right] \\ &= \sum_{n=1}^{\infty} \mathbf{E}\left[\sum_{v=1}^{N} A(v)^s \mathbb{1}_{\{N=n\}}\right]\mathbf{E}\left[1\wedge\theta\right] = \gamma\mathbf{E}[1\wedge\theta] \le \mathbf{E}[1\wedge\theta].\end{aligned}$$

Hence $\theta \le 1$ almost surely and $\mathbf{P}\{\theta > 0\} > 0$ only if $\gamma = 1$. This already shows that no flow is possible if $\gamma < 1$. In the case $\gamma = 1$ we get from (8.2) and independence, using that $\theta \le 1$, that

$$\text{ess sup}\,(\theta) = \text{ess sup}\left(\sum_{v=1}^{N} A(v)^s\right)\text{ess sup}\,(\theta)\,.$$

Hence, if ess sup $(\theta) > 0$ we have ess sup $(\sum_{v=1}^{N} A(v)^s) = 1$. As $\mathbf{E}[\sum_{v=1}^{N} A(v)^s] = \gamma = 1$ we must have $\sum_{v=1}^{N} A(v)^s = 1$, which is the excluded case. Hence $\theta = 0$ almost surely, which means that no flow is possible.

Proof of (b) We first look at a fixed (deterministic) tree T with weights $A(v)$ attached to the vertices. We introduce a family of random variables on this tree T as follows. Independently for every edge $e = (\bar{v}, v) \in E$ we let

$$X(e) = \begin{cases} 1 & \text{with probability } A(v)^s, \\ 0 & \text{with probability } 1 - A(v)^s. \end{cases}$$

The intuition is that an edge e is *open* if $X(e) = 1$ and otherwise closed. We consider the subtree $T^* \subset T$ consisting of all edges which are connected to the root by a path of open edges. Let $Q(T) = P\{T^* \text{ is infinite}\}$, where the probability P and the associated expectation E refer to the percolation process on the fixed tree T. For any cutset Π note that $\sum_{e\in\Pi} C_s(e)$ is the expected number of edges in Π, which are also in T^*. Hence

$$\sum_{e\in\Pi} C_s(e) \ge P\big\{e \in T^* \text{ for some } e \in \Pi\big\} \ge P\big\{T^* \text{ is infinite}\big\}.$$

If $\theta(T)$ is the maximal strength of a flow in T, then the last inequality together with the max-flow min-cut theorem shows that

$$Q(T) > 0 \implies \theta(T) > 0. \tag{8.3}$$

Now we use this result for a Galton–Watson tree, by performing a two-step experiment: first sampling the tree T and the reducing it to T^*. As a result of the experiment, T^* is another Galton–Watson tree. Denoting by $v_1, \ldots, v_n$ the children of the root, we get for the mean number of children in T^*,

$$\begin{aligned}
\mathbf{E}\left[\sum_{i=1}^{N} X\big((\rho, v_i)\big)\right] &= \mathbf{E}\left[\sum_{n=1}^{\infty}\sum_{i=1}^{n} X\big((\rho, v_i)\big)\mathbb{1}_{\{N=n\}}\right] \\
&= \sum_{n=1}^{\infty}\sum_{i=1}^{n} \mathbf{E}\left[X\big((\rho, v_i)\big)\mathbb{1}_{\{N=n\}}\right] \\
&= \sum_{n=1}^{\infty}\sum_{i=1}^{n} \mathbf{E}\left[A(v_i)^s\mathbb{1}_{\{N=n\}}\right] \\
&= \sum_{n=1}^{\infty} \mathbf{E}\left[\sum_{i=1}^{N} A(v_i)^s\mathbb{1}_{\{N=n\}}\right] = \mathbf{E}\left[\sum_{i=1}^{N} A(v_i)^s\right] = \gamma.
\end{aligned}$$

If $\gamma > 1$, by the criterion for Galton–Watson trees being infinite, we have

$$0 < \mathbf{P}\big\{T^* \text{ is infinite } \big\} = \mathbf{E}\big[Q(T)\big].$$

Hence $Q(T) > 0$ with positive probability, and by (8.3) we infer that $\theta(T) > 0$ with positive probability. In other words, $\mathbf{P}\{\theta(T) = 0\} < 1$. As the event $\{\theta(T) = 0\}$ is inherited, we infer from the Galton–Watson zero-one-law that we have $\theta(T) > 0$, almost surely on the tree being infinite. □

Up to some technicalities, the dimension formula for Galton–Watson fractals, found independently by Falconer (1986) and Mauldin and Williams (1986), now follows by combining Falconer's theorem and the dimension formula for tree representations, Theorem 8.1.

Theorem 8.4 (Hausdorff dimension of Galton–Watson fractals) *Suppose that $I(T)$ is a Galton–Watson fractal associated with a weighted Galton–Watson tree with generating variable $X = (N, A_1, \ldots, A_N)$. Then, almost surely on the event $\{I(T) \neq \emptyset\}$,*

$$\dim I(T) = \min\left\{s \,:\, E\left[\sum_{i=1}^{N} A_i^s\right] \le 1\right\}.$$

An interesting corollary comes from the fact that in the *critical case* $\gamma = 1$ flow is impossible unless we are in the excluded case $\sum_{i=1}^{N} A_i^s = 1$, in which flow is obviously possible.

Corollary 8.5 *If $\dim I(T) = s$ and $\sum_{i=1}^{N} A_i^s \neq 1$ with positive probability, then $\mathcal{H}_s(I(T)) = 0$ almost surely.*

We now exploit our main result by giving formulae for the Hausdorff dimension of a variety of sets. The main example, presented in some detail, is the zero set of a linear Brownian motion, which we study avoiding the use of local times.

Example 8.6 We define *percolation fractals*, or *percolation limit sets*. Fix the ambient dimension d, a parameter $p \in (0,1)$ and an integer $n \geq 2$. Divide $[0,1]^d$ into n^d nonoverlapping compact subcubes of equal sidelength. Keep each independently with probability p, and remove the rest. Apply the same procedure to the remaining cubes ad infinitum. The remaining set is a Galton–Watson fractal which has a generating random variable $(N, A_1, \ldots, A_N)$, where N is binomial with parameters n^d and p, and A_i deterministic with $A_i = 1/n$. The probability that it is nonempty is positive if and only if $p > 1/n^d$. Moreover,

$$E\left[\sum_{i=1}^{N} A_i^s\right] = \left(\frac{1}{n}\right)^s EN = \frac{n^d p}{n^s}.$$

This is ≤ 1 if and only if $s \geq d + \frac{\log p}{\log n}$. Hence, almost surely on $\{I(T) \neq \emptyset\}$,

$$\dim I(T) = d + \frac{\log p}{\log n}.$$

Example 8.7 We compare the following two random fractals: On the one hand a percolation fractal based on dividing the unit interval $[0,1]$ into three nonoverlapping intervals of length $1/3$ and keeping each with probability $p = 2/3$, on the other hand the random fractal obtained by dividing $[0,1]$ into three nonoverlapping intervals of length $1/3$ and keeping two randomly chosen intervals out of the three, proceeding like this ad infinitum.

In both cases we obtain fractals of Hausdorff dimension $s = \log 2/\log 3$. To see this in the second case just observe that the 3-adic coding tree of the fractal is the dyadic tree, exactly as in the case of the ordinary ternary Cantor set. Corollary 8.5 indicates a significant difference between the two examples. Whereas for the first case, by the corollary, the s-Hausdorff measure is zero, one can show that in the second case the s-Hausdorff measure is strictly positive. This can be seen from the fact that there exists a flow on the coding tree with capacities $C_s(\bar{v}, v) = |I_v|^s$ in the second example, whilst there is none in the first.

8.2.3 *The dimension of the zero-set of Brownian motion*

We now use the theory developed so far to calculate the dimension of the *zero-set* of a Brownian motion $W\colon [0,1] \to \mathbb{R}$. The idea of this proof is based on Galton–Watson fractals and is due to Graf *et al.* (1988).

A first step is to make the problem more symmetric by looking at a *Brownian bridge* instead of a Brownian motion. There are several ways of defining a Brownian bridge B from a Brownian motion W:

- The process $B(t) = W(t) - tW(1)$, for $t \in [0,1]$, is a Brownian bridge.
- Let $T = \sup\{t < 1\colon W(t) = 0\}$, then the process $C(t) = \sqrt{1/T}\, W(tT)$, for $t \in [0,1]$, is also a Brownian bridge.

Note that for a given sample path W of Brownian motion the two bridges B and C have quite different sample paths. From the second definition it is easy to see that the dimension of the zero set of a Brownian bridge and of a Brownian motion have the same law. An important property of the Brownian bridge is *symmetry*: If $\{B(t) \colon 0 \le t \le 1\}$ is a Brownian bridge, then so is the process $\{\tilde{B}(t) \colon 0 \le t \le 1\}$ defined by $\tilde{B}(t) = B(1-t)$.

To study the dimension of the zero set of a Brownian bridge, define

$$T_1 = \sup\{t \le 1/2 \,:\, B(t) = 0\} \text{ and } T_2 = \inf\{t \ge 1/2 \,:\, B(t) = 0\}.$$

By symmetry the random variables T_1 and $1 - T_2$ have the same distribution (but they are not independent). The interval (T_1, T_2) does not contain any zeros, and we remove it from $[0,1]$, which leaves us with two random intervals $[0, T_1]$ at the left and $[T_2, 1]$ on the right. Moreover, it is not hard to show that the process

$$\left\{\sqrt{1/T_1}\, B(tT_1) \,:\, 0 \le t \le 1\right\}$$

is a Brownian bridge, which is independent of $\{B(t) \,:\, t \ge T_1\}$.

Now we can represent the zero set of the Brownian bridge as a Galton–Watson fractal: we start with the interval $[0,1]$ and remove the interval (T_1, T_2). To the left of the removed interval, we have an independent Brownian bridge

$$\left\{\sqrt{1/T_1}\, B(tT_1) \,:\, 0 \le t \le 1\right\}.$$

By symmetry, we also have an independent Brownian bridge

$$\left\{\sqrt{1/(1-T_2)}\, B(1 - t(1-T_2)) \,:\, 0 \le t \le 1\right\}.$$

to the right of the removed interval. If we apply the same procedure on each of the remaining bridges, we iteratively construct the zero set of the Brownian bridge by removing all gaps. The essence of all this is the following:

Lemma 8.8 *The zero set of a Brownian bridge B is a Galton–Watson fractal with generating random variable $X = (2, T_1, 1 - T_2)$. Hence* $\dim\{t \in [0,1] : B(t) = 0\} = \alpha$, *where α is the unique solution of*

$$\mathbf{E}\left[T_1^\alpha + (1-T_2)^\alpha\right] = 1.$$

We can now calculate the dimension by evaluating this expectation for the right value of α.

Lemma 8.9

$$\mathbf{E}\left[\sqrt{T_1} + \sqrt{1-T_2}\right] = 1.$$

Proof By symmetry of the Brownian bridge, T_1 and $1 - T_2$ have the same distribution, hence it suffices, to show that $\mathbf{E}\sqrt{1-T_2} = 1/2$. We have, using the definition of the Brownian bridge and the time inversion property of Brownian motion,

$$\begin{aligned}
T_2 &= \inf\big\{1/2 \le t \le 1 \colon B(t) = 0\big\} \\
&= \inf\big\{1/2 \le t \le 1 : W(t) - tW(1) = 0\big\} \\
&\stackrel{d}{=} \inf\big\{1/2 \le t \le 1 : tW(1/t) - tW(1) = 0\big\} \\
&= \inf\big\{1/2 \le t \le 1 \colon W(1/t) - W(1) = 0\big\} \\
&= 1/\sup\big\{1 \le s \le 2 \colon W(s) - W(1) = 0\big\}.
\end{aligned}$$

As $\{W(s) - W(1) : s \ge 1\}$ has the same law as $\{W(s-1) : s \ge 1\}$, we have

$$T_2 \stackrel{d}{=} \frac{1}{1 + \sup\big\{0 \le t \le 1 : W(t) = 0\big\}}$$

and, in particular,

$$\mathbf{E}\sqrt{1 - T_2} = \int_0^1 \sqrt{\frac{x}{1+x}}\, f(x)\,\mathrm{d}x\,,$$

where f is the density of the random variable $L = \sup\big\{0 \le t \le 1 : W(t) = 0\big\}$. This random variable has the arcsine-distribution, which can be verified using the *reflection principle* of Brownian motion, see e.g. Mörters and Peres (2009). We get that

$$\begin{aligned}
\mathbf{E}\sqrt{1 - T_2} &= \frac{1}{\pi}\int_0^1 \sqrt{\frac{x}{1+x}}\, \frac{\mathrm{d}x}{\sqrt{x(1-x)}} \\
&= \frac{1}{\pi}\int_0^1 \frac{\mathrm{d}x}{\sqrt{1-x^2}} = \frac{1}{\pi}\arcsin(1) = \frac{1}{2},
\end{aligned}$$

which completes the proof of Lemma 8.9. □

We have thus proved the following result.

Theorem 8.10 *Almost surely,*

$$\dim\{t \in [0,1] \colon W(t) = 0\} = \dim\{t \in [0,1] \colon B(t) = 0\} = \tfrac{1}{2}.$$

8.3 Fine properties of stochastic processes

In this section we discuss two deeper results, which were solved using the tree approach. We also state an interesting open problem, which may be suitable for a treatment based on these ideas.

8.3.1 *Favourite points of planar Brownian motion*

Suppose $(W(t)\colon 0 \le t \le 1)$ is a planar Brownian motion, and denote by

$$T(A) = \int_0^1 \mathbb{1}_{\{W(s)\in A\}}\, \mathrm{d}s$$

the *occupation time* of the path in $A \subset \mathbb{R}^2$. A famous problem of Erdős and Taylor (stated in 1960 for the analogous random walk case) is to find the asymptotics of the occupation time around the favourite points,

$$T^*(\epsilon) = \max_{x\in\mathbb{R}^2} T\big(\mathcal{B}(x,\epsilon)\big) \qquad \text{as } \epsilon \downarrow 0 .$$

This problem was solved by Dembo, Peres, Rosen, and Zeitouni (2001) exploiting the deep self-similar structure of the Brownian path using tree ideas.

Theorem 8.11 *Almost surely, $T^*(\epsilon) \sim 2\epsilon^2 \log^2 \epsilon$ as $\epsilon \downarrow 0$.*

A detailed account of the proof of this and some closely related results can be found in Dembo (2005). Other than the original paper Dembo *et al.* (2001), this highly recommended source also discusses the tree analogy in depth. In our account we focus entirely on a rough sketch of this analogy. This captures the main idea of the proof, but neglects a lot of (often interesting) technical details.

Recall that a planar Brownian motion is neighbourhood recurrent, i.e. any ball is visited infinitely often as time goes to infinity. The main difficulty in the proof of Theorem 8.11 lies in the fact that the occupation time in a ball $\mathcal{B}(x,\epsilon)$ is accumulated during a large number of excursions from its boundary, whose lengths vary across a large range of scales. This leads to a complicated dependence between $T(\mathcal{B}(x,\epsilon))$ and $T(\mathcal{B}(y,\epsilon))$, even if x and y are relatively far away. The main merit of the tree picture is to organise this dependence structure in a natural fashion.

If a ball is visited often, by the law of large numbers, the time spent in the ball can be well approximated by the number of excursions from its boundary. To be more precise, let $x \in \mathbb{R}^2$ and consider a sequence of decreasing radii such that $\epsilon_k/\epsilon_{k-1} = k^{-3}$. Fix $a > 0$ and let N_k^x be the number of excursions from $\partial\mathcal{B}(x,\epsilon_{k-1})$ to $\partial\mathcal{B}(x,\epsilon_k)$ before time one. We call $x \in \mathbb{R}^s$ an *n-perfect* point if

$$N_k^x \approx 3ak^2 \log k \qquad \text{for all } k \in \{2,\dots,n\} .$$

During an excursion from $\partial\mathcal{B}(x,\epsilon_k)$ to $\partial\mathcal{B}(x,\epsilon_{k-1})$ the path spends on average about $3\epsilon_k^2 \log k$ time units in the ball $\mathcal{B}(x,\epsilon_k)$, and these times are all independent, so that a law of large numbers applies. As $\log(1/\epsilon_k) \approx 3k\log k$, we get that if x is n-perfect then it is *n-favourite* in the sense that

$$\frac{T\big(\mathcal{B}(x,\epsilon)\big)}{\epsilon^2 \log^2 \epsilon} \approx a \qquad \text{for all } \epsilon_2 \ge \epsilon \ge \epsilon_n . \tag{8.4}$$

Strictly speaking the n-perfect points are only a subset of the n-favourite points, but the difference is small enough for us to neglect this distinction from now on. Note that, by definition, if x is n-perfect, it is also m-perfect for all $m \leq n$.

We now focus on the favourite points inside a square S of sidelength ϵ_1. We partition S into $(\epsilon_n/\epsilon_1)^{-2} = (n!)^6$ non-overlapping squares $S(n,i)$ of sidelength ϵ_n with centres $x_{n,i}$. This decomposition yields a natural tree representation of the cube S, with squares $S(n,i)$ associated to the vertices in the nth generation, such that any vertex is offspring of another one, if its associated square is contained in that of the other. Observe that in this tree, denoted T, any vertex of the kth generation has exactly $(k+1)^6$ children.

In a rough approximation, which needs to be refined in the actual proof, we represent the set

$$\left\{ x \in S \colon \lim_{\epsilon \downarrow 0} \frac{T\big(\mathcal{B}(x,\epsilon)\big)}{\epsilon^2 \log^2 \epsilon} \approx a \right\}$$

by the tree T_a consisting of all vertices in the nth generation corresponding to squares with n-perfect centre. Here we *neglect* the fact that, because of the different centres, a square of sidelength ϵ_n with n-perfect centre may be contained in a square of sidelength ϵ_k, $k \leq n$, whose centre fails to be k-perfect. As most squares are sufficiently far away from the boundary of their parental square, this approximation turns out to be safe. We therefore have to show that, almost surely, the tree T_a is infinite if $a < 2$ and finite if $a > 2$.

To get hold of the squares with perfect centre, we fix a square $S(n,i)$ and map the planar Brownian curve onto a homogeneous Markov chain $(Z_k \colon k \in \mathbb{N})$ with values on the set $\{1, \ldots, n\}$. This Markov chain is started in $Z_0 = n$, and the transition probabilities of the Markov chain are given, for $j \geq 1$, as

$$\mathbf{P}\big\{ Z_j = \ell \,\big|\, Z_{j-1} = k \big\} = \begin{cases} 1 & \text{if } k = n,\ \ell = n-1, \\ p_k & \text{if } 1 < k < n,\ \ell = k-1, \\ 1 - p_k & \text{if } 1 < k < n \text{ and } \ell = k+1, \\ 1 & \text{if } k = \ell = 1, \end{cases}$$

for

$$p_k = \frac{\log \epsilon_{k+1} - \log \epsilon_{k+2}}{\log \epsilon_k - \log \epsilon_{k+2}}.$$

The rationale behind this choice is that, if $S = S_1 \supset S_2 \supset \cdots \supset S_n = S(n,i)$ is the sequence of construction squares containing $S(n,i)$, we follow the Brownian curve from the first time it hits the boundary of S_n and, as indicated in Fig. 8.2, whenever the motion moves from the boundary of S_k to the boundary of $S_{k\pm1}$, the chain moves from state k to $k \pm 1$. If squares are approximated by concentric balls of the same diameter, the probability that a Brownian motion started on the sphere of radius ϵ_{k+1} hits the sphere of radius ϵ_k before the sphere of radius ϵ_{k+2} is given by p_k, see e.g. Mörters and Peres (2009). The motion is stopped once it leaves S, which makes one an absorbing state.

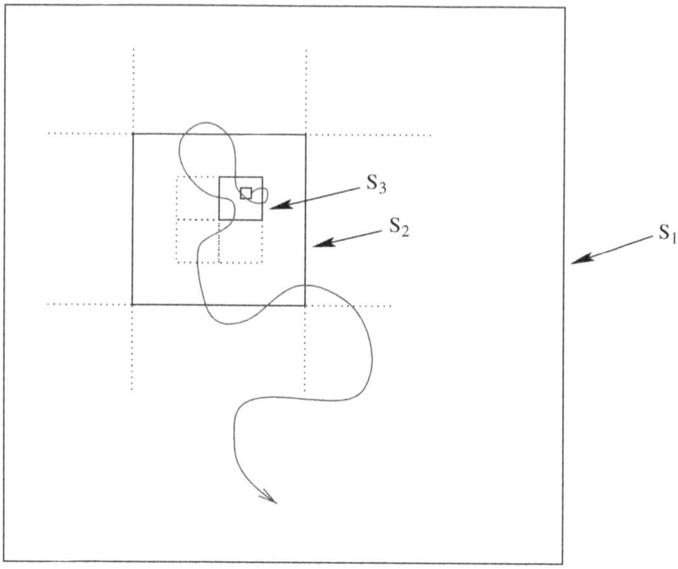

FIG. 8.2. Brownian motion moving between squares. In this picture $n = 4$ and the shown path yields the chain $4, 3, 4, 3, 2, 3, 2$.

Summarizing, the square $S(n, i)$ is kept in the construction if and only if the associated Markov chain satisfies

$$\sum_{\ell=1}^{\infty} \mathbb{1}\{Z_\ell = k-1, Z_{\ell+1} = k\} \approx 3ak^2 \log k \qquad \text{for all } k = 2, \dots, n.$$

The picture given so far suffices to show that $\partial T_a = \emptyset$ if $a > 2$. Indeed, using a Markov chain calculation, one can see that, for any vertex $v \in V$ with $|v| = n$ we have $\mathbf{P}\{v \in T_a\} \approx (n!)^{-3a}$, and hence, looking at the expected number of retained vertices,

$$\mathbf{E}\#\{v \in T_a \colon |v| = n\} \approx (n!)^{6-3a} \longrightarrow 0 \qquad \text{if } a > 2\,.$$

For the lower bound first moment arguments as above are insufficient and we need to look at the more complex picture arising when two squares are considered simultaneously. For this purpose we fix $a < 2$ and two vertices v, w from the nth generation of T, whose oldest common ancestor is in generation $0 < m < n$. To get hold of $\mathbf{P}\{v, w \in T_a\}$ we look at the Markov chain $(Z_n \colon n \in \mathbb{N})$ on the branching set $\{1, \dots, n\} \,\dot\cup\, \{m+1, \dots, n\}$ shown in Fig. 8.3. The chain can only change branch when it moves up from state m, and in this case each branch is chosen with the same probability. Otherwise the transition probabilities are the same as before, where we allow ourselves an abuse of notation by using the same symbol for the *distinct* states on the two branches of the state space that emerge from state m.

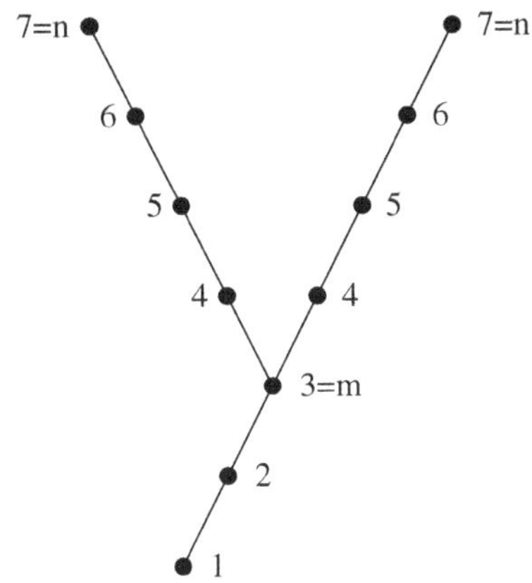

FIG. 8.3. The statespace of the Markov chain as a branching structure.

The rationale behind this chain is that the state j on the left branch represents the construction square of sidelength ϵ_j containing the square representing v, and the state $j > k$ on the right branch represents the construction square of sidelength ϵ_j containing the square representing w. The transition probabilities mimic the consecutive visits of the boundaries of these squares by the Brownian curve, though this mapping is imprecise about excursions between squares of radius ϵ_{m+1} and ϵ_m. This effect turns out to be negligible.

A Markov chain calculation shows that

$$\mathbf{P}\{v, w \in T_a\} \approx (n!)^{-6a}\,(k!)^{3a}\,,$$

and from this we obtain a constant $C > 0$ and a bound on the variance

$$\mathbf{E}\left[(\#\{v \in T_a \colon |v| = n\})^2\right] \le C\,\mathbf{E}\big[\#\{v \in T_a \colon |v| = n\}\big]^2.$$

We can therefore use the Paley–Zygmund inequality to derive, for any $0 < \lambda < 1$,

$$\begin{aligned}\mathbf{P}\Big\{\#\{v \in T_a : |v| = n\} &\ge (1-\lambda)\mathbf{E}\big[\#\{v \in T_a \colon |v| = n\}\big]\Big\} \\ &\ge \lambda^2\,\frac{\mathbf{E}\left[\#\{v \in T_a \colon |v| = n\}\right]^2}{\mathbf{E}\left[(\#\{v \in T_a \colon |v| = n\})^2\right]} \ge C^{-1}\,\lambda^2 > 0\,.\end{aligned}$$

Recall that, if $a < 2$, we have

$$\mathbf{E}\big[\#\{v \in T_a \colon |v| = n\}\big] \longrightarrow \infty,$$

and hence this argument shows that $\partial T_a \ne \emptyset$ with positive probability. A self-similarity argument (not unlike the Galton–Watson zero-one law) shows that this must therefore hold with probability one.

Let me emphasize the importance of the correct choice of the scales (ϵ_k) for the success of the tree approximation. If the ratio $\epsilon_{k-1}/\epsilon_k$ is chosen significantly smaller, the excursion counts typically do not reflect the occupation times at all radii and centres; observe that we need the equivalence analogous to (8.4) *simultaneously* for all squares of sidelength ϵ_k, $k \in \{2, \ldots, n\}$, so that a rigorous proof requires a much more quantitative approach to this part of the argument than our informal discussion suggests. Conversely, if the ratio $\epsilon_{k-1}/\epsilon_k$ is chosen significantly larger, we lose the necessary control over the occupation times for intermediate radii.

Finally, a note of caution: Turning this picture into a full proof of Theorem 8.11 still requires skill and a lot of work, as we oversimplified at many places. Nevertheless, the tree representation gives a neat organization of the complicated dependencies, which greatly helps understanding and solving this hard problem.

8.3.2 *The multifractal spectrum of intersection local time*

The *multifractal spectrum* is an important means of describing the fine structure of a fractal measure, see Mörters (2008) for a subjective discussion of its importance in the context of stochastic processes. For a precise definition, fix a locally finite measure μ, which may be random or non-random. The value $f(a)$ of the multifractal spectrum is the Hausdorff dimension of the set of points x with local dimension

$$\lim_{r \downarrow 0} \frac{\log \mu(\mathcal{B}(x,r))}{\log r} = a, \tag{8.5}$$

where $\mathcal{B}(x,r)$ denotes the open ball of radius r centred in x. In some cases of interest, the limit in (8.5) has to be replaced by liminf or limsup to obtain an interesting nontrivial spectrum.

Examples of multifractal spectra for measures arising in probability are the occupation measures of stable subordinators (Hu and Taylor, 1997), the states of super-Brownian motion (Perkins and Taylor, 1998), and the harmonic measure on a Brownian path (Lawler, 1997). The example we study here has some likeness with the first two examples, for which a similar tree analogy could be built, though details in the proof invariably differ considerably.

We look at two independent planar Brownian motions $\big(W_t^{(1)} : 0 \le t \le 1\big)$ and $\big(W_t^{(2)} : 0 \le t \le 1\big)$ and study the intersection set

$$\mathcal{S} = \big\{x \in \mathbb{R}^2 \colon \text{ there exist } 0 \le s, t \le 1 \text{ with } W_s^{(1)} = W_t^{(2)} = x\big\}.$$

The natural measure on $\mathcal{S}$ is the *intersection local time* μ defined symbolically by

$$\mu(A) = \int_A \mathrm{d}x \prod_{i=1}^{2} \int_0^1 \mathrm{d}s \; \delta_0\,(W_s^{(i)} - x).$$

Rigorous definitions of μ can be given by approximation of the 'delta-function' δ_0, but also as a suitable Hausdorff measure on $\mathcal{S}$. Technical details of the construction are not of interest to us here.

Theorem 8.12 *For every* $2 \le a \le \frac{70}{11}$ *we have, almost surely,*

$$\dim\Big\{x \in \mathcal{S} \colon \limsup_{r \downarrow 0} \frac{\log \mu(\mathcal{B}(x,r))}{\log r} = a\Big\} = \frac{1}{12}\Big(\frac{70}{a} - 11\Big).$$

Moreover, there are no points with local dimension $a < 2$ or $a > \frac{70}{11}$ in any sense (liminf, limsup, or lim). At least heuristically, all the results concerning values $a \ge 2$ can be read off a tree picture, which we describe below. The full proof of the result, which is inspired by this tree picture but does not make explicit use of it, can be found in Klenke and Mörters (2005).

As $\mathcal{S}$ is the intersection of two independent sets of full dimensions (the Brownian paths) it is not surprising that $\dim \mathcal{S} = 2$ and therefore $\mu(\mathcal{B}(x,r)) \approx r^2$ for typical points $x \in \mathcal{S}$. Fix $a > 2$ for the remainder of this section. For the points $x \in \mathcal{S}$ with

$$\mu(\mathcal{B}(x,r)) \approx r^a \ll r^2$$

we expect that

- the ball $\mathcal{B}(x,r)$ is visited only once by each Brownian motion,
- the intersection local time spent in $\mathcal{B}(x,r)$ during this visit is small.

Due to the first item, the recurrence effects that were so crucial in the proof of Theorem 8.11 do not play a rôle here. Indeed, here we can assume that for disjoint balls $\mathcal{B}(x,r)$ and $\mathcal{B}(y,r)$ the events $\{\mu(\mathcal{B}(x,r)) \approx r^a\}$ and $\{\mu(\mathcal{B}(y,r)) \approx r^a\}$ are essentially independent. This simplifies the informal discussion immensely, but making this argument rigorous is one of the main difficulties in the proof of Theorem 8.12, which we do not discuss here.

The remainder of our discussion of this example is based on this independence (or locality) assumption. Fix a square $S \subset \mathbb{R}^2$ of unit sidelength and pick a large integer m. Divide the square into m^2 squares of sidelength $1/m$, and keep a square if it contains a point of $\mathcal{S}$, then repeat this procedure with any square kept, and so on at infinitum. Identifying the squares kept in the procedure with vertices in a tree $T = (V, E)$, we obtain a tree representation of $\mathcal{S} \cap S$.

To connect the intersection local time μ to this tree representation, we recall a result of Le Gall (1986), which states that μ can be recovered from the volume of the Wiener sausages around the two Brownian paths, more precisely

$$\lim_{\epsilon \downarrow 0} \frac{(\log \epsilon)^2}{\pi^2} \operatorname{vol}(S_\epsilon^{(1)} \cap S_\epsilon^{(2)} \cap A) = \mu(A)\,,$$

where $S_\epsilon^{(i)} = \big\{x \in \mathbb{R}^2 \colon |W_t^{(i)} - x| \le \epsilon \text{ for some } 0 \le t \le 1\big\}$. This suggests that, given a square $v \in V$, we have that

$$\mu(v) \approx \lim_{n \to \infty} \frac{Z_n(v)}{m^{2n} n^{-2}}\,,$$

where $Z_n(v)$ is the number of offspring of v in the nth generation. Note that the mean number of children of a vertex in the nth generation is of order $\approx m^2\left(\frac{n-1}{n}\right)^2$ and hence is generation dependent.

Instead of looking for a strong analogy and discussing generation dependent offspring distributions, for this exposition we sacrifice precision in favour of simplicity and claim that in this analogous case the most interesting features of the original problem are still present. More precisely, we look at a Galton–Watson tree such that every vertex has a mean number m^2 of children, and discuss the multifractal spectrum of the *branching measure* $\tilde{\mu}$ on its boundary, defined by

$$\tilde{\mu}\big(\mathcal{B}(v)\big) = \lim_{n\to\infty} \frac{Z_n(v)}{m^{2n}},$$

where $\mathcal{B}(v)$ is the set of rays passing through the vertex v. Fixing some $b > 1$, we endow ∂T with the metric such that the distance of two rays is b^{-n}, where n is the generation of their last common ancestor. In this metric, the set $\mathcal{B}(v)$ is the ball centred in v of radius $b^{-|v|}$, so that for the choice of $b = m$ this corresponds to the sidelength and therefore, up to a constant, to the diameter of the represented square.

We state a general result for the multifractal spectrum of Galton–Watson trees with generating variable N and finite mean, which is taken from Mörters and Shieh (2004).

Theorem 8.13 *Suppose* $P\{N = 0\} = 0$ *and* $0 < P\{N = 1\} < 1$. *Define*

$$a = \log EN > 0 \quad \textit{and} \quad \tau = -\log P\{N = 1\}/\log EN > 0.$$

Then, for all $a \le \theta \le a\left(1 + \frac{1}{\tau}\right)$, *almost surely,*

$$\dim\left\{(v_0, v_1, \ldots) \in \partial T \,:\, \limsup_{n\to\infty} \frac{\log \tilde{\mu}(\mathcal{B}(v_n))}{-n} = \theta\right\} = \frac{a}{\log b}\left(\frac{a}{\theta}(1+\tau) - \tau\right).$$

Before looking at the structure of this result in more detail, let us adapt the parameters of our tree representation in good faith. We have already noted that $EN = m^2$ and by construction we have $N \ge 1$ so that the conditions of Theorem 8.13 are satisfied. For the metric we would like to choose $b = m$, and the remaining parameter is $P\{N = 1\}$, which we write as $m^{-\eta}$ for some $\eta > 0$, which we discuss later. We obtain $a = 2\log m$, $\tau = \eta/2$ and hence a predicted spectrum of

$$\dim\left\{x \in \mathcal{S}\colon \limsup_{r\downarrow 0} \frac{\log \mu(\mathcal{B}(x,r))}{\log r} = \theta\right\} = 2\left(\frac{2}{\theta}\left(1 + \frac{\eta}{2}\right) - \frac{\eta}{2}\right).$$

Note that neither side of this equation has any dependence on m, which gives us a handle on η, which we only have to determine asymptotically for $m \uparrow \infty$.

To do this we require knowledge of a *probability exponent*, roughly defined as the rate of decay (as $r \uparrow \infty$) of the probability of an increasingly unlikely event involving Brownian paths running until they exit the ball $\mathcal{B}(0,r)$. Various kinds of exponents can be defined and used in fractal geometry, see Lawler (1999).

In the present case we need an intersection exponent. To define these, suppose $k, m \geq 1$ are integers, and $(W_s^{(i)}: s \geq 0)$ for $i \in \{1, \ldots, k+m\}$ are independent Brownian motions started on the unit sphere $\partial\mathcal{B}(0,1)$, and stopped upon leaving $\mathcal{B}(0, \mathrm{e}^n)$, i.e. at times $T_n^{(i)} = \inf\{s > 0: |W_s^{(i)}| = \mathrm{e}^n\}$. We denote by

$$\mathfrak{B}_n^{(1)} = \bigcup_{i=1}^{k} \{W_s^{(i)} : 0 \leq s \leq T_n^{(i)}\},$$

$$\mathfrak{B}_n^{(2)} = \bigcup_{i=k+1}^{k+m} \{W_s^{(i)} : 0 \leq s \leq T_n^{(i)}\},$$

two packets of paths, and assume that the starting points in different packets are different. Denote by V_n the event that the two packets $\mathfrak{B}_n^{(1)}$ and $\mathfrak{B}_n^{(2)}$ do not intersect each other. The *intersection exponents* are defined by the requirement that there exist constants $0 < c < C$ such that

$$c \exp\{-n\,\xi(k,m)\} \leq \mathbf{P}(V_n) \leq C \exp\{-n\,\xi(k,m)\}.$$

Lawler (1996b) showed that the intersection exponents $\xi(k,m)$ are well-defined by this requirement, and some years later the (highly nontrivial) techniques of stochastic Loewner evolution (SLE) enabled Lawler, Schramm and Werner to give the explicit values

$$\xi(k,m) = \frac{\left(\sqrt{24k+1} + \sqrt{24m+1} - 2\right)^2 - 4}{48}.$$

For a short survey of the key steps in this development and some other early applications of the SLE technique, see Lawler *et al.* (2001).

Let us explain how intersection exponents help identifying $P\{N=1\}$ in our tree model. Suppose S is any square containing an intersection point (hence corresponding to a vertex in the tree). The event $\{N = 1\}$ means that the Brownian paths intersect in one of the m^2 congruent nonoverlapping subsquares which cover S, but nowhere else in S; see Fig. 8.4 for an illustration.

We now fix one f these subsquares, say S', and assume (without significant loss of generality, as in the previous section) that it is located sufficiently far away from the boundary of S. We split both motions at the first time they hit $\partial S'$ and apply time-reversal to the initial part of each motion. Though the reversed part are strictly speaking not Brownian motions, they are sufficiently similar to treat them as such. Then we are faced with *four* Brownian motions started at

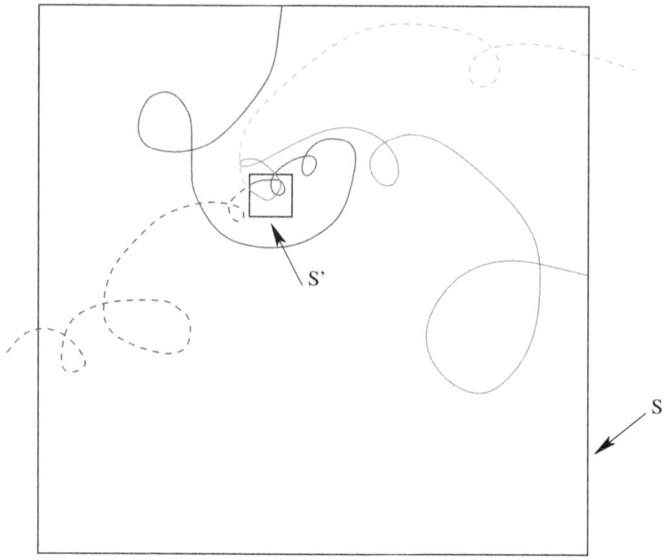

FIG. 8.4. Paths realizing the event $\{N = 1\}$, the initial parts of both paths are dashed.

$\partial S'$, which we divide in two packets of two, with each packet consisting of the (time-reversed) first part and the (non-reversed) second part of the same original motion. We consider all motions up to the first time when they hit ∂S, which is at distance of order m times the typical distance of the starting points. Hence, applying Brownian scaling, we get

$$\mathbf{P}\{ \text{ paths do not intersect outside } S'\} \approx m^{-\xi(2,2)}.$$

As these events are disjoint for the m^2 different squares $S' \subset S$ we can sum the probabilities and obtain

$$\mathbf{P}\{N = 1\} \approx m^{2-\xi(2,2)} \quad \text{as } m \uparrow \infty.$$

Hence $\eta = \xi(2,2) - 2$ and plugging this into the prediction yields

$$\dim\left\{x \in \mathcal{S}\colon \limsup_{r\downarrow 0} \frac{\log \mu(\mathcal{B}(x,r))}{\log r} = \theta\right\} = 2\,\frac{\xi(2,2)}{\theta} + 2 - \xi(2,2)\,.$$

Using the known value $\xi(2,2) = \frac{35}{12}$ of the intersection exponents gives the precise formula claimed in Theorem 8.12.

As in the previous example, a note of caution is necessary: The tree analogy is very suitable to develop an intuition for the problem and guess the right multifractal spectrum. However, in setting up the tree analogy, we have gone too far to prove Theorem 8.12 by justification of the steps undertaken in this simplification and it is preferable to start this proof from scratch. The original problem needs serious treatment before some form of the claimed locality assumption can be exploited, and it seems to be impossible to carry out the proof without using the full power of the strong Markov property of the two Brownian motions.

However, an inspection of the proof of Theorem 8.13 gives structural insight, which is directly applicable to the proof of Theorem 8.12. Indeed, given a vertex v with $|v| = n$, the event

$$\{\tilde{\mu}(\mathcal{B}(v)) \approx \mathrm{e}^{-n\theta}\}$$

is typically coming up when $Z_k(v) = 1$ for $k = n\theta/\log EN$, i.e. when the vertex v has just one offspring for k generations. This fact can be translated directly into the Brownian world. A point $x \in \mathcal{S}$ typically satisfies

$$\limsup_{r\downarrow 0} \frac{\log \mu(\mathcal{B}(x,r))}{\log r} \approx a$$

if there exists a sequence $r_n \downarrow 0$ of radii such that

$$(\mathcal{B}(x,r_n) \setminus \mathcal{B}(x,r_n^{a/2})) \cap \mathcal{S} = \emptyset\,.$$

The occurrence of large empty annuli at selected radii is also key to the understanding of the multifractal spectrum of super-Brownian motion (Perkins and Taylor, 1998). Hence, despite greatly oversimplifying the situation, the tree approach gives valuable insight into the original problem, which can be exploited directly in the proof.

8.3.3 *Points of infinite multiplicity*

In this section we turn our attention to an attractive *unsolved* problem. It is known for a long time that planar Brownian motion has points of multiplicity p, for any positive integer p. Moreover, Dvoretzky *et al.* (1958) have shown that, almost surely, there exist points of uncountably infinite multiplicity, see Le Gall (1992) or Mörters and Peres (2009) for modern proofs. These arguments can also be used to show that the Hausdorff dimension of the set of points of uncountably infinite multiplicity is still two.

How far can we go, before we see a reduction in the dimension? A natural way is to count the number of excursions from a point. To be explicit, let $(W_s\colon s \geq 0)$ be a planar Brownian motion and fix $x \in \mathbb{R}^2$ and $\epsilon > 0$. Let $S_{-1} = 0$ and, for any integer $j \geq 0$, let $T_j = \inf\{s > S_{j-1}\colon W_s = x\}$ and $S_j = \inf\{s > T_j\colon |W_s - x| \geq \epsilon\}$. Then define

$$N_\epsilon^x = \max\{j \geq 0\colon T_j < \infty\},$$

which is the number of excursions from x hitting $\partial\mathcal{B}(x,\epsilon)$. Observe that $N_\epsilon^x = 0$ for almost every point on the curve (with respect to the occupation time T introduced in Section 8.3.1) and that

$$\lim_{\epsilon\downarrow 0} N_\epsilon^x = \infty \qquad \Longleftrightarrow \qquad x \text{ has infinite multiplicity.}$$

It is therefore a natural question to ask how rapidly N_ϵ^x can go to infinity when $\epsilon \downarrow 0$. A partial answer is given in the following theorem of Bass *et al.* (1994).

Theorem 8.14

(a) Let $0 < a < \frac{1}{2}$. Then, almost surely,

$$\dim\left\{x \in \mathbb{R}^2 \colon \lim_{\epsilon\downarrow 0} \frac{N_\epsilon^x}{\log(1/\epsilon)} = a\right\} \geq 2 - a\,.$$

(b) Let $0 < a < 2\mathrm{e}$. Then, almost surely,

$$\dim\left\{x \in \mathbb{R}^2 \colon \lim_{\epsilon\downarrow 0} \frac{N_\epsilon^x}{\log(1/\epsilon)} = a\right\} \leq 2 - \frac{a}{\mathrm{e}}\,.$$

(c) Almost surely, for every $x \in \mathbb{R}^2$, we have

$$\limsup_{\epsilon\downarrow 0} \frac{N_\epsilon^x}{\log(1/\epsilon)} \leq 2\mathrm{e}.$$

Note, for comparison, that for a linear Brownian motion, almost surely, for every $x \in \mathbb{R}$, we have

$$\lim_{\epsilon\downarrow 0} (4\epsilon)\, M_\epsilon^x = L_t^x,$$

where M_ϵ^x is the number of excursions from x hitting $\{x-\epsilon, x+\epsilon\}$ before time t, and L_t^x is the local time at x, see e.g. Mörters and Peres (2009).

The proof of parts (b) and (c) of Theorem 8.14 is fairly straightforward, though the statements are certainly not optimal. The delicate part is the lower bound, given in (a). This argument is based on the construction of a local time, a nondegenerate measure on the set

$$\left\{x \in \mathbb{R}^2 \colon \lim_{\epsilon\downarrow 0} \frac{N_\epsilon^x}{\log(1/\epsilon)} = a\right\}.$$

The restriction to values $a < 1/2$ is due to the use of L^2-estimates and appears to be of a technical nature. It is believed that the following conjecture is true.

Conjecture 2 *Almost surely,*

$$\max_{x\in\mathbb{R}^2} \lim_{\epsilon\downarrow 0} \frac{N_\epsilon^x}{\log(1/\epsilon)} = 2.$$

Moreover, for any $0 < a < 2$, almost surely,

$$\dim\left\{x \in \mathbb{R}^2 \colon \lim_{\epsilon\downarrow 0} \frac{N_\epsilon^x}{\log(1/\epsilon)} = a\right\} = 2 - a\,.$$

This is still an open problem. Hope for its solvability comes from the fact that one can represent the dependence structure of the random variables N_ϵ^x in a tree picture similar to the one indicated in Section 8.3.1. However, because the Brownian path is required to return *exactly* to a given point, this problem has much less inherent continuity than the two previous ones, and therefore appears to be much harder.

8.4 More on the planar Brownian path

We have seen in the second example of the previous section that in some cases, once the tree technique has been exploited, there remains a serious challenge to identify the rate of decay of certain probabilities associated with the underlying process. This challenge can be formalised in the notion of *probability exponents*, and in this section we give further evidence of their use in fractal geometry, following ideas surveyed in Lawler (1999).

8.4.1 *The Mandelbrot conjecture*

We look at a famous example, the *Mandelbrot conjecture*: Let $(W_s : 0 \leq s \leq 1)$ be a planar Brownian motion running for one time unit and consider the complement of its path,

$$\{x \in \mathbb{R}^2 : x \neq W_s \text{ for any } 0 \leq s \leq 1\}.$$

This set is open and can be decomposed into connected components, exactly one of which is unbounded. We denote this component by U and define its boundary ∂U as the *frontier* of the Brownian path. The frontier can be seen as the set of points on the Brownian path which are accessible from infinity and is therefore also called the *outer boundary* of Brownian motion.

According to a frequently told legend, Mandelbrot, when presented with a simulation of the Brownian frontier, cast a brief glance at the picture and immediately identified its dimension as 4/3, see Mandelbrot (1982). However, a more rigorous confirmation of this conjecture took a long time. In the late nineties Bishop *et al.* (1997) showed that the frontier has Hausdorff dimension strictly larger than one, and about the same time Lawler (1996a) identified the Hausdorff dimension in terms of a disconnection exponent.

The disconnection exponents $\xi(k)$, $k \in \mathbb{N}$, can be defined as follows: Suppose $(W_s^{(i)} : s \geq 0)$ for $i \in \{1, \ldots, k\}$ are independent Brownian motions started on the unit sphere $\partial\mathcal{B}(0,1)$, and stopped upon leaving $\mathcal{B}(0, \mathrm{e}^n)$, i.e. at times $T_n^{(i)} = \inf\{s > 0 : |W_s^{(i)}| = \mathrm{e}^n\}$. We denote by

$$\mathfrak{B}_n = \bigcup_{i=1}^{k} \{W_s^{(i)} : 0 \leq s \leq T_n^{(i)}\}$$

the union of the paths, and by V_n the event that $\mathfrak{B}_n$ does not disconnect the origin from infinity, i.e. the origin is in the unbounded connected component of the complement of $\mathfrak{B}_n$. The *disconnection exponents* are defined by the requirement that there exist constants $0 < c < C$ such that

$$c \exp\{-n\,\xi(k)\} \le \mathbf{P}(V_n) \le C \exp\{-n\,\xi(k)\}\,.$$

Lawler (1996a) showed that the disconnection exponents $\xi(k)$ are well-defined by this requirement, and – just as in the case of intersection exponents – Lawler, Schramm and Werner found the explicit values

$$\xi(k) = \frac{\left(\sqrt{24k+1}-1\right)^2 - 4}{48}.$$

Note that this is in line with the intersection exponents as (formally, because of our requirement that m be an integer)

$$\lim_{m\downarrow 0} \xi(k,m) = \xi(k),$$

and this corresponds to the observation that if $\mathfrak{B}_n$ disconnects the origin from infinity, no further independent packet (no matter how slim, i.e. how small m) started at the origin can reach $\partial\mathcal{B}(0,\mathrm{e}^n)$ without intersecting $\mathfrak{B}_n$. This can be made rigorous by extending the definition of intersection exponents to noninteger arguments.

In Lawler (1996a) the dimension of the frontier was identified to be $2-\xi(2)$, so that Mandelbrot's conjecture follows.

Theorem 8.15 *Almost surely, the Hausdorff dimension of the frontier is*

$$\dim \partial U = \tfrac{4}{3}\,.$$

It is not hard to paint a tree picture that makes the connection of the disconnection exponents and the frontier clear. This time we prefer to work in the time domain and use the following striking result of Kaufman, for a proof see e.g. Mörters and Peres (2009).

Lemma 8.16 (Kaufman's lemma) *Suppose $d \ge 3$ and $(W_s\colon s \in [0,1])$ is a d-dimensional Brownian motion. Then, almost surely, for every $A \subset [0,1]$,*

$$\dim\big\{W_s\colon s \in A\big\} = 2\,\dim A\,.$$

Note that the 'dimension doubling' rule holds simultaneously for all sets $A \subset [0,1]$ with a single exceptional set of probability zero. It can therefore be applied to any random set A, which makes Kaufman's lemma a powerful tool.

We now look at the decomposition of the unit interval $[0,1]$ into 2^n nonoverlapping intervals of equal length. Any such interval $[j2^{-n},(j+1)2^{-n}]$ is associated to a vertex in a representing tree T if the set

$$\mathfrak{B}_n^{(j)} = \big\{W_s\colon 0 \le s \le j2^{-n} \text{ or } (j+1)2^{-n} \le s \le 1\big\}$$

does not disconnect $\{W_s : j2^{-n} \le s \le (j+1)2^{-n}\}$ from infinity. With the rule that a vertex v is an offspring of w if the interval associated to v is contained in that associated to w, this constitutes a tree representation of the set

$$I(T) = \{s \in [0,1] : W_s \in \partial U\}.$$

This representation does not make $I(T)$ a Galton–Watson fractal, but the following lemma taken from Lawler (1999, Lemma 1 and 2) indicates how, in a special situation, the independence conditions of Theorem 8.4 can be weakened.

Lemma 8.17 *Given a family of (not necessarily independent) zero-one valued random variables*

$$\{Y(j_1,\ldots,j_n) : j_i \in \{0,1\},\ n \in \mathbb{N}\},$$

we build a random fractal A iteratively. Let $\mathcal{S}_0 = \{[0,1]\}$ and, given a collection $\mathcal{S}_n$ of compact intervals of length 2^{-n}, construct a collection $\mathcal{S}_{n+1}$ by

- *splitting each interval in $\mathcal{S}_{n-1}$ into two nonoverlapping intervals of half the length,*
- *adding any interval thus constructed to the collection $\mathcal{S}_n$ if*

  $$Y(j_1,\ldots,j_n) = 1,$$

  where $\sum_{i=1}^n j_i 2^{-i}$ is the left endpoint of the interval.

Define the random fractal as

$$A = \bigcap_{n=1}^{\infty} \bigcup_{I \in \mathcal{S}_n} I.$$

Then

(i) If $\displaystyle\sum_{j_1,\ldots,j_n=0}^{1} \mathbf{E} \prod_{k=1}^{n} Y(j_1,\ldots,j_k) \le C\, 2^{(1-\alpha)n}$ *for some $C>0$, then*

$$\dim A \le 1-\alpha \quad \textit{almost surely.}$$

(ii) If, for some $0<c<C<\infty$ and $\epsilon>0$,

$$c\,2^{-\alpha n} \le \mathbf{E} \prod_{k=1}^{n} Y(j_1,\ldots,j_k) \le C\,2^{-\alpha n}$$

for all $\epsilon \le \sum_{i=1}^n j_i 2^{-i} \le 1-\epsilon$, and

$$\mathbf{E} \prod_{k=1}^{n} Y(j_1,\ldots,j_k) Y(i_1,\ldots,i_k) \le C\,2^{-2\alpha n} \left(\sum_{i=1}^{n} (i_i - j_i) 2^{-i}\right)^{-\alpha},$$

for all $\epsilon \le \sum_1^n j_i 2^{-i} < \sum_1^n i_i 2^{-i} \le 1-\epsilon$, then

$$\dim A \ge 1-\alpha \quad \textit{with positive probability.}$$

This lemma exploits a tree representation of A with the tree T given as a subtree of a binary tree with vertices in the nth generation canonically denoted by $(j_1, \dots, j_n)$. Such a vertex is contained in T if and only if

$$\prod_{k=1}^{n} Y(j_1, \dots, j_k) = 1 .$$

The set attached to the vertex $(j_1, \dots, j_n)$ is the closed interval of length 2^{-n} with left endpoint $\sum_1^n j_i 2^{-i}$, and the number of children of this vertex is

$$Y(j_1, \dots, j_n, 0) + Y(j_1, \dots, j_n, 1) .$$

Supposing that all the random variables $Y(j_1, \dots, j_k)$ are independent, we get

$$\begin{aligned} \min \Big\{ s \colon \mathbf{E}\big\{2^{-s} Y(j_1, \dots, j_n, 0) + 2^{-s} Y(j_1, \dots, j_n, 1)\big\} \le 1 \Big\} \\ = \min \big\{ s \colon 2^{-s-\alpha+1} \le 1 \big\} = 1 - \alpha, \end{aligned}$$

confirming that this generalises a special case of Theorem 8.4.

In our case, we let $Y(j_1, \dots, j_n) = 1$ if and only if the set

$$\big\{ W_s \colon 0 \le s \le \textstyle\sum_{i=1}^n j_i 2^{-i} \text{ or } \sum_{i=1}^n j_i 2^{-i} + 2^{-n} \le s \le 1 \big\}$$

does not disconnect

$$\Big\{ W_s \colon \sum_{i=1}^{n} j_i 2^{-i} \le s \le \sum_{i=1}^{n} j_i 2^{-i} + 2^{-n} \Big\}$$

from infinity. Using Brownian scaling and the definition of the disconnection exponents, one can show that the probability of this event is of order $2^{-\frac{n}{2}\xi(2)}$, and that the second condition in (ii) also holds. This argument therefore gives that

$$\dim \big\{ s \in [0,1] \colon W_s \in \partial U \big\} = 1 - \tfrac{1}{2}\,\xi(2) = \tfrac{2}{3},$$

with positive probability, and an application of Kaufman's lemma gives $\dim \partial U = \frac{4}{3}$ with positive probability. Some nontrivial extra work is required to show that this actually holds with probability one.

8.4.2 *More on the geometry of the Brownian frontier*

There are a variety of subsets of Brownian paths whose Hausdorff dimensions can be expressed in terms of different probability exponents. Examples with known exponents, like cutpoints, pioneer points and cone points of planar Brownian motion are given in Lawler (1999) and Lawler *et al.* (2001). Here we sketch two results, which reveal further details about the geometry of the frontier.

To begin with, it is easy to observe that the Brownian frontier contains double points of the Brownian motion. The argument, which is probably due to Paul Lévy, goes roughly like this: If it did not, then by construction the frontier would just be a stretch of the original Brownian path. This would however imply that it had double points, which is a contradiction.

Knowing that there are double points on the frontier, it is natural to ask, whether the frontier contains *triple points*. This problem was solved by Burdzy and Werner (1996).

Theorem 8.18 *Almost surely, there are* no *triple points on the frontier of a planar Brownian motion.*

A second natural question that comes up is *how many* double points one can find on the Brownian frontier. Surprisingly, it turns out that while the set

$$D = \big\{x \in \mathbb{R}^2 \colon x = W_s = W_t \text{ for distinct } s,t \in [0,1]\big\}$$

of double points has full dimension on the entire path, it does not have full dimension on the frontier. The following curious result is due to Kiefer and Mörters (2008).

Theorem 8.19 *Almost surely, the set of double points on the Brownian frontier satisfies*

$$\dim\big(D \cap \partial U\big) = \frac{\sqrt{97}+1}{24}.$$

In the proof of this result, a spatial approach is preferable. In this context it is natural to consider Brownian motion up to the first exit time τ from a big ball, rather than up to time one (it is not hard to see that this is equivalent). We fix a compact square S_0 of unit sidelength inside this ball, and a small $\epsilon > 0$. Let $\mathcal{S}_n$ be the collection containing those of the 2^{2n} nonoverlapping subsquares $S \subset S_0$ of sidelength 2^{-n}, which satisfy

- the Brownian motion $(W_s \colon s \geq 0)$ hits S, then moves to distance ϵ from S, and then hits S again before time τ;
- the union of the paths outside the square S does not disconnect its boundary ∂S from infinity.

Then we have

$$S_0 \cap D \cap \partial U = \bigcup_{\epsilon>0} \bigcap_{n=1}^{\infty} \bigcup_{S \in \mathcal{S}_n} S,$$

and the Hausdorff dimension can be determined (with positive probability) by verifying a first and second moment criterion analogous to (i), (ii) in Lemma 8.17.

For the first moment criterion, we have to show that the probability that a cube of sidelength 2^{-n} is in $\mathcal{S}_n$ is bounded from above and below by constant multiples of $2^{-n\xi(4)}$. Indeed, we may use three stopping times to split the path into four pieces: They are the first hitting time of S, the first time afterwards where the path has moved to distance ϵ from the square, the first hitting time of S after that. If we reverse the first and third part in time, the four pieces are sufficiently close to Brownian paths started on the boundary of the cube and running for one time unit, to infer that the probability of disconnection is, up to a factor which is polynomial in n, of order $2^{-n\xi(4)}$. At this place the proof is a bit more delicate than the arguments in Section 8.3.2, because a careful control of the polynomial factors is required.

This first moment argument, a slightly more sophisticated one for the second moment, and a tree framework similar to the one above, show that, with positive probability,

$$\dim\big(D \cap \partial U\big) = 2 - \xi(4) = \frac{\sqrt{97}+1}{24}.$$

Again, some more work is required to show that this holds almost surely.

For Theorem 8.18 the argument is easier, as no lower bound is needed. Using no more than the Borel–Cantelli lemma one can infer that for

$$T = \big\{x \in \mathbb{R}^2 \colon x = W_r = W_s = W_t \text{ for distinct } r, s, t \in [0,1]\big\}$$

we have $T \cap \partial U = \emptyset$ almost surely, if $\xi(6) > 2$. The merit of the paper by Burdzy and Werner (1996) is mostly in providing this estimate long before the SLE-technology allowed the precise calculation of this value.

Acknowledgements

I would like to thank Yuval Peres for teaching me the 'tree' point of view, and Richard Kiefer for permission to include our (yet unpublished) result. I would also like to acknowledge the support of EPSRC through an Advanced Research Fellowship.

References

Bass, R. F., K. Burdzy, and D. Khoshnevisan (1994). Intersection local time for points of infinite multiplicity. *Ann. Probab. 22*(2), 566–625.

Bishop, C. J., P. W. Jones, R. Pemantle, and Y. Peres (1997). The dimension of the Brownian frontier is greater than 1. *J. Funct. Anal. 143*(2), 309–336.

Burdzy, K. and W. Werner (1996). No triple point of planar Brownian motion is accessible. *Ann. Probab. 24*(1), 125–147.

Dembo, A. (2005). Favorite points, cover times and fractals. In *École d'Été de Probabilités de Saint-Flour XXXIII–2003*, Volume 1869 of *Lecture Notes in Math.*, pp. 1–101. Berlin: Springer.

Dembo, A., Y. Peres, J. Rosen, and O. Zeitouni (2001). Thick points for planar Brownian motion and the Erdős-Taylor conjecture on random walk. *Acta Math. 186*(2), 239–270.

Dvoretzky, A., P. Erdős, and S. Kakutani (1958). Points of multiplicity $\mathfrak{c}$ of plane Brownian paths. *Bull. Res. Council Israel Sect. F 7F*, 175–180 (1958).

Falconer, K. (1986). Random fractals. *Math. Proc. Cambridge Philos. Soc. 100*(3), 559–582.

Falconer, K. (2003). *Fractal Geometry Mathematical Foundations and Applications* (Second ed.). Hoboken, NJ: John Wiley & Sons Inc.

Ford, Jr., L. R. and D. R. Fulkerson (1962). *Flows in Networks*. Princeton, N.J.: Princeton University Press.

Graf, S., R. D. Mauldin, and S. C. Williams (1988). The exact Hausdorff dimension in random recursive constructions. *Mem. Amer. Math. Soc. 71*(381).

Hausdorff, F. (1918). Dimension und äußeres Maß. *Math. Ann. 79*(1-2), 157–179.

Hu, X. and S. J. Taylor (1997). The multifractal structure of stable occupation measure. *Stochastic Process. Appl. 66*(2), 283–299.

Kiefer, R. and P. Mörters (2008). Hausdorff dimension of the set of double points on the Brownian frontier. *Submitted for publication.*

Klenke, A. and P. Mörters (2005). The multifractal spectrum of Brownian intersection local times. *Ann. Probab. 33*(4), 1255–1301.

Lawler, G. (1996a). The dimension of the frontier of planar Brownian motion. *Electron. Comm. Probab. 1*, Paper 5, 29–47.

Lawler, G. (1996b). Hausdorff dimension of cut points for Brownian motion. *Electron. Journal Probab. 1*, Paper 2, 1–20.

Lawler, G. (1997). The frontier of a Brownian path is multifractal. *Unpublished manuscript*, 24pp.

Lawler, G. (1999). Geometric and fractal properties of Brownian motion and random walk paths in two and three dimensions. *Bolyai Society Mathematical Studies 9*, 219–258.

Lawler, G. F., O. Schramm, and W. Werner (2001). The dimension of the planar Brownian frontier is 4/3. *Math. Res. Lett. 8*(4), 401–411.

Le Gall, J.-F. (1986). Sur la saucisse de Wiener et les points multiples du mouvement brownien. *Ann. Probab. 14*(4), 1219–1244.

Le Gall, J.-F. (1992). Some properties of planar Brownian motion. In *École d'Été de Probabilités de Saint-Flour XX–1990*, Volume 1527 of *Lecture Notes in Math.*, pp. 111–235. Berlin: Springer.

Lyons, R. and Y. Peres (2009). *Probability on Trees and Networks*. Cambridge: Cambridge University Press. Forthcoming.

Mandelbrot, B. B. (1982). *The Fractal Geometry of Nature*. San Francisco, California: W. H. Freeman and Co.

Mauldin, R. D. and S. C. Williams (1986). Random recursive constructions: asymptotic geometric and topological properties. *Trans. Amer. Math. Soc. 295*(1), 325–346.

Mörters, P. (2008). Why study multifractal spectra? In *Trends in Stochastic Analysis: A Festschrift in Honour of Heinrich v. Weizsäcker*, Volume 353 of *London Mathematical Society Lecture Note Series*. Cambridge: Cambridge University Press.

Mörters, P. and Y. Peres (2009). *Brownian Motion.* Cambridge: Cambridge University Press. Forthcoming.

Mörters, P. and N.-R. Shieh (2004). On the multifractal spectrum of a Galton-Watson tree. *J. Appl. Prob. 41*, 1223–1229.

Perkins, E. A. and S. J. Taylor (1998). The multifractal structure of super-Brownian motion. *Ann. Inst. H. Poincaré Probab. Statist. 34*(1), 97–138.

Taylor, S. J. (1986). The measure theory of random fractals. *Math. Proc. Cambridge Philos. Soc. 100*(3), 383–406.

PART III

STATISTICS AND STOCHASTIC GEOMETRY

9

INFERENCE

Jesper Møller

Abstract

This contribution concerns *statistical inference* for *parametric models* used in stochastic geometry and based on quick and simple *simulation free procedures* as well as more comprehensive methods based on a *maximum likelihood* or *Bayesian* approach combined with *Markov chain Monte Carlo (MCMC)* techniques. Due to space limitations the focus is on *spatial point processes.*

9.1 Spatial point processes and other random closed set models

Recall that a spatial point process $\mathbf{X}$ considered as a random closed set in $\mathbb{R}^d$ is nothing but a random locally finite subset of $\mathbb{R}^d$. This means that the number of points $n(\mathbf{X} \cap B)$ falling in an arbitrary bounded Borel set $B \subset \mathbb{R}^d$ is a finite random variable. This extends to a marked point process where to each point $x_i \in \mathbf{X}$ there is associated a mark, that is a random variable K_i defined on some measurable space $\mathcal{M}$ (for details, see e.g. Stoyan *et al.* 1995 or Daley and Vere-Jones 2003).

Most theory and methodology for inference in stochastic geometry concern spatial point processes and to some extent marked point processes, cf. the monographs by Ripley (1981, 1988), Cressie (1993), Stoyan *et al.* (1995), Stoyan and Stoyan (1995), Van Lieshout (2000), Diggle (2003), Møller and Waagepetersen (2004), Baddeley *et al.* (2006), and Illian *et al.* (2008*b*). A particular important case is a *germ-grain model* (Hanisch, 1981), where the x_i are called germs and the K_i primary grains, the mark space $\mathcal{M} = \mathcal{K}$ is the set of compact subsets of $\mathbb{R}^d$, and the object of interest is the random closed U set given by the union of the translated primary grains $x_i + K_i = \{x_i + x : x \in K_i\}$. The most studied case is the Boolean model (Hall, 1988; Molchanov, 1997), i.e. when the germs form a Poisson process and the primary grains are mutually independent, identically distributed, and independent of the germs. Extensions to models with interacting grains have been studied in Kendall *et al.* (1999) and Møller and Helisova (2008, 2009).

9.2 Outline and some notation

In the sequel we mostly confine attention to planar point processes, but many concepts, methods, and results easily extend to $\mathbb{R}^d$ or a more general metric space, including many marked point process models. (For example, when discussing the Norwegian spruces in Fig. 9.6 below, this may be viewed as a marked point process of discs.) We concentrate on the two most important classes of spatial point process models, namely Cox (including Poisson) processes in Section 9.3 and Gibbs (or Markov) point processes in Section 9.4. We illustrate the statistical methodology with various application examples, where most are examples of inhomogeneous point patterns. We discuss the state of the art of simulation-based maximum likelihood inference as well as Bayesian inference for parametric spatial point process models, where fast computers and advances in computational statistics, particularly MCMC methods, have had a major impact. The MCMC algorithms are treated in some detail; for a general introduction to MCMC and for spatial point processes in particular, see Møller and Waagepetersen (2004) and the references therein. We also discuss more classical methods based on summary statistics, and more recent simulation-free inference procedures based on composite likelihoods, minimum contrast estimation, and pseudo likelihood. Often the `R` package `spatstat` (Baddeley and Turner 2005, 2006) has been used; some software in `R` and `C`, developed by my colleague Rasmus Waagepetersen in connection to our recent paper Møller and Waagepetersen (2007), is available at `www.math.aau.dk/~rw/sppcode`.

Throughout this contribution we use the following notation. Unless otherwise stated, $\mathbf{X}$ is a planar spatial point process and $\mathcal{B}$ is the class of Borel sets in $\mathbb{R}^2$. We let $W \in \mathcal{B}$ denote a bounded observation window of area $|W| > 0$, and assume that a realization $\mathbf{X} \cap W = \mathbf{x}$ is observed (most often only one such realization is observed). Here the number of points, denoted $n(\mathbf{x})$, is finite, and typically W is a rectangular region. Moreover, $\mathbf{1}[\cdot]$ (or $\mathbf{1}_{[\cdot]}$) is an indicator function.

9.3 Inference for Poisson and Cox process models

Poisson processes (Kingman, 1993) are models for point patterns with complete spatial independence, while Cox processes (Cox, 1955) typically model a positive dependence between the points. Both kinds of processes may model inhomogeneity (Diggle, 2003; Møller and Waagepetersen 2003, 2007). See also Karr (1991), Stoyan *et al.* (1995), and Daley and Vere-Jones (2003).

As an illustrative example of an inhomogeneous point pattern we refer to the rain forest trees in Fig. 9.1. These data have previously been analyzed in Waagepetersen (2007) and Møller and Waagepetersen (2007), and they are just a small part of a much larger data set comprising hundreds of thousands of trees belonging to hundreds of species (Hubbell and Foster, 1983; Condit *et al.*, 1996; Condit, 1998).

9.3.1 *Definition and fundamental properties of Poisson and Cox processes*

The most fundamental spatial point process model is the *Poisson process* in $\mathbb{R}^2$, see Chapter 1 of this volume. The Poisson process is defined in terms of a locally finite and diffuse measure μ on $\mathcal{B}$. These assumptions on μ allow us to view the Poisson process as a locally finite random closed set $\mathbf{X} \subseteq \mathbb{R}^2$ so

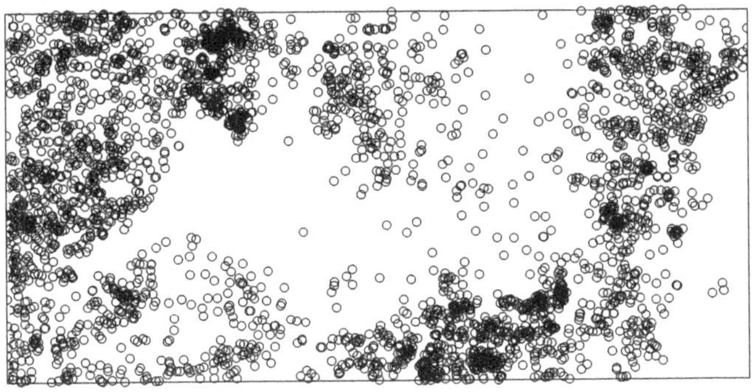

FIG. 9.1. Locations of 3605 *Beilschmiedia pendula Lauraceae* trees observed within a 500×1000 m region at Barro Colorado Island.

that its distribution is characterized by the avoidance probability (9.1) specified below; for more general definitions, relaxing the assumptions on μ, see e.g. Kingman (1993). Below we recall the definition of the Poisson process, and discuss why extensions to Cox processes are needed.

Characterizations of Poisson processes are described in Kingman (1993), Stoyan *et al.* (1995), and Daley and Vere-Jones (2003). One constructive description is that

(i) $N(B) = n(\mathbf{X} \cap B)$, the number of points falling within any bounded $B \in \mathcal{B}$, is Poisson distributed with mean $\mu(B)$,
(ii) conditional on $N(B)$, the points in $\mathbf{X} \cap B$ are mutually independent and identically distributed with a distribution given by the normalized restriction of μ to B,
(iii) $N(B_1), N(B_2), \ldots$ are independent if $B_1, B_2, \ldots \in \mathcal{B}$ are disjoint.

Clearly (i) implies that μ is the intensity measure of the Poisson process, (ii) implies that the locations of the points in $\mathbf{X} \cap B$ are independent of $N(B)$, and (iii) is called the independent scattering property. The Poisson process is easily constructed considering a partition of $\mathbb{R}^2$ into disjoint bounded Borel sets $B_1, B_2, \ldots$ and applying (i)–(iii). This establishes the existence of the process.

When verifying different properties of Poisson processes, it is often useful to recall that the distribution of a random closed set $\mathbf{X} \subseteq \mathbb{R}^2$ is uniquely determined by its avoidance probabilities $\mathbf{P}(\mathbf{X} \cap B = \emptyset)$, $B \in \mathcal{B}$, see e.g. Stoyan *et al.* (1995). By (i), for the Poisson process $\mathbf{X}$, we simply have that

$$\mathbf{P}(\mathbf{X} \cap B = \emptyset) = \exp(-\mu(B)), \quad B \in \mathcal{B}. \tag{9.1}$$

This clearly shows that the Poisson process is uniquely defined by its intensity measure.

Throughout this contribution, as in most statistical applications, μ is assumed to be absolutely continuous with respect to the Lebesgue measure, i.e., $\mu(B) = \int_B \rho(x)\mathrm{d}x$, where ρ is the intensity function. If $\rho(x) = c$ is constant for all x in a set $B \in \mathcal{B}$, then the Poisson process is said to be homogeneous on B with intensity c. Note that stationarity of the Poisson process (i.e. that its distribution is invariant under translations in $\mathbb{R}^2$ of the point process) is equivalent to the process being homogeneous on $\mathbb{R}^2$.

Although various choices of inhomogeneous intensity functions may generate different kinds of aggregated point patterns, the Poisson process is usually considered to be too simplistic for statistical applications because of the strong independence properties described in (ii)–(iii) above. A natural extension giving rise to a model for aggregated point patterns with more dependence structure is given by a doubly stochastic construction, where $\Lambda(x)$ is a random locally integrable function so that $\mathbf{X}$ conditional on a realization $\Lambda(x) = \rho(x)$, $x \in \mathbb{R}^2$, is a Poisson process with intensity function $\rho(x)$. Then $\mathbf{X}$ is said to be a *Cox process* (Cox, 1955) driven by $\mathbf{\Lambda}$ (or driven by the random measure given by $M(B) = \int_B \Lambda(x)\mathrm{d}x$). The simplest example is a *mixed Poisson process*, i.e. when $\Lambda(x) = \Lambda(o)$ does not depend on x where $\Lambda(o)$ is a non-negative random variable, see Grandell (1997). In practice more flexible models are needed, as discussed in the following section.

9.3.2 *Modelling intensity*

In order to model inhomogeneity, spatial heterogeneity or aggregation it is important to account for possible effects of *covariates* (also called explanatory variables), which we view as a non-random $(p+1)$-dimensional real function $z(x) = (z_0(x), \ldots, z_p(x))$, $x \in \mathbb{R}^2$, and which we assume is known at least within the observation window W. In the sequel, as in most applications, we assume that $z_0 \equiv 1$, while the covariates z_j are non-constant functions for $j > 0$.

In practice $z(x)$ may only be partially observed on a grid of points, and hence some interpolation technique may be needed (Rathbun, 1996; Rathbun *et al.*, 2007; Waagepetersen, 2008). For instance, Fig. 9.2 shows two kinds of covariates z_1 and z_2 for the rain forest trees in Fig. 9.1, letting z_1 and z_2 be constant on each of 100×200 squares of size 5×5 m.

Poisson processes with log-linear intensity The statistically most convenient model for a Poisson process is given by a *log-linear intensity function* (Cox, 1972)

$$\rho(x) = \exp(\beta \cdot z(x)) \tag{9.2}$$

where $\beta = (\beta_0, \ldots, \beta_p)$ is a real $(p+1)$-dimensional parameter and $\cdot$ denotes the usual inner product. We refer to $\beta_0, \ldots, \beta_p$ as regression parameters, and call β_0 the intercept since $z_0 \equiv 1$. The process is homogeneous if either $p = 0$ or $\beta_1 = \ldots = \beta_p = 0$, and it is in general inhomogeneous otherwise.

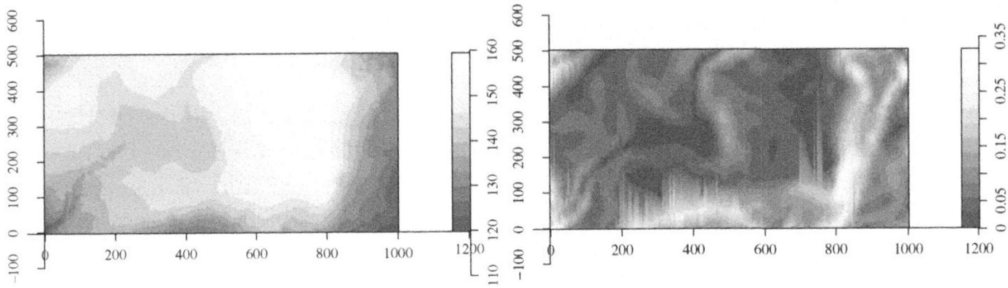

FIG. 9.2. Rain forest trees: the covariates z_1 (altitude; left panel) and z_2 (norm of altitude gradient; right panel) are recorded on a 5×5 m grid (the units on the axes are in metres).

Log-Gaussian Cox processes Let $\mathbf{\Phi} = \{\Phi(x) : x \in \mathbb{R}^2\}$ denote a stationary Gaussian process, and recall that its distribution is fully described by the mean of $\mathbf{\Phi}$ and the covariance function c, where by stationarity of $\mathbf{\Phi}$, $c(x - y) = \text{Cov}(\Phi(x), \Phi(y))$ (Adler, 1981). The log-linear Poisson process model extends naturally to a Cox process driven by

$$\Lambda(x) = \exp(\beta \cdot z(x) + \Phi(x)). \tag{9.3}$$

Such *log-Gaussian Cox process* models (Coles and Jones, 1991; Møller *et al.*, 1998) provide a lot of flexibility corresponding to different types of covariance functions c.

Henceforth, since we can always absorb the mean of $\mathbf{\Phi}$ into the intercept β_0, we assume without loss of generality that $\mathbf{E}\Phi(x) = -c(0)/2$ or equivalently that $\mathbf{E}\exp(\Phi(x)) = 1$. Thereby the log-Gaussian Cox process also has intensity function (9.2).

Shot-noise Cox processes Another very flexible class of models is that of the *shot-noise Cox processes* (Wolpert and Ickstadt, 1998; Brix, 1999; Møller, 2003). As an example of this, let

$$\Lambda(x) = \exp(\beta \cdot z(x)) \frac{1}{\omega \sigma^2} \sum_{y \in \mathbf{Y}} k((x - y)/\sigma) \tag{9.4}$$

where $\mathbf{Y}$ is a stationary Poisson process with intensity $\omega > 0$, $k(\cdot)$ is a density function with respect to Lebesgue measure, and $\sigma > 0$ is a scaling parameter. This form of the random intensity function can be much generalized, replacing $\mathbf{Y}$ with another kind of model and the kernel $k(x - y)$ by another density $k(\cdot; y)$ which may be multiplied by a positive random variable γ_y, leading to (generalized) shot-noise Cox processes (Møller and Torrisi, 2005). In analogy with the assumption made above for log Gaussian Cox processes, the random part following after the exponential term in (9.4) is a stationary stochastic process with mean one, so the shot-noise Cox process also has intensity function (9.2). Often a bivariate normal kernel $k(x) = \exp(-\|x\|^2/2)/(2\pi)$ is used, in which case we refer to (9.4) as a modified Thomas process (Thomas, 1949).

Related models Shot-noise Cox processes are a particular class of Poisson cluster processes, see e.g. Stoyan *et al.* (1995) and Section 9.3.3 below. Extensions of shot-noise Cox processes to Lévy driven Cox processes, and of log-Gaussian Cox processes and log shot-noise Cox processes to log Lévy driven Cox processes have recently been studied in Hellmund *et al.* (2008).

9.3.3 *Simulation*

Due to the complexity of spatial point process models, simulations are often needed when fitting a model and studying the properties of various statistics such as parameter estimates and summary statistics. For a Poisson process $\mathbf{X}$ with intensity function ρ, it is useful to notice that *independent thinning* $\{x \in \mathbf{X} : U(x) \leq p(x)\}$, where $p : \mathbb{R}^2 \to [0,1]$ is a Borel function and the $U(x)$ are mutually independent uniform random variables on $[0,1]$ which are independent of $\mathbf{Y}$, results in a Poisson process with intensity function $p\rho$.

Simulation of a stationary Poisson process $\mathbf{X}$ restricted to a bounded region B is straightforward, using (i)–(ii) in Section 9.3.1. If B is of irregular shape, we simulate $\mathbf{X}$ on a larger region $D \supset B$ such as a rectangle or a disc, and return $\mathbf{X} \cap B$ (corresponding to an independent thinning with retention probabilities $p(x) = \mathbf{1}[x \in B]$).

Usually the intensity function $\rho(x)$ of an inhomogeneous Poisson process $\mathbf{X}$ is bounded by some constant c for all $x \in B$. Then we can first simulate a homogeneous Poisson process on B with intensity c, and second make an independent thinning of this with retention probabilities $p(x) = \rho(x)/c$.

Thus the essential problem of simulating a Cox process on B is how to simulate its random intensity function $\Lambda(x)$ for $x \in B$. For the log-Gaussian Cox process driven by (9.3), there exist many ways of simulating the Gaussian process (Schlater, 1999; Lantuejoul, 2002; Møller and Waagepetersen, 2004). For the shot-noise Cox process $\mathbf{X}$ driven by (9.4), it is useful to notice that this is distributed as a Poisson cluster process obtained as follows. Conditional on $\mathbf{Y}$ in (9.4), let $\mathbf{X}_y$ for all $y \in \mathbf{Y}$ be mutually independent Poisson processes, where $\mathbf{X}_y$ has intensity function $\rho_y(x) = \exp(\beta \cdot z(x))k((x-y)/\sigma)/(\omega\sigma^2)$. Then $\mathbf{X} = \cup_{y\in\mathbf{Y}}\mathbf{X}_y$, that is, the superposition of 'clusters' $\mathbf{X}_y$ with 'centres' $y \in \mathbf{Y}$. We may easily simulate each cluster, but how do we simulate those centre points generating clusters with points in B? In fact such centre points form a Poisson process with intensity function $\rho(y) = \omega\mathbf{P}(\mathbf{X}_y \cap B \neq \emptyset)$, and the paper by Brix and Kendall (2002) shows how we can simulate this process as well as the non-empty sets $\mathbf{X}_y \cap B$. This paper provides indeed a neat example of a perfect simulation algorithm ('perfect' or 'exact' in the sense as defined later in Section 9.4.7), but it is easier to use an algorithm for approximate simulation, simulating only those centres which appear within a bounded region $D \supset B$ such that $\mathbf{P}(\mathbf{X}_y \cap B \neq \emptyset)$ is very small if $y \notin D$ (Møller, 2003; Møller and Waagepetersen, 2004).

9.3.4 *Maximum likelihood*

Consider a parametric point process model specified by an unknown parameter θ, e.g. $\theta = \beta$ in case of (9.2); many other examples will appear in the sequel. We assume that the point process $\mathbf{X}$ restricted to W has a density f_θ with respect to a stationary Poisson process $\mathbf{Y}$ with intensity one, i.e., $\mathbf{P}(\mathbf{X} \cap W \in F) = \mathbf{E}\left(f_\theta(\mathbf{Y} \cap W)\mathbf{1}[\mathbf{Y} \cap W \in F]\right)$ for all events F. Given the data $\mathbf{X} \cap W = \mathbf{x}$, the *likelihood function* $L(\theta; \mathbf{x})$ is any function of θ which is proportional to $f_\theta(\mathbf{x})$, and the *maximum likelihood estimate (MLE)* is the value of θ which maximizes the likelihood function, provided such a value exists and it is unique (apart from so-called exponential family models, existence and uniqueness of the MLE are not always easy to establish, see e.g. Geyer (1999)). Thus when we specify the likelihood function we may possibly (and without mentioning) exclude a multiplicative term of the density, if this term depends only on $\mathbf{x}$ (and not on θ). Obviously, the MLE does not depend on such a multiplicative term.

The log-likelihood function of the Poisson process with log-linear intensity function (9.2) is

$$l(\beta; \mathbf{x}) = \beta \cdot \sum_{x \in \mathbf{x}} z(x) - \int_W \exp\left(\beta \cdot z(x)\right) \mathrm{d}x. \tag{9.5}$$

Here the integral can be approximated by the Berman and Turner (1992) device, and the MLE can be obtained by `spatstat` (Baddeley and Turner (2005, 2006). Asymptotic properties of the MLE are studied in Rathbun and Cressie (1994) and Waagepetersen (2007).

In general, apart from various parametric models of mixed Poisson processes, the likelihood for Cox process models is intractable. For instance, for the log-Gaussian Cox process driven by (9.3), the covariance function $c(x)$ and hence the distribution of the underlying stationary Gaussian process $\mathbf{\Phi}$ may depend on some parameter γ, and the likelihood function is given by

$$L(\theta; \mathbf{x}) = \mathbf{E}_\theta \left[\exp\left(-\int_W \Lambda(x)\mathrm{d}x\right) \prod_{x \in \mathbf{x}} \Lambda(x)\right]$$

where the expectation is with respect to the distribution of $\mathbf{\Lambda}$ with parameter $\theta = (\beta, \gamma)$. This depends on the Gaussian process in such a complicated way that $L(\theta; \mathbf{x})$ is not expressible in closed form. Similarly, the likelihood function for the unknown parameter $\theta = (\beta, \omega, \sigma)$ of the shot-noise Cox process driven by (9.4) is not expressible in closed form.

Møller and Waagepetersen (2003, 2007) discuss various ways of performing maximum likelihood inference based on a missing data MCMC approach (see also Geyer (1999) though this reference is not specifically about Cox processes). Since these methods are rather technical, Section 9.3.7 discusses simpler ways of making inference for Cox processes, using moment results as given below.

9.3.5 *Moments*

A spatial point process $\mathbf{X}$ has *nth order product density* $\rho^{(n)}(x_1,\ldots,x_n)$ if for all non-negative Borel functions $q(x_1,\ldots,x_n)$ defined for $x_1,\ldots,x_n \in \mathbb{R}^2$,

$$\mathbf{E} \sum_{x_1,\ldots,x_n\in\mathbf{X}}^{\neq} q(x_1,\ldots,x_n) = \int\cdots\int q(x_1,\ldots,x_n)\rho^{(n)}(x_1,\ldots,x_n)\mathrm{d}x_1\cdots\mathrm{d}x_n$$

where the $\neq$ over the summation sign means that the points $x_1,\ldots,x_n$ are pairwise distinct. See e.g. Stoyan *et al.* (1995). For pairwise distinct $x_1,\ldots,x_n$, we may interpret $\rho^{(n)}(x_1,\ldots,x_n)\mathrm{d}x_1\cdots\mathrm{d}x_n$ as the probability of observing a point in each of n infinitesimally small regions containing $x_1,\ldots,x_n$ and of areas $\mathrm{d}x_1,\ldots,\mathrm{d}x_n$, respectively. Of most interest are the *intensity function* $\rho(x) = \rho^{(1)}(x)$ and the *pair–correlation function* $g(x,y) = \rho^{(2)}(x,y)/(\rho(x)\rho(y))$ (provided $\rho(x)\rho(y) > 0$).

For a Poisson process, $\rho^{(n)}(x_1,\ldots,x_n) = \rho(x_1)\cdots\rho(x_n)$ – this follows from the Slivnyak–Mecke formula (Mecke, 1967), which is later specified below (9.13) – and so $g(x,y) = 1$. For a general point process $\mathbf{X}$, at least for small distances $\|x-y\|$, we may expect $g(x,y) > 1$ in the case where realizations of $\mathbf{X}$ as compared to a Poisson process tend to be aggregated point patterns (e.g. as in a shot-noise Cox process or a Poisson cluster process), and $g(x,y) < 1$ in the case where realizations tend to be regular point patterns (e.g. caused by a repulsion between the points in a pairwise interaction point process as introduced later in Section 9.4.2).

For a Cox process driven by $\mathbf{\Lambda}$,

$$\rho^{(n)}(x_1,\ldots,x_n) = \mathbf{E}\left[\Lambda(x_1)\cdots\Lambda(x_n)\right]. \tag{9.6}$$

This takes a nice form for a log-Gaussian Cox process (Møller *et al.*, 1998) where in particular $\rho(x) = \exp(\beta\cdot z(x))$ (since the Gaussian process $\mathbf{\Phi}$ is assumed to be stationary with mean $-c(0)/2$) and $g(x,y) = \exp(c(x-y))$. This shows that the distribution of $(\mathbf{X},\mathbf{\Phi})$ is specified by (ρ, g), where $g(x,y) = g(x-y)$ (with a slight abuse of notation) is translation invariant and as one may expect, $g(x) > 1$ if and only if $c(x) > 0$. For the shot-noise Cox process given by (9.4), we obtain from (9.6) and Campbell's theorem (which is later specified below (9.13)) that also $\rho(x) = \exp(\beta\cdot z(x))$. For the modified Thomas process, we obtain from (9.6) and the Slivnyak–Mecke formula that $g(x,y) = g(r)$ depends only on the distance $r = \|x-y\|$, where

$$g(r) = 1 + \exp\left(-r^2/\left(4\sigma^2\right)\right)/\left(4\pi\omega\sigma^2\right) \tag{9.7}$$

showing that $g(r) > 1$ is a decreasing function.

9.3.6 *Summary statistics and residuals*

Exploratory analysis and validation of fitted spatial point process models are usually based on various summary statistics and residuals. In this section we focus mostly on summary statistics and residuals associated with the first and second order moment properties. These apply for many point process models which are not necessarily stationary, including the inhomogeneous Poisson and Cox process models discussed in this contribution, and the Gibbs point process models considered later in Section 9.4.

First order properties In the case of a spatial point process with constant intensity function on W, we naturally use the non-parametric estimate $\hat{\rho} = n(\mathbf{x})/|W|$. This is in fact the maximum likelihood estimate if $\mathbf{X}$ is a homogeneous Poisson process on W. In the inhomogeneous case,

$$\hat{\rho}(x) = \frac{1}{c(x)} \sum_{y \in \mathbf{x}} k_0((x-y)/b), \quad x \in W \tag{9.8}$$

is a non-parametric kernel estimate (Diggle, 1985), where k_0 is a density function with respect to Lebesgue measure, $b > 0$ is a user-specified parameter, and $c(x) = \int_W k_0((x-y)/b)\mathrm{d}y$ is an edge correction factor ensuring that $\int_W \hat{\rho}(x)\mathrm{d}x$ is an unbiased estimate of $\int_W \rho(x)\mathrm{d}x$. The kernel estimate is usually sensitive to the choice of the band width b, while the choice of kernel k_0 is less important.

Various kinds of *residuals* may be defined (Baddeley *et al.*, 2005; Waagepetersen, 2005). For the sake of brevity we mention only the smoothed raw residual field obtained as follows. By Campbell's theorem, if we replace $\mathbf{x}$ in (9.8) by $\mathbf{X} \cap W$, then $\hat{\rho}(x)$ has mean $\int_W \rho(y)k_0((x-y)/b)\mathrm{d}y/c(x)$. Suppose we have fitted a parametric model with parameter θ and estimate $\hat{\theta}$, where the intensity function ρ_θ may depend on θ. Then the smoothed raw residual field is given by

$$s(x) = \hat{\rho}(x) - \frac{1}{c(x)} \int_W \rho_{\hat{\theta}}(y)k_0((x-y)/b)\mathrm{d}y, \quad x \in W. \tag{9.9}$$

Positive/negative values of s suggest that the fitted model under/overestimates the intensity function. In `spatstat` the residual field is represented as a greyscale image and a contour plot. A somewhat similar type of plot but in a Bayesian setting is later shown in Fig. 9.5.

Second order properties Assume that the pair–correlation function is invariant under translations, i.e., $g(x, y) = g(x - y)$. This assumption is called *second order intensity re-weighted stationarity* (Baddeley *et al.*, 2000) and it is satisfied for the Poisson process, the log-Gaussian Cox, and shot-noise Cox processes studied in this contribution, cf. Section 9.3.5. It is also satisfied for a stationary point process.

The *inhomogeneous K and L-functions* (Baddeley *et al.*, 2000) are defined by

$$K(r) = \int_{\|x\| \le r} g(x)\mathrm{d}x, \quad L(r) = \sqrt{K(r)/\pi}, \quad r > 0.$$

In the stationary case we obtain *Ripley's K-function* (Ripley, 1976, 1977) with the interpretation that $\rho K(r)$ is the expected number of points within distance r from the origin o, where the expectation is with respect to the so-called reduced Palm distribution at o (intuitively, this is the conditional distribution given that $\mathbf{X}$ has a point at o, see Chapter 1). For a Poisson process, $L(r) = r$ is the identity, which is one reason why one often uses the L-function (see also Besag 1977). For a log-Gaussian Cox process, we may obtain $K(r)$ by numerical methods. For the modified Thomas process, (9.7) implies that

$$K(r) = \pi r^2 + \left(1 - \exp\left(-r^2/\left(4\sigma^2\right)\right)\right)/\omega. \tag{9.10}$$

A spatial point process is said to be *isotropic* if its distribution is invariant under rotations of the point process around the origin. In order to investigate for a possible anisotropy, directional K-functions have been constructed (Stoyan and Stoyan, 1995; Brix and Møller, 2001), while Mugglestone and Renshaw (1996) use a method as part of two-dimensional spectral analysis. Frequently, g is assumed to be invariant under rotations, i.e. $g(x, y) = g(\|x - y\|)$, in which case K and g are in a one-to-one correspondence, and it is usually easy to estimate and investigate a plot of g. Non-parametric kernel estimation of $g(r)$ (with $r = \|x - y\|$) is discussed in Stoyan and Stoyan (2000); it is sensitive to the choice of band width and a truncation for distances r near zero is needed.

An unbiased estimate of $K(r)$ is given by

$$\sum_{x,y \in \mathbf{X}}^{\neq} \frac{\mathbf{1}[\|x - y\| \le r]}{\rho(x)\rho(y)|W \cap (W + x - y)|} \tag{9.11}$$

where $W + x$ denotes W translated by x, and $|W \cap (W + x - y)|$ is an edge correction factor, which is needed since we sum over all pairs of points observed within W. In practice we need to plug in an estimate of $\rho(x)\rho(y)$, and denote the resulting estimate of K by $\hat{K}$, which then becomes biased. It may be given by a parametric estimate $\rho_{\hat{\theta}}(x)\rho_{\hat{\theta}}(y)$ or by a non-parametric estimate, e.g. $\hat{\rho}(x)\hat{\rho}(y)$ from (9.8) (but see also Stoyan and Stoyan (2000) in the stationary case and Baddeley *et al.* (2000) in the inhomogeneous case). Using the parametric estimate seems more reliable, since the degree of clustering exhibited by $\hat{K}$ may be very sensitive to the choice of band width if (9.8) is used.

The non-parametric estimate $\hat{K}$ for the tropical rain forest trees (Figs. 9.1 and 9.2) obtained with a parametric estimate of the intensity function is shown

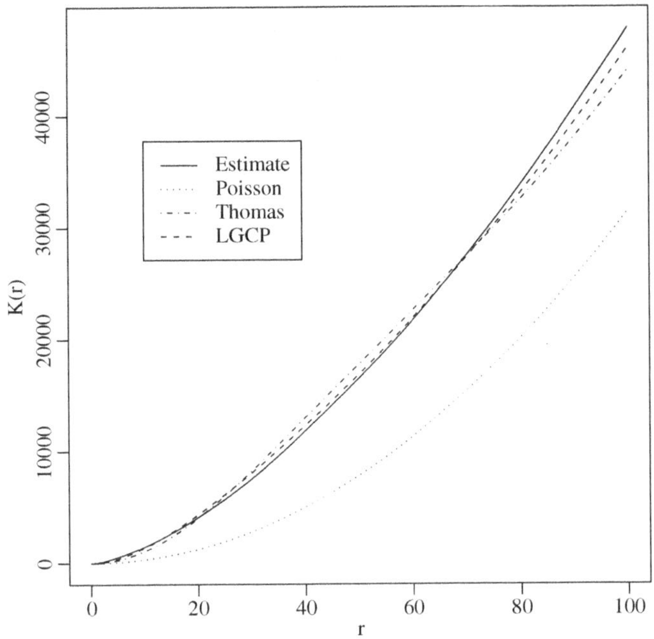

FIG. 9.3. Estimated K-function for tropical rain forest trees and theoretical K-functions for fitted Thomas, log Gaussian Cox, and Poisson processes.

in Fig. 9.3 (Section 9.3.7 describes the estimation procedure). The plot also shows theoretical K-functions for fitted log Gaussian Cox, modified Thomas, and Poisson processes, where all three processes share the same intensity function (see again Section 9.3.7). The trees seem to form a clustered point pattern, since $\hat{K}$ is markedly larger than the theoretical K-function for a Poisson process.

Further summaries and confidence bounds Other summary statistics than those based on first and second order moment properties usually assume stationarity of the point process $\mathbf{X}$. This includes the F, G, J-functions below, which are based on interpoint distances, and where variations for multivariate and marked point processes exist (see Diggle 2003, and the references therein). Also summary statistics based on higher-order moment properties exist (Stoyan and Stoyan, 1995; Møller *et al.*, 1998; Schladitz and Baddeley, 2000).

Suppose that $\mathbf{X}$ is stationary with intensity $\rho > 0$. The *empty space function* F and the *nearest-neighbour function* G are the cumulative distribution functions of the distance from respectively an arbitrary location to the nearest point in $\mathbf{X}$ and a 'typical' point in $\mathbf{X}$ (i.e., with respect to the reduced Palm distribution at o) to its nearest-neighbour point in $\mathbf{X}$. For a stationary Poisson process, $F(r) = G(r) = 1 - \exp(-\pi r^2)$. In general, at least for small distances, $F(r) < G(r)$ indicates aggregation and $F(r) > G(r)$ indicates regularity. Van Lieshout and Baddeley (1996) study the nice properties of the J-function defined by $J(r) = (1 - G(r))/(1 - F(r))$ for $F(r) < 1$. Non-parametric estimation of F and G accounting for edge effects is straightforward, see e.g. Stoyan *et al.* (1995), and combining the estimates we obtain an estimate of J.

Finally, estimates $\hat{g}(r), \hat{K}(r), \hat{F}(r), \ldots$ may be supplied with *confidence intervals* for each value of r obtained by simulations $\hat{g}_i(r), \hat{K}_i(r), \hat{F}_i(r), \ldots, \ i = 1, 2, \ldots$ under an estimated model, see e.g. Møller and Waagepetersen (2004). This is later illustrated in Figures 9.7, 9.9, and 9.10. The K-functions in Fig. 9.3 tell the same story as if we were plotting the corresponding L-functions, however, it is often preferable to consider L-functions when confidence intervals are wanted, as exemplified in Fig. 9.9.

9.3.7 *Composite likelihood and minimum contrast estimation*

The intensity functions of the log-linear Poisson process, the log-Gaussian Cox, and shot-noise Cox processes in Section 9.3.2 are of the same form, viz. $\rho_\beta(x) = \exp(\beta \cdot z(x))$, cf. Section 9.3.5. Their pair correlation functions are different, where g_ψ depends on a parameter $\psi \in \Psi$, which is usually assumed to be variation independent of $\beta \in \mathbb{R}^{p+1}$, i.e., $(\psi, \beta) \in \Psi \times \mathbb{R}^{p+1}$; e.g. $\psi = (\omega, \sigma)$ and $\Psi = (0, \infty)^2$ in the case of the modified Thomas process, cf. (9.7). Following Møller and Waagepetersen (2007) we may estimate β by maximizing a composite likelihood function based on ρ_β, and ψ by another method based on g_ψ or its corresponding K-function K_ψ given by e.g. (9.10).

The *composite likelihood function* is derived as a certain limit of a product of marginal likelihoods. Let C_i, $i \in I$, be a finite partitioning of W into disjoint cells C_i of small areas $|C_i|$, and define $N_i = \mathbf{1}[\mathbf{X} \cap C_i \neq \emptyset]$ and $p_i(\theta) = \mathbf{P}_\theta(N_i = 1) \approx \rho_\theta(u_i)|C_i|$, where u_i denotes a representative point in C_i. Under mild conditions the limit of $\log \prod_i p_i(\theta)^{N_i}(1-p_i(\theta))^{(1-N_i)}$ becomes (9.5); this is the log composite likelihood function.

For the rain forest trees (Figs. 9.1 and 9.2) we obtain the maximum composite likelihood estimate $(\hat{\beta}_0, \hat{\beta}_1, \hat{\beta}_2) = (-4.989, 0.021, 5.842)$ (under the Poisson model this is the MLE). Assuming asymptotic normality (Waagepetersen, 2007) 95% confidence intervals for β_1 and β_2 under the fitted shot-noise Cox process are $[-0.018, 0.061]$ and $[0.885, 10.797]$, respectively, while much more narrow intervals are obtained under the fitted Poisson process ($[0.017, 0.026]$ and $[5.340, 6.342]$).

For instance, ψ may be estimated by minimizing the contrast

$$\int_a^b \left(\hat{K}(r)^\alpha - K_\psi(r)^\alpha\right)^2 \mathrm{d}r$$

where typically $0 \le a < b < \infty$ and $\alpha > 0$ are chosen on an ad hoc basis, see e.g. Diggle (2003) and Møller and Waagepetersen (2004). For the rain forest trees, replacing $\rho(x)\rho(y)$ in (9.11) by the parametric estimate of $\rho(x)\rho(y)$ obtained in the previous paragraph, and taking $a = 0$, $b = 100$, and $\alpha = 1/4$, we obtain $(\hat{\omega}, \hat{\sigma}) = (8 \times 10^{-5}, 20)$. The estimated theoretical K-functions are shown in Fig. 9.3.

These *'simulation-free' estimation procedures* are fast and computationally easy, but the disadvantage is that we have to specify tuning parameters such as a, b, α. Theoretical properties of maximum composite likelihood estimators are investigated in Waagepetersen (2007) and Waagepetersen and Guan (2007), and of the minimum contrast estimators in Heinrich (1992).

9.3.8 *Simulation-based Bayesian inference*

A Bayesian approach often provides a flexible framework for incorporating prior information and analyzing Poisson and Cox process models.

As an example, consider Fig. 9.4 which shows the locations of just one species (called seeder 4 in Illian *et al.* 2009) from a much larger data set where the locations of 6378 plants from 67 species on a 22 m by 22 m observation window W in the south western area of Western Australia have been recorded (Armstrong, 1991). The plants have adapted to regular natural fires, where resprouting species survive the fire, while seeding species die in the fire but the fire triggers the shedding of seeds, which have been stored since the previous fire. As argued in more detail in Illian *et al.* (2009) it is therefore natural to model the locations of the reseeding plants conditionally on the locations of the resprouting plants. In the sequel we consider seeder 4 plants conditional on the 19 most dominant

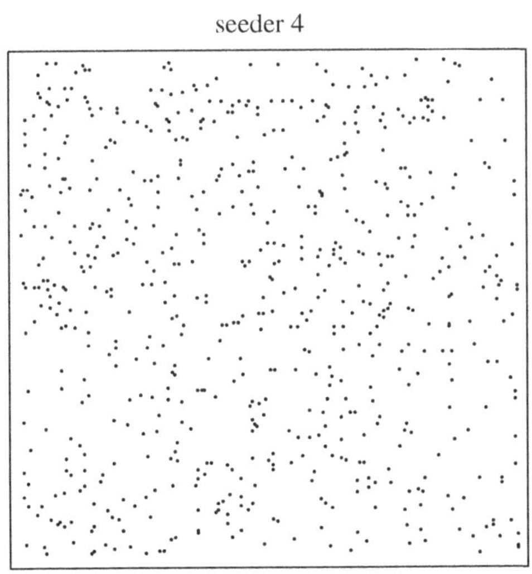

FIG. 9.4. Locations of 657 *Leucopogon conostephioides* plants observed within a 22 × 22 m window.

(influential) species of resprouters, with $\mathbf{y}_1, \ldots, \mathbf{y}_{19}$ the observed point patterns of resprouters as given in Fig. 1 in Illian *et al.* (2009). For a discussion of possible interaction with other seeder species, and the biological justification of the following covariates, we refer to Illian *et al.* (2009).

Define covariates by

$$z_i(x) = \sum_{y \in \mathbf{y}_i:\, \|x-y\| \le \kappa_{y,i}} \left(1 - (\|x - y\|/\kappa_{y,i})^2\right)^2, \quad i = 1, \ldots, 19$$

where $\kappa_{y,i} \geq 0$ is the radius of interaction of the ith resprouter at location y. Note that we suppress in the notation that z_i depends on all the $\kappa_{y,i}$ with $y \in \mathbf{y}_i$. As usual $\mathbf{X} \cap W = \mathbf{x}$ denotes the data (here the point pattern of seeder 4), where we assume that $\mathbf{X} \cap W$ is a Poisson process with log-linear intensity function (9.2) and parameter $\beta = (\beta_0, \ldots, \beta_{19})$, where β_0 is an intercept and β_i for $i > 0$ controls the influence of the ith resprouter. Thus the log-likelihood is of the form (9.5) but with unknown parameter $\theta = (\beta, \kappa)$, where κ denotes the collection of all the radii $\kappa_{y,i}$.

Using a fully Bayesian set up, we treat $\theta = (\beta, \kappa)$ as a random variable, where Table 1 in Illian *et al.* (2009) provides some prior information on κ. Specifically, following Illian *et al.* (2009), we assume a priori that the $\kappa_{y,i}$ and the β_i are mutually independent, each $\kappa_{y,i}$ follows the restriction of a normal distribution $\mathrm{N}(\mu_i, \sigma_i^2)$ to $[0, \infty)$, where (μ_i, σ_i^2) is chosen so that under the unrestricted normal distribution the range of the zone of influence is a central 95% interval, and each β_i is $\mathrm{N}(0, \sigma^2)$-distributed, where $\sigma = 8$. Combining these prior assumptions with the log-likelihood we obtain the posterior density

$$\pi(\beta, \kappa|\mathbf{x}) \propto \exp\left(-\beta_0^2/(2\sigma^2) - \sum_{i=1}^{19}\left\{\beta_i^2/(2\sigma^2) + \sum_{y \in \mathbf{y}_i}(\kappa_{y,i} - \mu_i)^2/(2\sigma_i^2)\right\}\right)$$
$$\times \exp\left(\beta \cdot \sum_{x \in \mathbf{x}} z(x) - \int_W \exp(\beta \cdot z(u)\,\mathrm{d}u)\right),$$
$$\beta_i \in \mathbb{R}, \kappa_{y,i} \geq 0 \tag{9.12}$$

(suppressing in the notation that z depends on κ and we have conditioned on $\mathbf{y}_1, \ldots, \mathbf{y}_{19}$ in the posterior distribution).

Monte Carlo estimates of various marginal posterior distributions are calculated using simulations from (9.12) obtained by a hybrid MCMC algorithm (see e.g. Robert and Casella 1999) where we alter between updating β and κ using random walk Metropolis updates (for details, see Illian *et al* 2009). For example, a large (small) value of $\mathbf{P}(\beta_i > 0|\mathbf{x})$ indicates a positive/attractive (negative/repulsive) association to the ith resprouter, see Fig. 2 in Illian *et al.*

(2009). The model can be checked following the idea of posterior predictive model assessment (Gelman *et al.*, 1996), comparing various summary statistics with their posterior predictive distributions. The *posterior predictive distribution* of statistics depending on $\mathbf{X}$ (and possibly also on (β, κ)) is obtained from simulations: we generate a posterior sample $(\beta^{(j)}, \kappa^{(j)})$, $j = 1, \ldots, m$, and for each j 'new data' $\mathbf{x}^{(j)}$ from the conditional distribution of $\mathbf{X}$ given $(\beta^{(j)}, \kappa^{(j)})$. For instance, the grey scale plot in Fig. 9.5 is a 'residual' plot based on quadrant counts. We divide the observation window into 100 equally sized quadrants and count the number of seeder 4 plants within each quadrant. The grey scales reflect the probabilities that counts drawn from the posterior predictive distribution are less or equal to the observed quadrant counts where dark means small probability. The stars mark quadrants where the observed counts are 'extreme' in the sense of being either below the 2.5% quantile or above the 97.5% quantile of the posterior predictive distribution. Figure 9.5 does not provide evidence against our model. A plot based on the L-function and the posterior predictive distribution is also given in Illian *et al.* (2009); again there is no evidence against our model.

Møller and Waagepetersen (2007) discuss another Bayesian analysis for the rain forest trees, using a shot-noise Cox process model. They consider the Poisson process $\mathbf{Y}$ of mother points from (9.4) as one part of the prior (more precisely, to obtain a finite process, $\mathbf{Y}$ is restricted to a bounded region $B \supset W$ as discussed in Section 9.3.3); remaining unknown parameters are denoted θ, and a certain prior is imposed on θ. Simulations from the posterior are again obtained by a hybrid MCMC algorithm, where one type of update involves conditional simulation of $\mathbf{Y}$ given the data and θ, using the Metropolis–Hastings birth–death algorithm in Section 9.4.7 (see also Møller 2003).

For simulation-based Bayesian inference for a log-Gaussian Cox process, we need among other things a procedure to make conditional simulation of the Gaussian process $\mathbf{\Phi}$ in (9.3) given the data and θ. This may be done by a Langevin–Hastings algorithm (Besag, 1994; Roberts and Tweedie, 1996) as detailed in Møller *et al.* (1998) and Møller and Waagepetersen (2004). Promising results in Rue and Martino (2005) suggest that it may be possible to compute accurate approximations of posterior distributions without MCMC.

9.4 Inference for Gibbs point processes

Briefly, a *Gibbs point process* models interaction between the points; the precise definition of a Gibbs point process is a bit technical and given later in this section. While Cox processes provide flexible models for aggregation or clustering in a point pattern, Gibbs point processes provide flexible models for regularity or repulsion (Van Lieshout, (2000); Møller and Waagepetersen, 2003). Though the density has an unknown normalizing constant (Section 9.4.1), likelihood inference based on MCMC methods is easier for parametric Gibbs point process models

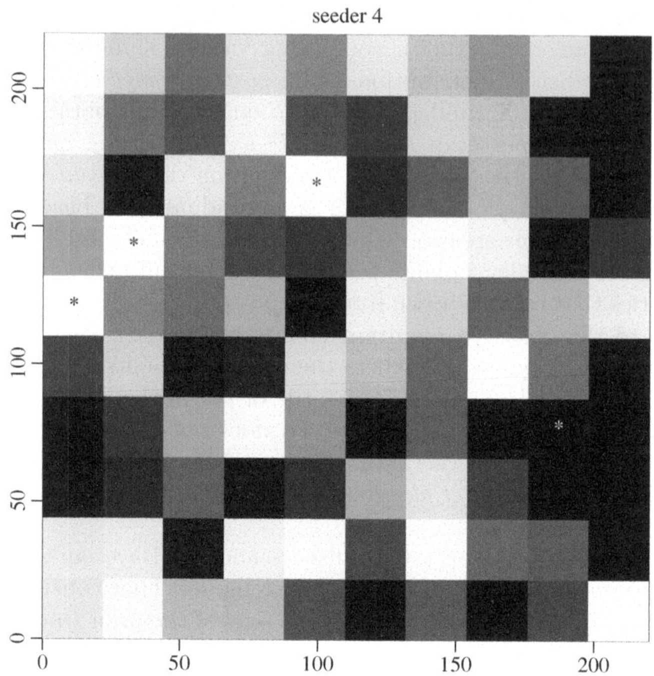

FIG. 9.5. Residual plots based on quadrant counts. Quadrants with a '*' are where the observed counts fall below the 2.5% quantile (white '*') or above the 97.5% quantile (black '*') of the posterior predictive distribution. The grey scales reflect the probabilities that counts drawn from the posterior predictive distribution are less or equal to the observed quadrant counts (dark means small probability).

(Sections 9.4.2 and 9.4.4) than for Cox process models. On the other hand, Bayesian inference is more complicated, involving an auxiliary variable MCMC method and perfect simulation based on a technique called dominated coupling from the past (Sections 9.4.6 and 9.4.7).

An example of a regular point pattern is shown in Fig. 9.6. The data set is actually a marked point pattern, with points given by the tree locations and marks by the stem diameters, where the discs may be thought of as 'influence zones' of the trees. Figure 9.7 shows estimates of F, G, J (as defined at the end of Section 9.3.6) based on the tree locations. The figure provides evidence of regularity in the point pattern, which in fact to a large extent is due to forest management but may also be caused by repulsion between the trees. The data set has been further analysed using point process methods by Fiksel (1984), Penttinen *et al.* (1992), Goulard *et al.* (1996), and Møller and Waagepetersen (2004, 2007).

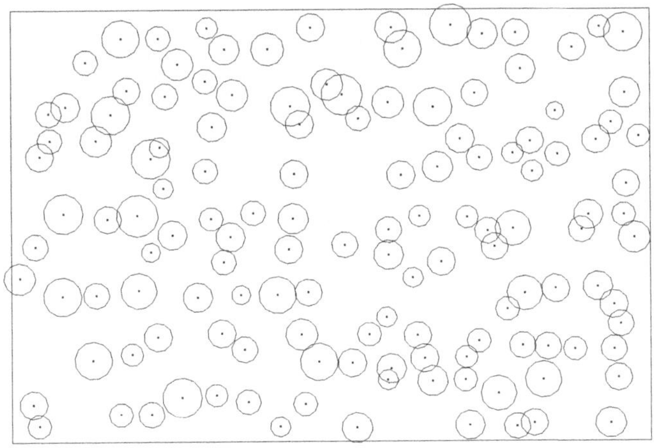

FIG. 9.6. Norwegian spruces observed in a 56×38 m window in Tharandter Forest, Germany. The radii of the 134 discs equal five times the stem diameters.

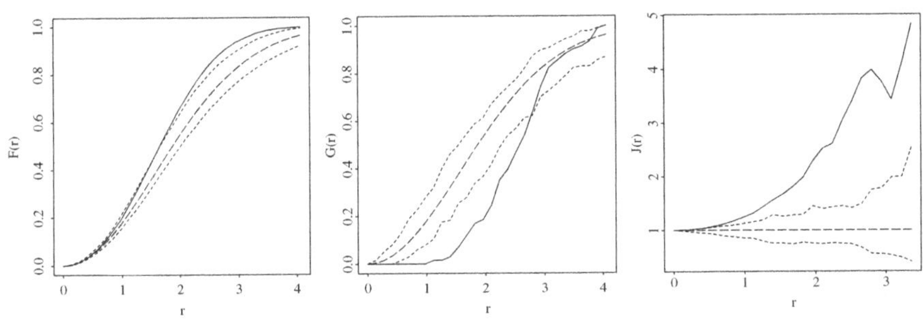

FIG. 9.7. Left to right: estimated F, G, J-functions for the Norwegian spruces (solid lines) and 95% envelopes calculated from simulations of a homogeneous Poisson process (dashed lines) with expected number of points equal to the observed number of points. The long-dashed curves show the theoretical values of F, G, J for a Poisson process.

9.4.1 *Gibbs point processes*

Gibbs point processes arose in statistical physics as models for interacting particle systems (see e.g. Ruelle 1969), and they have important applications in stochastic geometry and spatial statistics (see e.g. Baddeley and Møller 1989, Van Lieshout 2000, and Stoyan *et al.* 1995). They possess certain Markov properties which are useful for 'local computations' and to account for edge effects. Summary statistics for Gibbs point processes, including the intensity function ρ, are in general not expressible in closed form. Instead the Papangelou conditional intensity (Papangelou, 1974) becomes the appropriate tool.

Definition of conditional intensity Below we define the *Papangelou conditional intensity* $\lambda(x; \mathbf{x})$ for a general point process $\mathbf{X}$, assuming the process exists; the question of existence of Gibbs point processes is discussed later. Here $\mathbf{x} \subset \mathbb{R}^2$ denotes a locally finite point configuration and $x \in \mathbb{R}^2$.

By the *Georgii–Nguyen–Zessin (GNZ) formula* (Georgii, 1976; Nguyen and Zessin, 1979), $\mathbf{X}$ has Papangelou conditional intensity λ if this is a non-negative Borel function such that

$$\mathbf{E} \sum_{x \in \mathbf{X}} h(x, \mathbf{X} \setminus \{x\}) = \int \mathrm{E}[\lambda(x; \mathbf{X}) h(x, \mathbf{X})] \, \mathrm{d}x \tag{9.13}$$

for non-negative Borel functions h. The GNZ-formula (9.13) is a very general and useful but indirect definition/characterization of the conditional intensity. If $\mathbf{P}^!_x$ is the reduced Palm distribution at the point x and $\mathbf{P}^!_x$ is absolutely continuous with respect to the distribution $\mathbf{P}$ of $\mathbf{X}$, we may take $\lambda(x; \mathbf{x}) = \rho(x)(\mathrm{d}\mathbf{P}^!_x / \mathrm{d}\mathbf{P})(\mathbf{x})$ (Georgii, 1976). Thus, for $x \notin \mathbf{x}$, we may interpret $\lambda(x; \mathbf{x})\mathrm{d}x$ as the conditional probability that there is a point of the process in an infinitesimally small region containing x and of area $\mathrm{d}x$ given that the rest of the point process coincides with $\mathbf{x}$. For a Poisson process, $\lambda(x; \mathbf{x}) = \rho(x)$ and (9.13) becomes the *Slivnyak–Mecke formula* (Mecke, 1967). When $h(x, \mathbf{X} \setminus \{x\}) = h(x)$ only depends on its first argument, (9.13) reduces to *Campbell's theorem* (e.g. Stoyan *et al.* 1995) with $\rho(x) = \mathbf{E}\lambda(x; \mathbf{X})$.

Existence and uniqueness of λ for an infinite point process is discussed in Georgii (1976, 1978) and Preston (1976). In fact two infinite point processes may share the same Papangelou conditional intensity (even in the stationary case); this phenomenon is known in statistical physics as *phase transition*. By (9.13), for a Cox process driven by Λ, we can take $\lambda(x; \mathbf{X}) = \mathbf{E}[\Lambda(x) \mid \mathbf{X}]$, but this conditional expectation is in general not expressible in closed form. As explained later, the GNZ-formula is more useful in connection to Gibbs point processes.

A conditional intensity exists for a finite point process $\mathbf{X}$ defined on a bounded region $B \in \mathcal{B}$, assuming this has a density f with respect to the unit rate Poisson process on B, and f is hereditary, i.e.

$$f(\mathbf{y}) > 0 \text{ whenever } f(\mathbf{x}) > 0, \ \mathbf{y} \subset \mathbf{x} \subset B, \ \mathbf{x} \text{ finite.} \tag{9.14}$$

Then we can take

$$\lambda(x; \mathbf{x}) = f(\mathbf{x} \cup \{x\}) / f(\mathbf{x} \setminus \{x\}) \tag{9.15}$$

if $f(\mathbf{x} \setminus \{x\}) > 0$, and $\lambda(x; \mathbf{x}) = 0$ otherwise. The precise definition of $\lambda(x; \mathbf{x})$ when $x \in \mathbf{x}$ is not that important, and (9.15) just covers this case for completeness. In fact $\lambda(x; \mathbf{x})$ is unique up to the product of a Lebesgue null set of points x and a Poisson null set of point configurations $\mathbf{x}$. Note that $\lambda(x; \mathbf{x}) = \lambda(x; \mathbf{x} \setminus \{x\})$, and (9.14) implies that f and λ are in a one-to-one correspondence. Often we specify $f(\mathbf{x}) \propto h(\mathbf{x})$ only up to proportionality, and call h an unnormalized density. Here we assume that h is hereditary and integrable with respect to the unit rate Poisson process on B, in which case we can replace f by h in (9.15). These conditions are ensured by *local stability*, which means that for some finite constant K,

$$h(\mathbf{x} \cup \{\mathbf{x}\}) \leq Kh(\mathbf{x}) \quad \text{for all finite } \mathbf{x} \subset B \text{ and } x \in B \setminus \mathbf{x}. \tag{9.16}$$

Then $\lambda(x; \mathbf{x}) \leq K$ is uniformly bounded. A weaker condition ensuring integrability of h is Ruelle stability (Ruelle, 1969). We restrict attention to local stability, since this property is usually satisfied for Gibbs point processes used in stochastic geometry, and it implies many desirable properties as discussed in Section 9.4.7.

Definition and Markov properties of Gibbs point processes Suppose that $R > 0$ is a given parameter and $U(\mathbf{x}) \in (-\infty, \infty]$ is defined for all finite $\mathbf{x} \subset \mathbb{R}^2$ such that $U(\mathbf{x}){=}0$ if $\mathbf{x}$ contains two points at least R units apart. Then a point process $\mathbf{X}$ is a *Gibbs point process* (also called a canonical ensemble in statistical physics and a Markov point process in spatial statistics) with *potential* U if it has a Papangelou conditional intensity of the form

$$\lambda(x; \mathbf{x}) = \exp\left(- \sum_{\mathbf{y} \subseteq \mathbf{x} \cap B_R(x)} U(\mathbf{y} \cup \{x\}) \right)$$

for all locally finite $\mathbf{x} \subset \mathbb{R}^2$ and $x \in \mathbb{R}^2 \setminus \mathbf{x}$, where $B_R(x)$ is the closed disc of radius R centred at x. Note that the conditional intensity is zero if $U(\mathbf{y} \cup \{x\}) = \infty$. It satisfies a *local Markov property*: $\lambda(x; \mathbf{x})$ depends on $\mathbf{x}$ only through $\mathbf{x} \cap B_R(x)$, i.e., the R-close neighbours in $\mathbf{x}$ to x. Usually R is chosen as small as possible, in which case it is called the *range of interaction*. We say that a Gibbs point process has *interactions of order* k, if k is the smallest integer so that the potential $U(\mathbf{x})$ does not vanish if $n(\mathbf{x}) \leq k$. The order is often $k = 2$ as discussed later in Section 9.4.2.

An equivalent characterization of a Gibbs point process is given by a so-called *local specification* (Preston, 1976). This means that a *spatial Markov property* should be satisfied: for any bounded $B \in \mathcal{B}$, the conditional distribution of $\mathbf{X} \cap B$ given $\mathbf{X} \setminus B$ agrees with the conditional distribution of $\mathbf{X} \cap B$ given $\mathbf{X} \cap \partial B$, where $\partial B = \{x \in \mathbb{R}^2 \setminus B : B_R(x) \cap B \neq \emptyset\}$ is the R-close neighbourhood to B. It also means that the conditional density of $\mathbf{X} \cap B$ given $\mathbf{X} \cap \partial B = \partial \mathbf{x}$ is of the form

$$f_B(\mathbf{x} | \partial \mathbf{x}) = \frac{1}{c_B(\partial \mathbf{x})} \exp\left(- \sum_{\mathbf{y} \subseteq \mathbf{x} : \mathbf{y} \neq \emptyset} U(\mathbf{y} \cup \partial \mathbf{x}) \right) \tag{9.17}$$

with respect to the unit rate Poisson process on B. Here $c_B(\partial\mathbf{x})$ is a normalizing constant (called the partition function in statistical physics), which in general is not expressible in closed form.

Conditions ensuring existence of an *infinite Gibbs point process*, uniqueness or non-uniqueness (phase transition), and stationarity have been discussed in Georgii (1976, 1988) and Preston (1976). In general closed form expressions for $\rho^{(n)}$ and other summary statistics are unknown (even in the stationary case or in the finite case).

For a *finite Gibbs point process* $\mathbf{X}$ defined on a bounded region $B \in \mathcal{B}$, the density is of the form

$$f(\mathbf{x}) = \frac{1}{c}\exp\left(-\sum_{\mathbf{y}\subseteq\mathbf{x}:\mathbf{y}\neq\emptyset} U(\mathbf{y})\right) \tag{9.18}$$

corresponding to the case of (9.17) with 'free boundary condition' $\partial\mathbf{x} = \emptyset$. The celebrated *Hammersley–Clifford theorem* (Ripley and Kelly, 1977) states that for a finite point process on B with hereditary density f, the local Markov property is equivalent to (9.18).

The spatial Markov property can be used to account for *edge effects* when the Gibbs process is observed within a bounded window W but it extends outside W. Let $W_{\ominus R} = \{x \in W : B_R(x) \subseteq W\}$ be the R-clipped window. Conditional on $\mathbf{X}\cap\partial W_{\ominus R} = \partial x$, we have that $\mathbf{X}\cap W_{\ominus R}$ is independent of $\mathbf{X}\setminus W$. The conditional density of $\mathbf{X}\cap W_{\ominus R}|\mathbf{X}\cap\partial W_{\ominus R} = \partial x$ is given by (9.17) with $B = W_{\ominus R}$, and combining this with (9.15) we see that the conditional intensity $\lambda_{W_{\ominus R},\partial\mathbf{x}}(x;\mathbf{x})$ of $\mathbf{X}\cap W_{\ominus R}|\mathbf{X}\cap\partial W_{\ominus R} = \partial x$ agrees with λ, that is

$$\lambda_{W_{\ominus R},\partial\mathbf{x}}(x;\mathbf{x}) = \lambda(x;\mathbf{x}\cup\partial\mathbf{x}), \quad \mathbf{x}\subset W_{\ominus R} \text{ finite}, x \in W_{\ominus R}\setminus\mathbf{x}. \tag{9.19}$$

9.4.2 *Modelling conditional intensity*

Usually parametric models of a Gibbs point process with a fixed interaction radius R have a conditional intensity of a log-linear form

$$\log\lambda_\theta(x;\mathbf{x}) = \theta\cdot t(x;\mathbf{x}) = \sum_{i=1}^{k}\theta_i\cdot t_i(x;\mathbf{x}) \tag{9.20}$$

where $\theta = (\theta_1,\ldots,\theta_k)$ is a real parameter vector (possibly infinite parameter values of some θ_i are also included as exemplified below for a hard core process), $t(x;\mathbf{x}) = (t_1(x;\mathbf{x}),\ldots,t_k(x;\mathbf{x}))$ is a real vector function, where $t_i(x;\mathbf{x})$ is of the same dimension as θ_i, $i = 1,\ldots,k$, and $\cdot$ is the usual inner product. Equivalently, if the Gibbs point process has interactions of order k, we usually assume that the potential $U(\mathbf{x})$ is depending linearly on θ_i if $n(\mathbf{x}) = i \le k$, that is,

$$U(\mathbf{x}) = -\theta_i\cdot V(\mathbf{x}) \quad \text{if } n(\mathbf{x}) = i \le k \tag{9.21}$$

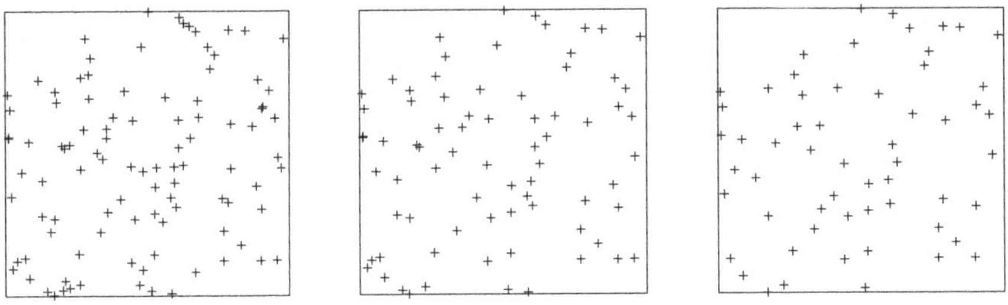

FIG. 9.8. Simulation of a Strauss process on the unit square (i.e., $V(\{x\}) = 1$ if $x \in [0,1]^2$ and $V(\{x\}) = 0$ otherwise) when $R = 0.05$, $\theta_1 = \log(100)$, and $\theta_2 = 0, -\log 2, -\infty$ (from left to right).

where $V(\mathbf{x})$ is a vector function with non-negative components; the connection between (9.20) and (9.21) is given by

$$t_i(x; \mathbf{x}) = \sum_{\mathbf{y} \subseteq \mathbf{x} \cap B_R(x):\, n(\mathbf{y}) = i-1} V(\mathbf{y} \cup \{x\}), \quad i = 1, \ldots, k. \tag{9.22}$$

Moreover, usually the first order term $V(\{x\})$ is constant one or it is given by a vector of covariates $V(\{x\}) = z(x)$, where $\theta_1 = \beta$ consists of regression parameters (Ogata and Tanemura, 1989), while the higher order terms $V(\mathbf{x})$ $(0 < n(\mathbf{x}) \leq k)$ specify the *interaction* between the points in $\mathbf{x}$.

We often consider *pairwise interaction processes* with repulsion between the points, i.e. $\theta_2 \leq 0$ and $\theta_n = 0$ whenever $n \geq 3$, where in many cases $V(\{x, y\}) = V(\|x - y\|)$ is a real decreasing function of $\|x - y\|$. A special case is the *Strauss process* (Strauss, 1975; Kelly and Ripley, 1976) where $V(\{x\}) = 1$, $V(\{x, y\}) = \mathbf{1}[\|x - y\| \leq R]$, and $(\theta_1, \theta_2) \in \mathbb{R} \times [-\infty, 0]$; if $\theta_2 = -\infty$, we use in (9.21) the convention that $-\infty \times 0 = 0$ and $-\infty \times 1 = -\infty$, and call R a hard core parameter since all points in the process are at least R-units apart from each other. Figure 9.8 shows simulations (in fact perfect simulations as later described in Section 9.4.7) of different Strauss processes restricted to the unit square.

The *Widom–Rowlinson model* (Widom and Rowlinson, 1970) (or *area-interaction process*, Baddeley and Van Lieshout (1995)) is given by

$$\log \lambda_\theta(x; \mathbf{x}) = \beta - \psi |B_x(R) \setminus \cup_{y \in \mathbf{x}} B_y(R)|$$

where $\beta, \psi \in \mathbb{R}$ and recalling that $|\cdot|$ denotes area. This is clearly of the form (9.20), and for a finite area-interaction process the corresponding density is proportional to

$$\exp\left(\beta n(\mathbf{x}) - \psi |\cup_{y \in \mathbf{x}} B_y(R)|\right).$$

It follows that this is a Markov point process with interaction range R, but the Hammersley–Clifford representation (9.18) is not so useful: using the inclusion-exclusion formula on $|\cup_{y\in\mathbf{x}} B_y(R)|$, we obtain this representation, but the expression for the potential is intricate. For $\psi \neq 0$, the process has interactions of all orders, since the $t_i(\mathbf{x})$-terms in (9.22) do not vanish for any $i \geq 1$ and the corresponding $\theta_i = \psi$ is non-zero for all $i \geq 2$. The model is said to be attractive if $\psi > 0$, since then $\lambda_\theta(x;\mathbf{x})$ is an increasing function of $\mathbf{x}$, and repulsive if $\psi < 0$, since then $\lambda_\theta(x;\mathbf{x})$ is a decreasing function of $\mathbf{x}$. Phase transition happens for all sufficiently large values of $\psi > 0$ (Ruelle, 1971; Häggström *et al.*, 1999).

Further specific examples of Gibbs point process models are given in Van Lieshout (2000) and the references therein.

Consider again Fig. 9.6. The conditional intensity for a Norwegian spruce with a certain influence zone should depend not only on the positions but also on the influence zones of the neighbouring trees. A tree with influence zone given by the disc $B_r(x)$, where x is the spatial location of the tree and r is the influence zone radius, is treated as a point (x, r) in $W \times (0,\infty)$, where W is the rectangular window in Fig. 9.6; similarly, we let now $\mathbf{x} \subset W \times (0,\infty)$ denote a finite configuration of such points (or discs). We assume that the influence zone radii belong to a bounded interval $M = [a, b]$, where a and b are estimated by the minimal and maximal observed influence zone radii, and we divide M into six disjoint subintervals of equal size. Confining ourselves to a pairwise interaction process with repulsion, we let

$$\lambda_\theta((x,r);\mathbf{x}) = \exp\left(\beta(r) + \psi \sum_{(x',r')\in\mathbf{x}} |B_r(x) \cap B_{r'}(x')|\right) \tag{9.23}$$

where $\beta(r) = \beta_k$ if r falls in the kth subinterval, and $\theta = (\beta_1, \ldots, \beta_6, \psi) \in \mathbb{R}^6 \times (-\infty, 0]$. This enables modelling the varying numbers of trees in the six different size classes, and the strength of the repulsion between two trees is given by ψ times the area of overlap between the influence zones of the two trees. However, the interpretation of the conditional intensity is not straightforward – it is for instance not in general a monotone function of r. On the other hand, for a fixed (x, r), the conditional intensity will always decrease if the neighbouring influence zones increase.

9.4.3 *Residuals and summary statistics*

Exploratory analysis and validation of fitted Gibbs point process models can be done using residuals and summary statistics, cf. Fig. 9.7. As mentioned theoretical closed form expressions for summary statistics such as ρ, g, K, L, F, G, J (see Sections 9.3.5-sec.sumstatres) are in general unknown for Gibbs point process models, so one has entirely to rely on simulations. Residuals are more analytical tractable due to the GNZ-formula (9.13), see Baddeley *et al.* (2005, 2008). These papers discuss in detail how various kinds of residuals can be defined, where `spatstat` can be used for the log-linear model (9.20). For example, the smoothed raw residual field is now given by

$$s(x) = \hat{\rho}(x) - \frac{1}{c(x)} \int_W \lambda_{\hat{\theta}}(y; \mathbf{x}) k_0((x-y)/b) \mathrm{d}y, \quad x \in W$$

with a similar interpretation of positive/negative values as in (9.9) but referring to the conditional intensity instead.

9.4.4 *Simulation-based maximum likelihood inference*

Ogata and Tanemura (1984) and Penttinen (1984) are pioneering works on simulation-based maximum likelihood inference for Markov point processes; see also Geyer and Møller (1994) and Geyer (1999). This section considers a parametric Gibbs point process model with finite interaction radius R, potential of the form (9.21), and finite interaction of order k (though other cases such as the area-interaction process are covered as well, cf. Section 9.4.2). We assume to begin with that R is known.

First, suppose that the process has no points outside the observation window W. By (9.18) and (9.21) the log-likelihood

$$l(\theta; \mathbf{x}) = \theta \cdot t(\mathbf{x}) - \log c_\theta, \quad t(\mathbf{x}) = (t_1(\mathbf{x}), \ldots, t_k(\mathbf{x})) \tag{9.24}$$

is of exponential family form (Barndorff-Nielsen, 1978), where

$$t_i(\mathbf{x}) = \sum_{\mathbf{y} \subseteq \mathbf{x} :\, n(\mathbf{y}) = i} V(\mathbf{y}), \quad i = 1, \ldots, k.$$

The first and second derivatives of l with respect to θ are called the *score function* $u(\theta)$ and *observed information* $j(\theta)$, respectively (suppressing in the notation the dependence of the data $\mathbf{x}$). By exponential family theory (Barndorff-Nielsen, 1978),

$$u(\theta) = t(\mathbf{x}) - \mathbf{E}_\theta t(\mathbf{X}), \quad j(\theta) = \mathrm{Var}_\theta t(\mathbf{X}),$$

where $\mathbf{E}_\theta$ and Var_θ denote expectation and variance with respect to $\mathbf{X}$ with parameter θ. Consider a fixed reference parameter value θ_0. The score function and observed information may then be evaluated using the *importance sampling formula*

$$\mathbf{E}_\theta q(\mathbf{X}) = \mathbf{E}_{\theta_0} \left[q(\mathbf{X}) \exp\left((\theta - \theta_0) \cdot t(\mathbf{X}) \right) \right] / (c_\theta / c_{\theta_0})$$

with $q(\mathbf{X})$ given by $t(\mathbf{X})$ or $t(\mathbf{X})^{\mathsf{T}} t(\mathbf{X})$. The importance sampling formula also yields

$$c_\theta / c_{\theta_0} = \mathbf{E}_{\theta_0} \left[\exp\left((\theta - \theta_0) \cdot t(\mathbf{X}) \right) \right].$$

Approximations of the log likelihood ratio $l(\theta;\mathbf{x}) - l(\theta_0;\mathbf{x})$, score, and observed information are then obtained by Monte Carlo approximation of the expectations $\mathbf{E}_{\theta_0}[\cdots]$ using MCMC samples from the process with parameter θ_0, see Section 9.4.7. The path sampling identity (Gelman and Meng, 1998) provides an alternative and often numerically more stable way of computing a ratio of normalizing constants.

Second, if $\mathbf{X}$ may have points outside W, we have to correct for edge effects. By (9.17) the log likelihood function based on the conditional distribution $\mathbf{X} \cap W_{\ominus R} | \mathbf{X} \cap \partial W_{\ominus R} = \partial x$ is of similar form as (9.24) but with c_θ depending on ∂x. Thus likelihood ratios, score, and observed information may be computed by analogy with the case above. Alternatively, if $\mathbf{X}$ is contained in a bounded region $S \supset W$, the likelihood function based on the distribution of $\mathbf{X}$ may be computed using a missing data approach, which is a more efficient and complicated approach, see Geyer (1999) and Møller and Waagepetersen (2004). Yet other approaches for handling edge effects are discussed in Møller and Waagepetersen (2004).

For a fixed R, the approximate (conditional) likelihood function can be maximized with respect to θ using Newton–Raphson updates (Geyer, 1999; Møller and Waagepetersen, 2004). Frequently the Newton–Raphson updates converge quickly, and the computing times for obtaining a MLE are modest. MLE's of R are often found using a profile likelihood approach, since the likelihood function is typically not differentiable and log concave as a function of R.

Asymptotic results for MLE's of Gibbs point process models are established in Mase (1991) and Jensen (1993) but under restrictive assumptions of stationarity and weak interaction. According to standard asymptotic results, the inverse observed information provides an approximate covariance matrix of the MLE, but if one is suspicious about the validity of this approach, an alternative is to use a parametric bootstrap, see e.g. Møller and Waagepetersen (2004).

For the overlap interaction model (9.23), Møller and Waagepetersen (2004) computed MLEs using both missing data and conditional likelihood approaches, where the conditional likelihood approach is based on the trees with locations in $W_{\ominus 2b}$, since trees with locations outside W do not interact with trees located inside $W_{\ominus 2b}$. The conditional MLE is given by

$$(\hat{\beta}_1, \ldots, \hat{\beta}_6) = (-1.02, -0.41, 0.60, -0.67, -0.58, -0.22), \quad \hat{\psi} = -1.13.$$

Confidence intervals for ψ obtained from the observed information and a parametric bootstrap are $[-1.61, -0.65]$ and $[-1.74, -0.79]$, respectively. The fitted overlap interaction process seems to capture well the second order characteristics for the point pattern of tree locations, see Fig. 9.9.

9.4.5 *Pseudo-likelihood*

The *maximum pseudo-likelihood estimate (MPLE)* is a simple and computationally fast but less efficient alternative to the MLE.

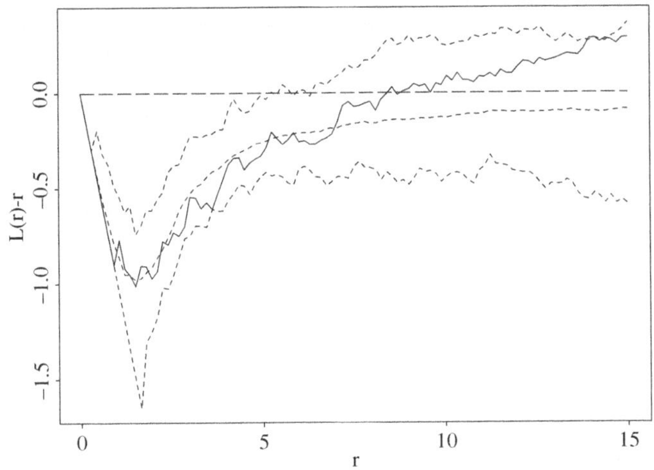

FIG. 9.9. Model assessment for Norwegian spruces: estimated $L(r)-r$ function for spruces (solid line), 95% envelopes (upper and lower dashed lines) and average (mid dashed line) computed from simulations of fitted overlap interaction model, and the theoretical curve for a Poisson process (long-dashed line).

First, consider a finite point process with no points outside the observation window W. Let C_i, $i \in I$, be a finite partitioning of W into disjoint cells C_i of small areas $|C_i|$, and define $N_i = \mathbf{1}[\mathbf{X} \cap C_i \neq \emptyset]$ and $p_i(\theta) = \mathbf{P}_\theta(N_i = 1|\mathbf{X} \setminus C_i) \approx \lambda_\theta(u_i, \mathbf{X} \setminus C_i)|C_i|$, where u_i denotes a representative point in C_i. Under mild conditions (Besag *et al.*, 1982; Jensen and Møller, 1991) the limit of $\log \prod_i (p_i(\theta)/|C_i|)^{N_i}(1 - p_i(\theta))^{(1-N_i)}$ becomes

$$\sum_{x \in \mathbf{x}} \lambda_\theta(x, \mathbf{x}) - \int_W \lambda_\theta(x, \mathbf{x}) \mathrm{d}x \tag{9.25}$$

which is known as the log *pseudo-likelihood function* (Besag, 1977*a*). By the GNZ formula (9.13), the pseudo score

$$s(\theta) = \sum_{x \in \mathbf{x}} \mathrm{d} \log \lambda_\theta(x, \mathbf{x})/\mathrm{d}\theta - \int_W (\mathrm{d} \log \lambda_\theta(x, \mathbf{x})/\mathrm{d}\theta) \lambda_\theta(x, \mathbf{x}) \mathrm{d}x \tag{9.26}$$

provides an unbiased estimating equation $s(\theta) = 0$ (assuming in (9.26) that $(\mathrm{d}/\mathrm{d}\theta) \int_W \cdots = \int_W (\mathrm{d}/\mathrm{d}\theta) \cdots$). This can be solved using `spatstat` if λ_θ is on a log linear form (Baddeley and Turner, 2000).

Second, suppose that $\mathbf{X}$ may have points outside W and $\mathbf{X}$ is Gibbs with interaction radius R. To account for edge effects we consider the conditional distribution $\mathbf{X} \cap W_{\ominus R} | \mathbf{X} \cap \partial W_{\ominus R} = \partial \mathbf{x}$. By (9.19) and (9.25) the log pseudo-likelihood function is then given by

$$\sum_{x \in \mathbf{x} \cap \partial W_{\ominus R}} \lambda_\theta(x, \mathbf{x}) - \int_{W_{\ominus R}} \lambda_\theta(x, \mathbf{x}) \mathrm{d}x.$$

Asymptotic results for MPLE's of Gibbs point process models are established in Jensen and Møller (1991), Jensen and Künsch (1994), and Mase (1995, 1999). Baddeley and Turner (2000) estimate the asymptotic variance by a parametric bootstrap.

9.4.6 *Simulation-based Bayesian inference*

Suppose we consider a parametric Gibbs point process model with a prior on the parameter θ. Then MCMC methods are needed for simulations of the posterior distribution of θ. The main difficulty in a 'conventional' MCMC algorithm (i.e. a Metropolis–Hastings algorithm) is that the so-called Hastings ratio (see Section 9.4.7) involves a ratio $c_\theta / c_{\theta'}$ of normalizing constants which we cannot compute (here θ denotes a current value of the Markov chain with invariant distribution equal to the posterior, and θ' denotes a proposal for the next value of the chain). Heikkinen and Penttinen (1999) avoided this problem by just estimating the posterior mode, while Berthelsen and Møller (2003) estimated each ratio $c_\theta / c_{\theta'}$ by path sampling (Section 9.4.4) which at each update of the Metropolis–Hastings algorithm involved several other MCMC chains. Recently, Møller *et al.* (2006) introduced an auxiliary variable method which avoids such approximations. The method has been used for semi- or non-parametric Gibbs point process models (Berthelsen and Møller, 2004, 2006, 2008).

This section provides a survey of Berthelsen and Møller (2008), where the cell data in the left panel in Fig. 9.10 was analysed by a pairwise interaction process,

$$f_\theta(\mathbf{x}) = \frac{1}{c_\theta} \prod_{x \in \mathbf{x}} \beta(x) \prod_{\{x,y\} \subseteq \mathbf{x}} \varphi(\|x - y\|), \tag{9.27}$$

ignoring edge effects and considering $\mathbf{x}$ as a finite subset of the rectangular observation window in Fig. 9.10; we shall later impose a very flexible prior for the first and second order interaction terms β and φ. The data set has also been analysed by Nielsen (2000) by transforming a Strauss point process model in order to account for inhomogeneity in the horizontal direction (Jensen and Nielsen, 2000; Nielsen and Jensen, 2004). The centre panel in Fig. 9.10 clearly shows that a Poisson process is an inadequate model for the data, where the low values of the estimated pair–correlation $\hat{g}(r)$ for distances $r < 0.01$ indicates repulsion between the points, so in the sequel we assume that $\varphi \leq 1$. The right panel in Fig. 9.10 shows simulated 95% envelopes under the fitted transformed Strauss point process, where $\hat{g}(r)$ is almost within the envelopes for small values of the distance r, suggesting that the transformed Strauss model captures the

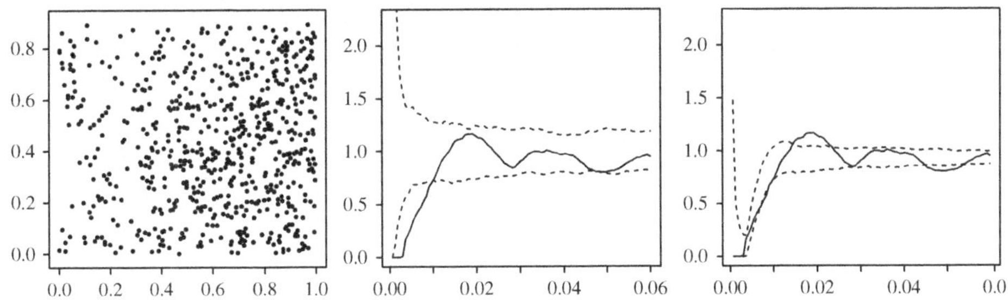

FIG. 9.10. Left panel: locations of 617 cells in a 2D section of the mucous membrane of the stomach of a healthy rat. Centre panel: non-parametric estimate of the pair-correlation function for the cell data (full line) and 95% envelopes calculated from 200 simulations of a fitted inhomogeneous Poisson process. Right panel: non-parametric estimate of the pair-correlation function for the cell data (full line) and 95% envelopes calculated from 200 simulations of fitted transformed Strauss process.

small scale inhibition in the data. Overall, $\hat{g}(r)$ follows the trend of the 95% envelopes, but it falls outside the envelopes for some values.

We assume a priori that $\beta(x) = \beta(x_1)$ is homogeneous in the second coordinate of $x = (x_1, x_2)$, where $\beta(x_1)$ is similar to the random intensity of a modified Thomas process (but with $p = 1$ and using a one-dimensional Gaussian kernel in (9.4)), and where various prior assumptions on the parameters of the shot-noise process are imposed (see Berthelsen and Møller 2008). The left panel in Fig. 9.11 shows five simulated realizations of β under its prior distribution. We also use the prior model in Berthelsen and Møller (2008) for the pairwise interaction function. First, $\varphi(r)$ is assumed to be a continuous and piecewise linear function which increases from zero to one, see the right panel in Fig. 9.11. The change points of $\varphi(r)$ are modelled by a homogeneous Poisson process on $[0, r_{\max}]$, where $r_{\max} = 0.02$ ($\hat{g}$ in Fig. 9.10 indicates that there is little interaction beyond an inter-point distance of 0.01, but to be on the safe side we let $r_{\max} = 0.02$). The step sizes of $\varphi(r)$ at the change points are modelled by a certain Markov chain to obtain some smoothness in φ, and various prior assumptions on the intensity of the Poisson process of change points and the parameters of the Markov chain for the step sizes are imposed (see Berthelsen and Møller 2008). We also assume independence between the collection of random variables θ_1 modelling β and the collection of random variables θ_2 modelling φ. Next, since a posterior analysis indicated the need for an explicit hard core parameter $h < r_{\max}$ in the model, we modify (9.27) by replacing $\varphi(r)$ throughout by

$$\tilde{\varphi}(r;h) = \begin{cases} 0 & \text{if } r < h \\ \varphi\left(\frac{(r-h)r_{\max}}{r_{\max}-h}\right) & \text{if } h \le r \le r_{\max} \\ 1 & \text{if } h > r_{\max}. \end{cases} \tag{9.28}$$

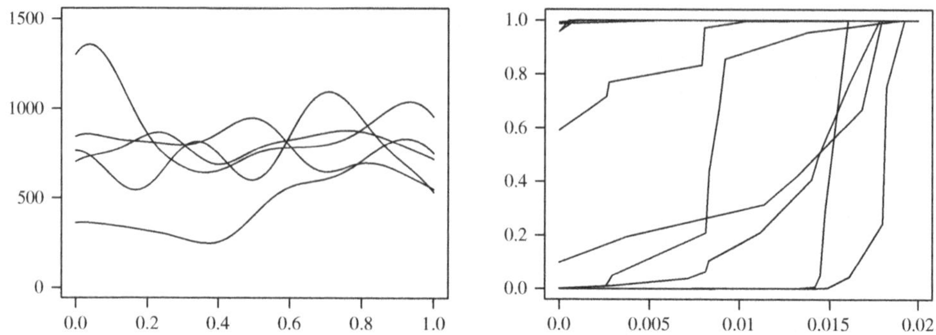

FIG. 9.11. Left panel: five independent realizations of β under its prior distribution. Right panel: Ten independent realizations of φ under its prior distribution.

Finally, h is assumed to be uniformly distributed on $[0, r_{\max}]$ and independent of (θ_1, θ_2), whereby the posterior density of $\theta = (\theta_1, \theta_2, h)$ can be obtained.

For the sake of brevity we omit the details on how to simulate from the posterior distribution, including how to use the auxiliary variable method from Møller *et al.* (2006). The details are given in Berthelsen and Møller (2008), and it should be noted that the auxiliary variable method requires perfect simulations from (9.28) as given in Section 9.4.7. The left panel of Fig. 9.12 shows the posterior mean $\mathbf{E}(\beta(x_1)|\mathbf{x})$ together with pointwise 95% central posterior intervals. Also the smooth estimate of the first order term obtained by (Nielsen, 2000) is shown, where the main difference compared with $\mathbf{E}(\beta(x_1)|\mathbf{x})$ is the abrupt change of $\mathbf{E}(\beta(x_1)|\mathbf{x})$ in the interval $[0.2, 0.4]$. For locations $x = (x_1, x_2)$ near the edges of W, $\mathbf{E}(\beta(x_1)|\mathbf{x})$ is 'pulled' towards its prior mean as a consequence of the smoothing prior. Apart from boundary effects, since $\beta(x_1, x_2)$ only depends on x_1, we may expect that the intensity $\rho_\theta(x_1, x_2)$ only slightly depends on x_2, i.e., $\rho_\theta(x_1, x_2) \approx \rho_\theta(x_1)$ where, if we let $W = [0, a] \times [0, b]$, $\rho_\theta(x_1)$ is given by $1/b$ times the integral of $\rho_\theta(x_1, x_2)$ over $x_2 \in [0, b]$. We therefore refer to $\rho_\theta(x_1)$ as the cell intensity, though it is more precisely the average cell intensity in W at $x_1 \in [0, a]$. A non-parametric estimate of $\rho_\theta(x_1)$ is given by

$$\hat{\rho}(x_1) = \frac{\sum_{(y_1,y_2)\in\mathbf{x}} \phi((x_1 - y_1)/\sigma_k)/\sigma_k}{b(\Phi((1 - x_1)/\sigma_k) - \Phi(x_1/\sigma_k))}$$

which is basically the one-dimensional edge-corrected kernel estimator of Diggle (1985) with bandwidth $\sigma_k = 0.075$ (here ϕ and Φ denote the density and cumulative distribution function of the standard normal distribution). The left panel of Fig. 9.12 also shows this estimate. The posterior mean of $\beta(x_1)$ is not unlike $\hat{\rho}(x_1)$ except that $\mathbf{E}(\beta(x_1)|\mathbf{x})$ is higher as would be expected due to the repulsion in the likelihood. The posterior mean of $\tilde{\varphi}$ is shown in the right panel of Fig. 9.12 together with pointwise 95% central posterior intervals. The figure

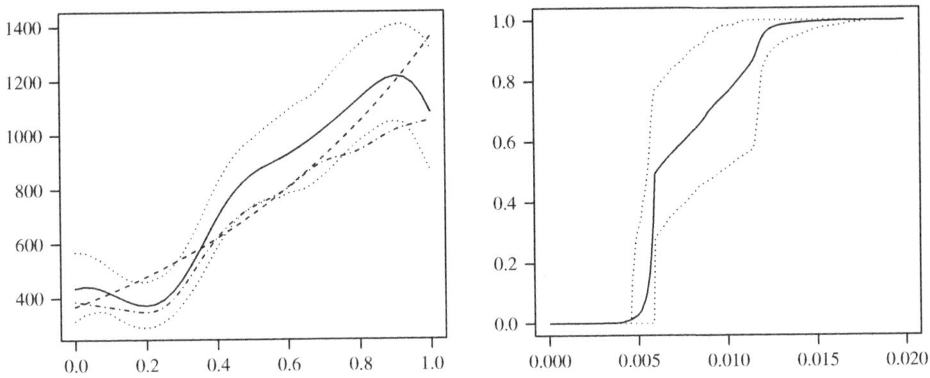

FIG. 9.12. Posterior mean (solid line) and pointwise 95% central posterior intervals (dotted lines) for β (left panel) and $\tilde{\varphi}$ (right panel). The left panel also shows the first order term (dashed line) estimated by a fitted transformed Strauss process, and an estimate of the cell intensity (dot-dashed line).

shows a distinct hard core on the interval from zero to the observed minimum inter-point distance d which is a little less than 0.006, and an effective interaction range which is no more than 0.015 (the posterior distribution of $\tilde{\varphi}(r)$ is concentrated close to one for $r > 0.015$). This further confirms that $r_{\max}$ was chosen sufficiently large. The corner at $r = d$ of the curve showing the posterior mean of $\tilde{\varphi}(r)$ is caused by that $\tilde{\varphi}(r)$ is often zero for $r < d$ (since the hard core is concentrated close to d), while $\tilde{\varphi}(r) > 0$ for $r > d$. Further posterior results, including a model checking based on the posterior predictive distribution, can be found in Berthelsen and Møller (2008).

9.4.7 *Simulation algorithms*

Consider a finite point process $\mathbf{X}$ with density $f \propto h$ with respect to the unit rate Poisson process defined on a bounded region $B \in \mathcal{B}$ of area $|B| > 0$, where h is a 'known' unnormalized density. The normalizing constant of the density is not assumed to be known. This section reviews algorithms for making simulated realizations of $\mathbf{X}$.

Birth–death algorithms Simulation conditional on the number of points $n(\mathbf{X})$ can be done using a variety of Metropolis–Hastings algorithms, e.g. using a Gibbs sampler (Ripley, 1977, 1979) or a Metropolis-within-Gibbs algorithm, where at each iteration a single point given the remaining points is updated, see Møller and Waagepetersen (2004). The standard algorithms (i.e. without conditioning on $n(\mathbf{X})$) are discrete or continuous time algorithms of the birth-death type, where each transition is either the addition of a new point (a birth) or the deletion of an existing point (a death). The algorithms can easily be extended to birth-death-move type algorithms, where, e.g. in the discrete time case the number of points is retained in a move by using a Metropolis–Hastings update (Norman and Filinov, 1969; Geyer and Møller, 1994).

In the discrete time case, the simplest version of the *Metropolis–Hastings birth–death algorithm* update a current state $\mathbf{X}_t = \mathbf{x}$ of the Markov chain as follows (other versions are discussed in Norman and Filinov (1969) and Geyer and Møller (1994)). Assume that h is hereditary, and define

$$r(x;\mathbf{x}) = \lambda(x;\mathbf{x})|B|/(n(\mathbf{x})+1)$$

where λ is the Papangelou conditional intensity. With probability 0.5 propose a birth, i.e. generate a uniform point x in B, and accept the proposal $\mathbf{X}_{t+1} = \mathbf{x}\cup\{x\}$ with probability $\min\{1, H_b(x;\mathbf{x})\}$, where $H_b(x;\mathbf{x}) = r(x;\mathbf{x})$ is the Hastings ratio. Otherwise propose a death, i.e. select a point $x \in \mathbf{x}$ uniformly at random, and accept the proposal $\mathbf{X}_{t+1} = \mathbf{x}\setminus\{x\}$ with probability $\min\{1, H_d(x;\mathbf{x})\}$, where $H_d(x;\mathbf{x}) = 1/r(x;\mathbf{x}\setminus\{x\})$ is the Hastings ratio. As usual in a Metropolis–Hastings algorithm, if the proposal is not accepted, $\mathbf{X}_{t+1} = \mathbf{x}$. This algorithm is irreducible, aperiodic, and time reversible with invariant distribution f. Thus the distribution of $\mathbf{X}_t$ converges towards f, and if h is locally stable, the rate of convergence is geometrically fast and a central limit theorem holds for Monte Carlo errors (Geyer and Møller, 1994; Geyer, 1999). If h is highly multimodal, e.g. in the case of a strong interaction like in a hard core model with a high packing density, the algorithm may be slowly mixing. Possibly it may then be worth considering a simulated tempering scheme as discussed in Geyer and Thompson (1995) and Mase *et al.* (2001).

An analogous continuous time algorithm is based on running a *spatial birth–death process* $\mathbf{X}_t$ with birth rate $\lambda(x,\mathbf{x})$ and death rate 1 (Preston, 1977; Ripley, 1977). This is also a reversible process with invariant density f, and convergence of $\mathbf{X}_t$ towards f holds under weak conditions (Preston, 1977), where local stability of h implies geometrically fast convergence (Møller, 1989). It can be extended to a coupling construction with a dominating spatial birth–death process $\mathbf{D}_t$, $t \geq 0$, which is easy to simulate (even in equilibrium) and from which we may easily obtain the spatial birth–death process $\mathbf{X}_t$ from above by a dependent thinning. Since the coupling construction is also used below in connection to perfect simulation we give here the details. Assume that h is locally stable with uniform upper bound $K \geq \lambda$. Initally, suppose that $\mathbf{X}_0 \subseteq \mathbf{D}_0$. Let $0 < \tau_1 < \tau_2 < \ldots$ denote the transition times in $\mathbf{D}_t$, and set $\tau_0 = 0$. For $i \geq 1$, if we condition on both $\tau_{i-1} = t$, $\mathbf{D}_t = \mathbf{d}$, $\mathbf{X}_t = \mathbf{x}$, and the history $(\mathbf{D}_s, \mathbf{X}_s)$, $0 \leq s < t$, then $\tau_i - \tau_{i-1}$ is exponentially distributed with mean $1/(K + n(\mathbf{d}))$. If we also condition on $\tau_i - \tau_{i-1}$, then with probability $K/(K + n(\mathbf{d}))$ a birth occurs so that $\mathbf{D}_{\tau_i} = \mathbf{d} \cup \{x\}$, where x is uniformly distributed on B, and otherwise a death happens so that both $\mathbf{D}_{\tau_i} = \mathbf{d} \setminus \{y\}$ and $\mathbf{X}_{\tau_i} = \mathbf{x} \setminus \{y\}$, where y is a uniformly selected point from $\mathbf{d}$. Furthermore, in case $\mathbf{D}_{\tau_i} = \mathbf{d} \cup \{x\}$, we generate a uniform number $R_i \in [0, 1]$ (which is independent of x and anything else so far generated), and set $\mathbf{X}_{\tau_i} = \mathbf{x} \cup \{x\}$ if $R_i \leq \lambda(x;\mathbf{x})/K$, and $\mathbf{X}_{\tau_i} = \mathbf{x}$ otherwise. It follows that $\mathbf{D}_t$ is dominating $\mathbf{X}_t$ in the sense that $\mathbf{X}_t \subseteq \mathbf{D}_t$ for all $t \geq 0$, and as required, $\mathbf{X}_t$ is a spatial birth–death process with birth rate λ and death rate 1. Moreover, $\mathbf{D}_t$ is a spatial birth–death process with birth rate K and death rate 1, so its equilibrium distribution is a homogeneous Poisson process with rate K.

Perfect simulation One of the most exciting recent developments in stochastic simulation is perfect (or exact) simulation, which turns out to be particular applicable for locally stable point processes. By this we mean an algorithm where the running time is a finite random variable and the output is a draw from a given target distribution (at least in theory – of course the use of pseudo random number generators and practical constraints of time imply that we cannot exactly return draws from the target distribution). For tutorials on perfect simulation, see Kendall and Thönnes (1999), Møller (2001), Thönnes (2000), Wilson (2000), and Dimakos (2001); the webpage `dimacs.rutgers.edu/~dbwilson/exact/` maintained by David Wilson contains much further information on perfect simulation. The most famous perfect simulation algorithm is due to Propp and Wilson (1996), but as discussed in the next paragraph the Propp–Wilson algorithm rarely works for spatial point processes. Another well-known perfect simulation algorithm is Fill's algorithm (Fill, 1998; Fill *et al.*, 2000), which in the context of spatial point processes has been discussed in Thönnes (1999) and Møller and Schladitz (1999). Various other perfect simulation algorithms related to spatial point processes have been proposed in Kendall (1998), Kendall and Møller (2000), Wilson (2000), Fernández *et al.* (2002), Berthelsen and Møller (2002*b*), and Huber (2007). For extensions to the case of marked point processes, see Van Lieshout and Stoica (2006). For application examples of perfect simulation for spatial point processes, see Lund and Thönnes (2004) and Berthelsen and Møller (2003, 2004, 2006, 2008).

Foss and Tweedie (1998) showed that the Propp–Wilson algorithm can in principle be applied to any uniformly ergodic Markov chain, though the details of 'in principle' may turn out to be very impractical. One problem with the Propp–Wilson algorithm is that it does not apply for most spatial point process algorithms, since they are only geometrically ergodic, see Kendall and Møller (2000), Møller and Waagepetersen (2004), and Kendall (2004). Another problem with the Propp–Wilson algorithm is that a kind of monotonicity property and existence of unique upper and lower bounds are needed with respect to a partial ordering, but the natural partial ordering of point processes is set inclusion, and for this ordering there is no maximal element (the minimal element is the empty point figuration $\emptyset$). Below we describe the simplest solution to these problems based on a technique called *dominating coupling from the past (DCFTP)*. DCFTP was first developed in Kendall (1998) and second extended in Kendall and Møller (2000) using spatial birth–death processes when considering spatial point processes; see also Huber (2007) who shows how an extra type of move given by a swap improves the DCFTP algorithm. As discussed in Kendall (2004), DCFTP can in principle be applied to any geometrically ergodic

Markov chain, though again the details of 'in principle' may turn out to be very impractical, but it works for many Gibbs point processes as explained below.

DCFTP uses the coupling construction for the spatial birth–death processes $\mathbf{D}_t$ and $\mathbf{X}_t$ described above, assuming that f is locally stable and that $\mathbf{D}_0$ follows its Poisson process equilibrium distribution. However, we generate $\mathbf{D}_t$ backwards in time $t \leq 0$, which by reversibility is the same stochastic construction as generating it forwards in time $t \geq 0$. One possibility, here called DCFTP1, would be to stop the backwards generation of $\mathbf{D}_t$ at time $\tau = \sup\{t \leq 0 : \mathbf{D}_t = \emptyset\}$, and then set $\mathbf{X}_\tau = \emptyset$ and use the same coupling construction as before when generating forwards $\mathbf{X}_t$, $\tau < t \leq 0$. In fact, with probability one, the algorithm terminates (i.e. $\tau > -\infty$) and $\mathbf{X}_0 \sim f$ is a perfect simulation; however, as demonstrated in Berthelsen and Møller (2002*a*), DCFTP1 is going to be far too slow in realistic cases. Note that we do not need to generate the exponential waiting times for transitions, since only the jumps of births and deaths are used, that is, we need only to generate $\mathbf{J}_0 = \mathbf{D}_0$ and if $\mathbf{J}_0 \neq \emptyset$, the jump chain $\mathbf{J}_1, \mathbf{J}_2, \ldots, \mathbf{J}_{M_1}$ corresponding to $\mathbf{D}_t$ backwards in time $0 > t \geq \tau$ (so that $\mathbf{J}_1 \neq \emptyset$, ..., $\mathbf{J}_{M_1-1} \neq \emptyset$, and $\mathbf{J}_{M_1} = \emptyset$) and the jump chain $\mathbf{I}_{1-M_1}, \mathbf{I}_{2-M_1}, \ldots, \mathbf{I}_0$ corresponding to $\mathbf{X}_t$ forwards in time $\tau < t \leq 0$ (so that $\mathbf{I}_0 = \mathbf{X}_0$). The infeasibility of DCFTP1 simply follows from that the number of jumps M_1 can be extremely large for spatial point process models with even a modest value of $K|B|$ (the expected number of points in $\mathbf{D}_0$).

In practice we use instead the DCFTP2-algorithm based on a sequence of discrete starting times $T_0 = 0 > T_1 > T_2 > \ldots$ for so-called upper and lower bounding processes $\mathbf{U}_t^n$ and $\mathbf{L}_t^n$, $t = T_n, T_n+1, \ldots, 0$, $n = 1, 2, \ldots$, constructed forwards in time as described below such that the following funnelling/sandwiching property is satisfied,

$$\mathbf{L}_t^{n+1} \subseteq \mathbf{L}_t^n \subseteq \mathbf{I}_t \subseteq \mathbf{U}_t^n \subseteq \mathbf{U}_t^{n+1} \subseteq \mathbf{J}_t, \quad t = T_n, T_n + 1, \ldots, 0, \; n = 1, 2, \ldots. \tag{9.29}$$

Here we imagine for the moment that $(\mathbf{D}_t, \mathbf{X}_t)$ is extended further back in time, which at least in theory is easily done, since the process regenerates each time $\mathbf{D}_t = \emptyset$. Thereby we see that $\mathbf{L}_0^n = \mathbf{U}_0^n = \mathbf{I}_0 = \mathbf{X}_0$ is a perfect simulation for all sufficiently large n, and we let M_2 denote the smallest possible n with $\mathbf{L}_0^n = \mathbf{U}_0^n$. The DCFTP2-algorithm consists in first generating $\mathbf{D}_0$ from its Poisson process equilibrium distribution, second if $\mathbf{D}_0 = \emptyset$ to return $\mathbf{X}_0 = \emptyset$, and else for $n = 1, \ldots, M_2$, generating $\mathbf{J}_{T_{n-1}+1}, \ldots, \mathbf{J}_{T_n}$ and $\left(\mathbf{U}_{T_n}^n, \mathbf{L}_{T_n}^n\right)$, ...,$(\mathbf{U}_0^n, \mathbf{L}_0^n)$, and returning $\mathbf{X}_0 = \mathbf{L}_0^{M_2}$. Note that only the dominating jump chain and the pairs of upper and lower processes are generated until we obtain coalescence $\mathbf{L}_0^n = \mathbf{U}_0^n$; it is not required to generate the 'target' jump chain $\mathbf{I}_t$. Empirical findings in Berthelsen and Møller (2002*a*) show that M_2 can be much smaller than M_1.

To specify the construction of upper and lower processes, recall that in case of a forwards birth (corresponding to a backwards death) of $\mathbf{D}_t$, we generated a uniform number used for determining whether a birth should occur or not in

$\mathbf{X}_t$. Let $R_0, R_1, \ldots$ be mutually independent uniform numbers in $[0,1]$, which are independent of the dominating jump chain $\mathbf{J}_t$. To obtain (9.29) we reuse these uniform numbers for all pairs of upper and lower processes as follows. For $n = 1, 2, \ldots$, initially, $\mathbf{U}^n_{T_n} = \mathbf{J}_{T_n}$ and $\mathbf{L}^n_{T_n} = \emptyset$. Further, for $t = T_n + 1, \ldots, 0$, if $\mathbf{J}_t = \mathbf{J}_{t-1} \cup \{x\}$, define

$$\max \lambda^n_t(x) = \max\left\{\lambda(x;\mathbf{x}) : \mathbf{L}^n_{t-1} \subseteq \mathbf{x} \subseteq \mathbf{U}^n_{t-1}\right\},$$

$$\min \lambda^n_t(x) = \min\left\{\lambda(x;\mathbf{x}) : \mathbf{L}^n_{t-1} \subseteq \mathbf{x} \subseteq \mathbf{U}^n_{t-1}\right\},$$

$$\mathbf{U}^n_t = \begin{cases} \mathbf{U}^n_{t-1} \cup \{x\} & \text{if } R_t \leq \max \lambda^n_t(x), \\ \mathbf{U}^n_{t-1} & \text{otherwise,} \end{cases}$$

$$\mathbf{L}^n_t = \begin{cases} \mathbf{L}^n_{t-1} \cup \{x\} & \text{if } R_t \leq \min \lambda^n_t(x), \\ \mathbf{L}^n_{t-1} & \text{otherwise.} \end{cases}$$

Furthermore, if $\mathbf{J}_t = \mathbf{J}_{t-1} \setminus \{x\}$, define $\mathbf{U}^n_t = \mathbf{U}_{t-1} \setminus \{x\}$ and $\mathbf{L}^n_t = \mathbf{L}_{t-1} \setminus \{x\}$. Hence by induction we obtain (9.29) if we are reusing the R_i when constructing $\mathbf{I}_t$ as in DCFTP1.

In practice we need some kind of monotonicity so that $\max \lambda^n_t$ and $\min \lambda^n_t$ can be quickly calculated. If $\lambda(x;\mathbf{x})$ is a non-decreasing function of $\mathbf{x}$ (the 'attractive case'), i.e., $\lambda(x;\mathbf{y}) \leq \lambda(x;\mathbf{x})$ whenever $\mathbf{y} \subset \mathbf{x}$, then $\max \lambda^n_t(x) = \lambda\left(x;\mathbf{U}^n_{t-1}\right)$ and $\min \lambda^n_t(x) = \lambda\left(x;\mathbf{L}^n_{t-1}\right)$. If instead $\lambda(x;\mathbf{x})$ is an non-increasing function of $\mathbf{x}$ (the 'repulsive case'), then $\max \lambda^n_t(x) = \lambda\left(x;\mathbf{L}^n_{t-1}\right)$ and $\min \lambda^n_t(x) = \lambda\left(x;\mathbf{U}^n_{t-1}\right)$.

As starting times T_i, we may use a doubling scheme $T_{i+1} = 2T_i$ as proposed in Propp and Wilson (1996). In Berthelsen and Møller (2002*a*) a random doubling scheme is used, where $T_1 = \sup\{t \leq 0 : \mathbf{J}_t \cap \mathbf{J}_0 = \emptyset\}$.

For instance, DCFTP2 has been used in connection to the pairwise interaction process fitting the cell data in Section 9.4.6. Figure 9.8 shows perfect simulations of different Strauss processes as defined in Section 9.4.2, with $\theta_2 = 0$ (the Poisson case), $\theta_2 = -\log 2$, and $\theta_2 = -\infty$ (the hard core case), using the same dominating process (and associated marks R_i) in all three cases, with $K = \exp(\theta_1) = 100$. Due to the dependent thinning procedure in the algorithm, the point pattern with $\theta_2 = 0$ contains the two others, but the point pattern with $\theta_2 = -\infty$ does not contain the point pattern with $\theta_2 = -\log 2$ because the Strauss process is repulsive.

Acknowledgments

Much of this contribution is based on previous work with my collaborators, particularly, Adrian Baddeley, Kasper K. Berthelsen, Janine Illian, Wilfrid Kendall, and last but not least Rasmus P. Waagepetersen. I am grateful to the three referees for their useful comments. Supported by the Danish Natural Science Research Council, grant no. 272-06-0442 ('Point process modelling and statistical inference').

References

Adler, R. (1981). *The Geometry of Random Fields.* Wiley, New York.

Armstrong, P. (1991). Species patterning in the heath vegetation of the northern sandplain. Honours thesis, University of Western Australia.

Baddeley, A., Gregori, P., Mateu, J., Stoica, R., and Stoyan, D. (ed.) (2006). *Case Studies in Spatial Point Process Modeling.* Springer Lecture Notes in Statistics 185, Springer-Verlag, New York.

Baddeley, A. and Møller, J. (1989). Nearest-neighbour Markov point processes and random sets. *International Statistical Review*, **2**, 89–121.

Baddeley, A., Møller, J., and Pakes, A. G. (2008). Properties of residuals for spatial point processes. *Annals of the Institute of Statistical Mathematics*, **60**, 627–649.

Baddeley, A., Møller, J., and Waagepetersen, R. (2000). Non- and semi-parametric estimation of interaction in inhomogeneous point patterns. *Statistica Neerlandica*, **54**, 329–350.

Baddeley, A. and Turner, R. (2000). Practical maximum pseudolikelihood for spatial point patterns. *Australian and New Zealand Journal of Statistics*, **42**, 283–322.

Baddeley, A. and Turner, R. (2005). Spatstat: an R package for analyzing spatial point patterns. *Journal of Statistical Software*, **12**, 1–42. URL: `www.jstatsoft.org`.

Baddeley, A. and Turner, R. (2006). Modelling spatial point patterns in R. In *Case Studies in Spatial Point Process Modeling* (ed. A. Baddeley, P. Gregori, J. Mateu, R. Stoica, and D. Stoyan), pp. 23–74. Springer Lecture Notes in Statistics 185, Springer-Verlag, New York.

Baddeley, A., Turner, R., Møller, J., and Hazelton, M. (2005). Residual analysis for spatial point processes (with discussion). *Journal of Royal Statistical Society Series B*, **67**, 617–666.

Baddeley, A. J. and van Lieshout, M. N. M. (1995). Area-interaction point processes. *Annals of the Institute of Statistical Mathematics*, **46**, 601–619.

Barndorff-Nielsen, O. E. (1978). *Information and Exponential Families in Statistical Theory.* Wiley, Chichester.

Berman, M. and Turner, R. (1992). Approximating point process likelihoods with GLIM. *Applied Statistics*, **41**, 31–38.

Berthelsen, K. K. and Møller, J. (2002*a*). A primer on perfect simulation for spatial point processes. *Bulletin of the Brazilian Mathematical Society*, **33**, 351–367.

Berthelsen, K. K. and Møller, J. (2002*b*). Spatial jump processes and perfect simulation. In *Morphology of Condensed Matter* (ed. K. Mecke and D. Stoyan), pp. 391–417. Lecture Notes in Physics, Vol. 600, Springer-Verlag.

Berthelsen, K. K. and Møller, J. (2003). Likelihood and non-parametric Bayesian MCMC inference for spatial point processes based on perfect simulation and path sampling. *Scandinavian Journal of Statistics*, **30**, 549–564.

Berthelsen, K. K. and Møller, J. (2004). An efficient MCMC method for Bayesian point process models with intractable normalising constants. In *Spatial Point Process Modelling and Its Applications* (ed. A. Baddeley, P. Gregori, J. Mateu, R. Stoica, and D. Stoyan). Publicacions de la Universitat Jaume I.

Berthelsen, K. K. and Møller, J. (2006). Bayesian analysis of Markov point processes. In *Case Studies in Spatial Point Process Modeling* (ed. A. Baddeley, P. Gregori, J. Mateu, R. Stoica, and D. Stoyan), pp. 85–97. Springer Lecture Notes in Statistics 185, Springer-Verlag, New York.

Berthelsen, K. K. and Møller, J. (2008). Non-parametric Bayesian inference for inhomogeneous Markov point processes. *Australian and New Zealand Journal of Statistics*, **50**, 257–272.

Besag, J. (1977*a*). Some methods of statistical analysis for spatial data. *Bulletin of the International Statistical Institute*, **47**, 77–92.

Besag, J. E. (1977*b*). Discussion of the paper by Ripley (1977). *Journal of the Royal Statistical Society Series B*, **39**, 193–195.

Besag, J. E. (1994). Discussion of the paper by Grenander and Miller. *Journal of the Royal Statistical Society Series B*, **56**, 591–592.

Besag, J., Milne, R. K., and Zachary, S. (1982). Point process limits of lattice processes. *Journal of Applied Probability*, **19**, 210–216.

Brix, A. (1999). Generalized gamma measures and shot-noise Cox processes. *Advances in Applied Probability*, **31**, 929–953.

Brix, A. and Kendall, W. S. (2002). Simulation of cluster point processes without edge effects. *Advances in Applied Probability*, **34**, 267–280.

Brix, A. and Møller, J. (2001). Space-time multitype log Gaussian Cox processes with a view to modelling weed data. *Scandinavian Journal of Statistics*, **28**, 471–488.

Coles, P. and Jones, B. (1991). A lognormal model for the cosmological mass distribution. *Monthly Notices of the Royal Astronomical Society*, **248**, 1–13.

Condit, R. (1998). *Tropical Forest Census Plots.* Springer-Verlag and R. G. Landes Company, Berlin, Germany and Georgetown, Texas.

Condit, R., Hubbell, S. P., and Foster, R. B. (1996). Changes in tree species abundance in a neotropical forest: impact of climate change. *Journal of Tropical Ecology*, **12**, 231–256.

Cox, D. R. (1955). Some statistical models related with series of events. *Journal of the Royal Statistical Society Series B*, **17**, 129–164.

Cox, D. R. (1972). The statistical analysis of dependicies in point processes. In *Stochastic Point Processes* (ed. P. A. W. Lewis), pp. 55–66. Wiley, New York.

Cressie, N. A. C. (1993). *Statistics for Spatial Data* (Second edn). Wiley, New York.

Daley, D. J. and Vere-Jones, D. (2003). *An Introduction to the Theory of Point Processes. Volume I: Elementary Theory and Methods* (Second edn). Springer-Verlag, New York.

Diggle, P. J. (1985). A kernel method for smoothing point process data. *Applied Statistics*, **34**, 138–147.

Diggle, P. J. (2003). *Statistical Analysis of Spatial Point Patterns* (second edn). Arnold, London.

Dimakos, X. K. (2001). A guide to exact simulation. *International Statistical Review*, **69**, 27–48.

Fernández, Roberto, Ferrari, Pablo A., and Garcia, Nancy L. (2002). Perfect simulation for interacting point processes, loss networks and Ising models. *Stochastic Processes and Their Applications*, **102**, 63–88.

Fiksel, T. (1984). Estimation of parameterized pair–potentials of marked and nonmarked Gibbsian point processes. *Elektronische Informationsverarbeitung und Kypernetik*, **20**, 270–278.

Fill, J. A. (1998). An interruptible algorithm for perfect sampling via Markov chains. *Annals of Applied Probability*, **8**, 131–162.

Fill, J. A., Machida, M., Murdoch, D. J., and Rosenthal, J. S. (2000). Extensions of Fill's perfect rejection sampling algorithm to general chains. *Random Structures and Algorithms*, **17**, 290–316.

Foss, S. G. and Tweedie, R. L. (1998). Perfect simulation and backward coupling. *Stochastic Models*, **14**, 187–203.

Gelman, A. and Meng, X.-L. (1998). Simulating normalizing constants: from importance sampling to bridge sampling to path sampling. *Statistical Science*, **13**, 163–185.

Gelman, A., Meng, X. L., and Stern, H. S. (1996). Posterior predictive assessment of model fitness via realized discrepancies (with discussion). *Statistica Sinica*, **6**, 733–807.

Georgii, H.-O. (1976). Canonical and grand canonical Gibbs states for continuum systems. *Communications of Mathematical Physics*, **48**, 31–51.

Georgii, H.-O. (1988). *Gibbs Measures and Phase Transition.* Walter de Gruyter, Berlin.

Geyer, C. J. (1999). Likelihood inference for spatial point processes. In *Stochastic Geometry: Likelihood and Computation* (ed. O. E. Barndorff-Nielsen, W. S. Kendall, and M. N. M. van Lieshout), Boca Raton, Florida, pp. 79–140. Chapman & Hall/CRC.

Geyer, C. J. and Møller, J. (1994). Simulation procedures and likelihood inference for spatial point processes. *Scandinavian Journal of Statistics*, **21**, 359–373.

Geyer, C. J. and Thompson, E. A. (1995). Annealing Markov chain Monte Carlo with applications to pedigree analysis. *Journal of the American Statistical Association*, **90**, 909–920.

Goulard, M., Särkkä, A., and Grabarnik, P. (1996). Parameter estimation for marked Gibbs point processes through the maximum pseudo-likelihood method. *Scandinavian Journal of Statistics*, **23**, 365–379.

Grandell, J. (1997). *Mixed Poisson Processes.* Chapman and Hall, London.

Häggström, O., van Lieshout, M. N. M., and Møller, J. (1999). Characterization results and Markov chain Monte Carlo algorithms including exact simulation for some spatial point processes. *Bernoulli*, **5**, 641–659.

Hall, P. (1988). *Introduction to the Theory of Covarage Processes.* Wiley, New York.

Hanisch, K.-H. (1981). On classes of random sets and point processes. *Serdica*, **7**, 160–167.

Heikkinen, J. and Penttinen, A. (1999). Bayesian smoothing in the estimation of the pair–potential function of Gibbs point processes. *Bernoulli*, **5**, 1119–1136.

Heinrich, L. (1992). Minimum contrast estimates for parameters of spatial ergodic point processes. In *Transactions of the 11th Prague Conference on Random Processes, Information Theory and Statistical Decision Functions*, Prague, pp. 479–492. Academic Publishing House.

Hellmund, G., Prokešová, M., and Jensen, E.B.V. (2008). Lévy based cox point processes. *Advances in Applied Probability*, **40**, 603–629.

Hubbell, S. P. and Foster, R. B. (1983). Diversity of canopy trees in a neotropical forest and implications for conservation. In *Tropical Rain Forest: Ecology and Management* (ed. S. L. Sutton, T. C. Whitmore, and A. C. Chadwick), pp. 25–41. Blackwell Scientific Publications.

Huber, M. (2007). Spatial birth-death-swap chains. Submitted for publication.

Illian, J., Penttinen, A., Stoyan, H., and Stoyan, D. (2008). *Statistical Analysis and Modelling of Spatial Point Patterns.* John Wiley and Sons, Chichester.

Illian, J. B., Møller, J., and Waagepetersen, R. P. (2009). Spatial point process analysis for a plant community with high biodiversity. *Environmental and Ecological Statistics.* (To appear).

Jensen, E. B. V. and Nielsen, L. S. (2000). Inhomogeneous Markov point processes by transformation. *Bernoulli*, **6**, 761–782.

Jensen, J. L. (1993). Asymptotic normality of estimates in spatial point processes. *Scandinavian Journal of Statistics*, **20**, 97–109.

Jensen, J. L. and Künsch, H. R. (1994). On asymptotic normality of pseudo likelihood estimates for pairwise interaction processes. *Annals of the Institute of Statistical Mathematics*, **46**, 475–486.

Jensen, J. L. and Møller, J. (1991). Pseudolikelihood for exponential family models of spatial point processes. *Annals of Applied Probability*, **1**, 445–461.

Karr, A. F. (1991). *Point Processes and Their Statistical Inference.* Marcel Dekker, New York.

Kelly, F. P. and Ripley, B. D. (1976). A note on Strauss' model for clustering. *Biometrika*, **63**, 357–360.

Kendall, W. S. (2004). Geometric ergodicity and perfect simulation. *Electronic Communications in Probability*, **9**, 140–151.

Kendall, W. S. and Thönnes, E. (1999). Perfect simulation in stochastic geometry. *Pattern Recognition*, **32**, 1569–1586.

Kendall, W. S., van Lieshout, M. N. M., and Baddeley, A. J. (1999). Quermass-interaction processes: conditions for stability. *Advances in Applied Probability*, **31**, 315–342.

Kendall, W. S. (1998). Perfect simulation for the area-interaction point process. In *Probability Towards 2000* (ed. L. Accardi and C. Heyde), pp. 218–234. Springer Lecture Notes in Statistics 128, Springer Verlag, New York.

Kendall, W. S. and Møller, J. (2000). Perfect simulation using dominating processes on ordered spaces, with application to locally stable point processes. *Advances in Applied Probability*, **32**, 844–865.

Kingman, J. F. C. (1993). *Poisson Processes.* Clarendon Press, Oxford.

Lantuejoul, C. (2002). *Geostatistical Simulation: Models and Algorithms.* Springer-Verlag, Berlin.

Lieshout, M. N. M. van (2000). *Markov Point Processes and Their Applications.* Imperial College Press, London.

Lieshout, M. N. M. van and Baddeley, A. J. (1996). A nonparametric measure of spatial interaction in point patterns. *Statistica Neerlandica*, **50**, 344–361.

Lieshout, M. N. M. van and Stoica, R. S. (2006). Perfect simulation for marked point processes. *Computational Statistics and Data Aanlysis*, **51**, 679–698.

Lund, J. and Thönnes, E. (2004). Perfect simulation for point processes given noisy observations. *Computational Statistics*, **19**, 317–336.

Mase, S. (1991). Asymptotic equivalence of grand canonical MLE and canonical MLE of pair–potential functions of Gibbsian point process models. Technical Report 292, Statistical Research Group, Hiroshima University.

Mase, S. (1995). Consistency of the maximum pseudo-likelihood estimator of continuous state space Gibbs processes. *Annals of Applied Probability*, **5**, 603–612.

Mase, S. (1999). Marked Gibbs processes and asymptotic normality of maximum pseudo-likelihood estimators. *Mathematische Nachrichten*, **209**, 151–169.

Mase, S., Møller, J., Stoyan, D., Waagepetersen, R. P., and Döge, G. (2001). Packing densities and simulated tempering for hard core Gibbs point processes. *Annals of the Institute of Statistical Mathematics*, **53**, 661–680.

Mecke, J. (1967). Stationäre zufällige Maße auf lokalkompakten Abelschen Gruppen. *Zeitschrift für Wahrscheinlichkeitstheorie und verwandte Gebiete*, **9**, 36–58.

Molchanov, I. (1997). *Statistics of the Boolean Model for Practitioners and Mathematicians.* Wiley, Chichester.

Møller, J. (1989). On the rate of convergence of spatial birth-and-death processes. *Annals of the Institute of Statistical Mathematics*, **3**, 565–581.

Møller, J. (2001). A review of perfect simulation in stochastic geometry. In *Selected Proceedings of the Symposium on Inference for Stochastic Processes* (ed. I. V. Basawa, C. C. Heyde, and R. L. Taylor), Volume 37, pp. 333–355. IMS Lecture Notes & Monographs Series, Beachwood, Ohio.

Møller, J. (2003). Shot noise Cox processes. *Advances in Applied Probability*, **35**, 4–26.

Møller, J. and Helisová, K. (2008). Power diagrams and interaction processes for unions of discs. *Advances in Applied Probability*, **40**, 321–347.

Møller, J. and Helisová, K. (2009). Likelihood inference for unions of interacting discs. Technical Report R-2008-18, Department of Mathematical Sciences, Aalborg University. To appear in *Scandinavian Journal of Statistics*.

Møller, J., Pettitt, A. N., Berthelsen, K. K., and Reeves, R. W. (2006). An efficient MCMC method for distributions with intractable normalising constants. *Biometrika*, **93**, 451–458.

Møller, J. and Schladitz, K. (1999). Extensions of Fill's algorithm for perfect simulation. *Journal of the Royal Statistical Society Series B*, **61**, 955–969.

Møller, J., Syversveen, A. R., and Waagepetersen, R. P. (1998). Log Gaussian Cox processes. *Scandinavian Journal of Statistics*, **25**, 451–482.

Møller, J. and Torrisi, G. L. (2005). Generalised shot noise Cox processes. *Advances in Applied Probability*, **37**, 48–74.

Møller, J. and Waagepetersen, R. P. (2004). *Statistical Inference and Simulation for Spatial Point Processes*. Chapman and Hall/CRC, Boca Raton.

Møller, J. and Waagepetersen, R. P. (2007). Modern spatial point process modelling and inference (with discussion). *Scandinavian Journal of Statistics*, **34**, 643–711.

Mugglestone, M. A. and Renshaw, E. (1996). A practical guide to the spectral analysis of spatial point processes. *Computational Statistics and Data Analysis*, **21**, 43–65.

Nguyen, X. X. and Zessin, H. (1979). Integral and differential characterizations of Gibbs processes. *Mathematische Nachrichten*, **88**, 105–115.

Nielsen, L. S. (2000). Modelling the position of cell profiles allowing for both inhomogeneity and interaction. *Image Analysis and Stereology*, **19**, 183–187.

Nielsen, L. S. and Jensen, E. B. V. (2004). Statistical inference for transformation inhomogeneous Markov point processes. *Scandinavian Journal of Statistics*, **31**, 131–142.

Norman, G. E. and Filinov, V. S. (1969). Investigations of phase transition by a Monte-Carlo method. *High Temperature*, **7**, 216–222.

Ogata, Y. and Tanemura, M. (1984). Likelihood analysis of spatial point patterns. *Journal of the Royal Statistical Society Series B*, **46**, 496–518.

Ogata, Y. and Tanemura, M. (1989). Likelihood estimation of soft-core interaction potentials for Gibbsian point patterns. *Annals of the Institute of Statistical Mathematics*, **41**, 583–600.

Papangelou, F. (1974). The conditional intensity of general point processes and an application to line processes. *Zeitschrift für Wahscheinlichkeitstheorie und werwandte Gebiete*, **28**, 207–226.

Penttinen, A. (1984). *Modelling Interaction in Spatial Point Patterns: Parameter Estimation by the Maximum Likelihood Method.* Number 7 in Jyväskylä

Studies in Computer Science, Economics, and Statistics, Univeristy of Jyväskylä.

Penttinen, A., Stoyan, D., and Henttonen, H. M. (1992). Marked point processes in forest statistics. *Forest Science*, **38**, 806–824.

Preston, C. (1976). *Random Fields.* Lecture Notes in Mathematics 534. Springer-Verlag, Berlin.

Preston, C. J. (1977). Spatial birth-and-death processes. *Bulletin of the International Statistical Institute*, **46**, 371–391.

Propp, J. G. and Wilson, D. B. (1996). Exact sampling with coupled Markov chains and applications to statistical mechanics. *Random Structures and Algorithms*, **9**, 223–252.

Rathbun, S. L. (1996). Estimation of Poisson intensity using partially observed concomitant variables. *Biometrics*, **52**, 226–242.

Rathbun, S. L. and Cressie, N. (1994). Asymptotic properties of estimators for the parameters of spatial inhomogeneous Poisson processes. *Advances in Applied Probability*, **26**, 122–154.

Rathbun, S. L., Shiffman, S., and Gwaltney, C. J. (2007). Modelling the effects of partially observed covariates on Poisson process intensity. *Biometrika*, **94**, 153–165.

Ripley, B. D. (1976). The second-order analysis of stationary point processes. *Journal of Applied Probability*, **13**, 255–266.

Ripley, B. D. (1977). Modelling spatial patterns (with discussion). *Journal of the Royal Statistical Society Series B*, **39**, 172–212.

Ripley, B. D. (1979). Simulating spatial patterns: dependent samples from a multivariate density. Algorithm AS 137. *Applied Statistics*, **28**, 109–112.

Ripley, B. D. (1981). *Spatial Statistics.* Wiley, New York.

Ripley, B. D. (1988). *Statistical Inference for Spatial Processes.* Cambridge University Press, Cambridge.

Ripley, B. D. and Kelly, F. P. (1977). Markov point processes. *Journal of the London Mathematical Society*, **15**, 188–192.

Robert, C. P. and Casella, G. (1999). *Monte Carlo Statistical Methods.* Springer-Verlag, New York.

Roberts, G. O. and Tweedie, R. L. (1996). Exponential convergence of Langevin diffusions and their discrete approximations. *Bernoulli*, **2**, 341–363.

Rue, H. and Martino, S. (2005). Approximate inference for hierarchical Gaussian Markov random fields models. Statistics Preprint 7/2005, Norwegian University of Science and Technology.

Ruelle, D. (1969). *Statistical Mechanics: Rigorous Results.* W.A. Benjamin, Reading, Massachusetts.

Ruelle, D. (1971). Existence of a phase transition in a continuous classsical system. *Physical Review Letters*, **27**, 1040–1041.

Schladitz, K. and Baddeley, A. J. (2000). A third-order point process characteristic. *Scandinavian Journal of Statistics*, **27**, 657–671.

Schlater, M. (1999). Introduction to positive definite functions and unconditional simulation of random fields. Technical Report ST 99-10, Lancaster University.

Stoyan, D., Kendall, W. S., and Mecke, J. (1995). *Stochastic Geometry and Its Applications* (Second edn). Wiley, Chichester.

Stoyan, D. and Stoyan, H. (1995). *Fractals, Random Shapes and Point Fields.* Wiley, Chichester.

Stoyan, D. and Stoyan, H. (2000). Improving ratio estimators of second order point process characteristics. *Scandinavian Journal of Statistics*, **27**, 641–656.

Strauss, D. J. (1975). A model for clustering. *Biometrika*, **63**, 467–475.

Thomas, M. (1949). A generalization of Poisson's binomial limit for use in ecology. *Biometrika*, **36**, 18–25.

Thönnes, E. (1999). Perfect simulation of some point processes for the impatient user. *Advances in Applied Probability*, **31**, 69–87.

Thönnes, E. (2000). A primer on perfect simulation. In *Statistical Physics and Spatial Statistics* (ed. K. R. Mecke and D. Stoyan), pp. 349–378. Lecture Notes in Physics, Springer, Berlin.

Waagepetersen, R. (2005). Discussion of the paper by Baddeley, Turner, Møller & Hazelton (2005). *Journal of the Royal Statistical Society Series B*, **67**, 662.

Waagepetersen, R. (2007). An estimating function approach to inference for inhomogeneous Neyman-Scott processes. *Biometrics*, **63**, 252–258.

Waagepetersen, R. (2008). Estimating functions for inhomogeneous spatial point processes with incomplete covariate data. *Biometrika*, **95**. To appear.

Waagepetersen, R. and Guan, Y. (2009). Two-step estimation for inhomogeneous spatial point processes. *Journal of the Royal Statistical Society, Series B*, **71**, 685–702.

Widom, B. and Rowlinson, J. S. (1970). A new model for the study of liquid-vapor phase transitions. *Journal of Chemical Physics*, **52**, 1670–1684.

Wilson, D. B. (2000). Layered multishift coupling for use in perfect sampling algorithms (with a primer to CFTP). In *Monte Carlo Methods* (ed. N. Madras), Volume 26, pp. 141–176. Fields Institute Communications Series, American Mathematical Society, Providence.

Wolpert, R. L. and Ickstadt, K. (1998). Poisson/gamma random field models for spatial statistics. *Biometrika*, **85**, 251–267.

10

STATISTICAL SHAPE THEORY

Wilfrid S. Kendall and Huiling Le

Abstract

There are many variations on what one may regard as statistical shape, depending on the application in mind. The focus of this chapter is the statistical analysis of the shapes determined by finite sequences of points in a Euclidean space. We shall draw together a range of ideas from statistical shape theory, including distributions, diffusions, estimations and computations, emphasizing the role played by the underlying geometry. Applications in selected areas of current interest will be discussed.

10.1 Motivations

One can trace the early origins of statistical shape theory back to the work of D'Arcy Thompson on growth and form (Thompson, 1917), who studied biological shape using a variety of empirical and quantitative methods. However shape theory in its modern form as a branch of geometrical statistics derives from two independent sources: (a) the work of D.G. Kendall, motivated by the desire to do statistics on 'what is left when the effects of location, size, and rotation have been filtered out' (e.g. D.G. Kendall 1984*b*); and (b) the work of F.L. Bookstein, motivated by biological and medical issues of morphology, and representing shapes and shape changes by deformations (e.g. Bookstein 1986). In particular, D.G. Kendall's approach diverges sharply from the more traditional concerns of classical statistics, which has largely concentrated on inference for location, scale, and orientation (precisely what is filtered out in the D.G. Kendall approach).

The seminal application for the D.G. Kendall approach concerns the Land's End dataset described in Broadbent (1980), recording locations of 52 standing stones in the extreme South West of England. The driving question was to devise means of assessing whether or not this dataset exhibited an excessive amount of collinear sub-structure (whether 'the layout of some of the sites reflects the action of factors favouring collinearities'; Kendall and Kendall 1980). One possible measure of collinear sub-structure is the number of triads of data points which are nearly aligned. This leads to considerations of the *shape* of triads of planar points, geometrically natural ways of representing such shapes, and associated questions of probability.

Typical applications for the Bookstein approach concern datasets generated by sagittal sections of skulls. Sections are converted into sequences of points by identifying biologically identified *landmarks*. Interest lies in what can be said about shapes of landmark sequences derived from different datasets.

From a mathematical perspective, in either case two landmark configurations $\underline{x} = (x_1, \ldots, x_k)$ and $\underline{y} = (y_1, \ldots, y_k)$, viewed as sequences of points, are said to have the same *shape* if one configuration can be derived from the other by a geometric similarity transform (a combination of translation, rotation, and dilation of scale). Differences between the approaches of D.G. Kendall and F.L. Bookstein derive from different ways of measuring dissimilarities between different shapes.

It is immediately apparent that there are many possible variations on the notion of shape. One may change the underlying symmetry group: thus replacement of the similarity group by the rigid motion group (dilations no longer allowed) results in the notion of 'size-and-shape' or 'form' (to use the succinct terminology of Stoyan and Stoyan 1994); while replacement by the affine group (for example, by allowing dilations to use different scales in different orthogonal directions) results in 'affine shape'. One may consider shapes deriving from points on higher-dimensional Euclidean spaces or even on spheres (the latter being a natural choice in shape problems arising from cosmology). For the sake of simplicity we will concentrate on shape as an equivalence class under similarities, for configurations lying in Euclidean space.

Before discussing shape theory in detail we first sketch an indicative approach to shape based on a Bayesian perspective, thus introducing shape concepts within an explicit statistical framework. Suppose one is provided with data in the form of n configurations of $k > 1$ landmarks in d-dimensional space $\mathbb{R}^d$. Suppose further that these are modelled statistically by

$$y_{ij} = g_i\, z_j + \varepsilon_{ij} \tag{10.1}$$

for $i = 1, \ldots, n$, $j = 1, \ldots, k$. Here $\underline{z} = (z_1, \ldots, z_k)$ is a configuration of idealized landmarks in $\mathbb{R}^d$, and $g_1, \ldots, g_n$ are members of Sim^d, the group of similarity transforms of $\mathbb{R}^d$ (a description of Sim^d as a matrix group can be found below). The ε_{ij} are random errors of known distribution; in the first instance we suppose them to be independent and identically distributed with common probability density $f(\cdot)$. For convenience we write $g\,\underline{z} = (g\,z_1, \ldots, g\,z_k)$; if the action $g\,z$ of the symmetry group can be expressed in matrix form then we can view $\underline{z}$ as a matrix constructed using the z_i as columns.

The parameters here are the nuisance parameters g_i and the landmark configuration $\underline{z}$. The idealized landmark configuration is not fully identifiable, so we use a Bayesian approach. We suppose the similarities $g_1, \ldots, g_n$ and the landmark configuration $\underline{z}$ to be independent, and stipulate that $\pi(\mathrm{d}\,\underline{z})$ is the prior probability measure for $\underline{z}$ while the g_i share the same diffuse prior. We suppose further that it is appropriate to replace the diffuse prior by the right-invariant Haar measure for Sim^d (details about left- and right-invariant Haar measures can be found in Loomis 1953, Chapter VI).

Berger (1985, Sections 3.3.2, 6.6.2) discusses the use of invariant priors in Bayesian statistics, and suggests an attractive structural justification for the use of the right-invariant Haar measure. We take a pragmatic view: from the point of view of shape theory we are interested in when we can draw shape-theoretic inferences without saying very much at all about the particular similarity transforms being employed (corresponding to D.G. Kendall's notion of filtering out the effects of location, rotation, and size). It is convenient to model this by using right-invariant Haar measure to approximate the prior for the similarity.

Here are some brief details on Sim^d and its Haar measure. The group Sim^d itself can be expressed as a semi-direct product of the *group of dilations* of multiplicative changes of scale $\{s : s > 0\}$ and the group Motion^d of rigid motions; Motion^d, denoted by G_d in Chapter 1, is a semi-direct product of the rotational group for $\mathbb{R}^d$ and the translation group for $\mathbb{R}^d$, with typical element (R, x). Motion^d and Sim^d can be expressed as matrix groups: if $\mathbb{R}^d$ is embedded into $\mathbb{R}^{d+1}$ using $y \mapsto \left(\begin{smallmatrix} y \\ 1 \end{smallmatrix}\right)$, then the general element of Motion^d can be viewed as a $(d+1) \times (d+1)$ partitioned matrix $\left(\begin{smallmatrix} R & x \\ 0 & 1 \end{smallmatrix}\right)$ where R is a $d \times d$ orthogonal matrix and the shift x is a d-vector, while the general element of Sim^d is the same except that the orthogonal matrix R is replaced by sR, where s is a positive real number.

Following Loomis (1953, Chapter VI, Section 30D), alternatively Schneider and Weil (2008, Theorem 13.2.10), Motion^d can be shown to be a unimodular group whose Haar measure $d\mu$ is the product of Lebesgue measure on the translation group and invariant Haar measure on the rotation group. (Recall, *unimodular* means Haar measure is both left- and right-invariant.) The right-invariant Haar measure $d\nu_R$ on Sim^d is the product of the measure ds/s on the group of dilations and the Haar measure $d\mu$ on Motion^d. The left-invariant Haar measure is given by $d\nu_L = d\nu_R/s^d$, since a scale-change by s changes the measure on Motion^d by a factor s^d. The presence of a positive power of s in the denominator for the expressions of both left-invariant $d\nu_L$ and right-invariant $d\nu_R$ means that both measures concentrate infinite mass near $s = 0$ even when the shift is bounded; their use as approximations to diffuse priors is legitimized only if the marginal posterior for the shape configuration $\underline{z}$ can be normalized to have finite total mass. If this is not the case then (speaking informally) posteriors resulting from diffuse priors are substantially affected by details of exactly what diffuse prior is being considered; in our context Bayesian inference about shape cannot then be separated from inference about location, orientation, and particularly size.

The improper posterior measure for $g_1, \ldots, g_n$ and $\underline{z}$ can be written formally (up to a constant of proportionality) as

$$\left(\prod_{i=1}^{n} \left(\prod_{j=1}^{k} f(y_{ij} - g_i\, z_j) \right) \nu_R(\mathrm{d}g_i) \right) \pi(\mathrm{d}\underline{z}) \,. \tag{10.2}$$

Integrating out $g_1, \ldots, g_n$ can lead to a proper posterior distribution for non-degenerate configurations $\underline{z}$ only if

$$\int_{\mathrm{Sim}^d} \left(\prod_{j=1}^{k} f(y_{ij} - g\, z_j) \right) \nu_R(\mathrm{d}g) < \infty \tag{10.3}$$

for each i. This criterion will fail if, for example, $\pi(\mathrm{d}\underline{z})$ and the error density f are continuous and positive everywhere, since ν_R will concentrate infinite mass in regions of nearly degenerate similarities g (those for example for which $g\, z_j$ is nearly a fixed vector c for all j): it will succeed if f has support in a ball of some fixed radius τ and the configuration data y_{ij} are such that for some $\varepsilon > 0$

$$\bigcap_{j=1}^{k} B_{\tau+\varepsilon}(y_{ij}) = \emptyset \qquad \text{for each } i = 1, \ldots, n\,. \tag{10.4}$$

(Here $B_r(x)$ refers to the ball of radius r centred at x.) Indeed in that case it can be shown that the integral is locally bounded as a function of non-degenerate $\underline{z}$ (not all z_j the same) and the marginalized posterior density $p(\underline{z}|y_{ij} : i = 1, \ldots, n; j = 1, \ldots, k)$ for $\underline{z}$ satisfies

$$\begin{aligned} & p(\underline{z}|y_{ij} : i = 1, \ldots, n;\ j = 1, \ldots, k)\pi(\mathrm{d}\underline{z}) \\ & \quad = \left(\prod_{i=1}^{n} \int_{\mathrm{Sim}^d} \left(\prod_{j=1}^{k} f(y_{ij} - g_i\, z_j) \right) \nu_R(\mathrm{d}g_i) \right) \pi(\mathrm{d}\underline{z})\,. \end{aligned}$$

Under mild conditions on π the posterior is then proper. Moreover the right-invariance of ν_R then means that the posterior density can be taken to be left-invariant: for any similarity $g \in \mathrm{Sim}^d$ we have

$$p(\underline{z}|y_{ij} : i = 1, \ldots, n;\ j = 1, \ldots, k) = p(g\, \underline{z}|y_{ij} : i = 1, \ldots, n;\ j = 1, \ldots, k). \tag{10.5}$$

Thus under the data constraint (10.4) we are led to view Bayesian inference of shape as being carried out not on individual configurations $\underline{z}$ but on entire similarity orbits $\mathrm{Sim}^d\, \underline{z} = \{g\, \underline{z} : g \in \mathrm{Sim}^d\}$.

The Bayesian approach outlined here does not seem satisfactory as a realistic inferential procedure, since the data constraint (10.4) depends on bounded support of the error distribution $f(\cdot)$. However it does suggest pertinent considerations for the general theory of statistical shape:

1. The data constraint (10.4) and the equivariance condition (10.5) focus attention on the *shape spaces*, made up of equivalence classes of non-degenerate configurations. Thus the Bayesian approach provides a clear motivation for excluding degenerate configurations from consideration

(in the original development of the theory this exclusion arose from the intuition that, were one to define the shape of a degenerate triad, then the shape would have to be viewed as infinitesimally close to all other shapes).

2. It is apparent that this analysis can be carried out for other symmetry groups, such as the group of affine transformations (leading to the notion of *affine shape*). In this case the natural data constraint varies from that of (10.4): using the notation of that constraint, the criterion (10.3) need not be satisfied if for some $i = 1, \ldots, n$ we find each of $y_{i1}, \ldots, y_{ik}$ is within $\tau + \varepsilon$ of the corresponding $\widetilde{y}_1, \ldots, \widetilde{y}_k$, for a configuration $(\widetilde{y}_1, \ldots, \widetilde{y}_k)$ which is degenerate in the sense that it lies on a hyperplane of $\mathbb{R}^d$.
3. No data constraint is required for the case when the underlying symmetry group is the rigid motion group Motion^d. In this case we obtain *size-and-shape space*, which may be expressed as a topological cone, namely the cartesian product of shape space Σ_d^k and size space $(0, \infty)$, with a single limiting point (representing all degenerate configurations $(z, \ldots, z)$) attached at the end corresponding to size 0.
4. The above Bayesian approach raises an interesting question: is there a natural error structure for the ε_{ij} which would lead to proper posteriors for the idealized configuration $\underline{z}$? Varying the model (10.1) to

$$y_{ij} = g_i\,(z_j + \varepsilon_{ij})$$

improves matters somewhat. In order to exploit symmetry we now need to use ν_L as improper prior for g_i, rather than ν_R; if $s(g)$ is the dilation parameter for the similarity transformation g then the criterion corresponding to (10.3) is

$$\begin{aligned}\int_{\text{Sim}^d} \left(\prod_{j=1}^{k} f(g^{-1}\, y_{ij} - z_j) \right) \frac{\nu_L(\mathrm{d}g)}{s(g)^{kd}} \\ = \int_{\text{Sim}^d} \left(\prod_{j=1}^{k} f(g\, y_{ij} - z_j) \right) s(g)^{kd}\, \nu_R(\mathrm{d}g) < \infty\,. \qquad (10.6)\end{aligned}$$

This criterion is less restrictive than (10.3). However we pay a price in that similarity invariance now pertains to the data rather than the parameter of interest; moreover the error structure $g_i\, \varepsilon_{ij}$ now depends on the unobserved scale component of the similarity g_i.

10.2 Concepts

Turning from the above Bayesian considerations, the classical approaches to statistical shape theory have led to a fruitful theory which is based on more empirical considerations and on development of appropriate geometries and distributions for the Euclidean shape space Σ_d^k (also called 'D.G. Kendall shape space'). We begin our discussion by making a formal definition of the Euclidean shape space Σ_d^k of k labelled points in d dimensions:

Definition 10.1 (Euclidean shape space) *The space Σ_d^k of shapes of k labelled points in $\mathbb{R}^d$ is defined to be the family of equivalence classes of configurations $\underline{z} = (z_1, \ldots, z_k)$ under the similarity group Sim^d generated by translations, rotations, and dilations, excluding the totally degenerate configurations $z_1 = \ldots = z_k$ for which all landmarks z_j are equal.*

It is useful to add to our discussion the notions of 'pre-shape' space (symmetry group is composed of translations and scaling) and size-and-shape space (symmetry group is the rigid motion group):

Definition 10.2 (Euclidean size-and-shape space) *The space $\mathrm{S}\Sigma_d^k$ of size-and-shapes of k labelled points in $\mathbb{R}^d$ is defined to be the family of equivalence classes of configurations $\underline{z} = (z_1, \ldots, z_k)$ under the group $Motion^d$ of rigid motions (generated by translations and rotations).*

Definition 10.3 (Euclidean pre-shape space) *The space $\mathcal{S}_d^k$ of pre-shapes of k labelled points in $\mathbb{R}^d$ is defined to be the family of equivalence classes of configurations $\underline{z} = (z_1, \ldots, z_k)$ under the semi-direct product group $(0, \infty) \ltimes \mathbb{R}^d$ generated by dilations and translations, excluding the totally degenerate configurations $z_1 = \ldots = z_k$ for which all landmarks z_j are equal.*

Note that there is no need to exclude degenerate configurations in Definition 10.2, as these simply lead to the unique trivial (zero-size) size-and-shape which we denote by o. Note also that the symmetry groups here do *not* include reflections.

There are natural maps which are helpful in describing the geometric features of Σ_d^k:

1. A map from configuration space to size-and-shape space (this factors out translations and rotations):
$$\left(\mathbb{R}^d\right)^k \to \mathrm{S}\Sigma_d^k .$$
2. A map from the space of non-trivial size-and-shapes to shape space (this factors out dilations):
$$\mathrm{S}\Sigma_d^k \setminus \{\mathrm{o}\} \to \Sigma_d^k .$$
3. A map from the space of non-trivial configurations to pre-shape space (this factors out translations and dilations):
$$\left(\mathbb{R}^d\right)^k \setminus \left\{(z, \ldots, z) \ : \ z \in \mathbb{R}^d\right\} \to \mathcal{S}_d^k .$$
4. A map from pre-shape space to shape space (this factors out rotations):
$$\mathcal{S}_d^k \to \Sigma_d^k .$$

10.3 Geometry of shape

A crucial issue in statistical shape theory is to decide on an appropriate geometry for Σ_d^k. One general approach is described in D.G. Kendall, Barden, Carne and Le ((1999), Section 6.4). This begins by representing the space $\mathbb{R}^{kd} = \mathbb{R}^d \times \ldots \times \mathbb{R}^d$ of configurations of k points in $\mathbb{R}^d$ as a space of $d \times k$ matrices furnished with a Euclidean metric based on the inner product $\langle \underline{x}, \underline{y} \rangle = \sum_i \sum_j x_{ij} y_{ij}$. The map from non-trivial configurations in $\mathbb{R}^{kd}$ to $\mathcal{S}_d^k$ can be viewed as a composition, first applying the orthogonal projection onto the subspace of configurations with centroid $(0, \ldots, 0)$, choosing new orthonormal coordinates to represent centered configurations X as belonging to a space $\mathbb{R}^{(k-1)d}$, then by normalizing each X by dividing through by the *Euclidean size*, the square of which is defined to be the sum of squared distances of vertices to the centroid of the configuration

$$\|X\|^2 = \operatorname{trace}\left(X^\top X\right) = \sqrt{\sum_i \sum_j X_{ij}^2} . \tag{10.7}$$

We denote the resulting *pre-shape* by $\underline{X}$; we will use underlined capital symbols to denote pre-shapes so as to distinguish them from general configurations $\underline{x}$.

Note that we have at this point chosen to use Euclidean size rather than some other measure of size: a different choice is made in the later discussion of simplex shape space and leads to different geometries.

Thus the space $\mathcal{S}_d^k$ of pre-shapes can be viewed as a unit sphere in a Euclidean space of dimension $d \times (k-1)$. This provides $\mathcal{S}_d^k$ with a metric as a hypersphere in $(\mathbb{R}^d)^{k-1}$, for example the chordal distance:

$$\widetilde{\operatorname{dist}}_{\mathcal{S}_d^k}(\underline{X}, \underline{Y}) = \|\underline{X} - \underline{Y}\| . \tag{10.8}$$

One can then define the distance between two shapes σ_1 and σ_2 in Σ_d^k as the smallest distance between two representing pre-shapes $\underline{X}$ and $\underline{Y}$. The resulting distance corresponds to a procrustean distance (Mardia, Kent and Bibby, 1979, Section 14.7) between the two configurations, if one bears in mind that the symmetry group used in this procrustes analysis does not contain reflections:

$$\widetilde{\operatorname{dist}}_{\Sigma_d^k}(\sigma_1, \sigma_2) = \min_{U \in SO_d} \|\underline{X} - U\,\underline{Y}\| . \tag{10.9}$$

It is convenient to transform this distance using arc-cosine to produce the distance associated with a Riemannian metric on Σ_d^k: bearing in mind that $\|\underline{X}\| = \|\underline{Y}\| = 1$, we obtain

$$\begin{aligned} \operatorname{dist}_{\Sigma_d^k}(\sigma_1, \sigma_2) &= \arccos \max_{U \in SO_d} \langle U\,\underline{Y}, \underline{X} \rangle \\ &= \arccos \max_{U \in SO_d} \operatorname{trace}\left(U \underline{Y}\,\underline{X}^\top\right). \end{aligned} \tag{10.10}$$

(We recommend Gallot, Hulin and Lafontaine, (1990), for those who need to revise this and further concepts from differential geometry.)

The matrix approach to $\mathcal{S}_d^k$ suggests a matrix parametrization of shape space Σ_d^k in case $k > d$. Introduce a 'pseudo-singular-values (PSV) decomposition' of a pre-shape $\underline{X}$:

$$\underline{X} = U(\Lambda, 0)V \tag{10.11}$$

where $U \in SO_d$, $V \in SO_{k-1}$, Λ is a $d\times d$ diagonal matrix, and 0 is a $d\times(k-1-d)$ matrix full of zero elements. We may remove some redundancy by supposing the diagonal entries of Λ satisfy

$$\lambda_1 \geqslant \lambda_2 \geqslant \ldots \geqslant \lambda_{d-1} \geqslant |\lambda_d| \geqslant 0\,,$$

so that all entries except perhaps the last are non-negative. The requirement of unit Euclidean shape translates into the requirement

$$\sum_i \lambda_i^2 = 1\,.$$

The shape σ representing $\underline{X}$ is the equivalence class $\{U\underline{X} : U \in SO_d\}$, so σ can be represented by the pair (Λ, V), in a generalization of polar coordinates for the sphere. There is more than one representing pair precisely when not all the λ_i are distinct.

The analogy of polar coordinates for the sphere turns out to be exact when considering Σ_2^3. In fact in this planar case D.G. Kendall (1984*a*) observed that considerable simplification is possible, since a general rotation of planar configurations can then be expressed using complex multiplication by unit modulus complex numbers, and a general transformation which rotates and changes scale can then be expressed using multiplication by non-zero complex numbers $\zeta \in \mathbb{C}^* = \{\zeta \in \mathbb{C} : \zeta \neq 0\}$. This means that in the planar case we can deal with just two maps:

1. The orthogonal projection to centred configurations:
$$(\mathbb{C})^k \to \left\{(z_1, \ldots, z_k) \in (\mathbb{C})^k \;:\; z_1 + \ldots + z_k = 0\right\} \cong (\mathbb{C})^{k-1}.$$
2. The map from the space of non-trivial centred configurations to shape space expressed as complex projective space:
$$(\mathbb{C})^{k-1} \setminus \{(0, \ldots, 0)\} \to \Sigma_2^k \cong \mathbb{C}P^{k-2}\,,$$
where $\mathbb{C}P^{k-2}$ is the space of equivalence classes $\{(\zeta z_1, \ldots, \zeta z_k) : \zeta \in \mathbb{C}^*\}$ under multiplication by non-zero complex numbers.

The special case of triads of labelled planar points Σ_2^3 is fundamental for several applications and is also particularly amenable to elegant treatment; the famous Hopf submersion

$$\mathbb{C}P^1 \cong \mathbb{S}^2(\tfrac{1}{2})$$

allows the representation of Σ_2^3 as a 2-sphere of radius $\frac{1}{2}$. (Note that this representation respects the natural procrustean metric structure of Σ_2^3, as does the more general representation of Σ_2^k as $\mathbb{C}P^{k-2}$.) This beautiful fact is exploited in D.G. Kendall (1984*a*) to construct the elegant data-analytic device known as the 'spherical blackboard', erasing the distinctions of reflection and labelling in order to represent shapes of triads as points on an equi-areal projection of a semi-lune of $\Sigma_2^3 \cong \mathbb{S}^2(\frac{1}{2})$.

Similar simplifications arise in the one-dimensional case: the shape spaces Σ_1^k can be viewed as standard spheres. Shape spaces Σ_d^k in higher dimension $d > 2$ present many challenges both mathematically and in visualization. If $k \leqslant d$ then the shape spaces acquire boundaries; otherwise for $d > 2$ the shape space has singularities and is not even a smooth manifold. Much detail on this can be found in D.G. Kendall *et al.* (1999), which includes a summary of a worked example on how to visualize Σ_3^4 (see also D.G. Kendall 1994*a*, 1994*b*, 1995).

The D.G. Kendall setting for shape is appropriate when the landmarks z_i of the configuration $(z_1, \ldots, z_k)$ are only weakly related; landmarks in biological applications delineate outlines which do not self-intersect and this stronger relationship already hints at the need for a different approach. F.L. Bookstein developed a different model for triangle shapes (Bookstein, 1986), based on measuring the difference between the shapes of two triangles by a multiple of the logarithm of the strain ratio of the affine transformation distorting one triangle into the other. This results in a representation of the space of shapes for triangles as the Poincaré half plane, a space of constant negative curvature, with the collinear triangles situated on the infinite horizon formed by the x-axis. Under this model, distortion of a non-collinear triangle into a collinear one is an extreme case and corresponds to moving the corresponding shape point out to the half-plane boundary.

Fix a reference non-degenerate triangular configuration $\underline{x} = (x_1, x_2, x_3)$. The distortion of this by a 2×2 matrix transformation g of positive determinant yields a new non-degenerate triangle $(g\,x_1, g\,x_2, g\,x_3)$, and by this means all possible non-degenerate triangle shapes may be obtained.

Consider geodesic polar coordinates (r, θ) of Bookstein's hyperbolic shape space, centred at the shape of $\underline{x}$. The coordinates (r, θ) for the shape of $(g\,x_1, g\,x_2, g\,x_3)$ can be related to the matrix g by

$$\sqrt{2}\,r = \log(\lambda_1/\lambda_2)\,; \quad \frac{1}{2}\sin\theta = \frac{\zeta}{\lambda_1/\lambda_2 - \lambda_2/\lambda_1},$$

where $\lambda_1 \geqslant \lambda_2$ are the non-negative square roots of the eigenvalues of $g^\top g$ and ζ is the non-diagonal element of $\det(g)^{-1} g^\top g$ (Kendall, 1998).

As noted by D.G. Kendall in the discussion of Bookstein (1986), there is a holomorphic correspondence (that is, complex-analytic and therefore conformal) between the structures of Bookstein's hyperbolic shape space and the upper hemisphere of Euclidean shape space Σ_2^3. If $(\rho(\sigma), \phi(\sigma))$ are geodesic polar coordinates for the shape σ of $\underline{x}$ viewed as a Euclidean shape and $(r(\sigma), \theta(\sigma))$ are geodesic polar coordinates when it is viewed as a Bookstein shape, with both coordinate systems centred at the shape σ_0 of equilateral triangles the pre-shape matrices of which have positive determinant, then the holomorphic diffeomorphism between the two systems of coordinates is given as follows (Kendall, 1998)

$$\phi = \theta + \text{constant}; \quad \sin(2\rho) = \tanh(\sqrt{2}\, r),$$

the second of which follows from

$$\cos\rho(\sigma) = \frac{\lambda_1 + \lambda_2}{\sqrt{2}\sqrt{\lambda_1^2 + \lambda_2^2}},$$

noting that $\rho(\sigma) = \operatorname{dist}_{\Sigma_2^3}(\sigma_0, \sigma)$.

Bookstein's triangle model is not able to treat $k > 3 = d + 1$ points in the plane $\mathbb{R}^2$; Bookstein (1989) deals with this by introducing a further analysis based on thin-plate splines which we will not discuss here. It *is* possible to generalize Bookstein's triangle model to non-degenerate (labelled) simplexes of $d + 1 > 3$ points in $\mathbb{R}^d$, and this is carried out by Small (1996) and Le and Small (1999). Here the shape of a simplex is its equivalence class under translations, rotations, dilations and reflections, where dilation is used to standardize the simplex volume, rather than the square root of the sum of the squared distances of vertices to the centroid of the simplex. The invariant distance between simplex shapes must then be a function of the principal strains of the affine transformation mapping one simplex to the other.

Representing a simplex in $\mathbb{R}^d$ by a $d \times (d+1)$ matrix formed by arranging vertex coordinates in columns, the simplex may be identified up to translation with an $d \times d$ non-singular matrix whose determinant is proportional to the simplex volume. Measuring size by simplex volume, we may identify the quotient space of non-degenerate simplexes in $\mathbb{R}^d$ modulo translations, reflections and dilations as

$$SL_d = \{X \,:\, X \text{ is a } d \times d \text{ matrix with } \det(X) = 1\}. \tag{10.12}$$

Accordingly the space of simplex shapes in $\mathbb{R}^d$ can be viewed as SL_d/SO_d (the quotient of SL_d by the left action of SO_d) and invariant metrics on the simplex shape space are then equivalent to right-SL_d-invariant metrics. Small's distance arises from requiring that the invariant metric be Riemannian, fixing it up to a scale factor. Moreover the resulting shape space for simplexes is an irreducible globally symmetric space of non-compact type (Le and Barden, 2001). The structure of such spaces is well understood by geometers. In particular, the symmetric space property means that for any point in the space there is an involution that reverses the geodesics through that point.

Using polar decomposition of matrices, it follows that simplex shape space can be coordinatized as

$$\begin{aligned}\mathbf{P}(d) &= \{Y \in SL_d \,:\, Y^\top = Y \text{ and } Y \text{ is positive definite}\}\\ &= \{Y = \sqrt{X^\top X} \,:\, X \in SL_d\},\end{aligned}$$

where the shape for regular simplexes is represented by the identity matrix I. Thus if $Y \in \mathbf{P}(d)$, the square of Small's distance between the simplex shapes corresponding to I and Y is proportional to

$$\sum_{i=1}^{d} (\log \lambda_i)^2 ,$$

where the λ_i are the eigenvalues of Y (noting the normalization $\lambda_1 \times \ldots \times \lambda_d = \det(Y) = 1$). When $d = 2$, and allowing for this normalization, this metric is indeed proportional to Bookstein's metric for triangle shapes. The SL_d-invariance of the distance implies that Small's distance between two simplex shapes of X_1 and X_2 with $X_i \in SL_d$ is equal to the distance between the simplex shapes of I and $X_1 X_2^{-1}$.

Finally, matrix calculus identifies the tangent space to $\mathbf{P}(d)$ and hence expressions for geodesic coordinate systems: the tangent space at I is formed by the vector space of traceless symmetric $d \times d$ matrices, and if V is such a matrix then the Riemannian exponential map $\text{Exp}_I(V)$ is simply the matrix exponential $\exp(V)$. In particular the geodesic on the simplex shape space starting from the shape corresponding to I and with initial tangent vector V is given by $\text{Exp}_I(tV)$ for $t \geqslant 0$.

Differences between the contexts for shape so far described arise from the motivating applications; essentially the difference comes from whether the landmarks are viewed as subject to individual perturbations (in which case the D.G. Kendall theory is relevant) or whether they are considered to be affected by more global distortions. In extreme contrast, applications from pattern recognition provide a motivation for treatments which seek to avoid any use of landmarks at all. The extensive literature deserves much more space than can be afforded here; we content ourselves with a brief description of just one example, namely *Mumford's shape space.* This is the manifold of simple closed smooth unparameterized curves in $\mathbb{R}^2$ (Michor and Mumford 2006, 2007). In this case the primary issue is not a question of factoring out similarities but of deparametrization: the configuration of interest is effectively the compact simply connected region in the plane bounded by the simple closed curve. The ensemble of simple closed smooth parametrized curves c can be described as the set $\text{Emb}(\mathbb{S}^1, \mathbb{R}^2)$ of smooth embeddings of $\mathbb{S}^1$ in the plane, and deparameterizing corresponds to quotienting out $\text{Diff}(\mathbb{S}^1)$, the group of diffeomorphisms of $\mathbb{S}^1$. Thus, Mumford's shape space is in fact

$$B_e(\mathbb{S}^1, \mathbb{R}^2) = \text{Emb}(\mathbb{S}^1, \mathbb{R}^2)/\text{Diff}(\mathbb{S}^1) .$$

Clearly Mumford's shape space is of infinite dimension. A tangent vector at a closed curve c is a vector field along c and therefore has also to be infinite-dimensional; the class of such tangent vectors is denoted by $\mathcal{C}^\infty(\mathbb{S}^1, \mathbb{R}^2)$. A Riemannian structure for Mumford's shape space is obtained by specifying a weighted inner product on the tangent space $T_c(\text{Emb}(\mathbb{S}^1, \mathbb{R}^2))$, which must be $\text{Diff}(\mathbb{S}^1)$-invariant. For example it can be of the form

$$G_c^{\Phi}(h,k) = \int_{\mathbb{S}^1} \Phi(\ell_c, \kappa_c(\theta)) \langle h(\theta), k(\theta) \rangle |\dot{c}(\theta)| \, \mathrm{d}\theta \,, \quad h,k \in \mathcal{C}^{\infty}(\mathbb{S}^1, \mathbb{R}^2)\,,$$

where ℓ_c is the length of c, $\kappa_c(\theta)$ is the curvature of c at $c(\theta)$ and Φ is a suitable auxiliary function. Tangent vectors of the particular form $g\,\dot{c}$, for $g \in \mathcal{C}^{\infty}(\mathbb{S}^1, \mathbb{R})$, run parallel to the re-parametrization symmetries induced by the diffeomorphism group $\text{Diff}(\mathbb{S}^1)$. Accordingly the tangent vectors relevant for $B_e(\mathbb{S}^1, \mathbb{R}^2)$ are those which are normal to the $\text{Diff}(\mathbb{S}^1)$-orbits, namely

$$\mathcal{N}_c = \{ b\, n_c \;:\; b \in \mathcal{C}^{\infty}(\mathbb{S}^1, \mathbb{R}) \}$$

where n_c is the normal unit field along c. This space is isometric with the tangent space to $B_e(\mathbb{S}^1, \mathbb{R}^2)$ at the shape of c, so the quotient Riemannian metric can be written as

$$G_c^{\Phi}(a\, n_c, b\, n_c) = \int_{\mathbb{S}^1} \Phi(\ell_c, \kappa_c(\theta))\, a(\theta)\, b(\theta)\, |\dot{c}(\theta)| \, \mathrm{d}\theta \,.$$

Michor and Mumford (2006, 2007) discuss geometric details at length and describe numerical examples: Younes *et al.* (2008) develop the theory for cases when translations, dilations and rotations are factored out, thus yielding a true shape theory.

10.4 Distributions of shape

Shape theory needs probability distributions both for inference and for modelling. There is a clear link to the theory of directional statistics (see for example Mardia and Jupp 2000), especially in the case of $\Sigma_2^3 \cong \mathbb{S}^2(\frac{1}{2})$. The most elementary shape distribution for Σ_2^3 first arose in Kendall and Kendall (1980); if three planar landmarks are independent and have identical symmetric Gaussian distribution then a simple symmetry argument shows that the resulting pre-shape is uniformly distributed on $\mathcal{S}_2^3$ and thus the resulting shape is uniformly distributed on the shape sphere Σ_2^3. (This uniform distribution plays a strong rôle in motivating the construction of the spherical blackboard.) If the common planar Gaussian distribution is *not* symmetric then the resulting shape distribution is a simple variation on the uniform distribution; namely it possesses the following probability density with respect to the uniform distribution on the shape sphere:

$$\frac{f}{\left(1 + (f^2 - 1)\sin^2(2\text{dist}_{\Sigma_2^3}(\sigma, \mathcal{L}))\right)^{3/2}} \,.$$

Here $f = \frac{1}{2}(\varsigma + \varsigma^{-1})$ if the parent Gaussian distribution has diagonal variance-covariance matrix $\text{diag}(\varsigma^2, \varsigma^{-2})$, while $\text{dist}_{\Sigma_2^3}(\sigma, \mathcal{L})$ is the distance from the shape

σ to the locus $\mathcal{L}$ of shapes of three aligned landmarks ($\mathcal{L}$ is in fact an equator of $\Sigma_2^3 = \mathbb{S}^2(\frac{1}{2})$).

Computations of Gaussian shape distributions for general Σ_d^k can be found in D.G. Kendall *et al.* (1999, Chapter 8). In the special planar case $d = 2$, symmetry considerations show that the symmetric planar Gaussian distribution leads to the uniform distribution on $\Sigma_2^k = \mathbb{C}P^{k-2}$ using the classic Fubini–Study metric, and there is an elegant expression for the density of the shape distribution corresponding to a general planar Gaussian distribution.

The natural next step is to consider shape distributions which derive from independent sampling of triads of points from uniform distributions over convex compact regions (more specifically, polygons). D.G. Kendall and Le have computed these algorithmically (D.G. Kendall *et al.* 1999, Chapter 8); this reference also describes the shape distribution on Σ_d^{d+1} arising by sampling tiles of Poisson–Delaunay tessellations in dimension d. For inferential purposes interest focusses more on shape distributions arising by sampling landmarks from symmetric planar Gaussian distributions with different means but the same variance. In the triad case this yields the celebrated *Mardia–Dryden* distribution on Σ_2^3, with density for the resulting shape σ of

$$(1 + \kappa(1 + \cos(2\mathrm{dist}_{\Sigma_2^3}(\sigma, \sigma_0)))) \exp\big(- \kappa(1 - \cos(2\mathrm{dist}_{\Sigma_2^3}(\sigma, \sigma_0)))\big). \qquad (10.13)$$

Here σ_0 is the shape determined by the means of the symmetric planar Gaussian distributions, while 4κ is the square of the Euclidean size determined by the means of the symmetric planar Gaussian distributions, divided by the common variance of the Gaussian distributions. Dryden and Mardia (1998) present derivations of this shape density together with many generalizations and variations.

There is a natural link here to the topic of *shape diffusions*; diffusion processes on shape spaces arising from randomly diffusing landmarks: small-time transition densities for shape diffusions (and their approximations) are natural candidates for useful shape distributions. The earliest discussion of shape diffusion dates back to D.G. Kendall (1977), studying the shape diffusion arising when three landmarks diffuse independently with Brownian motion in $\mathbb{R}^2$ (hence leading to shape diffusion on Σ_2^3) and $\mathbb{R}^3$ (hence leading to shape diffusion on Σ_3^3). In most cases of shape diffusion it is necessary to account for the way in which shape will change faster when the Euclidean size of the configuration is smaller; this is achieved by subjecting the shape process to a random time change which converts it into a genuine Itô diffusion. Under this time-change the two-dimensional case leads to a shape diffusion on Σ_2^3 which is Brownian motion on the sphere. The three-dimensional case is interesting because the shape space Σ_3^3 has a boundary, the locus $\mathcal{L}$ of shapes of three aligned landmarks (arising because in three-dimensional space there is an additional symmetry corresponding to picking up the triad and turning it over). The shape diffusion on Σ_3^3 avoids falling off the boundary because it is Brownian motion subject to a drift directed towards the equilateral shape (the north pole if $\mathcal{L}$ is the equator); this drift becomes singular near the boundary at a rate which prevents the shape diffusion from ever encountering the boundary. Kendall (1988) used computer algebra to show that this general picture extends to the case of Σ_d^3 for $d \geqslant 3$; this was further developed in Kendall (1990*a*) to account for the general diffusion of shape in Σ_d^k as a 'warped product' of a diffusion on eigenvalues of the SVD decomposition driving a rotation-matrix-valued process; finally D.G. Kendall (1991) and Le (1991) used stochastic calculus to derive the Mardia–Dryden formula and generalizations.

Motivated by a problem in cell biology, Ball, Dryden and Golalizadeh (2008) describe a modified shape diffusion Σ_2^k, which is subject to a drift directed towards a reference shape. Computations are demanding, but aided by the geometric insight arising because $\Sigma_2^k \cong \mathbb{C}P^{k-2}$ is a rank-1 symmetric space.

As we have noted, Euclidean shape space is not the only possible geometry for shapes. Kendall (1998) showed how to derive a shape diffusion appropriate for Bookstein's shape space for triangles, based on a natural Brownian motion on the space of 2×2 matrices of positive determinant. The resulting shape diffusion (with no time-change required) is a Brownian motion on the Poincaré half plane representation of Bookstein's shape space. Asymptotic approximations relate the transition density to a hyperbolic von Mises–Fisher distribution (Jensen, 1981). This work suggests the still-open question, how might one adapt shape diffusion treatments to provide stochastically motivated generalizations of Bookstein's shape space for general k-ads of points on the plane?

10.5 Some topics in shape statistics

We turn from shape geometry and shape distributions to some statistical matters.

10.5.1 *Mean shapes and regression*

One of the most important issues in the statistics of shape is estimation of mean shape. Shape spaces possess curved geometry, so the definition of average or mean shape is a non-trivial matter. Fortunately there is a relevant notion: the so-called *Fréchet mean* of a probability distribution μ on a general metric space $(\boldsymbol{M}, \text{dist})$. This is defined to be any point in $\boldsymbol{M}$ that achieves the *global* minimum of the function

$$F_\mu(x) = \frac{1}{2}\int_{\boldsymbol{M}} \text{dist}(x, y)^2 \mathrm{d}\mu(y)\,. \tag{10.14}$$

Variant definitions use increasing functions of dist (Kendall, 1998) or use the usual distance arising from a given embedding in Euclidean space (the so-called *extrinsic* Fréchet means, Bhattacharya and Patrangenaru 2003). Averaging a set of sample data can be defined using empirical measure.

In the case of Euclidean shape space, commonly used metrics are the Riemannian metric and the procrustean distance. For simplex shape spaces, the most common metric is the Riemannian distance dist, although $\sqrt{(\cosh(\sqrt{2}\,\text{dist}) - 1)}$

(equivalently, $\sinh(\text{dist}/\sqrt{2})$, Kume and Le 2000) can be convenient for the Bookstein shape space. Here we assume that $\boldsymbol{M}$ is a shape space of dimension m and focus on Fréchet mean shapes using Riemannian metrics.

Note that Stoyan and Molchanov (1997) present a related approach to mean shapes based on the theory of random compact sets. Random compact sets can be represented as random functions in Hilbert spaces in several different ways; each representation leads to a different notion of mean value.

Minimizers of (10.14) need not be unique, though uniqueness will hold for negatively curved and simply connected spaces (e.g. simplex shape space). Moreover there are no general closed forms for Fréchet means, so the evaluation of the Fréchet mean of data is a computational issue. Thus *local* minima of (10.14) are important; these are the so-called *Karcher means*. If x is a Karcher mean of μ, then $\text{grad}F_\mu(x)$ vanishes if defined at x: if Exp_x denotes the Riemannian exponential map at x and if $\text{Exp}_x^{-1}(y)$ is well-defined for μ-almost all y then

$$\text{grad}F_\mu(x) = -\int_{\boldsymbol{M}} \text{Exp}_x^{-1}(y)\,\text{d}\mu(y)) = -\int_{T_x(\boldsymbol{M})} v\,\text{d}(\mu \circ \text{Exp}_x)(v). \qquad (10.15)$$

(Here $T_x(\boldsymbol{M})$ is the tangent space to $\boldsymbol{M}$ at x.)

The sufficient conditions preceding (10.15) hold either when μ is a probability distribution on simplex shape space or when μ is defined on a small enough geodesically convex ball B of Euclidean shape space. It is possible to be explicit about *how* small: namely the ball must be contained in an (open) geodesic ball of radius $r < \bar{r}$, where $\bar{r} = \frac{1}{2}\min\{\text{inj}(\boldsymbol{M}), \pi/\sqrt{\varrho}\}$, $\varrho \geqslant 0$ being an upper bound of the curvature of $\boldsymbol{M}$ and $\text{inj}(\boldsymbol{M})$ being the lower bound of distances between pairs of points connected by more than one shortest geodesic (so $\bar{r} = \infty$ for spaces with non-positive curvature). For *any* μ supported on such a ball B, Kendall (1990*b*) showed there is just one Karcher mean within the ball and it is characterized by $\text{grad}F_\mu = 0$ (a direct consequence of the proof of Lemma 7.2); Le (2001, Lemma 1 and the proof of Theorem 1) then showed that this Karcher mean provides the Fréchet mean of μ. Thus, under the above sufficient conditions, (10.15) relates Fréchet mean shapes to Euclidean means: x is the Fréchet mean of μ if and only if the origin is the Euclidean mean of the distribution $\mu \circ \text{Exp}_x$ on the tangent space at x.

Under the condition that the support of μ is contained in B as above, this characterization of Fréchet means provides an iterative algorithm for computing the Fréchet mean shapes using the flow

$$\Phi(t, x) : x \mapsto \text{Exp}_x\left(-t\,\text{grad}F_\mu(x)\right).$$

If $\text{grad}F_\mu(x) = 0$, then x is a fixed point of the flow. If $\text{grad}F_\mu(x) \neq 0$ for an $x \in B$ then for small enough $t > 0$ (Le, 2004)

$$F_\mu(x) - F_\mu(\Phi(t, x)) \geqslant \frac{t}{2}\,\|\text{grad}F_\mu(x)\|^2.$$

For Euclidean shape space there is a $t_0 > 0$ such that

$$F_\mu(x) - F_\mu(\Phi(t_0, x)) \geqslant \frac{t_0}{2} \, \|\mathrm{grad} F_\mu(x)\|^2$$

for all $x \in B$. In particular, if the radius of B in Σ_2^k is bounded by $\bar{r}/3$ where $\bar{r} = \pi/4$, then $t_0 = 1$ suffices (Le, 2001). Thus an iterative algorithm follows: the sequence

$$x_0 \in B, \qquad x_{n+1} = \Phi(t_0, x_n),$$

converges to the unique Fréchet mean shape of μ.

A variant of the above iterative algorithm finds Fréchet sample mean shapes on simplex shape spaces (Kume and Le, 2003). It uses the matrix representation (10.12) to project sample shapes to the tangent space at the regular simplex shape I using Exp_I^{-1}. The idea is to show that $F_\mu(\gamma(t))$ is a decreasing function of small enough $t > 0$ and

$$F_\mu(I) - F_\mu(\gamma(t)) \geqslant \frac{t}{2} \|\hat{V}\|^3,$$

where $\hat{V}$ is the Euclidean mean of the projections of the samples onto the tangent space at I and γ is the geodesic starting at I defined by $\gamma(t) = \mathrm{Exp}_I(t\hat{V})$. Thus the algorithm is linked to computation of matrix means.

Symmetry considerations can permit exact computation of Fréchet means for certain classes of probability distributions. For example, for Euclidean planar shape space, if the density of a shape distribution is a non-increasing essentially non-constant function of the Riemannian distance to a fixed shape then the fixed shape is the unique Fréchet mean (Le, 1998). For simplex shape spaces, if the density function is invariant with respect to the involutive symmetry at a given point, then the Fréchet mean shape is located there if it exists at all. This was proved for the Bookstein shape space in Kume and Le (2000). However, the proof holds for all simplex shape spaces.

Given that they are defined, are Fréchet sample means strongly consistent estimators for the theoretical Fréchet sample mean? Euclidean shape space is compact, so strong consistency follows from results of Ziezold (1977). Simplex shape spaces are non-compact; however strong consistency then follows from the work of Bhattacharya and Patrangenaru (2003).

It is also possible to derive a central limit theorem for Fréchet mean shapes. This is a local result, so it is necessary to assume that the support of the measure μ in question is contained in a coordinate patch of the shape space $\boldsymbol{M}$. Actually this restriction is very mild: in the case of shape space any $x \in \boldsymbol{M}$ possesses a local coordinate chart $\phi : \boldsymbol{M} \setminus \mathcal{C}_x \to \mathbb{R}^m$ determined by $\mathrm{Exp}_x^{-1}(\cdot)$ and defined everywhere except the cut locus $\mathcal{C}_x$ of x (and the cut-locus has co-dimension at most one). Bhattacharya and Patrangenaru (2005) show that if ϕ is a coordinate chart defined on the support of μ, and if the Fréchet mean $\hat{x}_\mu$ of μ is unique and some further mild regularity conditions hold, then

$$\sqrt{n}\left\{\phi\left(\hat{\xi}_n\right)-\phi\left(\hat{x}_\mu\right)\right\} \xrightarrow{d} N(0, H^{-1}C\,(H^\top)^{-1})\,. \tag{10.16}$$

Here $\hat{\xi}_n$ is the Fréchet sample mean of an independent sample of n random variables ξ_i with distribution μ, C is the covariance matrix of the Euclidean gradient $\frac{1}{2}\text{grad}\left(\text{dist}(\phi(\xi),\phi(\hat{x}_\mu))\right)^2$, and H is the expected Euclidean Hessian matrix of $\frac{1}{2}\left(\text{dist}(\phi(\xi),\phi(\hat{x}_\mu))\right)^2$.

A two-sample test follows as an immediate application (Bhattacharya and Patrangenaru, 2005): for example, if $\hat{\xi}_n$ and $\hat{\eta}_n$ are independent Fréchet means of independent samples of size n from μ_1 and μ_2 respectively, then

$$\sqrt{n}\left\{\phi\left(\hat{\xi}_n\right)-\phi\left(\hat{x}_{\mu_1}\right)\right\}-\sqrt{n}\left\{\phi\left(\hat{\eta}_n\right)-\phi\left(\hat{x}_{\mu_2}\right)\right\} \xrightarrow{d} N(0,\Gamma)\,,$$

where $\Gamma = H_\xi^{-1}C_\xi\,(H_\xi^\top)^{-1} + H_\eta^{-1}C_\eta\,(H_\eta^\top)^{-1}$.

We indicate here the lines of an intrinsic treatment of the case when ϕ is chosen to be the inverse Riemannian exponential map $\text{Exp}_{\hat{x}_\mu}^{-1}(\cdot)$ located at the Fréchet mean $\hat{x}_\mu$. The result (10.16) then takes on a more geometric form using the covariant derivative D for the shape manifold: if the random sample in question is $\xi_1, \ldots, \xi_n$ then it can be deduced that

$$\frac{1}{\sqrt{n}}\sum_{i=1}^{n} D_{\text{Exp}_{\hat{x}_\mu}^{-1}(\hat{\xi}_n)}\left(\text{Exp}_{\hat{x}_\mu}^{-1}(\xi_i)\right) \xrightarrow{d} N(0,\Sigma)\,. \tag{10.17}$$

Here Σ is the self-adjoint linear operator on $T_{\hat{x}_\mu}(\boldsymbol{M})$ defined by

$$\text{Var}(\langle v, \text{Exp}_{\hat{x}_\mu}^{-1}(\xi)\rangle) = \langle v, \Sigma\, v\rangle \qquad \forall v \in T_{\hat{x}_\mu}(\boldsymbol{M})\,.$$

Define the linear operator R_x on $T_{\hat{x}_\mu}(\boldsymbol{M})$ by $R_x(v) = D_v(\text{Exp}_{\hat{x}_\mu}^{-1}(x))$. It can then be shown that

$$\sqrt{n}\text{Exp}_{\hat{x}_\mu}^{-1}(\hat{\xi}_n) \xrightarrow{d} N(0, (E[R_\xi])^{-1}\Sigma\,(E[R_\xi])^{-\top})\,. \tag{10.18}$$

Since

$$\sum_{i=1}^{n}\text{Exp}_{\hat{x}_\mu}^{-1}(\xi_i) \approx -\sum_{i=1}^{n} D_{\text{Exp}_{\hat{x}_\mu}^{-1}(\hat{\xi}_n)}\left(\text{Exp}_{\hat{x}_\mu}^{-1}(\xi_i)\right)$$

and $\text{Exp}_{\hat{x}_\mu}^{-1}(\xi_i) = \xi_i - \hat{x}_\mu$ in Euclidean space, the above version of the central limit theorem indeed generalizes the classical one on Euclidean space.

Statistical analysis of shape changes over time also involves geometry. The curvature of shape space means that the usual tools from a Euclidean context (regression curves and spline functions) will be suitable only if the total shape change is small: in this case one can use Exp_x^{-1} (for, say, the Fréchet mean x of the entire data set) to linearize the problem.

When variability is large enough that geometry must be taken into account, one possibility is to derive a function that minimizes the required objective functional on shape space. Unfortunately, the solution for such a minimizing problem in curved spaces is generally intractable. An alternative approach is to linearize using ideas of Cartan development of curves: unrolling/unwrapping onto appropriate tangent spaces where the standard Euclidean fitting procedure may be applied. This approach was first suggested by Jupp and Kent (1987) for fitting smooth curves to spherical data using spherical spline functions. Kume, Dryden and Le (2007) define smoothing splines for Euclidean shape space in a similar manner: 'the fitted path to the data is defined such that its unrolled version at the tangent space of the starting point is the cubic spline fitted to the unwrapped data with respect to that path'. Unrolling from the curved space to its flat tangent spaces preserves useful geometric features such as distances and angles. If regression lines are fitted in the tangent spaces then the resulting fitted curve for the shape dataset will be piecewise-geodesic.

Kume, Dryden and Le (2007) suggest an iterative algorithm for finding the Euclidean shape space smoothing spline for a given shape dataset; this is similar to the method of Jupp and Kent (1987), but lifts isometrically to the pre-shape sphere, rather than on the shape space itself. This requires derivation of technical results on the characterization of the lifting of the unrolling procedure for shape curves onto the pre-shape sphere (see also Le, 2003). For the Euclidean shapes of configurations with k landmarks in the plane, this lifting of the unrolling procedure along a geodesic from shape σ_0 to shape σ_1 can be specified in closed form. Representing the pre-shapes of σ_i by $(k-1)$-dimensional complex column vectors $\underline{X}_i$, $i = 0, 1$, such that $\underline{X}_0^* \underline{X}_1 > 0$ where $\underline{X}_0^*$ is the transposed complex conjugate (row) vector of $\underline{X}_0$, the lifting of the unrolling procedure is given by:

$$\Psi_{\underline{X}_0,\underline{X}_1} : \mathbf{H}_{\underline{X}_1}(\mathcal{S}_2^k) \to \mathbf{H}_{\underline{X}_0}(\mathcal{S}_2^k); \quad v \mapsto v - \frac{1-\cos s_0}{\sin^2 s_0} \underline{X}_0^* v\,(\underline{X}_0 + \underline{X}_1),$$

where $\mathbf{H}_{\underline{X}}(\mathcal{S}_2^k)$ denotes the horizontal tangent space to the pre-shape sphere $\mathcal{S}_2^k$ at $\underline{X}$ and $s_0 = \mathrm{dist}_{\Sigma_2^k}(\sigma_0, \sigma_1)$. The lift of the unwrapping of a shape point σ observed at time t_1 along this geodesic can be expressed as

$$s_0\, v_{\underline{X}_0,\underline{X}_1} + \Psi_{\underline{X}_0,\underline{X}_1}\left(\mathrm{Exp}_{\underline{X}_1}^{-1}(\underline{Y})\right),$$

where $v_{\underline{X}_0,\underline{X}_1}$ is the unit tangent vector at $\underline{X}_0$ of the horizontal lift of the geodesic and $\underline{Y}$ is the pre-shape of σ satisfying $\underline{X}_1^*\underline{Y} > 0$. The lifting of unrolling and unwrapping to a piecewise geodesic is obtained by applying Ψ recursively along each geodesic segment in turn.

10.5.2 *MCMC techniques*

Green and Mardia (2006) have introduced the use of Markov chain Monte Carlo in problems related to shape theory. Motivated by problems in bioinformatics, they consider partial matching problems linking two sets of partial observations $\{x_j\}$ and $\{y_l\}$ of unobserved ideal landmarks $\{z_i\}$. The observations are linked by the statistical model

$$\begin{aligned} x_j &= z_{\alpha_j} + \varepsilon_{1j} \\ \mathcal{A} y_l &= z_{\beta_l} + \varepsilon_{2l} \end{aligned} \tag{10.19}$$

where errors ε_{ij} are independent with common error density $f(\cdot)$ and the transformation $\mathcal{A}$ may be a rigid motion, or a more general transformation, and may be assumed known or unknown. In case $\mathcal{A}$ is an unknown rigid motion this is clearly related to size-and-shape analysis. The indexing arrays $\{\alpha_j\}$, $\{\beta_l\}$ determine the set of matched points together with a binary matching matrix M, and this matching matrix is of primary interest, as it indicates which x_j are linked (if at all) to which y_l.

On the face of it, the problem appears to be of varying dimension according to the number of matches and total number of landmarks. Green and Mardia (2006) escape this by an ingenious use of a Poisson process prior for the ideal landmarks z_i, which allows unobserved ideal landmarks to be integrated out of the problem. Other aspects of the prior include the following:

1. Each ideal landmark z_i independently may be observed in one, the other, both, or neither of the observed sets $\{x_j\}$, $\{y_l\}$.
2. The transformation is of affine form $\mathcal{A}(x) = Ax + \tau$, for an unknown shift τ and a matrix A which for planar size-and-shape analysis will be distributed as a rotation with a von Mises distribution.
3. The ideal landmarks are supposed to be confined to a large region, and a limiting form is derived for the resulting posterior neglecting boundary effects.
4. The error density $f(\cdot)$ is taken to be Gaussian of variance ς^2.
5. The parameters τ and ς^2 are given Gaussian-Inverse Gamma priors.

Much use is made of conjugacy to implement a Gibbs sampler: the matching matrix M is subject to Metropolis–Hastings moves (flipping entries between 0 and 1, switching entries in the same column or the same row).

It is striking that this treatment allows landmarks to be unobserved, or observed only in one or the other configuration.

Mardia *et al.* (2007) give several examples of this approach, combined with an initial graph-matching, in bioinformatics problems. Dryden *et al.* (2007) describe a variant approach to a similar problem in chemoinformatics, in particular using Gaussian distributions on tangent spaces to size-and-shape space to assess goodness of partial matchings.

10.6 Two recent applications

As examples of current developments in statistical shape theory, we discuss work on the random topology of linkages and on diffusions of projected shape. The work on random topology links suggestively with the foundational work on the topology of Σ_d^k described in D.G. Kendall *et al.* (1999).

10.6.1 *Random Betti numbers*

Recent work by Farber and co-workers is concerned with the topology of linkages of rods of specified lengths. The primary motivation arises from motion studies in robotics; there are also applications to the shapes of large bio-molecules. In mathematical essence, one studies a specified *linkage*, a sequence of positive numbers $L = (\ell_1, \ldots, \ell_k)$ encoding the consecutive distances between k nodes cyclically linked by inextensible rods of fixed lengths. Considering the planar case for simplicity, suppose that the directions of the rods are specified by unit vectors $u_1, \ldots, u_k \in \mathbb{R}^2$. The linkage is cyclic, so node $i-1$ is linked to node i by a rod of length ℓ_i in direction u_i, where node 0 is identified with node k. The cyclic nature implies that $\sum \ell_i u_i = 0$. The k nodes specify a planar shape in Σ_2^k; the region in size-and-shape space mapped out by varying directions u_i and fixed L can be viewed as a slice through a k-torus. Thus the region M_L in Σ_2^k mapped out by varying directions u_i and fixed $L = (\ell_1, \ldots, \ell_k)$ can be represented as

$$M_L = \left\{(u_1, \ldots, u_k) \in \mathbb{S}^1 \times \ldots \times \mathbb{S}^1 \;:\; \sum \ell_i u_i = 0\right\} \Big/ SO_2 . \qquad (10.20)$$

Previous work has shown that in the generic case (when $\sum \ell_i \varepsilon_i \neq 0$ for $\varepsilon_i = \pm 1$) M_L is a smooth manifold, and otherwise possesses just a finite number of singularities. Interest lies in understanding the random topology formed when $L = (\ell_1, \ldots, \ell_k)$ is chosen at random according to a suitable probability measure. Note that it is evident that the topology depends only on the *linkage simplex*, which is made up of the possible vectors $L = (\ell_1, \ldots, \ell_k)$ normalized so that $\sum_i \ell_i = 1$.

In order to deal with this random topology, one focuses on the *Betti numbers* $b_p(M_L)$ for $p = 0, 1, \ldots$. Formally these are the (non-negative integer) ranks of the homology groups $H^p(M_L)$; at a less formal level for example the first Betti number $b_1(M_L)$ may be thought of as counting the number of holes in M_L, while $b_0(M_L)$ counts the number of components. Thus they quantify various aspects of the topology of M_L. Farber and Kappeler (2008) show that if $L = (\ell_1, \ldots, \ell_k)$

is randomized according to one of two possible uniform probability models then for fixed p and large k

$$\mathbf{E}\left[b_p(M_L)\right] \sim \binom{k-1}{p}. \tag{10.21}$$

Farber and Schuetz (2007) relate the Betti numbers $b_p(M_L)$ to combinatorical considerations about $L = (\ell_1, \ldots, \ell_k)$, specifically the notion of *short* subsets $J \subseteq \{1, \ldots, k\}$ such that $\sum_{j \in J} \ell_j < \sum_{j \notin J} \ell_j$; Farber (2008) establishes the asymptotics of (10.21) for probability measures satisfying for example an exponential decay condition away from the centre of the linkage simplex.

10.6.2 *Radon shape diffusion*

Biophysicists use electron microscopes to image single particles moving in an aqueous environment to study the structure of biological macromolecules. The information provided by such a method is the projected structure of the particles. A simple model for the movement of such a particle would be to assume its location and rotation follow independent Brownian motions: one is led to consider the diffusion of the shape of the projected structure. Interest lies in relating the projected diffusion to the underlying particle shape.

Panaretos (2006, 2008) defines Radon shape diffusion as follows. Being interested only in the shape of the projection, it suffices to model the movement of the particle by assuming the location is fixed and the rotation follows a Brownian motion. When Brownian motion in SO_d acts on the k landmarks of a given configuration in $\mathbb{R}^d$, it leaves the shape of the configuration unchanged. However, the shape of its projection on a fixed hyperplane gives rise to a *Radon shape diffusion* on Σ^k_{d-1}.

Equivalently the Radon shape diffusion can be viewed as the shape of the projection of the given configuration onto a randomly moving hyperplane, where the motion of the hyperplane is such that it is normal to a Brownian motion u_t on $\mathbb{S}^{d-1}$. Such a projection can be expressed as $P_t = I - u_t u_t^\top$ where u_t satisfies the Itô stochastic differential equation

$$\mathrm{d}u_t = P_t \, \mathrm{d}B_t - \frac{d-1}{2} u_t \, \mathrm{d}t \,, \tag{10.22}$$

with B_t being a standard Brownian motion on $\mathbb{R}^d$. If X_0 is a given centred configuration, then $P_t X_0$ is also centred and so the Radon shape diffusion is the shape diffusion induced by the diffusion $P_t X_0$.

In the following we concentrate on the planar case $d = 2$. If the given configuration X_0 is degenerate, i.e. all its landmarks lie on a straight line or, equivalently, $\operatorname{rank}(X_0) = 1$, then the shape of $P_t X_0$ (when not itself degenerate) is either the shape of X_0 or the shape of its reflection. Hence, we further assume that the matrix X_0 is of full rank. Since $d = 2$ we write $P_t = v_t v_t^\top$, where $v_t = J\, u_t$ and J is anticlockwise rotation by $\pi/2$. Thus, v_t is also a Brownian motion on $\mathbb{S}^1$ and $v_t^\top X_0$ comprises the coordinates of the projection of the vertices of X_0 on v_t.

Since SO_1 is trivial, $\tilde{\sigma}_t = v_t^\top X_0$ and $\sigma_t = v_t^\top X_0/\|v_t^\top X_0\|$ are the size-and-shape and shape of $v_t^\top X_0$ respectively. It now follows from (10.22) that

$$\mathrm{d}v_t = J\,v_t\,\mathrm{d}B_t^1 - \frac{1}{2}v_t\,\mathrm{d}t\,,$$

where B_t^1 is a Brownian motion on $\mathbb{R}$, and so the size-and-shape of the projection satisfies the Itô stochastic differential equation

$$\mathrm{d}\tilde{\sigma}_t = \tilde{\sigma}_t\,X_0^\top\,(X_0X_0^\top)^{-1}J^\top X_0\,\mathrm{d}B_t^1 - \frac{1}{2}\tilde{\sigma}_t\,\mathrm{d}t\,.$$

Note that, if $R \in SO_2$, then $R^\top JR = J$ so that $X_0^\top\,(X_0X_0^\top)^{-1}J^\top X_0$ is a function of the shape of X_0.

To obtain the Itô equation for the shape σ_t, we note that the path $\tilde{\sigma}_t$ lies on the ellipse $\{x \in \mathbb{R}^{k-1}\ :\ y^\top X_0 = x,\ y \in \mathbb{S}^1\}$. This implies that the Radon shape σ_t lies on a circle, a one-dimensional subspace of the shape space $\Sigma_1^k = \mathbb{S}^{k-1}$.

Up to a rotation matrix on the left which is of no interest to us, we can write $X_0 = \Lambda R$, where R is a $2 \times (k-1)$ matrix with orthonormal rows, $\Lambda = \mathrm{diag}\,(\lambda_1, \lambda_2)$ and $\lambda_i > 0$. Since $\|v_t^\top \Lambda\| = \|\tilde{\sigma}_t\|$, applying the Itô formula to the function $(x, y) \mapsto \arg(x/y)$ shows that $\vartheta_t = \arg(v_t^\top \Lambda)$ satisfies

$$\mathrm{d}\vartheta_t = \frac{\det(\Lambda)}{\|\tilde{\sigma}_t\|^2}\left\{\mathrm{d}B_t^1 + u(\vartheta_t)^\top \Lambda J\Lambda^{-1}u(\vartheta_t)\,\mathrm{d}t\right\}.$$

Now $\Lambda J\Lambda^{-1} = \Lambda^2 J/\det(\Lambda)$, $u(\vartheta_t) = (\cos\vartheta_t, \sin\vartheta_t)^\top$; therefore

$$\begin{aligned}\mathrm{d}u(\vartheta_t)^\top &= -u(\vartheta_t)^\top J\,\mathrm{d}\vartheta_t - \frac{1}{2}\left(\frac{\det(\Lambda)}{\|\tilde{\sigma}_t\|^2}\right)^2 u(\vartheta_t)^\top\,\mathrm{d}t\\ &= -\frac{\det(\Lambda)}{\|\tilde{\sigma}_t\|^2}\left\{u(\vartheta_t)^\top J\,\mathrm{d}B_t^1 + w_t\,\mathrm{d}t\right\}\end{aligned}$$

where

$$w_t = \frac{1}{\det(\Lambda)}u(\vartheta_t)^\top \Lambda^2 J u(\vartheta_t)u(\vartheta_t)^\top J + \frac{\det(\Lambda)}{2\|\tilde{\sigma}_t\|^2}\,u(\vartheta_t)^\top\,.$$

However, since $u(\vartheta_t)^\top = v_t^\top \Lambda/\|\tilde{\sigma}_t\|$, we have $\sigma_t = u(\vartheta_t)^\top R$. Thus,

$$\begin{aligned}\mathrm{d}\sigma_t = -\frac{\det(\Lambda)}{\|\tilde{\sigma}_t\|^2}\Bigg\{&\sigma_t R^\top JR\,\mathrm{d}B_t^1 + \left(\frac{1}{\det(\Lambda)}\sigma_t X_0^\top X_0 R^\top JR\sigma_t^\top \sigma_t R^\top JR\right.\\ &\left.+\frac{\det(\Lambda)}{2\|\tilde{\sigma}_t\|^2}\sigma_t\right)\mathrm{d}t\Bigg\}.\end{aligned} \qquad (10.23)$$

Note that, when $k = 3$, $R^\top JR = J$.

The stochastic differential equation (10.23) thus suggests ways to draw inferences about X_0 from observations of the evolution of the Radon shape diffusion σ.

When the dimension d is greater than two, the complexity of Euclidean shape spaces increases. Panaretos' (2008) results on stochastic differential equations for Radon shape diffusion in this situation are expressed using unoriented shapes and unoriented size-and-shapes, as they have convenient global representations in terms of inner products of landmarks (see also Kendall 1990*a*).

References

Ball, F. G., Dryden, I. L., and Golalizadeh, M. (2008). Brownian motion and Ornstein–Uhlenbeck processes in planar shape space. *Methodol. Comput. Appl.*, **10**, 1–22.

Berger, J. O. (1985). *Statistical Decision Theory and Bayesian Analysis* (2nd edn). Springer Series in Statistics. Springer-Verlag, New York.

Bhattacharya, R. and Patrangenaru, V. (2003). Large sample theory of intrinsic and extrinsic sample means on manifolds. I. *Ann. Statist.*, **31**(1), 1–29.

Bhattacharya, R. and Patrangenaru, V. (2005). Large sample theory of intrinsic and extrinsic sample means on manifolds. II. *Ann. Statist.*, **33**(3), 1225–1259.

Bookstein, F. L. (1986). Size and shape spaces for landmark data in two dimensions. *Stat. Sci.*, **1**(2), 181–242.

Bookstein, F. L. (1989). Principal warps: thin-plate splines and the decomposition of deformations. *IEEE T. Pattern Anal.*, **11**(6), 567–585.

Broadbent, S. R. (1980). Simulating the Ley-Hunter. *J. Roy. Stat. Soc. Ser. A-G*, **143**, 109–140.

Dryden, I. L., Hirst, J. D., and Melville, J. L. (2007). Statistical analysis of unlabeled point sets: comparing molecules in chemoinformatics. *Biometrics*, **63**(1), 237–251, 315.

Dryden, I. L. and Mardia, K. V. (1998). *Statistical Shape Analysis.* Wiley Series in Probability and Statistics: Probability and Statistics. John Wiley & Sons Ltd., Chichester.

Farber, M. (2008). Topology of random linkages. *Algebraic and Geometric Topology*, **8**, 155–172.

Farber, M. and Kappeler, T. (2008). Betti numbers of random manifolds. *Homology, Homotopy and Applications*, **10**, 205–222.

Farber, M. and Schuetz, D. (2007). Homology of planar polygon spaces. *Geometria Dedicata*, **25**, 75–92.

Gallot, S., Hulin, D., and Lafontaine, J. (1990). *Riemannian Geometry* (2nd edn). Universitext. Springer-Verlag, New York.

Green, P. J. and Mardia, K. V. (2006). Bayesian alignment using hierarchical models, with applications in protein bioinformatics. *Biometrika*, **93**(2), 235–254.

Jensen, J. L. (1981). On the hyperboloid distribution. *Scand. J. Statist.*, **8**(4), 193–206.

Jupp, P. E. and Kent, J. T. (1987). Fitting smooth paths to spherical data. *Appl. Statist.*, **36**, 34–46.

Kendall, D. G. (1977). The diffusion of shape. *Adv. in Appl. Probab.*, **9**, 428–430.

Kendall, D. G. (1984*a*). Shape manifolds, procrustean metrics, and complex projective spaces. *Bulletin of the London Mathematical Society*, **16**(2), 81–121.

Kendall, D. G. (1984*b*). Statistics, geometry and the cosmos. (The Milne lecture, 1983). *Quart. J. R. Astron. Soc.*, **25**, 147–156.

Kendall, D. G. (1991). The Mardia–Dryden shape distribution for triangles: a stochastic calculus approach. *J. Appl. Probab.*, **28**(1), 225–230.

Kendall, D. G. (1994*a*). How to look at a 5-dimensional shape space. I: Looking at distributions. *Teor. Ver. Prim.*, **39**, 242–247.

Kendall, D. G. (1994*b*). How to look at a 5-dimensional shape space. II: Looking at diffusions. In *Probability, Statistics and Optimization: A Tribute to Peter Whittle* (ed. F. Kelly), Chichester and New York, pp. 315–324. John Wiley & Sons Ltd.

Kendall, D. G. (1995). How to look at a 5-dimensional shape space. III: Looking at geodesics. *Adv. in Appl. Probab.*, **27**, 35–43.

Kendall, D. G., Barden, D., Carne, T. K., and Le, H. (1999). *Shape and Shape Theory*. Wiley Series in Probability and Statistics. John Wiley & Sons Ltd., Chichester.

Kendall, D. G. and Kendall, W. S. (1980). Alignments in two-dimensional random sets of points. *Adv. in Appl. Probab.*, **12**(2), 380–424.

Kendall, W. S. (1988). Symbolic computation and the diffusion of shapes of triads. *Adv. in Appl. Probab.*, **20**(4), 775–797.

Kendall, W. S. (1990*a*). The diffusion of Euclidean shape. In *Disorder in Physical Systems*, Oxford Sci. Publ., pp. 203–217. Oxford Univ. Press, New York.

Kendall, W. S. (1990*b*). Probability, convexity, and harmonic maps with small image. I. Uniqueness and fine existence. *Proc. London Math. Soc. (3)*, **61**(2), 371–406.

Kendall, W. S. (1998). A diffusion model for Bookstein triangle shape. *Adv. in Appl. Probab.*, **30**(2), 317–334.

Kume, A., Dryden, I. L., and Le, H. (2007). Shape-space smoothing splines for planar landmark data. *Biometrika*, **94**, 513–526.

Kume, A. and Le, H. (2000). Estimating Fréchet means in Bookstein's shape space. *Adv. in Appl. Probab.*, **32**(3), 663–674.

Kume, A. and Le, H. (2003). On Fréchet means in simplex shape spaces. *Adv. in Appl. Probab.*, **35**(4), 885–897.

Le, H. (1991). A stochastic calculus approach to the shape distribution induced by a complex normal model. *Math. Proc. Cambridge Philos. Soc.*, **109**(1), 221–228.

Le, H. (1998). On the consistency of procrustean mean shapes. *Adv. in Appl. Probab.*, **30**(1), 53–63.

Le, H. (2001). Locating Fréchet means with application to shape spaces. *Adv. in Appl. Probab.*, **33**(2), 324–338.

Le, H. (2003). Unrolling shape curves. *J. London Math. Soc. (2)*, **68**(2), 511–526.

Le, H. (2004). Estimation of Riemannian barycentres. *LMS J. Comput. Math.*, **7**, 193–200 (electronic).

Le, H. and Barden, D. (2001). On simplex shape spaces. *J. London Math. Soc. (2)*, **64**(2), 501–512.

Le, H. and Small, C. G. (1999). Multidimensional scaling of simplex shapes. *Pattern Recogn.*, **32**(9), 1601–1613.

Loomis, L. H. (1953). *An Introduction to Abstract Harmonic Analysis*. D. Van Nostrand Company, Inc., Toronto-New York-London.

Mardia, K. V. and Jupp, P. E. (2000). *Directional Statistics*. Wiley Series in Probability and Statistics. John Wiley & Sons Ltd., Chichester.

Mardia, K. V., Kent, J. T., and Bibby, J. M. (1979). *Multivariate Analysis*. Academic Press [Harcourt Brace Jovanovich Publishers], London.

Mardia, K. V., Nyirongo, V. B., Green, P. J., Gold, N. D., and Westhead, D. R. (2007). Bayesian refinement of protein functional site matching. *BMC Bioinformatics*, **8**(257), 18 pp.

Michor, P. W. and Mumford, D. (2006). Riemannian geometries on spaces of plane curves. *J. Eur. Math. Soc. (JEMS)*, **8**(1), 1–48.

Michor, P. W. and Mumford, D. (2007). An overview of the Riemannian metrics on spaces of curves using the Hamiltonian approach. *Appl. Comput. Harmon. Anal.*, **23**(1), 74–113.

Panaretos, V. M. (2006). The diffusion of Radon shape. *Adv. in Appl. Probab.*, **38**(2), 320–335.

Panaretos, V. M. (2008). Representation of Radon shape diffusions via hyperspherical Brownian motion. *Math. Prob. Cambridge*, **145**(2), 457–470.

Schneider, R. and Weil, W. (2008). *Stochastic and Integral Geometry*. Springer, Berlin, Heidelberg.

Small, C. G. (1996). *The Statistical Theory of Shape*. Springer Series in Statistics. Springer-Verlag, New York.

Stoyan, D. and Molchanov, I. S. (1997). Set-valued means of random particles. *J. Math. Imaging Vision*, **7**(2), 111–121.

Stoyan, D. and Stoyan, H. (1994). *Fractals, Random Shapes and Point Fields*. Wiley Series in Probability and Mathematical Statistics: Applied Probability and Statistics. John Wiley & Sons Ltd., Chichester.

Thompson, D'A. W. (1917). *On Growth and Form*. Cambridge University Press, Cambridge.

Younes, L., Michor, P., Shah, J., and Mumford, D. (2008). A metric on shape space with explicit geodesics. *Rend. Lincei Mat. Appl.*, **9**, 25–57.

Ziezold, H. (1977). On expected figures and a strong law of large numbers for random elements in quasi-metric spaces. In *Trans. Seventh Prague Conference on Information Theory, Statistical Decision Functions, Random Processes*, Volume A, Prague, pp. 591–602. Academia.

11

SET ESTIMATION

Antonio Cuevas and Ricardo Fraiman

Abstract

The problem of estimating a set from a random sample of points is considered from different points of view, including the estimation of supports, boundaries and level sets. The literature on this topic is surveyed and the main applications, as well as some typical mathematical tools, are commented on.

11.1 Introduction

The goal of set estimation is to approximate (i.e. 'to estimate', in statistical terms) a compact set S in the Euclidean space $\mathbb{R}^d$ from a sample of randomly chosen points whose distribution is closely related to S; for example S could be the support of the distribution which generates the points. The term 'set' is maybe too general for our purposes since some geometrical assumptions on S are usually needed in order to properly carry out the estimation of S. In other words, the family of Euclidean compact sets is indeed enormous and includes very strange members. The effective reconstruction of a set using statistical methods is not always possible in practice, unless some geometrical restrictions, oriented to ruling out the 'monster sets', are imposed. As a result, set estimation, as it will be presented here, deals with problems in the interplay between statistics and finite-dimensional geometry.

Of course there are other various statistical fields in which the sets or bodies in the Euclidean space are primary objects of interest; this is the case of stereology (see Chapter 15 in this book) or some topics in image reconstruction; see e.g. Baddeley and Molchanov (1998, Sec. 6). However, these theories differ from set estimation either in the emphasis (which is not mainly focussed on estimating a set) or in the structure of the sampling data (which is not necessarily made of randomly chosen points).

In a way, set estimation theory can be seen as a branch of nonparametric functional estimation (see e.g. Simonoff 1996) where the emphasis is placed on sets (distribution supports, level sets, boundaries, ...) rather than on functions. So set estimation has a geometrical flavor which is not that present in nonparametric functional estimation.

The aim here is to present an up-to-date survey of set estimation theory. There are, at least, two very common obstacles, besides the space limitations, in this type of surveys. First, the difficulty to establish the subject scope, sometimes close to the borders of other related issues; second, the need to find a classification criterion, reasonably simple and complete (see Section 11.2), to present the existing literature in a systematic, insightful way. In both respects we have made our choices. This overview is more inclined to the results in the nonparametrics line, with the emphasis on asymptotic properties (consistency, convergence rates, etc.), or smoothing methods. Among them we give priority to those involving methods or ideas closer to stochastic geometry. A rough classification for the literature on set estimation is followed in Section 11.2, according to the main target of interest (the support, its boundary or a level set) in the estimation problem and/or the available sample information (with or without auxiliary covariates).

In the selection of the material to be commented on there is a slight deliberate bias in favour of the more recent references, within say the last 10 years, over the older ones.

There is a wide branch of image analysis dealing with image compression and reconstruction which also uses statistical procedures but it is more related to harmonic analysis combined with numerical methods than to stochastic geometry. We do not consider here this important field which would require, at least, a survey as the present one. The paper by Donoho (1999) is an example of this line of work.

11.2 The problems under study

The following subsections define a tentative classification criterion aimed to present the set estimation problems in a systematic way. The symbol $\to$ will denote convergence as $n \to \infty$, unless otherwise stated.

11.2.1 *Support estimation*

Statement of the problem Given a random variable X with values in $\mathbb{R}^d$, denote by P_X and S, respectively, the distribution and the support of X. The problem is to find a suitable estimator $S_n = S_n(X_1, \ldots, X_n)$ of S based on a random sample $X_1, \ldots, X_n$ from P_X. We will refer to this sampling model as the *iid inner model*, just to indicate that the sample information comes only from P_X, hence all of the sample points belong to S. This is the most direct version of the set estimation problem.

It is often assumed that the support S is compact and X is absolutely continuous but this assumption is not necessary for some basic results.

If S is assumed to be convex there is a natural estimator for S, the convex hull of the sample points, $\operatorname{conv}(X_1, \ldots, X_n)$. Two early references by Rényi and Sulanke (1963, 1964) deal with the convex case which however will not be considered here since it is analysed in detail in Chapter 2 of this book.

The performance of a set estimator S_n is usually evaluated through either the Hausdorff distance $\rho_H(S_n, S)$, see Chapter 1, or the distance in measure $d_\mu(S_n, S)$ between S_n and S, defined by

$$\rho_H(S_n, S) = \inf\{\varepsilon > 0 : S_n \subset S^\varepsilon \text{ and } S \subset S_n^\varepsilon\} \text{ and } d_\mu(S_n, S) = \mu(S_n \Delta S),$$

where A^ε denotes the ε-parallel set (see Chapter 1) of A and μ stands for a reference measure. Often μ is either a probability measure or λ_d, the Lebesgue measure on $\mathbb{R}^d$. Anyway, in order to ensure $\mu(S_n \Delta S) < \infty$ for compact S_n and S we should obviously assume $\mu(B) < \infty$ for any bounded measurable B.

Constructing set estimators Similarly to the classical theory of parametric inference, in set estimation there are also different general methods to construct estimators. We should mention at least two general principles which, in various versions, are used in the different set estimation problems. First, *the plug-in approach*, based on the general principle of estimating the target set S by an analogous sampling version S_n. For example, if S is the (compact) support of a density f, S_n could be the support of a suitable compact supported density estimator f_n. A second usual approach is the *minimization of empirical functionals.* It arises when the target set S is found to be the minimizer of a certain functional $H(C)$ defined on some class $\mathcal{C}$ of sets to which S is supposed to belong. Then, a natural estimator can be constructed by minimizing over the class $\mathcal{C}$ an empirical estimator $H_n(C)$ of such functional. For example, the functional $H(C)$ could be defined just in terms of a distance, or discrepancy measure, $d(C, S)$ between C and S but other more sophisticated alternatives, as the so-called *excess mass approach*, will be considered below. In general terms, the plug-in estimates are more convenient for the practical use but the minimization of empirical functionals is more suitable to incorporate shape restrictions (through the corresponding choice of the minimization class $\mathcal{C}$) as well as to obtain optimality results (in particular, optimal convergence rates).

However, not every estimator is explicitly constructed according to these principles. In some cases the definition is given after some elementary intuitive considerations or using the *hull principle* which leads to define the set estimator as the 'smaller' set fulfilling some shape restriction. This is the case of the above mentioned *convex hull* but we will also comment other examples.

Some simple estimators The simplest support estimator, see Chevalier (1976), is maybe

$$S_n = \{X_1, \dots, X_n\}^{\varepsilon_n} = \bigcup_{i=1}^{n} B_{\varepsilon_n}(X_i). \tag{11.1}$$

A thorough study of the consistency properties of this natural estimator is due to Devroye and Wise (1980) who proved (Th. 1) the following result. If $\varepsilon_n \to 0$ and $n\varepsilon_n^d \to \infty$ then $d_\mu(S_n, S) \to 0$ in probability for all probability measure μ whose restriction to S is absolutely continuous with respect to the common distribution of the X_i.

The generality of this result is remarkable. No assumption at all is imposed on the distribution of the X_i. It is interesting to point out that the assumptions imposed on the sequence ε_n of smoothing parameters are the same, see Devroye (1983), needed to guarantee the universal L_1-consistency of kernel density estimators. More precisely, define

$$f_n(t) = \frac{1}{n} \sum_i K_h(t - X_i), \tag{11.2}$$

where $K_h(z) = (1/h^d)K(z/h)$, $K : \mathbb{R}^d \to \mathbb{R}$ is a fixed function called *kernel*, that is chosen by the user (we assume that K, and hence f_n, is a probability density) and $h = h_n$ a sequence of smoothing parameters, see, e.g., Simonoff (1996) for details. Devroye (1983) proved that a necessary and sufficient condition for the almost sure (a.s.) universal L_1-consistency, $\int |f_n - f| \mathrm{d}\lambda_d \to 0$, is precisely $h_n \to 0$ and $nh_n^d \to \infty$. Note that the estimator (11.1) can be seen as the support of a density estimator f_n whose kernel K is the uniform distribution on the unit ball. As a consequence, if P_X is assumed to be absolutely continuous with density f and compact support S, we can obtain also the strong consistency $d_\mu(S_n, S) \to 0$ a.s. by just assuming $\varepsilon_n \to 0$ and $n\varepsilon_n^d \to \infty$, since, with probability one

$$\mu(S_n \setminus S) \leq \mu(S^{\varepsilon_n}) \text{ and } P_X(S \setminus S_n) \leq \int |f_n - f| \mathrm{d}\lambda_d. \tag{11.3}$$

In general, the condition on ε_n required to ensure the strong universal consistency $d_\mu(S_n, S) \to 0$ a.s. (see Th. 2 in Devroye and Wise 1980) is $\varepsilon_n \to 0$ and $\sum_n \exp\left(-\alpha n \varepsilon_n^d\right) < \infty$, for all $\alpha > 0$.

Estimator (11.1) is also studied in the book by Grenander (1981) as an example of a 'regularization method' in 'abstract' (infinite-dimensional) inference.

The book by Korostelev and Tsybakov (1993a) is a further landmark in the development of set estimation theory. It includes a compilation of the authors' work on the subject together with a survey of the most relevant theory. The estimator (11.1) is considered in Section 7.2 of that book, under the title 'A simple and rough support estimator'. Assuming that the support S belongs to a certain class $\mathcal{G}$ of sets with piecewise Lipschitz boundaries and the distribution of the X_i is uniform on S, it is proved there (Th. 7.2.1) that for suitable ϵ_n

$$\sup_{S \in \mathcal{G}} \mathbf{E} d_{\lambda_d}(S_n, S) = \mathcal{O}\left(\frac{\log n}{n}\right)^{1/d}.$$

Another simple computationally efficient estimator is essentially based on the following idea. Assume that $S \subset [0,1]^d$. Consider a sequence $\mathcal{P}_n$ of partitions (grids) of $[0,1]^d$ that become finer as n increases. For example, $\mathcal{P}_n$ could be made of 'pixels' or rectangles whose size decreases as n grows. The estimator S_n is defined as the union of the elements of $\mathcal{P}_n$ containing at least one sample observation. This idea, as well as some derivations of it, has been analyzed in Ray Chaudhuri, Chaudhuri and Parui (1997) and Ray Chaudhuri *et al.* (1999) with a motivation on image analysis. In these papers the name s-shape is proposed for this estimator, s being the grid-side length. Further developments on the same idea have been independently proposed in Chaudhuri *et al.* (2004), which includes interesting smoothed versions of the estimator, still with a view toward image analysis, and in Baíllo and Cuevas (2006).

On r-convexity and estimation under shape restrictions If a shape restriction is imposed on the target set S it seems reasonable to define an estimator fulfilling this condition. The best known example is convexity, mentioned above. However, other milder shape-restrictions appear also in a quite natural way in set estimation. This is the case of *r-convexity*. A closed set A is said to be r-convex when it it coincides with its *r-convex hull*,

$$C_r(A) = \bigcap_{\text{int}(B_r(x))\cap A=\emptyset} (\text{int}(B_r(x)))^c.$$

Thus, a r-convex set can be expressed as an intersection of the complement sets of a family of balls of radius r. Letting r tend to infinity we see that $C_r(A)$ approaches the convex hull of A. Note that any closed convex set is r-convex for all $r > 0$.

Moreover, r-convexity is also closely related to the notions of erosion and dilation from mathematical morphology. It is not difficult to show that $C_r(A) = (A \oplus r(\text{int}(B))) \ominus r(\text{int}(B))$, where $\oplus$ and $\ominus$ denote, respectively, the Minkowski sum and difference (see Chapter 1), B is the unit ball $B_1(0)$, $A \oplus C = \{a + c : a \in A, c \in C\}$, $A \ominus C = \{x : \{x\} \oplus C \subset A\}$ and $rC = \{rc : c \in C\}$.

If we assume that the compact support S is r-convex and denote $\mathcal{X}_n = \{X_1, \ldots, X_n\}$, a natural estimator of S is $S_n = C_r(\mathcal{X}_n)$. The convergence rates of S_n have been studied in Rodríguez-Casal (2007) where the following general result (Th. 2) is given (see also Cuevas and Rodríguez-Casal 2004 for related ideas). Assume that S is r-convex and *standard* with respect to P_X, which means that there exist $\varepsilon_0 > 0$ and $\delta > 0$ such that for all $x \in S$ and $\varepsilon \leq \varepsilon_0$

$$P_X(B_\varepsilon(x) \cap S) \geq \delta\lambda_d(B_\varepsilon(x)), \tag{11.4}$$

then, with probability one,

$$\rho_H(S_n, S) = \mathcal{O}\left(\frac{\log n}{n}\right)^{1/d}.$$

Interestingly, this rate coincides with that obtained in Dümbgen and Walther (1996) for $\rho_H(S, \text{conv}(\mathcal{X}_n))$ when S is convex. Faster rates, of order $(\log n/n)^{2/(d+1)}$, are obtained in Rodríguez-Casal (2007) under additional assumptions on P_X and S which, in particular, entail also the r-convexity of $\text{cl}(S^c)$. Moreover, these rates hold for $\rho_H(\partial S_n, \partial S)$ and $d_{\lambda_d}(S_n, S)$ as well.

A more radical shape restriction arises when the set S is assumed to be the hypograph under a smooth curve. These sets are sometimes called 'boundary fragments' and will be considered again below in the context of boundary estimation (in connection with the problem of efficient boundary in productivity analysis) and level set estimation. With regard to support estimation under this restriction, let us mention for example the results in Härdle, Park, and Tsybakov (1995) where the rate of a 'piecewise polynomial' estimator of S is obtained. This rate depends on the smoothness of ∂S and the rate of decay of the density f near ∂S. A more recent reference is Girard and Jacob (2003) where the considered estimator relies both on Haar series and extreme values of the point process.

11.2.2 *Boundary estimation*

A further natural goal is estimating the boundary (or frontier) ∂S of the compact support S. This problem is clearly different from (and more difficult than) support estimation, as S_n could be close to S (in the Hausdorff metric) without the respective boundaries being close to each other. However, when set estimation is viewed from the perspective of image analysis, the identification of the boundary is maybe the most important issue.

In Cuevas and Rodríguez-Casal (2004) this problem is considered in a general context for the estimator (11.1), using the iid inner model. It is shown that $\rho_H(\partial S_n, \partial S) \to 0$ a.s. provided that $S \subset S_n$ eventually a.s. and $\varepsilon_n \to 0$ a.s. Under some shape restrictions on the support, sharp a.s. convergence rates of type $\mathcal{O}\left(\left(n^{-1}\log n\right)^{1/d}\right)$ are also obtained. See also Rodríguez-Casal (2007) for the r-convex case.

The Poisson model In boundary estimation the iid inner model is often replaced with the *Poisson model*, where the sample is given by a random number N of observations from a Poisson process whose intensity function f has a compact support S. Again the target is estimating ∂S. The asymptotic results are typically obtained by assuming that the intensity function becomes νf with $\nu \to \infty$. As pointed out in Hall, Park and Turlach (2002), first-order asymptotic results are generally the same in both contexts. A classical reference is Ripley and Rasson (1977) where the case of convex S is considered and the estimator is essentially a dilation of the convex hull of the sample. Some deep results have been obtained by Khmaladze and Weil (2007) concerning the local behaviour of Poisson processes in the vicinity $(\partial K)^\varepsilon$ of the boundary of a body K. Although these authors consider only the case of convex K, they also suggest that some of their results could be translated to more general settings.

When less attention is paid to geometrical considerations concerning the support of f, the problem shifts to the wide field of estimation of the intensity function, see Kutoyants (1998), in the same way as set estimation is related to nonparametric density estimation in the inner iid model.

The Poisson model is also followed in a number of papers, by Peter Hall and several co-authors, where a 'tracking method' is used to iteratively reconstruct a boundary or a fault line in a response surface. This method is proposed in Hall, Peng and Rau (2001) and Hall and Rau (2000). According to Cheng and Hall (2006) the basic idea behind this method is *(...) to estimate the boundary by steadily following a univariate track generated by the estimate itself. The calculations involve reaching a bandwidth radius into the plane from the current point estimate, gathering the data within that radius, and using this information to compute the next point estimate.* For a related 'rolling ball approach' see Hall, Park and Turlach (2002).

The efficient frontier problem An important part of the literature on boundary estimation is motivated by a problem in economics, in the field of productivity analysis. The efficiency of a company which transforms some inputs x (capital investments, human resources, etc.) into an output y (capital gains) can be measured by the difference $g(x) - y$ between y and the 'best attainable output' associated with the input x, which is defined by some function $g(x)$. In practice, $g(x)$ is not exactly known so that it must be estimated from a sample $(x_1, y_1), \ldots, (x_n, y_n)$ giving the performances of n randomly selected companies. This is, in short, the statement of the so-called 'efficient frontier problem'. To see how this problem is related to boundary estimation note that the sample points belong to a *hypograph set*,

$$S = \{(x, y) : x \in \mathcal{I},\, y \leq g(x)\}\,,$$

where $\mathcal{I}$ is the set of all possible inputs. The aim of estimating the function g is equivalent to the estimation of the upper boundary of the hypograph S. The classical approach to this problem, started by Farrell (1957), imposes a sort of monotonicity property on the boundary called the 'free disposal assumption' (that is, if $(x, y) \in S$ then $(x', y') \in S$ for any $x' \geq x$, $x' \in \mathcal{I}$, and $y' \leq y$) as well as the convexity of S. The standard estimator in this setup, often denoted DEA (Data Envelopment Analysis), has been extensively analyzed in the literature on the subject. It is essentially defined as the lowest function $\widehat{g}$ whose hypograph $\hat{S}_n$ includes the sample points, is convex and satisfies the free disposal property, see, e.g. Gijbels *et al.* (1999). A more general estimator, relying only in the free disposal assumption thus dropping convexity requirements, is the FDH (Free Disposal Hull). It is defined, see Deprins, Simar and Tulkens (1984), by an analogous principle to that of DEA. Some asymptotic theory on this estimator can be found in Korostelev, Simar and Tsybakov (1995a), Korostelev, Simar and Tsybakov (1995b) and Park, Simar and Weiner (2000). A more recent reference is Aragon, Daouia, and Thomas-Agnan (2005). A survey of these topics is given in Simar and Wilson (2000).

The true target in the efficient frontier problem here is the upper bound of an hypograph, defined by a function g. So in some sense this theory is closer to the classical functional estimation setup than to the more geometrically oriented framework of set estimation. However, there are other features in this problem which allow us to include it in the set estimation field. For example the structure of the data, which are drawn from a set whose upper border is defined by g and the construction of the estimators by the 'hull principle'. On the other hand, as mentioned in Korostelev and Tsybakov (1993a), the hypographs have some independent interest when considered as 'building blocks' for other more complicated structures, see Section 5.3.2 in the book by Korostelev and Tsybakov (1993a).

11.2.3 *Level set estimation*

In some situations where the underlying distribution is absolutely continuous with huge 'almost empty' areas of very low probability density, the support is not a so relevant feature; the level sets $L(c) = \{x : f(x) \geq c\} = \{f \geq c\}$ (f being the underlying density and $c > 0$ a given constant) arise as a more interesting target since they represent the 'substantial support' where most probability is concentrated.

The plug-in approach Maybe the most direct idea for estimating $L(c)$ is the *plug-in* method, consisting on replacing the unknown underlying density f with a suitable non-parametric density estimator f_n, for example of kernel type. This leads to an estimator of type $L_n = \{f_n \geq c\}$. When compared with other procedures (see below) this plug-in approach is more direct and easy to motivate but it is also less suitable in order to incorporate prior assumptions about the shape of $L(c)$.

A recent interesting contribution to the study of plug-in estimators is due to Cadre (2006) who provides (Th. 2.1) sufficient conditions to ensure the convergence in probability of $\sqrt{nh^d}\lambda_d(L_n \Delta L)$ to a constant. In particular, this shows that, under appropriate conditions, the exact rate of convergence of the plug-in estimator L_n with respect to the above defined distance 'in measure' d_{λ_d}, is $(nh^d)^{-1/2}$. Taking into account the conditions imposed on the smoothing parameter h, this rate turns out to be slower than $\mathcal{O}(n^{-2/(d+4)})$ which is in turn slower than those, of type $\mathcal{O}\left(n^{-1}\log n\right)^{2/(d+1)}$ for $d > 3$ and $\mathcal{O}\left(n^{-1/2}\log n\right)$ for $d \leq 3$, obtained in Walther (1997), Th. 4, for another different estimator (see below) under assumptions that essentially entail that f has a jump along $\partial L(c)$. On the contrary, Cadre's result applies to the case of 'smooth densities' with a possibly slow decay to zero. In any case, the condition that $\|\nabla f\|$ is bounded away from zero on $f^{-1}(c)$ is imposed. This is aimed to exclude the existence of a plateau at level c in the graph of f. Such a condition appears often, under different forms, in level set estimation, see Section 11.3.

Cuevas, González-Manteiga and Rodríguez-Casal (2006) and Molchanov (1998) coincide in considering plug-in estimators $\{h_n \geq c\}$ for general level sets of type $\{h \geq c\}$, where h is a real function of interest (not necessarily a density) defined on a metric space M (not necessarily $\mathbb{R}^d$). This is interesting not only for the sake of generality but also for practical reasons. For example, h could be a regression function, $h(x) = \mathbf{E}(Y|X = x)$ and M could be the unit sphere S^2, which is the space where the data live in the problems with directional data; see Mardia and Jupp (2000). The paper by Molchanov (1998), Th. 3.1, provides an exact convergence rate for the $\rho_H(L_n, L)$. In Cuevas, González-Manteiga and Rodríguez-Casal (2006, Th. 2), the problem of estimating the boundary ∂L is considered. It is shown that, under certain conditions which in particular exclude the existence of 'flat density regions',

$$\rho_H(\partial L_n, \partial L) = \mathcal{O}(\|h_n - h\|_\infty) \text{ a.s.},$$

so the rate (in the supremum norm) of h_n as an estimator of h is inherited for the estimator of the boundary ∂L.

The obvious connection between level sets and probability support can be used to construct a plug-in sequence of support estimators based on level set estimators. The basic idea, as developed in Cuevas and Fraiman (1997), is as follows. In most regular cases, the support S (i.e. the minimal closed set with probability one) essentially coincides (unless a λ_d-null set) with the set $\{f > 0\}$. Then we could think of estimating S by just replacing f in $\{f > 0\}$ by an estimator f_n. Note that if f_n is of kernel type (11.2) and the function K is strictly positive everywhere, the estimator $\{f_n > 0\}$ would be useless as it would always coincide with $\mathbb{R}^d$; on the other hand, if the support of K is compact, $\{f_n > 0\}$ would be just modified version of (11.1) with the balls replaced by another structural element. However, the plug-in approach could be still used by estimating the support S through a sequence $S_n = \{f_n > c_n\}$ of 'approximating level sets' where c_n is a sequence of tuning parameters with $c_n \downarrow 0$. The asymptotic behaviour of S_n is studied in Cuevas and Fraiman (1997). Thus, it is shown (Th. 1) that the L_1 convergence rates of the estimator f_n are translated into d_{λ_d}-convergence rates for S_n. Also, under some assumptions concerning boundedness of f and standardness (11.4) of S it is shown that, with a suitable choice of h_n, $\mathbf{E}d_{\lambda_d}(S_n, S)$ can achieve any rate of type $\mathcal{O}(n^{-s})$ slower than $n^{-1/d}$, (that is with $0 < s < 1/d$) and also the rates for $\rho_H(S_n, S)$ are shown to be, under quite general conditions, not faster than $\left(n^{-1}\log n\right)^{1/d}$. Note that, unlike the estimator (11.1), those of type $\{f_n > c_n\}$ can have smooth boundary.

The excess mass methodology This method has proved to be a particularly fruitful alternative idea for the estimation of $L(c)$. It is based (see Hartigan 1987, Müller and Sawitzki 1991) on the simple observation that the functional

$$H_c(B) = \int_B (f(x) - c)dx = P(B) - c\lambda_d(B)$$

is maximized by the level set $L(c)$. Then if $\mathcal{B}$ is a given class of sets, a natural estimator $L_n(c)$ of $L(c)$ under the shape restriction $L(c) \in \mathcal{B}$ would be the maximizer on $\mathcal{B}$ of the empirical excess mass $H_{c,n}(B) = P_n(B) - c\lambda_d(B)$. Hartigan (1987) considered the case where $\mathcal{B}$ is the class of convex sets and proposed and algorithm involving $\mathcal{O}(n^3)$ steps to obtain $L_n(c)$. Asymptotic results for the estimator restricted to much more general classes $\mathcal{B}$ are given by Polonik (1995) using the empirical process theory. Müller and Sawitzki (1991) showed that the function $c \mapsto E(c)$, where $E(c) = \sup_B\{H_c(B)\}$ can be useful to provide a methodology for the difficult problem of testing multimodality where the usual notion of mode as a local maximum is replaced with the more intrinsic concept of c-cluster, defined as a connected component of $L(c)$. Tsybakov (1997) uses a 'local' version of the excess mass methodology to deal with the problem of level set estimation under smoothness assumptions on the boundary. He gets piecewise polynomial estimators which fulfil some optimality properties of asymptotically minimax type.

It can be seen that the excess mass approach is a natural way to incorporate shape restrictions to the level set estimation problem. In fact, these restrictions are required in order to guarantee the applicability of this method. If no assumption is imposed on the class $\mathcal{B}$ the maximizer would be just $\{X_1, \ldots, X_n\}$. In general, the maximization on $\mathcal{B}$ will not be an easy task. An algorithm for the case where $\mathcal{B}$ is the class of ellipsoids in $\mathbb{R}^d$ is proposed in Nolan (1991).

The estimation of regression level sets $\{x : \mathbf{E}(Y|X = x) \geq c\}$ using the excess mass approach has been considered in Polonik and Wang (2005) and Cavalier (1997).

If the underlying density is assumed to be uniform (or at lest bounded away from below) and c is small enough, the problems of estimating the level set $L(c)$ and the support coincide. In this setting Klemelä (2004) has used a 'penalized' version of the excess mass method to get support estimators S_n built from a class $\mathcal{P}_n$ (increasing with n) of dyadic partitions of a d-dimensional rectangle. A penalization term is added to the excess mass functional in order to take into account the size of the class $\mathcal{P}_n$. The support S is assumed to be a so-called 'boundary fragment' with a smooth upper boundary given by a Hölder function $h : [0,1]^{d-1} \to \mathbb{R}$ of order $s \in (0,1]$. The corresponding rates for $\mathbf{E}d_{\lambda_d}(S_n, S)$ are, up to a log factor, of order $\mathcal{O}(n^{-s/(s+d-1)})$ which has been proved to be the minimax rate by Korostelev and Tsybakov (1993a, Th. 7.3.1).

Granulometric smoothing. The rolling property Besides the plug-in and the excess-mass estimators, we should mention the method of 'granulometric smoothing' proposed by Walther (1997). It is based on nice geometrical ideas and has good computational and statistical properties. It relies on the assumption that the level set $L = L(c)$ satisfies the following *rolling property.* For some $r_0 > 0$, a ball of radius r 'rolls freely' inside each path-connected component of L and $\text{cl}(L^c)$ for all $0 \leq r \leq r_0$ (we say that a ball of radius r rolls freely inside a set A if for all $x_0 \in \partial A$ there exists $x \in \mathbb{R}^d$ such that $x_0 \in B_r(x) \subset A$).

The rolling condition is a sort of smoothness property with an intuitive, geometrical character. It is closely related to the r-convexity considered above. More precisely, see Th. 1 in Walther (1997) and Walther (1999), a compact non-empty set $S \subset \mathbb{R}^d$ satisfies the above r_0-rolling property (for both S and $\text{cl}(S^c)$) if and only if both S and $\text{cl}(S^c)$ are r_0*-convex*, and each path-connected component of S has non-empty interior. This is in turn equivalent to $\Psi_r(S) = S$ for all $r \in (-r_0, r_0]$, where for $r \geq 0$,

$$\Psi_r(S) = (S \ominus rB) \oplus rB = \bigcup_{B_r(x) \subset S} B_r(x), \quad \Psi_{-r}(S) = (S \oplus rB) \ominus rB,$$

B being the unit ball of radius 1 centred at 0.

The transformation $r \mapsto \Psi_r(S)$, called the *granulometry of S* by Matheron (1975), provides information on the changes in S when it is dilated and eroded with the 'structural element' B. The invariance $\Psi_r(S) = S$ can be seen as a smoothness property of the set S.

Walther's (1997) proposal for estimating a level set $L = L(c)$ that satisfies the r_0-rolling property is based on two steps. First, we need an auxiliary density estimator f_n, for example of kernel type (11.2), based on the sample $\mathcal{X}_n = \{X_1, \ldots, X_n\}$. Then define

$$\mathcal{X}_n^+(c) = \{X \in \mathcal{X}_n : f_n(X) \geq c\}, \quad \mathcal{X}_n^-(c) = \mathcal{X}_n \setminus \mathcal{X}_n^+(c).$$

Second, the estimator of $L(c)$ is defined as an empirical approximation of $\Psi_r(L)$ given by

$$L_n(c) = \left(\mathcal{X}_n^+(c) \cap (\mathcal{X}_n^-(c) \oplus r_n B)^c\right) \oplus r_n B,$$

where r_n is a numerical sequence converging to zero. The estimator $L_n(c)$ is computationally feasible. It consists of the union of balls around those points in $\mathcal{X}_n^+$ that have a distance of at least r_n from each point in $\mathcal{X}_n^-$. The required number of computation steps is $\mathcal{O}(dn^2)$, so it is linear in the dimension.

The corresponding convergence rates in case of smooth f are faster than those obtained with the excess mass approach in Polonik (1995). In the non-smooth case the rates obtained from both approaches coincide up to log factors.

The estimation of level sets with a given probability content arises in connection with statistical quality control, see Section 11.5. Walther's (1997) granulometric approach can be also used in this case and the convergence rates agree with those obtained for the regular level set estimation problem. A plug-in approach to this problem, based on kernel estimates, can be found in Baíllo (2003) where convergence rates are obtained with respect to the measure distance based on the underlying distribution. These rates depend on the existence of moments and on the probability contents near the boundary of the target level set.

A Bayesian approach to level set estimation is considered in Gayraud and Rousseau (2005).

11.2.4 *Estimation in problems with covariates*

We consider here a relatively heterogeneous class of papers having as a common feature the fact that the sample data come from the observation of two, possibly related, random variables X and Y.

Image analysis As a first example, let us mention the paper by Carlstein and Krishnamoorty (1992), where the target is estimating the boundary of a compact set $S \subset [0,1]^d$ from the information provided by the observation of real random variables indexed by the nodes of a finite, deterministic, d-dimensional grid $\mathcal{J} \subset [0,1]^d$. The random variables X_i whose node i belongs to S have a cumulative distribution function F, while those falling in S^c have cumulative distribution $H \neq F$. There is no further information. If $d = 1$ and i indicates time, we are on the classical change point problem, or in an epidemic change model. For $d = 2$ if we think of $[0,1]^d$ as a geographical area, several applications have been considered. For example, the observations X_i could represent heights of trees on a forest, and F is the distribution of a healthy stand, while G is the distribution of a diseased stand. Thus ∂S will correspond to the boundary between both regions on the forest. An estimator of this boundary is proposed by Carlstein and Krishnamoorty (1992) based on the idea of maximizing (along the possible partitions (J_1, J_2) of the grid of nodes $\mathbb{J}$) the Kolmogorov–Smirnov distance between the empirical distributions F_{n1}, F_{n2} calculated from the observations with nodes in J_1 and J_2, respectively. For closely related ideas see Ferger (2004).

Other early references, regarding also boundary estimation, are Rudemo and Stryhn (1994) and Tsybakov (1994).

Mammen and Tsybakov (1995) study several sampling models which include, as particular cases, the main models previously analysed in the literature. Thus they consider the case where $S \subset [0,1]^d$ and the sampling observations are (X_i, Y_i), $i = 1, \ldots, n$ with $Y_i = (2\mathbf{1}_{\{X_i \in S\}} - 1)\xi_i$, where the ξ_i are iid random variables, independent from the X_i, taking values 1 and -1 with probabilities $1/2 + a_n$ and $1/2 - a_n$, respectively, a_n being a sequence with $0 < a_n < 1/2$. The value Y_i is interpreted as the image level at the point X_i, where $Y_i = 1$ stands for black and $Y_i = -1$ for white. Assuming that ∂S has a smooth parametrization these authors derive the expression of the asymptotically optimal d_{λ_d}-rates for the estimation of S; these rates depend on the smoothness degree of the boundary parametrization. These conclusions are very general and useful in order to know what are the best performance to be expected; they are focused on existence results rather than on computational or practical aspects. It should be noted that the requirement of smooth parametrization does not entail that the boundary itself is smooth. The results generalize those obtained in Korostelev and Tsybakov (1993a) for classes of smooth sets of hypograph type (boundary fragments).

Efficient frontier, again Most papers dealing with the problem of estimation of the efficient frontier, considered above in the subsection devoted to boundary estimation, could also be included here as they involve the use of covariates. Some examples are Girard and Jacob (2003), Hall, Nussbaum and Stern (1997) and Hall, Park and Stern (1998). A similar remark holds for those references, e.g. Hall, Peng and Rau (2001), Hall and Rau (2000), concerning the estimation of fault lines in regression surfaces.

Machine learning Another set-estimation problem with covariates, of increasing importance, in the field of machine learning, arises in nonparametric supervised classification where the aim is to predict the class $Y \in \{0, 1\}$ (for example, in a medical study 0 could correspond to 'healthy' and 1 to 'ill') to which an individual belongs, using the information provided by a $\mathbb{R}^d$-valued random variable X (for example, the result of a blood analysis) observed on this individual. It is assumed that we have a 'training sample' (X_i, Y_i), $i = 1, \ldots, n$ of correctly classified individuals (i.e. the value Y_i is known for them). In formal terms, the aim is to define an appropriate classifier $\hat{Y}(X)$, where $\hat{Y} : \mathbb{R}^d \longrightarrow \{0,1\}$, and $\hat{Y}(X) = \hat{Y}(X; X_1, \ldots, X_n)$ to predict the class Y of a new coming individual (X, Y) from which only X is observed. It is well-known (see Devroye, Györfi and Lugosi 1996 for details) that the optimal classifier Y^* which minimizes the expected risk $\mathbf{P}\{\hat{Y} \neq Y\}$ is given by the indicator function $\hat{Y}^*$ of the level set $G^* = \{x : \eta(x) \geq 1/2\}$, where $\eta(x) = \mathbf{P}(Y = 1|X = x)$. In the paper by Tsybakov (2004) classification is 'considered as a special case of nonparametric estimation of sets'. A related paper is Audibert and Tsybakov (2007) which includes also valuable, somewhat surprising, results questioning the alleged superiority of the classifiers obtained by empirical risk minimization over those constructed by the plug-in methodology which amounts to estimate the function η in the expression $\hat{Y}^*$ of the optimal classifier.

A recent contributions on the interplay between set estimation and the so-called *active learning* (a version of the supervised classification problems where there is a possibility to choose a sampling design for the observations X_i) is Castro and Novak (2008).

11.2.5 *Estimation of functionals and hypothesis testing problems related to the support*

A natural aim associated with set estimation is the inference about some real (or vectorial) functionals $\phi(S)$ defined from the set of interest S. For example if S is the (compact) support of the underlying distribution, $\phi(S)$ could be its Lebesgue measure, the gravity center or the boundary measure. The work by Korostelev and Tsybakov (1993b) is an earlier reference on the estimation of smooth functionals of the support. Also Chapter 8 in Korostelev and Tsybakov (1993a) is devoted to the problem of estimating the support area in the bi-dimensional case.

Under some smoothness assumptions on the support S, a test for the null hypothesis $H_0 : S = S_0$ versus an alternative of type $H_1 : |\int_S \varphi - \int_{S_0} \varphi| \geq \varepsilon$, where φ is a given function on $\mathbb{R}^d$, is proposed in Gayraud (2001). The problem of testing whether the contour of an image belongs to some given parametric family is analysed in Gayraud and Tsybakov (2002).

The paper by Cuevas, Fraiman and Rodríguez-Casal (2007) focuses on the topic of estimation of the boundary measure $L_0 = L_0(S)$ of a compact support S, as given for the $(d-1)$-dimensional Minkowski measure

$$L_0(S) = \lim_{\varepsilon \to 0} \frac{\lambda_d((\partial S)^\varepsilon)}{2\varepsilon}$$

which coincides with the better-known Haussdorf measure in regular cases. The method requires an inner-outer sampling model. In fact, the data consist of a random sample of points, taken on a rectangle containing S, in which we are able to identify whether every point is inside or outside S. The proposed estimator L_n is proved to fulfil $\mathbf{E}|L_n - L_0| \leq \mathcal{O}(n^{-1/2d})$ under a positive reach condition on S and S^c. This condition is conceptually related to the rolling property which has been considered above see e.g. Walther (1997). The *reach* (Federer 1959) of a closed set $G \subset \mathbb{R}^d$ is defined as the largest (possibly ∞) value r_0 such that if $x \in \mathbb{R}^d$ and the distance from x to G is smaller than r_0, then G contains a unique point nearest to x. Under additional shape restrictions, a related estimator $\tilde{L}_n$ is proposed in Pateiro-López and Rodríguez-Casal (2008) with a faster rate

$$|\tilde{L}_n - L_0| = \mathcal{O}\left(\left(\frac{\log n}{n}\right)^{1/(d+1)}\right) \quad a.s.$$

11.3 On optimal convergence rates

A major chapter in set estimation theory is the study of optimal convergence rates. To fix ideas take the problem of estimating a level set $L = \{h \geq c\}$, where h could be either a density or a regression function (similar ideas apply to support estimation). Given the nonparametric nature of set estimation problems it is quite unusual to have the expression of the optimal estimator (with respect to a distance d) for a given sample size, unless very strong assumptions are imposed on S. Instead, the optimality claims adopt often the following pattern: *assuming some conditions, say* (Ch), *for the function* h *and* (CL) *on the level set* L, *the best attainable estimation error for* L *fulfils* $d(L_n, L) = O(r_n)$ *(in probability or almost surely), where* $r_n \downarrow 0$, *and this rate is achieved for an estimator* L_n *belonging to a class* $\mathcal{S}$. As indicated there, the optimal estimators are not always easy to construct explicitly but still the knowledge of the optimal rate is very useful as a benchmark to assess the performance of any other estimator at hand (maybe easier to construct). Thus we could have some kind of informal trade-off between simplicity and efficiency.

A typical (Ch) condition is the so-called *margin assumption* which is established through an inequality of type

$$\mu\{x : |h(x) - c| \leq \varepsilon\} \leq C_0 \varepsilon^\gamma \tag{11.5}$$

for some constants $C_0 > 0$, $\gamma > 0$ and a measure μ (often the Lebesgue measure). Condition (11.5) appears (under different versions) in many papers, for example, Tsybakov (1997), Steinwart, Hush and Scovel (2005), Rigollet and Vert (2008), Audibert and Tsybakov (2007), Singh, Scott and Novak (2007). It prevents the function h from approaching too slowly the border of the level set $\{h \geq c\}$ thus avoiding the existence of a 'quasi-plateau' near $\{h = c\}$. The above commented condition that $\|\nabla h\|$ is bounded away from zero in $h^{-1}(c)$ can be seen as another formulation of this idea. Among the assumptions (CL) on the level set we could cite the paper by Tsybakov (1997) where a star-shape condition is imposed on the level sets, together with the existence of a smooth parameterization, expressed in polar coordinates, for their borders. The smoothness assumption for sets introduced by Dudley (1974) has been used by Mammen and Tsybakov (1995). Another interesting condition aimed to ensure that the level set 'is not arbitrarily narrow anywhere' is introduced in Singh, Scott and Novak (2007). It is similar to the *expandability assumption* introduced in Cuevas and Rodríguez-Casal (2004).

It is not easy to summarize the specific optimal convergence rates obtained in the recent literature since they are established for different classes of sets which depend on different parameters. As an example, let us mention that the optimal d_{λ_d}-rate obtained in Tsybakov (1997) for density level sets is of type $(1/n)^{((2\alpha+1)\beta+d-1)}$ where α and β are positive parameters controlling the increase rate of f around $f^{-1}(c)$ (through a margin condition) and the regularity of the level set. The ρ_H-rate is of the same type with $1/n$ replaced with $\log n/n$. Fast rates, arbitrarily close to n^{-1} can be attained for the optimal estimators under appropriate conditions, see Rigollet and Vert (2008). The same rates (or even *super fast rates*, that is rates faster than n^{-1}) have been obtained by Audibert and Tsybakov (2007) for the error in supervised classification problems. Optimal ρ_H-rates of type $(\log n/n)^{1/(d+2\alpha)}$ are obtained for plug-in estimators of level sets under less restrictive assumptions in Singh, Scott and Novak (2007); here α is the analogous of γ in (11.5) for a suitable version of the margin condition. These rates generalize several other results, see for example Cuevas and Fraiman (1997), with rates of type $\mathcal{O}(\log n/n)^{1/d}$ obtained for support estimation problems in the 'sharp', easier case $\alpha = 0$.

Further results on optimal convergence rates can be found, among others, in Korostelev and Tsybakov (1993a), Härdle, Park, and Tsybakov (1995), Mammen and Tsybakov (1995), Cavalier (1997), Scovel, Hush and Steinwart (2005), Steinwart, Hush and Scovel (2005). A recent result on asymptotic normality is given in Mason and Polonik (2009).

11.4 Some mathematical tools

The discussion in the previous sections provides an overview of set estimation, divided in several topics. We have thus reviewed different methods to construct estimators, including the simplest methodologies considered in Subsection 11.2.1, excess mass, plug-in approaches, hull-type estimators, etc. For each of them the relevant mathematical results concern consistency, convergence rates, optimality (often in minimax sense) and limit distributions, under different sampling models. As we cannot dwell on technical aspects, we will try at least to present a few typical tools in order to convey the mathematical flavor of the subject.

Shape restrictions As we have seen, the shape restrictions on the target set S (connectedness, convexity and r-convexity, reach and rolling conditions, boundary fragments, etc.) are used in two different ways. First they can be incorporated into the estimators, for example by using a 'hull principle', second they reduce the huge class of candidates to be considered thus allowing the use of tools leading to get convergence rates and, in general, sharper asymptotic results beyond simple consistency.

Most shape restrictions stem from the concept of convexity, one of the deepest and richest mathematical ideas which is at the origin of set estimation. A usual pattern is to generalize the property of convexity by imposing just one of the properties derived from it. This is the case with r-convexity, rolling property, etc. Another relevant example is the assumption of star-shape. A set S in the Euclidean space is said to be *star-shaped* if there exists a point $x_0 \in S$ such that for all $x \in S$ the segment joining x_0 and x is included in S. The set of all points x_0 fulfilling this property is called the 'kernel' of S. Some references on estimation of star-shaped sets are Baíllo and Cuevas (2001), De Haan and Resnick (1994), Korostelev and Tsybakov (1993a), Rudemo and Stryhn (1994) and Tsybakov (1997).

Empirical processes In general terms, empirical processes theory deals with the approximation of the underlying probability P from the empirical measure P_n based on a sample of size n. Typical results in this theory concern uniform convergence of Glivenko–Cantelli type such as $\sup_{B \in \mathcal{B}} |P_n(B) - P(B)| \to 0$ a.s., and weak convergence towards a Gaussian process G, $\sqrt{n}(P_n(\cdot) - P(\cdot)) \xrightarrow{w} G$. In both cases the 'size' or 'complexity' of the class $\mathcal{B}$ where the sets B live must be limited in order to ensure the existence of the limit. This is often done through some entropy conditions. A classical reference is Pollard (1984). In set estimation, the use of empirical processes arises, for example, in the above-mentioned problem of excess mass approach to level set estimation. More precisely, let $\mathcal{B} \subset \mathbb{R}^d$ be a class of measurable sets to which the level sets $L(c)$ are assumed to belong. Let $E_{\mathcal{B}}(c) = \sup_{B \in \mathcal{B}}\{H_c(B)\}$ and let $E_{\mathcal{B},n}(c) = \sup_{B \in \mathcal{B}}\{H_{c,n}(B)\}$ be the empirical version. Since

$$\begin{aligned} |\sup_{B \in \mathcal{B}}\{H_{c,n}(B)\} - \sup_{B \in \mathcal{B}}\{H_c(B)\}| &\leq \sup_{B \in \mathcal{B}} |H_{c,n}(B) - H_c(B)| \\ &= \sup_{B \in \mathcal{B}} |P_n(B) - P(B)|, \end{aligned} \tag{11.6}$$

the strong consistency of $E_{\mathcal{B},n}(.)$ to $E_{\mathcal{B}}(.)$ follows from that of the right-hand side of (11.6). The classes $\mathcal{B}$ for which this convergence holds are called 'Glivenko–Cantelli (GC) classes'. An example is the class of all d-dimensional closed ellipsoids. The class of all closed convex sets in $\mathbb{R}^d$, $d \geq 2$, is a GC-class for all distributions P which have a bounded Lebesgue density. See Polonik (1995) for details. In that paper the empirical processes methodology is also used to analyze conditions under which, assuming $L(c) \in \mathcal{B}$, the limit process of $\sqrt{n}(E_{\mathcal{B},n}(.) - E_{\mathcal{B}}(.))$ is a Brownian bridge.

Therefore, we could say that empirical process theory is an important tool in set estimation. More examples of its use can be found in Baíllo (2003), Mammen and Tsybakov (1995) and Tsybakov (1997), among others.

Some tools from stochastic geometry and graph theory Penrose (1999) has proved the following result. Suppose that $X_1, \ldots, X_n$ are iid observations in $\mathbb{R}^d$, $d \geq 2$, with common absolutely continuous distribution P_X whose density f has a connected and compact support S, with smooth boundary ∂S and $f|_{\partial S}$ is continuous. Let M_n denote the smallest r such that the union of balls of diameter r centred at the first n sample points is connected. Let ω_d denote the volume of the unit ball. Then as $n \to \infty$,

$$\frac{n\omega_d M_n^d}{\log n} \longrightarrow \max\left\{\left(\min_S f\right)^{-1}, 2\left(1-\frac{1}{d}\right)\left(\min_{\partial S} f\right)\right)^{-1}\right\} \text{ a.s.} \tag{11.7}$$

Note that, in the terminology of graph theory, M_n is the longest edge of the minimal spanning tree with vertices in the sample points.

From the point of view of set estimation, (11.7) provides the exact order, including the constant, for $\bar{\varepsilon}_n = M_n/2$ which is the 'connectivity radius', i.e. the minimal value of ε_n which makes connected the union-of-balls estimator (11.1). So, if we denote by C_0 the constant in the right-hand side of (11.7), the support estimator (11.1) with $\varepsilon_n = C\left(n^{-1}\log n\right)^{1/d}$ for any $C > \left(2^{-d}C_0/\omega_d\right)^{1/d}$ is eventually connected a.s. Then if μ is a measure with $\mu|_S \ll P_X$ and $\mu(T) < \infty$ for any bounded T, the weak consistency $d_\mu(S_n, S) \to 0$ (in prob.) can be obtained from Th. 1 in Devroye and Wise (1980) and the strong (a.s.) consistency follows from the bounds in (11.3); see Baíllo and Cuevas (2001).

The recent paper by Penrose (2007) (see also Penrose and Yukich 2003) gives additional valuable insights on the connection between random graphs and set estimation.

In Edelsbrunner, Kirkpatrick and Seidel (1983) a generalization (closely linked to the definition of r-convexity) is proposed for the convex hull of a sample of points. The related concept of α-shape is also defined. For $\alpha > 0$ a sample point X_i is said to be α-extreme if it lies in the boundary of a closed ball of radius $1/\alpha$ which contains all the points. Two points are called α-neighbours when they both lie in the boundary of a closed ball of radius $1/\alpha$ which contains all other points. For $\alpha < 0$ these notions work replacing the closed $1/\alpha$-ball with the complement of an open $-(1/\alpha)$-ball. For $\alpha = 0$ the balls are replaced with half-spaces. Given a set $\mathcal{X}_n = \{X_1, \ldots, X_n\}$ of points in the plane and an arbitrary real α, the *α-shape* of $\mathcal{X}_n$ is the straight line graph whose vertices are the α-extreme points and whose edges connect the respective α-neighbours. The algorithmic aspects of these ideas, when extended to three-dimensional points, as well as some related geometric concepts, are discussed by Edelsbrunner and Mücke (1994).

Spacings The 'multivariate spacings' are a tool of stochastic geometry that generalizes to the multivariate setting the classical notion of 'spacing' between two consecutive order statistics in a sample of real random observations.

The paper by Deheuvels (1983) is a pioneering reference. Given a sample $\mathcal{X}_n = \{X_1, \ldots, X_n\} \subset S = [0,1]^d$, the maximal spacing can be defined as the radius of the largest ball which can be included in S without including any sample point in its interior. This is formally defined by

$$\Delta_n = \sup\{r : \exists x \text{ with } B(x,r) \subset S \setminus \mathcal{X}_n\},$$

and from the results in Janson (1987) it follows that, if the X_i are uniformly distributed,

$$\lim_{n\to\infty} \left(\frac{n}{\log n}\right)^{\frac{1}{d}} \Delta_n = \left(\frac{1}{\omega_d}\right)^{\frac{1}{d}} \quad \text{a.s.}$$

where ω_d denotes the Lebesgue measure of the unit ball in $\mathbb{R}^d$. Using this, it is not difficult to prove (see Cuevas and Rodríguez-Casal 2004 for details) that, if S_n is an estimator of type (11.1) then, with probability one,

$$\liminf_{n\to\infty} \left(\frac{n}{\log n}\right)^{\frac{1}{d}} \rho_H(S_n, S) \geq \frac{1}{2}\left(\frac{1}{\omega_d}\right)^{\frac{1}{d}},$$

which means that the rate cannot be improved in general even if S is assumed to be standard. For a related use of spacings in set estimation see Th. 3 in Cuevas and Fraiman (1997).

Exponential inequalities In general terms, probabilistic exponential inequalities deal with the probability of deviations larger than a positive t between a random variable Y and its expectation $\mathbf{E}(Y)$. They establish exponentially small bounds (as t increases) for such deviations. Of course the precise expression of the bound and the assumptions on the variable Y, vary for the different inequalities. These inequalities are used as concentration results around the expectation which enable us to replace the random variable Y with its expectation in some reasonings. Important examples are Bernstein's inequalities, for which Y is a sum of iid random variables, and McDiarmid's inequality, for which $Y = f(X_1, \ldots, X_n)$ is a general function of n independent variables with the property of bounded differences in each variable, that is $|f(x_1, \ldots, x_i, \ldots, x_n) - f(x_1, \ldots, x_i', \ldots, x_n)| \leq c_i$, for some constants c_i. The book by Devroye, Györfi and Lugosi (1996) provides useful and complete information on exponential inequalities and its practical application in statistics.

Some examples of use of Bernstein's inequalities in set estimation can be found in Cheng and Hall (2006), Cuevas and Fraiman (1997), Mammen and Tsybakov (1995), Polonik (1995) and Tsybakov (1997). As for McDiarmid's inequality see, for example, Baíllo (2003).

Differential geometry Just to cite a few examples, the papers by Walther (1997, 1999) include some interesting arguments relating the rolling condition with the differential properties of the boundary and the so-called 'Serra's regular model' for sets; see Serra (1982).

The paper by Cadre (2006) shows an example of the use of $(d-1)$-dimensional Hausdorff measures. In Cuevas, Fraiman and Rodríguez-Casal (2007) the $(d-1)$-dimensional surface measures (in the more restrictive Minkowski version) arise again, together with an application of Federer's (1959) extension of Steiner formula on the measure of convex parallel sets.

Finally, the paper by Khmaladze and Weil (2007) includes a wealth of tools from differential geometry.

Tools from random set theory The paper by Molchanov (1998) provides an example of the use of central limit theorem for random sets in the framework of set estimation. Also, the distances proposed in Baddeley and Molchanov (1998) are likely to provide useful tools in this setting.

11.5 Further applications

In the previous sections we have already mentioned different applications (e.g. productivity analysis) of set estimation as well as connections with other topics.

Image analysis We have also mentioned several references whose aim or motivation was partially related to the field of image analysis. These include Baíllo and Cuevas (2006), Carlstein and Krishnamoorty (1992), Cheng and Hall (2006), Cuevas, Fraiman and Rodríguez-Casal (2007), Korostelev and Tsybakov (1993a), Ray Chaudhuri, Chaudhuri and Parui (1997), Ray Chaudhuri *et al.* (1999), Ray Chaudhuri *et al.* (2004), and Rudemo and Stryhn (1994).

Anomaly detection Another practical aspect which deserves some attention arises in connection with the following *statistical quality control* problem. Consider a process yielding independent observations $X_1, X_2, \ldots$ of a d-dimensional random variable X, for example, certain quality characteristics of a manufactured item. Initially, when the process is 'in control', the observations follow a distribution F, with density f. Let us assume that we have a 'pilot' (pre-run) sample $X_1, \ldots, X_n$ from F taken during a monitoring (in control) period. At some stage, the process may run out of control and the distribution of the X_i's changes to some unknown G. The goal is to detect a real change in the distribution of subsequent observations X_{n+k}, $k \geq 1$, as quickly as possible. Devroye and Wise (1980) proposed to use set (support) estimation in this problem. The basic idea would be just decide that a change has occurred at the stage $n+k$, based on a single new observation, when $X_{n+\mathrm{k}} \notin S_n$ where $S_n = S_n(X_1, \ldots, X_n)$ is an appropriate estimator of the support S of the 'in control' distribution F. This approach is reminiscent of the classical methodology based on Shewhart's control charts in a multivariate and non-parametric version. Different theoretical and practical aspects of this idea have been developed in Baíllo (2003) and Baíllo, Cuevas and Justel (2000).

Clustering Cluster analysis (or unsupervised classification) provides another application for level set estimation. The starting point is Hartigan's notion which identifies a (population) c-cluster with a connected component of the level set $L(c) = \{f \geq c\}$. Thus, a natural proposal is to estimate the c-clusters by the empirical clusters, defined as the connected components of $L_n(c) = \{f_n \geq c\}$, f_n being a non-parametric estimator of f. Finally, the data clusters in which the sample $\{X_1, \ldots, X_n\}$ is divided could be defined by just classifying the observations according to the empirical cluster they belong. The theoretical and practical problems involved in these ideas are addressed in Cuevas, Febrero and Fraiman (2000). Note that this approach to clustering takes into account the geometrical shape of the clusters, unlike the classical k-means algorithm which tend to provide 'globular' clusters. This flexibility is important in applications to astronomy where, clearly, the clusters of stars or galaxies need not be globular. An interesting application in astronomy of this density-based algorithm (as well as some computational improvements of it) can be found in Jang and Hendry (2007). For a related application in astronomy see Cuevas, González-Manteiga and Rodríguez-Casal (2006).

Acknowledgements

We are very grateful for the suggestions and criticisms from the editors and an anonymous reviewer who brought to our attention several important ideas and references.

We are also indebted to our co-authors in set estimation subjects, Amparo Baíllo, Manuel Febrero and Alberto Rodríguez-Casal who have influenced decisively our view of this topic.

This research has been partially supported by grant MTM2007-66632 from the Spanish Ministry of Education.

References

Aragon, Y., Daouia, A. and Thomas-Agnan, C. (2005). Nonparametric frontier estimation: A conditional quantile-based approach. *Economet. Theor.*, **21**, 358–389.

Audibert, J.Y. and Tsybakov, A.B. (2007). Fast learning rates for plug-in classifiers. *Ann. Statist.*, **35**, 608–633.

Baddeley, A. and Molchanov, I. (1998). Averaging of random sets based on their distance functions. *J. Math. Imaging Vision*, **8**, 79–92.

Baíllo, A. (2003). Total error in a plug-in estimator of level sets. *Statist. Probab. Lett.*, **65**, 411–417.

Baíllo, A. and Cuevas, A. (2001). On the estimation of a star-shaped set. *Adv. in Appl. Probab.*, **33**, 1–10.

Baíllo, A. and Cuevas, A. (2006). Image estimators based on marked bins. *Statistics*, **40**, 277–288.

Baíllo, A., Cuevas, A. and Justel, A. (2000). Set estimation and nonparametric detection. *Canad. J. Statist.*, **28**, 765–782.

Cadre, B. (2006). Kernel estimation of density level sets. *J. Multivariate Anal.*, **97**, 999–1023.

Carlstein, E. and Krishnamoorty, C. (1992). Boundary estimation. *J. Amer. Statist. Assoc.*, **87**, 430–438.

Castro, R.M. and Nowak, R.D. (2008). Minimax bounds for active learning. *IEEE T. Inform. Theory*, **5**, 2339–2353.

Cavalier, L. (1997). Nonparametric estimation of regression level sets. *Statistics*, **29**, 131–160.

Cheng, M.Y. and Hall, P. (2006). Methods for tracking support boundaries with corners. *J. Multivariate Anal.*, **97**, 1870–93.

Chevalier, J. (1976). Estimation du support et du contour de support d'une loi de probabilité. *Ann. Inst. H. Poincaré B*, **12**, 339–364.

Cuevas, A. and Fraiman, R. (1997). A plug-in approach to support estimation. *Ann. Statist.*, **25**, 2300–2312.

Cuevas, A., Febrero, M. and Fraiman, R. (2000). Estimating the number of clusters. *Canad. J. Statist.*, **28**, 367–382.

Cuevas, A., Fraiman, R. and Rodríguez-Casal, A. (2007). A nonparametric approach to the estimation of lengths and surface areas. *Ann. Statist.*, **35**, 1031–1051.

Cuevas, A., González-Manteiga, W. and Rodríguez-Casal, A. (2006). Plug-in estimation of general level sets. *Aust. N. Z. J. Stat.*, **48**, 7–19.

Cuevas, A. and Rodríguez-Casal, A. (2004). On boundary estimation. *Adv. in Appl. Probab.*, **36**, 340–354.

Deheuvels, P. (1983). Strong bounds for multidimensional spacings. *Z. Wahrsch. verw. Gebiete*, **64**, 411–424.

Deprins, D., Simar, L. and Tulkens, H. (1984). Measuring labor efficiency in post offices. In *The Performance of Public Enterprises: Concepts and Measurements* (ed. M. Marchand, P. Pestieau and H. Tulkens), pp. 243–267. North-Holland, Amsterdam.

Devroye, L. (1983). The equivalence of weak, strong and complete convergence in L_1 for kernel density estimates. *Ann. Statist.*, **11**, 896–904.

Devroye, L., Györfi, L. and Lugosi, G. (1996). *A Probabilistic Theory of Pattern Recognition*. Springer-Verlag, New York.

Devroye, L. and Wise, G. (1980). Detection of abnormal behaviour via nonparametric, estimation of the support. *SIAM J. Appl. Math.*, **3**, 480–488.

Donoho, D.L. (1999). Wedgelets: nearly minimax estimation of edges. *Ann. Statist.*, **27**, 859–897.

Dudley, R.M. (1974). Metric entropy of some classes of sets with differentiable boundaries. *J. Approx. Theory*, **10**, 227–236.

Dümbgen, L. and Walther, G. (1996). Rates of convergence for random approximations of convex sets. *Adv. in Appl. Probab.*, **28**, 384–393.

Edelsbrunner, H., Kirkpatrick, D.G. and Seidel, R. (1983). On the shape of a set of points in the plane. *IEEE T. Inform. Theory*, **IT-29**, 551–559.

Edelsbrunner, H. and Mücke, E.P. (1994). Three dimensional α-shapes. *ACM T. Graphic*, **13**, 43–72.

Farrell, M.J. (1957). The measurement of productive efficiency. *J. Roy. Statist. Soc. Ser. A*, **120**, 253-281.

Federer, H. (1959). Curvature measures. *Trans. Amer. Math. Soc.*, **93**, 418–491.

Ferger, D. (2004). Boundary estimation based on set-indexed empirical processes. *J. Nonparametr. Stat.*, **16**, 245–260.

Gayraud, G. (2001). Minimax hypothesis testing about the density support. *Bernoulli*, **7**, 507–525.

Gayraud, G. and Rousseau, J. (2005). Rates of convergence for a Bayesian level set estimation. *Scand. J. Statist.*, **32**, 639–660.

Gayraud, G. and Tsybakov, A.B. (2002). Testing hypotheses about contours in images. *J. Nonparametr. Stat.*, **14**, 67–85. 62G10 (62G20).

Gijbels, I., Mammen, E., Park, B.U. and Simar, L. (1999). On estimation of monotones and concave frontier funtions. *J. Amer. Statist. Assoc.*, **94**, 220–228.

Girard, S. and Jacob, P. (2003). Extreme values and Haar series estimates of point process boundaries. *Scand. J. Statist.*, **30**, 369–384.

Grenander, U. (1981). *Abstract Inference.* Wiley, New York.

De Haan, L. and Resnick, S. (1994). Estimating the home range. *J. Appl. Probab.*, **31**, 700–720.

Hall, P., Nussbaum, M. and Stern, S.E. (1997). On the estimation of a support curve of indeterminate sharpness. *J. Multivariate Anal.*, **62**, 204–232.

Hall, P., Park, B.U. and Stern, S.E. (1998). On polynomial estimators of frontiers and boundaries. *J. Multivariate Anal.*, **66**, 71–98.

Hall, P., Park, B. and Turlach, B. (2002). Rolling-ball method for estimating the boundary of the support of a point-process intensity. *Ann. Inst. H. Poincaré Probab. Statist.*, **38**, 959–971.

Hall, P., Peng, L. and Rau, C. (2001). Local likelihood tracking of fault lines and boundaries. *J. Roy. Statist. Soc. Ser. B*, **63**, 569–582.

Hall, P. and Rau, C. (2000). Tracking a smooth fault line in a response surface. *Ann. Statist.*, **28**, 713–733.

Härdle, W., Park, B.U. and Tsybakov, A.B. (1995). Estimation of non-sharp support boundaries. *J. Multivariate Anal.*, **55**, 205–218.

Hartigan, J.A. (1987). Estimation of a convex density contour in two dimensions. *J. Amer. Statist. Assoc.*, **82**, 267–270.

Janson, S. (1987). Maximal spacings in several dimensions. *Ann. Probab.*, **15**, 274–280.

Jang, W. and Hendry, M. (2007). Cluster analysis of massive datasets in astronomy. *Stat. Comput.*, **17**, 253–262.

Khmaladze, E. and Weil, W. (2007). Local empirical processes near boundaries of convex bodies. *Ann. Inst. Stat. Math.*, **60**, 813–842.

Klemelä, J. (2004). Complexity penalized support estimation. *J. Multivariate Anal.*, **88**, 274–297.

Korostelev, A.P., Simar, L., and Tsybakov, A.B. (1995a) On estimation of monotone and convex boundaries. *Publ. Inst. Statist. Univ. Paris*, **39**, 3–18.

Korostelev, A. P., Simar, L., and Tsybakov, A. B. (1995b) Efficient estimation of monotone boundaries. *Ann. Statist.*, **23**, 476–489.

Korostelev, A.P. and Tsybakov, A.B. (1993). *Minimax Theory of Image Reconstruction.* Lecture Notes in Statistics 82. Springer-Verlag, New York.

Korostelev, A.P. and Tsybakov, A.B. (1993). Estimating the support of a density and functionals of it. *Problems Inform. Transmission*, **29**, 1–15.

Kutoyants, Y. (1998). *Statistical Inference for Spatial Poisson Processes.* Lecture Notes in Statistics, 134. Springer-Verlag, New York.

Mammen, E. and Tsybakov, A.B. (1995). Asymptotical minimax recovery of sets with smooth boundaries. *Ann. Statist.*, **23**, 502–524.

Mardia, K.V. and Jupp, P.E. (2000). *Directional Statistics.* Wiley, Chichester.

Mason, D.M. and Polonik, W. (2009). Asymptotic normality of plug-in level set estimates. *Ann. Appl. Probab.* **19**, 1108–1142.

Matheron, G. (1975). *Random Sets and Integral Geometry.* Wiley, New York.

Molchanov, I. (1998). A limit theorem for solutions of inequalities. *Scand. J. Statist.*, **25**, 235–242.

Müller, D.W. and Sawitzki, G. (1991). Excess mass estimates and tests of multimodality. *J. Amer. Statist. Assoc.*, **86**, 738–746.

Nolan, D. (1991). The excess-mass ellipsoid. *J. Multivariate Anal.*, **39**, 348–371.

Park, B.U., Simar, L. and Weiner, C. (2000). The FDH estimator for productivity efficiency scores. *Economet. Theor.*, **16**, 855–877.

Pateiro-López, B. and Rodríguez-Casal, A. (2008). Length and surface area estimation under convexity type restrictions. *Adv. in Appl. Probab.*, **40**, 348–358.

Penrose, M.D. (1999). A strong law for the longest edge of the minimal spanning tree. *Ann. Probab.*, **27**, 246–260.

Penrose, M.D. (2007). Laws of large numbers in stochastic geometry with statistical applications. *Bernoulli*, **13**, 1124–1150.

Penrose, M.D. and Yukich, J.E. (2003). Weak laws of large numbers in geometric probability. *Ann. Appl. Probab.*, **13**, 277–303.

Pollard, D. (1984). *Convergence of Stochastic Processes.* Springer, New York.

Polonik, W. (1995). Measuring mass concentration and estimating density contour clusters-an excess mass approach. *Ann. Statist.*, **23**, 855–881.

Polonik, W. and Wang, Z. (2005). Estimation of regression contour clusters. An application of the excess mass approach to regression. *J. Multivariate Anal.*, **94**, 227–249.

Ray Chaudhuri, A., Chaudhuri, B. B. and Parui, S. K. (1997). A novel approach to computation of the shape of a dot pattern and extraction of its perceptual border. *Comput. Vis. Image Und.*, **68**, 257–275.

Ray Chaudhuri, A., Basu, A., Bhandari, S.K. and Chaudhuri, B.B. (1999). An efficient approach to consistent set estimation. *Sankhya Ser. B*, **61**, 496–513.

Ray Chaudhuri, A., Basu, Tan, K., Bhandari, S.K. and Chaudhuri, B.B. (2004). An efficient set estimator in high dimensions: consistency and applications to fast data visualization. *Comput. Vis. Image Und.*, **93**, 260–287.

Rényi, A. and Sulanke, R. (1963). Über die konvexe Hülle von n zufällig gewählten Punkten. *Z. Wahrsch. verw. Gebiete*, **2**, 75–84.

Rényi, A. and Sulanke, R. (1964). Über die konvexe Hülle von n zufällig gewählten Punkten II. *Z. Wahrsch. verw. Gebiete*, **3**, 138–147.

Rigollet, P. and Vert, R. (2008). Fast rates for plug-in estimators of density level sets. *Manuscript.*

Ripley, B.D. and Rasson, J.P. (1977). Finding the edge of a Poisson forest. *J. Appl. Probab.*, **14**, 483–491.

Rodríguez-Casal, A. (2007). Set estimation under convexity-type assumptions. *Ann. Inst. H. Poincaré Probab. Statist.*, **43**, 763–774.

Rudemo, M. and Stryhn, H. (1994). Approximating the distribution of maximum likelihood contour estimators in two-region images. *Scand. J. Statist.*, **21**, 41–55.

Scovel, C., Hush, D. and Steinwart, I. (2005). Learning rates for density level detection. *Analysis and Applications*, **3**, 356–371.

Serra, J. (1982). *Image Analysis and Mathematical Morphology.* Academic Press, London.

Simar, L. and Wilson, P. (2000). Statistical inference in nonparametric frontier models: The state of the art. *J. Prod. Anal.*, **13**, 49–78.

Simonoff, J. S. (1996). *Smoothing Methods in Statistics.* Springer-Verlag, New York.

Singh, A., Scott, C. and Nowak, R. (2007) Adaptive Hausdorff estimation of density level sets. *Manuscript.*

Steinwart, I., Hush, D. and Scovel, C. (2005). A classification framework for anomaly detection. *J. Machine Learn. Res.*, **6**, 211-232.

Tsybakov, A.B. (1994). Multidimensional change-point problems and boundary estimation. In *Change-Point Problems* (ed. M. Carlstein, H.G.. Müller and D. Siegmund), IMS Lecture Notes Monograph Series, 23, pp. 317–329.

Tsybakov, A.B. (1997). On nonparametric estimation of density level sets. *Ann. Statist.*, **25**, 948–969.

Tsybakov, A.B. (2004). Optimal aggregation of classifiers in statistical learning. *Ann. Statist.*, **32**, 135–166.

Walther, G. (1997). Granulometric smoothing. *Ann. Statist.*, **25**, 2273–2299.

Walther, G. (1999). On a generalization of Blaschke's Rolling Theorem and the smoothing of surfaces. *Math. Methods Appl. Sci.*, **22**, 301–316.

12

DATA DEPTH: MULTIVARIATE STATISTICS AND GEOMETRY

Ignacio Cascos

Abstract

This chapter presents several ways to measure the degree of centrality of a point with respect to a multivariate probability distribution or a data cloud. Such degree of centrality is called *depth*, and it can be used to extend a wide range of univariate techniques that are based on the natural order on the real line to the multivariate setting.

12.1 Introduction, background and history of data depth

The real line has many convenient properties that simplify the task of performing the analysis of univariate data sets. Many of those properties are shared by the multivariate Euclidean space, but unfortunately, the existence of a natural order is not among them. The usual order relation $\leq$ on $\mathbb{R}$ is a total order relation and the sorting of data points (with respect to it) allows us to rank them and is in the basis of many data analysis techniques (though not always in an obvious manner). The lack of a natural order on $\mathbb{R}^d$ for $d \geq 2$ is an obstacle when translating those techniques to the multivariate setting. Of course, there are several extensions of the relation $\leq$ to $\mathbb{R}^d$, among them the componentwise order. Unfortunately, such multivariate extensions lack of some of the properties of the univariate order. In particular, the componentwise order is not a total order.

As statisticians, we develop procedures based on data, and thus the ordering on $\mathbb{R}^d$ that we use may be data-based, too. In addition to multivariate ordering, we are interested in statistical notions that capture the particular features of a given data set. A *depth function* assigns each point in $\mathbb{R}^d$ its degree of centrality with respect to a probability distribution or a data cloud, in the sense that for the more central points (the *deepest* ones) the depth is higher, and the depth decreases as we move towards the tails of the distribution. As a consequence, a depth function enables us to rank the data points with respect to the centre-outward ordering that it defines. Further, by construction, the data depths do not favour any direction, and moreover, the rankings they provide us with are independent of the chosen system of coordinates.

In order to capture information about the geometrical configuration of a data cloud, some geometrical objects built from a random vector can be put into play. Evidently, a random vector can be treated as a random singleton, but, as a singleton, it has no structure and contains little geometrical information. Nevertheless, we can build a random segment from a fixed point to the random vector and compute its selection expectation. Alternatively, we can consider the random set defined as the convex hull of a given number of independent copies of the random vector (and maybe a fixed point) and study its coverage function, its selection expectation or its average volume. From $d-1$ observations of a random vector and a fixed point, we obtain a random hyperplane. All these random sets appear in the construction of classical depth functions, some of which will be considered in the present chapter.

Associated with each notion of data depth, we can build the level sets of the depth function. Those level sets constitute a family of *depth-trimmed regions* or *central regions*, a family of nested sets formed by all points whose depth is, at least, some given value. These depth-trimmed regions capture the shape of a data cloud and visualize valuable information about the location, scatter and relations between the variables.

We commence by summarizing the history of these methods. The first notion of *data depth* was proposed by Tukey (1975) in a data analysis context. J.W. Tukey wanted to obtain a bivariate analog of the univariate rank statistic and, further, develop some contour plots that would help exploring a bivariate sample. In this work, J.W. Tukey coined the term *depth* to refer to a function that depends on the data and assigns the highest values to central points and decreases as we move away from the center. Independently, Barnett (1976) published an influential work describing extensions of the concepts of median, range and quantiles to the multivariate setting. V. Barnett claimed that, although there is no natural order for multivariate data, most univariate statistics that are naturally defined as transformations of order statistics, can be extended to higher dimensions.

One of the first exploratory data analysis techniques that induced an ordering in a multivariate data set is the method of *convex hull peeling*. Given the data set, its first step consists in obtaining the convex hull of all data points, and in a second step, the data points which are extreme values of that convex hull are deleted. These steps are iterated until no data point remains. The name 'convex hull peeling' arises because subsequent layers of data points are peeled off, like onion layers. Such a procedure is useful in data analysis but lacks a distributional counterpart. Consequently, limit theorems for convex hull peeling, like those of Hueter (2005), are scarce and describe particular features of the layers.

In Fig. 12.1 we present the contour plots of the convex hull peeling layers of some real data and the numbers of the convex hull peeling layer that corresponds to each data point. The data represents the results of the 30 athletes that competed in the 2004 Olympics Decathlon in long jump (in meters, axis X) and in the 100 m race (in seconds, axis Y).

The papers of Tukey and Barnett rapidly influenced other authors like Green and Silverman (1979) and Green (1981) who further improved the technique of convex hull peeling, constructed efficient algorithms and described applications. Eddy and Gale (1981) obtained limit theorems for the convex hull (the outermost layer of the convex hull peeling) of a random sample. Later, Eddy (1982)

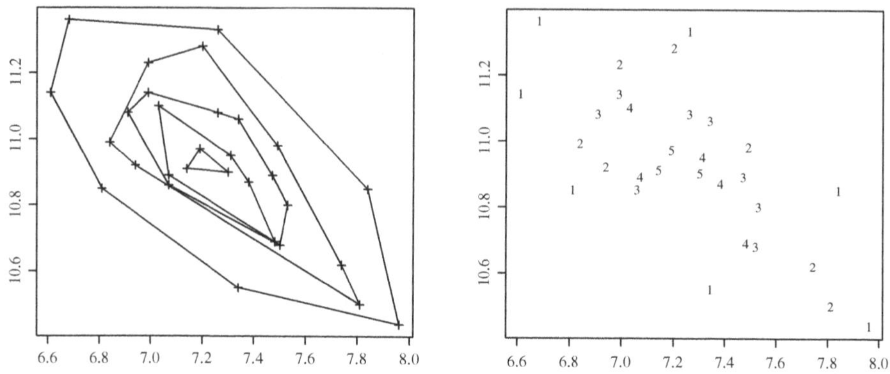

FIG. 12.1. Convex hull peeling for the Decathlon data.

obtained the first limit theorems for Tukey's depth and Eddy (1984) constructed set-valued order statistics for a multivariate sample, with ordering determined by the inclusion relation.

Recent results about convex hull peeling (also considered in Chapter 2) concern the number of vertices in the convex hull peeling layers, see Finch and Hueter (2004) and Hueter (2005).

Chronologically, the second notion of data depth appeared in 1990, when R. Liu described her simplicial depth. During the 1990s extensive work was performed in relation with data depth. Theoretical studies about the halfspace depth in data sets were developed by Donoho and Gasko (1992), for probability distributions by Rousseeuw and Ruts (1999) and about its depth-trimmed regions by Massé and Theodorescu (1994). During this decade Koshevoy and Mosler (1996, 1997*a*, 1997*b*, 1998) performed extensive work on a new notion of data depth, the zonoid depth, see also Mosler (2002). Algorithms for the efficient computation of the different notions of data depth and their trimmed regions were developed, and finally, Liu, Parelius and Singh (1999) reviewed the statistical applications of the different notions of data depth that had been developed during those years.

More recently, Zuo and Serfling (2000*a*) came up with a comprehensive study comparing the theoretical properties of several notions of data depth and their trimmed regions (Zuo and Serfling, 2000*c*). Other excellent surveys on data depth are Liu (1992), Chapters 3 and 4 in Mosler (2002) and Serfling (2006) which is included in Liu, Serfling and Souvaine (2006), a collection of essays about data depth that covers foundational aspects, applications and algorithms.

However, probably the most remarkable current fact about depth is that it has been extended to the regression setting, see Rousseeuw and Hubert (1999), laying the ground for a general notion of *parameter depth*, which was explored in the location-scale model by Mizera and Müller (2004).

The structure of the chapter is as follows. Section 12.2 is devoted to the introduction of several classical notions of data depth, in Section 12.3 we briefly describe some applications of depth functions and depth-trimmed regions, Section 12.4 is devoted to algorithms, in Section 12.5 we present stochastic orderings for random vectors and random sets related to data depth. Finally, in Section 12.6 we describe extensions of depth to a general parametric model.

12.2 Notions of data depth

A *(statistical) depth function* measures how central a point is relative to a probability distribution or a data cloud. Given a random vector X in $\mathbb{R}^d$, we will denote by P_X the probability distribution of X.

The properties of the simplicial depth studied by Liu (1990) have been accepted as common requirements for a depth function. Here we will follow the formulation of Dyckerhoff (2004). A *depth function* is a mapping $\mathrm{D}(\cdot; P_X) : \mathbb{R}^d \mapsto [0,1]$ that satisfies the properties of:

- *Affine invariance.* $\mathrm{D}(Ax+b; P_{AX+b}) = \mathrm{D}(x; P_X)$ for all $b \in \mathbb{R}^d$ and each $d \times d$ nonsingular matrix A.
- *Vanishing at infinity.* $\lim_{\|x\|\to\infty} \mathrm{D}(x; P_X) = 0$.
- *Upper semicontinuity.* The level set of a depth function at any level α, $\{x \in \mathbb{R}^d : \mathrm{D}(x; P_X) \geq \alpha\}$ is closed.
- *Monotonicity with respect to deepest point.* If $\theta = \arg\max_x \mathrm{D}(x; P_X)$, then $\mathrm{D}(x; P_X) \leq \mathrm{D}(\theta + \alpha(x-\theta); P_X)$ for any $0 \leq \alpha \leq 1$.

Related to the concept of depth, is the one of *centre* of a probability distribution, which is a point that satisfies a kind of multivariate symmetry condition. A common notion of symmetry is the one of *angular symmetry*, see Rousseeuw and Struyf (2004). A probability distribution P is *angularly symmetric* about θ if for any Borel cone $K \subseteq \mathbb{R}^d$ (i.e. $rK = K$ for $r > 0$ and K Borel), it holds

$$P(\theta + K) = P(\theta - K).$$

Zuo and Serfling (2000*a*) considered another requirement for a depth function.

- *Maximality at centre.* $\mathrm{D}(\theta; P_X) = \sup_{x\in\mathbb{R}^d} \mathrm{D}(x; P_X)$ if P_X is centred at θ.

In a data analysis context, a depth function measures the degree of centrality of a point with respect to a data cloud. Given a d-dimensional data set $X_1, X_2, \ldots, X_n$, let P_n denote the empirical distribution on these data and $\mathrm{D}_n(x)$, unless it is explicitly stated otherwise, is the depth of x with respect to P_n.

For any chosen depth function $\mathrm{D}(\cdot; P_X)$, the *depth-trimmed region* (or *central region*) of level α, denoted by $\mathrm{D}^\alpha(P_X)$, is the set of all points whose depth is, at least, α, i.e. depth-trimmed regions are the level sets of a depth function,

$$\mathrm{D}^\alpha(P_X) = \{x \in \mathbb{R}^d : \mathrm{D}(x; P_X) \geq \alpha\}.$$

Depth-trimmed regions satisfy various properties, some of which depend on the notion of data depth they were built from. From their construction as level sets, they constitute a nested family of sets.

- *Nesting.* If $\alpha \geq \beta$, then $\mathrm{D}^\alpha(P_X) \subseteq \mathrm{D}^\beta(P_X)$.

  As a consequence of the general properties of depth functions, see Dyckerhoff (2004), the central regions satisfy the properties of:

- *Affine equivariance.* $\mathrm{D}^\alpha(P_{AX+b}) = A\mathrm{D}^\alpha(P_X) + b$ for all random vector X in $\mathbb{R}^d$ and all $d \times d$ nonsingular matrix A and $b \in \mathbb{R}^d$.
- *Compactness.* $\mathrm{D}^\alpha(P_X)$ is compact.
- *Starshapedness.* If $x \in \mathrm{D}^\alpha(P_X)$ for all α such that $\mathrm{D}^\alpha(P_X)$ is nonempty, then each $\mathrm{D}^\alpha(P_X)$ is starshaped with respect to x.

  Some authors, see Zuo and Serfling (2000*c*), have relaxed the starshapedness condition on central regions and require only connectedness. On the other hand, if the depth function is not only decreasing on rays from the deepest point, but also quasiconcave, then the regions are convex.

  Other interesting properties of depth-trimmed regions that do not follow from the ones that were proposed for depth functions are:

- *Monotonicity.* If $Y \leq X$ (interpreted componentwise) a.s., then for all $x \in \mathrm{D}^\alpha(P_X)$ there is $y \in \mathrm{D}^\alpha(P_Y)$ with $y \leq x$ (componentwise).
- *Subadditivity.* $\mathrm{D}^\alpha(P_{X+Y}) \subseteq \mathrm{D}^\alpha(P_X) + \mathrm{D}^\alpha(P_Y)$, where the summation of central regions is in the Minkowski (elementwise) sense.

Since depth-trimmed regions depend on probability distributions and not on realizations of random vectors, the monotonicity condition can be written in terms of a stochastic order, see Section 12.5. Further, once the Minkowski addition has been introduced, the monotonicity can be expressed as, if $Y \leq X$ a.s., then $\mathrm{D}^\alpha(P_X) + \mathbb{R}^d_+ \subseteq \mathrm{D}^\alpha(P_Y) + \mathbb{R}^d_+$, where $\mathbb{R}^d_+$ is the positive quadrant.

Observe that a combination of the affine equivariance and monotonicity conditions implies that the componentwise $\leq$ order can be substituted by the order defined by any polyhedral cone. That is, if both properties are satisfied, given X, Y two random vectors and a nonsingular matrix A such that $AY \leq AX$ a.s. (or equivalently $X + A^{-1}\mathbb{R}^d_+ \subseteq Y + A^{-1}\mathbb{R}^d_+$ holds *a.s.*), then $A\mathrm{D}^\alpha(P_X) + \mathbb{R}^d_+ \subseteq A\mathrm{D}^\alpha(P_Y) + \mathbb{R}^d_+$, which amounts to $\mathrm{D}^\alpha(P_X) + A^{-1}\mathbb{R}^d_+ \subseteq \mathrm{D}^\alpha(P_Y) + A^{-1}\mathbb{R}^d_+$.

12.2.1 *Halfspace depth*

The *halfspace depth* (or *Tukey depth*) of a fixed point $x \in \mathbb{R}^d$ with respect to a probability distribution is the infimum of the probabilities of all closed halfspaces that contain x, see Rousseeuw and Ruts (1999),

$$\mathrm{HD}(x; P) = \inf\{P(H) : H \text{ closed halfspace}, x \in H\}. \tag{12.1}$$

Expression (12.1) is a natural extension of Tukey's (1975) original construction, where the notion of halfspace depth was introduced in a data analysis context. Given a data set $X_1, X_2, \ldots, X_n$ in $\mathbb{R}^d$, the halfspace depth of a point $x \in \mathbb{R}^d$ is the smallest fraction of data points in a closed halfspace containing

x, or alternatively, the smallest fraction of data points that must be deleted so that x lies outside the convex hull of the remaining data points. That is, if we denote the halfspace depth of x with respect to the previous sample by $\mathrm{HD}_n(x)$, then

$$\mathrm{HD}_n(x) = n^{-1} \min_{u \in \mathbb{R}^d} \#\{i : \langle X_i, u \rangle \geq \langle x, u \rangle\}.$$

The original aim of Tukey was to rank points with respect to a finite bivariate data set in order to obtain multivariate order statistics. The halfspace depth, determines a center-outward ranking in the data set through the ordering,

$$\mathrm{HD}_n(X_{[1]}) \leq \mathrm{HD}_n(X_{[2]}) \leq \ldots \leq \mathrm{HD}_n(X_{[n]}).$$

In the univariate case, the halfspace depth of a point $x \in \mathbb{R}$ with respect to a continuous probability distribution is $\mathrm{HD}(x; P_X) = \min\{F_X(x), 1 - F_X(x)\}$, where F_X is the cdf of the random variable X. That is, if x is the α-quantile of X, then it is the minimum of α and $1 - \alpha$.

The halfspace depth is affine invariant, vanishes at infinity, is upper semicontinuous and decreases on the rays from the deepest point (in fact it is quasiconcave). Further, it is maximal at the point of angular symmetry (whenever such a point exists) and is a continuous function of x if P is absolutely continuous. Note, however, that the deepest point is not necessarily unique.

The maximal possible halfspace depth on an absolutely continuous distribution is 2^{-1} and it is attained at the point of symmetry of angularly symmetric distributions, see Rousseeuw and Struyf (2004). Mizera (2002) showed that for any probability distribution P, there exists at least one point whose halfspace depth with respect to P is, at least, $(1 + d)^{-1}$.

The *halfspace trimmed region* of level α, $\mathrm{HD}^\alpha(P)$, is the intersection of all closed halfspaces whose probability content is greater than $1 - \alpha$,

$$\mathrm{HD}^\alpha(P) = \bigcap \{H : H \text{ closed halfspace}, P(H) > 1 - \alpha\}. \tag{12.2}$$

In the univariate setting, the halfspace depth-trimmed regions amount to the classical quantile trimming,

$$\mathrm{HD}^\alpha(P_X) = \left[\inf\{x : \mathbf{P}\{X \leq x\} \geq \alpha\}, \inf\{x : \mathbf{P}\{X \leq x\} > 1 - \alpha\}\right]. \tag{12.3}$$

The lower extreme in (12.3) is the (left-continuous) α-quantile of X, and the upper extreme is the negative of the α-quantile of $-X$.

The halfspace trimmed regions are nested, affine equivariant, compact and convex. However, in general they fail to be subadditive and even monotonic, because some directions might not count in their construction. That is, given three halfspaces from (12.2), the intersection of two of them might be contained in the interior of the third one, see also (12.4).

Massé and Theodorescu (1994) considered the trimmed regions given by the intersection of all closed halfspaces whose probability content is, at least,

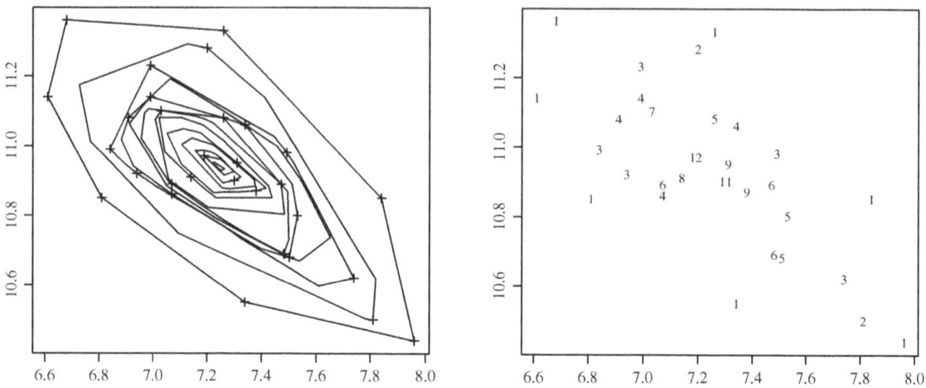

FIG. 12.2. Halfspace depth for the Decathlon data.

$1 - \alpha$. That is, those satisfying the nonstrict inequality in (12.2). Such regions have been shown to be more appropriate for certain applications. The difference between them and the classical regions is similar to the difference between the right-continuous and the left-continuous quantile functions. In fact, they would transform (12.3) by interchanging left-continuous and right-continuous quantiles.

In relation with the affine equivariance, for a general $p \times d$ matrix A, we have $\mathrm{HD}^\alpha(P_{AX}) \supseteq A\mathrm{HD}^\alpha(P_X)$ which implies that the support function on any $u \in \mathbb{R}^d$ is bounded by the right-continuous $(1-\alpha)$-quantile of $\langle X, u\rangle$,

$$h(\mathrm{HD}^\alpha(P_X), u) \leq \inf\{x : \mathbf{P}\{\langle X, u\rangle \leq x\} > 1 - \alpha\}. \tag{12.4}$$

Figure 12.2 represents the contours of the halfspace trimmed regions of the Decathlon data and the halfspace depths of each data point, multiplied by 30.

Massé (2004) described the asymptotic behaviour of the empirical halfspace depth process $n^{1/2}(\mathrm{HD}_n(\cdot) - \mathrm{HD}(\cdot; P))$, showing that it may not converge weakly, but for any given $x \in \mathbb{R}^d$, $n^{1/2}(\mathrm{HD}_n(x) - \mathrm{HD}(x; P))$ converges and, under some smoothness condition on P at x, it is asymptotically normal.

12.2.2 *Simplicial depth*

Liu (1990) proposed a notion of depth based on random simplices. The *simplicial depth* of a point $x \in \mathbb{R}^d$ with respect to the distribution of a random vector X is the probability that it lies inside the random simplex whose vertices are $d+1$ independent copies of X, that is,

$$\mathrm{SD}(x; P_X) = \mathbf{P}\{x \in \mathrm{conv}\{X_1, X_2, \ldots, X_{d+1}\}\},$$

where $X_1, X_2, \ldots, X_{d+1}$ are independent copies of X.

In the univariate case, the simplicial depth of a point $x \in \mathbb{R}$ with respect to a continuous probability distribution amounts to $\mathrm{SD}(x; P_X) = 2F_X(x)(1 - F_X(x))$

which clearly implies that the maximal simplicial depth on the real line is 2^{-1} and it is attained at the median.

The simplicial depth is affine invariant, upper semicontinuous and vanishes at infinity. Further, on an absolutely continuous distribution, it is a continuous function of x and is monotonic with respect to the point of angular symmetry. After Wendel (1962), it follows that the simplicial depth at the point of angular symmetry of a distribution that assigns probability zero to the hyperplanes through it is 2^{-d}. Moreover, the results of Wagner and Welzl (2001) imply that the maximal simplicial depth of any absolutely continuous distribution is 2^{-d}.

The *simplicial trimmed regions* are nested, affine equivariant, compact and starshaped for angularly symmetric and absolutely continuous distributions. If the distribution is not absolutely continuous, they might even fail to be connected. A simple coupling argument shows that they are monotonic.

The depth-trimmed regions of a continuous univariate distribution are

$$\mathrm{SD}^{\alpha}(P_X) = \mathrm{HD}^{\beta}(P_X)\,, \qquad \text{where } \beta = \frac{1-\sqrt{1-2\alpha}}{2}\,, \qquad \text{for } \alpha \in (0, 1/2]\,.$$

The sample version of the simplicial depth is a U-statistic. Consider all simplices whose vertices are $d+1$ distinct data points, the depth of x is the proportion of simplices that contain it. Given a d-dimensional data set $X_1, X_2, \ldots, X_n$, let

$$\mathrm{SD}_n(x) = \binom{n}{d+1}^{-1} \sum_{1 \le i_1 < i_2 < \ldots < i_{d+1} \le n} \mathbf{1}(x \in \mathrm{conv}\{X_{i_1}, X_{i_2}, \ldots, X_{i_{d+1}}\})\,,$$

where $\mathbf{1}(\cdot)$ is the indicator function.

Figure 12.3 represents a grey-scale image of the sample simplicial depth of the Decathlon data and the simplicial depths of each data point.

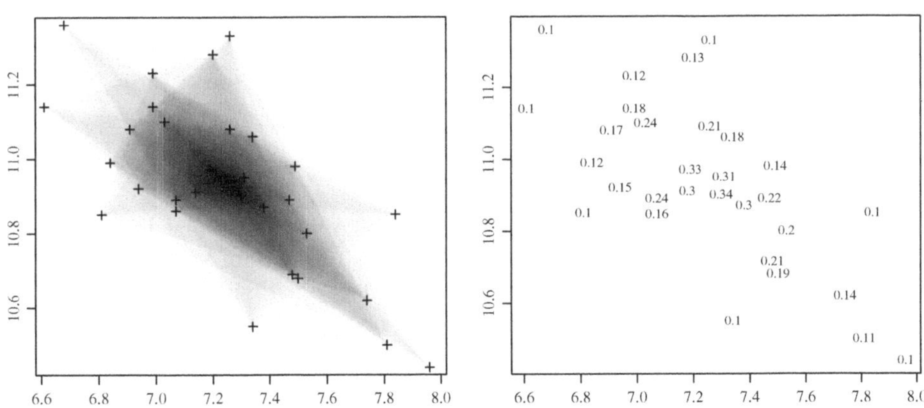

FIG. 12.3. Simplicial depth for the Decathlon data.

Liu (1990) proved the uniform consistency of $\mathrm{SD}_n(\cdot)$ and Dümbgen (1992) showed that the process $n^{1/2}(\mathrm{SD}_n(\cdot) - \mathrm{SD}(\cdot; P))$ is asymptotically Gaussian.

12.2.3 *Zonoid depth*

G. Koshevoy and K. Mosler developed the *lift zonoid theory* using techniques from multivariate statistics, stochastic geometry, stochastic orders and the study of income inequality, see Mosler (2002). The basic concepts are the zonoid of a probability distribution and its lifting (the lift zonoid).

Zonoids are a family of centrally symmetric convex bodies, see Chapter 1. The *zonoid* (of moments) of an integrable probability distribution P_X is

$$\mathrm{Z}(P_X) = \left\{\mathbf{E}Xg(X) : g : \mathbb{R}^d \mapsto [0,1] \text{ measurable}\right\} = \mathbf{E}\,\mathrm{conv}\{o, X\},$$

the selection expectation of the random segment that joins the origin and X, see Chapter 1 or Molchanov (2005). It is centrally symmetric about the point $\mathbf{E}X/2$ and if P_X is a discrete probability, then its zonoid is a zonotope (Minkowski sum of segments).

The *lift zonoid* of P_X is the zonoid of the probability distribution of the random vector $(1, X)$ in $\mathbb{R}^{d+1}$ that is obtained by *lifting* X with one coordinate equal to 1,

$$\widehat{\mathrm{Z}}(P_X) = \mathrm{Z}(P_{(1,X)}) = \mathbf{E}\mathrm{conv}\{o, (1, X)\}.$$

An immediate interpretation of the lift zonoid can be given in terms of the Lorenz curve, see Lorenz (1905). Given a random variable X, we define

$$L_X(t) = \int_0^t F_X^{-1}(s)\,\mathrm{d}s; \quad \overline{L}_X(t) = \int_{1-t}^1 F_X^{-1}(s)\,\mathrm{d}s, \tag{12.5}$$

where F_X^{-1} is the quantile function of X, see for example the lower extreme in (12.3). If X is *a.s.* nonnegative and $\mathbf{E}X > 0$, then its Lorenz curve is $L_X(t)/\mathbf{E}X$. The classical interpretation of the Lorenz curve arises when X determines the distribution of the wealth in a population, so that $L_X(t)/\mathbf{E}X$ is the fraction of the total wealth that is possessed by the fraction t of the poorest individuals.

The lift zonoid of a univariate probability distribution can be expressed as

$$\widehat{\mathrm{Z}}(P_X) = \left\{(t, s) : 0 \le t \le 1\,,\ L_X(t) \le s \le \overline{L}_X(t)\right\}, \tag{12.6}$$

i.e. the *lower boundary* of the lift zonoid, $L_X(t)$, is a non-normalized generalized *Lorenz curve*. It is generalized because it is built for random variables with no sign restrictions, and it is non-normalized since it is not divided by $\mathbf{E}X$.

Koshevoy and Mosler (1996) noticed the relation between the Lorenz curve and the lift zonoid of a univariate distribution and transformed the lift zonoid into the Lorenz zonoid, a multivariate generalization of the Lorenz curve.

The lift zonoid uniquely characterizes integrable probability distributions. Given an integrable univariate probability distribution, its lift zonoid, determines

$L_X(t)$ and, as a consequence, it also determines its derivative, the quantile function, or equivalently, the distribution of X. For $u \in \mathbb{R}^d$, let A be the $2 \times (d+1)$ matrix whose first row is $(1, o)$, where o is the origin in $\mathbb{R}^d$ and its second row $(1, u)$. From the linearity of the selection expectation, we have

$$A\widehat{Z}(P_X) = \mathbf{E}\mathrm{conv}\{o, (1, \langle X, u\rangle)\} = \widehat{Z}(P_{\langle X,u\rangle}),$$

i.e. the lift zonoid of P_X determines the lift zonoid of $P_{\langle X,u\rangle}$ for all $u \in \mathbb{R}^d$. Finally, by the Crámer–Wold Theorem, since the lift zonoid of P_X characterizes all its univariate projections, it also characterizes P_X.

The *zonoid trimmed regions* of the probability distribution of a random vector X with finite first moment are built, from the lift zonoid, as

$$\mathrm{ZD}^\alpha(P_X) = \alpha^{-1}\mathrm{proj}_\alpha(\widehat{Z}(P_X)), \qquad \text{for} \quad \alpha \in (0,1],$$

where $\mathrm{proj}_\alpha(\widehat{Z}(P_X))$ is the projection of the intersection of $\widehat{Z}(P_X)$ with the hyperplane $\{(\alpha, x) : x \in \mathbb{R}^d\}$ to the last d coordinates. Alternatively, we can write

$$\mathrm{ZD}^\alpha(P_X) = \left\{\mathbf{E}Xg(X) : g : \mathbb{R}^d \mapsto [0, \alpha^{-1}] \text{ measurable and } \mathbf{E}g(X) = 1\right\},$$

which makes clear the difference between the zonoid trimmed regions and the halfspace and simplicial ones. The zonoid trimming does not generalize the concept of quantile to the multivariate setting, but the concept of *expectile* instead.

From equations (12.5) and (12.6), the zonoid trimmed regions of a univariate probability distribution can be written as

$$\mathrm{ZD}^\alpha(P_X) = \left[\frac{1}{\alpha}\int_0^\alpha F_X^{-1}(t)\,\mathrm{d}t, \frac{1}{\alpha}\int_{1-\alpha}^1 F_X^{-1}(t)\,\mathrm{d}t\right]. \tag{12.7}$$

The zonoid trimmed regions are nested, affine equivariant, compact, convex, monotonic and subadditive. For any $\alpha \in (0,1]$, the trimmed region $\mathrm{ZD}^\alpha(P_X)$ is nonempty and, for $\alpha = 1$, we obtain $\mathrm{ZD}^1(P_X) = \{\mathbf{E}X\}$. Finally, they are continuous, in the Hausdorff distance, with respect to the underlying probability.

As an extension of the affine equivariance, $\mathrm{ZD}^\alpha(P_{AX}) = A\mathrm{ZD}^\alpha(P_X)$ for any $p \times d$ matrix A. Therefore, the projections of the zonoid trimmed regions are the zonoid trimmed regions of the projected data and for any $u \in \mathbb{R}^d$, we have

$$h(\mathrm{ZD}^\alpha(P_X), u) = \frac{1}{\alpha}\int_{1-\alpha}^1 F_{\langle X,u\rangle}^{-1}(t)\,\mathrm{d}t.$$

The *zonoid depth* of a point x is the largest value α such that x is contained in the zonoid trimmed region of level α,

$$\mathrm{ZD}(x; P) = \sup\{\alpha : x \in \mathrm{ZD}^\alpha(P)\},$$

if x does not belong to any zonoid trimmed region of P, then $\mathrm{ZD}(x; P) = 0$. The zonoid depth satisfies the usual properties of a depth function, it is affine

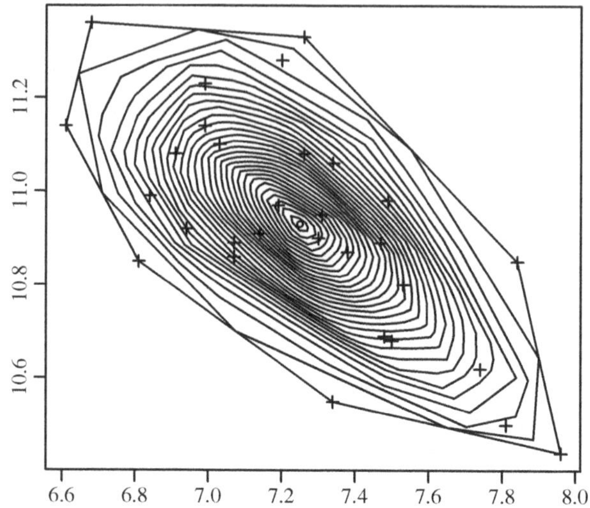

FIG. 12.4. Zonoid trimming for the Decathlon data.

invariant, vanishes at infinity, is continuous on x and P, monotonic with respect to the deepest point (in fact it is quasiconcave), and further, it is uniquely maximal at the expectation. Observe that the expectation might not be the point of symmetry of an angularly symmetric distribution.

The d-dimensional sample $X_1, X_2, \ldots, X_n$ induces empirical zonoid trimmed regions denoted by ZD_n^α. If $\alpha \in (k/n, (k+1)/n]$ for $k \in \{0, 1, \ldots, n-1\}$, then

$$\mathrm{ZD}_n^\alpha = \mathrm{conv}\left\{\frac{1}{\alpha n}\sum_{j=1}^{k} X_{i_j} + \left(1 - \frac{k}{\alpha n}\right) X_{i_{k+1}} : 1 \leq i_1, \ldots, i_{k+1} \leq n\right\}. \quad (12.8)$$

Figure 12.4 represents the contour plots of the zonoid trimmed regions of the Decathlon data.

A law of large number for the zonoid trimmed regions was obtained by Koshevoy and Mosler (1997*b*).

12.2.4 *Other notions of data depth*

In the statistical literature, there are many other notions of data depth apart from the three above. However, we will only focus our attention on two other families of depth-trimmed regions.

Integral trimmed regions Given a probability distribution P and $\alpha \in (0, 1]$, we define the α-trimming of P, denoted by P^α, as

$$P^\alpha = \{Q \in \mathbb{P} : Q(B) \leq \alpha^{-1} P(B) \text{ for all Borel sets } B \subseteq \mathbb{R}^d\}.$$

where $\mathbb{P}$ is the set of all probabilities on the Borel sets of $\mathbb{R}^d$. That is, P^α is the set of probability measures that are bounded above by $\alpha^{-1}P$. While $P^1 = \{P\}$, the family P^α grows if α is getting smaller.

The *integral trimmed regions*, see Cascos and López-Díaz (2005), of P of level α generated by the set of measurable functions $\mathfrak{F}$ form the family

$$\mathrm{D}^\alpha_{\mathfrak{F}}(P) = \left\{ x \in \mathbb{R}^d : \text{ for some } Q \in P^\alpha,\, f(x) \le \int f \, \mathrm{d}Q \,, \text{ for all } f \in \mathfrak{F} \right\}.$$

The family $\mathfrak{F}$ defines a location parameter as $L_{\mathfrak{F}}(P) = \bigcap_{f \in \mathfrak{F}} f^{-1}((-\infty, \int f \, \mathrm{d}P])$ and the integral trimmed region of level α is the union of all those $\mathfrak{F}$-location parameters applied to the probabilities in P^α.

If we define the depth of a point x as the minimum value α such that x belongs to the integral trimmed region of level α, then this depth could be interpreted as the smallest fraction of probability that has to be relocated (still inside the support of P) in order to make x (belong to) an $\mathfrak{F}$-location parameter.

For the appropriate family of functions (for example linear or convex), the integral trimmed regions generate the zonoid trimming. Alternatively, we could directly apply a general location parameter (not necessarily built from a family of functions) to all the probabilities from P^α. If this location parameter is the expectation, we obtain the zonoid trimming.

Further, it is possible to consider a generalization of the integral trimmed regions by considering a family of sets of functions and the intersection of all trimmed regions generated by them. In this framework, we can obtain the half-space trimmed regions as integral ones.

The properties of the integral trimmed regions follow directly from the properties of their generating sets of functions, except for the nesting property, which is derived from the nesting of the family $\{P^\alpha\}_{\alpha \in (0,1]}$.

Cascos and López-Díaz (2008) showed that the sequence of α-trimmings of empirical probabilities converge, in the Painlevè–Kuratowski sense, to the α-trimming of the population probability. Such result is useful in the study of the consistency of the empirical integral trimmed regions.

Expected convex hull trimmed regions The *expected convex hull trimmed region* (Cascos, 2007*a*) of level $1/k$ of the probability distribution of a random vector with finite first moment X is defined as the selection expectation of the convex hull of k independent copies of X, i.e.

$$\mathrm{CD}^{1/k}(P_X) = \mathbf{E}\, \mathrm{conv}\{X_1, X_2, \ldots, X_k\} \,,$$

where $X_1, X_2, \ldots, X_k$ are independent copies of X.

Vitale (1987) proved that, if X has finite first moment, then the sequence $\{\mathrm{CD}^{1/k}(P_X)\}_k$ characterizes its distribution.

In the univariate case, we obtain a segment whose extreme values are the expectations of extreme order statistics,

$$\mathrm{CD}^{1/k}(P_X) = [\mathbf{E}\min\{X_1, X_2, \ldots, X_k\}, \mathbf{E}\max\{X_1, X_2, \ldots, X_k\}] \,. \tag{12.9}$$

These regions are related to the zonoid trimmed regions in the sense that they do not generalize the concept of quantiles to the multivariate setting, but integrated quantiles (or expectiles) instead.

The expected convex hull trimmed regions are nested, affine equivariant, compact, convex, monotonic, and subadditive. As it is the case for the zonoid trimmed regions, the most central set is a singleton formed by the mean of the random vector, $\text{CD}^1(P_X) = \{\mathbf{E}X\}$. Further, $\text{CD}^{1/k}(P_{AX}) = A\text{CD}^{1/k}(P_X)$ for a general $p \times d$ matrix A, in particular for any $u \in \mathbb{R}^d$,

$$h(\text{CD}^{1/k}(P_X), u) = \mathbf{E}\max\{\langle X_1, u\rangle, \langle X_2, u\rangle, \ldots, \langle X_k, u\rangle\}.$$

The depth of a point x is the inverse of the minimum number of observations of a random vector that must be taken so that x belongs to the expectation of their convex hull, i.e.

$$\text{CD}(x; P_X) = \max\{1/k : x \in \text{CD}^{1/k}(P_X)\}.$$

The expected convex hull depth is affine invariant, decreases in rays from the deepest point (in fact it is quasiconcave), vanishes at infinity and is maximal at the expectation of the random vector.

The sample version of these regions is given in the form of a U-statistic of degree k. For the sample $X_1, X_2, \ldots, X_n$, the empirical expected convex hull trimmed region of level k, denoted by $\text{CD}_n^{1/k}$, is the Minkowski sum of all convex hulls of all subsamples of size k,

$$\text{CD}_n^{1/k} = \binom{n}{k}^{-1} \sum_{1 \le i_1 < i_2 < \ldots < i_k \le n} \text{conv}\{X_{i_1}, X_{i_2}, \ldots, X_{i_k}\}.$$

Clearly enough, we have $\text{CD}_n^{1/n} = \text{conv}\{X_1, X_2, \ldots, X_n\}$.

Figure 12.5 represents the contour plots of the expected convex hull regions of the Decathlon data.

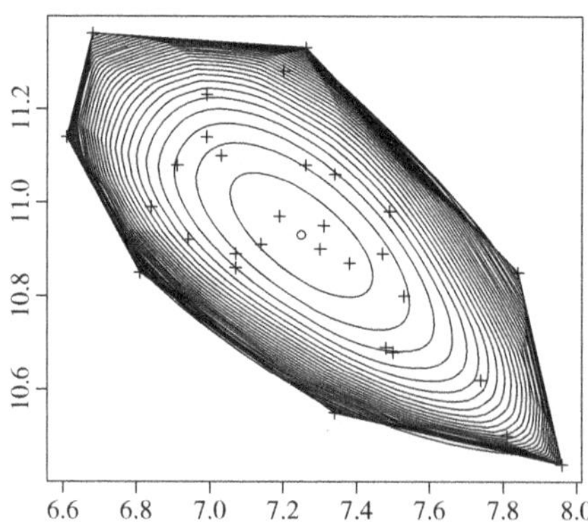

FIG. 12.5. Expected convex hull trimming for the Decathlon data.

Notice that the region $\mathrm{CD}_n^{1/2}$ is a Minkowski sum of segments and, thus, it is a zonotope. As a consequence, it is centrally symmetric about the sample average value of the data points. However, the regions of other levels might neither be zonotopes, nor even centrally symmetric, but they will capture the shape of the data cloud with greater detail instead.

The consistency of these central regions follows from the results on U-statistics in Banach spaces of Borovskikh (1996) applied to the support functions of the convex hulls of subsets of the data points.

12.3 Applications

The depth-trimmed regions of a data set contain information about the location, scatter, correlation, skewness, and tails of the data. Therefore, they can be considered as set-valued descriptive statistics. In the following pages, we will describe some simple applications of data depths and trimmed regions.

12.3.1 *Multivariate goodness-of-fit*

A useful graphical tool to check whether two multivariate data sets were drawn from the same distribution is the *DD-plot* (depth vs. depth plot), see Liu, Parelius and Singh (1999). It is a multivariate generalization of the PP-plot and is obtained the following way: take every point from each of the two data sets to be compared and plot its depth with respect to the first data set versus its depth with respect to the second data set. If both data samples were drawn from the same distribution, then our collection of points should concentrate around the line through the origin with slope 1.

In Fig. 12.6 we have plotted three DD-plots, all of them obtained using the simplicial data depth.

12.3.2 *Bagplot*

Rousseeuw, Ruts and Tukey (1999) developed the *bagplot*, a bivariate generalization of Tukey's boxplot based on the halfspace depth. In a bagplot, we first

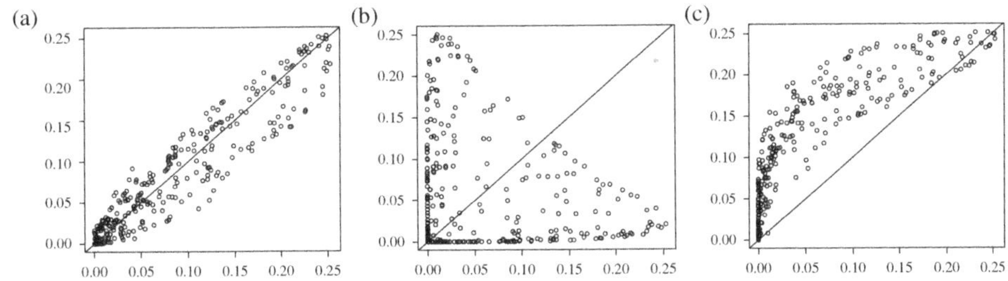

FIG. 12.6. DD-plots of (a) two identical distributions, (b) two distributions with a location shift and (c) two distributions with different scale parameter.

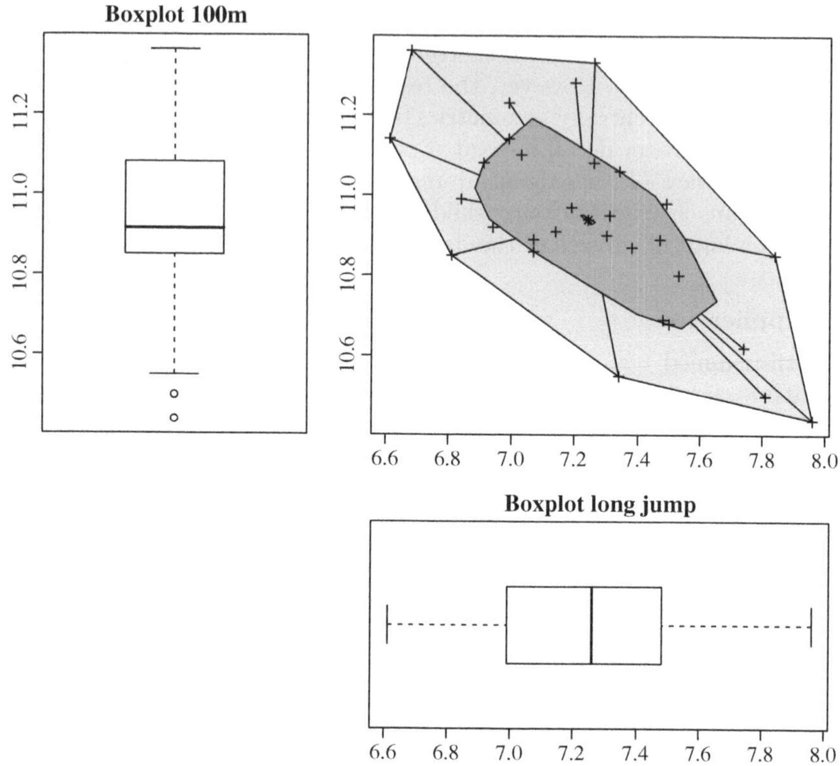

FIG. 12.7. Bagplot and boxplots of the Decathlon data.

paint the *bag* which is the halfspace trimmed region that contains the deeper half of the observations. On a second step, this bag is enlarged radially from the centre of gravity of the deepest region, by a factor of 3, in order to obtain an unplotted *fence*. Finally, the convex hull of the observations out of the bag, but inside the fence is constructed, and the points outside the fence are marked as outliers.

In Fig. 12.7 we present the bagplot of the Decathlon data and the boxplots of its marginals. Clearly enough, the bagplot captures the shape of the data cloud and it detects outliers more efficiently than the univariate boxplots.

12.3.3 *Location estimates*

The point of maximal depth is a multivariate location estimate. Since the point of maximal zonoid and expected convex hull depth is the mean, it does not bring anything new. But for the halfspace depth and the simplicial depth, those points are called the *halfspace median* and the *simplicial median*. They satisfy standard requirements for a multivariate location estimate (Oja, 1983), like being affine equivariant and were already considered by Small (1990) in his review of multidimensional medians.

If the halfspace depth does not have a unique maximum, the halfspace median is defined as the center of gravity of the innermost halfspace trimmed region. The halfspace median has interesting robustness properties, being its breakdown point, at least, $(d+1)^{-1}$, see Donoho and Gasko (1992).

Alternatively, we can obtain pointwise multivariate location estimates by means of depth-weighted L-statistics. That is, L-statistics of the type

$$L = \frac{\sum_i X_i W(\mathrm{D}_n(X_i))}{\sum_i W(\mathrm{D}_n(X_i))}$$

for a suitable weight function $W : [0,1] \mapsto [0,1]$. Dümbgen (1992) has studied limit theorems for L-statistics built from the simplicial depth and Massé (2004) limit theorems for L-statistics built from the halfspace depth.

12.3.4 *Scatter estimates*

Since more scattered data produce larger central regions, the volume of the depth-trimmed regions can be used as a multivariate scatter estimate. By the affine equivariance of the central regions, their volumes $\lambda_d(\cdot)$ satisfy

$$\lambda_d(\mathrm{D}^\alpha(P_{AX+b})) = |\det(A)|\lambda_d(\mathrm{D}^\alpha(P_X)),$$

for any $d \times d$ nonsingular matrix A and any $b \in \mathbb{R}^d$, which was already observed by Zuo and Serfling (2000*b*) and is a classical requirement for a multivariate location estimate, see Oja (1983).

It can be shown that the volume of the expected convex hull trimmed region of level 1/2 is a multivariate generalization of the classical Gini index, see Cascos (2007*a*). Further, from the relation between the lift zonoid and the Lorenz curve, it follows that the volume of the lift zonoid, studied by Koshevoy and Mosler (1997*a*), is another variant of multivariate Gini index.

12.3.5 *Risk measurement*

Risk measures are used in finance to assess the risk of a random portfolio, which is classically modeled as a random variable X that represents the financial gain associated to an investment. The most widely used risk measure is the *value-at-risk* (the opposite of a fixed quantile of X) and other common risk measures are the *expected shortfall* and the *expected minimum*. These three risk measures are, up to a change of sign, the smallest extreme values of the halfspace, zonoid, and expected convex hull trimmed regions of a univariate probability distribution, respectively, see (12.3), (12.7) and (12.9).

Those relations enable us to build risk measures for multivariate portfolios (modeled as random vectors). The basic idea, which is better explained in

Chapter 17 and developed in detail by Cascos and Molchanov (2007) is the following. A random multivariate portfolio X is acceptable if its depth-trimmed region of some level lies inside the positive quadrant, $\mathbb{R}^d_+$, and a set-valued risk measure can be defined as the set of all deterministic portfolios that, although they are added to X, cannot make it acceptable. This way, the risk of X is assessed by the set

$$\rho(X) = -\left(\mathrm{D}^\alpha(P_X) + \mathbb{R}^d_+\right) ,$$

i.e. the reflected set of $\mathrm{D}^\alpha(P_X) + \mathbb{R}^d_+$. In this framework the subadditivity property of depth-trimmed regions has an interpretation in terms of risk diversification.

12.4 Algorithms

A naive algorithm to compute either the halfspace depth or the simplicial depth of a point x with respect to a d-dimensional data set of size n has complexity $\mathcal{O}(n^{d+1})$. For the halfspace depth, such an algorithm involves considering each of the $\binom{n}{d}$ hyperplanes determined by x and any d points from the data set and counting the number of data points on each of the halfspaces it determines. For the simplicial depth, we should consider all $\binom{n}{d+1}$ simplices whose vertices are $d+1$ data points. Clearly, these algorithms are too demanding and extensive work has been done to obtain faster implementations.

Rousseeuw and Ruts (1996) described algorithms of complexity $\mathcal{O}(n \log n)$ to compute the halfspace depth and the simplicial depth of a point with respect to a bivariate data cloud. The basic idea is to obtain, for each data point x_i, the angle defined by the line through x and x_i and the horizontal axis and to obtain the halfspace and simplicial depths in terms of these angles. A similar argument was already used by Jewell and Romano (1982) in order to analytically compute the coverage probability of the convex hull of a set of identically distributed random points in the plane. The complexity of the algorithms of Rousseeuw and Ruts (1996) was shown to be optimal by Aloupis, Cortés, Gómez, Soss and Toussaint (2002).

The computation of the zonoid depth requires solving a linear program, see (12.8). Dyckerhoff, Koshevoy and Mosler (2000) proposed an algorithm for computing the zonoid depth on any dimension based on a Dantzig–Wolfe decomposition, whose (simulated) complexity appears to be below $\mathcal{O}(n^2 \log n)$.

The fastest algorithm to compute all convex hull peeling contours of a bivariate data cloud was built by Chazelle (1985) and has complexity $\mathcal{O}(n \log n)$. The algorithms for depth-trimmed regions of planar sets are not that fast.

The first efficient algorithms to compute the halfspace (Ruts and Rousseeuw, 1996), zonoid (Dyckerhoff, 2000) and expected convex hull (Cascos, 2007*b*) contours of a bivariate data cloud are based on the *circular sequence* argument. The basic idea is to consider all possible orderings of the univariate projections of the data points and find the extreme points of the univariate trimmed regions on that direction. This way we obtain a set of halfplanes and the desired trimmed region is their intersection. For a data set of size n, the univariate projections of the points can be ordered, at most, in $2\binom{n}{2}$ possible ways, determined by the projection of the data cloud in a normal line to the one defined by each two data points (direct order and reverse order). Finally, the complexity of those algorithms is $\mathcal{O}(n^2 \log n)$.

Using techniques of computational geometry, in particular, the topological sweep of the dual arrangement of lines, algorithms for computing all halfspace contours (Miller, Ramaswami, Rousseeuw, Sellarès, Souvaine, Streinu and Struyf, 2003) and zonoid contours for $\alpha = 1/n, 2/n, \ldots, 1$ (Gopala and Morin, 2008) of complexity $\mathcal{O}(n^2)$ have been constructed. After obtaining these contours, the halfspace and zonoid depth of any point can be computed in a fast way. However, these techniques cannot be applied to the expected convex hull contours directly because each of their extreme values is a convex linear combination of the data points with a different weight on each of them.

12.5 Stochastic orderings

Stochastic orders are partial order relations between probability distributions of random elements, see Müller and Stoyan (2002). They have plenty of applications in economics, actuarial sciences, queueing theory and other areas.

There is a wide range of stochastic orders between random variables and random vectors, but not so many between random sets.

12.5.1 *Quantile orderings*

We say that X is stochastically smaller than Y in the *usual stochastic order*, and write $X \leq_{\mathrm{st}} Y$, if their quantile functions, see for example the lower extreme in (12.3), are ordered as $F_X^{-1}(t) \leq F_Y^{-1}(t)$ for all t, or equivalently, if there exist two random variables X' and Y' distributed as X and Y respectively such that $\mathbf{P}\{X' \leq Y'\} = 1$.

The multivariate version of the usual stochastic ordering is given by applying the above coupling construction to the componentwise $\leq$ order. The monotonicity of the zonoid, expected convex hull and simplicial trimmed regions puts them in relation with the usual multivariate stochastic order.

Since halfspaces that are normal to all possible directions are involved in the construction of the halfspace trimmed regions, see (12.2), these regions fail to be monotonic. However, it is possible to define a one-sided version of the halfspace trimmed regions by considering only *upper* halfspaces, i.e. those halfspaces $H \subset \mathbb{R}^d$ such that if $x \in H$ and $x \leq y$, then $y \in H$

$$\mathrm{HD}_+^\alpha(P) = \bigcap \{H : H \text{ closed upper halfspace}, P(H) > 1 - \alpha\}.$$

Clearly enough, if $X \leq_{\text{st}} Y$, then $\text{HD}_+^\alpha(P_Y) \subseteq \text{HD}_+^\alpha(P_X)$ for all α. However, the reverse implication is not true in general.

Scatter orderings Zuo and Serfling (2000*b*) defined multivariate stochastic orderings built as $X \leq_{\text{sc}} Y$ if the volumes of their corresponding halfspace trimmed regions satisfy $\lambda_d(\text{HD}^\alpha(P_X)) \leq \lambda_d(\text{HD}^\alpha(P_Y))$ for all α.

Although the notions of dispersion and scatter are closely related, the term *dispersion* in the literature about stochastic orders is traditionally deferred to more widely separated quantiles and, in the univariate case $X \leq_{\text{disp}} Y$ if $F_X^{-1}(t) - F_X^{-1}(s) \leq F_Y^{-1}(t) - F_Y^{-1}(s)$ for all $0 < s < t < 1$. In the multivariate case Massé and Theodorescu (1994) defined $X \leq_{\text{disp}} Y$ if

$$\lambda_d(\text{HD}^\alpha(P_X)) - \lambda_d(\text{HD}^\beta(P_X)) \leq \lambda_d(\text{HD}^\alpha(P_Y)) - \lambda_d(\text{HD}^\beta(P_Y))$$

for all $\alpha < \beta$.

12.5.2 *Variability orderings*

The usual stochastic order is commonly defined either in terms of a relation between quantile functions or a coupling construction. However, it is also characterized by

$$\mathbf{E}f(X) \leq \mathbf{E}f(Y) \tag{12.10}$$

for all increasing functions f such that both expectations exist. Stochastic orders defined by imposing (12.10) for all functions from a certain family are named *integral stochastic orders*, and the family of functions is called the set of *generators* of the stochastic order. These integral stochastic orders are usually named after the description of the families of functions that generate them. When the functions are convex, we usually talk about variability orderings or, more simply, convex orderings. Among them, we have the following orders:

- *Convex:* $X \leq_{\text{cx}} Y$ if $\mathbf{E}f(X) \leq \mathbf{E}f(Y)$ for all convex function f.
- *Increasing convex:* $X \leq_{\text{icx}} Y$ if $\mathbf{E}f(X) \leq \mathbf{E}f(Y)$ for all increasing convex function f.
- *Linear convex:* $X \leq_{\text{lcx}} Y$ if $\langle X, u\rangle \leq_{\text{cx}} \langle Y, u\rangle$ for all $u \in \mathbb{R}^d$, or alternatively $\mathbf{E}f \circ l(X) \leq \mathbf{E}f \circ l(Y)$ for all convex function $f : \mathbb{R} \mapsto \mathbb{R}$ and linear $l : \mathbb{R}^d \mapsto \mathbb{R}$.
- *Increasing positive linear convex:* $X \leq_{\text{iplcx}} Y$ if $\langle X, u\rangle \leq_{\text{cx}} \langle Y, u\rangle$ for all $u \in \mathbb{R}_+^d$.

Only random vectors with the same expectation are comparable by means of the convex and linear convex stochastic orders, while $X \leq_{\text{cx}} Y$ if and only if $X \leq_{\text{icx}} Y$ and $-X \leq_{\text{icx}} -Y$, which naturally leads to $\mathbf{E}X = \mathbf{E}Y$. The set of all stop-loss functions $f_t(x) = (x - t)_+$, for $t \in \mathbb{R}$ is also a generator of the univariate increasing convex order.

Another common stochastic order is generated by all increasing concave functions. However, it can be characterized in terms of the increasing convex order and X is smaller than Y in the *increasing concave stochastic order* if and only if $-Y \leq_{\text{icx}} -X$.

Lift zonoid The inclusion of lift zonoids, or equivalently the inclusion of the zonoid trimmed regions of all levels, characterizes the linear convex stochastic order, i.e. we have $X \leq_{\text{lcx}} Y$ if and only if $\widehat{\mathrm{Z}}(P_X) \subseteq \widehat{\mathrm{Z}}(P_Y)$, see Koshevoy and Mosler (1998). In fact, by the representation of the lift zonoid as the selection expectation of a random segment and the fact that the support function of the selection expectation of a random set on any $u \in \mathbb{R}^d$ equals the expectation of the support function of the random set on u, the inclusion $\widehat{\mathrm{Z}}(P_X) \subseteq \widehat{\mathrm{Z}}(P_Y)$ implies that $\mathbf{E}(\langle X, u\rangle - t)_+ \leq \mathbf{E}(\langle Y, u\rangle - t)_+$ for all $u \in \mathbb{R}^d$ and $t \in \mathbb{R}$, whence $\langle X, u\rangle \leq_{\text{icx}} \langle Y, u\rangle$ for all $u \in \mathbb{R}^d$.

The increasing positive linear convex stochastic order, can also be characterized in terms of the lift zonoid or the zonoid trimmed regions. We have $X \leq_{\text{iplcx}} Y$ if $\widehat{\mathrm{Z}}(P_X)+(\{o\}\times\mathbb{R}^d_-) \subseteq \widehat{\mathrm{Z}}(P_Y)+(\{o\}\times\mathbb{R}^d_-)$, where o is here the origin in $\mathbb{R}$ and $\mathbb{R}^d_-$ is the negative quadrant, which is equivalent to $\mathrm{ZD}^\alpha(P_X)+\mathbb{R}^d_- \subseteq \mathrm{ZD}^\alpha(P_Y)+\mathbb{R}^d_-$ for all $\alpha \in (0, 1]$.

A final characterization of the increasing positive linear convex order $X \leq_{\text{iplcx}} Y$ can be given in terms of the selection expectation and closed halfspaces as $\mathbf{E}(\{X\} \cup H\}) \subseteq \mathbf{E}(\{Y\} \cup H)$ for all *lower* halfspace H (the halfspace H is lower, if its reflection, $-H$, is upper). By taking upper halfspaces, we would obtain $-X \leq_{\text{iplcx}} -Y$, which, once the roles of X and Y are reversed, is equivalent to a notion of increasing positive linear concave ordering.

Expected convex hull By inclusion of the expected convex hull trimmed regions we define the *convex hull order*. Let $X \leq_{\text{ch}} Y$ if $\mathrm{CD}^{1/k}(P_X) \subseteq \mathrm{CD}^{1/k}(P_Y)$ for all k, that is, the selection expectation of the convex hull of any given number of independent copies of X is contained in the equivalent set for Y.

In the univariate case we can define $X \leq_{\max} Y$ if $\mathbf{E}\max\{X_1, \dots, X_k\} \leq \mathbf{E}\max\{Y_1, \dots, Y_k\}$ for all k and it is possible to characterize the convex hull order by $X \leq_{\text{ch}} Y$ if and only if $X \leq_{\max} Y$ and $-X \leq_{\max} -Y$. For this reason, we can establish a parallel with the convex and increasing convex stochastic orders. However, there is hierarchy between these four orderings, since the ordering of the maximum is weaker than the increasing convex, and thus, the convex hull order is weaker than the linear convex order. The role of the increasing concave stochastic order would be played now by a minimum stochastic order, that could be characterized in terms of the expectations of the minimum order statistics.

The multivariate expected convex hull order is a linear stochastic order, in the sense that $X \leq_{\text{ch}} Y$ if $\langle X, u\rangle \leq_{\text{ch}} \langle X, u\rangle$ for all $u \in \mathbb{R}^d$. This ordering is again weaker than the linear convex stochastic order.

It is possible to build a multivariate generalization of the maximal order, the *increasing convex hull order* $X \leq_{\text{ich}} Y$ if $\langle X, u\rangle \leq_{\max} \langle Y, u\rangle$ for all $u \in \mathbb{R}^d_+$ or alternatively $\mathrm{CD}^{1/k}(P_X) + \mathbb{R}^d_- \subseteq \mathrm{CD}^{1/k}(P_Y) + \mathbb{R}^d_-$ for all k.

12.5.3 *Stochastic orderings for random sets*

Koshevoy (2002) and Cascos and Molchanov (2003) built generalizations of the linear convex stochastic order to random sets. Given a compact convex random set $\boldsymbol{X}$ with capacity functional $T_{\boldsymbol{X}}$, we define its *lift pseudo-zonoid* as

$$\widehat{\mathrm{Z}}(T_{\boldsymbol{X}}) = \mathbf{E}\,\mathrm{conv}\{\{o\} \cup (1, \boldsymbol{X})\}\,. \tag{12.11}$$

Unlike the lift zonoid of a probability distribution, the lift pseudo-zonoid of a capacity functional, does not characterize it and neither is it a zonoid.

The *set linear convex* stochastic order for random sets can be characterized by inclusion of lift pseudo-zonoids as $\boldsymbol{X} \leq_{\text{slcx}} \boldsymbol{Y}$ if $\widehat{\mathrm{Z}}(T_{\boldsymbol{X}}) \subseteq \widehat{\mathrm{Z}}(T_{\boldsymbol{Y}})$. Clearly this is not an order relation between the distributions of random sets, but only a preorder since lift pseudo-zonoids do not determine capacity functionals.

Apart from the similarity of the constructions, the reason we call this order set linear convex is for its relation with the linear convex order between the selections of the random sets (random vectors that belong *a.s.* to the random sets). Given $\boldsymbol{X}, \boldsymbol{Y}$ two random sets, if either for every selection of $\boldsymbol{X}$ there exists a selection of $\boldsymbol{Y}$ that is greater than the first one in the linear convex order or $\mathbf{E}\boldsymbol{X} \subseteq \mathbf{E}\boldsymbol{Y}$ and for every selection of $\boldsymbol{Y}$ there exists a selection of $\boldsymbol{X}$ that is smaller than the first one in the linear convex order, then $\boldsymbol{X} \leq_{\text{slcx}} \boldsymbol{Y}$.

An alternative characterization of the set linear convex stochastic order can be given in terms of the support function. We have $\boldsymbol{X} \leq_{\text{slcx}} \boldsymbol{Y}$ if $h(\boldsymbol{X}, u) \leq_{\text{icx}} h(\boldsymbol{Y}, u)$ holds for all $u \in \mathbb{R}^d$.

Koshevoy (2002) characterized this ordering between random sets as an integral stochastic order for the Choquet integration with respect to the capacity functionals of the random sets and their containment functionals.

It is also possible to extend the positive linear convex stochastic order to random sets, as well as the convex hull stochastic order and related concepts.

12.6 Parameter depth and other concepts of depth

In Section 12.2, the halfspace depth of a fixed point with respect to a sample was presented, among other equivalent constructions, as the minimum fraction of data points that must be deleted from the sample in order to make the fixed point lie outside of the convex hull of the remaining data points. This would make the fixed point a *nonfit*, i.e. a not suitable candidate for a location parameter.

Instead of the d-dimensional Euclidean space of points, we can consider any parameter space and define the depth of a fixed parameter as the minimal fraction of points that must be deleted from a sample in order to make the fixed parameter a nonfit. In the location setting (location parameter) the original definition of halfspace depth given by Tukey fits easily in this framework, only by considering that a nonfit is a point that lies outside the convex hull of a data set. This idea was exploited by Mizera (2002) in order to construct a general theory of depth, applied to linear regression by Rousseeuw and Hubert (1999), and to location-scale estimation by Mizera and Müller (2004).

Regression depth Rousseeuw and Hubert (1999) generalized the concept of depth to a linear regression problem. For a regression line, a nonfit is a line that does not contain any data point and whose residuals change sign only once. Alternatively, it is a line that can be rotated at some point into a vertical line, without touching any data point. The regression depth of a line would be the smallest fraction of data points that have to be removed in order to make it a nonfit, or the smallest fraction of data points that are touched when the line is rotated into a vertical one. The deepest regression line is a robust regression line.

Location-scale depth Mizera and Müller (2004) extended the notion of depth to location-scale parameters. In their construction, a different notion of location-scale depth appears for each distribution model.

Functional data depth López-Pintado and Romo (2009) defined data depth for functional data. The observations they work with are real continuous functions defined on some compact interval, $\{f_i(t)\}_{i=1}^n$ for $t \in I$ compact. For a subset of those functions $\{f_i\}_{i\in J}$, they build their band, the set of all functions f such that $\min_{i\in J} f_i(t) \leq f(t) \leq \max_{i\in J} f_i(t)$ for all $t \in I$. Depth functions for functional data assign to a fixed function f a degree of centrality obtained in terms of the proportion of sets of a fixed number of functions from $\{f_i(t)\}_{i=1}^n$ that contain f in their band.

Random sets In Section 12.5 stochastic orders for random sets were considered. In order to define the set linear convex stochastic order, we have built in (12.11) the lift pseudo-zonoid of a random set and, from it, we can obtain a family of (pseudo-)zonoid trimmed regions. The expected convex hull trimmed regions can also be immediately translated to random sets. Those central regions would be set-valued location estimates of a random set, satisfying properties similar to those presented in Section 12.2.

Acknowledgements

This work has been partially supported by the Spanish Ministry of Education and Science under grant MTM2005-02254 and by the Region of Madrid and Universidad Carlos III de Madrid under grant CCG07-UC3M/HUM-3260.

The manuscript has benefited from wise comments and suggestions of a referee and the editors. The author is grateful to his coauthors Miguel López-Díaz and Ilya Molchanov for their constant guidance, support and patience.

References

Aloupis, G., Cortés, C., Gómez, F., Soss, M., and Toussaint, G. (2002). Lower bounds for computing statistical depth. *Comput. Statist. Data Anal.*, **40**, 223–229.

Barnett, V. (1976). The ordering of multivariate data (with discussion). *JRSS, Series A*, **139**, 318–354.

Borovskikh, Y.V. (1996). *U-statistics in Banach spaces.* VSP, Utrecht.

Cascos, I. (2007*a*). Depth functions based on a number of observations of a random vector. WP 07-29. Statistics and Econometrics Series, Universidad Carlos III de Madrid. Available from `http://econpapers.repec.org/paper/ctewsrepe/`.

Cascos, I. (2007*b*). The expected convex hull trimmed regions of a sample. *Comput. Statist.*, **22**, 557–569.

Cascos, I. and López-Díaz, M. (2005). Integral trimmed regions. *J. Multivariate Anal.*, **96**, 404–424.

Cascos, I. and López-Díaz, M. (2008). Consistency of the α-trimming of a probability. Applications to central regions. *Bernoulli*, **14**, 580–592.

Cascos, I. and Molchanov, I. (2003). A stochastic order for random vectors and random sets based on the Aumann expectation. *Statist. Probab. Lett.*, **63**, 295–305.

Cascos, I. and Molchanov, I. (2007). Multivariate risks and depth-trimmed regions. *Finance Stoch.*, **11**, 373–397.

Chazelle, B. (1985). On the convex layers of a planar set. *IEEE Trans. Inform. Theory IT*, **31**, 509–517.

Donoho, D. and Gasko, M. (1992). Breakdown properties of location estimates based on halfspace depth and projected outlyingness. *Ann. Statist.*, **20**, 1803–1827.

Dümbgen, L. (1992). Limit theorems for the simplicial depth. *Statist. Probab. Lett.*, **14**, 119–128.

Dyckerhoff, R. (2000). Computing zonoid trimmed regions of bivariate data sets. In *COMPSTAT 2000 – Proceedings in Computational Statistics* (ed. J. Bethlehem and P. Heijden), pp. 295–300. Physica-Verlag, Heidelberg.

Dyckerhoff, R. (2004). Data depths satisfying the projection property. *Allg. Stat. Arch.*, **88**, 163–190.

Dyckerhoff, R., Koshevoy, G., and Mosler, K. (2000). Zonoid data depth: theory and computation. In *COMPSTAT 1996 – Proceedings in Computational Statistics* (ed. A. Prat), pp. 235–240. Physica-Verlag, Heidelberg.

Eddy, W.F. (1982). Convex hull peeling. In *COMSPTAT 1982 – Proceedings in Computational Statistics* (ed. H. Cassinus, P. Ettinger, and R. Tomassone), pp. 42–47. Physica-Verlag, Vienna.

Eddy, W.F. (1984). Set-valued orderings for bivariate data. In *Stochastic Geometry, Geometric Statistics, Stereology (Oberwolfach, 1983)* (ed. R. Ambartzumian and W. Weil), Volume 65 of *Teubner-Texte Math.*, pp. 79–90. Teubner, Leipzig.

Eddy, W.F. and Gale, J.D. (1981). The convex hull of a spherically symmetric sample. *Adv. Appl. Probab.*, **13**, 751–763.

Finch, S. and Hueter, I. (2004). Random convex hulls: a variance revisited. *Adv. Appl. Probab.*, **36**, 981–986.

Gopala, H. and Morin, P. (2008). Algorithms for bivariate zonoid depth. *Comput. Geom.*, **39**, 2–13.

Green, P.J. (1981). Peeling bivariate data. In *Interpreting Multivariate Data* (ed. V. Barnett), pp. 3–19. Wiley, Chichester, England.

Green, P.J. and Silverman, B.W. (1979). A comparative study of algorithms for convex hulls of bivariate data sets, with applications. *Computer J.*, **22**, 262–266.

Hueter, I. (2005). Limit theorems for convex hull peels. Available from `http://faculty.baruch.cuny.edu/ihueter/`.

Jewell, N.P. and Romano, J.P. (1982). Coverage problems and random convex hulls. *J. Appl. Probab.*, **19**, 546–561.

Koshevoy, G. (2002). Orderings of random sets. Unpublished.

Koshevoy, G. and Mosler, K. (1996). The Lorenz zonoid of a multivariate distribution. *J. Amer. Statist. Assoc.*, **91**, 873–882.

Koshevoy, G. and Mosler, K. (1997*a*). Multivariate Gini indices. *J. Multivariate Anal.*, **60**, 252–276.

Koshevoy, G. and Mosler, K. (1997*b*). Zonoid trimming for multivariate distributions. *Ann. Statist.*, **25**, 1998–2017.

Koshevoy, G. and Mosler, K. (1998). Lift zonoids, random convex hulls and the variability of random vectors. *Bernoulli*, **4**, 377–399.

Liu, R. (1990). On a notion of data depth based on random simplices. *Ann. Statist.*, **18**, 405–414.

Liu, R. (1992). Data depth and multivariate rank tests. In *L_1-statistical analysis and related methods. Proceedings of the Second International Conference on Statistical Data Analysis Based on the L_1-norm and Related Methods (Neuchâtel, 1992)* (ed. Y. Dodge), pp. 279–294. North-Holland, Amsterdam.

Liu, R., Parelius, J.M., and Singh, K. (1999). Multivariate analysis by data depth: Descriptive statistics, graphics and inference. With discussion and a rejoinder by Liu and Singh. *Ann. Statist.*, **27**, 783–858.

Liu, R., Serfling, R., and Souvaine, D.L. (ed.) (2006). *Data Depth: Robust Multivariate Analysis, Computational Geometry and Applications*, Volume 72 of *DIMACS Series*. American Mathematical Society.

López-Pintado, S. and Romo, J. (2009). On the concept of depth for functional data. *J. Amer. Statist. Assoc.*, **104**, 718–734.

Lorenz, M.O. (1905). Methods of measuring the concentration of wealth. *Amer. Stat. Assoc.*, **9**, 209–219.

Massé, J.-C. (2004). Asymptotics for the Tukey depth process, with an application to a multivariate trimmed mean. *Bernoulli*, **10**, 397–419.

Massé, J.-C. and Theodorescu, R. (1994). Halfplane trimming for bivariate distributions. *J. Multivariate Anal.*, **48**, 188–202.

Miller, K., Ramaswami, S., Rousseeuw, P.J., Sellarès, J.A., Souvaine, D., Streinu, I., and Struyf, A. (2003). Efficient computation of location depth contours by methods of computation geometry. *Stat. Comput.*, **13**, 153–162.

Mizera, I. (2002). On depth and deep points: a calculus. *Ann. Statist.*, **30**, 1681–1736.

Mizera, I. and Müller, C.H. (2004). Location-scale depth. With comments and a rejoinder by the authors. *J. Amer. Statist. Assoc.*, **99**, 949–989.

Molchanov, I. (2005). *Theory of Random Sets.* Springer, London.

Mosler, K. (2002). *Multivariate Dispersion, Central Regions and Depth. The Lift Zonoid Approach*, Volume 165 of *Lecture Notes in Statistics.* Springer, Berlin.

Müller, A. and Stoyan, D. (2002). *Comparison Methods for Stochastic Models and Risks.* Wiley Series in Probability and Statistics. Wiley, New York.

Oja, H. (1983). Descriptive statistics for multivariate distributions. *Statist. Probab. Lett.*, **1**, 327–332.

Rousseeuw, P.J. and Hubert, M. (1999). Regression depth. With discussion and a reply by the authors and Stefan Van Aelst. *J. Amer. Statist. Assoc.*, **94**, 388–433.

Rousseeuw, P.J. and Ruts, I. (1996). Bivariate location depth. *Appl. Statist.*, **45**, 516–526.

Rousseeuw, P.J. and Ruts, I. (1999). The depth function of a population distribution. *Metrika*, **49**, 213–244.

Rousseeuw, P.J., Ruts, I., and Tukey, J.W. (1999). The Bagplot: a bivariate Boxplot. *Amer. Statist.*, **53**, 382–387.

Rousseeuw, P.J. and Struyf, A. (2004). Characterizing angular symmetry and regression symmetry. *J. Stat. Plann. Inference*, **122**, 161–173.

Ruts, I. and Rousseeuw, P.J. (1996). Computing depth contours of bivariate point clouds. *Comput. Statist. Data Anal.*, **23**, 153–168.

Serfling, R. (2006). Depth functions in nonparametric multivariate inference. In *Data Depth: Robust Multivariate Analysis, Computational Geometry and Applications* (ed. R. Liu, R. Serfling, and D. Souvaine), Volume 72 of *DIMACS Series*, pp. 1–16. American Mathematical Society.

Small, C.G. (1990). A survey of multidimensional medians. *Internat. Statist. Rev.*, **58**, 263–277.

Tukey, J.W. (1975). Mathematics and the picturing of data. In *Proceedings of the International Congress of Mathematics (Vancouver, 1974)* (ed. R. James), Volume 2, pp. 523–531.

Vitale, R. (1987). Expected convex hulls, order statistics, and Banach space probabilities. *Acta Appl. Math.*, **9**, 97–102.

Wagner, U. and Welzl, E. (2001). A continuous analogue of the Upper Bound Theorem. *Discrete Comput. Geom.*, **26**, 205–219.

Wendel, J.G. (1962). A problem in geometric probability. *Math. Scand.*, **11**, 109–111.

Zuo, Y. and Serfling, R. (2000*a*). General notions of statistical depth function. *Ann. Statist.*, **28**, 461–482.

Zuo, Y. and Serfling, R. (2000*b*). Nonparametric notions of multivariate 'scatter measure' and 'more scattered' based on statistical depth functions. *J. Multivariate Anal.*, **75**, 62–78.

Zuo, Y. and Serfling, R. (2000*c*). Structural properties and convergence results for contours of sample statistical depth functions. *Ann. Statist.*, **28**, 483–499.

PART IV

APPLICATIONS

13

APPLICATIONS OF STOCHASTIC GEOMETRY IN IMAGE ANALYSIS

Marie-Colette N.M. van Lieshout

Abstract

A discussion is given of various stochastic geometry models (random fields, sequential object processes, polygonal field models) which can be used in intermediate- and high-level image analysis. Two examples are presented of actual image analysis problems (motion tracking in video, foreground/background separation) to which these ideas can be applied.

13.1 Introduction

The new millennium has opened with what can only be described as a data explosion following advances in digital technology, a development that has led to a strong demand for tools to analyse digital data such as still images, video streams, audio signals, and text.

As a field, statistical image analysis took off during the 1980s with the seminal work by Besag (1986) and Geman and Geman (1984) on the restoration of pictures degraded by noise. Much of the work in this period is focussed on 'low level' tasks, that is, it aims to de-noise, sharpen, segment, or classify the image. A good overview can be found in the supplement to *Journal of Applied Statistics* edited by Mardia and Kanji (1993) which includes reprints of the seminal papers mentioned above and a list of early references.

With the improvement in image quality, in the course of the 1990s the emphasis shifted towards the 'high level' task of describing an image by its content, for example in terms of the objects it contains and the spatio-temporal relations between them. Early work in this direction includes Baddeley and Van Lieshout (1992, 1993), Molina and Ripley (1989), Ripley and Sutherland (1990), as well as the work by Grenander and co-authors – although usually couched in pattern theoretic language (Grenander 1976, 1978, 1981).

Given the difference in goals, it is hardly surprising that different stochastic models are used in low and high level vision. For example, classification at the lowest conceptual level calls for a pixel based model. Even in this context, though, due to the high dimension of image data, fitting a model is a far from trivial task that cannot be done explicitly except for the simplest models. Instead, an algorithmic approach is often taken in which small changes – involving only a few pixels – are proposed iteratively. The crucial ingredient of many such algorithms is the difference in log likelihood between the new and old states. For computational reasons, it would be highly desirable if this ratio would be 'local', a concept that can be formalized by a Markov property. Further details can be found in Geman (1990) or Winkler (2003).

At the other extreme, the focus of attention are the objects in the image as described by their location, shape, and colour parameters. In this context, it is natural to use marked point processes (Daley and Vere-Jones 2003), in particular those satisfying a Markov property (Van Lieshout 2000a). Which mark to use depends on the application, ranging from simple geometric shapes (Baddeley and Van Lieshout 1992, Van Lieshout 1994, 1995) through deformable template models (Amit *et al.* 1991, Hansen *et al.* 2002, Hurn 1998, Mardia *et al.* 1997, Pievatolo and Green 1998, Rue and Hurn 1999, Rue and Husby 1998) via the complex ensembles of simple shapes studied by Lacoste *et al.* (2005), Stoica *et al.* (2002, 2004, 2007), and Ortner *et al.* (2007) to the spatio-temporal models of Van Lieshout (2007).

Intermediate level modelling tries to avoid a full scene description while preserving a global approach. It focusses therefore on image regions, either conceived as a conglomerate of pixels (Møller and Waagepetersen 1998, Tjelmeland and Holden 1993) or as a (continuous) tessellation of space (Clifford and Middleton 1989, Kluszczyński *et al.* 2007, Møller and Skare 2001, Nicholls 1998).

The present chapter is organized as follows. In Section 13.2, we present random fields, (sequential) object processes and polygonal field models. In Section 13.3, we discuss relations between the various Markov properties discussed in Section 13.2, indicate how a random field is affected by the choice of the number of class labels, and show that certain polygonal field models can be seen as non-overlapping object processes. In the last section, we present two applications. The first one concerns tracking of a variable number of moving objects through a video sequence, the other one a foreground/background separation problem. For the first example, a ground truth is known to which the outcome can be compared; in the second application, visual inspection must validate the results.

The emphasis in this chapter is on stochastic geometric modelling at the expense of a detailed discussion of the Monte Carlo algorithms employed for inference. The interested reader is referred to, e.g. Chapter 9 or Winkler (2003). We also refrain from discussing mathematical morphology, a branch of image analysis with similar roots to stochastic geometry (Matheron 1975). Instead we refer to Serra (1982) or Dougherty (1992).

13.2 Stochastic geometric models: From random fields to object processes

13.2.1 *Random fields*

A digital image is simply a finite array of 'colour' or 'intensity' labels. The entries of the array S, typically a rectangle of raster points in $\mathbb{Z}^2$, are referred to as pixels. We shall use the notation $S = (s_1, \ldots, s_m)$, $m \in \mathbb{N}$. For the moment, assume the set of labels is finite as well, and denoted by Λ.

Let $X = (X_1, \ldots, X_m)$ be a random field with values in Λ^S. Thus, X_i is the label in Λ assigned to pixel s_i. The labels can be categorical, for instance foreground/background, or be related to the intensity values of the image. The distribution of X is given by the joint probability mass function $\mathbf{P}\{X_1 = x_1, \ldots, X_m = x_m\}$ for $x = (x_1, \ldots, x_m) \in \Lambda^S$, which we assume to be strictly positive so that conditional distributions are well-defined. The condition may be relaxed a little (Besag 1974, Clifford 1990) to include models such as the lattice gas discussed below.

Let $\sim$ be a symmetric, reflexive relation on S. For instance, on $\mathbb{Z}^2$ one may define $s_i \sim s_j$ to be neighbours if and only if they are directly or diagonally adjacent, that is, if $||s_i - s_j|| \leq \sqrt{2}$. In graph theoretical terms, the pixels are the vertices and an edge is drawn between $s_i \neq s_j$ if and only if $s_i \sim s_j$. The random field X is said to be Markov with respect to $\sim$ if for all $i = 1, \ldots, m$ the conditional distribution

$$\mathbf{P}\{X_i = x_i \mid X_j = x_j, j \neq i\} = \mathbf{P}\{X_i = x_i \mid X_j = x_j, s_i \sim s_j, j \neq i\} \qquad (13.1)$$

depends only on x_i and the labels at those pixels s_j that share an edge with s_i (see e.g. Besag 1974).

Example: Lattice gas (Lebowitz and Gallavotti 1971) Let S be a finite square of raster points $\{s_1, \ldots, s_m\} \subseteq \mathbb{Z}^2$ equipped with the neighbourhood relation $s_i \sim s_j$ if and only if $||s_i - s_j|| \leq \sqrt{2}$. Let Λ be the set $\{0, 1, 2\}$, fix the parameter $\alpha > 0$, and consider the random field X on Λ^S whose joint probability mass function

$$\mathbf{P}\{X_1 = x_1, \ldots, X_m = x_m\} \propto \prod_{i=1}^{m} \alpha^{\mathbf{1}_{x_i \neq 0}} \prod_{s_i \sim s_j, i<j} \mathbf{1}_{x_i = x_j \text{ or } x_i x_j = 0}$$

is proportional to the right hand side product. In words, labels are assigned independently with probabilities $1/(1+2\alpha)$ for 0 and $\alpha/(1+2\alpha)$ otherwise, conditional on the event that for every pair of neighbour pixels $s_i \sim s_j$, $i < j$, either one of the labels is zero ($x_i x_j = 0$) or they are identical ($x_i = x_j$). Clearly, some elements of Λ^S occur with probability zero. Nevertheless, whenever the conditioning event on the left hand side of (13.1) has strictly positive probability, the equation (13.1) holds, and we say that $\mathbf{P}$ is Markov with respect to $\sim$, with labels 1 and 2 'repelling' each other. A realization with $\alpha = 0.75$ is shown in Fig. 13.1.

From a computational point of view, the Markov property implies that iteratively sampling X_i from the right hand side of (13.1) leaves $\mathbf{P}$ invariant, an observation that is crucial for Monte Carlo inference (see e.g. Geman 1990).

Assume that X is a random field with strictly positive joint probability mass function $\mathbf{P}$. Then, by the Hammersley–Clifford theorem (see the historical account by Clifford 1990), X is a Markov random field if and only if $\mathbf{P}$ can be written as

$$\mathbf{P}\{X_1 = x_1, \ldots, X_m = x_m\} = \prod_{C \in \mathbf{C}} \varphi_C(x_c, c \in C) \qquad (13.2)$$

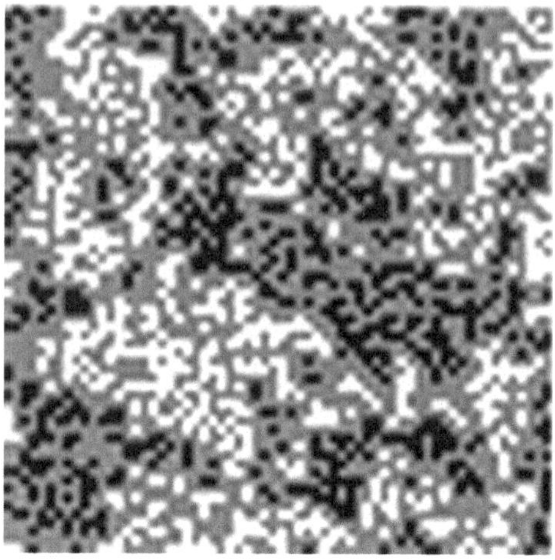

FIG. 13.1. Realization of a lattice gas model with $\alpha = 0.75$. Black represents colour label '1', grey '0', and white '2'.

for some interaction functions $\varphi_C : \Lambda^C \to \mathbb{R}^+$ defined for $C \in \mathbf{C}$, the family of sets consisting of pairwise neighbours. By convention, singletons and the empty set are included in $\mathbf{C}$. Indeed, $\varphi_\emptyset$ can be regarded as a normalizing constant.

Pairwise interaction models for which $\varphi_C \equiv 1$ whenever the cardinality of C exceeds two are the most convenient, and widely used. The lattice gas model falls in this class. Indeed, $\varphi_{\{s_i\}}(x_i) = \alpha^{\mathbf{1}_{x_i \neq 0}}$, while $\varphi_{\{s_i,s_j\}}(x_i, x_j) = 0$ if and only if $s_i \sim s_j$ and $x_i = 1$, $x_j = 2$ or vice versa, one otherwise. Such discouragement for neighbouring pixels to have different labels is typical for image analysis. Another famous example is the Potts interaction function $\varphi_{\{s_i,s_j\}}(x_i, x_j) = \exp\left\{-\beta\, \mathbf{1}_{x_i \neq x_j}\right\}$ for some $\beta > 0$. The reader may find it useful to think of φ_C as a regularization term for the C-local behaviour of desirable realizations x of the random field X.

The global behaviour of a pairwise interaction Markov random field cannot be expected to reflect the global appearance of complex pictures, which is the reason why greedy ascent type algorithms tend to give better results than methods aimed at a global optimum (Winkler 2003). If global characteristics are the object of interest, it is advisable to use larger neighbourhoods and non-trivial interaction functions for a rich subfamily of $\mathbf{C}$ (Tjelmeland and Besag 1998). Note that (13.2) can be augmented with interactions on the dual graph (Geman *et al.* 1990, Geman and Reynolds 1992) when preservation of the natural edges in an image is important.

A stronger form of conditional local dependence is that of the Markov mesh models proposed by Abend *et al.* (1965). Let S be a finite rectangle of raster points in $\mathbb{Z}^2$ and call pixel s_j a predecessor of s_i if it either lies above or to the left of s_i. Then X is a Markov mesh model if the conditional distribution of X_i,

$i = 1, \ldots, m$, given the values x_j at all predecessors s_j of s_i depends only on a limited set of s_j, say the three nearest ones. These conditional distributions are often assumed to be given in closed form and easy to simulate from. In this case, the joint probability mass function factorizes as the product over such conditional distributions, and – in contrast to most Markov random field models – is analytically tractable and amenable to sequential procedures. However, due to their causal non-symmetric nature, Markov mesh models tend to produce realizations with striping effects not apparent in many natural images (Cressie and Davidson 1998, Lacroix 1987, Qian and Titterington 1991).

To conclude this subsection, Gaussian Markov random fields deserve special mention. So far, for convenience, we have assumed that Λ is finite. However, the theory discussed here remains valid in the context of real-valued labels if we replace probability mass functions by densities. When X is normally distributed, the precision matrix Q captures the spatial dependence. Typically in applications, Q will be sparse, so that Monte Carlo methods can often be avoided and replaced by numerical ones (Rue and Held 2005). Note that the class of Gaussian fields includes classic conditional autoregression models for smoothly varying images (Ripley 1988, see also Besag and Kooperberg 1995, Besag *et al.* 1991, Kűnsch 1987).

13.2.2 *Intermediate level modelling*

Intermediate level image analysis aims to describe an image in terms of its homogeneous regions. There are two main strands. In the first, one stays within the random field framework but considers a more global interaction structure; the second approach is based on the concept of a spatial tessellation (chapter 5 in this volume).

Random fields Pairwise interaction Markov random fields like the Potts model with moderate neighbourhood sizes can be used successfully as regularization terms for low level tasks such as the restoration of noisy pictures (Section 13.2.1). However, for large values of β, realizations of the Potts model tend to be dominated by a single colour.

In order to construct models that tend to produce realizations containing compact regions, Tjelmeland and Besag (1998) introduced higher order interaction functions φ_C that express the relative likelihood of various types of labellings on C. For example, if $\Lambda = \{0, 1\}$, the C could be monochrome, contain convex/concave corners or edges between the two labels. See also Gimel'farb (1999).

With the same objective in mind, Møller and Waagepetersen (1998) focussed on image regions directly and proposed Markov connected component fields defined by a joint probability mass function of the form

$$\mathbf{P}\{X_1 = x_1, \ldots, X_m = x_m\} \propto \prod_{K \in \mathbf{K}(x)} \Psi_K(l(x_K)), \qquad (13.3)$$

where the product ranges over the maximal $\sim$-connected components of identically labelled pixels in $x = (x_1, \ldots, x_m)$, $l(x_K)$ denotes the common label in the component $x_K = (x_k, k \in K)$, and $\Psi_K : \Lambda \to \mathbb{R}^+$ is the interaction function. The functions Ψ_K may be based on fundamental geometric quantities such as the area, perimeter, and Euler–Poincaré characteristic of K, as well as on corner features similar to the ones considered by Tjelmeland and Besag (1998). Inter-component interaction can be taken into account, leading to a joint probability mass function proportional to

$$\prod_{K \in \mathbf{K}(x)} \Psi_K(l(x_K)) \prod_{\{K,L\} \subseteq \mathbf{K}(x)} \Phi_{K,L}(l(x_K), l(x_L)), \tag{13.4}$$

the semi-Markov random fields introduced by Tjelmeland and Holden (1993). Here, $\Phi_{K,L} \geq 0$ are the symmetric inter-component interaction functions between components K and L in $\mathbf{K}(x)$. Note that the Potts model of Section 13.2.1 permits a factorization of the form (13.3)–(13.4). In general, though, Markov random fields and connected components fields or semi-Markov random fields with respect to the same neighbourhood relation $\sim$ are not comparable in the sense that neither class is contained in the other.

Spatial tessellations A coloured spatial tessellation is a partition of the image window $D \subseteq \mathbb{R}^2$, for example the convex hull of the pixels, into a finite number of disjoint, usually polygonal, regions whose united closures fill the window (Stoyan *et al.* 1995) and where each polygon carries a label. The labels may be assigned uniformly under the condition that no adjacent regions share the same label, or, if the polygons are seen as building blocks for larger regions, by a conditional Markov random field model with respect to the adjacency relation that encourages similar labels between polygons separated by a common boundary segment (for example a Potts or Gaussian model).

A random tessellation can be obtained in various ways. We shall focus here on models derived from the Poisson line process L. Note that a line $\ell_{\theta,p}$ in the plane may be parametrized by the pair (θ, p), where p is the signed distance of the line to the origin and $\theta \in [0, \pi)$ is the angle between the normal to the line and the x-axis as illustrated in Fig. 13.2. The measure $\mathrm{d}\theta\,\mathrm{d}p$ can be shown to be invariant under translations and rotations, so that the random set of lines L parametrized by a Poisson process on $[0, \pi) \times \mathbb{R}$ with intensity measure $\mathrm{d}\theta\,\mathrm{d}p$ is stationary and isotropic. Clearly, the lines of L that intersect D form a partition of the image window. Moreover, the angle between two 'randomly chosen' lines of L is distributed according to the probability density $\sin(\phi)/2$ on $[0, \pi)$.

More refined tessellations can be obtained by using the lines of L as a skeleton (Arak 1982). Suppose that D is a bounded, non-empty, open convex set with a piecewise smooth border ∂D, for example a rectangle. Define the family Γ_D of admissible tessellations of D as the set of all planar graphs γ in $D \cup \partial D$ with non-intersecting straight line segments as edges such that no two edges are co-linear, all vertices in D have degree two, and $\gamma \cap \partial D$ is either empty or consists of

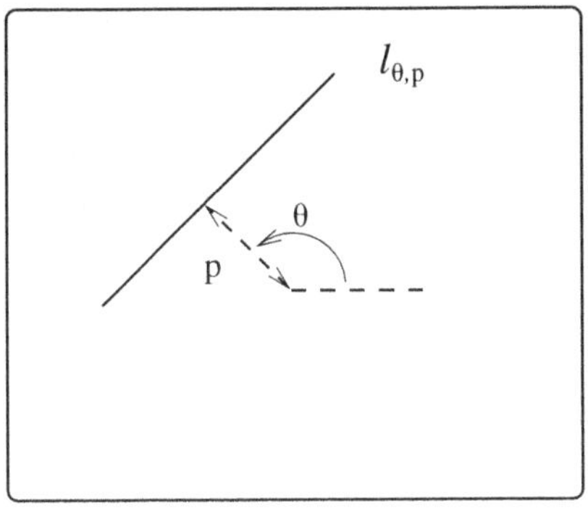

FIG. 13.2. Parametrization of lines in the plane.

vertices of degree 1. In statistical physics, the latter is referred to as an 'empty' or 'free' boundary condition, respectively. For a realization $l_D = \{l_i\}_{i=1}^n$ of the Poisson line process L_D restricted to D, let $\Gamma_D(l_D)$ be the family of $\gamma \in \Gamma_D$ such that the graph $\gamma \subseteq \bigcup_{i=1}^n l_i$ is constructed on lines of l_D and $\gamma \cap l_i$, $i = 1, \ldots, n$, contains a single interval of strictly positive length. Now, the polygonal Arak field A_D on D is defined by

$$\mathbf{P}\{A_D \in G\} \propto \mathbf{E}\left[\sum_{\gamma \in \Gamma_D(L_D) \cap G} \exp\{-2\,l(\gamma)\}\right], \tag{13.5}$$

where $l(\gamma)$ is the total edge length of γ and the expectation is with respect to the distribution of L_D. A realization of (13.5) with free boundary condition is shown in Fig. 13.3. The Arak field is Markovian in the following sense: the conditional distribution of A_D within a piecewise smooth closed curve C depends only on the intersection points and directions of A_D with C.

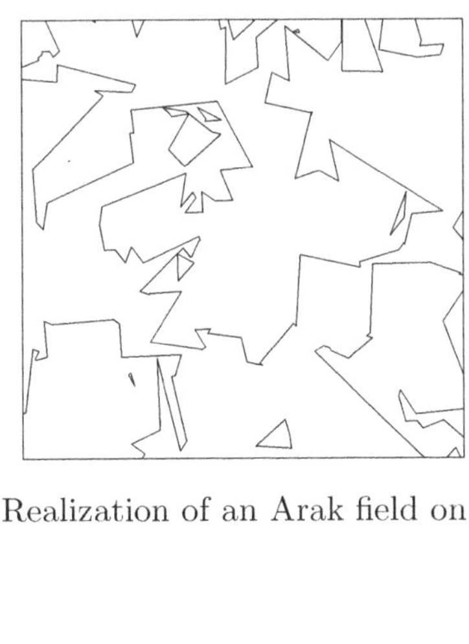

FIG. 13.3. Realization of an Arak field on $[0, 4] \times [0, 4]$.

Several variations have been considered in the literature. Firstly, the probability distribution (13.5) may be used as a dominating measure to define further random tessellations. For instance, for $\beta > 0$, a length-interacting polygonal Markov field has probability density $f(\gamma) \propto \exp\{-\beta\, l(\gamma)\}$ with respect to the law of A_D (Arak 1982, Van Lieshout and Schreiber 2007, Schreiber 2005), thus favouring tessellations with small total edge lengths. Secondly, the class of admissable tessellations may be modified. In this vein, Nicholls (1998) restricts himself to triangles, Arak and Surgailis (1989) allow vertices of degrees 3 and 4, and Arak *et al.* (1993) consider point rather than line based constructions. Finally, Voronoi tessellations (Okabe *et al.* 2000) have been used as alternatives for polygonal fields (Blackwell and Møller 2003, Green 1995, Heikkinen and Arjas 1998, Møller and Skare 2001, Skare *et al.* 2007).

13.2.3 *Object processes*

As in the previous section, let D be the image window. Let Q be a Polish space that captures object attributes such as size, shape and colour. We shall define an object process as a marked point process (see Chapter 1 of this volume) specified by its probability density f with respect to the distribution of a unit rate Poisson process on $\bar{D} = D \cup \partial D$ marked in an i.i.d. fashion according to some mark distribution $\mathbb{Q}$ on Q. Realizations are denoted by $\mathbf{x} = \{x_1, \ldots, x_n\}$, where $n \in \mathbb{N}_0$, and $x_i \in \bar{D} \times Q$, $i = 1, \ldots, n$.

Example: Penetrable spheres (Widom and Rowlinson 1970) Let $Q = \{1, 2\}$. In analogy to the lattice gas model discussed in Section 13.2.1, the penetrable spheres mixture model is defined as a homogeneous Poisson process of rate $2\beta > 0$ in which points are independently assigned each of the two types with probability 1/2, conditional on the event that points of different type keep at least a distance $R > 0$ away from each other. A realization with $R = 0.1$ and $\beta = 100$ is shown in Fig. 13.4. Write X_1 for the ensemble of points with mark 1, X_2 for those marked 2. Then the model has joint probability density

$$f(\mathbf{x}_1, \mathbf{x}_2) \propto \beta^{n(\mathbf{x}_1)+n(\mathbf{x}_2)}\, \mathbf{1}_{d(\mathbf{x}_1, \mathbf{x}_2) > R}$$

with respect to the product measure of two independent unit rate Poisson processes on $\bar{D}$. Here $n(\mathbf{x}_i)$ denotes the cardinality of $\mathbf{x}_i$, and $d(\mathbf{x}_1, \mathbf{x}_2)$ is the shortest distance from a point in $\mathbf{x}_1$ to one in $\mathbf{x}_2$. The marginal distributions of X_i, $i \in \{1, 2\}$, are called area-interaction processes (Baddeley and Van Lieshout 1995, Häggström *et al.* 1999) and have probability density proportional to

$$\beta^{n(\mathbf{x}_i)} \exp\{-\beta\, |U_{\mathbf{x}_i} \cap \bar{D}|\}$$

with respect to a unit rate Poisson process on $\bar{D}$, where $|\cdot|$ denotes area, and $U_{\mathbf{x}_i}$ is the union of closed balls of radius R centred at the points of $\mathbf{x}_i$, $i = 1, 2$. Note that points of the same type prefer to be close.

FIG. 13.4. Realization of a penetrable spheres model on $[0,1] \times [0,1]$ with $\beta = 100$ and $R = 1/10$. Points of type 1 are represented by triangles, points of type 2 by crosses.

Let $\sim$ be a reflexive, symmetric neighbourhood relation on $\bar{D} \times Q$. For example, the objects parametrised by x and y may be neighbours whenever they overlap (allowing for blur and shading where necessary). An object process X with probability density f is said to be Markov with respect to $\sim$ if f is hereditary in the sense that whenever $f(\mathbf{x}) > 0$ for some configuration of objects, also $f(\mathbf{y}) > 0$ for all $\mathbf{y} \subseteq \mathbf{x}$, and, for all $u \in (\bar{D} \times Q) \setminus \mathbf{x}$, the ratio

$$\lambda(u \mid \mathbf{x}) = \frac{f(\mathbf{x} \cup \{u\})}{f(\mathbf{x})}, \tag{13.6}$$

known as the conditional intensity (Papangelou 1974), depends only on u and $\{x_i \in \mathbf{x} : u \sim x_i\}$ (Ripley and Kelly 1977). A factorization is provided by the Ripley–Kelly theorem stating that a marked point process with probability density f is a Markov object process if and only if its probability density can be written as

$$f(\mathbf{x}) = \prod_{\text{cliques } \mathbf{y} \subseteq \mathbf{x}} \varphi(\mathbf{y}) \tag{13.7}$$

for some non-negative interaction function φ. The product ranges over object configurations $\mathbf{y} \subseteq \mathbf{x}$ that consist of pairwise neighbours (including singletons and the empty configuration). The resemblance to (13.1) is obvious. To see that the area-interaction process is Markov, define $x \sim y$ if and only if $||x - y|| \leq 2R$, and observe that the conditional intensity $\lambda(u \mid \mathbf{x}) = \beta \exp\{-\beta|(B(u,R) \setminus U_{\mathbf{x}}) \cap \bar{D}|\}$ depends only on u and those $x_i \in \mathbf{x}$ that are neighbours of u. The interaction function is given by

$$\varphi(\{x_1, \ldots, x_k\}) = \exp\left\{\beta(-1)^k \left| \left(\bigcap_{i=1}^{k} B(x_i, R) \right) \cap \bar{D} \right| \right\}$$

for $k \geq 1$.

If $\varphi(\mathbf{x}) \equiv 1$ whenever $\mathbf{x}$ contains more than two objects, f defines a pairwise interaction process. Note that some care has to be taken to ensure that a model defined by some φ is integrable with respect to the dominating measure and hence can be normalized to unity. A sufficient condition is that $\varphi(\mathbf{x}) \leq 1$ for all cliques $\mathbf{x} \neq \emptyset$. Fortunately, such a condition is not restrictive in the context of image analysis, as it is usually undesirable for a scene to contain many overlapping objects. Forbidding overlap altogether, that is, setting $\phi(\mathbf{x}) = 0$ whenever $x_1 \sim x_2$ for some $x_1 \neq x_2 \in \mathbf{x}$, results in a hard core model. For further details as well as a wide range of examples, the reader is referred to Van Lieshout (2000a).

Continuous analogues of the Markov mesh models and connected component fields discussed in Section 13.2.1 exist (Baddeley and Møller 1989, Baddeley *et al.* 1996, Cressie *et al.* 2000, Häggström *et al.* 1999) but to date seem not to have found many applications in high level image analysis.

Markov marked point processes are useful modelling tools for scenes composed of objects that do not overlap or are of similar appearance. However, they do not take into account the relative depth – distance to the camera – of objects in the image (Mardia *et al.* 1997) due to the invariance under permutations inherent in their definition, nor can they cope with non-symmetric neighbour relations (Ortner *et al.* 2007). In such cases, finite sequential spatial processes (Van Lieshout 2006a, 2006b) can be used. The realizations of such a process, denoted by $\vec{\mathbf{x}} = (x_1, \ldots, x_n)$ for $n \in \mathbb{N}_0$ and $x_i \in \bar{D} \times Q$, $i = 1, \ldots, n$, consist of vectors of arbitrary length – in contrast to the sets that arise as realizations of marked point processes. An example is given in Fig. 13.6, where for $i < j$, the coloured square x_i lies in the foreground compared to square x_j. The distribution of a sequential spatial process on $\bar{D}$ with marks in Q may be specified by its probability density with respect to the distribution of a random sequence of Poisson length with independent components that are uniformly distributed over $\bar{D}$ and marked in an i.i.d. fashion according to $\mathbb{Q}$. See also Daley and Vere-Jones (2003).

Let $\sim$ be a reflexive relation on $\bar{D} \times Q$, not necessarily symmetric. If $y \sim z$, the object parametrized by the marked point z is said to be a directed neighbour of y. Now, a sequential spatial process Y with probability density f is said to be Markov with respect to $\sim$ if $f(\vec{\mathbf{y}}) > 0$ implies $f(\vec{\mathbf{z}}) > 0$ for all subsequences $\vec{\mathbf{z}}$ of $\vec{\mathbf{y}}$, and, for $u \notin \vec{\mathbf{y}}$, the ratio $f((\vec{\mathbf{y}}, u))/f(\vec{\mathbf{y}})$ depends only on u and its directed neighbours $\{y_i \in \vec{\mathbf{y}} : u \sim y_i\}$ in $\vec{\mathbf{y}}$. Usually the set of neighbours is small compared to the global configuration, a fact that can be employed to advantage in the design of Monte Carlo algorithms. The ratio is related to the sequential conditional intensity

$$\lambda_i(u \mid \vec{\mathbf{y}}) = \frac{f(s_i(\vec{\mathbf{y}}, u))}{(n+1)\, f(\vec{\mathbf{y}})}, \quad f(\vec{\mathbf{y}}) > 0, \tag{13.8}$$

for inserting $u \notin \vec{\mathbf{y}}$ at position $i \in \{1, \ldots, n+1\}$ of $\vec{\mathbf{y}} = (y_1, \ldots, y_n)$. Here $s_i(\vec{\mathbf{y}}, u) = (y_1, \ldots, y_{i-1}, u, y_i, \ldots, y_n)$. The overall conditional intensity for finding a marked point at u in any position in the vector given that the remainder of the sequence equals $\vec{\mathbf{y}}$ is given by $\sum_{i=1}^{n+1} \lambda_i(u \mid \vec{\mathbf{y}})$. The expression should be compared to its classic counterpart (13.6).

An analogue of (13.7) holds in the sequential setting. Indeed, for $u \in \bar{D} \times Q$, the sequence $\vec{\mathbf{y}}$ is said to be a u-clique with respect to $\sim$ if $\vec{\mathbf{y}}$ either has length zero or all its components y_i satisfy $u \sim y_i$. The definition is u-directed but otherwise permutation invariant, so we may map $\vec{\mathbf{y}}$ onto the set $\mathbf{y}$ by ignoring the permutation. Similarly, write $\mathbf{y}_{<i} = \{y_1, \ldots, y_{i-1}\}$. Then, a sequential spatial process with probability density f is Markov with respect to $\sim$ if and only if it can be factorized as

$$f(y_1, \ldots, y_n) = f(\emptyset) \prod_{i=1}^{n} \prod_{\mathbf{z} \subseteq \mathbf{y}_{<i}} \varphi(y_i, \mathbf{z}) \tag{13.9}$$

for some non-negative interaction function φ such that $\varphi(u, \mathbf{z}) = 1$ if $\mathbf{z}$ is no u-clique with respect to $\sim$. As for marked point processes, when defining a Markov density by its interaction function, integrability must be verified.

Clearly, any sequential spatial process Y immediately defines a classic object process by ignoring the permutation. Alternatively, provided it is integrable, one may define a marked point process X by geometric averaging, that is, by defining a probability density f_X with respect to a unit rate Poisson process on $\bar{D}$ marked in an i.i.d. fashion according to $\mathbb{Q}$ by $f_X(\mathbf{x}) \propto \left(\prod_{\vec{\mathbf{x}}} f(\vec{\mathbf{x}})\right)^{1/n(\mathbf{x})!}$ where the product ranges over all permutations $\vec{\mathbf{x}}$ of object configuration $\mathbf{x}$, and f is a probability density of Y in the sequential setting. In this case, if f admits a Hammersley–Clifford factorisation of the form (13.9), the probability density of X factorizes as

$$f_X(\mathbf{x}) \propto \prod_{\emptyset \neq \mathbf{y} \subseteq \mathbf{x}} \tilde{\varphi}(\mathbf{y})$$

where $\tilde{\varphi}(\mathbf{y}) = \{\prod_{y \in \mathbf{y}} \varphi(y, \mathbf{y} \setminus \{y\})\}^{1/n(\mathbf{y})}$. It follows that if Y is a pairwise interaction Markov sequential spatial process with respect to some relation $\sim$ having probability density $f(\emptyset) \prod_{i=1}^{n} \prod_{j<i} \varphi(y_i, y_j)$, X is a pairwise interaction marked point process with respect to the symmetric relation $\approx$ for which $x \approx y$ if and only if $x \sim y$ or $y \sim x$. When interactions of higher order occur, $\tilde{\varphi}(\mathbf{y}) \neq 1$ implies that some $y \in \mathbf{y}$ can be found such that for all $z \in \mathbf{y} \setminus \{y\}$, $y \sim z$. The interesting dual property that any finite sequential spatial process can be derived as the time-ordered vector of points in a classic spatio-temporal marked point process can be shown to hold as well. For further details, see Van Lieshout (2006a, 2006b).

13.3 Properties and connections

The success of Markov random fields in low level vision has given a boost to the development of Markov marked point processes. Indeed, as we saw, mesh or pairwise interaction models have their analogues in point process theory. Similar remarks hold for the intermediate level. For example, Markov connected component fields (13.3) are close in spirit to quermass interaction processes (Kendall *et al.* 1999, Møller and Helisova 2009). Reversely, methods designed for continuous polygonal fields have inspired algorithms for Markov random fields (Schreiber and Van Lieshout 2007).

In this section we consider the intermediate level models in closer detail and study some interesting properties.

13.3.1 *Merging of labels*

In image segmentation, one has to face the model selection problem of choosing the number of categorical labels, that is, the cardinality of the set Λ. It would be desirable if the local dependence structure of the model would not change with the number of labels. Indeed, the class of Markov connected component fields is closed under merging of labels, but the class of Markov random fields is not (Van Lieshout and Stoica 2009).

To be specific, let Λ be the set $\{1,\dots,k\}$, $k \geq 2$, and X a k-label Markov connected component field (13.3) with respect to some reflexive, symmetric relation $\sim$ on S. Define the random field Y with labels in $\{0,\dots,k-2\}$ by $Y_i = X_i\,\mathbf{1}_{X_i \leq k-2}$. Then Y is a $(k-1)$-label Markov connected component field with respect to $\sim$. To see this, fix $y = (y_1,\dots,y_m) \in \{0,\dots,k-2\}^S$. For $x \in \{z \in \Lambda^S : y_i = z_i\,\mathbf{1}_{z_i \leq k-2}\}$ and $j \in \Lambda$, write $\mathbf{K}_j(x)$ for the set of maximal connected components in x labelled j. Note that the maximal j-labelled connected components in x and y are identical for $j = 1,\dots,k-2$ and that each $(k-1)$- or k-component is part of a single 0-component in y. Hence, $\mathbf{P}\{Y_1 = y_1,\dots,Y_m = y_m\}$ is proportional to

$$\left\{\prod_{j=1}^{k-2}\prod_{K\in\mathbf{K}_j(y)}\Psi_K(j)\right\}\times\prod_{K\in\mathbf{K}_0(y)}\left\{\sum_{x\in\{k-1,k\}^K}\prod_{j=k-1}^{k}\prod_{L\in\mathbf{K}_j(x)}\Psi_L(j)\right\},$$

which proves the claim and gives an expression for $\Psi_K(0)$. It is interesting to note that the above result can be seen as the discrete analogue of the fact that the class of Markov connected component point processes (Baddeley and Møller 1989) is closed under independent superposition (Chin and Baddeley 1999).

For Markov random fields, the situation is more complicated, as could be expected from the behaviour of Markov point processes with respect to superposition (Van Lieshout 2000b). By the above remarks, for Markov random fields such as the Potts model that can also be factorized as in (13.3), merging labels $k-1$ and k leads to a Markov connected component field Y. However, Y is not necessarily a Markov random field with respect to the given relation. A counterexample is the Potts model with three colours, $\beta = 1$, and the horizontal and vertical neighbours relation: in configurations such as $\begin{pmatrix} 0 & ? \\ 0/1 & 0 \end{pmatrix}$, the conditional distribution of the top right pixel depends on whether Y takes value 0 or 1 at the bottom left pixel. Furthermore, since the class of Markov random fields is not contained in that of Markov connected component fields with respect to the same relation, merging labels does not always lead to a Markov connected component field. A counterexample is the 6-label Geman and Reynolds (1992) field X on $S = \{s_1, s_2\}$ with $s_1 \sim s_2$ defined by its joint probability mass function $\mathbf{P}\{X_1 = x_1, X_2 = x_2\} \propto \exp\{1/(|x_1 - x_2| + 1)\}$. Then clearly X is a pairwise interaction Markov random field, but, by arguments similar to those in Example A2 in Møller and Waagepetersen (1998), the joint probability mass function of Y does not factorize over maximal connected components.

13.3.2 *The length-weighted Arak field is a hard core object process*

In this section, we will consider the length-weighted Arak field with empty boundary condition introduced in Section 13.2.2. One can show that for β large enough the model is a Markov object process in the sense of Section 13.2.3 and give a convenient algorithmic description for the mark distribution.

To do so, let C_n be the set of all closed polygons in $[-n, n]^2$ which do not touch the boundary and set $\mathbf{C} = \cup_n C_n$. To each polygon θ, attach a unique $r(\theta)$, say the extreme lower left point, to serve as location parameter. Each polygon lying in the image window D as in the discussion surrounding (13.5), including the empty one, can than be described as $x + \theta$ with $x \in D$ and $\theta \in Q$ defined as $Q = \{\theta \in \mathbf{C} \mid r(\theta) = 0\} \cup \{\emptyset\}$.

For fixed $\beta \geq 2$, define a mark distribution $\mathbb{Q}$ by the continuous time random walk representation of Schreiber (2006) as follows. Fix Z_0 at 0, pick an initial direction uniformly in $(0, 2\pi)$, and, between direction updates, move in a constant direction with unit speed. Directions are updated with rate 4; the angle between the old and new direction is chosen according to the probability density $p(\phi) = |\sin(\phi)|/4$ on $(0, 2\pi)$, mirroring properties of the Poisson line process L discussed in Section 13.2.2. The resulting random walk $Z_t, t \geq 0$, is killed with rate $\beta - 2$ and whenever it hits its past trajectory to obtain $\tilde{Z}_t$, $t \geq 0$. Also draw a loop-closing half-line l^* through 0 whose angle with the initial segment of $(Z_t)_{t \geq 0}$ is distributed according to $p(\phi)$. Finally, if $\tilde{Z}_t$ hits l^* before being killed and the contour θ^* formed by l^* and the trajectory of $\tilde{Z}_t$ up to the moment of hitting has lower left extreme point 0, let e^* be the segment along l^* from the origin up to its intersection with $\tilde{Z}_t$, and generate a contour as follows:

- with probability $\exp\{-(\beta + 2)\, l(e^*)\}$, output θ^*;
- otherwise, output $\emptyset$.

In all remaining cases, the empty polygon is outputted. The procedure is illustrated in Fig. 13.5.

The above algorithmic definition of $\mathbb{Q}$ is easy to implement. Moreover, the union of polygonal contours of the marked point process defined by its conditional intensity

$$\lambda\left((u, \theta) \mid \{(x_i, \theta_i)\}_{i=1}^{n}\right) = \begin{cases} 4\pi, & \text{if } (u + \theta) \cap \left(\bigcup_{i=1}^{n} (x_i + \theta_i)\right) = \emptyset, \ u + \theta \subseteq D; \\ 0, & \text{otherwise,} \end{cases}$$

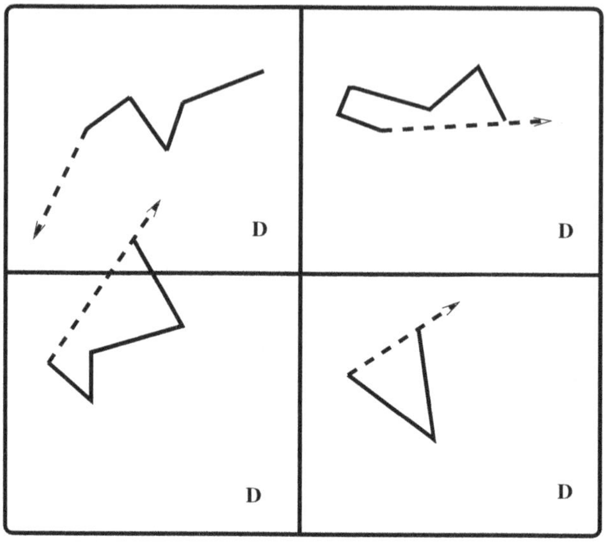

FIG. 13.5. Random walk construction. The dashed line is the loop-closing half-line, the solid one $\tilde{Z}_t$. The bottom pictures may output a non-empty polygon. The conditional intensity for the bottom left polygon is zero regardless of the remainder of the configuration since it does not lie entirely in D.

with respect to the product of Lebesgue measure on D and $\mathbb{Q}$ on Q coincides in distribution with the length-weighted Arak field (Van Lieshout and Schreiber 2007). Hence, one may regard it as a hard core object process of intensity 4π.

13.4 Examples

To illustrate the theory reviewed above, two image analysis problems are presented. The first example concerns the task of tracking a variable number of moving objects across a video sequence, the second one is devoted to foreground/background separation. In both cases, a regularization framework is chosen in which a goodness of fit term between the data and its semantic description is balanced by terms that favour smooth, sparse descriptors.

13.4.1 *Motion tracking*

Object tracking is important, as motion is a major source of semantic information in domains such as surveillance, robotics, and depth estimation. The aim is to find the objects that appear in a video sequence, and to follow their movements over time. Clearly, this is a high level task, hence we place ourselves in the framework of Section 13.2.3.

Consider the synthetic data given in Fig. 13.6. The objects are squares of various sizes and colours. Hence $Q = [s_{\min}, s_{\max}] \times \{0, \ldots, 255\}$ where the side

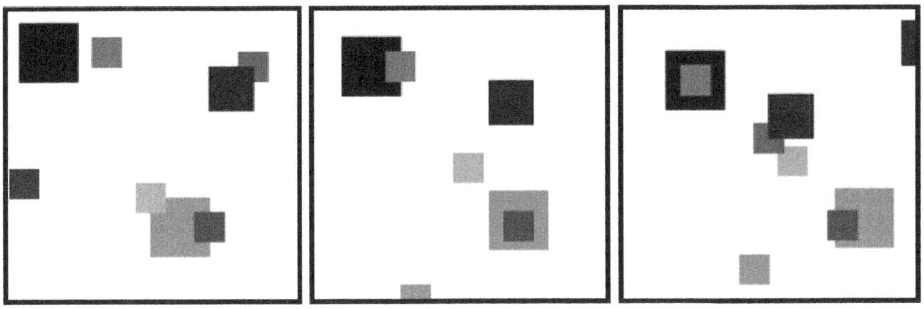

FIG. 13.6. Video sequence of moving squares.

length lies in some interval with $0 < s_{\min} < s_{\max} < \infty$, and the colour space is the set of 8-bit grey levels. We use the top left corner as our location parameter. Each square x occupies a set of pixels $R(x)$ in the digital image. As the object colour is not constant and there is significant overlap, we take vectors of squares to describe a scene. The components with a low index are those close to the camera. Doing so, we may define the signal $\theta_t(\vec{\mathbf{x}}^i)$ of configuration $\vec{\mathbf{x}}^i$ in video frame i at pixel t as the colour of the object closest to the camera if $t \in \cup R(x^i_j)$, and as the background colour otherwise.

In order to formulate object tracking as a statistical inference problem, we shall define a probability density $f(\mathbf{x}) \propto \exp\{-U(\mathbf{x})\}$ at the sequence of object configurations $\mathbf{x} = (\vec{\mathbf{x}}^1, \vec{\mathbf{x}}^2, \vec{\mathbf{x}}^3)$ with respect to the product distribution of independent unit rate sequential Poisson processes marked by a colour chosen according to the data histogram and a uniformly chosen side length. For notational convenience, we suppress the dependence on the data. The unknown configuration $\mathbf{x}$ is now treated as a parameter to be estimated, in other words, the aim is to find $\mathbf{x}$ that minimise the energy $U(\mathbf{x})$.

Upon observation of the video sequence $\mathbf{y} = (\mathbf{y}^1, \mathbf{y}^2, \mathbf{y}^3) = (y^i_t)$, indexed by pixels $t \in T$ and image frames $i = 1, 2, 3$, the goodness of fit is taken to be proportional to

$$\sum_{i=1}^{3} L_1(\mathbf{y}^i, \theta(\vec{\mathbf{x}}^i)) = \sum_{i=1}^{3} \sum_{t \in T} |y^i_t - \theta_t(\vec{\mathbf{x}}^i)|. \tag{13.10}$$

Note that the probability density function f defined by energy function U given by (13.10) has independent frame marginals that are Markov sequential spatial processes with respect to the overlapping objects relation $u \sim v$ if and only if $R(u) \cap R(v) \neq \emptyset$. Optimisation of (13.10) over $\mathbf{x}$ amounts to a least absolute deviation regression. However, a minimum is not guaranteed to exist, nor, if it does exist, to be unique. Indeed, small spurious squares behind the signal of those closer to the camera (having a lower index) do not affect the goodness of fit.

To overcome this problem, we add terms to the energy function that penalize overlap, favour temporal cohesion between squares in subsequent frames, and include matchings to keep track of a square as it moves across the video frames. To prevent spurious overlap, we add a positive penalty for each object and each pair of overlapping squares (known as the Strauss potential). Note that doing so results in a Markov overlapping objects process. Furthermore, we say that x_k^i and x_l^{i+1} are matched if they are instances of the same square in adjacent frames. The quality of a match is described by a weighted sum of the absolute difference in side lengths and colour and the squared distance between the top left points of the two squares involved. In order to quantify coherence between consecutive frames, we add a positive penalty term for each missing match, offset by the quality of matches that are present. Finally, in order to propagate relative depth information gathered when two squares overlap over time, a positive penalty is added to U for each overlapping pair of matched squares such that the relative depth is not preserved in the next or previous frame. Again, the dependence structure is local in the sense that only pairs of overlapping objects and consecutive video frames are involved.

To find an optimal configuration sequence with associated matchings, we use simulated annealing (Haario and Saksman 1991) within the Metropolis–Hastings framework (see Chapter 9). Inspired by Lund *et al.* (1999), we allow addition and deletion of matched or unmatched squares, modification of the permutation order, addition and deletion of a match, changes in location, colour, and side length, splitting a square in two, and merging two close squares into a single one. The synthetic data presented in Fig. 13.6 were chosen to include squares entering and leaving the image window or passing each other, a square leaving one connected component to join another, complete occlusion, as well as varying contrast. The signal of the near optimal sequence of configurations coincides with that of the data. Not visible in the signal but present in the near optimal configuration is a square hidden behind the dark one in the top right quadrant of the middle frame, see the listed sequence of object configurations with associated matchings in Table 13.1. The matches are correctly reproduced as well.

In order to quantify depth, we estimate the probabilities p_{kl}^i of square k in frame i lying closer to the camera than square $l > k$ in the same frame. Table 13.2 lists the result for $i = 2$. By standard combinatorial arguments, the correctness of these empirical values is verified. Other examples and further details are given in Van Lieshout (2008).

13.4.2 *Foreground/background separation*

Consider an image such as the one depicted in Fig. 13.7 and suppose the goal is to separate the foreground from the background. For this task, a large variety of models and methods is available, ranging from elementary thresholding through contour extraction methods to scene modelling. For a comprehensive overview, see for example Chapter 10 in Rosenfeld and Kak (1982). More recent material

TABLE 13.1. Result after simulated annealing (see text) for Fig. 13.6. The columns contain the square parameters (x and y coordinates of top left corner, side length, and grey level) for frames $i = 1, 2, 3$. For matched objects, the index of the associated square in the next frame is listed in the columns labelled 'm'.

$i = 1$				m	$i = 2$				m	$i = 3$			
89	119	20	180	7	119	49	30	30	1	99	59	30	30
129	139	20	80	6	49	29	20	127	4	89	79	20	105
1	109	20	55		59	189	20	155	9	190	9	30	30
99	129	40	155	8	19	19	40	0	8	39	39	21	127
139	39	30	30	1	125	57	20	105	2	139	139	20	80
59	19	20	127	2	129	139	20	80	5	144	124	40	155
9	9	40	0	4	95	99	20	180	7	105	95	20	180
159	29	20	105	5	119	125	40	155	6	29	29	40	0
										79	169	20	155

can be found in Vincent (1999), which treats morphological methods, or in the volume edited by Osher and Paragios (2003) which contains contributions on level set approaches, active contour models and variational methods, and more. In this chapter on stochastic geometric methods, we place ourselves at the intermediate conceptual level that regards the sought after segmentation as a coloured tessellation. The approach is a compromise between the full scene modelling feasible in restricted application domains only and the Markov random field models of Section 13.2.1.

As discussed in Section 13.2.2, at the intermediate level one may either choose sets of pixels or tessellations as the focus of interest. Since reality is continuous rather than discrete, we find the latter approach the most appealing and will use polygonal field models. The idea to use such a model as a prior distribution for image segmentation is due to Clifford and Middleton (1989). A Metropolis–Hastings style sampler was developed by Clifford and Nicholls (1994) and applied

FIG. 13.7. Segmented image of a cat.

TABLE 13.2. Pairwise probabilities p^2_{kl} of object k having a lower sequence index than object l after annealing (see text) for frame 2 and data as given in Fig. 13.6.

–	0.66	0.79	0.93	1.00	0.75	1.00	1.00
	–	0.67	1.00	0.62	0.59	0.82	0.95
		–	0.67	0.42	0.42	0.62	0.83
			–	0.22	0.24	0.43	0.71
				–	0.50	1.00	1.00
					–	0.75	1.00
						–	1.00
							–

to an image reconstruction problem. A modification was suggested by Paskin and Thrun (2005) for use in robotic mapping.

In order to formulate foreground extraction as a statistical inference problem, we shall define a probability density $f(\hat{\gamma}) \propto \exp\{-U(\hat{\gamma})\}$ on labelled tessellations $\hat{\gamma} \in \hat{\Gamma}_D$ with respect to the coloured Arak field $\hat{A}_D$ with free boundary conditions. The notation $\hat{A}_D$ is used to distinguish the dominating measure from its colourless counterpart A_D, cf. (13.5), with a similar remark for $\hat{\Gamma}_D$. The labels are binary: 1 indicates the foreground, 0 the background. Since adjacent regions must have different labels, each admissable γ can be coloured in two ways only. In the reference field $\hat{A}_D$, each possible colouring is given probability 1/2. Note that the energy may and will depend on the data image. Having formulated a probability density f, the unknown configuration $\hat{\gamma}$ is treated as a parameter to be estimated, in other words, the aim is to minimize the energy $U(\hat{\gamma})$ over the family of admissible labelled tessellations.

Write $\mathbf{y} = (y_t)$, $t \in T$, for the data image. Here, each pixel value $y_t \in \{0, \ldots, 255\}$ is a grey level. Given $\mathbf{y}$, the energy function U is the sum of regression and regularization terms. In analogy to (13.10), the goodness of fit is described in terms of $L_1(\mathbf{y}, \hat{\gamma})$. Optimization of the goodness of fit criterion alone suffers from similar problems as those discussed in the context of motion tracking: In general, the minimum will not be unique and tend to result in an over-segmentation. To overcome these problems, we add a regularization term proportional to the total edge length $\ell(\gamma)$.

To find the optimal configuration, we again use simulated annealing within the Metropolis–Hastings framework. The update proposals are based on the dynamic representation of Arak (1982), an equivalent definition of A_D in terms of the evolution of a one-dimensional particle system. Briefly, the edges of the field that separate foreground and background components are interpreted as the traces of particles travelling in time and one-dimensional space. Particles are born in D in pairs according to a homogeneous Poisson point process, while at the boundary ∂D, single particles are generated according to a Poisson point process with an appropriate intensity measure. Particles change their velocity according to a pure jump process with a suitably chosen transition kernel, which show up as the vertices of γ in the trace. When particles collide or reach the boundary of D, they die. For details, see Arak and Surgailis (1989). Schreiber (2005) observed that when a new particle birth site $u \in D$ is added to configuration γ, the symmetric difference with the old configuration is a closed polygonal curve that may be self-intersecting or chopped off at the boundary. The addition of a birth site $u \in \partial D$ gives rise to a single self-avoiding polygonal curve that may be chopped off at the the boundary. Hence, a Metropolis–Hastings sampler can be set up that proposes to add or delete birth sites combined with swaps between foreground and background. The resulting moves are easy to implement and sufficiently flexible for an efficient exploration of the state space (Kluszczyński *et al.* 2007, Schreiber 2005).

The result for an image from the Pascal Network of Excellence challenge 2006 `http://www.pascal-network.org/challenges/VOC/thumbs/VOC2006` is overlaid on the data in Fig. 13.7. The misclassification rate, calculated manually due to the lack of a ground truth, is about 3%. Further details can be found in Kluszczyński *et al.* (2006).

Acknowledgements

Much of the work presented here was done in collaboration with others. Special thanks are due to all co-authors, as well as to B. Lisser and A. Steenbeek for programming assistance.

References

Abend, K., Harley, T. J., and Kanal, L. N. (1965). Classification of binary random patterns. *IEEE Trans. Information Theory*, **11**, 538–544.

Amit, Y., Grenander, U., and Piccioni, M. (1991). Structural image restoration through deformable templates. *J. Amer. Statist. Assoc.*, **86**, 376–387.

Arak, T. (1982). On Markovian random fields with finite number of values. *4th USSR-Japan Symposium on Probability Theory and Mathematical Statistics, Abstracts of Communications*, Tbilisi.

Arak, T. and Surgailis, D. (1989). Markov fields with polygonal realizations. *Probab. Theory Related Fields*, **80**, 543–579.

Arak, T., Clifford, P., and Surgailis, D. (1993). Point-based polygonal models for random graphs. *Adv. in Appl. Probab.*, **25**, 348–372.

Baddeley, A. J. and Lieshout, M. N. M. van (1992). Object recognition using Markov spatial processes. In *Proceedings 11th IAPR International Conference on Pattern Recognition*, **B**, 136–139. IEEE Computer Society Press, Los Alamitos.

Baddeley, A. J. and Lieshout, M. N. M. van (1993). Stochastic geometry models in high-level vision. In *Statistics and Images, Volume 1*, K. V. Mardia and G. K. Kanji (Eds.), *Advances in Applied Statistics*, a supplement to *Journal of Applied Statistics*, **20**, 231–256. Carfax, Abingdon.

Baddeley, A. J. and Lieshout, M. N. M. van (1995). Area-interaction point processes. *Ann. Inst. Statist. Math.*, **46**, 601–619.

Baddeley, A. J. and Møller, J. (1989). Nearest-neighbour Markov point processes and random sets. *Internat. Statist. Rev.*, **57**, 89–121.

Baddeley, A. J., Lieshout, M. N. M. van, and Møller, J. (1996). Markov properties of cluster processes. *Adv. in Appl. Probab.*, **28**, 346–355.

Besag, J. (1974). Spatial interaction and the statistical analysis of lattice systems (with discussion). *J. Roy. Statist. Soc. Ser. B*, **36**, 192–236.

Besag, J. (1986). On the statistical analysis of dirty pictures (with discussion). *J. Roy. Statist. Soc. Ser. B*, **48**, 259–302.

Besag, J. and Kooperberg, C. (1995). On conditional and intrinsic autoregressions. *Biometrika*, **82**, 733–746.

Besag, J., York, J., and Mollié, A. (1991). Bayesian image restoration, with two applications in spatial statistics (with discussion). *Ann. Inst. Statist. Math.*, **43**, 1–59.

Blackwell, P. G. and Møller, J. (2003). Bayesian analysis of deformed tessellation models. *Adv. in Appl. Probab.*, **35**, 4–26.

Chin, Y. C. and Baddeley, A. J. (1999). On connected component Markov point processes. *Adv. in Appl. Probab.*, **31**, 279–282.

Clifford, P. (1990). Markov random fields in statistics. In *Disorder in Physical Systems. A Volume in Honour of J. M. Hammersley*, G. R. Grimmett and D. J. A. Welsh (Eds.), pages 19–32. Oxford University Press, New York.

Clifford, P. and Middleton, R. D. (1989). Reconstruction of polygonal images. *J. Appl. Stat.*, **16**, 409–422.

Clifford, P. and Nicholls, G. (1994). A Metropolis sampler for polygonal image reconstruction. *Available at*: `http://www.stats.ox.ac.uk/~clifford/papers/met_poly.html`

Cressie, N. and Davidson, J. L. (1998). Image analysis with partially ordered Markov models. *Comput. Statist. Data Anal.*, **29**, 1–26.

Cressie, N., Zhu, J., Baddeley, A. J., and Nair, M. G. (2000). Directed Markov point processes as limits of partially ordered Markov models. *Methodol. Comput. Appl. Probab.*, **2**, 5–21.

Daley, D. J. and Vere–Jones, D. (2003). *An Introduction to the Theory of Point Processes. Volume I. Elementary Theory and Methods* (2nd edn). Springer-Verlag, New York.

Dougherty, E. R. (1992). *An Introduction to Morphological Image Processing.* SPIE Press, Bellingham.

Geman, D. (1990). Random fields and inverse problems in imaging. *École d'été de Probabilités de Saint-Flour XVIII – 1988, Lecture Notes in Mathematics*, **1427**, 113–193. Springer-Verlag, Berlin.

Geman, S. and Geman, D. (1984). Stochastic relaxation, Gibbs distributions and the Bayesian restoration of images. *IEEE Transactions on Pattern Analysis and Machine Intelligence*, **6**, 721–741.

Geman, D. and Reynolds, G. (1992). Constrained restoration and the recovery of discontinuities. *IEEE Transactions on Pattern Analysis and Machine Intelligence*, **14**, 367–383.

Geman, D., Geman, S., Graffigne, C., and Dong, P. (1990). Boundary detection by constrained optimization. *IEEE Transactions on Pattern Analysis and Machine Intelligence*, **12**, 609–628.

Gimel'farb, G. L. (1999). *Image Textures and Gibbs Random Fields*. Kluwer, Dordrecht.

Green, P. J. (1995). Reversible jump Markov chain Monte Carlo computation and Bayesian model determination. *Biometrika*, **82**, 711–732.

Grenander, U. (1976). *Lectures on Pattern Theory, Vol. 1: Pattern Synthesis*. Applied Mathematical Sciences vol. 18. Springer-Verlag, New York.

Grenander, U. (1978). *Lectures on Pattern Theory, Vol. 2: Pattern Analysis*. Applied Mathematical Sciences vol. 24. Springer-Verlag, New York.

Grenander, U. (1981). *Lectures on Pattern Theory, Vol. 3: Regular Structures*. Applied Mathematical Sciences vol. 33. Springer-Verlag, New York.

Haario, H. and Saksman, E. (1991). Simulated annealing process in general state space. *Adv. in Appl. Probab.*, **23**, 866–893.

Häggström, O., Lieshout, M. N. M. van, and Møller, J. (1999). Characterization results and Markov chain Monte Carlo algorithms including exact simulation for some spatial point processes. *Bernoulli*, **5**, 641–658.

Hansen, M. B., Møller, J., and Tøgersen, F. A. (2002). Bayesian contour detection in a time series of ultrasound images through dynamic deformable template models. *Biostatistics*, **3**, 213–228.

Heikkinen, J. and Arjas, E. (1998). Non-parametric Bayesian estimation of a spatial Poisson intensity. *Scand. J. Statist.*, **25**, 435–450.

Hurn, M. A. (1998). Confocal fluorescence microscopy of leaf cells: an application of Bayesian image analysis. *J. Roy. Statist. Soc. Ser. C*, **47**, 361–377.

Kendall, W. S., Lieshout, M. N. M. van, and Baddeley, A. J. (1999). Quermass-interaction processes: Conditions for stability. *Adv. in Appl. Probab.*, **31**, 315–342.

Kluszczyński, R., Lieshout, M. N. M. van, and Schreiber, T. (2007). Image segmentation by polygonal Markov fields. *Ann. Inst. Statist. Math.*, **59**, 465–486.

Künsch, H. R. (1987). Intrinsic autoregressions and related models on the two-dimensional lattice. *Biometrika*, **74**, 517–524.

Lacoste, C., Descombes, X., and Zerubia, J. (2005). Point processes for unsupervised line network extraction in remote sensing. *IEEE Transactions on Pattern Analysis and Machine Intelligence*, **27**, 1568–1579.

Lacroix, V. (1987). Pixel labelling in a second-order Markov mesh. *Signal Processing*, **12**, 59–82.

Lebowitz, J. L. and Gallavotti, G. (1971). Phase transitions in binary lattice gases. *J. Math. Phys.*, **12**, 1129–1133.

Lieshout, M. N. M. van. (1994). Stochastic annealing for nearest-neighbour point processes with application to object recognition. *Adv. in Appl. Probab.*, **26**, 281–300.

Lieshout, M. N. M. van. (1995). *Stochastic Geometry Models in Image Analysis and Spatial Statistics, CWI Tract*, **108**. CWI, Amsterdam.

Lieshout, M. N. M. van. (2000a). *Markov Point Processes and their Applications.* Imperial College Press/World Scientific Publishing, London/Singapore.

Lieshout, M. N. M. van. (2000b). Propagation of spatial interaction under superposition. In *Accuracy 2000, Proceedings of the 4th International Symposium on Spatial Accuracy Assessment in Natural Resources and Environmental Sciences*, G. B. M. Heuvelink and M. J. P. M. Lemmens (Eds.), pp. 687–694. Delft University Press, Delft.

Lieshout, M. N. M. van. (2006a). Markovianity in space and time. In *Dynamics and Stochastics: Festschrift in Honour of Michael Keane*, D. Denteneer, F. den Hollander, and E. Verbitskiy (Eds.), *Lecture Notes Monograph Series*, **48**, 154–167. Institute for Mathematical Statistics, Beachwood.

Lieshout, M. N. M. van. (2006b). Campbell and moment measures for finite sequential spatial processes. In *Proceedings Prague Stochastics 2006*, M. Hušková and M. Janžura (Eds.), pages 215–224. Matfyzpress, Prague.

Lieshout, M. N. M. van. (2008). Depth map calculation for a variable number of moving objects using Markov sequential object processes. *IEEE Transactions on Pattern Analysis and Machine Intelligence*, **30**, 1308–1312.

Lieshout, M. N. M. van and Schreiber, T. (2007). Perfect simulation for length-interacting polygonal Markov fields in the plane. *Scand. J. Statist.*, **34**, 615–625.

Lieshout, M. N. M. van and Stoica, R. S. (2009). A note on pooling of labels in random fields. Research Report PNA-E0906, CWI, Amsterdam.

Lund, J., Penttinen, A., and Rudemo, M. (1999). Bayesian analysis of spatial point patterns from noisy observations. Research Report, Department of Mathematics and Physics, The Royal Veterinary and Agricultural University, Copenhagen.

Mardia, K. V. and Kanji, G. K. (Eds.) (1993). *Statistics and Images, Volume 1, Advances in Applied Statistics*, a supplement to *Journal of Applied Statistics*, **20**. Carfax, Abingdon.

Mardia, K. V., Qian, W., Shah, D., and De Souza, K. M. A. (1997). Deformable template recognition of multiple occluded objects. *IEEE Transactions on Pattern Analysis and Machine Intelligence*, **19**, 1036–1042.

Matheron, G. (1975). *Random Sets and Integral Geometry.* John Wiley and Sons, New York.

Molina, R. and Ripley, B. D. (1989). Using spatial models as priors in astronomical image analysis. *J. Appl. Stat.*, **16**, 193–206.

Møller, J. and Helisova, K. (2008). Power diagrams and interaction processes for unions of discs. *Adv. in Appl. Probab.*, **40**, 321–347.

Møller, J. and Skare, Ø. (2001). Bayesian image analysis with coloured Voronoi tessellations and a view to applications in reservoir modelling. *Statistical Modelling*, **1**, 213–232.

Møller, J. and Waagepetersen, R. P. (1998). Markov connected component fields. *Adv. in Appl. Probab.*, **30**, 1–35.

Nicholls, G. K. (1998). Bayesian image analysis with Markov chain Monte Carlo and coloured continuum triangulation models. *J. Roy. Statist. Soc. Ser. B*, **60**, 643–659.

Okabe, A., Boots, B., Sugihara, K., and Chiu, S. N. (2000). *Spatial Tessellations: Concepts and Applications of Voronoi Diagrams* (2nd edn). John Wiley and Sons, Chichester.

Ortner, M., Descombes, X., and Zerubia, J. (2007). Building outline extraction from digital elevation models using marked point processes. *International Journal of Computer Vision*, **72**, 107–132.

Osher, S. and Paragios, N. (Eds.) (2003). *Geometric Level Set Methods in Imaging, Vision, and Graphics.* Springer, New York.

Papangelou, F. (1974). The conditional intensity of general point processes and an application to line processes. *Z. Wahrscheinlichkeitstheorie und Verw. Gebiete*, **28**, 207–226.

Paskin, M. A. and Thrun, S. (2005). Robotic mapping with polygonal random fields. In *Proceedings in Artificial Intelligence UAI-05.*

Pievatolo, A. and Green, P. J. (1998). Boundary detection through dynamic polygons. *J. Roy. Statist. Soc. Ser. B*, **60**, 609–626.

Qian, W. and Titterington, D. M. (1991). Multidimensional Markov chain models for image textures. *J. Roy. Statist. Soc. Ser. B*, **53**, 661–674.

Ripley, B. D. (1988). *Statistical Inference for Spatial Processes.* Cambridge University Press, Cambridge.

Ripley, B. D. and Kelly, F. P. (1977). Markov point processes. *J. London Math. Soc.*, **15**, 188–192.

Ripley, B. D. and Sutherland, A. I. (1990). Finding spiral structures in images of galaxies. *Philosophical Transactions of the Royal Society of London, Series A*, **332**, 477–485.

Rosenfeld, A. and Kak, A. C. (1982). *Digital Picture Processing* (2nd edn). Academic Press, Orlando.

Rue, H. and Held, L. (2005). *Gaussian Markov Random Fields: Theory and Applications, Monographs on Statistics and Applied Probability*, **104**. Chapman & Hall/CRC, Boca Raton.

Rue, H. and Hurn, M. A. (1999). Bayesian object identification. *Biometrika*, **86**, 649–660.

Rue, H. and Husby, O. K. (1998). Identification of partly destroyed objects using deformable templates. *Statistics and Computing*, **8**, 221–228.

Schreiber, T. (2005). Random dynamics and thermodynamic limits for polygonal Markov fields in the plane. *Adv. in Appl. Probab.*, **37**, 884–907.

Schreiber, T. (2006). Dobrushin–Kotecký–Schlosman theorem for polygonal Markov fields in the plane. *J. Statist. Phys.*, **123**, 631–684.

Schreiber, T. and Lieshout, M. N. M. van. (2007). Disagreement loop and path creation/annihilation algorithms for planar Markov fields with applications to image segmentation. EURANDOM Report 2007-045.

Serra, J. (1982). *Image Analysis and Mathematical Morphology.* Academic Press, London.

Skare, Ø., Møller, J., and Jensen, E. B. V. (2007). Bayesian analysis of spatial point processes in the neighbourhood of Voronoi networks. *Statistics and Computing*, **17**, 369–379.

Stoica, R., Descombes, X., and Zerubia, J. (2004). A Gibbs point process for road extraction in remotely sensed images. *International Journal of Computer Vision*, **57**, 121–136.

Stoica, R. S., Descombes, X., Lieshout, M. N. M. van, and Zerubia, J. (2002). An application of marked point processes to the extraction of linear networks from images. In *Spatial Statistics: Case Studies*, J. Mateu and F. Montes (Eds.), pages 287–312. WIT Press, Southampton.

Stoica, R. S., Martinez, V., and Saar, E. (2007). A three dimensional object point process for detection of cosmic filaments. *J. Roy. Statist. Soc. Ser. C*, **56**, 459–477.

Stoyan, D., Kendall, W. S., and Mecke, J. (1995). *Stochastic Geometry and its Applications* (2nd edn). John Wiley and Sons, Chichester.

Tjelmeland, H. and Besag, J. (1998). Markov random fields with higher-order interactions. *Scand. J. Statist.*, **25**, 415–433.

Tjelmeland, H. and Holden, L. (1993). Semi-Markov random fields. In *Geostatistics Troia '92*, A. Soares (Ed.). Kluwer, Amsterdam.

Vincent, L. (1999). Current topics in applied morphological image analysis. In *Stochastic Geometry, Likelihood and Computation*, O. Barndorff-Nielsen, W. S. Kendall and M. N. M. van Lieshout (Eds.), pages 199–283. CRC Press/Chapman and Hall, Boca Raton.

Widom, B. and Rowlinson, J. S. (1970). A new model for the study of liquid-vapor phase transitions. *J. Chem. Phys.*, **52**, 1670–1684.

Winkler, G. (2003). *Image Analysis, Random Fields and Markov Chain Monte Carlo Methods. A Mathematical Introduction* (2nd ed), *Applications of Mathematics, Stochastic Modelling and Applied Probability*, **27**. Springer-Verlag, Berlin.

14

STEREOLOGY

Werner Nagel

Abstract

In the present chapter some essential mathematical principles of stereology are presented. Considering that several excellent new books and surveys appeared over the past years, it is not aimed to review the recent development in the field. The emphasis is on rather general quantitative as well as qualitative basic principles and a few new results from the literature.

14.1 Motivation

Why are we still dealing with stereology, the science of statistical inference on features of a geometric structure from information that is available on sections with lower-dimensional sets? During the last two decades enormous technical achievements – confocal microscopy and tomography (e.g. CT, MRT) – came into the labs and become more and more affordable. Torquato (2002, p. 294) apostrophizes stereology as a 'poor man's' tomography. At first glance, this could frustrate those who dedicated a lot of their energy and time to stereology. But a second look makes clear that – even if some of the classical problems can now be solved more easily, and sometimes indeed without stereological methods – the new techniques also yield new challenges for theoretical research, e.g. concerning 3D image analysis, and higher demands concerning the resolution or precision of the results.

There are both practical as well as statistical and mathematical issues. In applications, any equipment that provides images has a finite resolution, i.e. a *sample* of the structure of interest is available only. Furthermore, there can be significant differences in the resolution between 2D and 3D imaging. Thus today's CT can deliver a resolution of about 1 micron, but there are other microscopical methods and equipment which perform much better, with a resolution in the nano-scale for 2D images. And indeed, there are structures, e.g. in materials science, where the particles are of such a small size that only their two-dimensional section profiles are observable with much higher resolution than it is available in 3D, and hence the only way to gain information about the 3D objects is to study 2D sections of them. Moreover, the images are digitized to be stored, processed and analysed. Summarizing the procedure of generating image data of a real structure, the image can be interpreted as a sample, and this can be modelled as a *section* of the structure with a 2D or 3D point lattice or with a line or with a plane – depending on the size of the objects compared to resolution and to digitization effects. This is discussed and illustrated by Ohser and Mücklich (2000,

p. 67ff). Thus, also in connection with modern imaging techniques there remain or newly occur intrinsic stereological problems: To infer information about a set of interest (structure/object/background) from a lower-dimensional sample of this set.

It has to be emphasized that it is not the purpose of stereology to reconstruct a set completely in a geometrical sense from the available lower-dimensional observations but to estimate some of its parameters or characteristics.

There is a series of recent and modern books written by the top researchers in the field. With a focus on a clear and sound theoretical foundation and the mathematical background, the most important books are those ones by Baddeley and Jensen (2005), Jensen (1998), Beneš and Rataj (2004), Schneider and Weil (2008), Ohser and Mücklich (2000). There are further books concerning several fields of application which are referred to in the mentioned ones. These books and the numerous original papers indicate that stereology, even if already well established and solidly founded, is also a current research field with interesting open problems.

What else can be written after these new and very good books which, taken all together, yield an exhaustive presentation of the theory and the state of the art? How do we gain any 'new perspective'?

In the present paper we take the liberty of a drastic reduction to a few mathematical principles and formulae in order to suggest a hopefully wide and clear view to the main ideas, approaches, principles and problems in stereology. This can probably allow the reader to have a more distanced view to it and thus to find new or alternative approaches or to see links to other mathematical fields. A drawback of this presentation is that interesting deep and detailed results cannot be provided.

In Chapter 1 of the present volume the reader can find historical remarks as well as the most important and famous formulae of stereology. Therefore, we abstain here from starting with the classical stereological and integral geometric formulae. Here the main emphasis is on formulae (14.1), (14.2), and the Crofton formulae come along later.

A purpose of this contribution is also to explain and summarize open problems, primarily from a mathematical point of view. A considerable part of the material is cited from the above mentioned books.

The chapter is organized as follows. A presentation of the setting in Section 14.2 is followed by a description of a general integral equation in 14.3. Then, in Section 14.4 several quantitative results are treated, whereas 14.5 is devoted to qualitative or structural issues. Finally, 14.7 contains a few remarks on statistics.

14.2 From higher to lower dimension – sections and induced mappings

A stereological problem can arise when a set which is embedded in a d-dimensional Euclidean space is intersected by a k-dimensional set with $0 \leq k \leq$

$d-1$. If the target is an inference on certain features of the original set, the first step is always an analysis of the mapping which is induced by this section. The actual stereological problem is then mostly the investigation of an inverse of this mapping. But sometimes the first step is already sufficient and an explicit calculation or inversion is not always necessary, e.g. for model classification or a goodness-of-fit test.

Let $X \subset \mathbb{R}^d$ denote the set of interest. In applications this can be the 'object' as well as the 'background'. It is assumed that a *lower-dimensional* sample of X and of its complement X^c is available, namely

$$X \cap T \quad \text{and hence also} \quad X^c \cap T.$$

A large variety of sets T that are used as section sets is studied. The choice of T can be both motivated by practical restrictions or demands and by theoretical results. The most important choices for T are

(1) a k-dimensional plane, $0 \leq k \leq d-1$.
(2) a lattice or grid of parallel k-planes. Of special practical value: serial planar sections, i.e. $k = 2$ in $\mathbb{R}^3$; and a point lattice with $k = 0$ in 2D and 3D image analysis.
(3) a circle or a cycloid or a higher-dimensional analogue which represents, due to the distributions of the tangential directions, sections with different directions.
(4) a wedge that is generated by two or more non-parallel k-planes.
(5) a thin section with parallel section planes. Usually it is assumed that an orthogonal projection of this thin layer can be observed. Sometimes a thin section can allow to extract extra information, e.g. a measurement of section angles.

The sections can be performed physically in the sense that a probe is cut with a knife or another tool. But there exists also equipment, like a confocal microscope or a tomograph, that allows to perform so-called optical sections. And also a combination of a physical section and an overlaid system of test curves, also referred to as two-stage sampling, is applied, e.g. with cycloidal test curves on vertical planar sections, see Baddeley and Jensen (2005), Beneš and Rataj (2004).

The mapping $(X, T) \mapsto X \cap T$ causes a loss of information about X. *It is a key issue of stereology to understand which part of the information is really lost and what is somehow comprised or implied in $X \cap T$.*

It is obvious that without further assumptions the set $X \cap T$ can be continued outside T in an arbitrary way to a set X and thus $X \cap T$ remains the only information available about X.

In order to perform an inference concerning X it has to be assumed that $X \cap T$ can be considered as a sample of X that is 'representative' and informative w.r.t. certain features of X. Therefore stochastic models are introduced for either X

or for T or for both (X, T). This yields the model-based, the design-based or the hybrid approaches, respectively.

The model-based approach Consider X as a realization of a random set Ξ. This can be a non-parametric model of a random closed set (RACS) as well as a semi-parametric or a parametric one like a Boolean model, a particle process (or germ-grain model), a random tessellation, a surface or fibre process, a process of ℓ-dimensional planes, $\ell = 1, \ldots, d-1$, a random fractal set.

For a thorough presentation of the most important models for random sets see Chapter 1 of the present volume or Schneider and Weil (2008) or Molchanov (2005). Also here, we use the notation $\mathcal{F}$ for the set of all closed sets in $\mathbb{R}^d$, and $\mathcal{F}$ is endowed with the hit-and-miss topology and the Borel σ-algebra generated by it.

Formally, the section is a mapping

$$(\Xi, T) \mapsto \Xi \cap T.$$

The design-based approach The set X is considered as non-random and fixed. Choose a prototype of a section set T and a probability distribution on the set of possible positions of T – i.e. on the set of images of T under rigid motions – and generate a random section set τ accordingly. The choice of a design depends on the target parameters of X to be estimated and on the treasury of formulae one has at one's disposal. Furthermore, there are restrictions imposed by the material of the probe and by the equipment for making sections. The main theoretical sources stereology lives from are integral geometry and statistical sampling theory. The application of integral geometric formulae requires some invariance under a group of motions, and that invariance has to go into the distribution of τ. Two typical statistical problems to be considered are the compensation of a sampling bias, i.e. that larger particles are hit by a section set with a higher probability than smaller ones and, on the other hand, the estimation of ratios.

The design-based approach is systematically and exhaustively treated in Baddeley and Jensen (2005), Jensen (1998).

The section scheme is

$$(X, \tau) \mapsto X \cap \tau.$$

The hybrid approach Interpret X as a realization of a random set Ξ and generate a random section set τ. This yields

$$(\Xi, \tau) \mapsto \Xi \cap \tau.$$

A standard example is a spatially stationary (homogeneous) but anisotropic random closed set Ξ and random isotropic section plane (linear subspace) τ which 'compensates' the anisotropy of Ξ.

In all these approaches the intersection set, also referred to as section profile, is a random set. Formally, the hybrid approach can be considered as the general one with model-based and design-based approaches as particular cases, where the distributions of Ξ or of τ are Dirac δ-measures, respectively. We will write (Ξ, τ) even if one of the parts is non-random.

14.3 A general integral equation

Given the distribution for (Ξ, τ), the mapping $(X, T) \mapsto X \cap T$ induces a probability distribution on the set of section profiles. If the distribution $P_{\Xi,\tau}$ of (Ξ, τ) is concentrated on $\mathcal{F}^2$, i.e. if Ξ and τ are random closed sets in $\mathbb{R}^d$, then the intersection induces a distribution $P_{\Xi\cap\tau}$ on $\mathcal{F}$. This link can be expressed by the transformation formula for integrals. For any non-negative and measurable function $h : \mathcal{F} \to [0, \infty)$

$$\int P_{\Xi\cap\tau}(\mathrm{d}Y)\, h(Y) = \int P_{\Xi,\tau}(\mathrm{d}(X, T))\, h(X \cap T). \tag{14.1}$$

If Ξ and τ are independent then

$$\int P_{\Xi\cap\tau}(\mathrm{d}Y)\, h(Y) = \int P_{\Xi}(\mathrm{d}X) \int P_{\tau}(\mathrm{d}T)\, h(X \cap T). \tag{14.2}$$

Notice that Ξ and τ will not always be independent if assumptions on the mutual position of Ξ and τ are made. This can be the case in local stereology, see Jensen (1998). As already mentioned, (14.2) concerns the model-based approach if $P_\tau = \delta_T$ and the design-based approach if $P_\Xi = \delta_X$.

The formulae (14.1) and (14.2) can be interpreted as 'basic section formulae', when reading them from left to right: The distribution of the section profile $\Xi \cap \tau$ is described by an integral w.r.t. the distribution of Ξ and of τ. On the other hand, when reading the formulae from right to left, one has the general form of the so-called inversion formulae, i.e. one studies which features of the set Ξ (or, more precisely, of the distribution P_Ξ) can be expressed as an integral on the right-hand side with an appropriate choice of the distribution P_τ and of the function h.

A further application of these formulae is to consider the right-hand side for $P_\tau = \delta_T$ (which corresponds to the model-based approach) and this as a function of T. When T varies within a certain domain of section sets, e.g. the k-dimensional linear subspaces of $\mathbb{R}^d$, then this function $h(\Xi \cap \cdot)$ can be used for an inference on Ξ. An important example is the 'the rose of intersections', where the intensity of a section (point) process on T is analysed as a function of T. This is applied to the estimation of the orientation distribution of the surface of Ξ or of the directional distribution of a fibre system Ξ.

It is quite natural that stereologists concentrate their efforts on finding P_τ and h in order to provide inversion formulae for a given model Ξ, e.g. a Boolean model or a Poisson–Voronoi tessellation. But it is at least also of theoretical interest to search for new models for Ξ which are 'well tailored' for stereological purposes, i.e. such models for which essential features can easily be expressed by integrals on the right side of (14.2). This may also concern the statistical and the numerical stability of solutions of integral equations.

Examples for such models can be parametric ones, where the parameters can directly be inferred from sections. A classical result of the type is the Rayleigh distribution in Wicksell's corpuscle problem: If for a stationary particle process of spheres in the three-dimensional space the sphere diameter distribution is a Rayleigh distribution, then the diameter distribution of the circles appearing on a planar section is the same Rayleigh distribution, see below.

A search for such models that behave particularly well in stereology could be considered as l'art-pour-l'art and far from practical demands. But one should not forget that many models, also in stochastic geometry, are often favoured due to their mathematical feasibility and elegance, and compromises are made concerning the goodness of fit to the data.

A complete review of stereological methods could be orientated on (14.1) and (14.2) by specifying Ξ, τ and h respectively. Here we only choose and discuss some cases of interest without claiming any completeness.

14.4 Some perspectives

14.4.1 *The capacity functional*

If Ξ as well as τ are random closed sets, then the distributions of Ξ and of the intersection $\Xi \cap \tau$ are uniquely determined by the respective capacity functionals T_Ξ and $T_{\Xi\cap\tau}$ that are defined on $\mathcal{C}$, the set of all compact subsets of $\mathbb{R}^d$. Put $h(Y) = \mathbf{1}_{Y\cap C=\emptyset}$, where $\mathbf{1}_M$ denotes the indicator function which is 1 if the condition M is fulfilled and 0 otherwise. Here we make use of the functional $Q_\Xi = 1 - T_\Xi$. Then (14.1) yields for $C \in \mathcal{C}$

$$Q_{\Xi\cap\tau}(C) = \mathbf{P}((\Xi \cap \tau) \cap C = \emptyset) = \mathbf{P}(\Xi \cap (\tau \cap C) = \emptyset).$$

If Ξ and τ are independent this can be written as

$$Q_{\Xi\cap\tau}(C) = \int P_\tau(\mathrm{d}T)\, Q_\Xi(C \cap T) \tag{14.3}$$

and, particularly in the model-based approach for a fixed T,

$$Q_{\Xi\cap T}(C) = Q_\Xi(C \cap T). \tag{14.4}$$

This shows in an abstract but direct way the limitation of stereological inference which is due to a certain choice of T or P_τ. The functional Q_Ξ can be expressed by $Q_{\Xi\cap T}$ only for sets that can be written as $C \cap T$, $C \in \mathcal{C}$. This information has to be supplemented by prior assumptions on the model Ξ. Thus, even if not the complete distribution of Ξ, at least some features of interest can be inferred from sections.

For a fixed k-dimensional linear subspace T in $\mathbb{R}^d$, $1 \le k \le d-1$, let $B \subset T$ be a k-dimensional convex compact set in T that contains the origin o. For a stationary random closed set Ξ and $C \in \{r\,B, r \geq 0\}$ formula (14.4) yields a formula for the contact distribution functions of Ξ and of $\Xi \cap T$ w.r.t. the so-called structuring element B, cf. Stoyan, Kendall and Mecke (1995), Schneider and Weil (2008).

$$H_{\Xi}^{B}(r) = 1 - \frac{Q_{\Xi}(r\,B)}{Q_{\Xi}(\{o\})} = 1 - \frac{Q_{\Xi\cap T}(r\,B)}{Q_{\Xi\cap T}(\{o\})} = H_{\Xi\cap T}^{B}(r), \quad r \geq 0. \tag{14.5}$$

If $B = [o, u]$ is a segment with u being a unit vector in T, then these *linear* contact distribution functions can be studied as functions of both r and of u. If B is a k-dimensional unit ball in T then one obtains a k-spherical contact distribution function. Even if all k-dimensional subspaces T are taken into account this does not provide the full-dimensional d-spherical contact distribution function.

Analogously, another immediate relation holds for the covariance functions of stationary random closed sets, cf. Stoyan, Kendall and Mecke (1995), Ohser and Mücklich (2000), Schneider and Weil (2008).

$$\mathrm{cov}_{\Xi}(ru) = Q_{\Xi^c}(\{0, ru\}) - [Q_{\Xi^c}(\{0\})]^2 = \mathrm{cov}_{\Xi\cap T}(ru) \tag{14.6}$$

for $r \geq 0$ and a unit vector $u \in T$. This can be generalized to Q_{Ξ^c} for sets of more than two points, and thus to a description of higher order moments of the volume measure of Ξ. For a k-dimensional subspace T the number of points should be less or equal $k+1$ such that they can be arranged in a general position.

Notice that for an anisotropic but stationary random closed set Ξ the linear contact distribution function as well as the covariance function can be determined for all directions $u \in \mathbb{S}^{d-1}$ already from a system of *vertical sections* T, see this subsection below.

Such functions provide an insight into certain features of a set concerning the issues of spatial arrangement (such as clustering vs. hard-core distances, regular vs. 'completely random'), orientation (when linear segments with different orientations are used as structuring elements), shape and size. On the other hand, they can provide more complete or detailed information if a parametric model is chosen for Ξ, e.g. a Boolean model with a parametric distribution of the grains or for tessellations. For some models it is possible to calculate the contact distribution function for certain structuring elements explicitly, and this can be used in statistics for goodness-of-fit tests as well as for parameter estimation, see Ohser and Mücklich (2000).

It could be fruitful to study contact distributions also for other structuring elements C, in particular when T is not a single linear subspace but a more complex section set.

These considerations for the capacity functional are also meaningful for the design-based approach but apparently not used there.

14.4.2 *Intrinsic volumes and mixed functionals – the application of integral geometry*

Even if the capacity functional is the fundamental tool to deal with a random closed set and it also immediately provides the contact distribution and covariance functions as particular cases, it would be very intricate to extract many further features of interest of Ξ only from 'hit-or-miss-observations' $\Xi \cap \tau \cap C \neq \emptyset$ or $\Xi \cap \tau \cap C = \emptyset$. Other observations and measurements are available, and the purpose of inference can often be described more conveniently in terms of functionals or of measures (e.g. surface or curvature measures). This means that in (14.1) the function h should not be restricted to the class of indicator functions.

The main use of (14.2) (or also (14.1)) is to plug in for h an intrinsic volume and to apply integral geometric formulae – the Crofton formula, see (1.18), the principal kinematic formula, translative integral formulae respectively, see Schneider and Weil (2008), Beneš and Rataj (2004), Baddeley and Jensen (2005) and Chapter 1 of the present volume. This yields stereological relations – even inversion formulae – for intrinsic volumes (or Minkowski functionals, quermassintegrals, respectively). In the design-based approach one considers the total value of such functionals for a bounded set while in the model-based approach their intensities, i.e. their mean total values per unit volume, are of interest.

In these formulae there appear integrals like

$$\int_M \mu(\mathrm{d}m)\, f(X \cap mT) = F(X,T)\,, \tag{14.7}$$

where M is a subgroup of the group G_d of rigid motions and μ a σ-finite invariant measure on M; for details on the topology and the corresponding Borel σ-algebra see Schneider and Weil (2008). This means that the relative position of X and T has to be moved and the value of the function f has to be integrated. Mostly, M is the group of translations or the group of rotations or of rigid motions, i.e. of translations and rotations. This formula can be understood as a particular case of formula (1.1) in Chapter 1.

Formula (14.2) (but also (14.1) can be considered) can be adapted to these integral geometric formulae by exploiting the motion-invariance properties of the distribution of Ξ. If there is no or not sufficient invariance in the model Ξ then one has to choose a distribution P_τ with amenable invariance properties. It is a crucial idea in the design-based and in the hybrid approaches that a random sampling design compensates a lack of invariance in the model. Given an invariance property of a distribution, an integration on a group of motions can be brought into (14.2) as follows.

Let Ξ be a random closed set which is invariant under a subgroup $M \subseteq G_d$ of motions, i.e. its distribution fulfils $P_\Xi = P_\Xi \circ m^{-1}$ for all $m \in M$. Further, let μ be a σ-finite invariant measure on M and $g : M \to [0,\infty)$ a measurable function with $\int \mu(\mathrm{d}m)\, g(m) = 1$. Then (14.2) yields for non-negative measurable h

$$\begin{aligned}\int P_{\Xi\cap\mathcal{T}}(\mathrm{d}Y)\, h(Y) &= \int P_{\Xi}(\mathrm{d}X) \int P_{\mathcal{T}}(\mathrm{d}T)\, h(X \cap T) \\ &= \int \mu(\mathrm{d}m)\, g(m) \int P_{\Xi} \circ m^{-1}(\mathrm{d}X) \int P_{\mathcal{T}}(\mathrm{d}T)\, h(X \cap T) \\ &= \int P_{\Xi}(\mathrm{d}X) \int P_{\mathcal{T}}(\mathrm{d}T) \int \mu(\mathrm{d}m)\, g(m)\, h((mX) \cap T). \end{aligned} \tag{14.8}$$

If h is invariant under the motions $m \in M$, as it holds for the intrinsic volumes, then $h((mX) \cap T) = h(X \cap m^{-1}T)$. The choice of the innermost integral, i.e. of $P_{\mathcal{T}}$, g and h, is decisive for the potential stereological interpretation of the section profiles in the sense of (14.7). If μ is finite then g can be chosen as the normalizing constant. Otherwise it is often chosen as the normalized indicator function of a set of motions which corresponds to the possible sections or observations to be made if a bounded observation window is available.

In proofs of particular formulae the above equation usually does not appear explicitly but in a more involved form. Also, if the set of interest is not modelled as a random closed set but as a particle process (i.e. as a point process on a space of compact sets) its distribution is described differently. But the basic idea behind the derivation of stereological formulae is analogous.

If, moreover, $P_{\mathcal{T}}$ is invariant under a subgroup $M' \subseteq G_d$ of motions such that $P_{\mathcal{T}} = P_{\mathcal{T}} \circ m'^{-1}$ for all $m' \in M'$, then this can be used to continue the derivation of (14.8) in an analogous way as it is shown for P_{Ξ}. An example is given in (14.10). Thus, by an appropriate choice of the subgroups of motions, a formula of the type (14.7) can be applied.

Two standard cases are that either Ξ is invariant under the group M of all rigid motions and thus $P_{\mathcal{T}}$ can be chosen as a δ-measure – this is the model-based approach – or that Ξ has no invariance property and thus M contains only the identity. The latter case is referred to as the design-based approach since all the necessary or desirable invariance has to be provided by $P_{\mathcal{T}}$, i.e. by an appropriate choice of a random section set.

The classical stereological formula is A. Delesse's (1847) equation $V_V = A_A$. The formula expresses that the (mean) fraction of the volume – volume intensity – that is occupied by a three-dimensional set Ξ equals the mean fraction of the area of its section profile on a two-dimensional plane. A historical review is given in Chapter 1.

We will illustrate here only a few particular results, namely an application of a translative formula as well as the hybrid approach for stationary anisotropic random closed sets and also the idea of vertical sections.

Assume that Ξ is stationary, i.e. invariant under translations in $\mathbb{R}^d$. Then M is the group of translations which can be identified with $\mathbb{R}^d$ itself, and the Lebesgue measure with the element $\mathrm{d}x$ is an invariant measure. In this case for

(14.7) a translative integral geometric formula has to be applied. Here we cite a part of Schneider and Weil (2008, Th. 9.4.7), see also Chapter 1.

A set $X \subset \mathbb{R}^d$ is locally polyconvex if it is a union of compact convex sets such that any compact convex set $B \subset \mathbb{R}^d$ is intersected by only finitely many of those convex sets, see Chapter 1. Let $N(X \cap B)$ denote the minimal number of compact convex sets that are necessary to generate $X \cap B$. A random closed set Ξ is called a *standard random set* if it is stationary and its realizations are a.s. locally polyconvex and $\mathbf{E}2^{N(\Xi\cap[0,1]^d)} < \infty$. Then the application of the translative integral formula to (14.8) yields the following assertion.

Theorem 14.1 *Let Ξ be a standard random closed set in $\mathbb{R}^d$, T a k-dimensional plane, $B_T \subset T$ a ball of k-volume 1 and $j \in \{0, \dots k\}$. Then $\Xi \cap T$ is a standard random set in T and*

$$\bar{V}_j(\Xi \cap T) = \bar{V}^{(j)}_{d-k+j,k}(\Xi, B_T)\,, \tag{14.9}$$

where $\bar{V}_j(\Xi \cap T)$ denotes the specific j-th intrinsic volume of the section profile $\Xi \cap T$, and $\bar{V}^{(j)}_{d-k+j,k}(\Xi, B_T)$ is the specific $(j, d-k+j)$-th mixed functional of Ξ and B_T, see Chapter 1.

Such translative formulae are not yet widely used for stereological purposes, and the interpretation of mixed functionals is more intricate than the understanding of intrinsic volumes (which coincide up to multiplicative constants and index shift with Minkowski functionals or quermassintegrals, respectively). In order to gain relations for intrinsic volumes a kinematic formula or the classical Crofton formula – this is *the* integral geometric formula for stereology – is applied. To do so one has to introduce isotropy, i.e. invariance under rotations.

Let SO_d denote the group of rotations of $\mathbb{R}^d$ around the origin, $\mathrm{d}\rho$ the element of the invariant probability measure on SO_d. If $P_\mathcal{T} = P_\mathcal{T} \circ \rho^{-1}$ for all $\rho \in SO_d$ then, analogously to the calculations in (14.8) one can augment an integration $\int \mathrm{d}\rho$. If h is invariant under rotations this yields

$$\int P_{\Xi\cap\mathcal{T}}(\mathrm{d}Y)\, h(Y) = \int P_\Xi(\mathrm{d}X) \int P_\mathcal{T}(\mathrm{d}T) \int \mathrm{d}\rho \int \mathrm{d}x\; g(x)\, h(X \cap \rho(T + x)). \tag{14.10}$$

The integrals $\int \mathrm{d}\rho \int \mathrm{d}x$ describe the integration w.r.t. an invariant measure on the group of all euclidean motions in $\mathbb{R}^d$ and thus integral geometric formulae of type (14.7) with $M = G_d$ can be applied. Often $P_\mathcal{T}$ is concentrated on a set $\{\rho T : \rho \in G\}$ for a fixed section set T and G a subgroup of SO_d. Then $P_\mathcal{T}$ can be generated as the image of the invariant probability measure on the group G.

It is a thoroughly investigated problem how isotropic random sections can be generated. Practically, it is often impossible to cut a probe physically one or more times in random independent directions. A key result was provided by A.J. Baddeley for so-called *vertical sections*, cf. Baddeley and Jensen (2005), Beneš and Rataj (2004). The basic idea is that one-dimensional isotropic sections in $\mathbb{R}^3$ can be generated by random lines on two-dimensional planar sections which are all 'vertical', i.e. which are all parallel to a fixed line. Given a vertical direction $v \in \mathbb{S}^2$, and thus all planar sections have normals orthogonal to v, it is obvious that any line in $\mathbb{R}^3$ can be embedded in a vertical plane. It is assumed that the distribution of the vertical planes is invariant under rotations around v. Then, in order to generate an isotropic line in $\mathbb{R}^3$, the directional distribution of the random line within a given vertical plane has to have the non-isotropic density $\sin\angle(w, v)$, where $\angle(w, v)$ denotes the angle between the direction w parallel to the line and v.

This result can be formulated for arbitrary dimensions d and vertical hyperplanes as follows, see Beneš and Rataj (2004, Lemma 4.3), which is proved there with the help of the coarea theorem.

Theorem 14.2 *Let $v \in \mathbb{S}^{d-1}$ be a fixed direction, $\mathfrak{V}$ the set of all hyperplanes ($(d-1)$-linear subspaces) which contain v, and U_v the rotation invariant distribution on $\mathfrak{V}$. For any $\mathcal{V} \in \mathfrak{V}$ the unit sphere is denoted by $\mathbb{S}^{d-2}(\mathcal{V})$ and the uniform distribution on it by $U_{\mathcal{V}}$. Then for the uniform distribution U on $\mathbb{S}^{d-1}$ and for any non-negative measurable function $f : \mathbb{S}^{d-1} \to [0, \infty)$ we have*

$$\int_{\mathbb{S}^{d-1}} U(\mathrm{d}u)\, f(u) = c_d \int_{\mathfrak{V}} U_v(\mathrm{d}\mathcal{V}) \int_{\mathbb{S}^{d-2}(\mathcal{V})} U_{\mathcal{V}}(\mathrm{d}w)\, |\sin\angle(w, v)|\, f(w) \quad (14.11)$$

with

$$c_d = \frac{(d-1)\Gamma(\frac{d}{2})^2}{(d-2)\Gamma(\frac{d-1}{2})\Gamma(\frac{d+1}{2})}.$$

In this formula only the directions are taken into account. The result can be used for a reformulation of the Crofton formula for sections with lines that are located on vertical section (hyper-)planes. The assertion of the theorem can also be extended by integrals over translations of section lines.

Here we cite a result which represents an application of the theorem for the design based approach, see Baddeley and Jensen (2005). Let $X \subset \mathbb{R}^3$ be a polyconvex set, i.e. a finite union of convex bodies, and $S(X)$ its surface area. Denote by $T_{\mathcal{V},w,q}$ the line that is contained in the vertical plane $\mathcal{V}$ and has the direction w (i.e. w is parallel to the line) and the distance q to the origin. Further, $\#(X \cap (T_{\mathcal{V},w,q} + x))$ denotes the number of intersection points of the shifted line with the boundary of X.

$$S(X) = 2V_{d-1} =$$

$$\frac{1}{\pi} \int_{\mathfrak{V}} U_v(\mathrm{d}\mathcal{V}) \int_{\mathcal{V}^{\perp}} \mathrm{d}x \int_{\mathbb{S}^1(\mathcal{V})} U_{\mathcal{V}}(\mathrm{d}w) \int_{(0,\infty)} \mathrm{d}q\, |\sin\angle(w, v)|\, \#(X \cap (T_{\mathcal{V},w,q} + x)). \quad (14.12)$$

The formulae for vertical sections with the sine-densities can be interpreted as a rule for generating a random section line on a vertical section. But it can also be understood that one has to weight the observations made on an isotropic line on a vertical section by this sine.

In order to gain observations on isotropic linear sections one can even go a step further by using a curved section line, where the distribution of tangent directions 'emulates' a certain directional distribution for straight lines. For the case of vertical sections this is realized by cycloidal curves on the planes, see Baddeley and Jensen (2005) and Beneš and Rataj (2004) for the proof based on geometric measure theory.

While the Crofton formula for sections with k-dimensional planes T is widely known and used, the potential of an application in stereology of the principal kinematic formula for other section sets seems not yet be fully recognized.

A further important tool is the Blaschke–Petkantschin formula that yields a decomposition of the volume measure to measures on sets of linear subspaces and volume measures on these subspaces. This is worked out in detail and applied to local stereology by Jensen (1998).

The coarea formula, which comes from geometric measure theory, also provides a rather general decomposition for Hausdorff measures, see e.g. Beneš and Rataj (2004). Such decompositions of measures comprise stereological applications if they refer to section sets and there to measures which can be determined on these section sets.

In integral geometry there are also local versions of the classical formulae, see Schneider and Weil (2008). They yield relations for curvature measures for the surface of a set. A survey on far reaching generalizations of integral geometric formulae of Crofton type is given by Schneider and Hug (2002). This concerns the classes of sets which are intersected as well as the functionals, measures or functions that are considered for these sets. Also in such formulae can be a potential for stereology which is not yet exploited.

14.4.3 *Normal measures and orientation analysis*

The functionals considered in the previous paragraph are global ones, i.e. they describe single real parameters of a set as a whole. Moreover, the intrinsic volumes are invariant under Euclidean motions of the set. For a characterization of the orientation or of the anisotropy of a set the mixed functionals could be used but they are rather intricate for this purpose. The preferential entity is the distribution of the normal direction of the surface of the set in a 'typical' point of its surface. For lower-dimensional sets like ensembles of manifolds, in particular systems of fibres, this is applied to the points of the set itself. The stereological methods for a determination of a directional distribution, also referred to as 'rose of directions', are based on a so-called 'rose of intersections'. The latter is that the sections of the set with k-planes of different directions are observed and analyzed as a function of that directions.

We illustrate this here for the model-based approach with a standard random closed set (stationary with realizations in the extended convex ring), in order to have the measures be defined properly.

Here we review a representation by Kiderlen (2008), where references are given also to previous work by other authors, like P. Goodey, J. Rataj,

R. Schneider, W. Weil and M. Zähle. For a polyconvex set $X \subset \mathbb{R}^d$ there are in a point $x \in \partial X$ either exactly one outer (unit) normal vector $u(X,x)$ or exactly two normal vectors $u(X,x)$ and $-u(X,x)$ or infinitely many ones. Denote the sets of boundary points of the first two types by $\partial^1 X$ and $\partial^2 X$ respectively. The $(d-1)$-Haudorff measure of the boundary is then given as $\mathcal{H}^{d-1}(\partial X) = \mathcal{H}^{d-1}(\partial^1 X \cup \partial^2 X)$. The *surface area measure* $S(X,\cdot)$ of X is a measure on $\mathbb{S}^{d-1}$ which is defined by

$$S(X,B) = \int_{\partial^1 X \cup \partial^2 X} \mathbf{1}_{u(X,x)\in B}\, \mathcal{H}^{d-1}(\mathrm{d}x) + \int_{\partial^2 X} \mathbf{1}_{-u(X,x)\in B}\, \mathcal{H}^{d-1}(\mathrm{d}x) \tag{14.13}$$

for measurable $B \subset \mathbb{S}^{d-1}$. If X is topological regular, i.e. it is the closure of its interior, then $\partial^2 X = \emptyset$.

For a standard random closed set Ξ the *oriented mean normal measure* $\bar{S}(\Xi,\cdot)$ can be defined on $\mathbb{S}^{d-1}$ by

$$\bar{S}(\Xi,\cdot) = \lim_{r\to\infty} \frac{1}{\lambda_d(rK)} \mathbf{E} S(\Xi \cap rK, \cdot)\,, \tag{14.14}$$

where λ_d is the d-dimensional Lebesgue measure in $\mathbb{R}^d$ and K is a convex body with positive volume. This limit does not depend on the particular choice of K. Notice that the 'rose of directions' is the normalized and symmetrized (i.e. unoriented) $\bar{S}(\Xi,\cdot)$.

For the section profile $\Xi \cap T$ on a linear k-subspace T the oriented mean normal measure is defined accordingly and denoted by $\bar{S}'(\Xi \cap T,\cdot)$. For $u \in \mathbb{S}^{d-1}$ the orthogonal projection onto T is denoted by $u|T$.

Theorem 14.3 (Weil 1997) *Let Ξ be a standard random set in $\mathbb{R}^d$ and T a fixed k-dimensional subspace with the unit sphere $\mathbb{S}^{k-1}(T)$. For any non-negative measurable function $f : \mathbb{S}^{k-1}(T) \to [0,\infty)$*

$$\int \bar{S}'(\Xi \cap T, \mathrm{d}u')\, f(u') = \int \bar{S}(\Xi, \mathrm{d}u)\, f\left(\frac{u|T}{\|u|T\|}\right) \|u|T\|. \tag{14.15}$$

It is a genuine stereological problem to characterize sets $\mathcal{L}$ of k-dimensional subspaces that are sufficiently large in the sense that $\bar{S}(\Xi,\cdot)$ is uniquely determined by $\{\bar{S}'(\Xi \cap T,\cdot) : T \in \mathcal{L}\}$. Here we cite only a most recent result by M. Kiderlen. Denote by $\mathcal{L}_k^d$ the set of all k-dimensional linear subspaces of $\mathbb{R}^d$ and by ν_k the rotation invariant probability measure on $\mathcal{L}_k^d$. A stationary random closed set Ξ is called simple polyhedral if $\bar{S}(\Xi,\cdot)$ is discrete.

Theorem 14.4 (Kiderlen 2008)

(a) For any $1 \le k \le d-1$ and any finite $\mathcal{L} \subset \mathcal{L}_k^d$ there are two stationary simple polyhedral random sets Ξ and Ξ' with different mean normal measures, but such that $\bar{S}'(\Xi \cap T,\cdot) = \bar{S}'(\Xi' \cap T,\cdot)$ for all $T \in \mathcal{L}$.

(b) *Let* $2 \leq k \leq d-1$, $m = \lfloor d/k \rfloor + 1$, *and* Ξ *be a stationary simple polyhedral random set with* $\mathbf{E}\mathcal{H}^{d-1}(\partial\Xi \cap K) < \infty$ *for all convex bodies* K *with interior points. Then, for* $\nu_k^{\otimes m}$*-almost all* $(T_1, ..., T_m) \in (\mathcal{L}_k^d)^m$ *and all stationary polyhedral random sets* Ξ' *the equation*

$$\bar{S}'(\Xi \cap T_i, \cdot) = \bar{S}'(\Xi' \cap T_i, \cdot) \quad \text{for all } i = 1, \ldots, m,$$

implies $\bar{S}(\Xi, \cdot) = \bar{S}(\Xi', \cdot)$.

The mean normal measure, oriented or non-oriented, bears essential information about the form or shape and the orientation of the set or of the particles which generate it.

There are also stereological inversion formulae and estimation methods for the directional distribution of fibre and surface processes when planar sections can be observed. A comprehensive presentation of that is given by Beneš and Rataj (2004). Recently, for the orientation analysis of particles in a particle process, an approach that uses a tensor – the mean area moment tensor – was introduced by Schneider and Schuster (2006), and a stereological formula for sections with hyperplanes was proved.

Besides the non-parametric approach, also parametric models are studied such as circular plates or particle processes of spheroids. Results for the multivariate joint distribution of size, shape and orientation are presented by Beneš and Rataj (2004).

14.4.4 *Size and shape distribution of particles*

It can be appropriate to model a geometric structure as a particle process rather than as a random closed set. The formal difference of the models allows the treatment of features of the individual particles which constitute the random set. Let Φ denote a particle process in $\mathbb{R}^d$, i.e. Φ is a simple point process in $\mathcal{F}$, where the distribution is concentrated on $\mathcal{C}'$, the family of non-empty compact sets. As usual, we interpret a simple point process as a random set as well as a random measure. Thus Φ describes a random ensemble of compact sets, and these sets will be considered individually – irrespective of overlaps.

Let Φ be a stationary particle process with distribution P, finite positive number intensity N_V and distribution of the typical particle (primary grain distribution) P_0. The process of section profiles on a k-dimensional subspace T is a particle process $\Phi \cap T = \{K \cap T : K \in \Phi, K \cap T \neq \emptyset\}$, and this process is stationary in T, i.e. its distribution $P^{(T)}$ is invariant under translations in $\mathbb{R}^d$ which leave T invariant. Denote by $N_A^{(T)}$ its intensity and by $P_0^{(T)}$ the distribution of the typical particle section profile. In order to relate the two particle processes consider the mapping $K \mapsto K \cap T$ for a fixed plane T and $K \in \mathcal{C}'$ and the respective mapping $\varphi \mapsto \varphi \cap T$ for particle ensembles φ. Thus the refined Campbell theorem for stationary particle processes (see Chapter 1) and the transformation theorem for integrals yield a general stereological integral equation for non-negative and measurable $f : \mathcal{C}' \to [0, \infty)$. Let $\mathrm{d}x^{(T)}$ denote the element of the Lebesgue measure on T.

$$\begin{aligned}
N_A^{(T)} \int_T \mathrm{d}x^{(T)} \int P_0^{(T)}(\mathrm{d}K')\, f(K' + x^{(T)}) \\
&= \int P^{(T)}(\mathrm{d}\varphi^{(T)}) \int \varphi^{(T)}(\mathrm{d}K')\, f(K') \\
&= \int P(\mathrm{d}\varphi) \int (\varphi \cap T)(\mathrm{d}K')\, f(K') \qquad (14.16)\\
&= \int P(\mathrm{d}\varphi) \int \varphi(\mathrm{d}K)\, f(K \cap T)\, \mathbf{1}_{K \cap T \neq \emptyset} \\
&= N_V \int_{\mathbb{R}^d} \mathrm{d}x \int P_0(\mathrm{d}K)\, f((K + x) \cap T)\, \mathbf{1}_{(K+x) \cap T \neq \emptyset}.
\end{aligned}$$

In particular, for $f(K) = \mathbf{1}_{c(K) \in B}$, where $B \subset T$ is a Borel set with $(d-k)$-volume 1, and $c(K)$ the center of the $(d-k)$-circumball of $K \subset T$ in T, this yields

$$N_A^{(T)} = N_V \int P_0(\mathrm{d}K)\, V^{(d-k)}(K|T^{\perp}), \qquad (14.17)$$

where $K|T^{\perp}$ denotes the orthogonal projection of K onto the orthogonal complement of T, and $V^{(d-k)}$ the $(d-k)$-volume.

These formulae can be specified for stationary and isotropic models or for randomized sections T by augmenting integrals over rotations. Further, if models with all homothetic particles are considered, or – more general – if the particles can be described by a parameter vector, e.g. spheroids, then the distributions of the typical particles can be defined on the respective parameter spaces. Then they are referred to as size distributions. *Wicksell's corpuscle problem* for a process of balls is the classical and most prominent example of this kind, and can be found in almost every review on stereology. The aim is to infer the diameter distribution of three-dimensional balls from the observation of the diameters of planar section circles. A series of results is also developed for spheroids and their two-dimensional distributions of axes lengths or even trivariate ones of axes lengths and orientation, see Cruz-Orive (1976), Cruz-Orive (1978), Beneš and Rataj (2004), Ohser and Mücklich (2000).

Here we restrict the presentation to processes of random balls. An application of formula (14.16) yields the famous integral equation that expresses the link between the distribution functions H of the ball diameters and G of the section circles respectively, when a three-dimensional stationary particle process of balls with random radii is intersected with a two-dimensional plane T. This reads as

$$N_A^{(T)}\, G(y) = 2N_V \int_y^{\infty} (x - \sqrt{x^2 - y^2})\, \mathrm{d}H(x), \quad y \geq 0. \qquad (14.18)$$

A closer examination shows that the Rayleigh distribution with the density

$$f_a(r) = \frac{r}{a^2} \exp\left(-\frac{r^2}{2a^2}\right), \quad r > 0, \tag{14.19}$$

$a > 0$ a parameter, is a fixed point under the transformation $H \mapsto G$ of distributions that is described by (14.18). Drees and Reiss (1992) showed that these are the only distributions that remain invariant.

Besides the complete distribution and their moments, recently interesting results were derived for the tail behaviour of the particle size distributions. This is of interest if certain quantiles or extremes are studied, and it is of practical relevance, e.g. in materials science. In the mentioned paper by H. Drees and R.-D. Reiss (cf. Beneš and Rataj 2004) the following result was shown which indicates how certain types of extremal size distributions are inherited by sections.

In statistical extreme value theory the following notions are introduced for the maximum of i.i.d. random variables. A univariate distribution function G belongs to the domain of attraction of a distribution function $\mathfrak{K}$, symbolically $G \in \mathcal{D}(\mathfrak{K})$, if there exist sequences (a_n), (b_n) of real numbers such that for all $x \in \mathbb{R}$

$$\lim_{n\to\infty} G^n(a_n x + b_n) = \mathfrak{K}(x). \tag{14.20}$$

The three classes of distribution functions which are max-stable and arise as limits are

$$\mathfrak{K}_{i,\gamma}(x) = \begin{cases} \exp(-x^{-\gamma}), & x \geq 0, \quad i = 1 \quad \text{(Fréchet)} \\ \exp(-(-x)^{\gamma}), & x \leq 0, \quad i = 2 \quad \text{(Weibull)} \\ \exp(-\mathrm{e}^{-x}), & x \in \mathbb{R}, \quad i = 3 \quad \text{(Gumbel)} \end{cases} \tag{14.21}$$

for $\gamma > 0$.

Theorem 14.5 (Drees/Reiss 1992) *Let Φ be a stationary particle process of balls in $\mathbb{R}^3$ with the parameters introduced above, and consider the stationary particle process of section circles on a two-dimensional plane. Then for $i = 1$ and $\beta = \gamma - 1 > 0$, or $i = 2$ and $\beta = \gamma + \frac{1}{2}$, $\gamma > 0$, or for $i = 3$:*

If $H \in \mathcal{D}(\mathfrak{K}_{i,\gamma})$ then $G \in \mathcal{D}(\mathfrak{K}_{i,\beta})$.

Analogous relations for the shapes of the lower tails of the diameter distributions of balls in $\mathbb{R}^d$ and their k-dimensional sections were derived by Kötzer and Molchanov (2006). The approach by H. Drees and R.-D. Reiss was also successfully applied to the extremal shape factor of spheroidal particles and worked out by Hlubinka (2006), see also Beneš and Rataj (2004).

14.4.5 *Arrangement of particles – second order quantities*

If a set is modeled as an ensemble of particles, also their mutual arrangement is of interest. An aim is to quantify intuitive conceptions like interaction or non-interaction, attraction or repulsion, regularity or non-regularity, clustering and also a dependence of the arrangement on the spatial direction if the structure is anisotropic. A comprehensive presentation of respective entities is given by Ohser and Mücklich (2000). It comprises also the covariance functions and the contact distribution functions which are considered in Section 14.4.1. An important second order characteristic is the pair-correlation function. For a stationary and isotropic point process Ψ in $\mathbb{R}^d$ with intensity γ and the Palm distribution P_0 (see Chapter 1 and Chapter 3, respectively) it is defined by

$$g(r) = \frac{1}{d\kappa_d r^{d-1}} \frac{\mathrm{d}}{\mathrm{d}r}\mathcal{K}(B_r(0)) \quad \text{for } r \geq 0, \tag{14.22}$$

where

$$\gamma\,\mathcal{K}(B_r(0)) = \int P_0(\mathrm{d}\psi)\,\psi(B_r(0)) - 1$$

is the mean number of points in a ball of radius r around a typical point of Ψ which is not counted itself.

The stereological problem of determining the pair-correlation function of a particle process from a planar section was studied for processes of spheres in $\mathbb{R}^3$. Let Φ be a stationary and isotropic particle process of balls in $\mathbb{R}^3$ and g_V the pair-correlation function of the point process Ψ of their centers. Denote by g_A the pair-correlation function of the stationary and isotropic point process of centers of section circles in $\Phi \cap T$, where T is a two-dimensional section plane. There are section formulae, expressing g_A as an integral over g_V for sphere processes with several assumptions on the diameters of the spheres. For the particular case of a constant sphere diameter D even an inversion formula was derived by Kalmykov and Shepilov (2000), which is presented here. The section formula is

$$g_A(r) = 2D^{-2} \int_r^{\sqrt{r^2+D^2}} \mathrm{d}t \left(\tfrac{D}{\sqrt{t^2-r^2}} - 1\right) t\, g_V(t), \quad r > 0. \tag{14.23}$$

Under the assumption that there is an $R_1 > 0$ such that $g_V(t) = 1$ for $t \geq R_1$ (interpretation: behaviour like a Poisson process for distances $\geq R_1$), and that R_1 is the smallest of these numbers, a solution is

$$\begin{aligned} g_V(t) &= \frac{1}{\pi} g^*(t) - \frac{D}{\pi t} \int_t^{R_1} \mathrm{d}x\, \frac{x g^{*\prime}(x)}{\sqrt{x^2-t^2}} + \frac{1}{\pi} \int_t^{R_1} \mathrm{d}x\, g^*(x) \\ &\quad \times \left[\frac{2x}{\pi D^2} \exp\left(\frac{x^2-t^2}{\pi D^2}\right) \left(1 + \operatorname{erf}\sqrt{\frac{x^2-t^2}{\pi D^2}}\right) \right. \\ &\quad \left. + \frac{2x}{\pi D \sqrt{x^2-t^2}} - \frac{D}{(x+t)\sqrt{x^2-t^2}} \right] \end{aligned} \tag{14.24}$$

for $t \in [\sqrt{R_1^2 - D^2}, R_1]$ and

$$g^*(r) = g_A(r) - 1 - \frac{R_1^2 - r^2}{D^2} + \frac{2}{D}\sqrt{R_1^2 - r^2}.$$

Kalmykov and Shepilov (2000) give a generalized version of this formula, where the knowledge of the exact value of R_1 is not necessary.

14.4.6 *Tessellations*

Random tessellations – also called mosaics – can be described as particle processes, with the cells as particles, see Chapter 5. Alternatively, they can also be described as the particle processes of the $(d-1)$-faces of the cells or also as the random closed set consisting of the union of the boundaries of the cells. Hence, all the general formulae for those models can be applied.

Here we restrict to random tessellations in $\mathbb{R}^d$ which have compact convex polytopes as cells. Obviously, a section with a k-plane T, $1 \leq k \leq d-1$ generates a random tessellation in T.

If the tessellation in $\mathbb{R}^d$ and its sections with k-planes belong to the same parametric model class then the stereological problem reduces to one or a few intensity parameters and – in anisotropic models – to the determination of directional distributions. Such an inheritance property can also be an aspect in the development of new tessellation models.

Concerning the important models the following inheritance properties are known. For hyperplane tessellations it is obvious that sections are also hyperplane tessellations and, in particular, if one has a Poisson hyperplane tessellation then the section profiles are Poisson too (see the remarks on infinite divisibility below in Section 14.5). And also the d-dimensional stationary STIT (i.e. stable with respect to iteration of tessellations) tessellations inherit the STIT property to sections with k-planes.

For Poisson–Voronoi tessellations the problem is much more involved. A planar section of a stationary three-dimensional Poisson–Voronoi tessellation yields a tessellation that is not a Voronoi tessellation in the sense that there is no point process that generates a Voronoi tessellation which coincides (in distribution) with this planar section, see Stoyan, Kendall and Mecke (1995, p. 375) and the reference given there. But the Voronoi tessellations can be embedded into the class of Laguerre tessellations or power diagrams. A planar section of a Laguerre tessellation is a Laguerre tessellation again. Thus also the planar section of a Poisson–Voronoi tessellation is a Laguerre tessellation. For fixed realizations this is shown by an explicit calculation in Okabe, Boots, Sugihara and Chiu (2000). But feasible quantitative formulae for the corresponding random entities, namely the weights or radii for the Laguerre section tessellation, are not yet known. For the stationary Poisson–Voronoi tessellation in $\mathbb{R}^d$ and k-dimensional planar sections formulae were developed by Muche and by Schlather for the edge length distribution of the section profile and for further characteristics of the neighbourhood of edges, see Muche (2005), and the references therein.

When dealing with general random stationary tessellations one is confronted with the problem that sections of random convex polytopes are hardly handled stereologically. This effect occurs already for the chord length distributions that are generated by random linear sections. Therefore, for many problems concerning tessellations, simulation studies play a relevant role too, cf. Okabe, Boots, Sugihara and Chiu (2000), Ohser and Mücklich (2000).

For stationary random tessellations it is also of interest to study the distribution of the typical cell or of the zero-cell (i.e. the cell that contains the origin) of the section profile. Since a section with a k-dimensional plane of a Poisson hyperplane or a STIT tessellation (see Chapter 5) in $\mathbb{R}^d$ is Poisson hyperplane or STIT, respectively, again, the stereological relations are relatively easy to describe. Essentially, formulae for the surface intensities or related intensities and the directional distributions have to be used (which more generally hold for any stationary tessellation). Since the distributions of Poisson hyperplane or STIT tessellations are uniquely determined by their surface intensities and the directional distributions, this uniqueness also holds for the distributions of the typical cell or of the zero cell. But it remains the problem to describe the distribution for cells more explicitly.

For Poisson–Voronoi tessellations or for Laguerre tessellations there are two problems: to study the distribution of the cells of the tessellations themselves and of the section profile cells.

14.5 Properties that are inherited by section profiles

As it was already mentioned in the previous subsections, there are parametric as well as non-parametric stereological relations. We will concentrate now on some 'qualitative' properties, and restrict considerations to random closed sets. The type of the assertions is: If Ξ has a property p, then $\Xi \cap \tau$ has the property $\tilde{p}$. A typical example for this is stationarity or isotropy. If Ξ is a stationary (or stationary and isotropic) random closed set in $\mathbb{R}^d$ and T is a k-dimensional plane then $\Xi \cap T$ is stationary (or stationary and isotropic) in T too. More general, if the distribution of Ξ is invariant under a group M of motions and M_T is a subgroup of motions which leave the section set T invariant then the distribution of the profile $\Xi \cap T$ is invariant under M_T. This can e.g. also be applied to systems of parallel section planes.

On the one hand, considerations of this kind may allow us to identify the essential parameters of a section profile to deal with. On the other hand, it can be useful to classify the model of Ξ or to exclude certain models if the section profile does not exhibit certain features.

Much more complicated is the inverse problem, namely whether one can also conclude: If $\Xi \cap \tau$ has a property $\tilde{p}$, then Ξ has the property p. Or in a stronger formulation: If for all section sets $T \in \mathcal{T}$ of an appropriate set $\mathcal{T}$ the profile $\Xi \cap T$ has a property $\tilde{p}$, does then have Ξ the property p? An example for this is: What can be asserted about a random stationary tessellation if all linear sections of it are Poisson point processes?

Now we give a list of some selected stochastic properties that are inherited by sections.

Ergodicity and mixing Intuitively, it seems to be clear that the random section profiles on a k-plane T of a stationary and ergodic random closed set are not necessarily ergodic. To see this, consider a stationary random closed set Ξ which consists of stripes of width 1, say, that are parallel to T. Thus the realizations of $\Xi \cap T$ are either T or $\emptyset$, both with positive probability, and hence $\Xi \cap T$ is not ergodic. But ergodicity is not yet studied profoundly in stochastic geometry. A standard reference for ergodicity of spatial processes is still the paper by Nguyen and Zessin (1979).

More emphasis is placed on the stronger property of being mixing. This property is bequeathed by planar sections. Let Ξ be a stationary and mixing random closed set in $\mathbb{R}^d$ and T a k-dimensional linear subspace. Then for $C_1, C_2 \in \mathcal{C}$ and $C_1, C_2 \subset T$ and with (Schneider and Weil, 2008, Th. 9.3.2)

$$\begin{aligned}\lim_{\substack{\|x\|\to\infty\\ x\in T}} Q_{\Xi\cap T}(C_1 \cup (C_2 + x)) &= \lim_{\substack{\|x\|\to\infty\\ x\in T}} Q_{\Xi}((C_1 \cap T) \cup ((C_2 \cap T) + x)) \\ &= Q_{\Xi}(C_1 \cap T)\, Q_{\Xi}(C_2 \cap T) \qquad (14.25) \\ &= Q_{\Xi\cap T}(C_1)\, Q_{\Xi\cap T}(C_2).\end{aligned}$$

This equation yields that $\Xi \cap T$ is mixing.

Infinite divisibility and stability w.r.t. an operation (union, iteration/nesting) The general stochastic concepts of infinite divisibility and stability can also be applied to random closed sets and appropriate operations. Matheron (1975) studied infinite divisibility w.r.t. the operation of union of sets; another operation, where this aspect was considered is the iteration (also referred to as nesting) of tessellations, if a tessellation is interpreted as the random closed set of the union of cell boundaries, see the assertions and references on STIT tessellations in Chapter 5 of the present volume.

Here we consider the union of sets. The equality in distributions is denoted by $\overset{D}{=}$. A random closed set is called infinitely divisible w.r.t. union if for any natural n there exists a sequence Ξ_i, $i = 1, \ldots, n$, of i.i.d. random closed sets such that

$$\Xi \overset{D}{=} \bigcup_{i=1}^{n} \Xi_i. \qquad (14.26)$$

Obviously, this equation implies $\Xi \cap T \overset{D}{=} \bigcup_{i=1}^{n} (\Xi_i \cap T)$ for any closed section set T, and the $\Xi_i \cap T$ are still i.i.d. and thus $\Xi \cap T$ is infinitely divisible too.

A random closed set is called stable w.r.t. union if for any natural n there exist a real number a_n and a sequence Ξ_i, $i = 1, \ldots, n$, of i.i.d. random closed sets with $\Xi \overset{D}{=} \Xi_i$ such that

$$\Xi \overset{D}{=} a_n \bigcup_{i=1}^{n} \Xi_i. \qquad (14.27)$$

Thus, if Ξ is a random closed set that is stable w.r.t. union this yields

$$\begin{aligned}\Xi \cap T &\overset{D}{=} \left(a_n \bigcup_{i=1}^{n} \Xi_i\right) \cap T \\ &= a_n \bigcup_{i=1}^{n} (\Xi_i \cap T) \qquad \text{if} \quad a_n T = T.\end{aligned} \tag{14.28}$$

Roughly speaking, if the section set T is a linear subspace and if the considered operation (here it is the union) commutes with the intersection with T, then stability is bequeathed from Ξ to $\Xi \cap T$. It will be fruitful to search for further operations and random closed sets that are stable w.r.t. them.

Semi-markovian property Matheron (1975) introduced a definition of semi-Markov random closed sets. As for the mixing property, it is easy to investigate for sections. Let Ξ be a random closed set in $\mathbb{R}^d$, T a k-dimensional affine subspace and $C, D, C' \in \mathcal{C}$, $C, D, C' \subset T$ and C, C' separated by D in the sense that any linear segment $[x, x']$ with $x \in C$, $x' \in C'$ hits D. If Ξ is semi-Markov, then

$$\begin{aligned} &Q_{\Xi\cap T}(C \cup D \cup C')\, Q_{\Xi\cap T}(D) \\ &= Q_{\Xi}((C \cup D \cup C') \cap T)\, Q_{\Xi}(D \cap T) \\ &= Q_{\Xi}((C \cap T) \cup (D \cap T))\, Q_{\Xi}((D \cap T) \cup (C' \cap T)) \\ &= Q_{\Xi\cap T}(C \cup D)\, Q_{\Xi\cap T}(D \cup C'). \end{aligned} \tag{14.29}$$

This means that $\Xi \cap T$ is semi-Markov too. The main argument is that two sets $C, C' \subset T$ which are separated within T by D are also separated by D if all the sets are considered as subsets of $\mathbb{R}^d$.

14.6 Uniqueness problems

Up to now we were focused upon the transformation of distributions that is induced by a section, i.e. we mainly 'read formulae (14.1) and (14.2) from left to right'. The original request of stereology is to find expressions for certain entities of the set Ξ itself, i.e. to find appropriate solutions of these integral equations. Obviously, it includes the problems of existence and uniqueness of solutions. This is referred to as an inverse problem or the search for stereological inversion formulae.

As it was illustrated in Section 14.4, formula (14.2) may yield immediate results on Ξ for certain contact distribution functions, for the covariance function for certain mean values (intensities) of functionals like volume, surface, or mixed functionals.

For other problems, formulae (14.1) or (14.2) appear as an integral equation, where the feature of interest is implicit, e.g. particle size distribution, orientation distribution. In some cases explicit solutions of these integral equations can be given.

The problem of uniqueness of a solution concerns the question whether the section profiles – or, more precisely, the observations performed on the section profiles – contain sufficient information to retrieve the feature of interest, at least theoretically. Here we pick out a few examples of results.

Cruz–Orive (1976, 1978) showed an important and famous result that for planar sections of stationary and isotropic processes of three-dimensional spheroids there are models with different mixtures of oblate and prolate spheroids which yield the same distribution of section profiles.

There are several uniqueness results concerning the mean oriented normal measure of a random closed set, see Section 14.4 and Theorem 14.4.

Goodey and Weil (2006) studied star-shaped bodies $X \subset \mathbb{R}^d$ (w.r.t. the origin) and intersections with k-dimensional half-spaces T (through the origin). For any direction $u \in \mathbb{S}^{d-1}$ denote by $s_k(X,u)$ the average of the intersection volume over all k-dimensional half-spaces T containing u (orthogonal to the boundary of T). It is proved that the function $s_k(X,\cdot)$ determines the star-shaped set X uniquely for the cases $k = 2$ and $d = 3$ or 4, as well as $k \leq (d+2)/2$ and $k > (2d+1)/3$ for $d \geq 5$. The authors also show that there are infinitely many pairs (k,d) for which uniqueness fails.

Finally we emphasize that sometimes the inverse problem does not play a major role. For instance, certain classification problems for the model of Ξ can be treated immediately with the section profiles. Another example appears for the stationary Poisson-Voronoi tessellations. Their distribution is determined by one real parameter, the intensity of the Poisson point process. This can be determined stereologically relatively easy, and there is no need to invert the rather involved section formulae, like the above mentioned one by M. Schlather and L. Muche for the edge length distribution of planar sections.

14.7 Statistical aspects

Sections are considered as samples. A main purpose of stereology is a statistical inference for certain features of the set or of its distribution if it is modelled as a random set. Even if this is a major topic in stereology we mention the statistical problems only very briefly here. Profound presentations are given in the books by Baddeley and Jensen (2005), with an emphasis on the design-based approach and a thorough treatise of the sampling theory that is relevant there, by Beneš and Rataj (2004) and Ohser and Mücklich (2000).

A general problem is that typically stereological inverse problems are ill-posed ones. The solutions of the arising integral equations are statistically and numerically unstable. An indicator of statistical instability can be an infinite variance, e.g. the infinite variance of an unbiased estimator of the mean sphere diameter in

Wicksell's corpuscle problem. Also, errors of observation or those ones due to an approximation or a discretization of the integral may yield inadmissible results, e.g. estimated probability densities with negative values.

These effects are due to a loss of information which is caused by the intersection of Ξ with a lower-dimensional section set τ. This loss can sometimes be partly compensated for by an appropriately large sample in terms of the size of the observation window or in the number of sections. But this is not always possible, in particular if the applied estimator has an infinite variance.

The assessment of stereological estimators or tests is a current issue. In particular, the determination of the variance of estimators is a main problem, see the list of open problems at the end of the book Baddeley and Jensen (2005).

Progress is made for the asymptotic distributions of stereological estimators, in particular for case of infinite moments, see Heinrich (2007).

A further aspect concerns numerical methods which are applied to solve inverse problems, in particular of integral equation type. A wide and thorough view to this is given by Ohser and Mücklich (2000).

Image analysis At first glance it may be surprising, but also in the analysis of three-dimensional (3D) images a stereological and statistical problem arises. Such a 3D image is generated by equipment with a finite resolution and thus yields only incomplete information about the set of interest. There are different approaches to model the physical process of generating an image of an object. The information is represented as a lattice of voxels, i.e. 3D 'pixels'. Thus, for the analysis the data can be presented in a 3D point lattice. One approach is that this point lattice is considered as the section set T and the available information is $\Xi \cap T$. For an estimation of the specific volume a simple count of those lattice points that belong to Ξ is sufficient. But for the specific surface area, the Euler–Poincaré characteristic or other intrinsic volumes one has to take into consideration sets of adjacent lattice points. Mostly, a cubic point lattice is assumed. For each 'elementary cube' of this lattice the intersection of the $2 \times 2 \times 2$ vertices with Ξ yields a so-called point configuration. These configurations are used as a base for the estimation of entities of Ξ. Thus the problem is to define weights for the $2^8 = 256$ possible configurations when each point either belongs to Ξ or to Ξ^c respectively. These weights express the local contribution of the configuration to the estimation of the entity of interest. The estimator is then – up to a correction of edge effects – the sum over all the weights of the configurations (on the vertices of 'elementary cubes') that appear in an image of Ξ, i.e. in the intersection of Ξ with the whole cubic lattice.

Several ideas were developed to find weights that yield good estimators. There are papers by Kiderlen and coauthors, also containing references to related work, see Ziegel and Kiderlen (2009). A stereological method was developed by Schladitz, Ohser and Nagel (2006), where the Crofton formula from integral geometry was adapted to point lattices. This way the estimation of intrinsic volumes can be traced back to an estimation of the Euler–Poincaré characteristic on lower-dimensional sections of the point lattice.

Acknowledgments I thank Viktor Beneš, Lothar Heinrich, Claudia Lautensack, Lutz Muche, Joachim Ohser, Matthias Reitzner, Rolf Schneider, Dietrich Stoyan, Wolfgang Weil for their helpful hints. I am most grateful to the anonymous referee for a very careful reading and a long list of helpful hints and corrections of several errors.

References

Baddeley, A. and Jensen, E.B. Vedel (2005). *Stereology for Statisticians.* Chapman & Hall.

Beneš, V. and Rataj, J. (2004). *Stochastic Geometry: Selected Topics.* Kluwer Academic Publishers.

Cruz-Orive, L.M. (1976). Particle size-shape distributions: the general spheroid problem. I. *J. Microsc.*, **107**, 235–253.

Cruz-Orive, L.M. (1978). Particle size-shape distributions: the general spheroid problem. II. *J. Microsc.*, **112**, 153–167.

Drees, H. and Reiss, R.D. (1992). Tail behaviour in Wicksell's corpuscle problem. In *Probability Theory and Applications* (ed. J. Galambos and J. Kátai), pp. 205–220. Kluwer, Dordrecht.

Goodey, P. and Weil, W. (2006). Average section functions for star-shaped sets. *Ad. Appl. Math.*, **36**, 70–84.

Heinrich, L. (2007). Limit distributions of some stereological estimators in Wicksell's corpuscle problem. *Image Analysis & Stereology*, **26**, 63–71.

Hlubinka, D. (2006). Size and shape factor extremes of spheroids. *Image Analysis & Stereology*, **25**, 145–154.

Hug, D. and Schneider, R. (2002). Kinematic and Crofton formulae of integral geometry: recent variants and extensions. In *Homenatge al professor Lluís Santaló i Sors* (ed. C. Barceló i Vidal), pp. 51–80. Universitat de Girona, Girona.

Jensen, E.B. Vedel (1998). *Local Stereology.* World Scientific, Singapore.

Kalmykov, A.E. and Shepilov, M.P. (2000). Analytical solution to the equation for pair–correlation function of particles formed in the course of phase separation in a glass. *Glass Physics and Chemistry*, **26**, 143–147.

Kiderlen, M. (2008). Estimation of the mean normal measure from flat sections. *Adv. Appl. Prob.*, **40**, 31–48.

Kötzer, S. and Molchanov, I. (2006). On the domain of attraction for the lower tail in Wicksell's corpuscle problem. In *Proceedings S4G* (ed. R. Lechnerová, I. Saxl, and V. Beneš), Prague, pp. 91–96. Union of Czech Mathematicians and Physisists.

Matheron, G. (1975). *Random Sets and Integral Geometry.* John Wiley & Sons, New York, London.

Molchanov, I. (2005). *Theory of Random Sets.* Springer, London.

Muche, L. (2005). The Poisson–Voronoi tessellation: Relationships for edges. *Adv. Appl. Prob. (SGSA)*, **37**, 279–296.

Nguyen, X.X. and Zessin, H. (1979). Ergodic theorems for spatial processes. *Z. Wahrscheinlichkeitsth. verw. Geb.*, **48**, 133–158.

Ohser, J. and Mücklich, F. (2000). *Statistical Analysis of Microstructures in Materials Science.* John Wiley & Sons, Chichester.

Okabe, A., Boots, B., Sugihara, K., and Chiu, S.N. (2000). *Spatial Tessellations. Concepts and Applications of Voronoi Diagrams* (2nd edn). John Wiley & Sons, Chichester.

Schladitz, K., Ohser, J., and Nagel, W. (2006). Measurement of intrinsic volumes of sets observed on lattices. In *13th International Conference on Discrete Geometry for Computer Imagery* (ed. A. Kuba, L. G. Nyul, and K. Palagyi), LNCS, Berlin, Heidelberg, New York, pp. 247–258. DGCI, Szeged: Springer.

Schneider, R. and Schuster, R. (2006). Particle orientation from section stereology. *Rendiconti del Circolo Matematico di Palermo*, **Suppl. 77**, 623–633.

Schneider, R. and Weil, W. (2008). *Stochastic and Integral Geometry.* Springer, Berlin Heidelberg.

Stoyan, D., Kendall, W.S., and Mecke, J. (1995). *Stochastic and Integral Geometry* (2nd edn). John Wiley & Sons, Chichester.

Torquato, S. (2002). *Random Heterogeneous Materials. Microstructure and Macroscopic Properties.* Springer, New York.

Ziegel, J. and Kiderlen, M. (2009). Estimation of surface area and surface area measure of three-dimensional sets from digitizations. *Image and Vision Computing.* `http://dx.doi.org/10.1016/j.imavis.2009.04.013`

15

PHYSICS OF SPATIALLY STRUCTURED MATERIALS

Klaus Mecke

Abstract

Modern imaging techniques with nanometre resolution open up the possibility to study the relationship between physical properties and geometric features of spatially structured materials.

15.1 Introduction

This chapter gives a brief overview of the current imaging and image analysis tools, and demonstrates how integral geometric methods connect spatial characteristics to physical properties. The aim of Section 15.2 is to describe the main experimental methods. X-ray computed tomography provides a three-dimensional porous network of nut shells with micrometer resolution. Fluorescence confocal microscopy allows us to study living cells or fluctuating liquid interfaces. Scanning force microscopy maps the structure of nanometer thin films and X-ray scattering techniques resolve the atomic structure of metal alloys, see Fig. 15.1. Image analysis and segmentation algorithms are crucial geometric tools for a quantitative analysis of noisy experimental data.

Section 15.3 introduces stochastic geometric models which can be applied to characterize the experimentally accessible information of material structures. Special emphasise is given on Gaussian random fields and germ-grain models. Stochastic geometry provides expectation values and variances for valuations of random sets, especially for intrinsic volumes. They can be used to quantify dewetting processes in thin films and to reconstruct porous materials.

In Section 15.4 stochastic geometric ideas are applied to derive structure–property relations of materials. As an example, it is shown that thermodynamic properties are not only *extensive* but *additive* quantities. This implies that the dependence, for instance, of Gibbs free energies and surface tensions on the size and shape of confining boundaries of the thermodynamic system is fully characterized by the contained intrinsic volumes. Also transport properties of fluids in porous media can be related to intrinsic volumes of the pore space.

15.2 Imaging of spatially structured materials: measuring stochastic geometries

Nature exhibits a breathtaking variety of different shapes and functions of materials (Hyde, Larsson, Blum, Landh, Lidin, Ninham and Andersson, 1996). Spatially

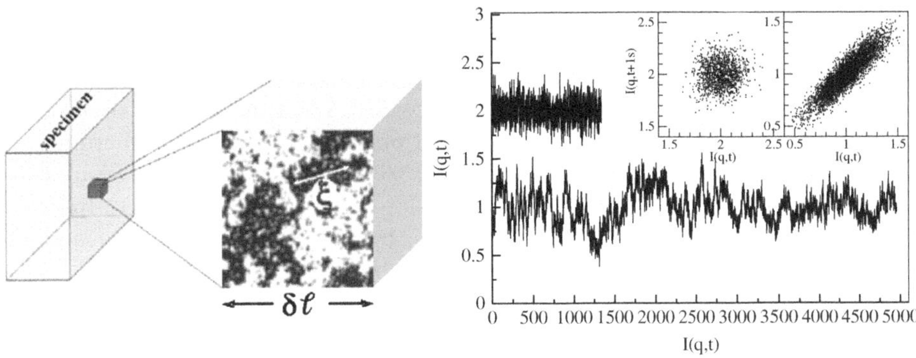

FIG. 15.1. Principle of a diffraction microscope: A microbeam illuminates a small volume sample V in the crystal Fe_3Al producing a fluctuating diffraction intensity $I(q,t)$ which reflects the spatial geometry of ordered domains. Close at a phase transition temperature T_c the typical size ξ of spatial domains becomes macroscopically large. An example of statistical analysis methods of the time-resolved diffraction data are lag plots, i.e. $I(t+\Delta t)$ versus $I(t)$ with $\Delta t = 1s$ which are shown for two selected temperatures $T \gg T_c$ and $T \approx T_c$, respectively. Absence of temporal correlations in an isotropic distribution of data points associated with $T > T_c$, while the ellipse-shaped distribution associated with data $T \approx T_c$ is characteristic for strong temporal correlations in the measured intensity fluctuations.

complex disordered matter such as foams, gels or polymer phases are of increasing technological importance due to their shape-dependent material properties. But the shape of disordered structures is a remarkably incoherent concept and cannot be captured by correlation functions alone which were almost a synonym for structural analysis in statistical physics since the very first X-ray scattering experiments. However, in the last 20 years numerous methods such as scanning probe microscopy and computed tomography have been developed which allow quantitative measurements of the shape of complex structures directly in real space. Integral geometry furnishes a suitable family of morphological descriptors, known as *intrinsic volumes* (Hadwiger, 1957; Santalò, 1976), which are related to curvature integrals and do not only characterize connectivity (topology) but also size and shape of disordered structures. Furthermore, intrinsic volumes are related to the spectrum of the Laplace operator, so that structure–property relations can be derived for complex materials. Percolation thresholds and fluid flow in porous media, for instance, can be predicted by measuring the intrinsic volumes of the pore space alone. Also, evidence was found in hard sphere fluids that the shape dependence of thermodynamic potentials in finite systems can be expressed solely in terms of intrinsic volumes. Finally, a density functional theory can be constructed on the basis of intrinsic volumes which allows an accurate calculation of correlation functions and phase behaviour of mesoscopic complex fluids such as microemulsions and colloids.

In the following, the imaging techniques are illustrated with one example each in order of decreasing resolution: from X-ray scattering at nanometre-sized order fluctuations in crystals (Section 15.2.1), via optical microscopy of micrometre-sized interface fluctuations (Section 15.2.2) towards computed tomography of millimetre-sized spatial structures in biomaterials (Section 15.2.4). In each step standard techniques are described to characterize the structure and to relate the measured signal to material properties. In Section 15.3 a new technique is introduced for the reconstruction of heterogeneous materials which utilizes intrinsic volumes and Boolean models from stochastic geometry. In Section 15.4 the additivity of these valuations and Hadwiger's characterization theorem is then the basis for a fundamental relationship for spatial structure of materials and their thermodynamic properties. It is shown that a thermodynamic potential of a fluid bounded by an arbitrarily shaped convex container can be calculated fully from the knowledge of the intrinsic volumes of the container.

15.2.1 *X-ray scattering: measuring correlation functions*

A prime objective of condensed matter research is to understand and predict on a microscopic level, how a material interacts with external fields, such as temperature, pressure or electric fields. More precisely, how does a macroscopic material property, for instance, the magnetization $m(h)$ depend on an external magnetic field $h(r)$ which itself may be inhomogeneous and depend on the position r in the sample. Usually, a heterogeneous material is modelled by a continuous real-valued function $\rho(r) \in \mathbb{R}$ which assigns a fluctuating density value of a physical quantity to each point r in space. Statistical Physics tells us that the macroscopic response functions, so-called 'susceptibilities' $\chi(r) = \frac{\delta m}{\delta h}$, of a given many-body system are intimately related to the set of 'structure fluctuations' $\rho(r)$ which are exploring all possible configurations of the material. More rigorously, the average static susceptibility

$$\chi = \lim_{q \to 0} S(q)$$

is related to the long-wavelength limit of the power spectrum

$$S(q) = \mathbb{E}\left[\tilde{\rho}(q)\tilde{\rho}^*(q)\right] = \frac{1}{V}\int\int \mathbb{E}\left[\rho(r)\rho(r')\right] \mathrm{e}^{-iq\cdot(r-r')}\mathrm{d}r\mathrm{d}r' \ .$$

with the Fourier transformed density

$$\tilde{\rho}(q) = \frac{1}{\sqrt{V}}\int \mathrm{e}^{-iq\cdot r}\hat{\rho}(r)\mathrm{d}^3r$$

inside a domain (observation window) of volume V. Thus, $S(q)$ is the Fourier transform of the well-known two-point correlation function $g_2(r, r') = \mathbb{E}\rho(r)\rho(r')$

of the physical quantity ρ, see Chapter 1 for a detailed discussion. For instance, the correlations of the particle density ρ determine the compressibility χ of the material.

Starting with the successful diffraction of X-rays at crystals in 1911 by Max von Laue, Paul Knipping and Walter Friedrich, scattering experiments became the most prominent tool to measure the spectrum and determine material properties on a microscopic basis. Scattering is a general physical process where light, sound or moving particles, for instance, interact with the material through which they pass. From the deviation of their radiation from a straight trajectory due to non-uniformities in the medium, one can extract useful spatial information on non-uniformities that can cause scattering, for example, of atoms in solids, bubbles or density fluctuations in fluids, defects in crystalline materials, roughness of surfaces, or biological cells in organisms; for a review on X-ray diffraction see Als-Nielsen and McMorrow (2001) and Warren (1990).

Although scattering with neutrons and electrons are important techniques, in the following we concentrate one X-ray scattering as an example to illustrate the obtained structure information and the relevance of stochastic noise. For elastic scattering without energy loss in the medium and for weak interactions such as light in typical condensed matter one may apply the so-called Born approximation to obtain the scattered intensity of the radiation

$$I(q) = I_0 \left| \int \rho(r) \mathrm{e}^{-iq \cdot r} \mathrm{d}r \right|$$

where the prefactor I_0 depends on interaction parameters of the radiation with the material.

Unfortunately, in conventional X-ray or neutron scattering experiments, the observed intensity $I(q)$ is always an average over large time intervals and not a realistic snapshot. More precisely, during the 'slow' data acquisition the scattered intensity $I(q)$ is averaged over the fast microscopic motions of the materials leading to a static two-point correlation function $\mathbb{E}\rho(r)\rho(r')$ which does not depend on time anymore. Thus, scattering techniques usually provide 'only' second order statistics of material fluctuations. This information is exactly what is needed if one is interested in determining static material properties such as the static averaged susceptibility χ. By now, the entire body of X-ray and neutron scattering experiments has led to a rather comprehensive and consistent picture of the static material properties of crystalline matter, both in the bulk as well as at the surface of matter.

However, local fluctuations are dynamic in nature, i.e. the density $\rho(r, t)$ is a function of space and time. Thus, the microscopic understanding of stochastic fluctuations is fundamentally incomplete as long as one does not have experimental access to the temporal behaviour of such spatial fluctuations on a microscopic scale.

In recent years, a new X-ray diffraction scheme has been developed which exploits the fully coherent part of the X-ray beam provided by modern synchrotron radiation sources. In this scheme the sample is illuminated by a coherent x-ray beam and produces a 'speckle' diffraction pattern which is uniquely related to the exact spatial arrangement of the disorder which fluctuates in time. Thus, through a proper time resolution, this speckle pattern carries the information on the dynamic part of the fluctuations. At the moment the coherent flux of current synchrotron radiation sources are still limited, but the X-ray free-electron laser XFEL, which is currently under construction at DESY (Hamburg), will open up new experimental possibilities in materials research that are inaccessible today. The temporal and spatial resolution of such future X-ray lasers will be several orders of magnitude higher than the existing ones, so that one can investigate chemical reactions on a femtosecond time scale. It will be possible to record 'movies' of molecular motions in liquids and to reveal the dynamic structure of macromolecules such as proteins which are important for a biological understanding of living systems.

A first glimpse of such dynamical fluctuations and the statistical analysis which is necessary can be obtained already today. By focusing a highly brilliant synchrotron X-ray beam to a micrometer spot on a crystalline material, it has been possible to measure in real time the intensity fluctuations $I(q,t)$ associated with order fluctuations in crystalline structure on a microscopic scale. If the spot size δl is reduced to a microscopic small size, than only a few fluctuations are tested by the X-ray beam and the diffraction experiment does no longer perform an average, but rather exhibits a fluctuating diffraction intensity $I(q,t)$ which directly reflects the dynamic behaviour of thermal fluctuation.

Materials which undergo continuous phase transformations such as liquid-vapour or ordered-disordered are of particular interest. Close to the phase transition the fluctuations grow without bound leading to a two-point correlation function which is approximately given by the Ornstein–Zernicke form

$$g_2(r,r') \sim \frac{1}{|r-r'|^{1+\eta}} e^{-\frac{|r-r'|}{\xi}} \quad .$$

The system-inherent correlation length $\xi = \xi_0 |\frac{T}{T_c} - 1|^{-\nu}$ diverges as the associated critical temperature T_c is approached with critical exponents $\nu \approx 0.67$ and $\eta \approx 0.036$. The resulting unlimited growth of the range ξ of fluctuations leads to a 'quasi-long-ranged' power law behaviour, and by this, to universal macroscopic response functions which depend only on the dimension d and the symmetry of the system and which render the system ultra-sensitive to external fields near the critical temperature.

If $\xi \approx \delta l$ is large enough and comparable with the micrometer spot size δl of the X-ray beam dynamical fluctuations of the density $\rho(r)$ become visible. A typical set of experimental results is shown in Fig. 15.1 where time-resolved X-ray intensities $I(q,t)$ are shown for various temperatures far away and close to the critical temperature. Interestingly, the intensities exhibit strong fluctuations in time which are most pronounced for $T = T_c$. In order to test the nature of the observed intensity fluctuations a detailed statistical analysis of the data based on the use of time-time correlation functions is necessary. A direct and unambiguous way to test the existence of correlations between the fluctuations as observed in the time-resolved intensity measurements are so-called lag plots, where $I(t + \Delta t)$ is plotted as a function of $I(t)$. For a given time lag Δt strong correlations manifest themselves in a highly anisotropic distribution for $T \approx T_c$, while a random signal as found for $T > T_c$ exhibits an isotropic distribution. For all temperatures the intensities are consistent with a Gaussian stochastic process with no signature of nonlinearity in the time series. Since the measured intensity $I(q, t)$ is already a two-point function of the order parameter $\rho(r, t)$ the observed correlation is governed by the four-point correlation function.

A theoretical understanding of these fourth order statistics is still missing. With the development of highly brilliant micro- and nanobeams at synchrotron radiation and free electron laser facilities such experiments will become more and more standard and other types of fluctuations will become accessible in the time and space – calling for detailed stochastic modelling. The X-ray free-electron laser XFEL will produce spatial speckle pattern which fluctuates in time, so that not a time series analysis as above but a stochastic geometric modelling is needed to gain insight in material properties on a microscopic level. With a X-ray laser even time-resolved holography is possible, which offers the unique opportunity to study even the dynamics of the three-dimensional structure, see Chapman and et al. (2007).

Unfortunately, scattering techniques do not provide spatial information directly, but Fourier transforms of real space structures. In conventional techniques, only the absolute value of the Fourier transform is measured and the phase information get lost. This phase problem makes it difficult to extract real space structure information directly by an inverse transformation. Moreover, as discussed, only two-point statistics can be obtained due to long observation times, so that the shape of fluctuations in materials are not accessible by scattering techniques. To overcome this uncomfortable situation several microscopic techniques in real space have been developed, most prominently scanning probe microscopy and fluorescence confocal microscopy, which will be discussed in the following sections.

15.2.2 *Fluorescence confocal microscopy: spatial information in real space*

With the development of powerful imaging techniques in real space such as scanning force microscopy, confocal microscopy and computed tomography, the analysis of spatial structures becomes more and more important in contemporary material science. These techniques provide the opportunity to experimentally measure the complex morphology of a range of materials in three dimensions at resolutions down to several micrometres and lower.

Confocal microscopy is an optical imaging technique for obtaining high-resolution data of three-dimensional spatial structures; for a review on confocal microscopy see Masters (2006) and Wilson (1990). In a conventional microscope

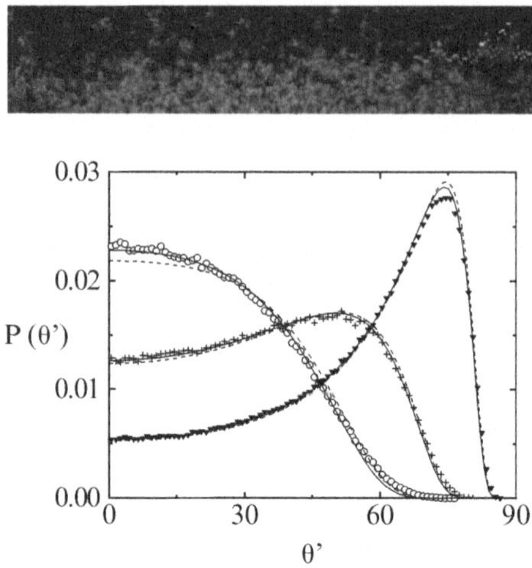

FIG. 15.2. Confocal scanning laser microscopy image (size $17.5 \times 85\mu m^2$) of a phase separating colloid-polymer mixture (Aarts, Schmidt and Lekkerkerker, 2004; Aarts, Schmidt, Lekkerkerker and Mecke, 2005); courtesy Dirk Aarts and Henk Lekkerkerker. Thermally excited capillary waves corrugate the interface and are related to important physical quantities such as surface tension and bending rigidities of fluid interfaces. The bright dots at the right indicate the surface location $\rho(x)$ which allows the determination of local angles θ'. For three different phase states the angle distribution $P(\theta')$ is shown, which is in good agreement with a Gaussian random field given by equation (15.2).

a spatial pinhole is used to eliminate out-of-focus light. Due to this pinhole only one point is illuminated at a time, so that a whole image requires scanning over a regular grid of the material. In laser scanning confocal microscopy the material is made fluorescent and illuminated by a point laser source.

To illustrate the importance of statistical modelling which emerged with this imaging technique, an image of a phase separating colloid-polymer mixture is shown in Fig. 15.2. In such fluid dispersions it is possible to tune length- and time-scales so that the thermal fluctuations of the interface between two phases can be studied directly by optical means (Aarts, Schmidt and Lekkerkerker, 2004) – in contrast to the nanoscopic noise in alloys visible in the X-ray scattering as discussed in the previous section. Also in molecular fluids the interface roughness is of a few nanometres, which is only accessible through scattering techniques. however, going from molecules to mesoscopic colloidal particles of size 140 nm the lengths are scaled up, so that capillary waves, which are an important material property of interfaces, can directly be observe in real space. Adding polymer to a colloidal suspension may induce a fluid-fluid demixing transition that is widely accepted to be the mesoscopic analogue of the liquid-gas phase transition in atomic substances (Aarts, Schmidt and Lekkerkerker, 2004; Aarts, Schmidt, Lekkerkerker and Mecke, 2005). The coexisting phases are a colloidal liquid (rich in fluorescent labelled colloid and poor in polymer) and a colloidal gas (poor in colloid and rich in polymer). The origin of the phase separation lies in the entropy-driven attraction between the colloids, which is mediated by the polymers.

Pictures, such as the one in Fig. 15.2, represent an intensity distribution of fluorescent light, $I(x, z, t)$, at a certain time t with x the horizontal (along the interface) and z the vertical (opposite to gravity) components of the space vector. The microscope records the fluorescence of excited dye within the colloids, hence the colloid-rich (liquid) phase appears bright and the colloid-poor (gas) phase appears dark. $I(x, z, t)$ is a direct measure of the local and instantaneous distribution of colloidal particles and provides the starting point for a statistical analysis. From the fluorescence intensity difference between the two phases the interface can easily be located and the time-dependent height function $\rho(x, t)$ are constructed. By analysing the height fluctuations of the interface position one can see that the interface roughness can be described within a Gaussian random field model . For instance, to describe the thermal noise contributions to the height function the two-point correlation functions

$$g_2(x, x') = \langle \rho(x, t')\ \rho(x', t') \rangle = \frac{k_B T}{2\pi\gamma_0} K_0 \left(\frac{x' - x}{L_{\|}} \right) \tag{15.1}$$

can be calculated and compared with the measured correlations at temperature T. Material properties of interfaces, such as surface tension γ_0, the capillary length $L_{\|} = \sqrt{\gamma_0/(G\Delta\rho)}$ and viscosity can directly be inferred from the images by comparing two-point correlations of these fluctuations; but now in real space and not in Fourier transform as in the scattering techniques described above. Moreover, one can easily probe higher order statistics in these real space images which is not possible by scattering.

The presented technique enables us also to measure the probability distribution of the tilt angle of the local interface normal and the vertical direction. To this end the derivatives $\rho_x = \tan\theta$ of $\rho(x, t)$ are numerically calculated because they stand in direct connection to the (projected) angle θ' normal to the interface. In Fig. 15.2 histograms are plotted of the absolute value of θ' for three different state-points. For a Gaussian random field model it can be shown that the distribution in one dimension is given by Mecke and Dietrich (2005)

$$P(\theta') = \frac{2}{\sqrt{2\pi\sigma'^2}} \frac{e^{-\frac{1}{2}\frac{\tan^2\theta'}{\sigma'^2}}}{\cos^2\theta'} \tag{15.2}$$

with the variance $\sigma'^2 \equiv \langle \tan^2\theta \rangle$ given by

$$\sigma'^2 = \frac{1}{2} \int \frac{k_B T q^2}{\Delta\rho G + \gamma_0 q^2} \frac{d^2 q}{(2\pi)^2} .$$

The particular form of this equation is a result of the Jacobian of the transformation of ρ_x to θ using $\partial\rho_x(\theta)/\partial\theta = 1/\cos^2\theta$. One can either use (15.2) to fit to the data with the variance as fitting parameter or obtain the variance directly from the experiment. In Fig. 15.2 both methods are plotted. The agreement is yet another confirmation that the probability of fluctuations of the local interface position around its equilibrium value can be described by a Gaussian random field model. As the variance σ'^2 is extremely sensitive to the molecular interactions, it is in principle possible to obtain the interfacial tensionγ_0 more accurately than by equation (15.1) and to determine the microscopic length scales directly in real space (Mecke and Dietrich, 2005). It could be interesting if on the particle level a wave-vector dependent surface tension $\gamma(q)$ can be detected due to molecular interactions, as predicted for simple liquids (Mecke and Dietrich, 1999) and observed in various liquids (Fradin, Braslau, Luzet, Smilgies, Alba, Boudet, Mecke and Daillant, 2000; Mora, Daillant, Mecke, Luzet, Braslau, Alba and Struth, 2003). The value of $\gamma(q)$ on nanometre scales is still a matter of intense discussions and any technique to measure $\gamma(q)$ is highly welcome. With the advance of real space techniques one may be able to find more sophisticated relations between $\gamma(q)$ and spatial features of $\rho(x)$ based on stochastic geometry.

Irregular spatial-temporal patterns occur in many systems, including chemical reaction-diffusion systems, hydrodynamic convection, or dewetting processes, for instance. The possibility of quantitative measurements of complex structures directly in real space, calls for a progression in image analysis methods and stochastic geometric modelling. Whilst the human eye can ascertain the similarity in a qualitative manner easily, mathematical descriptors for structures that are not perfectly symmetric are not trivial as we will see in the following section.

15.2.3 *Scanning probe microscopy: two-dimensional images*

In contrast to optical (confocal) microscopy the image of phonograph-like techniques such as scanning tunnelling microscopy or atomic force microscopy (AFM) is obtained by scanning with an atomic tip over a surface (for a review on scanning probe microscopy, see Meyer, Hug and Bennewitz 2005 and Bonnell 2001). Scanning probe microscopy covers several related technologies for imaging and measuring surfaces on a fine scale, down to the level of molecules and groups of atoms. At the other end of the scale, a scan may cover a distance of over 100 micrometres in the x and y directions and 4 micrometres in the z direction. This is an enormous range and it can truly be said that the development of this technology is a major achievement, for it has profound effects on many areas of science and engineering.

These microscopy technologies share the concept of scanning an extremely sharp tip (3–50 nm radius of curvature) across the object surface. The tip is mounted on a flexible cantilever, allowing the tip to follow the surface profile.

When the tip moves in proximity to the investigated object, forces of interaction between the tip and the surface influence the movement of the cantilever. These movements are detected by selective sensors. Various interactions can be studied depending on the mechanics of the probe. The three most common scanning probe techniques are:

(i) Atomic Force Microscopy (AFM) measures the interaction force between the tip and surface. The tip may be dragged across the surface, or may vibrate as it moves. The interaction force will depend on the nature of the sample, the probe tip and the distance between them.
(ii) Scanning Tunnelling Microscopy measures a weak electrical current flowing between tip and sample as they are held a very distance apart.
(iii) Near-Field Scanning Optical Microscopy scans a very small light source very close to the sample and can provide resolution below that of the conventional light microscope. The detection of this light energy forms the image.

Here, the focus is on AFM measurements of dewetting morphologies in thin films. The importance of stochastic geometry is demonstrated by applying Gaussian random field and intrinsic volumes to determine physical properties of thin films.

Dewetting of thin films In the course of miniaturization of electronic and microfluidic devices reliable predictions of the stability of ultra-thin films gain an important role for design purposes. The thickness $\rho(x)$ of, e.g. insulating layers or photo resists for the fabrication of electronic chips now reaches the order of nanometres, resulting in new challenges to guarantee stability during production and use of the device. Given sufficient knowledge of system parameters such as dielectric constants ϵ and the Hamaker constants A of the participating media, the principal question of stability or instability of a given thin film can be answered by an effective interface potential $V(\rho) = \epsilon/\rho^8 - A/(12\pi\rho^2)$ which covers intermolecular forces beyond capillarity. Figure 15.3 displays atomic force microscopy (AFM) scans taken in situ of liquid polystyrene films beading off an oxidised Si wafer with $\epsilon = 6.3 \cdot 10^{-76}$ Jm^6 and $A = 2.2 \cdot 10^{-20}$ J leading to an equilibrium thickness of $\rho = 1.3$ nm.

Theoretically the dynamics of the thin film is described by the Navier–Stokes equation in the lubrication approximation,

$$\eta \frac{\partial \rho}{\partial t} = \nabla \cdot \left\{ \frac{\rho^3}{3} \nabla \left[V'(\rho) - \gamma_0 \nabla^2 \rho \right] \right\} \tag{15.3}$$

with the viscosity η. To solve this equation numerically, one can use a finite element method on a simplicial triangulation of the substrate (Becker *et al.*, 2003). The comparison of the simulations with the AFM experiment is shown in Fig. 15.3. Both are in close qualitative and quantitative agreement. This can already be seen by looking at the time-scales involved. Taking the formation of

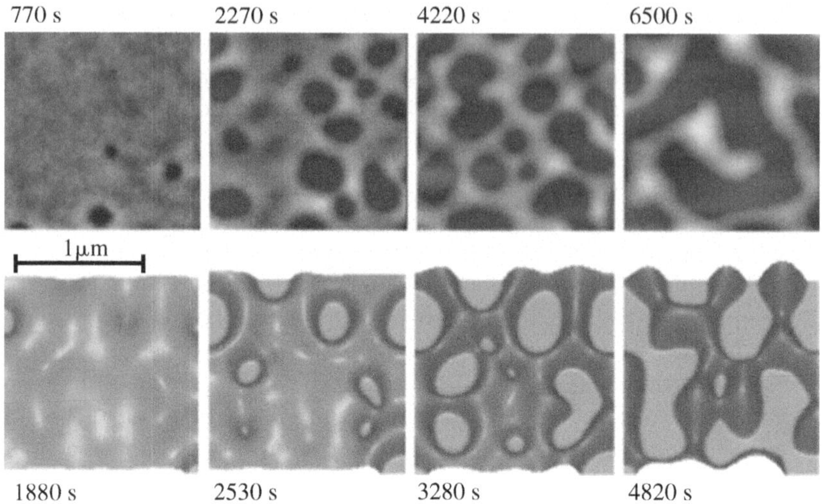

FIG. 15.3. A 3.9 nm polystyrene film beads off an oxidised Si wafer: temporal series of experimental scanning force microscopy of the dewetting process (top) can be simulated by the Navier–Stokes equation in lubrication approximation, equation (15.3), with identical system parameters (Becker *et al.* 2003). The temporal evolution of the dewetting morphology is characterized by intrinsic volumes in Fig. 15.5 and can be modelled quantitatively by a Gaussian random field.

the first hole as the origin of the time axis in simulation and experiment, the absolute time-scale for the appearance of, e.g. connected holes is of the same order of magnitude. While the accordance of the time-scales of experiment and simulation are a first indication of the quantitative agreement, the precise nature of the simulation can be demonstrated by pattern analysis based on intrinsic volumes and a Gaussian random field model which provides accurate comparison tools beyond the visual inspection of the patterns, see Section 15.3.1. However, a morphometric analysis also revealed that thermal noise can strongly influence characteristic time-scales of fluid flow and cause coarsening of typical length-scales. This lead to an extension of the Navier–Stokes equation and equation (15.3) by a stochastic noise term which finally lead to a fully quantitative agreement between experiments and theory (Mecke and Rauscher, 2005; Grün, Mecke and Rauscher, 2006; Fetzer, Rauscher, Seemann, Jacobs and Mecke, 2007).

The three examples shown in Figs. 15.1–15.3 illustrate, that on length-scales smaller than micrometers thermal fluctuations are an important general phenomenon in materials. The spatial structures of these materials are stochastic in nature which calls for appropriate geometric models to characterize the observed disorder.

Scanning probe microscopy is usually restricted to two-dimensional images. In recent years it became possible to use scanning probe techniques also for nanotomography, i.e. for imaging three-dimensional structures (Magerle, 2000). Unfortunately, for this a layer-by-layer removal of the material is needed, so that non-destructive techniques such as computed tomography are more promising three-dimensional imaging methods.

15.2.4 *Computed tomography: three-dimensional images*

Since the discovery of X-rays by Wilhelm Roentgen in 1895 this electromagnetic radiation has been used to study materials, see Roentgen (1995). In the beginning only two-dimensional projections of objects were taken. The images obtained distinguish strong absorbing regions like bones from weakly absorbing regions like muscles. The two-dimensional projection, which is the intensity $I(x, y)$ of the transmitted X-ray beam, obtained from a classic X-ray image can be interpreted as a Radon transform of the materials absorption coefficient $\mu(r)$, see Bronnikov (2002). If enough projections from different angles are taken, this transformation can be inverted to obtain $\mu(r)$, i.e. the full three-dimensional function. Since the most important interaction of photons in a material at typical X-ray energies of 10–100 keV is the Compton scattering from electrons, $\mu(r)$ can be related to the local electron density which is connected to the local density of the material.

Direct measurement of a three-dimensional structure of materials is now available with a pixel resolution of 0.3 µm via the non-destructive method of synchrotron-based X-ray micro-tomography. Based on the use of electron microscopes it recently became even possible to image with molecular resolutions (Frank, 2006; Downing, Sui and Auer, 2007). This nano-tomography of molecules and nanoparticles is becoming increasingly popular in science and it will have a huge impact on biology, chemistry and nanotechnology.

Electron and X-ray computed tomography offers the opportunity to investigate the shape of a spatial structure which determines the environment for all physical, chemical and biological processes on micrometre scales; for a review on computed tomography, see Thorsten (2008), Kalender (2005), Oppelt (2005) and Hsieh (2003). Therefore, visualization and geometric quantification of the 3D architecture is of great importance and should be achieved at different scales from the macro to micro. For instance, the understanding of soil structure is of crucial importance for microbial processes, the sequestration of organic carbon, water storage and preferential flow. This is greatly associated with processes evolving at the larger scale and there is a need to merge observations from different scales for understanding soil functioning. Linking 3-D quantified data to its function is therefore vital for advancing in environmental protection, for instance.

For strongly absorbing materials like metals or rocks, X-ray micro-tomography has been extensively used to study structures qualitatively as well as quantitatively (Kinney and Nichols, 1992; Ashby and et al., 2000; Sakellariou, Sawkins, Senden and Limaye, 2004). However, for weakly absorbing materials, like biological materials, absorption tomography is not suitable. Also materials that are strongly absorbing but which have a very homogeneous absorption coefficient μ can not be imaged using absorption tomography. For the analysis of these classes of material, phase contrast tomography has recently opened a opportunity to study them (Wilkins, Gureyev, Gao, Pogany and Stevenson, 1996; Cloetens, Barrett, Baruchel and Guigay, 1996; Cloetens, Ludwig, Pateyron-Salome, Buffiere, Peix, Baruchel, Peyrin and Schlenker, 1997; Cloetens *et al.*, 1999; Cloetens, Ludwig, Boller, Peyrin, Schlenker and Baruchel, 2002). Imaging the three-dimensional structure of foams (Lambert, Cantat, Delannay, Renault, Graner, Glazier, Veretennikov and Cloetens, 2005; Hutzler and Weaire, 2001) or bones (Jones, Sheppard, Sok, Arns, Limaye, Averdunk, Brandwood, Sakellariou, Senden, Milthorpe and Knackstedt, 2004), for instance, is the first necessary step to understand stability and function of these important biomaterials.

Here, two examples are given at the microscale, the topology and connectivity of the cellular pore network in nutshells (see Fig. 15.4) and in sandstones (see Fig. 15.6 below). It is a great opportunity for stochastic geometry in making progress of image analysis techniques and advances in software specifically for 3-D analysis that may provide solutions to present difficulties with data processing. Image analysis of stochastic data is becoming increasingly demanding for computing and programming resources.

Pore space in sandstone by absorption contrast tomography Fontainebleau sandstone is very homogeneous and considered a ‘benchmark’ of a homogeneous rock in the petroleum industry. However, a plot of the porosity variations shows considerable variability on the local grain scale. Four 4.52 mm diameter cylindrical core samples of the Fountainebleau sandstone have been used for X-ray computed microtomography with a resolution of 5.68 μm, see Fig. 15.6. X-ray computed tomographic images of porous media are grey scale images, usually with a bimodal population apparent, one mode corresponding to the signal from the void space (pores), the second to the signal from the grain space. Because this absorption contrast between pores and grains, i.e. because of their large differences in X-ray attenuation, a simple thresholding leads to well segmented data. From each of the binarized cylindrical plugs of the Fountainebleau sandstone with bulk porosities $\phi = 7.5\%$, 13%, 15%, and 22% a centred 480^3 cubic subset was extracted for analysis corresponding to a volume of $20.3\,\mathrm{mm}^3$ (see Fig. 15.6). The samples show the variability of shapes and structures typical for sandstones. In Section 15.3.2 intrinsic volumes of the pore space are used to reconstruct these samples by a Boolean model. The basic physical properties of such porous and cellular materials are elasticity and transport which are related in Section 15.4 to stochastic and geometric features of the observed structure.

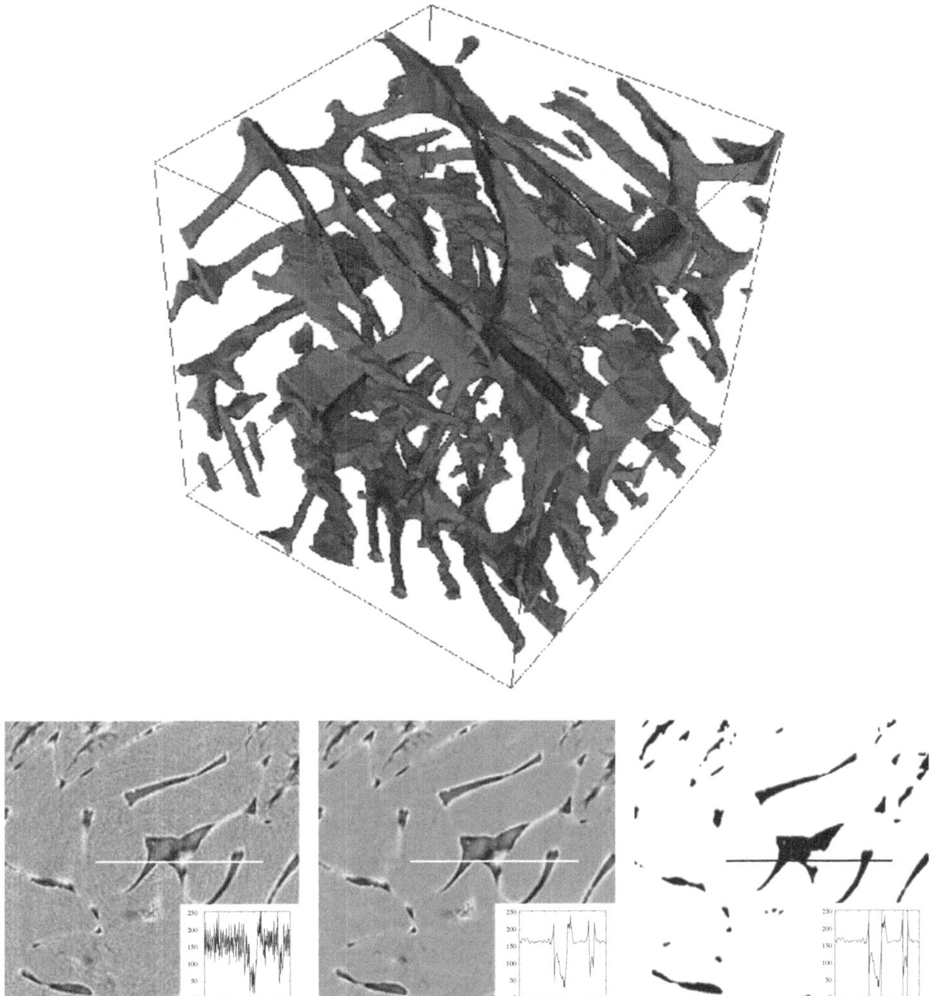

(a) (b) (c)

FIG. 15.4. Pore space inside the shell of a coconut measured by high resolution X-ray tomography at the ESRF (Grenoble, beamline ID19; images from Breidenbach (2007), Breidenbach, Sheppard, Wegst, Cloetens and Mecke (2009)). The pore space is highly connected (negative Euler number V_0) and seems to have a preferred orientation. Shown is also a slice through a 500^3 subset (a) with a line-scan in the inset. It is smoothed using anisotropic diffusion (b). The noise is clearly reduced, while pore features are kept. Finally the smoothed image is segmented by a region growing algorithm (c). Vertical lines in the line-scan indicate the position of the pore boundaries after segmentation.

Pore space in nutshells by phase contrast tomography Since man-made and mineralized materials such as sandstones have either high attenuation coefficients or great differences in the X-ray attenuation coefficients of their components, their structure is easily accessible through an investigation by X-ray tomography with absorption contrast. Unfortunately, the same cannot be said about polymeric and non-mineralized biological materials. They absorb X-rays only weakly and the attenuation coefficients of their components are very similar. Nowadays this problem can be overcome, however, when phase contrast is used for imaging. Phase contrast is available at third generation synchrotron radiation sources such as the ESRF in Grenoble, France, due to the spatial coherence, homogeneity and brilliance of its X-ray beam. It is a technique based on Fresnel diffraction, which leads to phase jumps at the interface between two phases or components. The interface corresponds to singularities in the refractive index and appears as a white-black contrast line in the reconstructed images (Cloetens, Ludwig, Pateyron-Salome, Buffiere, Peix, Baruchel, Peyrin and Schlenker, 1997). In phase contrast images, the intensity $\rho(r)$ is a non-local function of the material density. This effect makes it easier to distinguish pores and matrix in soft materials, i.e. materials with low absorption coefficients. The disadvantage of phase contrast datasets is that the standard segmentation procedures available in commercial tools cannot be applied, as the grey values of a porous material, for example, are the same within the material and the pore, see line scans in Fig. 15.4(a)–(c). While the eye can distinguish one from the other due to the Fresnel rings, the software cannot. It is, therefore, necessary to develop new algorithms and tools to overcome this problem and to make automatic segmentation and morphometric analysis possible also for phase contrast datasets.

Image analysis and segmentation Tomographic images of two-phase materials are grey-scale images, usually with a bimodal population. Quantitative investigation of the geometry of the phase space requires a voxel-by-voxel determination of the phase type, a process known as *segmentation*. This transformation of the grey level data into a black-and-white image, i.e. the discrimination of the two phases, material and pore space, can be numerically quite difficult. A simple thresholding based on matching a predetermined bulk measurement (phase fraction) is often used to segment a tomographic image. In practice this may not be a reasonable method due to the peak overlap in the intensity histogram. For these reasons, to segment the original tomographic sophisticated algorithms has been developed such as an edge-finding (kriging-based) algorithm or a watershed region-growing algorithm based on a converging active contours algorithm (Malladi, Sethian and Vemuri, 1995; Caselles, Kimmel and Sapiro, 1997). The Mango image analysis framework of the Applied Mathematics Department of the Australian National University provides this implementation (Sheppard and Knackstedt, 2006).

The images are also corrected for noise: before segmentation by filtering and after segmentation by re-identifying the phase type of all isolated grain and void voxel clusters. These isolated voxels have a strong effect on many of the morphological measures. For adsorption contrast images simple Gauss filtering may be appropriate to remove noise and artefacts. For phase contrast images instead edge-preserving anisotropic diffusion filtering is more efficient but also numerically more involved (Frangakis and Hegerl, 2001). A typical example of the different steps in image analysis is shown in Fig. 15.4, which illustrates the smoothing and segmentation process for a slice through the coconut data set. After the segmentation the structures can be analysed by methods from stochastic geometry, for examples by intrinsic volumes and correlation functions. Alternative approaches of image analysis to extract structure from random geometries are described in Bilodeau, Meyer and Schmitt (2005), Rosenfeld (1976), Serra (1992) and Stoyan, Kendall and Mecke (1995).

Due to the large size of such high-resolution tomographic datasets, their analysis is a demanding task. Typical image sizes of up to 2048 voxels in every space direction pose large demands on software and computers. In the last 10 years Adrian Sheppard (ANU, Canberra) has developed and perfected the software Mango for the segmentation and quantitative analysis of tomographic datasets. Mango is the only code to date that can perform image analysis on 8 GB data sets. This is possible because, in contrast to commercial packages, it is written to run on a parallel computer and thus is not limited by available RAM, but by available computing time. An additional advantage of this tool is that the code is not working as a black box, but readily controllable and extensible. Figure 15.4 gives an impressive example of the capability of the code. The output of this procedure is a binary image of the material in which the pore space voxels carry the value 0 and the matrix voxels carry the value 1. This dataset can now be used to analyse the structure of the material (Section 15.3) and serve as input for property calculations such as conductivity calculations or finite element calculations of elasticities (Section 15.4).

15.3 Characterizing and modelling: utilizing intrinsic volumes for image analysis

Many statistical methods have been developed for the analysis of two- and three-dimensional patterns, their power and potential has been explored, and there are many successful applications, in particular in biometry, physics and medicine (for reviews see Bilodeau, Meyer and Schmitt 2005; Serra 1992; Stoyan, Kendall and Mecke 1995; Matheron 1967; Matheron 1975; Cressie 1993; Lohmann 1998; Ohser and Mücklich 2000; Torquato 2002; Mecke and Stoyan 2000; Mecke and Stoyan 2002). Many scientists consider second-order characteristics such as the pair correlation function as particularly powerful and prefer to use variances to measure roughness of thin films, for instance. In the previous Section 15.2.1 scattering techniques have been shown to be method to measure second-order characteristics via the spectrum $S(q)$ of the scattered intensity. Whereas for regular patterns these measures provide useful information about characteristic length scales and orientational order, they are incapable of distinguishing irregular structures of different topology. It can be shown that commonly used analysis methods like pair correlation functions and Fourier transforms are inapplicable for describing spatial features in quite a number of complex systems (Mecke, 2000; Arns, Knackstedt and Mecke, 2003). They can only measure a distribution of distances, other morphological peculiarities remain undetected. Hence, extended topological measures have to be considered in these cases. Statistical methods that are sensitive to the morphology or shape of structures (curvatures and connectivity) have been investigated extensively in other fields such as image analysis and pattern recognition (Lohmann, 1998; Rosenfeld, 1976; Serra, 1992; Ohser and Mücklich, 2000). But also in the understanding of condensed matter it became evident that curvature of the spatial structures play an important role (Hyde, Larsson, Blum, Landh, Lidin, Ninham and Andersson, 1996). Nowadays, the measurement of interfacial curvatures from three-dimensional digital images (Mecke, 1994; Nishikawa, Jinnai, Koga, Hashimoto and Hyde, 1998; Arns, Knackstedt, Pinczewski and Mecke, 2001; Arns, Knackstedt and Mecke, 2004; Parra-Denis, Moulin and Jeulin, 2007) is a standard procedure in physics, chemistry and biology to characterize materials, for example, polymer blends (Jinnai, Koga, Nishikawa, Hashimoto and Hyde, 1997; Rehse, Mecke and Magerle, 2008), porous sandstones (Arns, Knackstedt, Pinczewski and Mecke, 2001), zeolites (Blum, Hyde and Ninham, 1993), complex fluids (Likos, Mecke and Wagner, 1995; Roth, Harano and Kinoshita, 2006; Hansen-Goos, Roth, Mecke and Dietrich, 2007), metallic particles (Parra-Denis, Moulin and Jeulin, 2007) or biomaterials such as nutshells (Breidenbach, 2007).

With the advance of modern imaging techniques in physics as described above, it became possible to go beyond second-order characteristics by applying modern image analysis technique for digitized grey-scale images. Here, the focus is on *intrinsic volumes* $V_\nu(A)$ introduced in Chapter 1 to characterize both, morphology and shape of domains A in two-dimensional Euclidean space. Such domains are obtained from greyscale images by thresholding, so that the functionals are actually functions $V_\nu(\rho)$ of the grey-scale ρ. To illustrate the technique the surface topologies of a dewetting film are studied, which are shown in Fig. 15.3. Differences between experiments and simulations can be detected and modelled by stochastic geometry which finally lead to a correction of the hydrodynamic equations by stochastic noise terms. Thus, stochastic geometric modelling is important to characterize physical processes and infer material properties. Various geometric processes for random sets are described in Chapter 1. In material science, the excursion sets of Gaussian random fields and the Boolean model became most popular (for reviews see Bilodeau, Meyer and Schmitt 2005; Matheron 1967; Matheron 1975; Molchanov 1997; Jeulin 1997; Mecke and Stoyan 2000; Mecke and Stoyan 2002; Torquato 2002). Applications of various Boolean models in physics are reviewed by Mecke (2000). In particular for the reconstruction of porous materials the Boolean model is successful as shown in Section 15.3.2, which finally allows excellent predictions of the shape dependence of thermodynamic quantities (see Section 15.4.1 and König, Roth and Mecke 2004) and transport properties (see Section 15.4.3 and Arns, Knackstedt and Mecke 2003) in porous media. To demonstrate the importance of Gaussian random field the physical properties of thin films are studied in the following section by means of integral geometry, see Hadwiger (1957) and Santalò (1976).

15.3.1 *Characterizing thin films by Gaussian random field*

In Fig. 15.3 the film surface during dewetting is seen to develop a correlated pattern of indentations and finally ruptures. Numerical solutions of equation (15.3) and experimental data are in close qualitative and quantitative agreement which can, e.g. already be seen by looking at the time-scales involved. However, the quantitative agreement of the simulations has supported by quantitative pattern analysis which provides accurate comparison tools beyond the visual inspection of the patterns (Mecke, 2000). This is of particular relevance as the morphology of the rupturing film has two contributions of different character: while the holes are easy to identify by eye, a description of the film in between the holes requires a closer inspection. Becker *et al.* (2003) used intrinsic volumes V_ν for a quantitative analysis of the evolution of dewetting patterns in both experiment and simulation. It was shown that the complex spatial and temporal evolution of the rupture of ultra thin films can be modelled by a Gaussian random field in quantitative agreement with both experiments and simulations. A Gaussian random field is an excellent stochastic model for many geometric features observed in physical data, for the cosmic background radiation field, for capillary waves, or for spinodal dewetting in the linear dynamic regime (Mecke, 2000; Mecke and Stoyan, 2002). The quantitative agreement with a stochastic model is essential for the development of efficient tools capable to manage thin film flow in technical systems.

The grey-scale image may be parameterized by a single-valued density field $\hat{\rho}(x)$ of the spatial coordinates $x \in \Omega \subset \mathbb{R}^2$ inside a domain (observation window) Ω of area V. As already mentioned in Sections 15.2.1 and 15.2.2 usually the mean density $\bar{\rho} = \mathbb{E}\rho(r)$ and the spectrum $\mathbb{E}\tilde{\rho}(q)\tilde{\rho}(q') = (2\pi)^2\delta(q+q')S(q)$ is used to characterize a grey-scale image, where $S(q)$ is the Fourier transformation of the two-point correlation function $g_2(r,r') = \mathbb{E}\hat{\rho}(r)\hat{\rho}(r')$. Of particular interest are integrals of the spectrum, i.e. the variance of the grey-values and of their derivatives,

$$\sigma^2 = \mathbb{E}\hat{\rho}(r)^2 = g(r=0) = \int_{\mathbb{R}^2} S(q)\frac{\mathrm{d}q}{(2\pi)^2}$$

$$k^2 = \frac{\mathbb{E}(\nabla\hat{\rho}(r))^2}{2\pi\mathbb{E}\hat{\rho}(r)^2} = -\frac{g''(r=0)}{2\pi\sigma^2} = \frac{1}{2\pi}\frac{\int q^2 S(q)\mathrm{d}q}{\int S(q)\mathrm{d}q}$$

which are the most convenient and prominent measures to characterize the distribution of the grey-scales $\hat{\rho}(r)$ in an image. For instance in Section 15.2.2 the slope variance $\sigma'^2 = 2\pi\sigma^2 k^2$ was used to determine the surface tension γ_0 of a thermally fluctuating interface – assuming a Gaussian random field. A main problem in using the spectrum $S(q)$ to characterise a grey-scale image is its strong dependence on boundaries, i.e. on the finite size of the observation windows, and on stochastic noise which can be dominant in small images usually measured by AFM or optical microscopy. In addition, the data analysis of the dewetting process shown in Fig. 15.3 is hindered by the emergence of growing holes in the thin film which limits the film regions where linear dynamics can be studied. The variances σ^2 and k^2 cannot be determined accurately from the Fourier transform of the AFM images as moments of the spectrum. The measured spectrum is convoluted with the Fourier transform of the observation window, which decays as $\sim q^{-2}$ for a sharp cutoff. Thus, the short-wavelength behaviour of the spectrum is for finite images dominated by the edges of the AFM image and not by the thermal fluctuations. Moreover, the exponentially damped spectrum $S(q,t) \sim \mathrm{e}^{-2t|\omega(q)|}$, characteristic for the deterministic dynamics given by equation (15.3), is masked by this convolution yielding a spectrum similar to the noise-induced algebraic decay $S_{CW}(q) \sim q^{-2}$ (Mecke and Dietrich, 1999, 2005). Since the 'bare' spectrum $S(q)$ is experimentally not accessible, one has to determine variances σ^2 and k^2 from the real space images directly – in contrast to the Fourier space employed by scattering techniques discussed in Section 15.2.1. Whereas determining σ^2 is straightforward, calculating k^2 from pixelized images using numerical differentiation is often too inaccurate – in contrast to Section 15.2.2 where $k^2 = \sigma'^2/(2\pi\sigma^2)$ could be determined directly.

However, one can apply an analysis technique based on Minkowski functionals, i.e. intrinsic volumes V_ν (Becker, Grün, Seemann, Mantz, Jacobs, Mecke and Blossey, 2003). The images are threshold at a certain film height ρ and measure the area $V_2(\rho)$, boundary length $V_1(\rho)$, and Euler number $V_0(\rho)$ as function of ρ. These functions are sensitive to the geometry of the film surface and measure spatial features which are not visible to the eye. The excursion set D_ρ over a given threshold ρ is defined as the binary image $D_\rho = \Theta(\hat{\rho}(x) - \rho)$ with the Heaviside step function $\Theta(x)$. Then the averaged Minkowski functions can be defined by

$$v_\nu(\rho) = \frac{1}{V} V_\nu(D_\rho) \; , \tag{15.4}$$

which depend on the threshold ρ. Although the functional dependence of the intrinsic volumes on the threshold ρ can be quite complex, one can find explicit expressions for standard models of stochastic geometries, which depend only on a finite number of model parameters, see contributions in this volume and Mecke (1998*a*, 2000). Comparing analytic results with experimental data one may numerically determine these parameters which finally can be used for statistical tests.

For a Gaussian random field model the mean intrinsic volumes of the excursion set are given by

$$v_\nu^{(GF)}(\rho) = k^{d-\nu} \frac{\omega_d}{\omega_{d-\nu}\omega_\nu} H_{d-\nu-1}\left(\frac{\rho - \bar{\rho}}{\sigma}\right) \tag{15.5}$$

with the Hermite functions

$$H_\nu(x) = \frac{1}{\sqrt{2\pi}} \left(-\frac{\partial}{\partial x}\right)^\nu \mathrm{e}^{-\frac{1}{2}x^2}$$

and the volume ω_d of a d-dimensional unit ball. Thus, no matter what the two-point correlation function $g_2(r, r')$ looks like, the mean values of the intrinsic volumes are given by only three parameters $\bar{\rho}$, σ and k, e.g. in two dimensions

$$\begin{aligned} v_2^{(GF)}(\rho) &= \tfrac{1}{2} - \tfrac{1}{\sqrt{2\pi}\,\sigma} \int_0^{\rho-\bar{\rho}} \mathrm{e}^{-\frac{1}{2}\frac{t^2}{\sigma^2}} \mathrm{d}t \;= \tfrac{1}{2}\left(1 - \mathrm{erf}\left(\tfrac{\rho-\bar{\rho}}{\sqrt{2}\sigma}\right)\right) \\ v_1^{(GF)}(\rho) &= \tfrac{k}{\sqrt{8\pi}} \mathrm{e}^{-\frac{1}{2}\frac{(\rho-\bar{\rho})^2}{\sigma^2}} \\ v_0^{(GF)}(\rho) &= \tfrac{k^2}{\sqrt{2\pi^3\sigma^2}} \, (\rho - \bar{\rho}) \, \mathrm{e}^{-\frac{1}{2}\frac{(\rho-\bar{\rho})^2}{\sigma^2}} \;. \end{aligned} \tag{15.6}$$

Although mean values of intrinsic volumes V_ν can often be calculated analytically for a wide range of stochastic models for greyscale images such as the Gaussian random field model, higher order moments such as variances are difficult to calculate explicitly. Thus, numerical algorithms such as the one presented by Mantz *et al.* (Mantz, Jacobs and Mecke, 2008) are necessary to gain insight into the geometry of such stochastic patterns shown in Fig. 15.3.

In some cases it may be more convenient to define *effective* measures by an alternative normalization of these functions, namely

$$\begin{aligned} s(\rho) &= -\frac{\partial v_2(\rho)}{\partial \rho} \frac{1}{2\pi v_1(\rho)} \\ u(\rho) &= -\log\left(2\pi v_1(\rho)\right) \\ \kappa(\rho) &= \frac{v_0(\rho)}{2 v_1(\rho)} \;. \end{aligned} \tag{15.7}$$

The advantage of these effective measures for spatial patterns compared to the original ones defined in equation (15.4) are the simplicity of the expression for a Gaussian random field model, where the effective measures read

$$\begin{aligned} s^{(GF)}(\rho) &= s_0 \\ u^{(GF)}(\rho) &= u_0 + u_2(\rho - \bar{\rho})^2 \\ \kappa^{(GF)}(\rho) &= \kappa_1(\rho - \bar{\rho}) \end{aligned} \tag{15.8}$$

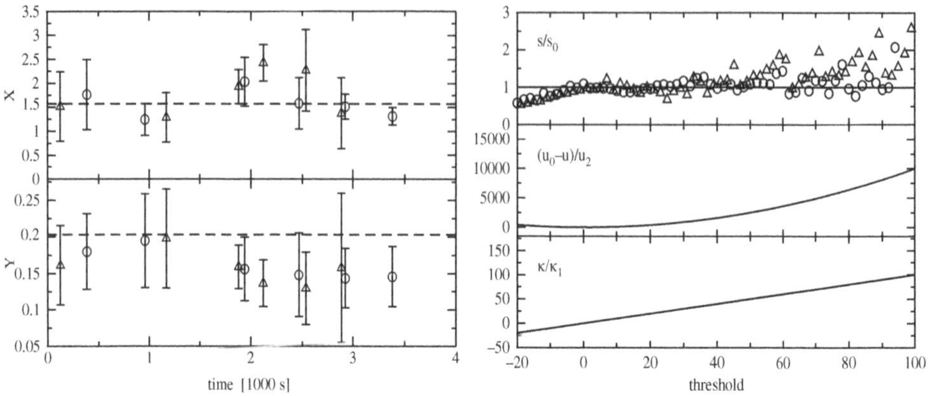

FIG. 15.5. The temporal evolution of the dewetting morphology of a thin film (see Fig. 15.3) can be analysed by intrinsic volumes $V_\nu(\rho)$ as function of threshold ρ; both in experimental AFM scans recorded *in situ* and in simulated dewetting morphology based on equation (15.3) with the identical system parameters as in the experiment. The effective measures $s(\rho)$, $u(\rho)$ and $\kappa(\rho)$ defined in equation (15.7) and fitted with a constant s_0, a parabola $u_0 + u_2\rho^2$ and a linear function $\kappa\rho$, respectively. From the fit values $s_0 \approx 5.0$, $u_0 \approx 1.14$, $u_2 \approx 9.67$ and $\kappa_1 \approx 0.4$ one obtains the ratio $Y = \frac{\kappa_1 s_0}{u_2} \approx 0.21$ which is in good agreement with the value $Y \approx 0.203$ for a Gaussian random field, see Becker, Grün, Seemann, Mantz, Jacobs, Mecke and Blossey (2003).

with $s_0 = \frac{1}{\pi\sigma k}$, $u_0 = -\log k\sqrt{\frac{\pi}{2}}$, $u_2 = \frac{1}{2\sigma^2}$ and $\kappa_1 = \frac{k}{\pi\sigma}$. Thus, one finds polynomials up to second order, i.e. a constant $s(\rho)$, a linear $\kappa(\rho)$ and a quadratic $u(\rho)$, with only five scalar and positive coefficients.

In order to test a grey-scale image for features of a Gaussian random field model one may measure the quantities

$$X := \frac{e^{-2u_0} s_0}{\kappa_1} = \frac{\pi}{2} \quad \text{and} \quad Y := \frac{\kappa_1 s_0}{u_2} = \frac{2}{\pi^2} \tag{15.9}$$

which are independent of the specific correlation function, i.e. do not depend on the details of a Gaussian random field, but are sensitive to non-Gaussian features as illustrated by Mantz, Jacobs and Mecke (2008) for various experimental data and stochastic geometric models. For instance, the effective Minkowski functions are not polynomials for a lattice model where pixels are occupied randomly which is often used to describe noise in digitized data. However, for the dewetting process of thin films in the early regime (see Figs. 15.3 and 15.5) not only polynomials are found but also the Gaussian values for X and Y given by equation (15.9). Such a test on Gaussian behaviour may be important also in astro-particle physics and cosmology, where the search for non-Gaussianity in the cosmic background radiation field is one of the most important tasks

(Kerscher, Mecke, Schmalzing, Beisbart, Buchert and Wagner, 2001*a*; Kerscher, Mecke, Schücker and collaboration, 2001*b*).

Surprisingly, the morphometric analysis of the dewetting process shows that experiments and simulations shown in Fig. 15.3 follow both a Gaussian random field model for contours above the average film thickness $\bar{\rho} \approx 0$, namely a constant $s(\rho) = s_0$, a parabolic $u(\rho) = u_2(\rho - \bar{\rho})^2$ and a linear behaviour in the curvature $\kappa(\rho) = \kappa_1(\rho - \bar{\rho})$ (compare equations 15.6–15.8). In Fig. 15.5 only the time-averaged normalized data for $\kappa(\rho)$ are shown for clarity but a similar good agreement between model expectation and data are found also for the other measures and for each snapshot at any given time. Only the parameters $s_0 = t^{-\nu_s}$, $u_2 = t^{-\nu_u}$ and $\kappa_1 = t^{-\nu_\kappa}$ depend on time t over at least two decades. An excellent consistency check is again provided by the ratios $X = \frac{e^{-2u_0} s_0}{\kappa_1} \approx 1.57$ and $Y = s_0\kappa_1/u_2 \approx 0.203$ (see equation 15.9 and Fig. 15.5) which nevertheless remains constant for Gaussian random fields (solid line). Moreover, the expected zeros $u(\bar{\rho}) \approx 0$ and $\kappa(\bar{\rho}) \approx 0$ are matched by experiments and simulations for all times.

It is interesting to note that in the simulations, the stadium until first holes are generated, is reached much later than in the experiments. A detailed study focusing on the time scale revealed that thermal fluctuations help in the experiments to speed up the initial stadium (Fetzer, Rauscher, Seemann, Jacobs and Mecke, 2007). Thus, the deterministic Navier–Stokes equation given by equation (15.3) in the lubrication approximation has to be extended by a stochastic noise term to match the experimental data (Mecke and Rauscher, 2005; Grün, Mecke and Rauscher, 2006; Fetzer, Rauscher, Seemann, Jacobs and Mecke, 2007). In the linear dynamics regime the fluctuations should follow a Gaussian distribution as just shown. Then, the mean values $v_\nu(\rho)$ depend only on the variances σ^2 and k^2, which can be determined as fit parameters to the functions $v_\nu(\rho)$. Besides providing accurate estimations of these values, this morphometric analysis allows to test whether the analysed film region is still in the linear regime. As soon as the non-linear dynamics sets in, the Gaussian model fails to fit the measured functions $v_\nu(\rho)$. Thus, introducing thresholds is an important tool to restrict the data analysis to those regions in the film which belong to the regime of linear dynamics. This morphometric image analysis reveals that the spinodally dewetting liquid film evolves as a correlated Gaussian random field with a time-dependent variance. As a consequence of the good quantitative agreement of the stochastic geometric model, experiment and simulation of thin film rupture over a time-interval exceeding the initial rupture event by far, the full dynamical evolution of complex film patterns becomes accessible, having impact on many advances in thin film technologies at nanometer scales.

15.3.2 *Modelling complex materials by Boolean models*

Predicting the relationship between morphology and transport properties of porous media, e.g. conductivity and permeability, is a long-standing problem of interest and important to a range of applications from geophysics to materials science (Matheron, 1967; Dullien, 1992; Sahimi, 1995; Sahimi, 2003; Torquato, 2002; Jeulin, 2005). Fundamental issues to be addressed include defining a set of morphological measures which allow one to quantitatively characterize morphology, optimally reconstruct model morphologies and finally accurately predict material properties, see Section 15.4.3.

The availability of 3D images through X-ray tomography data (see Section 15.2.4) has accelerated the development of computational tools to directly measure the stochastic nature of the porous materials and to construct realistic representations of the complex space. The method illustrated here, applies the Boolean model. for definition and the explicit expressions for the densities $\bar{V}_i$ of the intrinsic volumes, see Chapter 1. The method described by Mecke and Arns (2005) allows the definition of an effective local shape for an inclusion from any complex system made up of a distribution of arbitrary shaped constituents. The method requires no prior knowledge of the original ensemble of inclusion size and shape.

The reconstruction method is based on the intrinsic volumes which describe the *global* morphology of a two phase complex material in terms of the densities $\bar{V}_i$ of the intrinsic volumes, i.e. the porosity $\phi = \bar{V}_0$, and the surface to volume ratio $S = 2\bar{V}_1$, the integral mean curvature $\bar{H} = 4\pi\bar{V}_2$, and the genus $\chi = \bar{V}_3$. The method is also based on the result that the global morphology, i.e. measured values of V_i can be related to densities $\bar{V}_i$ of a Boolean model at a density ρ defined by local grains of volume $V = V_3$, surface area $A = 2V_2$ and integral mean curvature $C = 4\pi V_2$; for grains made up as unions of cubic voxels, it is more convenient to normalize differently, namely $V_3^{\square} = V_3$, $V_2^{\square} = V_2/3$, $V_1^{\square} = \frac{4}{3}V_1$, $V_0^{\square} = V_0$ so that their values are one for unit cubes. Instead of the equation given in Chapter 1 for a Boolean model in Euclidean space one obtains on a cubic lattice the densities (Mecke, 1998*a*, 2000)

$$\begin{aligned}
\bar{V}_0 &= 1 - \mathrm{e}^{-\rho V_3^{\square}} \qquad (15.10)\\
\bar{V}_2 &= \left(1 - \mathrm{e}^{-\rho V_2^{\square}}\right) \mathrm{e}^{-\rho V_3^{\square}}\\
\bar{V}_1 &= \left(-1 + 2\mathrm{e}^{-\rho V_2^{\square}} - \mathrm{e}^{-\rho(2V_2^{\square}+V_1^{\square})}\right) \mathrm{e}^{-\rho V_3^{\square}}\\
\bar{V}_0 &= \left(1 - 3\mathrm{e}^{-\rho V_2^{\square}} + 3\mathrm{e}^{-\rho(2V_2^{\square}+V_1^{\square})} - \mathrm{e}^{-\rho(3V_2^{\square}+3V_1^{\square}+1)}\right) \mathrm{e}^{-\rho V_3^{\square}} .
\end{aligned}$$

This result holds for any complex mixture of grains; $V_\nu^{\square}$ for an ensemble of n grains is given by averaged values $V_\nu^{\square} = \sum_{j=1}^{n} p_j V_{\nu j}^{\square}$, weighted by the probability of their occurrence. It also holds if the grains are correlated (hard-shell, soft shell models) where $V_i^{\square}$ depend on all n-point correlation functions. From a single 3D tomographic image or from a pair of two-dimensional serial sections of a complex material one can directly measure the global morphological parameters ϕ, S, H, χ by counting the numbers of voxels, faces, edges and vertices on the image (Arns, Knackstedt, Pinczewski and Mecke, 2001). The measurement of

TABLE 15.1. Densities of the local intrinsic volumes $V_i^\square$ ($V_0^\square = 1$ for convex grains) for the four Fontainebleau sandstone samples of size 480^3 with porosities 7.5%, 13%, 15% and 22%. The lengths a, b, c of the half-axes for randomly oriented spheroids with two different types of probability p. The best match is always achieved when one of the particles is very small and has a width of the order of a few voxels.

Sample	$V_3^\square$	$V_2^\square$	$V_1^\square$	p	a [μm]	b [μm]	c [μm]
7.5%	0.4019	3.994	0.0473	0.848	12.5	11.9	8.52
				0.152	83.5	80.7	77.2
13%	0.3619	3.752	0.0466	0.863	15.9	11.4	9.09
				0.137	84.6	76.7	76.1
15%	0.4506	4.520	0.0523	0.794	15.3	12.5	8.52
				0.206	84.6	80.1	76.7
22%	0.3361	3.965	0.0509	0.751	16.5	10.8	8.52
				0.249	74.4	67.0	63.6

intrinsic volumes, in particular of the Euler number, is biased on grids and subject to edge effects, which was studied in detail in Rosenfeld (1976), Serra (1992) and taken into account in Arns, Knackstedt, Pinczewski and Mecke (2001). From this measurement and equation (15.10) one can determine the local shape of an equivalent grain ensemble, $V_\nu^\square$, $\nu = 0, 1, 2$, and its density ρ. The results are summarized in Table 15.1 for the Fontainebleau sandstone samples. The error in the prediction of the equivalent grain ensemble V_ν^{im} from the image, $E_\nu = |V_\nu - V_\nu^{im}|/V_\nu$ is always smaller than 10^{-5}. The error is due to the use of discrete particle shapes and the finite system size. As techniques regularly produce 3D images at scales of $\geq 1000^3$ voxels, an accurate equivalent ensemble of grains can be derived from a single experimental image.

This technique is called 'method of intensities' (see Stoyan, Kendall and Mecke 1995, p. 89) and statisticians consider it as the best estimation method. Of course, the microstructure of a sandstone is a result of a complex physical process, which can include consolidation, compaction and cementation of an original grain packing and more realistic models of sandstones have been derived. These methods require however, the simulation of the generating process including primary grain sedimentation followed by a diagenetic process such as compaction and cementation. This process is both computationally expensive and requires several fitting parameters. Reconstructing the microstructure of sandstones by the Boolean model is simple and lead to an excellent match. The original sandstone microstructure for the sample at 15% and reconstructions via the Boolean model with a mixture of spheroids with two different sizes (ROS$^{(2)}$-model, half axis are given in Table 15.1) are illustrated in Fig. 15.6. Visual inspection suggests that the ROS$^{(2)}$-model closely resemble the original microtomographic image, but quantitative measures are definitely

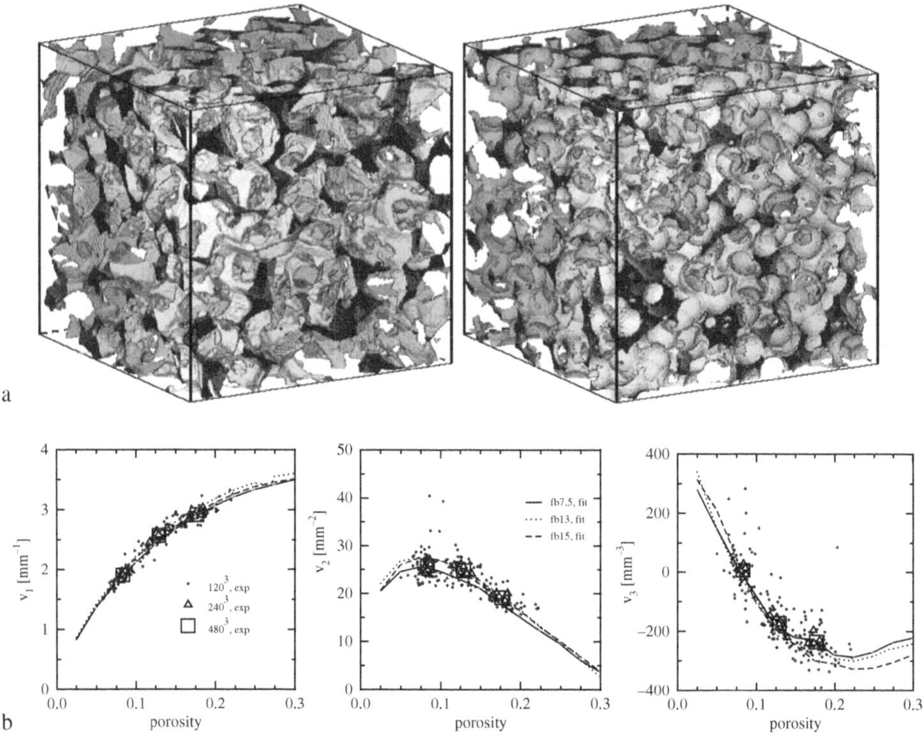

FIG. 15.6. (a) Visual comparisons of a 240^3 subset of the original Fountainebleau sandstone sample ($\phi = 15\%$) and its reconstruction by randomly oriented spheroids ROS$^{(2)}$. (b) Comparison of the intrinsic volumes over fraction ϕ for each of the four Fontainebleau sandstone samples (symbols) to the stochastic reconstruction, i.e. the ROS$^{(2)}$ model, see Arns, Knackstedt and Mecke (2002), Mecke and Arns (2005). The lines show the intrinsic volumes of the reconstructed systems predicted from a single image at porosities indicated by the circle (solid: $\phi = 7.5$, dotted: $\phi = 13$, dashed: $\phi = 15$, long-dashed: $\phi = 22$). Although the correlation functions are similar in the models the morphometric measures differ considerably, indicating that the ROS$^{(2)}$-model matches best the sandstone structure.

needed to verify the quality of the reconstruction. First, let us mention, that the Boolean model ROS$^{(2)}$ exhibit a remarkable agreement with the correlation function $g_2(r)$ of the sandstone sample, although the model does not use this information.

Let us now compare the intensities $\bar{V}_i$ of the global intrinsic volumes for the ROS$^{(2)}$-model and for the sandstone samples, see right column of Fig. 15.6. An analysis of the full 480^3 cubic subsets would only give a single value for the porosity of each of the four samples and would provide little data to compare to stochastic models. However, the samples are reasonably heterogeneous in the pore volume fraction ϕ. Due to this natural heterogeneity and by appropriately choosing different window sizes on the image it is possible to measure morphological parameters for the sandstone images across a range of pore volume fractions ϕ. This gives us a more comprehensive data set with which to compare experimental images to equivalent models. For the Fontainebleau samples cubic blocks of 480^3, 240^3 and 120^3 are considered. This provided in all cases a good spread of porosities across different sampling volumes. The measured morphological properties $S, H, \chi(\phi)$ resulting from the sampling window at 120^3 are summarised in Fig. 15.6. It is interesting that the fourth Fontainebleau data set at 22% exhibits very different measures to the first three sets, what could indicate the potential presence of heterogeneity in the sandstone.

The Boolean model do reasonably well to describe the structure across a range of porosities. This is important because transport and mechanical processes depend on specific morphological measures. For example, single phase flow and conductivity in clean sandstones is strongly affected by surface-to-volume ratio and the topology. Multiphase flow properties depend crucially on the curvature of the surfaces where immiscible phases meet. A reconstruction model which accurately describes these measures are expected to yield good agreement with experiment.

15.4 Structure–property relations: physics based on intrinsic volumes

One important goal of material science is the prediction of physical properties from the knowledge of the material structure. In particular, for cellular and porous materials the knowledge of the spatial structure is mandatory for an understanding of the material properties (Gibson and Ashby, 1997).

Here, the success of geometric ideas to derive structure–property relations is illustrate for the thermodynamic properties of a fluid bounded by an arbitrarily shaped convex container, see Fig. 15.7. Although thermodynamics is built on extremely general assumptions its implications are far reaching and powerful. One basic building block is geometry which has a long history in thermodynamics and statistical physics of condensed matter. A few examples are the formulation of thermodynamics in terms of differential forms (Frankel, 1997), the scaled-particle theory for fluids, depletion forces of colloids in biological cells, and the density functional theory based on fundamental geometric functionals known as intrinsic volumes (Rosenfeld, 1989). Reviews on application of integral geometry and stochastic geometry in physics can be found in Mecke (1998*a*, 2000), Mecke and Stoyan (2002). Here, as an example it is shown that the free energy of a fluid in a convex container can be calculated in the thermodynamic limit fully from the knowledge of only four morphometric shape descriptors, namely the intrinsic volumes of the container. This result is based on the assumption that a thermodynamic potential is an 'additive' functional of the container which can be

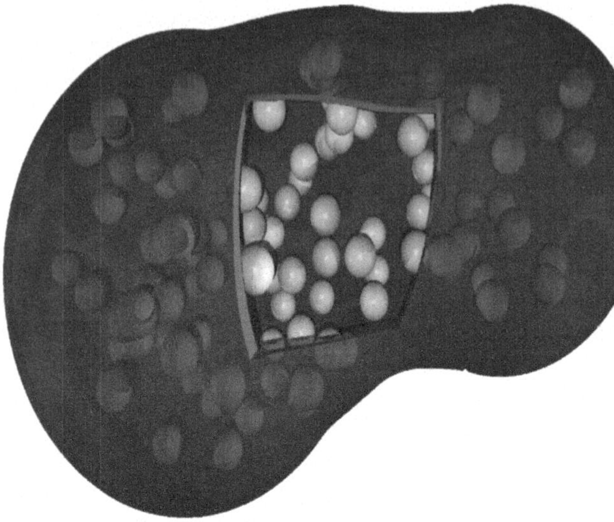

FIG. 15.7. Sketch of fluid particles in a finite container. The Gibbs free energy $\Omega(T,\mu;V_\nu)$ (see equation 15.14 and Fig. 15.9), as well as the surface energy $\gamma(T,\mu;V_\nu)$ (see equation 15.15) and the contact density $\rho_c(T,\mu;V_\nu)$ (see equation 15.17 and Fig. 15.9) of the fluid at the wall depend besides temperature T and chemical potential μ only on the 'additive' morphological functionals of the container, i.e. the volume V_3, surface area V_2, integral mean curvature V_1 and integral Gaussian curvature V_0. Further details of the shape does not enter in thermodynamic quantities.

understood as a more precise definition for the conventional term 'extensive'. As a consequence, the surface energy and other thermodynamic quantities should contain, beside a constant term, only contributions linear in the mean and Gaussian curvature of the container. This can be tested numerically in the entropic system of hard spheres bounded by a curved wall.

15.4.1 *Thermodynamic properties of confined fluids*

A thermodynamic potential Φ is considered to be an extensive quantity which means that it scales linearly with the 'size' of the system. By partitioning a large system into identical small subsystems one normally concludes that extensive quantities are proportional to the volume of the container. For instance, the Gibbs free energy $\Omega(T,\mu,V)$ which is a function of temperature T, chemical potential μ and volume V of the system is given by the pressure $p(T,\mu)$,

$$\Omega(T,\mu,V) = -p(T,\mu)\,V\,.$$

This is the well-known Gibbs–Duhem relation found in every textbook on thermodynamics. This argument however is only true for infinite bulk systems as it ignores that a *physical* partitioning of a finite system induces in general changes in the extensive quantity Φ due to the influence of the dividing wall, so that Φ depends on the shape of the chosen subsystems. Therefore, as a rule, an extensive quantity for a finite container depends not only on the volume but also on the shape and possibly on the connectivity of the enclosing container. For instance, in Fig. 15.9 the contribution to the Gibbs free energy of a colloidal fluid are shown which are due to the area and curvature of a finite container. This finding is in particular important for systems on a microscale such as porous media or biological cells where fluids are confined to complex shaped compartments and where the dependence of thermodynamic quantities and transport properties on the shape of pores or cells have significant functional and biological consequences.

Let us be more precise and consider a physical medium (condensed matter) located within a closed set $D \subset \mathbb{R}^d$ in the d-dimensional Euclidean space. Let us demand the set to be *closed* to emphasise that any wall bounding the system is also considered to be part of the set D under consideration. A physical realization of such a setup may be a fluid completely confined in a finite container or a semi-infinite system in contact with a cavity/blob. A thermodynamic potential $\Phi[D]$ is formally a mapping $\Phi : R \to \mathbb{R}$ of the convex ring R of polyconvex sets $D \in R$ onto real numbers, for instance, the Gibbs free energy of a system. Notice, that such a mapping depends in general on the actual type and state of the condensed matter system under consideration but here let us focus solely on the dependence on the shape of the container D.

'Extensivity' of a quantity in the usual meaning implies that for two disjunct containers D_1 and D_2 with $D_1 \cap D_2 = \emptyset$ the thermodynamic potential is strictly additive: $\Phi[D_1 \cup D_2] = \Phi[D_1] + \Phi[D_2]$. Notice that this assumption is only valid if interactions between the two domains D_1 and D_2 are excluded. If however the containers have points in common, a more precise definition of extensivity is needed. Let us therefore define three general properties a thermodynamic potential $\Phi : R \to \mathbb{R}$ should possess in order to be an extensive thermodynamic quantity:

(i) Motion invariance: The thermodynamics of a physical system is independent of its location and orientation in space if no external field is applied. Let G_d be the group of rigid motions in Euclidean space, namely translations and rotations in $\mathbb{R}^d$, see Chapter 1. The transitive action of $g \in G$ on a domain $D \in R$ is denoted by gD. Then, the thermodynamic potential should fulfil

$$\Phi[gA] = \Phi[A] .$$

(ii) Continuity: Small distortions of the boundary should not effect the physical properties of a system. Otherwise digital imaging would be of no use and discretizations in modelling would not be possible. If a sequence of

convex sets $K_n \to K$ for $n \to \infty$, converges towards the convex set K, then one needs

$$\Phi[K_n] \to \Phi[K].$$

Intuitively, this continuity property expresses the fact that an approximation of a convex domain by, e.g. convex polyhedra also yields an approximation of the thermodynamic potential $\Phi[K]$ by $\Phi[K_n]$.

(iii) *Additivity:* In the thermodynamic limit physical properties should be additive: the functional of the union $D_1 \cup D_2$ of two domains $D_i \in R$ is the sum of the functional of the single domains subtracted by the intersection

$$\Phi[D_1 \cup D_2] = \Phi[D_1] + \Phi[D_2] - \Phi[D_1 \cap D_2]\,. \qquad (15.11)$$

This relation generalizes the common rule for the addition of an extensive quantity of two disjunct domains to the case of overlapping domains by subtracting the value of the thermodynamic quantity of the double-counted intersection. This property is the key feature of an extensive thermodynamic quantity and makes the notion of 'extensive' more precise. It is important to notice, that for instance the intersection of two neighboured cubes is not empty if they have a face in common. Although the volume of such a face is zero and does not contribute to the additivity relation (15.11) for the volume, this does not need to be the case for other geometric measures such as the surface area and therefore in general not for a thermodynamic potential. However, one can expect the additivity to hold only if all intrinsic length-scales of the system such as the correlation length of the interaction range are small compared to the system size. for finite ratios of these lengths one expect exponentially small corrections to the additivity relation (15.11).

In $d = 3$ dimensions the volume V of the container D features properties (i)–(iii). The same is trivially also the case for any container-independent (i.e. constant) multiple of the volume V. This leads to the well-known ansatz $\Phi[D] = \phi\, V[D]$ that expresses the extensive potential Φ as a product of an intensive quantity ϕ and the volume V. The pressure $p = -\phi$ exclusively depends on the type and state of the condensed matter system whereas the volume V exclusively depends on the shape of the container D.

The surface area A of the container D also obeys properties (i)–(iii). Therefore a thermodynamic potential $\Phi[D] = \phi V[D] + \gamma_0 A[D]$ has in general an additional contribution that is proportional to the surface A and capture the influence of the boundary ∂D. The proportionality factor γ_0 is normally referred to as surface tension.

Naturally the question arises about the most general form for the mapping $\Phi : R \to \mathbb{R}$ that obeys the conditions (i)–(iii). Hadwiger's characterization theorem (discussed in Chapter 1) gives the answer: any motion-invariant, conditional continuous, and additive thermodynamic potential $\Phi[D]$ on closed convex system K in d dimensions can be written as a linear combination of the $d+1$ intrinsic volumes (or Minkowski functionals)

$$\Phi[D] = \sum_{\nu=0}^{d} \phi_\nu V_\nu[D], \tag{15.12}$$

with real coefficients ϕ_ν independent of D, see Chapter 1. In an abstract picture, every extensive quantity is an element from a $(d+1)$-dimensional vector space and the intrinsic volumes form a basis set for this space. In $d = 3$ dimensions the four intrinsic volumes can be written with convenient normalizations as

$$V_3 = V, \quad 2V_2 = A, \quad 4\pi V_1 = C, \quad 4\pi V_0 = X, \tag{15.13}$$

with the volume V, the total surface area $A = \int_{\partial D} \mathrm{d}S$, the integral mean curvature $C = \int_{\partial D} H \, \mathrm{d}S$, and the integrated Gaussian curvature (Euler characteristic) $X = \int_{\partial D} K \, \mathrm{d}S$.

Equation (15.12) recovers the findings described above that extensive quantities feature contributions proportional to the volume V and the surface A of the container D. A complete description of a thermodynamic potential however must include additional contributions: The integrated mean curvature C can be seen to describe the effect induced by bending the wall whereas the Euler characteristic X can be seen to describe the contributions due to the connectivity of the system. Apart from such a physical interpretation we emphasise that equation (15.12) is exact for every quantity which is extensive in the sense of respecting properties (i)–(iii). For instance, the Gibbs free energy Ω should feature all properties (i)–(iii) such that one can apply equation (15.12) yielding the explicit expression

$$\begin{aligned} \Omega(T, \mu; D) = &-p(T, \mu)V[D] + \gamma_0(T, \mu)A[D] \\ &+\kappa(T, \mu)C[D] + \bar{\kappa}(T, \mu)X[D] \end{aligned} \tag{15.14}$$

with the pressure $p = -\Omega_V$, the surface tension $\gamma_0 = \Omega_A$, and the bending rigidities $\kappa = \Omega_C$ and $\bar{\kappa} = \Omega_X$, which are properties of the fluid and do not depend on the shape of the container. Hadwigers theorem (15.12) asserts that $\delta\Omega = \Omega + pV - \gamma_0 A - \kappa C - \bar{\kappa} X = 0$ vanishes for all containers D in the thermodynamic limit – irrespectively of its size and shape. Thus, the dependence of the Gibbs potential Ω on the possibly complex shape of a container D enters only via four numbers. In particular, no higher derivatives of the boundary of D are necessary, nor higher powers of the mean and Gaussian curvature. In other words, the intrinsic volumes given by equation (15.13) are sufficient to quantify the the influence of the shape on thermodynamic variables.

The simple form of equation (15.12) raises the question whether property (i)–(iii) are fulfilled for a thermodynamic potential of every condensed matter system and every shape of the container D. From a physical point of view, motion invariance (i) demands the isotropy and homogeneity of (phase) space and will be assumed to be always valid if no external field is present.

Conditional continuity (ii) however fails, e.g. for containers with the size of a particle of the condensed matter system. When approximated by a polyhedra, no particles can enter the system and therefore the approximated container has substantially different thermodynamic properties. Let us exclude such and similar 'microscopic' effects and demand that the system is in the thermodynamic limit in the sense that the number of particles included in the container is large.

The additivity (iii) of a system cannot be fulfilled (or only approximative) in presence of long-ranged forces. As mentioned earlier, for two disjunct containers the potential must not depend on the mutual separation (this would be a violation of equation 15.11). This forbids wall-particle and inter-particle potentials with a range exceeding the spacing between two containers. In addition to this, long range correlations must also be excluded. This can be seen if a smaller container is excluded from a large one. The change of the extensive quantity must not depend on where the smaller container is excluded. If long-ranged correlations are present, induced, e.g. by long-ranged particle-particle or substrate-particle interactions, by critical phenomena or by wetting or drying (wetting by the gas phase) phenomena, this condition can be violated.

These constraints can be combined to one, namely that a container must be large compared to typical internal length-scales (particle sizes, potential ranges, correlation length, wetting layers, etc.). These arguments can only give a vague idea about when equation (15.12) holds and a more rigorous condition should be sought which is an important research problem in current theoretical statistical physics.

Surface energy and Steiner's formula The shape dependent interface energy $\gamma = \frac{1}{A}(\Omega + pV)$ can be defined as the excess grand potential per unit area which, by using equation (15.14), can be written as

$$\gamma(T, \mu; D) = \gamma_0 + \kappa\bar{H} + \bar{\kappa}\bar{K}, \tag{15.15}$$

where $\bar{H} = C/A$ and $\bar{K} = X/A$ are the average mean and average Gaussian curvature respectively. This relation follows directly from equation (15.14) and allows the calculation of the interface energy for arbitrarily shaped object with the knowledge of only three expansion coefficients. The definition of γ requires a particular choice of the volume V of the system and the surface area A of the dividing interface. One choice is the surface in which the density profile is discontinuous. The curvatures $\bar{H}$ and $\bar{K}$ of the surface are also measured in this dividing interface. If we were to choose a different parallel surface as dividing interface and container border, the qualitative forms of equations (15.14) and (15.15) would remain unchanged, as can be easily shown by applying Steiner's formula, see Chapter 1. The coefficients Ω_ξ, with $\xi = V, A, C, X$ however change quantitatively as $\bar{H}$ and $\bar{K}$ are different for another parallel surface.

Contact density The form of equation (15.15) has implications for other physical quantities and therefore can be tested in several ways. Let us expand the cavity by an infinitesimal amount at each point of the surface in the direction of the surface normal (pointing inwards) and analyse the change ∂_n of some quantity in linear order of infinitesimal shifts. The right hand side of the equation $\Omega = -pV + \gamma A$ changes according to the geometric relations $\partial_n V = -A$ and $\partial_n A = 2C$ (see Steiner's formula for convex sets), while the change of the left hand side is calculated within the framework of density-functional theory. This connects the thermodynamics with the microscopic (equilibrium) density distribution $\rho(r)$. In density-functional theory the equilibrium density distribution of a system minimizes the functional of the grand-canonical potential $\Omega[\rho(r)]$, i.e. it can be obtained from a variational principle (Evans, 1979). For $\rho(r)$ the functional $\Omega[\rho(r)]$ reduces to the grand potential of the system, hence the grand potential can be expressed as

$$\Omega = \Omega[\rho(r)] = \int_D \rho(r)(V_{\text{ext}}(r)\mathrm{d}r - \mu) + F[\rho(r)]$$

where V_{ext} denotes the external potential exert on the fluid by the wall. The quantity $F[\rho(r)]$ is the functional of the *intrinsic* Helmholtz free energy and does not depend explicitly on the external potential. For the external potential a hard wall potential

$$V_{\text{ext}}(r) = \begin{cases} \infty, & r \notin D \\ 0, & r \in D \end{cases}$$

is used. One obtains $\partial_n \Omega = \int_{\partial D} \rho(r)\mathrm{d}^2 r$ by exploiting that the equilibrium density profile minimizes the grand canonical functional Ω. The range of the integration of the density is the border of the container, therefore only the densities at the border (contact densities) contribute to the derivative. Thus, the right hand side reads $A\bar{\rho}^c$ where $\bar{\rho}^c$ denotes the average contact density at the container D. Combing these results one obtains the sum rule

$$\bar{\rho}^c = p + 2\bar{H}\gamma + \partial_u \gamma \tag{15.16}$$

which relates the equation of state p and the surface energy γ to the average contact density $\bar{\rho}^c$ for arbitrarily shaped objects. Combining equation (15.16) and the morphometric relation equation (15.15) leads to

$$\bar{\rho}^c(T, \mu; D) = p + 2\gamma\bar{H} + \kappa\bar{K} \; . \tag{15.17}$$

This shows that similar to the interface energy in equation (15.15), also the average contact density features this particular simple dependency on the shape of the container D.

15.4.2 *Numerical test of additivity for a hard-sphere fluid*

It is a remarkable feature of equations (15.15) and (15.17) that the interface energy as well as the average contact density of the fluid can be expressed in a form that separates the geometry of the wall and properties of the fluid and that only four expansion coefficients, namely p, γ, κ, and $\bar{\kappa}$, are required.

In order to test these predictions of morphometry, one can study a one-component fluid of hard spheres with radius R in the grand-canonical ensemble within the framework of density functional theory close to planar, cylindrical and spherical walls. The chemical potential μ is kept fixed and determines the bulk number density of particles ρ_{bulk} or equivalently the bulk packing fraction $\eta_{\text{bulk}} = \frac{4\pi}{3}R^3\rho_{\text{bulk}}$. The hard-sphere system remains fluid for all packing fractions $\eta_{\text{bulk}} < 0.494$ and does not undergo any fluid-fluid phase separation. Hence all subtleties introduced through long-ranged correlation by wetting or drying or critical phenomena do not apply for a hard-sphere fluid and therefore the conditions on the grand potential, (i)–(iii), can be assumed to be fulfilled.

Unfortunately, for arbitrarily shaped wall geometries, there exists no reliable and accurate *direct* method to find the interface energy γ and the average contact density $\bar{\rho}^c$ of the hard sphere fluid. This would require to calculate a three-dimensional density distribution of the fluid close to the arbitrarily shaped wall, which is extremely time consuming and in addition gives usually rather inaccurate results.

However, there are three different shapes for the cavity, which are in particular simple and for which the interface energy and the contact density can be solved: an infinitely large planar wall, an infinitely long cylindrical cavity with radius R_c and and a spherical cavity with radius R_s. By exploiting the spatial symmetry of these cavities, all thermodynamic quantities can be calculated very reliably using density functional theory. For instance, Rosenfeld's fundamental measure theory (Rosenfeld, 1989) has proven to describe properties of inhomogeneous hard-sphere fluids very accurately.

Minimizing the density functional $\Omega[\rho(r)$ for these simple shapes one obtains the equilibrium density profile of the hard-sphere fluid $\rho(r)$ and the Gibbs free energy $\Omega = \Omega[\rho(r)]$. Applying a curvature expansion to the whole density profile

$$\begin{aligned}\rho(r) &= \rho_P(u) + \rho_H(u)H + \rho_K(u)K + \rho_{H^2}(u)H^2 \\ &\quad + \rho_{H^3}(u)H^3 + \rho_{HK}(u)HK + \mathcal{O}(R^{-4})\end{aligned}$$

the expansion coefficients $\rho_\xi(u)$ are functions of the normal distance u. The contact density can be obtained by evaluating the density profile at $u = 0$. The result is shown in Fig. 15.8. In the left panel the coefficient functions are shown of the planar wall contribution and those proportional to H and K which correspond to additive functionals and hence display a non-vanishing contact value at $u = 0$. Higher order contributions are shown in the right panel up to third order in curvature. These coefficient functions have a vanishing contact density but away from the wall they display structure with a similar amplitude than the

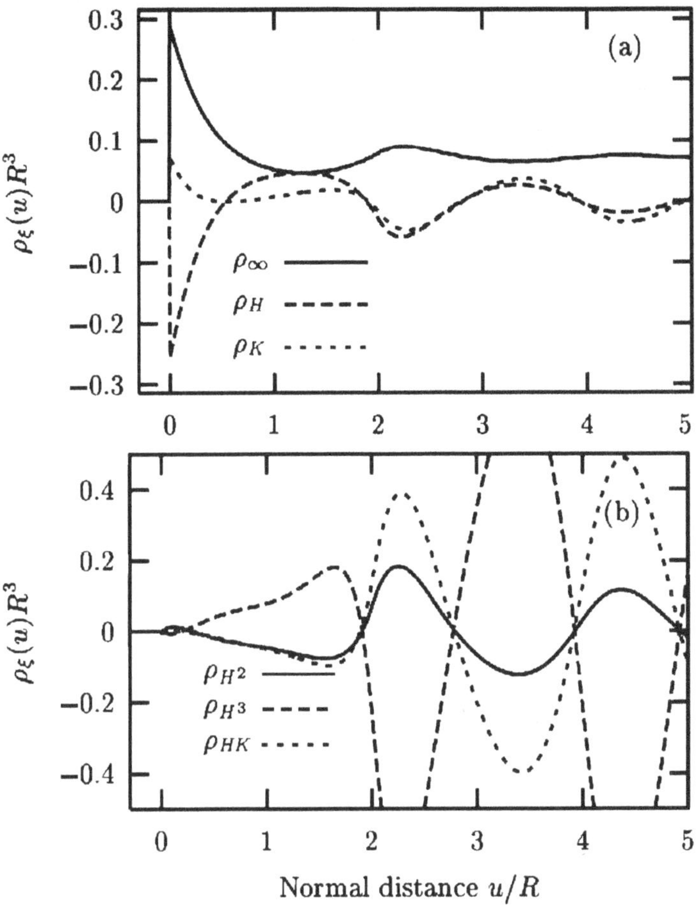

FIG. 15.8. Curvature expansion coefficient functions of the density profile for a hard sphere fluids with $\eta_{\mathrm{bulk}} = 0.3$ modelled with the Rosenfeld functional (from König, Bryk Roth and Mecke 2004). In (a) the additive contributions are shown while (b) shows the residual coefficient functions. In (a) the finite contact value at $u = 0$ is clearly visible.

additive functionals. Notice, that in contrast to the contact position ($u = 0$), there is no reason that a linear expansion in terms of H and K will be sufficient to capture the influence of a curved boundary on the density distribution at arbitrary normal distance u. Although K and H^2 have the same order in curvature, The Gaussian curvature K contributes to the contact value as an additive functional, whereas H^2 does not.

In Fig. 15.9 the contact densities ρ^c_Ξ of these expansion profiles are shown as function of the bulk packing fraction η of the fluid. The contact density

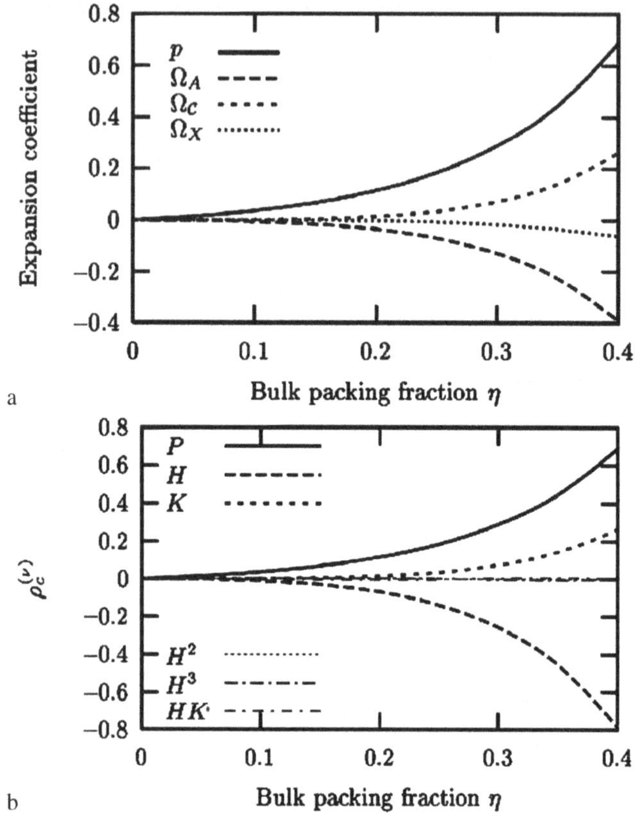

FIG. 15.9. The Gibbs free energy Ω is strictly composed of only four terms, where the first one, $-pV$, is proportional to the volume but the remaining depend on area A, integral mean curvature C and integral Gaussian curvature X of the confining wall (from König, Roth and Mecke 2004). (a) Expansion coefficients p, $\gamma_0 = \Omega_A$, $\kappa = \Omega_C$, and $\bar{\kappa} = \Omega_X$ of the grand potential Ω depend on the bulk packing fraction η_{bulk} of the fluid but not on container shape; (b) Curvature expansion coefficients for the contact density for different bulk packing fractions η_{bulk}. Notice, that only the contributions ρ_P^c, ρ_H^c, and ρ_K^c contribute to the contact value.

obviously has only contributions form additive functionals A, C and X. From the behaviour of Ω as function of η_{bulk} and the $\bar{H}$ and $\bar{K}$ in these geometries, one can extract also the expansion coefficients Ω_ξ for the Gibbs free energy shown in Fig. 15.9. Again in agreement with morphometry, only those contributions are non-vanishing which correspond to additive functionals, whereas higher powers in the mean and Gaussian curvature have contribution to the contact density which vanish numerically. Notice that K (which is additive) and

H^2 (which is not additive) are the same order $\sim R_i^2$ in curvature but nevertheless contribute quite differently to thermodynamic quantities $\Phi[D]$, since K enters in the general form given by equation (15.12), but H^2 does not. In this sense, the general expression (15.12) for a thermodynamic potential is not a mere expansion in curvatures but a unique representation in terms of *additive* functionals. These results for the hard-sphere fluid can presumably be generalised to different fluids with arbitrary short ranged molecular interactions as long as 'intrinsic' length scales are small compares to typical features of the container.

The idea of morphological thermodynamics goes back to applications of *intrinsic volumes* in statistical physics, in particular on composite materials (Mecke, 1994, 1996, 1998*b*). For instance, studying microemulsions and other complex fluids one may assume that the energy, i.e. the Hamiltonian of the oil-water mixture is given by additive functionals on scales larger than the persistence length where the bending rigidity κ of the oil/surfactant/water interface vanishes. This assumption was used in previous work heuristically to justify a Hamiltonian linear in intrinsic volumes yielding quit rich phase behaviour and thermodynamic properties (Mecke, 1994; Likos, Mecke and Wagner, 1995; Brodatzki and Mecke, 2001; Brodatzki and Mecke, 2002).

The relations (15.14), (15.15) and (15.17) for interface energy and contact density of fluids should not be understood as an expansion in powers of the inverse radius of curvature, because the series exactly truncates. However, since the concept of parallel surfaces and curvature expansion are very common and important tools in condensed matter physics, these results may inspire interesting applications, even where the series does not truncate. It is numerically not feasible to calculate physical properties for any possible shape of a container or a particle. Truncating curvature expansion at the Gaussian curvature may be the best option to derive at least approximate expressions for such physical properties. In the following section a method is proposed to predict physical properties of porous media (percolation thresholds, conductivities, elasticity, phase transitions, capillary condensation) which rests essentially on parallel surfaces and the expansion in curvatures (Mecke, 2002; Arns, Knackstedt and Mecke, 2002, 2003, 2004).

15.4.3 *Transport properties and Boolean model*

Relating transport properties to the structure of porous media is a long-standing task in science (Matheron, 1967; Dullien, 1992; Hilfer, 1998; Sahimi, 1995, 2003). In Section 15.3.2 it was shown that an accurate stochastic reconstruction of a complex material made up of discrete pores or grains can be derived from a single 3D snapshot at any porosity. The method is based on determining the intrinsic volume of the pore space and determining a Boolean model with grain shapes $V_\nu^\square$, so that the densities $\bar{V}_\nu$ match the measured values V_ν.

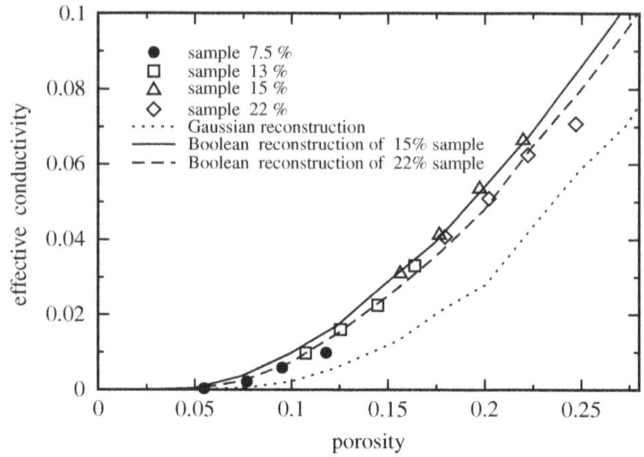

FIG. 15.10. Comparison between the prediction for conductance of the matching Boolean model ROS$^{(2)}$-model and the Fontainebleau sandstone data: [a] 7.5%, [b] 13%, [c] 15%, and [d] 22% (data from Mecke and Arns 2005). Reconstruction with a Gaussian random model gives a poor match to the conductivity data over the full fractional range although 2-point correlation function is matched precisely.

From the previous section we know, that thermodynamic quantities of a fluid inside the pore space should be well captured by the reconstructed Boolean model. Moreover, Kac (1966) showed that the short time behaviour of the diffusion equation in a complex porous medium is governed by the intrinsic volumes V_i defined by integral geometry. Naturally, one may assume that a reconstructed porous medium based on the equivalent shape which honours these additive measures exhibits similar transport properties to the original system.

One may numerically determine physical properties such as the conductance of the original sandstone and of the reconstructed model. The conductivity calculation is based on a solution of the Laplace equation with periodic boundary conditions. In Fig. 15.10 the effective conductivity σ_{eff} of the Fontainebleau sandstone at a scale of 120^3 is shown and compared with its equivalent stochastic model ROS$^{(2)}$ at infinite contrasts $\sigma_{\text{pore}} : \sigma_{\text{grain}} = 1 : 0$. The match of the equivalent single spheroidal grain ensemble is excellent in all cases and superior to a prediction based on spherical grains or reconstructions based on Gaussian models. It is remarkable that one can generate the conductance curve of a complex porous system *across all porosities from a single image at a given porosity.*

15.4.4 *Outlook: elastic properties and tensor valuations*

Many attempts have been made to predict elastic properties to structural features; some of them are based on Boolean model and random set models (Jeulin and Ostoja-Starzewski, 2001; Jeulin, 2005; Ostoja-Starzewski, 2007). In the previous section it was shown that the use of equivalent shapes in terms of intrinsic volumes leads to excellent predictions of thermodynamic and transport properties. But does a reconstructed material which honours the additive measures exhibits also similar elastic properties to the original system. A finite element method can be used to estimate the elastic properties of the original and equivalent granular systems. First results indicate, that it is also possible to predict elastic properties (Mecke and Arns, 2005). However, in contrast to the scalar conductivity the elasticity is a tensorial quantity. Thus, tensorial valuations have to be introduced to analyse tomographic data such as the sandstone in Fig. 15.6.

Tensorial valuations are also promising tools to characterize anisotropic features which are important in biomaterials such as the nutshell shown in Fig. 15.4. The structure and mechanical performance of natural cellular materials, such as nutshells, are interesting both from a biological and a materials science point of view. In nature, the nutshell's microstructure and shape result in excellent material properties at low density and provide an efficient trade-off between the protection of the seed against herbivory: they make it difficult to 'get in', while still ensuring that the seed can 'get out' during germination. In industry, ground nutshells are valued as soft-grit abrasives. They cleanse precision parts without scratching and pitting and can be reused because their resistance to rupture, deformation and breakdown is high. Phase contrast X-ray microtomography made it recently possible to analysis quantitatively the microstructure (Breidenbach, 2007; Breidenbach, P., Wegst, Cloetens and Mecke, 2009). To date, little research has been carried out in order to correlate the microstructure of nutshells to their macroscopical mechanical performance both in their natural habitat and as an industrial material. First results on the characterization of the shape of the pore space by tensorial Minkowski valuations indicate that the structure of nut shells is non-trivial and varies enormously between different species. The application of tensorial valuations on anisotropic biomaterials, as well as medial axis transforms, periodic minimal surfaces and distribution of curvatures in anisotropic Gaussian random fields would be exciting applications of stochastic geometry in material science.

References

Aarts, D. G. A., Schmidt, L. M., and Lekkerkerker, H. N. W. (2004). Direct visual observation of thermal capillary waves. *Science*, **304**, 847.

Aarts, D. G. A., Schmidt, L. M., Lekkerkerker, H. N. W., and Mecke, K. (2005). Microscopy on thermal capillary waves in demixed colloid-polymer systems. *Advances in Solid State Physics*, **45**, 15–27.

Als-Nielsen, Jens and McMorrow, Des (2001). *Elements of Modern X-ray Physics*. John Wiley & Sons.

Arns, C. H., Knackstedt, M. A. and Mecke, K. (2002). *Characterising the morphology of disordered materials, p. 40-78 in K. Mecke and D. Stoyan, Morphology of Condensed Matter, LNP 600.* Springer, Berlin.

Arns, C. H., Knackstedt, M. A., and Mecke, K. (2003). Reconstructing complex materials via effective grain shapes. *Phys. Rev. Lett.*, **91**, 215506.

Arns, C. H., Knackstedt, M. A., and Mecke, K. (2004). Characterisation of irregular spatial structures by parallel sets and integral geometric measures. *Colloids and Surfaces A*, **241**(1–3), 351–372.

Arns, C. H., Knackstedt, M. A., Pinczewski, W. V., and Mecke, K. (2001). Euler–Poincaré characteristics of classes of disordered media. *Phys. Rev. E*, **63**(31112), 1–13.

Ashby, M. and et al. (2000). *Metal Foams: A Design Guide.* Butterworth-Heinemann, Oxford.

Becker, J., Grün, G., Seemann, R., Mantz, H., Jacobs, K., Mecke, K., and Blossey, R. (2003). Complex dewetting scenarios captured by thin film models. *Nature Materials*, **2**, 59–64.

Bilodeau, M., Meyer, F., and Schmitt, M. (Eds.) (2005). *Structure and Randomness, Contributions in Honor of Georges Matheron in the Fields of Geostatistics, Random Sets, and Mathematical Morphology; Series: Lecture Notes in Statistics, Vol. 183, XIV.* Springer-Verlag, New York.

Blum, Z., Hyde, S. T., and Ninham, B. W. (1993). Adsorption in zeolites, dispersion self-energy and gaussian curvature. *J. Phys. Chem.*, **97**, 661–665.

Bonnell, Dawn A. (2001). *Scanning Probe Microscopy and Spectroscopy: Theory, Techniques, and Applications.* Wiley, New York.

Breidenbach, B. (2007). *Scalar and tensor-valued Minkowski functionals of spatially complex structures.* PhD thesis, Universität Erlangen-Nürnberg, Erlangen.

Breidenbach, B. P., Sheppard A., Wegst, U., Cloetens, P., and Mecke, K. (2009). Phase contrast x-ray tomography and morphometry of nut and seed shells. *in preparation for Nature Materials* .

Brodatzki, U. and Mecke, K. (2001). Morphological model for colloidal suspensions. *cond-mat*, 0112009.

Brodatzki, U. and Mecke, K. (2002). Simulating stochastic geometries: morphology of overlapping grains. *Computer Physics Communications*, **147**, 218–221.

Bronnikov, A. (2002). Theory of quantitative phase-contrast computed tomography. *J. Opt. Soc. Am. A*, **19**, 472–480.

Caselles, V., Kimmel, R., and Sapiro, G. (1997). Geodesic active contours. *Int. J. Comp. Vis.*, **22**, 61–79.

Chapman, Henry N. and et al. (2007). Femtosecond time-delay x-ray holography. *Nature*, **448**, 676–680.

Cloetens, P., Barrett, R., Baruchel, J., and Guigay, J. P. (1996). Phase objects in synchrotron radiation hard x-ray imaging. *J. Phys. D. Appl. Phys.*, **29**, 133–146.

Cloetens, P., Ludwig, W., Baruchel, J., Guigay, J., Rejmankova-Pernot, P., Salome, M., Schlenker, M., Buffiere, J., Maire, E. and G. Peix (1999). Hard x-ray phase imaging using simple propagation of a coherent synchrotron radiation beam. *J. Phys. D: Appl. Phys*, **32**, A145–A151.

Cloetens, P., Ludwig, W., Boller, E., Peyrin, F., Schlenker, M., and Baruchel, J. (2002). 3D imaging using coherent synchrotron radiation. *Image Anal. Stereol.*, **21**(Suppl. 1), S75–S85.

Cloetens, P., Ludwig, W., Pateyron-Salome, M., Buffiere, J. Y., Peix, G., Baruchel, J., Peyrin, F., and Schlenker, M. (1997). Observation of microstructure and damage in materials by phase sensitive radiography and tomography. *J. Appl. Phys.*, **81**(12), 5878–5886.

Cressie, Noel A. C. (1993). *Statistics for Spatial Data*. Wiley, New York.

Downing, Kenneth H., Sui, Haixin, and Auer, Manfred (2007). Electron tomography: a 3D view of the subcellar world. *Analytical Chemistry*, **79**(21), 7949–7957.

Dullien, F. A. L. (1992). *Porous Media, Fluid Transport and Pore Structure.* Academic Press, San Diego.

Evans, R. (1979). The nature of the liquid-vapour interface and other topics in the statistical mechanics of non-uniform, classical fluids. *Adv. Phys.*, **28**, 143.

Fetzer, R., Rauscher, M., Seemann, R., Jacobs, K., and Mecke, K. (2007). Thermal noise influences fluid flow in thin films during spinodal dewetting. *Physical Review Letters*, **99**, 114503.

Fradin, C, Braslau, A., Luzet, D., Smilgies, D., Alba, M., Boudet, N., Mecke, K., and Daillant, J. (2000). Reduction in the surface energy of liquid interfaces at short length scales. *Nature*, **403**, 871.

Frangakis, A. S. and Hegerl, R. (2001). Noise reduction in electron tomographic reconstructions using nonlinear anisotropic diffusion. *Journal of Structural Biology*, **135**, 239–250.

Frank, Joachim (2006). *Electron Tomography. Methods for Three-Dimensional Visualization of Structures in the Cell, 2nd ed.* Springer.

Frankel, Th. (1997). *The Geometry of Physics: An Introduction.* Cambridge University Press, Cambridge.

Gibson, L. J. and Ashby, M. F. (1997). *Cellular Solids – Structure and Properties.* Cambridge University Press, Cambridge.

Grün, G., Mecke, K., and Rauscher, M. (2006). Thin-film flow influenced by thermal noise. *J. Stat. Phys.*, **122**, 1261–1291.

Hadwiger, H. (1957). *Vorlesungen über Inhalt, Oberfläche und Isoperimetrie.* Springer, Berlin.

Hansen-Goos, Hendrik, Roth, Roland, Mecke, Klaus, and Dietrich, S. (2007). Solvation of proteins: linking thermodynamics to geometry. *Physical Review Letters*, **99**, 128101.

Hilfer, R. (1998). *Transport and Relaxation Phenomena in Porous Media, Vol. XCII of Advances in Chemical Physics, pp. 299-424.* John Wiley & Sons, Inc, New York.

Hsieh, Jiang (2003). *Principles, Design, Artifacts, and Recent Advances Computed Tomography.* SPIE Press.

Hutzler, S. and Weaire, D. (2001). *The Physics of Foams.* Oxford University Press, Oxford.

Hyde, S., Larsson, K., Blum, Z., Landh, T., Lidin, S., Ninham, B.W., and Andersson, S. (1996). *The Language of Shape: The Role of Curvature in Condensed Matter: Physics, Chemistry and Biology.* Elsevier, Amsterdam.

Jeulin, D. (2005). Random Structures in Physics, in: *Structure and Randomness* edited by M. Bilodeau, F. Meyer and M. Schmitt, pp. 183–222. Springer-Verlag, New York.

Jeulin, D. and Ostoja-Starzewski, M. (Eds) (2001). *Mechanics of Random and Multiscale Microstructures, CISM Lecture Notes N 430.* Springer Verlag, New York.

Jeulin, D. (Ed) (1997). *Proceedings of the Symposium on the Advances in the Theory and Applications of Random Sets, Fountainebleau, 9-11 October 1996.* World Scientific Publishing Company.

Jinnai, H., Koga, T., Nishikawa, Y., Hashimoto, T., and Hyde, S. T. (1997). Curvature determination of interfaces in spinodally phase-separated structures of a polymer blend. *Phys. Rev. Lett.*, **78**, 2248–2251.

Jones, A. C., Sheppard, A. P., Sok, R. M., Arns, C. H., Limaye, A., Averdunk, H., Brandwood, A., Sakellariou, A., Senden, T. J., Milthorpe, B. K., and Knackstedt, M. A. (2004). Three-dimensional analysis of cortical bone structure using X-ray micro-computed tomography. *Physica A*, **339**, 125–130.

Kac, M. (1966). Can one hear the shape of a drum? *Amer. Math. Monthly*, **73**, 1.

Kalender, Willi A. (2005). *Computed Tomography: Fundamentals, System Technology, Image Quality, Applications.* Publicis, Erlangen.

Kerscher, M., Mecke, K., Schmalzing, J., Beisbart, C., Buchert, T., and Wagner, H. (2001*a*). Morphological fluctuations of large-scale structure: the pscz survey. *Astronomy & Astrophysics*, **373**, 1–11.

Kerscher, M., Mecke, K., Schücker, P., and collaboration, REFLEX (2001*b*). Non-gaussian morphology on large scales: Minkowski functionals of the reflex cluster catalogue. *Astronomy & Astrophysics*, **377**, 1–16.

Kinney, H. and Nichols, M. C. (1992). X-ray tomographic microscopy (XTM) using synchrotron radiation. *Annual Review of Materials Science*, **83**, 121–152.

König, P.-M., Roth, R., and Mecke, K. (2004). Morphological thermodynamics of fluids: shape dependence of free energies. *Physical Review Letters*, **93**, 160601.

Lambert, J., Cantat, I., Delannay, R., Renault, A., Graner, F., Glazier, J. A., Veretennikov, I., and Cloetens, P. (2005). Extraction of relevant physical parameters from 3Dimages of foams obtained by X-ray tomography. *Colloids Surf., A Physicochem. Eng. Asp.*, **263**, 295–302.

Likos, C. N., Mecke, K., and Wagner, H. (1995). Statistical morphology of random interfaces in microemulsions. *J. Chem. Phys.*, **102**, 9350–9361.

Lohmann, G. (1998). *Volumetric Image Analysis.* Wiley-Teubner, Chichester, New York.

Magerle, R. (2000). Nanotomography. *Phys. Rev. Lett*, **85**, 2749.

Malladi, R., Sethian, J., and Vemuri, B. (1995). Shape modeling with front propagation. *IEEE Trans. Pattern Anal. Mach. Intell.*, **17**, 158–175.

Mantz, H., Jacobs, K., and Mecke, K. (2008). Utilising minkowski functionals for image analysis. *JSTAT*, P12015.

Masters, Barry R. (2006). *Confocal microscopy and multiphoton excitation microscopy: the genesis of live cell imaging.* SPIE Press, Bellingham, WA.

Matheron, G. (1967). *Eléments pour une théorie des milieux poreux.* Masson, Paris.

Matheron, G. (1975). *Random Sets and Integral Geometry.* Wiley, New York.

Mecke, K. (1994). *Integralgeometrie in der Statistischen Physik.* Verlag Harri Deutsch, Frankfurt.

Mecke, K. (1996). Morphological model for complex fluids. *J. Phys.: Condensed Matter*, **8**, 9663–9667.

Mecke, K. (1998*a*). Integral geometry and statistical physics. *International Journal of Modern Physics B*, **12**, 861–899.

Mecke, K. (1998*b*). Morphological thermodynamics of composite media. *Fluid Phase Equilibria*, **150/151**, 591–598.

Mecke, K. (2000). Additivity, convexity, and beyond: Applications of Minkowski functionals in statistical physics. In: K.Mecke and D. Stoyan (Eds.), *Statistical Physics and Spatial Statistics*, Lecture Notes in Physics, Vol. 554. Springer, Berlin, pp. 111–184.

Mecke, K. (2002). The shapes of parallel surfaces: porous media, fluctuating interfaces and complex fluids. *Physica A*, **314**, 655–662.

Mecke, K. and Arns, C. H. (2005). Fluids in porous media: a morphometric approach. *J. Phys.: Condens. Matter*, **17**, S503–S534.

Mecke, K. and Dietrich, S. (1999). Effective hamiltonian for liquid-vapor interfaces. *Phys. Rev. E*, **59**, 6766–6784.

Mecke, K. and Dietrich, S. (2005). Local orientations of fluctuating fluid interfaces. *Journal of Chemical Physics*, **123**, 204723.

Mecke, K. and Rauscher, M. (2005). On thermal fluctuations in thin film flow. *J. Phys.: Condens. Matter*, **17**, S3515–S3522.

Mecke, K. and Stoyan, D. (2000). *Statistical Physics and Spatial Statistics - The Art of Analyzing and Modeling Spatial Structures and Pattern Formation, Lecture Notes in Physics, Vol. 554.* Springer, Berlin.

Mecke, K. and Stoyan, D. (2002). *Morphology of Condensed Matter – Physics and Geometry of Spatially Complex Systems, Lecture Notes in Physics, Vol. 600.* Springer, Berlin.

Meyer, Ernst, Hug, Hans Josef, and Bennewitz, Roland (2005). *Scanning Probe Microscopy: The Lab on a Tip.* Springer-Verlag, Berlin, Heidelberg.

Molchanov, I. (1997). *Statistics of the Boolean Model for Practitionners and Mathematicians.* Wiley, Chichester.

Mora, S., Daillant, J., Mecke, K., Luzet, D., Braslau, A., Alba, M., and Struth, B. (2003). Determination of the structure and surface energy of liquid-vapor interfaces at short-lengthscales by synchrotron x-ray scattering. *Phys. Rev. Lett.*, **90**, 216101.

Nishikawa, Y., Jinnai, H., Koga, T., Hashimoto, T., and Hyde, S. T. (1998). Measurements of interfacial curvatures of bicontinuous structure from three-dimensional digital images. 1. a parallel surface method. *Langmuir*, **14**, 1242–1249.

Ohser, J. and Mücklich, F. (2000). *Statistical Analysis of Materials Structures.* John Wiley & Sons, Chichester, New York.

Oppelt, A. (2005). *Imaging Systems for Medical Diagnostics.* Publicis Corporate Publishing, Erlangen.

Ostoja-Starzewski, M. (2007). *Microstructural Randomness and Scaling in Mechanics of Materials.* Modern Mechanics and Mathematics Series. Chapman & Hall, Taylor & Francis, Boca Raton.

Parra-Denis, E., Moulin, N., and Jeulin, D. (2007). Three dimensional complex shapes analysis from 3d local curvature measurements: applications to intermetallic particles in aluminium alloy. *Image Analysis and Stereology*, **26**, 157–164.

Rehse, S., Mecke, K., and Magerle, R. (2008). Characterization of the dynamics of block copolymer microdomains with local morphological measures. *Physical Review E*, **77**, 051805.

Roentgen, W. C. (1995). über eine neue Art von Strahlen. *Sitzungsberichte der Physikalisch-medizinischen Gesellschaft zu Wřzburg,* **9**.

Rosenfeld, Azriel (1976). *Digital Picture Processing.* Academic Press, New York.

Rosenfeld, Y. (1989). Free-energy model for the inhomogeneous hard-sphere fluid mixture and density-functional theory of freezing. *Phys. Rev. Lett*, **63**, 980.

Roth, Roland, Harano, Yuichi, and Kinoshita, Masahiro (2006). Morphometric approach to the solvation free energy of complex molecules. *Physical Review Letters*, **97**, 078101.

Sahimi, M. (1995). *Flow and Transport in Porous Media and Fractured Rock.* VCH, Weinheim, Germany.

Sahimi, M. (2003). *Heterogeneous Materials, vol I.* Springer-Verlag, New York.

Sakellariou, A., Sawkins, T. J., Senden, T. J., and Limaye, A. (2004). X-ray tomography for mesoscale physics applications. *Physica A*, **339**, 152–158.

Santalò, L. A. (1976). *Integral Geometry and Geometric Probability.* Addison-Wesley, Reading.

Serra, J. (1992). *Image Analysis and Mathematical Morphology, Vol. 2,: Theoretical Advances.* Academic Press, New York.

Sheppard, A. P. and Knackstedt, M. A. (2006). Mango. *http://xct.anu.edu.au/mango.*

Stoyan, D., Kendall, W. S., and Mecke, J. (1995). *Stochastic Geometry and its Applications.* John Wiley & Sons, Chichester.

Thorsten, M. Buzug (2008). *Computed Tomography. From Photon Statistics to Modern Cone-Beam CT.* Springer, Berlin.

Torquato, S. (2002). *Random Heterogeneous Materials: Microstructure and Macroscopic Properties.* Springer, New York.

Warren, Bertram Eugene (1990). *X-Ray Diffraction.* Courier, Dover.

Wilkins, W., Gureyev, T. E., Gao, D., Pogany, A., and Stevenson, A. W. (1996). Phase-contrast imaging using polychromatic hard X-rays. *Nature*, **384**, 335–338.

Wilson, Tony (1990). *Confocal microscopy.* Academic Press, London.

16

STOCHASTIC GEOMETRY AND TELECOMMUNICATIONS NETWORKS

Sergei Zuyev

To my dearest friend, my love and my wife: Ania

Abstract

Just as queueing theory revolutionized the study of circuit switched telephony in the twentieth century, stochastic geometry is gradually becoming a necessary theoretical tool for modelling and analysis of modern telecommunications systems, in which spatial arrangement is typically a crucial consideration in their performance evaluation, optimization or future development.

In this survey we aim to summarize the main stochastic geometry models and tools currently used in studying modern telecommunications. We outline specifics of wired, wireless fixed and ad hoc systems and show how stochastic geometry modelling helps in their analysis and optimization. Point and line processes, Palm theory, shot-noise processes, random tessellations, Boolean models, percolation, random graphs and networks, spatial statistics and optimization: this is a far from exhaustive list of techniques used in studying contemporary telecommunications systems and which we shall briefly discuss.

16.1 Crash course on telecommunications

It is impossible to imagine our lives today without telecommunications, be it picking up a fixed line or a mobile phone to call a friend, watching digital terrestrial or satellite TV broadcasts, browsing the web with your PDA when taking a bus home, listening to an internet radio streaming your favourite rumba music from the other side of the world or downloading the latest distribution of your Linux OS via a P2P network (sure, you never download illegal copies of blockbusters with bittorrent!)

But there is also a great number of other communications activities going without you ever noticing them: remote environment monitoring, road traffic management, GPS, security cameras streaming, etc.

The multitude of communications tasks reflect a multitude of existing communications systems. But broadly we can distinguish a few main classes based on the way they are structured and the way their information flow is organized. All that a communication system is about is to provide a means to pass information from a point A (person, computer, camera, etc.) to a point B. If the distance between A and B is small enough to shout and be heard, there is, perhaps, no need for any intermediary, but if the two parties willing to communicate are far from each other or require larger information exchange than a voice can provide, the information must be relayed through a *communications system (network)* basically consisting of *nodes* (stations, antennas, servers, etc.) linked between each other in some way (by copper cables, optical fibres, radio waves, laser beams or beacons on two adjacent hilltops). According to the nature of the nodes, of the links and the way the information is relayed, a few major classes of communication systems can be identified: fixed line telephony and the Internet, cellular communications, ad hoc networks. It should be noted that all kinds of intermediate and mixed architectures also abound. So let us focus our attention on these main classes of telecommunications systems.

16.1.1 *Circuit switched telephony*

The Public Switched Telephone Networks (PSTN[1]) are good old telephone systems. Their name originates from the times when a human operator was manually connecting two wires: between the caller and the respondent. This created a circuit of a 'twisted pair' of cables, enabling the electrical impulses to modulate the human voice, pass it through the wires, demodulate on the other end and be heard. In these systems, the media carrying communications is the copper wires and/or optical fibres. The nodes (accumulation points, switching centres, etc.) are the fixed points where these cables join or split and where the switching between the destinations is performed (much like a human operator a hundred years ago did).

Nowadays, such networks possess a very complex topology consisting of many hierarchical layers, both physical (i.e. the physical layout of the cables) and logical (i.e. how the circuits are interconnected). It is common, however, to distinguish between its two main parts: *distribution* and *transport* networks. The distribution or the 'Local Loop' or the 'Last Mile' network starts at the socket you plug your telephone in, then goes to a concentration point in your building or the floor where the wires from your neighbouring flats meet and join in a cable, then proceeds to the next concentration point, perhaps, on a corner of your street where these cables are assembled into yet a bigger cable. The next level concentration point could be an optical line terminal where the electrical signals are transformed into a laser beam and sent via an optical fibre. This kind of aggregation may yet proceed further, until the distribution cables reach *access nodes* which not only assemble the cables but also carry out the *switching*

[1]Once my wife, a mathematician by education, was playing Lego® blocks with our 2 year old son. When his grandma exclaimed: 'Why do you speak such complicated words to the child: tetrahedron, parallelogram?', my wife argued that at this age all words are complicated, so let's learn the right terms from the start! It is the author's intention to familiarize the reader with 'complicated' acronyms which are commonly used in the telecommunications community. Knowing these not only helps to understand what the specialists are talking about, but also provide valuable keywords to research further information. Ohh ... did I mention that our son graduated in mathematics from the University of Cambridge?

tasks between different routes. They are so called because they provide access to a higher level structure of a transport network whose main goal is to route information inside the network or to other networks.

While the distribution network topology looks very much like a forest, i.e. a collection of trees (loopless graphs) rooted at the access points and branching out to the consumers' telephones, the transport network topology can vary significantly. The main factors are the geography of the area, its urban type, density of the population, the types of services provided, development history, cost factors, redundancy, quality of service (QoS) requirements, etc. A rather popular topology in metropolitan areas is a *ring network* when a dozen of access points are organized in a ring connected by means of an optical fibre of a large capacity. The ring itself (or a system of rings) is connected through egress nodes to the highest speed *backbone* network which is the top hierarchy responsible for carrying information to large distances (inter-city, inter-state and trans-continental). The ring topology has the main advantage of being robust to link failures: if a link between two nodes on a ring breaks, it is still possible to route the information going round the ring in the opposite direction.

It should be noted that two topologically different links (or, as they are often called, 'logical links') may physically be a bunch of cables in the same trench. So different logical links failures may well happen dependently on each other, not to speak of cascade failures which sometimes also happen. The physical layout of the network carries a footprint of the underlying urban factors: for instance, cables are usually laid along the streets. This provides easy access to them and also minimizes the civil engineering cost.

The following features characterize the PSTN. The topology of the network is fixed: the nodes and the telephones occupy known geographical locations and the links are permanently present (provided they are not broken).

16.1.2 *Internet and P2P networks*

Closely related to the fixed telephone networks is the *Internet* which is a system of interconnected computers. Its predecessor was the US Department of Defence funded ARPANET which was superseded in the 1980s by NSFNET, the National Science Foundation series of networks that already included many of today's main structural components: the TCP/IP protocol responsible for routing of messages over the Internet, computer naming conventions, and commonly used Internet services such as e-mail and hypertext. Physically the structure of the Internet resembles and, to a large extent, is integrated with the fixed telephone networks. As in the case of PSTN, two layers: can be distinguished the Transport high speed and capacity network called the *Backbone* responsible long distance information transmission supplied by Network Service Providers, and the Access layer delivering a physical link to the transport network and managed by the Internet Service Providers. According to `www.internetworldstats.com` for summer 2008 there was an estimated 1.5 billion Internet users, i.e. just over

1/5 of the whole world population. Although specialized optical fibre Internet access to households is no longer exotic, the most common way to provide the Internet is still through an existing telephone line, therefore the PSTN 'local loop' is still a crucial part of the Internet structure.

For many research questions it is not important to know how physically the computers are connected, but rather the logical stricture in the form of domains with dedicated servers providing routing of the information between the computers inside and between domains. Here one can observe huge differences between servers of the local networks 'responsible' for just a few connected computers and large clusters receiving gigantic information flows, like the popular search websites. Fascinating graph representations of the Internet can be found on the website of the OPTE project: `www.opte.org/maps`.

The highly connected structure of the Internet and high access speed brought to life different collaborative initiatives: IP telephony, file sharing, remote storage, distributed computing, peer-to-peer or P2P networks, to name a few. In a 'pure' P2P network all participating computers play an equal role being 'servers' and 'clients' at the same time effectively sharing the resources and routing information. Since the information is not kept at one known place or can even be available only by pieces from different sources, additional infrastructure is employed to assist in files and hosts search. It should be noted that P2P is not something that concerns just download of pirate films nor just IT techies: the all-popular ICQ® and Skype™ use P2P protocols!

16.1.3 *Cellular telephony*

The last decades have witnessed a real boom in cellular telephony: at the start of the new millennium the number of cellular phones has already exceeded the number of landline phones! Compared to the fixed networks described above, the 'Last Mile' distribution network is replaced in the cellular telephone networks with a wireless link via radio channels between the antenna: the *base station* (BS), and the user's mobile phone. The BSs serve as an entry point to the network for the mobiles, the stations themselves are (usually) a part of a wired network. First generation systems used analogue radio transmissions which were replaced by the second generation (2G) digital systems in the beginning of the 1990s. 2G systems still employ the *circuit switching*, i.e. for each call, a set of connections is established and remains fixed for the whole duration of the call. The third generation (3G) systems that are a product of this millennium already employ *packet switching*, i.e. the data packets are routed, queued and buffered subjected to random delays, the same way as information is transmitted in the Internet. At the same time 3G also support circuit switched channels, in particular, for high bandwidth applications like video. The most widely used standard in 2G systems is the Global System Mobile (GSM), while the Universal Mobile Telecommunication System (UMTS) is the standard for 3G. A gradual change to 3G gave rise to a few intermediate architectures, namely: 2.5G – GPRS

(General Packet Radio Service) with GSM and 2.75G – EDGE (Enhanced Data rate for GSM Evolution).

Since the contemporary systems use digital transmissions when the voice is first encoded into a digital stream, then compressed and multiplexed to save bandwidth and then passed over the radio channel,[2] communication is possible only when the received signal can be decoded successfully. Physics and information theory tell us that to make this possible the signal must be strong enough compared to the background noise. This quantity is measured by the Signal-to-Noise ratio (SNR). In order to communicate at a given rate, SNR must be above a certain threshold which is a function of this rate. By the noise one means not only random radio waves coming from all kinds of sources around us, but also the emissions from other BSs and other cellular phones which may interfere with the transmission thus preventing the communication to be decoded successfully. Hence the Signal-to-Interference plus Noise ratio (SINR) is more often used to distinguish these two main components: random background noise and the interference from other emitters; we shall come to details on this later in Section 16.2.2.

With every BS there is associated its *service zone* which represents the locations where a reliable communication with the BS is possible (this is the very *cell* in a Cellular Network). Surely, the strongest signal is near a BS, the signal strength then falls down with the distance (*fading*) and is also prone to different phenomena complicating the reception. The main such phenomenon is arguably multiple path fading or *multipath*, for short, when the signal is received together with its in-phase and reversed phase copies coming after reflections from buildings and the ground. When the opposite phase copies are summing up with the main wave, it effectively weakens the received signal. Other phenomena include random fading and motion fading. All this makes the geometry of the service zones dynamically changing with somewhat fuzzy boundaries which also depend on the number and locations of the on-going transmissions.

The radio transmission protocols may vary: FDMA, TDMA, CDMA standing for Frequency, Time or Code Division Multiple Access and define the way multiple cellular phones share the medium to communicate successfully with the BS. The main reason for existence of complex protocols is that the radio waves medium with a finite frequency range can only carry a finite number of simultaneous communications: their number is limited by physical phenomena, the main being interference of radio transmissions on close frequencies. Thus a BS may only have a finite, and well separated, set of radio channels for communications with the mobiles. The neighbouring BSs can not use the same or close frequencies as this would create interference on the boundaries between their service zones.

Early analogue radio systems were FDMA when each call simply used a different frequency. With emergence of digital transmission techniques, TDMA

[2] Actually, *two* different radio channels are usually used for sending and receiving: the uplink might be, for instance, 879.360 Mhz and the downlink might be 834.360 Mhz.

became the technology of choice when several calls are transmitted on the same channel. This is obtained by chopping up the call data into small pieces and then transmitting them in the right order at the right time: each user occupies a radio frequency for a predetermined amount of time in an assigned time slot. TDMA is used in GSM networks. The 3G systems, in contrast, use its successor technique CDMA to transmit.

The CDMA initially developed by Qualcomm uses orthogonal codes to encode the call data and the BS is then able to decode all the calls it receives simultaneously. The CDMA provides much more efficient spectral efficiency by accommodating more users per MHz of bandwidth. Though putting more and more users on the same frequency is theoretically possible (this is dubbed the *Soft Capacity* of the CDMA), it also makes error free decoding more difficult for all users. This manifests in audio quality degradation or longer data downloads.

Modern systems use a technique called *Frequency Hopping* in which a mobile will switch in order through a sequence of available frequencies in order to avoid a co-channel interference should the current frequency be clobbered by other conversations or to reduce receiving problems due to multipaths. But the most important technique to achieve the maximal wireless system capacity is *Power Control* by which a cellular phone continuously adapts its transmitting power in accordance with Transmit Power Control (TPC) commands received from the serving cell(s) antennas. The cells estimate the current SINR on the basis of the received uplink signals and dictate to the mobile whether to increase or to decrease its transmission power to the least acceptable for good quality communication level. The transmission power of the downlink is also controlled by the network, modern BSs are capable of beaming different power in different directions compared to previous generation static omni-directional antennas. Effective power control reduces interference with the other active network users by reducing the noise they receive and still maintaining the necessary SINR for a successful decoding. No less important, keeping the power to a minimum reduces exposure of the cell phone user to RF radiation and also saves the phone's battery life.

What also makes the cellular systems radically different from the fixed switched networks is the mobility of the users. This mobility creates new challenges for engineers, notably, *Paging* and *Handover* (or *Handoff* in American English, the term *Handover* is standardized by the 3rd Generation Partnership Project: 3GPP). Contrary to a telephone socket which is always at the same place, a mobile telephone user may take a car or plane and soon appear in another city or another country. Therefore the system should be able to trace the current position of its subscribers and make necessary updates in the users' location registry in order to effectively relay the calls to the part of the network or to another network which provides the service in the area. Organization of this task is what comprises the paging.

A handover is a procedure to organize a smooth transition of all the transmission tasks from one BS to another when an active cellular phone leaves its current service zone. As we have already mentioned, the association of a mobile with a particular BS is based on the strength of the signal, more exactly, on the SINR. But the signal is subject to random fluctuations so that when a mobile is at about the same distance from two or more BSs, the strongest signal station may change many times. Different mechanisms of handover were devised to intelligently handle passing control over the mobile from one BS to another and prevent frequent swapping of the stations in such boundary situations.

16.1.4 *Wireless ad hoc networks*

The network systems considered previously have a fixed hierarchical structure of their transport layer which allows operators to effectively plan routing of information in the network. Since the topology of the network is known, the operator can pre-define rules for the information flow. These rules are often called the *routing matrix* which defines a path a communication between nodes A and B should take. In contrast, the *ad hoc networks* do not have a centralized control, do not possess (at least, initially) a hierarchical structure and, typically, its elements have only a limited knowledge of the system as a whole.

Ad hoc networks consist of a set of fixed or mobile communication devices or *nodes*. Since there is no centralized control, they have to cooperate and self-organize in order to provide efficient information transmission. The topology of ad hoc networks may be constantly changing due to mobility, failures of nodes or, possibly, specific protocols prescribing some nodes to temporarily shut down to preserve their energy or not to cause interferences. MANET is the acronym used for mobile ad hoc networks.

A typical example of ad hoc networks provide *sensor networks* which consist of similar devices capable of collecting, transmitting and sometimes also of a preliminary processing of the information they have gathered or received from other sensors to be relayed further. The sensors may be mobile, e.g. health monitoring sensors attached to herd, or fixed scattered over the area to be monitored.

With advances in miniaturization and the constantly dropping price of mass production, the creation of small cheap radio transmission devices has become possible which could be deployed in mass quantities in order to monitor and transmit information about the environment they are put in. Such devices usually consist only of a sensor gathering the information they are intended for, a tiny processing unit, a transmitter/receiver and a source of power – a battery. In military applications it is intended that these sensors are hard to detect due to their tiny size of just a few millimetres, so the term *smart dust* became common. Because of the size limitations, the sensors can only transmit to relatively short distances and because of memory and processing limitations they only have some knowledge of their local neighbourhood.

A common scenario of usage: sensors are distributed over a territory (for example, dropped from a plane). The sensors then self-organize in a network based on the other sensors they 'hear'. The collected information is then sent across the network (possibly, with some processing) from sensor to sensor in, generally, a multihop fashion. To collect this information, a dedicated device (or devices) dubbed a *sink* is also deployed in the field. A sink usually has a larger communication range allowing it to transmit the gathered and processed information when requested to, for instance, a plane passing over it. A range of tasks include weather, ecology and seismic activity monitoring and, of course, military applications.

A real challenge in sensor networks is how to organize efficiently the information flow. Each activity of a sensor, especially the radio transmission, cost the sensor its battery life. A good protocol should guarantee transmission of information from any sensor in the network to the sink for as long as possible, so it must consume as little as possible of the sensors' energy still maintaining connectivity of the network.

Another yet futuristic example is the 4th Generation (4G) telecommunication systems when heterogeneous devices: network access devices, roaming mobile users, cars, computers, even fridges and microwaves will 'sense' each other and self-organize to provide seamless fast access to the network resources.

16.1.5 *Further reading*

The area of telecommunications is arguably the most dynamic area of technology development and research. So for the most up-to-date information one needs to go to the web resources. An excellent site with constantly updated information and wiki-possibilities is `www.privateline.com`. Comprehensive web resources on all aspects of design of wireless systems are `www.wirelessnetdesignline.com`, `www.umtsworld.com` and `www.techonline.com`. The books by Bedell (1999), Glover and Grant (2004), Shepard (2005) and Hill Associates (2001) provide a comprehensive description of telecommunications systems and the technologies behind. A very interesting monograph, Webb (2007), gives many experts' forecasts on the future of wireless communications: from technology progress to mobile cellular systems to military applications. Future 4G systems are discussed in Wisely (2007) especially in the view of emerging non-cellular systems, such as WiMAX and WLAN. A broad account of the state-of-the-art in sensor networks is given in the book by Karl and Willing (2007). See also another excellent book by Callaway (2003) and the collection in Phoha *et al.* (2007) on the subject.

International programme COST 279: Action, Analysis and Design of Advanced Multiservice Networks supporting Mobility, Multimedia, and Internetworking involved over 40 research institutions from the academic, industrial, and telecom operator worlds. The results of four years of its functioning covering the areas of wireless, optical networks and peer-to-peer systems are summarized in the Final Report: Brazio *et al.* (2005).

What fate is my tomorrow brewing?[3] What does the future hold for us? It is clear that telecommunications are already becoming ubiquitous. Coming 4G networks will definitely be packet switched and are planned to integrate and dynamically utilise various available network resources to provide communications from anywhere to anywhere at speeds allowing all existing digital services functioning smoothly: voice, multimedia messaging, data services, digital video broadcasts, etc. For this, a fast moving user should get no less than 100Mb/sec and a stationary user, no less than 1Gb/sec access speed to the network. New technologies like WiMAX (IEEE standard 802.16d, see `www.wimaxforum.org`) are thought to provide these goals. Working groups include the Wireless World Research Forum: `www.wireless-world-research.org` and the WINNER project which is a consortium co-ordinated by Nokia Siemens Networks.

For a deeper dive into how the things work one necessarily has to get down to the engineering nitty gritty: the standards. A comprehensive source of information on all aspects of telecommunications technologies is the UN agency International Telecommunication Union, see `www.itu.int`. For existing technical specifications and new proposals for current and future wireless systems, point your browser to the 3rd Generation Partnership Project's web page: `www.3gpp.org`.

The Internet in its basic principle is a conglomerate of independent networks, but in order to be inter-operable they follow commonly defined standards historically called RFC: Request For Comment. For all aspects of the Internet's standards and organization, see the Internet Engineering Task Force website: `www.ietf.org`.

16.2 Stochastic geometry modelling

The reader has already realized that the contemporary network systems possess a complicated multi-layer and multi-level architecture. They are described by a huge number of parameters and it is generally not clear in advance which of these many parameters are predominant for what we are going to achieve, be it the cost, service characteristics or future planning. It is sometimes possible to put the detailed network description into a computer and simulate its functioning as a whole. But even ignoring vast scales of memory, storage and calculation time requirements, the same problem inevitably shows up: among these millions of parameters which main few would influence the desirable properties of the system? So for the sake of analysis of the system as a whole, identification of the determinant factors and sensible strategic planning, development of reliable models is imperative.

Any model building is certainly based on what we trying to model. When considering telecommunications systems, it becomes clear readily that their spatial structure has direct bearing on their functioning, cost and performance characteristics. Randomness is also a key feature: the service requests are mainly

[3] *Eugene Onegin: A Novel in Verse* by Alexandre Pushkin, Ch. 6.XXI, Translated by Charles Johnston.

random, but often there is also an intrinsic randomness in the network system structure itself: the positions of the cell phone users and their trajectories are, to a large extent, unpredictable, data packets may take different routes to the destination, even the topology of the network itself can be random, as it is the case for ad hoc systems (think of the sensors scattered from the plane).

All these factors represent a big challenge for a modeller, and place stochastic geometry at the centre of necessary tools the modeller should master. In this section we give a non-representative sample of examples on how stochastic geometry tools could be used for analysis and development of telecommunications networks with the ultimate goal to provide a better and cheaper service to their clients – to all of us.

16.2.1 *Fixed line networks: the last mile to go*

Challenges: Cost evaluation, optimization, strategic planning
Techniques: Point processes, tessellations, random graphs

An early example is the paper by Sallai (1988) which concerned the modelling of the distribution network ('last-mile') in the fixed telephone networks: PSTN. The idea is to represent the positions of the nodes: fixed telephone terminals (or the households) and the concentration points of different levels as a realization of a stochastic point process. The other characteristics of the network: its topology, the cost, etc. then become functionals of the point process defined for each realization by some 'natural' rules. For instance, the basic model of the distribution network proposed in Baccelli *et al.* (1995) is based on realizations of two (stochastically) independent point processes: Π_0 representing subscribers, and Π_1 representing the first level nodes (concentration points). It is natural to assume that the subscribers are connected to their closest concentration points, so the wires are represented by the segments drawn from the points of Π_0 to their *closest* points of the process Π_1. More generally, a multi-level distribution network is represented by a *hierarchical model* with processes Π_i, $i = 1, \dots, n$ describing the ith level concentration points and each such point (apart from the highest level n representing the access points to the transport network) is connected to its closest concentration point of the next level $i+1$. These models reduce the number of the parameters of the system to just a few easily statistically identifiable ones: the telephone subscribers' density and the density of different level nodes. A good survey of point process techniques used in telecommunications modelling with emphasis on parameters fitting and simulations can be found in Frey and Schmidt (1998*a*, 1998*b*).

Even in the simplest case when homogeneous Poisson processes are used to represent subscribers and the nodes, the model gives rather accurate results. Consider, for instance, the problem of estimation of the total lengths of cables in the distribution network. At the first glance it seems strange that such a problem arises at all: is the network operator supposed to know its own network? In fact, yes, it does know, but in order to get this information many people should dig deep into the detailed engineering plans which often are not kept in the same place but scattered over the local directions and spend many hours with a ruler collecting this information. Sometimes, it is simply not possible, for instance, if it concerns a competitor's network or when just planning deployment of a new network. Also it may be not that critical to have the *exact* figures, but rather a good estimate which would perfectly do if one is to choose between different strategic development scenarios, for example.

A frequently raised critique: why *stochastic* models if nothing is random in the nodes position? Certainly true. But it is a commonly accepted scientific method to treat complex statistically homogeneous systems as realisations of stochastic processes. And it is often the case that the typical configurations of the network would not differ much for, say, two cities with a similar population density and a similar road system. After all, there is nothing random in the distribution of prime numbers, but probabilistic methods are powerful tools in number theory. It is reassuring that even the basic model described above gives errors of only about 2–3% when applied to estimation of the cable lengths in rural areas. For large cities the error could reach 10%, so the model needs refinements, and below we shall outline approaches taken in this direction.

To give a flavour of the techniques used, we present an idea of the cost computation of a three-level hierarchical model with independent homogeneous Poisson processes Π_0, Π_1, Π_2 representing the subscribers, the first and the second level concentration points, respectively. The cost G of the sub-tree rooted at a second level accumulation point z is the sum of the cost of the twisted pairs connecting z to each of its subscriber in the tree, the cost of civil engineering (e.g. trenches) and the cost C of the concentration points y_i of level 1 attached to z. Note that the twisted pairs are grouped together in one cable at points y_i so they share the same civil engineering cost. Let λ_0, λ_1 and λ_2 denote the intensities of the processes, let a be the cost of a unit length of a cable and b_{ij} be the unit cost of the civil engineering for connections between the concentrations points of level i to level $j = i + 1$, $i = 0, 1$. The total cost of the network in a large area W can be well approximated by the intensity λ_2 of the second level nodes times the area of W times the expected cost of the tree rooted at a typical z, which is formally the Palm distribution of G with respect to Π_2. Using the fact that under the Palm distribution the rest of the process is still Poisson (the Slivnyak–Mecke theorem), we set z at the origin and write

$$\mathbf{E}_2^0 G = \mathbf{E}_2^0 \Big[\sum_{y_i \in \Pi_1 \cap V_0(\Pi_2)} C + b_{12}|y_i| + a|y_i|\Pi_0(V_{y_i}(\Pi_1)) \\ + \sum_{x_j \in \Pi_0 \cap V_{y_i}(\Pi_1)} b_{01}|x_j - y_i| + a|x_j - y_i| \Big].$$

In the above equation, $V_x(\Pi)$ stands for the Voronoi cell centred at $x \in \Pi$ constructed with respect to Π. Indeed, Π_1 points are connected to their closest points among Π_2, so effectively, to the nuclei of the Π_2-Voronoi cells they belong to. The same is true for the connection between Π_0 and Π_1 points. Applying Neveu's exchange formula relating the Palm distributions of two processes, see Theorem 1.10 in Chapter 1 and the original paper by Neveu (1976), the above expression simplifies to

$$\frac{\lambda_1}{\lambda_2}C + \frac{\lambda_1}{\lambda_2}\mathbf{E}_1^0\left[b_{12}|z_0| + a|z_0|\Pi_0(V_0(\Pi_1)) + \sum_{x_j \in \Pi_0 \cap V_0(\Pi_1)} (a + b_{01})|x_j|\right],$$

where z_0 is the closest to the origin point of Π_2. Now, denoting by $B_r(x)$ the closed ball of radius r centred at x,

$$\begin{aligned}
\mathbf{E}_1^0|z_0| &= \mathbf{E}_1^0 \int |z| \mathbf{1}\{\Pi_2(B_{|z|}(0)) = 0\}\, \Pi_2(\mathrm{d}z) \\
&= \mathbf{E} \int |z| \mathbf{1}\{\Pi_2(B_{|z|}(0)) = 0\}\, \Pi_2(\mathrm{d}z) \\
&= \lambda_2 \int |z| \mathbf{P}_2^0\{\Pi_2(B_{|z|}(-z)) = 0\}\, \mathrm{d}z \\
&= 2\pi\lambda_2 \int_0^\infty r^2 e^{-\lambda_2 \pi r^2}\, \mathrm{d}r = \frac{1}{2\sqrt{\lambda_2}}.
\end{aligned}$$

The second line follows from the first one by the independence of Π_1 and Π_2 and the next line follows from the refined Campbell theorem, see, e.g. Stoyan *et al.* (1995). Similarly, the other terms can be computed, these types of functionals were extensively studied in Foss and Zuyev (1996), where the first two moments as well as their distribution's tail decay were obtained. These results, in particular, suggest fitting a statistical Weibull-type model for the Palm distribution of the cost G.

The overall expression for $\mathbf{E}_2^0 G$ above appears to be a convex function in λ_1. So there is an optimal (for the given costs parameters) value of the density of the first level concentration points which minimizes the costs of connections. It is easy to understand: when λ_1 is large, we gain in the cost of the connections from Π_0 to Π_1, but lose in the civil engineering cost: we practically have to dig a separate trench for each twisted pair. If λ_1 is small, subscribers have to reach far to their closest Π_1-point which is not optimal neither. Explicit expressions for the optimal cost and further details can be found in Baccelli and Zuyev (1999), where more general hierarchical models are also considered. For instance, direct connections from subscribers to second level points are possible to account for. Note that cables are almost never laid in a straight line but rather follow existing infrastructure, e.g. the roads. Nonlinear cost functions of the distance considered in Baccelli and Zuyev (1999) can take care of this.

To each station $z^{(k)}$ (node) of level k there corresponds its *service zone* $Z(z^{(k)}; \Pi_1, \ldots, \Pi_k)$ consisting of those locations in the plane which are attached to $z^{(k)}$ (served by $z_{(k)}$) through a system of intermediate nodes of lower levels.

Since in the hierarchical model each subscriber is attached to its unique node on every hierarchical level, the service zones represent a tessellation of the plane. If one accepts the general principle by which the nodes are connected to the closest node of the next level in the hierarchy, the service zone for a node $z^{(1)}$ of level $k = 1$, $Z(z^{(1)}; \Pi_1) = V(z^{(1)}; \Pi_1)$ is the Voronoi cell with nucleus z_1 constructed with respect to the first level nodes Π_1. The zone of a second level node $Z(z^{(2)})$ consists of those Voronoi cells $V(z_i^{(1)}; \Pi_1)$ whose nuclei $z_i^{(1)}$ are closer to $z^{(2)}$ than to any other points of Π_2, i.e. which belong to the Voronoi cell $V(z^{(2)}; \Pi_2)$, etc. These types of *aggregative tessellations* were studies in Tchoumatchenko and Zuyev (2001), in particular, the coverage probability is of a special interest for the communications modelling.

The models described above are based on homogeneous point processes and they apply successfully to either large scale networks or the networks with a more or less homogeneous distribution of the population of users. Generally, it is also possible to work with non-homogeneous point processes, notably, Poisson. For example, a Poisson process Π_0 in the basic model has intensity measure $\mu_0(dx)$ related to the population distribution, and we seek for the intensity measure $\mu_1(dx)$ for the process of the first level concentration points Π_1 which minimizes the expected cost of connections. Although the processes are assumed to be statistically independent, they are related through the choice of their intensity measures, so we expect to put more concentration points where the density of the population is higher. Since the parameters of the model now include the intensity measures of the non-homogeneous Poisson processes, optimization goal functions are now functions of measures. Specific variation analysis tools for such functionals were developed in Molchanov and Zuyev (2000) which allowed to obtain the following interesting result: the asymptotically optimal density of the concentration points is proportional to power $2/3$ of the population density, see also Molchanov and Zuyev (2002) for the computational aspects of optimizations of functionals of measures. The author learnt with surprise that a similar power law has already been empirically established by the network planners.

For more accurate estimates of the last-mile cost associated with the connections, especially when dealing with urban areas, one has to take into account the topology of the road system which the cables usually follow. For this, it is necessary to develop reliable models for the road infrastructure which may vary significantly from area to area. Indeed, Manhattan as many other cities in the United States has a rectangular road grid of streets and avenues. In contrast, Moscow has a concentric circle road structure intersected by radial chaussés while central Paris is characterized by a number of junctions where five and more roads meet. It was suggested to use the *nested* or *iterated tessellations* to describe the road network. One tessellation, X, describes the main roads system. Then, given a realization of X, independent realisations of another tessellation, Y, are drawn in each of the cells of X to produce the secondary roads web which all end up in the main roads: boundaries of the X-cells. The popular choices for X and Y are

Voronoi, Delaunay, Poisson lines and regular tessellations, see Gloaguen *et al.* (2006).

Considerable work has been done recently to express the main geometrical characteristics of these tessellations for the sake of statistical fit of these parameters to the road system under consideration, see Maier and Schmidt (2003), Maier *et al.* (2004), Heinrich *et al.* (2005). Another challenge is taking into account spatial inhomogeneity of the real networks on the global scale. Tessellations based on different types of Cox point processes prove adequate for this purpose. Such processes are doubly stochastic Poisson process driven by the intensity measure which is itself randomly generated. In Fleischer *et al.* (2008*a*, 2008*b*) the authors study Voronoi tessellations with respect to network components represented by a Cox process with intensity supported by random Poisson lines and the edges of another Voronoi tessellation representing the roads in order to characterise the mean length of subscriber lines. Gloaguen *et al.* (2008) consider a Cox process with the intensity defined by a realization of a Boolean model of balls with possibly random radii: it has different constant values on the balls and outside of them. An ultimate goal is to have an automated procedure which would take a territory road map as input and give estimates of the required network's performance and cost factors as output. And according to a recent private communication this goal is not far away.

Another important cost function relates to the teletraffic, in particular, to the distribution of the proportion of stations of different levels engaged in routing information in the net. This depends on the the distance between the two talking parties: if the call is local, the communication route will only involve regional exchanges, in contrast, a long-distance call will use most of the hierarchy in the network topology. Gloaguen (2001) uses aggregate tessellations to obtain analytic expression for traffic production functions which are based on the call patterns distribution and the number of hierarchical levels engaged in the communications. The case of two concurrent operators is considered in Baccelli *et al.* (2000) which leads to studying the tessellations obtained by superposition of two Poisson–Voronoi tessellations, see also Gloaguen (2006). Modelling the spatial telecommunication traffic is an area of research in its own right. A wide range of publications addresses these issues, to name a few, Tutschku and Tran-Gia (2005) and the references therein, or Chapter 3 in Brazio *et al.* (2005). Since the teletraffic is related to the population, some publications estimate the cost directly based on the demographic density or on its models. For instance, Appleby (1995 and 1996) and Norros (2005) make use of multifractals to model the population distribution in order to characterize telecommunications demand and estimate the corresponding system cost.

Finally we should mention the emergence of hybrid wireless-optical networks (WOBAN) which replace the lower part of the wired access network with wireless link between the end users (or low hierarchy distribution points) to the end-points of the high-speed fibre network: Optical Network Units (ONU). WiMAX is currently the most often used technology behind the wireless link. In WiMAX, the user first negotiates access with the wireless access point competing with other users. Once this initialization process is completed, the user gets scheduled access time-slots which guarantee the negotiated quality of service. Currently, WiMAX can provide up to 70Mb/sec speed in the vicinity of a base station and can reach up to 50 km (at much lower rates). These hybrid networks become a viable cost-effective alternative to wired networks in the regions with difficult geography or in the new areas without developed infrastructure, see, e.g., Sarkar *et al.* (2007). Keeping the right balance between how far from the local loop to the end users the optical links in the form of the Passive Optical Network (PON) should go before going wireless at ONUs for the last reach is the main question for optimizing the cost of such networks. The problem of dimensioning and planning of WOBAN was addressed, e.g. in Elsayed *et al.* (2001), Farinetto and Zuyev (2004) and in Sarkar *et al.* (2008).

16.2.2 *Modelling cellular mobile networks*

> **Challenges:** Service coverage, power management, paging, handover
> **Techniques:** Coverage processes (Boolean and germ-grain models, incomplete and SINR-based tessellations), Spatial queueing systems, Gibbs processes and Gibbs sampler

One of the first applications of stochastic geometry in studying telecommunications was the paper by Gilbert (1961) who considered a network of stations represented by a Poisson point process in which each pair of stations is linked if the distance between them is less than a fixed R. The paper studies connectivity of thus defined network which is equivalent to percolation properties of the Boolean model of spheres. It was just shortly before the turn of the millennium that Gilbert's ideas made their way into engineering literature, i.e. before massive popularization of wireless communications. Since then, the number of publications using some form of stochastic geometry increased dramatically driven by attempt to account for specific spatial aspects of wireless communications and this is one of the most important recent trends in networks research.

Conceptually, cellular mobile networks differ from PSTN networks in two main aspects: radio wave connections replacing the wired access network and mobility of the users. Inevitably, the main directions of research are related to these two areas: studying and modelling of the wireless medium, its coverage characteristics, development of wireless multiple access protocols, congestion and power control. Then there comes the users' mobility modelling, development of handover protocols, user location and tracking algorithms. Integral activities to all these are the economy of the networks, tariffication and planning according to the projected future demand, capacity and the quality of service (QoS) requirements.

Since the capability to receive and decode a digital transmission by a mobile terminal in CDMA and UMTS networks is determined by the signal to interference plus noise ratio SINR at the position of the mobile, the service zone of an antenna can be defined as the set of locations where the SINR level exceeds a given threshold determined by the required minimum communication rate. A commonly accepted theoretical model for the signal strength is the *Gaussian radio channel* by which the power received at a point at the distance r from the emitting station with power P is given by $Pl(r)$, where $l(r)$ is the *attenuation* or *path loss* at distance r. In vacuum, the function $l(r)$ decays at infinity as r^{-2}, but in physical environments it is usually assumed that $l(r) \sim r^{-\alpha}$ with some α in the range from 2 to 4 which is called the *path loss exponent*. In complex environments the signal comes together with many of its reflections (multipath phenomenon) which interfere with each other effectively weakening the received signal. So the path loss exponent may be even up to 6.[4] The most frequently used in modelling attenuation function is $l(r) = (Ar)^{-\alpha}$ which is fine as an approximation for moderate to large values of r but which explodes at $r = 0$. Hence other popular choices are $l(r) = (1 + Ar)^{-\alpha}$ and $l(r) = A(\max\{r_0, r\})^{-\alpha}$ for some $r_0 > 0$.

In a system with only one emitter the maximal bit-rate W of a reception at the distance r in the presence of a background (or thermal) noise is given by *Shannon's capacity formula*: $W = H\log(1+\mathrm{SNR}(r)/H)$, where $\mathrm{SNR} = Pl(r)/N$ with the bandwidth H of the channel in Hz and the noise spectral density $N/2$ in watts/Hz. Different applications usually require different minimal rates so both the emitter and the receiver must be capable to adapt their coding scheme to the perceived power in order to communicate at different maximal bit-rates. In many systems today though the necessary bit-rate is fixed, dictated by the QoS requirements, so that there is a threshold β such that the communication at this rate is possible if and only if $\mathrm{SNR} \geq \beta$.

In the presence of many emitters at positions $\Phi = \{x_i : i = 1, 2, \dots\}$ emitting at respective powers $\{P_i\}$, their signals contribute to the noise received by the terminals. Thus successful decoding at position y of the signal emitted at power P_0 from a station at position x_0 is possible if the signal to interference plus noise level is at least β:

$$\mathrm{SINR}(y, \Phi) = \frac{P_0\, l(\|y - x_0\|)}{N_0 + \gamma I(y, \Phi)} \geq \beta, \qquad (16.1)$$

where

$$I(y, \Phi) = \sum_{i \geq 1} P_i l(\|y - x_i\|)$$

is the sum of powers from all other emitters. A stochastic geometer may recognize in $I(y, \Phi)$ a *shot-noise* process evaluated at y, see, e.g., Daley (1971) or Møller (2003) and the references therein.

The coefficient γ which is between 0 and 1 stems from imperfect orthogonality of the received codes: the value 0 corresponds to a perfect orthogonality: the signal is decoded perfectly so no interference is received from the other emitters.

[4]In contrast, tunnels may produce coherent paths amplifying the signal thus making the exponent α less than 2.

The maximal interference with $\gamma = 1$ is typical for narrowband systems while for wideband systems one typically has $0 < \gamma < 1$ determined by the radio wave spreading, the processing gain, etc., see, e.g. Viswanath and Anantharam (1999).

Consider an example of the calculation of the probability that a mobile at position y can talk to a station x_0 at the origin (on the Palm space of the Poisson process Φ of base stations), i.e. it has a sufficient for decoding SINR level β as in (16.1). Assume that the emitting powers P_i are i. i. d. exponentially $\mathrm{Exp}(1/\mu)$ distributed with mean $1/\mu$ and that $\mathcal{L}_{N_0}(z) = \mathbf{E}\exp\{-zN_0\}$ is the Laplace transform of the background noise at the position of the mobile. These are usual assumptions dubbed *Rayleigh fading channels*. We then have that

$$
\begin{aligned}
\mathbf{P}\{\mathrm{SINR}(y,\Phi) > \beta\} &= \mathbf{P}\Big\{\frac{\beta(N_0 + \gamma I(y,\Phi))}{l(\|y\|)}\Big\}\\
&= \int_0^\infty e^{-z\mu\beta/l(\|y\|)} \mathrm{d}\mathbf{P}\{N_0 + \gamma I(y,\Phi) \le z\}\\
&= \mathcal{L}_{N_0}\Big(\frac{\mu\beta}{l(\|y\|)}\Big)\,\mathcal{L}_{I(y,\Phi)}\Big(\frac{\gamma\mu\beta}{l(\|y\|)}\Big).
\end{aligned}
\tag{16.2}
$$

The Laplace transform of the shot-noise process is just a particular form of the probability generating functional for the marked (by the powers P_i) point process:

$$
\begin{aligned}
\mathcal{L}_{I(y,\Phi)}(z) &= \mathbf{E}e^{-zI(y,\Phi)} = \mathbf{E}\prod_{x_i\in\Phi} e^{-zP_i l(\|y-x_i\|)}\\
&= \exp\Big\{-\lambda\int_{\mathbb{R}^2}\big(1 - \mathbf{E}e^{-zPl(\|x\|)}\big)\,\mathrm{d}x\Big\},
\end{aligned}
$$

where λ is the stations' density. Since $P \sim \mathrm{Exp}(1/\mu)$, this further simplifies to

$$
\begin{aligned}
\mathcal{L}_{I(y,\Phi)} &= \exp\left\{-\lambda\int_{\mathbb{R}^2}\Big(1 - \frac{\mu}{zl(\|x\|)+\mu}\Big)\,\mathrm{d}x\right\}\\
&= \exp\left\{-2\pi\lambda\int_0^\infty \frac{r\,\mathrm{d}r}{1+\mu/(zl(r))}\right\}
\end{aligned}
$$

and leads to the final expression:

$$
\mathbf{P}\{\mathrm{SINR}(y,\Phi) > \beta\} = \mathcal{L}_{N_0}\Big(\frac{\mu\beta}{l(\|y\|)}\Big)\,\exp\Big\{-2\pi\lambda\int_0^\infty \frac{r\,\mathrm{d}r}{1 + l(\|y\|)/(\gamma\beta l(r))}\Big\}.
$$

The nice decomposition (16.2) of the probability into two terms corresponding to the noise and interference is specific to exponential distribution assumption for received power, unfortunately. When the distribution of P is not exponential, an expression for the above probability can still be obtained by contour integration in the complex plane, see Baccelli and Błaszczyszyn (2001) for details.

When a mobile is switched on, it listens to a pilot signal periodically emitted by all the base stations (BSs) and on the basis of the SINR level of these signals the mobile determines which station cell (or cells) it is in. The geometry of the cells is thus based on (16.1) and has a rather complex structure, especially taking into account that the background noise N_0 is a time varying random quantity. The cells may intersect, i.e. more than one BS can provide a sufficient for decoding level of the SINR. When a few stations are participating in a call by sending the same signal to the mobile, it is known as soft-handover opposed to hard-handover case when only one BS (usually with the strongest signal) is talking to the mobile. Soft-handover improves quality of the received signal on the boundaries of the cells. An overview of different techniques used for organization of handover is presented in Ekiz *et al.* (2005).

Strictly speaking, the BS cell concerns only the possibility of the *mobile* to decode messages emitted by the BS, i.e. the *downlink*. Similarly, the *uplinkuplink* concerns capability of a station to decode the bit stream from the mobile. To reduce interference and thus to increase the capacity of the network, a *power control* mechanism is employed: while initializing a call, the stations participating in the soft handover negotiate with the mobile the signal level they want to receive from it based on the type of call and the QoS demands. The call is established if the mobile can adapt its transmission power to these requirements. Once the call is active, the phone is constantly adjusting its transmission power on the basis of Transmit Power Control (TPC) messages exchanged with the controlling BS. An effective power control reduces interferences and allows for more simultaneous communications (capacity) but also saves the telephone's battery, so it can be used longer without recharging.

Modelling the geometry of the cells is thus a very important issue since the key performance characteristics of the network depend on it. Early studies of cellular telephone systems used models where the BSs are arranged in a regular triangular grid and the antenna cells are the regular hexagons of the same size centred at the stations and thus forming a honeycomb lattice. Though many interesting results were obtained in this framework, it has been realized that for more accurate analysis one should take into account spatial variability of the cells. A natural approach is to assume that the antennae locations are given by a realization of a random point process and the cells are modelled by a tessellation constructed with respect to this point process. The performance characteristics now become functionals of the point process and the tessellation. If the process and the tessellation are stationary ergodic, then the expectations of the characteristics could be interpreted as the spatial averages of these characteristics for different locations. For instance, the probability that there are three stations in the r-neighbourhood of the origin equals the proportion of those space locations which have three stations at the distance less than r. The simplest model of this kind uses homogeneous Poisson process as a model for the stations' positions and the corresponding Voronoi tessellation as a model for the cells. It is also a model corresponding to an isotropic monotonically decreasing attenuation function in the hard handover and the antennas emitting at same power. Here the mobile users are always associated with the closest BS irrespectively of the received signal level. Imposing the minimal power level requirements in the hard handover case leads to incomplete Voronoi tessellations (or rather coverage processes) studied in Muche (1993). When the BSs emit at different powers, the Voronoi tessellation is replaced by the Laguerre tessellation, see Lautensack (2007) and Lautensack and Zuyev (2008).

Generally, in the soft handover case the cells based on the SINR levels do not form a tessellation: there may be places for which the relation (16.1) holds for more than one BS as well as there are places where SINR levels are below the threshold β for all BSs. The coverage process of the SINR cells was first considered in Baccelli and Błaszczyszyn (2001) where the main coverage characteristics: volume and shape of a typical cell, proportion of uncovered space and multiply covered space were obtained. The model has been further developed by the authors in Baccelli *et al.* (2002) where the cells are defined by the levels of a linear function defined through the path loss functions to all nuclei. This model also includes the Boolean model and the Voronoi tessellation as special cases.

Knowing the parameters of the handovers is of a primary importance for planning mobile communication systems, see e.g. Tran-Gia *et al.* (2000). Since they relate to mobiles in communication crossing the boundaries of BS cells, these parameters rely on the the mobility patterns of the users. The majority of handovers comes from relatively fast moving users in cars or in public transport so modelling the road traffic is the first thing to do. There is a huge traffic science literature, see, e.g. Marcotte and Nguyen (1998) and the references therein. Perhaps, the simplest model is to look at the mobile users as points moving along the lines representing the roads. The speed could be independently drawn for each mobile from a known distribution (which is a good model for multi-lane roads) or given by a velocity field which all the users follow (which is appropriate for narrow or congested roads), see e.g. Leung *et al.* (1994). Baccelli and Zuyev (1997) used the above Poisson–Voronoi model for the cells and the Poisson line process for the road system to obtain expressions for the intensity of the handover process.

Paging is another important issue arising due to mobility. In the GSM networks, each mobile number has an associated Home Location Register (HLR) which defines the set of BSs, the *location zone*, where the mobile is sought in the first instance when a call is coming for it. When a mobile user is out of his/her home area, the current location is stored in the Visitor Location Register (VLR) and also communicated to the mobile's HLR. The registers are usually kept at the Mobile Services switching Centres (MSC). Each register has a number of BSs attached to it spanning a few location zones. Since the position of a mobile is known up to a location zone only, all the BSs in the location zone broadcast a paging (search) signal when the call for the mobile is coming. So the fewer BS that are in the location zone the smaller is the cost of the search. But this cost should be offset by the cost of the register updates which soars when the locations zones are too small: the smaller the zones the higher the number of updates because of the users' mobility. This parametric optimization was considered in Baccelli and Zuyev (1999), where the location zones were modelled by the Voronoi cells.

The power or load control is another challenging complex optimization problem: given a configuration of the mobiles and their QoS requirements find a minimal allocation of powers at which all the mobiles and the BSs transmit so that the required SINRs for all the communications are provided. Solving this problem exactly requires complete knowledge of all active calls in the whole system which is hardly practically achievable, see e.g. Gunnarsson (1998) and the references therein. This is why telecom operators employ decentralized control schemes which depend on the characteristics local to the cell in which the calls are serviced. The problem becomes even more complicated if one wants to take mobility of the users and the associated handovers into account. The feasibility of the global power allocation was shown in Zander (1992) to depend on the required bit rates and the spectral radius of a matrix of normalised attenuation functions from user i to the BS j. Based on the Poisson model for the BSs and users' locations, Baccelli *et al.* (2004) establish feasibility and examine the performance of a decentralized power control in the case of unlimited powers. In the sequel, in his PhD thesis Karray (2007) proposes decentralized admission/congestion control protocols in the practical case of limited powers also.

Considering the instances of call initiation as 'births', the duration of the calls until hangup or forced termination as 'deaths' and the trajectories of the mobiles along which the transmission conditions should be maintained as 'service' leads to a new interesting class of spatial birth-mobility-and-death processes. They can be seen, on one hand, as generalizations of the spatial birth-and-death processes, see e.g. Preston (1977), and, on the other hand, as generalizations of the spatial queueing systems considered, for instance, in Serfozo (1999). Like the finite birth-and-death processes, they also possess a stationary distribution which is a Gibbs measure, and like the spatial Whittle networks, see Huang and Serfozo (1999), they are, under certain conditions, also time ergodic, see Baccelli *et al.* (2005, 2007) and Karray (2007). These results allowed the authors to evaluate or approximate call blocking probabilities for CDMA systems (when a new call cannot be accepted because of the capacity constraints) and the communication cut probabilities (when a call is dropped out because of, for instance, displacement to a non-serviced area).

To provide the 4G goal of seamless integration of various dynamic network users, people research into self-organizing networks. This calls for development of efficient distributed algorithms for resource allocation, in particular of the shared wireless bandwidth in WLAN. Paper Kauffmann *et al.* (2007) proposes a few algorithms of fair allocation of the bandwidth by wireless access nodes based on the Gibbs sampler where the potential function depends on the transmitting powers and the SINRs. Gibbs random fields look particularly promising for designing and analysing self-configuring wireless networks.

A comprehensive account of the state-of-the-art in the stochastic analysis of the wireless systems should soon be available in Baccelli and Błaszczyszyn (2008).

16.2.3 *Modelling ad hoc networks*

Challenges: Coverage, connectivity, cost, power management, medium access and sharing, routing, distributed information processing, self-organization and self-management

Techniques: Coverage processes, tessellations, random graphs, percolation, Matérn processes

There is a great number of challenges that ad hoc networks represent both for their physical engineering and for development of efficient protocols of their functioning. Leaving aside physical construction issues: size, battery capacity, antenna range, microchip design, onboard channel coding, modulation, etc., the most common points to be addressed on higher functional levels concern connectivity of the network, organization of effective gathering, prioritisation and pre-processing of the information, power control of individual nodes, distributed self-organization providing efficient routing, robustness to failures and topology changes, adaptation to dynamic environments and changing roles. Depending on applications, many other requirements and constraints can be added: real-time processing or transmission of information (video-streaming in ad hoc networks, time-critical monitoring), data sharing and privacy (multicast and VPN), provision of self-healing, longevity, cost, etc.

A variety of usage scenarios implies variety of models and research activities, but where the stochastic geometry techniques have been effectively used so far concerns such issues as the capacity, connectivity, optimal coverage, power management, media access control, energy aware multi-hop routing, hierarchical clustering for the data transmission and precessing, tracking and optimal positioning of the nodes.

The term 'ad hoc' assumes randomness, so widely used are the models where the nodes are represented by a realization of a stochastic point process on a line or on a plane. One-dimensional models are relevant for, say, a network of cars moving along the road and communicating with each other and are also often used as a testbed for new methods. The wireless connections between the nodes are typically either supposed to be based on the SINR levels as in (16.1), or just on the power of the signal. The first type of models are referred to as *physical models* (also known as *information theoretic connectivity* models). The second type is dubbed the *protocol models*. In the case of the protocol model, reception of a signal emitted at power P_i by a node x_i is possible at a node x_j if the power of the transmitted signal at x_j is above a fixed threshold (determined by the required data rate). Usually, the signal strength at position z emitted at power P from the origin is modelled by $Pl(z)$ with a path-loss function l as in the previous section. Then x_j can receive from x_i if $x_j \in S(y_i)$, where $S(y_i) = \{x : P_i l(x - y_i) \geq \beta$. Geometrically, the ensemble of sets $\{S(y_i)\}$ is a Boolean model, hence the term *Boolean connectivity* is also used. In the simplest case, the function l depends only on the distance and all the signals are emitted at the same powers. Then the corresponding Boolean model consists of balls of the same radius r and the two nodes can communicate if $\|x_i - x_j\| \leq r$. A few variations of this approach are possible. For directional antennas one replaces the balls with elongated shapes, like ellipses, or their unions (multi-beam antennas). Some accounting for the radio interferences in the framework of the protocol model can be made by assuming that the reception at x_j from x_i is only possible if, in addition, $\|x_k - x_j\| \geq (1 + \Delta)\|x_k - x_l\|$ for any other node x_l transmitting at the same time at the same channel to another node x_l.

The interference is the main issue which bounds the carrying capacity of ad hoc networks as it was first shown in Gupta and Kumar (2000). This work represents an example of how stochastic geometry techniques yield important results that should be undeniably taken into consideration when planning ad hoc networks. It raises serious issues about scalability of the ad hoc systems and it is, perhaps, the most cited paper on telecommunications which uses stochastic geometry hitting over 2000 citations after seven years since its publication. The paper considers a maximum throughput obtainable by each node when the nodes can transmit at a maximum rate of W bits per second to a randomly chosen destination. The throughput is the number of bits per second transmitted by the node when the relayed bits are only counted once. So when a k-bit long sequence is transmitted by m nodes in one second, the average throughput per node is k/m although each node effectively relayed k bits. The authors shows that in a large network consisting of n nodes, the maximum throughput per node is only of order $O(W/\sqrt{n})$. This fact is due to limitations in the transmission capacity of the nodes and due to the radio wave interference which prevents a node from receiving two different communications at the same time on the same frequency (like one cannot really listen to two radio stations on the same radioband). Although the formal proof in the paper is quite involved using a combination of percolation, random tessellations and Vapnik–Chernovenkis dimension techniques, why this result should be expected is easy to explain in geometric terms using the protocol model.

Imagine a network of n nodes randomly placed in a square $[0, \sqrt{n}]^2$ so that their density is kept constant for all n. Assume that each node wants to send some information to a randomly chosen node so that the average distance to the destination is of order $O(\sqrt{n})$. Assume also that there is only one radio frequency available: if there is a finite range of these the throughput just changes proportionally and this does not affect the asymptotics. An ongoing transmission from a node x to a range r makes it impossible for all other nodes in the ball of the radius r centred in the receiving node to transmit due to the interference. So the footprint of all currently going communications is a set of disjoint circles representing the 'silence zones' around the active senders. In order for the information to pass, each such circle should contain a receiving node, so that the average radius of a circle is at least of order of a constant. Thus the maximal number of concurrent transmissions has the maximal order of $\mathcal{O}(n)$. But a typical destination is at the distance $\mathcal{O}(\sqrt{n})$, so it takes $\mathcal{O}(\sqrt{n})$ hops to transmit each bit, hence there is only a maximum of $\mathcal{O}(\sqrt{n})$ source-destination routes supported in total by a maximum of n nodes and each transmitting maximum of W bits/second that gives the throughput of $\mathcal{O}(W/\sqrt{n})$ bits per second per node. The authors also demonstrate the same phenomenon in the physical model based on the SINR connectivity, see also Lévêque and Telatar (2004) and Xue and Kumar (2006).

An apparent strength of the paper is that it shows that the rate of $\mathcal{O}(W/\sqrt{n})$ is actually achievable by the choice of appropriate transmission algorithms and careful positioning of the nodes though this may not always be possible to guarantee in practice. For randomly placed nodes a $\mathcal{O}(W/\sqrt{n \log n})$ lower bound for the throughput per node was obtained. It was believed that the additional $\log n$ was a necessary price to pay if one wants to have a full connectivity of the network: this is the order of the longest edge in the shortest spanning tree of n randomly distributed nodes, see Penrose and Pisztora (1996). Hence, to reach these nodes one has to transmit at the power growing with n that leads to all kinds of undesirable phenomena. However, as it was shown in Franceschetti *et al.* (2007), even in the case of randomly placed nodes the throughput of (the strict) order $\mathcal{O}(W/\sqrt{n})$ is achievable with a TDMA scheme.

The argument is to exploit percolation properties which are central for ad hoc systems. Not only the existence of a giant percolating cluster means that every part of the system can communicate with any other part, but this cluster presents preferential pathways for information transmission. The reason for this is twofold: first, it is known from the percolation theory that the (necessarily unique) infinite cluster in 2D topologically resembles a mesh which rather homogeneously spans the whole plane, so for most of the points this cluster is not far away. Second, no matter how far two nodes are in the same cluster, they can exchange information at a constant guaranteed rate. To see this, we follow Dousse *et al.* (2006*a*).

Assume all the nodes transmit at the same power P consider two nodes in the same cluster of balls of radii $r/2$ centred in the nodes and the shortest path between them. Each ball in the shortest path, apart from the first and the last ones, intersect with exactly two other balls on the path: 'predecessor' and 'successor', otherwise a shortcut would be possible avoiding one of the balls. Now colour the balls in three colours, say, red, blue and green, starting from the initial node and repeating this sequence of colours along the path. Now schedule TDMA transmission scheme so that each colour nodes transmit at the same time and at the next time slot the next colour nodes transmit, etc. Assume red balls are transmitting and blue are receiving. Since the distances between red and blue balls are at at most r, any blue ball x_0 receives transmission from its predecessor red ball at power at least $Pl(r)$ (as above, we assume that the path loss function

$l(r)$ is decreasing). All the other transmitting red nodes contributing to the interference are at the distances at least r from the receiving blue node. This interference is no more than interference from the balls arranged in the densest packing, i.e. placed in the vertices of a regular honeycomb lattice. Since in the regular hexagon lattice there are six neighbours at distance r, six at distance $\sqrt{3}r$, etc., the powers received from these other red balls at x_0 does not exceed

$$\sum_{i=1}^{\infty} P(6l(r) + 6l(\sqrt{3}r) + 6l(2r) + \dots) \le P \sum_{i=1}^{\infty} 6(i+1)l(ir).$$

The sum above, call it $K(r)$, is finite because $\int rl(r)\,\mathrm{d}r < \infty$. Hence SINR at each receiver is at least a constant:

$$\mathrm{SINR} \ge \frac{Pl(r)}{N_0 + PK(r)}.$$

Now the idea is to use the percolating cluster of nodes as the 'backbone' information routes. By appropriately choosing the transmission radius, this cluster can cover a share of the infinite space arbitrarily close to 100%. The throughput of order $\mathcal{O}(W/\sqrt{n})$ along the backbone is achievable, the factor $1/\sqrt{n}$ appears here only because of the collision avoidance as explained earlier. The distance from the rest of the nodes to the backbone has order $\mathcal{O}(\log n)$ which accounts only for the first or the last hop in the route and thus contributes the same throughput rate. The maximal achievable throughput in a large wireless network from the information theoretical point is considered in Liu and Srikant (2008).

The problem of connectivity and coverage in ad hoc networks is an important subject on its own right. One usually wants a full connectivity in order to be able to pass information from any node to any node (or to a dedicated information collection point – the *sink*). In case of sensor networks, it is often required to provide a full coverage of the area so that no event happening in a 'hole' passes undetected. In some cases, it is just sufficient that the 'hole' space is not 1-connected: for instance, a vehicle will necessarily cross the area covered by sensors even if there are holes in the coverage provided they are bounded, see e.g. Biswas and Phoha (2007), Ahmed *et al.* (2005), Kesidis and Phoha (2003).

Depending on the model chosen for the wireless links, one studies the percolation properties of the specific Boolean model corresponding to the protocol connectivity, or connectivity properties of the random graphs related to the SINR-based links.

The Boolean connectivity of the net of n nodes with communication radius $r(n)$ is equivalent to the longest edge in the nodes' nearest-neighbour graph to be less than $r(n)$. Penrose (1997) finds the asymptotic distribution of the longest edge length for the case of nodes uniformly and independently distributed in a cube in a general dimension (see also Gupta and Kumar (1999) who studied asymptotic Boolean connectivity for uniform nodes in a 2D disk). These results give the order of the size of a largest 'void' in the network of many uniform nodes and establish a criterion for an asymptotic Boolean connectivity: in the planar case the probability that the network is connected approaches 1 iff $\pi n r^2(n) - \log n \to \infty$. It is interesting that this is similar to the connectivity of very different Bernoulli graphs, see Bollobàs (1985, Th. VII.3).

Percolation properties of the information theoretic connectivity graph depend not only on the density of the nodes but also on the orthogonality factor γ in (16.1), the parameter that measures the interference. When $\gamma = 0$, the SINR connectivity is equivalent to the Boolean connectivity when the intensity N_0 of the background noise is constant. For a large class of attenuation functions l, percolation in the physical model is, basically, the same as in the Boolean model: for the density λ of nodes high enough there is a critical value $\gamma'(\lambda)$ such that there is an infinite connected component for all $\gamma < \gamma'(\lambda)$, see Dousse *et al.* (2005, 2006*b*).

Routing in ad hoc systems is one of the most complex problems. Because of absence of a centralized control and constantly changing environment, application of the classical minimal path algorithms may not be feasible. Routing schemes should thus allow for construction of multihop paths from the source to destination using the limited information available to nodes. One class of such schemes is deterministic routing which takes into account only the geometry of the nodes. Examples are provided by the Voronoi Cell (VC) routing and the k-nearest neighbour routing. In VC-routing, the network region is divided into Voronoi cells with respect to a smaller number of elected *clusterhead* nodes. The source node first transits information to its closest clusterhead (hence Voronoi in the name) and next this clusterhead passes this information to its neighbouring clusterheads along the line to the destination node. The properties of such a path were studied in Baccelli, Tchoumatchenko and Zuyev (1998). In the other scheme, information is transmitted on each hop to the k-nearest node of the transmitting node which is closer to the destination. In the case $k = 1$ the routes emanating from the origin form the so-called *radial spanning tree* (RST) extensively studied in Baccelli and Bordenave (2007). Bordenave (2008) makes a further step considering an abstract *navigation* process on Poisson points: an iterative procedure which eventually leads to the 'destination' and establishes its important properties like asymptotic progress speed and limit theorems.

A serious limitation of the deterministic routing is that it does not take into account the remaining energy nor the bandwidth restrictions of the individual nodes which may lead to a congestion. Therefore alternative approaches were suggested in the literature by which multiple paths to the destination are dynamically constructed and used to prevent fast energy depletion along the preferential routes in energy constrained systems, see e.g. Gupta and Kumar (1997), Chang and Tassiulas (2000), Servetto and Barrenechea (2002). Farinetto *et al.* (2005) argue that homogenization of the residual energy is the key factor to prolonging the lifetime of energy constrained systems and suggested a few routing algorithms aiming at this. Baek and de Veciana (2005) use stochastic geometry techniques to evaluate spatial energy burdens under proactive multipath routing. Borkar and Kumar (2001) suggest a distributed asynchronous routing protocol inspired by the Traffic Science which aims at attaining a Wardrop-type equilibrium: the transmission delays over the utilized routes should be equal and smaller than those over the unutilised routes. Papers Haenggi (2005) and Zheng and Li (2008) compare performances of different deterministic and opportunistic routing schemes both in light teletraffic load when Poisson assumption on the distribution of the nodes is appropriate and in a saturated regime where it is more appropriate to use Matérn type of process to model the set of active nodes. The underlying reason is that due to randomness of demands and mobility of the nodes it can happen that too many nodes require transmission in one place, so to prevent the collisions associated with interferences, networks employ Media Access Control mechanisms (MAC) to prevent this from happening. This is the purpose of the Carrier Sense Multiple Access (CSMA) technique which bans origination of a new transmission in the vicinity of an on-going transmission. Under this scheme, the set of sending nodes at any given time zone represents a type of a hard-core process with non-overlapping 'silence zones' alluded above. Practically, this can be organized by sending a signalling burst before sending a data packet which defines the priority of the transmission. This mechanism is realized, for example, in HiPERLAN type 1 MAC. Considering this burst length as a mark of the node requesting a transmission, observe the apparent similarity with the elimination procedure in the construction of Matérn hard-core processes. This is why the Matérn-type processes are natural models for the time snapshot of the system of transmitting nodes in dense wireless ad hoc networks, see Nguyen *et al.* (2007).

A predecessor to CSMA MAC scheme is a spatial variant of ALOHA which has been a technique widely used in networks for a few decades now. It consists in resending the packets which were not acknowledged as been received at the destination at a later time. The slotted ALOHA allows every transmission to start only at discrete times lots. The probability of re-sending the lost packet by the transmitter at every start of the slot is a parameter of the protocol called the Media Access Probability or MAP. Baccelli *et al.* (2006) analysed the performance of ALOHA in wireless ad hoc settings and showed that ALOHA schemes with an optimized MAP can achieve the spatial reuse factor (defined as the distance to the receiver divided by the average distance between successive emitters on the communication path) to be just marginally worse than that in the CSMA. See the survey by Kumar *et al.* (2008) for more engineering details on popular MAC schemes employed in wireless networks.

Designing efficient information collection, processing and flow often calls for hierarchical organisation of the network. We have already mentioned the higher hierarchy *clusterheads*, which are responsible for gathering information from their cluster nodes and then transmitting it via shortcut long-distance links. This type of topology is attracting much attention and is known as *small-world* models. The name is inspired by the Internet which has developed into a real world-wide web so highly connected that it rarely takes more than a dozen of hops to reach any computer in the world from any place, see e.g. Reittu and Norros (2004), Ganesh and Xue (2007) and the references therein.

A range of algorithms were suggested for building hierarchy in wireless ad hoc networks, see e.g. Boukerche (2005), Yang and de Veciana (2005), Younis *et al.* (2006). A clusterheads election algorithm was proposed in Bandyopadhyay and Coyle (2003) and analysed by means of stochastic geometry. A closely related problem concerns an optimal sink(s) placement in the sensor networks. These are specialized nodes that the information gathered by the sensors should eventually reach, see e.g. Schmitt *et al.* (2006), Poe and Schmitt (2007). Information submission to the sinks could be considered as the task, in a sense, opposite to the *multicast*, when a server needs to disseminate information to a variety nodes, see e.g. Nandi *et al.* (2007). Multicast can be efficiently done by relaying the information through a series of local servers which will disseminate it further to their local neighbourhoods. This is why the stochastic methods developed for multicast, see Infocom (1999), Błaszczyszyn and Tchoumatchenko (2004), should also be applicable for ad hoc networks.

Self-organization of sensor networks is a hot subject since the sensors are often deployed randomly and it is often beneficial to put some sensors to sleeping state if the neighbours can provide adequate coverage of their sensing area. The sleeping nodes are scheduled to awake periodically and replace the nodes whose energy is depleted. In some applications the nodes have the ability to move in order to maximise coverage or optimize the message traffic. One of the algorithms used is based on the Voronoi tessellation when a node moves towards its farthest Voronoi neighbour vertex to fix the coverage or routing hole most likely to happen in the 'centre' of the cell, see a survey paper Younis and Akkaya (2008) and the references therein.

For a node in a ad hoc system, finding the destination or, more generally, finding the required information distributed over the nodes may be a non-trivial task. It is often a consequence of a small amount of information which may be available to the nodes, because, for instance, of constantly changing environment or due to the energy or storage or processing constraints at the individual nodes. A widely used destination search method is the Expanding Ring Search (ERS) by which a destination search packet is sent first to some neighbourhood of the sending node (the first ring). If the search produced no result, the search is then performed at a larger neighbourhood (the second ring) etc., eventually finishing by *flooding* when all the nodes in the network are requested. The paper Chang and Liu (2004) discusses a dynamic programming algorithm that optimises the ERS and which assumes complete knowledge of the system by the transmitting node. Deng and Zuyev (2008) show that with a small overhead a search close to the optimal may be performed with at most three rings before opting for the flooding.

It is worth mentioning in this respect various activities related to peer-to-peer applications in ad hoc systems. Since the future networks are to be heterogeneous, decentralized and dynamic, there is no choice for the network nodes (peers) but to cooperate in order to perform various functions: the information transmission, distributed storage and computing, search, authentication, etc. Structured P2P systems use Distributed Hash Table (DHT) to store data in the network and provide the nodes with their unique ID. Dozens of DHT lookup algorithms and routing protocols were proposed for the networks organised in loops, see e.g. Stoica *et al.* (2003), Leong *et al.* (2004). Non-DST P2P systems have also been proposed like ontology-based search Rostami *et al.* (2006). However, suitable P2P algorithms for ad hoc networks are yet to be developed.

16.3 Conclusion

An evident conclusion of this short survey is that during the last decade stochastic geometry has gained well deserved recognition in the telecommunication community as a necessary and vital tool for studying present and, more importantly, future telecommunication systems. There is a wide range of vibrant applications for an enthusiastic stochastic geometer to choose from and to make his or her own contribution to this important dynamic area of research. In the author's view, the most promising area in the nearest future will be (and actually is already becoming now) the ad hoc systems where a competence in point processes, random tessellations, percolation, random graphs, optimisation, contact processes, stochastic modelling will be in high demand. *May the road rise up to meet you!*[5]

References

Ahmed, N., Kanhere, S. S., and Jha, S. (2005). The holes problem in wireless sensor networks: a survey. *ACM SIGMOBILE Review*, **9**(2), 4–18.

Appleby, S. (1995). Estimating the cost of a telecommunications network using the fractal structure of the human population distribution. *IEE Proceedings: Communications*, **142**(3), 172–178.

Appleby, S. (1996). Multifractal characterization of the distribution pattern of the human population. *Geographic Analysis*, **28**(2), 147–160.

Baccelli, F. and Błaszczyszyn, B. (2001). On a coverage process ranging from the Boolean model to the Poisson Voronoi tessellation, with applications to wireless communications. *Adv. Appl. Probab.*, **33**, 293–323.

Baccelli, F. and Błaszczyszyn, B. (2008). *Spatial Modelling of Wireless Communications, A Stochastic Geometry Approach.* NOW Publishers.

Baccelli, F., Błaszczyszyn, B., and Karray, M. (2004). Up and downlink admission/congestion control and maximal load in large homogeneous CDMA networks. *MONET*, **9**, 605–617.

Baccelli, F., Błaszczyszyn, B., and Karray, M. (2005, March). Blocking rates in large CDMA networks via spatial Erlang formula. In *Proc. of IEEE INFOCOM*, Miami.

[5]A traditional Gaelic blessing.

Baccelli, F., Błaszczyszyn, B., and Karray, M. (2007). A spatial Markov queueing process and its applications to wireless loss systems. Technical Report 5517, INRIA.

Baccelli, F., Błaszczyszyn, B., and Mühlethaler, P. (2006). An Aloha protocol for multihop mobile wireless networks. *IEEE Trans. Information Theory*, **52**(2), 421–436.

Baccelli, F., Błaszczyszyn, B., and Tournois, F. (2002). Spatial averages of coverage characteristics in large CDMA networks. *ACM Wireless Networks*, **8**, 569–586.

Baccelli, F. and Bordenave, C. (2007). The radial spanning tree of a poisson point process. *Ann. Applied Probab.*, **17**(1), 305–359.

Baccelli, F., Gloaguen, C., and Zuyev, S. (2000). Superposition of planar Voronoi tessellations. *Comm. Statist. Stoch. Models*, **16**(1), 69–98.

Baccelli, F., Klein, M., Lebourges, M., and Zuyev, S. (1995). Géométrie aléatoire et architecture de réseaux de communications. Technical Report 2542, INRIA, May.

Baccelli, F., Tchoumatchenko, K., and Zuyev, S. (1998, May). Markov paths on the Poisson-Delaunay graph with applications to routing in mobile networks. Technical Report 3420, INRIA, Sophia-Antipolis, France.

Baccelli, F. and Zuyev, S. (1997). Stochastic geometry models of mobile communication networks. In *Frontiers in Queueing. Models and Applications in Science and Engineering.* (ed. J. Dshalalow), pp. 227–244. CRC Press.

Baccelli, F. and Zuyev, S. (1999). Poisson–Voronoi spanning trees with applications to the optimization of communication networks. *Oper. Research*, **47**(4), 619–631.

Baek, S. J. and de Veciana, G. (2005). Spatial energy balancing large-scale wireless multihop networks. In *Proc. IEEE INFOCOM*, pp. 1–12.

Bandyopadhyay, S. and Coyle, E. (2003). An energy efficient hierarchical clustering algorithm for wireless sensor networks. In *Proceedings of the 22nd Annual Joint Conference of the IEEE Computer and Communications Societies (Infocom 2003).*

Bedell, P. (1999). *Cellular/PCs Management. A Real World Perspective.* McGraw-Hill Professional Publishing.

Biswas, P. K. and Phoha, S. (2007). Hybrid sensor network test bed for reinforced target tracking. In *Sensor Network Operations* (ed. S. Phoha, T. LaPorta, and C. Griffin), Chapter 13, pp. 689–704. Wiley.

Błaszczyszyn, B. and Tchoumatchenko, K. (2004). Performance characteristics on multicast flows on random trees. *Stochastic Models*, **20**, 341–361.

Bollobàs, B. (1985). *Random Graphs.* Academic Press, Orlando, FL.

Bordenave, C. (2008). Navigation on a Poisson point process. *Ann. Applied Probab/*, **18**(2), 708–746.

Borkar, V. and Kumar, P. R. (2001). Dynamic Cesaro–Wardrop equilibration in networks. *IEEE Trans. on Automatic Control*, **48**(3), 382–396.

Boukerche, A. (2005). *Handbook of Algorithms for Wireless Networking and Mobile Computing*, Volume 8 of *Computer & Information Science Series*. Chapman & Hall/CRC.

Brazio, J., Tran-Gia, P., Akar, N., Beben, A., Burakowski, W., Fiedler, M., Karasan, E., Menth, M., Olivier, P., Tutschku, K., and Wittevrongel, S. (ed.) (2005). *Analysis and Design of Advanced Multiservice Networks Supporting Mobility, Multimedia, and Internetworking: COST Action 279 Final Report*. Springer.

Callaway, E. H. (2003). *Wireless Sensor Networks: Architectures and Protocols*. CRC Press.

Chang, J.-H. and Tassiulas, L. (2000). Fast approximate algorithms for maximum lifetime routing in wireless ad-hoc networks. In *Proceedings of the IFIP-TC6/European Commission International Conference on Broadband Communications, High Performance Networking, and Performance of Communication Networks*, Volume 1815 of *Lecture Notes In Computer Science*, pp. 702–713. Springer-Verlag.

Chang, N. and Liu, M. (2004, September 26 – October 1). Revisiting the TTL-based controlled flooding search: Optimality and randomization. In *Proc. of the 10th Annual ACM/IEEE Internation Conference on Mobile Computing and Networking (MobiCom '04)*, Philadelphia, PA, USA, pp. 85–99.

Daley, D. J. (1971). The definition of a multidimansional generalization of a short noise. *J. Appl. Probab.*, **8**, 128–135.

Deng, J. and Zuyev, S. (2008). On search sets of expanding ring search in wireless networks. *Ad Hoc Networks*, **8**(7), 1168–1181.

Dousse, O., Baccelli, F., and Thiran, P. (2005). Impact of interferences on connectivity in ad-hoc networks. *IEEE/ACM Trans. Networking*, **13**, 425–436.

Dousse, O., Franceschetti, M., Macris, N., Meester, R., and Thiran, P. (2006*b*). Percolation in the signal to interference ration graph. *J. Appl. Probab.*, **43**, 552–562.

Dousse, O., Franceschetti, M., and Thiran, P. (2006*a*). On the throughput scaling of wireless relay networks. *IEEE Trans. on Information Theory*, **52**(6), 2756–2761.

Ekiz, N., Küçüköner, S., and Fidanboylu, K. (2005, June). An overview of handoff techniques in cellular networks. *Proc. of World Academy of Science, Engineering and Technology*, **6**, 1–4.

Elsayed, K. M. F., Gerlich, N., and Tran-Gia, P. (2001). Efficient design of voice carrying fixed-network links in cdma mobile communication systems. *Telecommunication Systems*, **17**, 9–29.

Farinetto, C., Harle, D., Tachtatzis, C., and Zuyev, S. (2005). Efficient routing for the extension of lifetime and quality of energy constrained Ad Hoc networks. In *Performance challenges for efficient next generation networks* (ed. X. Liang, Z. Xin, V. Inversen, and G. Kuo), pp. 759–770. Publishing House,

BUPT. Proc. of the 19th International Teletrafic Congress, Beijing China, August 29–September 2, 2005.

Farinetto, C. and Zuyev, S. (2004). Stochastic geometry modelling of hybrid wireless/optical networks. *Perf. Eval.*, **57**, 441–452.

Fleischer, F., Gloaguen, C., Schmidt, H., Schmidt, V., and Schweiggert, F. (2008*b*). Simulation of typical modulated Poisson–Voronoi cells and their application to telecommunication network modelling. *Japan Journal of Industrial and Applied Mathematics*, **25**, 305–330.

Fleischer, F., Gloaguen, C., Schmidt, V., and Voss, F. (2008*a*). Simulation of the typical Poisson–Voronoi–Cox–Voronoi cell. *J. Statistical Computation and Simulation*.

Foss, S. and Zuyev, S. (1996). On a certain Voronoi aggregative process related to a bivariate Poisson process. *Adv. in Appl. Probab.*, **28**, 965–981.

Franceschetti, M., Dousse, O., Tse, D., and Thiran, P. (2007). Closing the gap in the capacity of random wireless networks via percolation theory. *IEEE Trans. on Information Theory*, **53**(3), 1009–1018.

Frey, A. and Schmidt, V. (1998*a*). Marked point processes in the plane I – A survey with applications to spatial modeling of communication networks. *Advances in Performance Analysis*, **1**(1), 65–110.

Frey, A. and Schmidt, V. (1998*b*). Marked point processes in the plane II – A survey with applications to spatial modeling of communication networks. *Advances in Performance Analysis*, **1**(2), 171–214.

Ganesh, A. and Xue, F. (2007). On the connectivity and diameter of Small-World networks. *Adv. Appl. Probab.*, **39**(4), 853–863.

Gilbert, E. (1961). Random plane networks. *J. Soc. Indust. Appl. Math.*, **9**(4), 533–543.

Gloaguen, C. (2001). Prise en compte de la dépendance spatiale du trafic dans un modèle hiérarchique stochastique du réseau. *Annales des Télécommunications*, **56**(3–4), 113–139.

Gloaguen, C. (2006). Conditional length distributions induced by the coverage of two points by a Poisson Voronoi tessellation: application to a telecommunication model. *Applied Stochastic Models in Business and Industry*, **22**(4), 335–350.

Gloaguen, C., Fleischer, F., Schmidt, H., and Schmidt, V. (2006). Fitting of stochastic telecommunication network models via distance measures and Monte-Carlo tests. *Telecommunication Systems*, **31**(4), 353–377.

Gloaguen, C., Fleischer, F., Schmidt, H., and Schmidt, V. (2008). Analysis of shortest paths and subscriber line lengths in telecommunication access networks. *Networks and Spatial Economics* (in print).

Glover, I. A. and Grant, P. M. (2004). *Digital Communications* (2nd edn). Pearson Education Ltd.

Gunnarsson, F. (1998). *Power control in cellular radio systems: Analysis, design and estimation.* Ph.D. thesis, Linkpings Universitet.

Gupta, P. and Kumar, P. R. (1997). A system and traffic dependent adaptive routing algorithm for ad hoc networks. In *Proceedings of the 36th IEEE Conference on Decision and Control*, San Diego, pp. 2375–2380.

Gupta, P. and Kumar, P. R. (1998). Critical power for asymptotic connectivity in wireless networks. In *Stochastic Analysis, Control, Optimisation and Applications* (ed. W. M. McEneany, G. Yin and Q. Zhang, *et al.*), pp. 547–566. Birkhäuser, Boston.

Gupta, P. and Kumar, P. R. (2000). The capacity of wireless networks. *IEEE Trans. Information Theory*, **IT-46**(2), 388–404.

Haenggi, M. (2005). On routing in random Rayleigh fading networks. *IEEE Trans. Wireless Commun.*, **4**(4), 1553–1562.

Heinrich, L., Schmidt, H., and Schmidt, V. (2005). Limit theorems for stationary tessellations with random inner cell structures. *Adv. Appl. Prob.*, **37**, 25–47.

Hill Associates, Inc. (2001). *Telecommunications: A Beginner's Guide (Network Professional's Library).* McGraw-Hill Osborne.

Huang, X. and Serfozo, R. (1999). Spatial queueing processes. *Math. Oper. Res.*, **24**, 865–886.

Infocom (1999). *Self Organizing Hierarchical Multicast Trees and their Optimization.* Infocom.

Karl, H. and Willing, A. (2007). *Protocols and Architectures for Wireless Sensor Networks.* Wiley.

Karray, M. (2007). *Analytic evaluation of wireless cellular networks performance by a spatial Markov process accounting for their geometry, dynamics and control schemes.* Ph.D. thesis, École Nationale Supérioire des Télécommunications, Paris.

Kauffmann, B., Baccelli, F., Chaintreau, A., Mhatre, V., Papagiannaki, K., and Diot, C. (2007). Measurement-based self organization of interfering 802.11 wireless access networks. In *INFOCOM 2007. 26th IEEE International Conference on Computer Communications. IEEE*, pp. 1451–1459.

Kesidis, G., Konstantopoulos T. and Phoha, S. (2003, October). Surveillance coverage of sensor networks under a random mobility strategy. In *IEEE Sensors Conference*, Toronto.

Kumar, S., Raghavan, V. S., and Deng, J. (2008). Medium access control protocols for ad hoc wireless networks: a survey. *Ad Hoc Networks*, **4**, 326–358.

Lautensack, C. (2007). *Random Laguerre Tessellations.* Ph.D. thesis, Universität Karlsruhe. Verlag Lautensack, Weiler bei Bingen, `http://www.itwm.fhg.de/bv/theses/works/lautensack`.

Lautensack, C. and Zuyev, S. (2008). Random Laguerre tessellations. *Adv. Appl. Probab.*, **40**(3), 630–650.

Leong, B., Liskov, B., and Demaine, E. (2004, November). Epichord: Parallelizing the chord lookup algorithm with reactive routing state management. In *Proc. of the IEEE International Conference on Networks (ICON 2004)*, Volume 1, Singapore, pp. 270–276.

Leung, K. K., Massey, W. A., and Whitt, W. (1994). Traffic models for wireless communication networks. *IEEE J. on Selected Areas in Commun.*, **12**(8), 1353–1364.

Lévêque, O. and Telatar, E. (2004). Upper bounds on the capacity of ad-hoc wireless networks. In *Proc. Inf. Theory Symp. (ISIT)*, Chicago, IL.

Liu, X. and Srikant, R. (2008). Asymptotic uniform data-rate guarantees in large wireless networks. *Ad Hoc Networks*, **6**, 325–343.

Maier, R., Mayer, J., and Schmidt, V. (2004). Distributional properties of the typical cell of stationary iterated tessellations. *Mathematical Methods of Operations Research*, **59**(2), 287–302.

Maier, R. and Schmidt, V. (2003). Stationary iterated tessellations. *Adv. Appl. Prob.*, **35**(2), 337–353.

Marcotte, P. and Nguyen, S. (ed.) (1998). *Equilibrium and Advanced Traffic Modelling.* Kluwer, Boston.

Molchanov, I. and Zuyev, S. (2000). Variational analysis of functionals of Poisson processes. *Math. Oper. Research.*, **25**(3), 485–508.

Molchanov, I. and Zuyev, S. (2002). Steepest descent algorithms in space of measures. *Statistics and Computing*, **12**, 115–123.

Møller, J. (2003). Shot noise Cox processes. *Adv. in Appl. Probab.*, **35**(3), 614–640.

Muche, L. (1993). An incomplete Voronoi tessellation. *Applicationes Mathematicae*, **22**, 45–53.

Nandi, A., Ganjam, A., Druschel, P., Stoica, I., Zhang, H., and Bhattacharjee, B. (2007, April). Saar: A shared control plane for overlay multicast. In *Proceedings of NSDI'07*, Cambridge, MA.

Neveu, J. (1976). Sur les mesures de Palm de deux processus ponctuels stationnaires. *Z. Wahrsch. verw. Gebiete*, **34**, 199–203.

Nguyen, H. Q., Baccelli, F., and Kofman, D. (2007, March). A stochastic geometry analysis of dense IEEE 802.11 networks. In *Proc. of IEEE Infocom'07.*

Norros, I. (2005). Teletraffic as a stochastic playground. *Scandinavian Journal of Statistics*, **32**(2), 201–215.

Penrose, M. D. and Pisztora, A. (1996). Large deviations for discrete and contionuous percolation. *Adv. Appl. Probab.*, **28**, 29–52.

Penrose, M. D. (1997). The longest edge of the random minimal spanning tree. *Ann. Appl. Prob.*, **7**(2), 340–361.

Phoha, S., LaPorta, T., and Griffin, C. (ed.) (2007). *Sensor Network Operations.* Wiley.

Poe, W. Yi. and Schmitt, J. B. (2007, July). Minimizing the Maximum Delay in Wireless Sensor Networks by Intelligent Sink Placement. Technical Report 362/07, University of Kaiserslautern, Germany.

Preston, C. (1977). Spatial birth-and-death processes. *Bull. Int. Statist. Inst.*, **46**(2), 371–391.

Reittu, H. and Norros, I. (2004). On the power-law random graph model of massive data networks. *Perform. Eval.*, **55**(1-2), 3–23.

Rostami, H., Habibi, J., Abolhassani, H., Amirkhani, M., and Rahnama, A. (2006). An ontology based local index in P2P networks. In *The 2nd International Conference on Semantics, Knowledge, and Grid.*

Sallai, G. (1988, June). Optimal network structure with randomly distributed nodes. In *12th International Teletraffic Congress*, Torino, pp. 2.1B.4.

Sarkar, S., Dixit, S., and Mukherjee, B. (2007). Hybrid wireless-optical broadband-access network (woban): A review of relevant challenges. *J. Lightwave Tech.*, **25**(11), 3329–3340.

Sarkar, S., Yen, H.-H., Dixit, S., and Mukherjee, B. (2008). Hybrid wireless-optical broadband access network (WOBAN): network planning and setup. *IEEE Journal on Selected Areas in Communications.* (to appear).

Schmitt, J. B., Zdarsky, F., and Roedig, U. (2006, May). Sensor network calculus with multiple sinks. In *Performance Control in Wireless Sensor Networks Workshop at the 2006 IFIP Networking Conference*, pp. 6–13.

Serfozo, R. (1999). *Introduction to Stochastic Networks.* Springer.

Servetto, S. D. and Barrenechea, G. (2002). Constrained random walks on random graphs: routing algorithms for large scale wireless sensor networks. In *Proc. of the 1st ACM international workshop on Wireless sensor networks and applications*, pp. 12–21.

Shepard, S. (2005). *Telecom Crash Course* (2nd edn). McGraw-Hill Professional.

Stoica, I., Morris, R., Liben-Nowell, D., Karger, D. R., Kaashoek, M. F., Dabek, F., and Balakrishnan, H. (2003). Chord: A scalable peer-to-peer lookup protocol for internet applications. *IEEE/ACM Transactions on Networking*, **11**(1), 36–43.

Stoyan, D., Kendall, W., and Mecke, J. (1995). *Stochastic Geometry and its Applications* (second edn). Wiley, Chichester.

Tchoumatchenko, K. and Zuyev, S. (2001). Aggregate and fractal tessellations. *Prob. Theory and Rel. Fields*, **121**(2), 198–218.

Tran-Gia, P., Leibnitz, K., and Tutschku, K. (2000). Teletraffic issues in mobile communication network planning. *Telecommunication Systems*, **15**.

Tutschku, K. and Tran-Gia, P. (2005). Traffic characteristics and performance evaluation of peer-to-peer systems. In *Peer-to-Peer-Systems and Applications* (ed. K. W. Ralf Steinmetz), Springer.

Viswanath, P. and Anantharam, V. (1999). Optimal sequences and sum capacity of syncronous CDMA systems. *IEEE Trans. Information Theory*, **45**, 1984–1991.

Webb, W. (ed.) (2007). *Wireless Communications: The Future.* Wiley.

Wisely, D. (2007). Cellular mobile: the generation game. *BT Technology Journal*, **25**(2), 27–41.

Xue, F. and Kumar, P. R. (2006). Scaling laws for ad hoc wireless networks: An information theoretic approach. Volume 1 of *Foundations and Trends in Networking*, pp. 145–270. NOW Publishers, Delft, The Netherlands.

Yang, X. and de Veciana, G. (2005). Inducing spatial clustering in mac contention for spread spectrum ad hoc networks. In *Proc. Sixth ACM International Symposium on Mobile Ad Hoc Networking and Computing (MobiHoc)*, pp. 121–132.

Younis, M. and Akkaya, K. (2008). Strategies and techniques for node placement in wireless sensor networks: a survey. *Ad Hoc Networks*, **6**, 621–655.

Younis, O., Krunz, M., and Ramasubramanian, S. (2006). Node clustering in wireless sensor networks: Recent developments and deployment challenges. *IEEE Network*, **20**(3), 20–25. (special issue on wireless sensor networking).

Zander, J. (1992). Performance of optimum transmitter power control in cellular radio systems. *IEEE Trans. Vehic. Technol.*, **41**(1), 57–62.

Zheng, Rong and Li, Chengzhi (2008). How good is opportunistic routing? – a reality check under Rayleigh fading channels. In *MSWiM'08: Proceedings of the 11th international symposium on modeling, analysis and simulation of wireless and mobile systems*, New York, NY, USA, pp. 260–267. ACM.

17

RANDOM SETS IN FINANCE AND ECONOMETRICS

Ilya Molchanov

Abstract

This chapter surveys several examples where random sets appear in mathematical finance and econometrics: trading with transaction costs, risk measures, option prices, and partially identified econometric models.

17.1 Introduction

Although the monetary wealth of individuals or financial institutions is additive and a merge or cash flow corresponds to the arithmetic sum of financial assets, a number of financial phenomena exhibits non-additivity. From the consumption (or consumption-related utility) viewpoints the combined utility for two persons each winning one million dollars may not be the same as for one person who has won two millions. The overall loss for 1000 lottery players each losing on 1 pound bet is not perceived to have the same effect as for one person who has lost on a 1000 pounds stake. In finance this effect is widely used in diversifying investments in order to reduce the overall risk. The linearity assumption is often replaced by subadditivity (superadditivity) or convexity (concavity) properties.

A quite general (if not ultimate) mathematical source of convex functionals is the supremum applied to a collection of linear functionals. While linear functionals on a linear space become points in the dual space, their suprema in many cases correspond to convex sets. This fact sometimes calls for the use of geometric arguments and tools that come from the theory of random sets in order to study random non-linear phenomena. As we will see, the key technique is provided by various selection theorems that establish the existence of (random) linear functionals dominating (or dominated by) some non-linear ones. In the language of random sets these results manifest in the existence of random vectors (selections) that have some useful properties and take values from particular random sets.

The studies of random sets, their distributions, expectations and limit theorems form an essential part of stochastic geometry, see Matheron (1975), Stoyan, Kendall and Mecke (1995) and Section 1.4 of Chapter 1. In many cases they bring forward new relationships between convex geometry and multivariate statistics, see Mosler (2002). The modern theory of random sets is presented in Molchanov (2005). In this chapter we demonstrate how the notions of random sets and associated concepts from convex geometry play a significant role in mathematical finance.

The remaining part of this introduction goes through several examples discussed in this survey.

Transaction costs The major part of classical mathematical finance deals with assets having well-defined prices at which they can be bought and sold. In the presence of transaction costs this classical model is no longer applicable. The transaction costs result in the appearance of two prices for each tradeable asset: the ask price at which the asset can be bought and the bid price at which it can be sold. The transaction costs can be clearly seen in currencies trading, even at the level of money exchange by physical persons. The non-linearity effect is apparent here: simultaneous purchase and sale of 1 Euro at the UK money exchange booth is worse than making no transaction at all.

Section 17.2 begins with discussion of Kabanov's model of transaction costs in currency markets following Kabanov (1999) and Schachermayer (2004). The trading operations in this model lead to currency portfolios that belong to certain polyhedral (possibly random) cones. The key issue is the closedness (in L^0) of cones formed by random vectors that almost surely belong to these cones. The closedness property makes it possible to utilize the separation argument to show that the only non-negative portfolio resulting from trades with no initial investment is the zero one. Section 17.2 then surveys various variants and generalizations of Kabanov's model for not necessarily polyhedral cones. The absence of arbitrage is reduced to the existence of a martingale selection for a certain set-valued process, i.e. the existence of a martingale that at each time point almost surely belongs to the random set being the value of this set-valued process. Finally, some conditions for the existence of a martingale selection of a set-valued process are presented. It should be noted that set-valued processes also naturally appear if transaction costs are absent, but some parameters of the model (most importantly, the volatility) are not exactly specified, see Mykland (2000).

Risk measures Measuring risk is one of the central topics in modern mathematical finance. The current government regulations require that financial institutions provide quantitative information about the level of risk involved in their operations and accordingly set aside necessary capital reserves. The psychological perception of risk is highly non-linear, and also the pure quantitative assessment of risk shows a natural non-linearity. For instance, if X and Y represent investments in two different currencies, then the risk of the total investment is less than the sum of individual risks, since the fluctuations of the exchange rate might move the values of X and Y in the opposite directions.

Section 17.3 surveys the basic results concerning univariate risk measures, see Artzner, Delbaen, Eber and Heath (1999) and Delbaen (2002). Although it is not possible to relate a general risk measure to a random set, in some cases this relationship leads to a new interpretation of risk measures. The obtained risk measures coincide with so-called spectral risk measures studied by Acerbi (2002).

The major part of the developing theory of risk measures concerns the univariate case, where the aim is to assess the risk of a single investment. However, in most practical situations the institutions operate with portfolios that consist of a number of different assets or other financial products. The dependency between these components becomes an important issue and may influence the risk assessment of the overall financial position. The commonly used approach is to aggregate the individual risks using their monetary equivalents and then assess the overall risk. This practice however does not reflect the financial realities, since losses in one or several components of a large portfolio are often offset by gains in other components. Section 17.3.3 describes first steps towards defining multivariate risk measures emphasising their relationships with multivariate statistics, see Cascos and Molchanov (2007). The mere nature of the multivariate setting calls for the use of geometric concepts, so that risk measures become set-valued.

Option prices Options are extremely important financial instruments that give the holder the right to perform certain financial transactions. In other words, the option pays the maximum of the net result of this transaction and zero, that corresponds to no transaction at all, i.e. the case when the option is worthless. The simplest case is provided by European options that can be exercised at some fixed (maturity) time T and pay the holder $(S_T - k)_+$ (call option) or $(k - S_T)_+$ (put option). Here S_T is the asset price at time T and k is a parameter of the option called the strike price. In a perfect market, the value of an option can be found as the expectation of the payoff with respect to the equivalent martingale measure $\mathbf{Q}$, so that the process $(S_t)_{t=0}^T$ is a $\mathbf{Q}$-martingale.

Section 17.4 highlights some relationships between the values of options and support functions of certain convex sets called lift zonoids, see Mosler (2002) and Chapter 13 in the present volume. It is based on the work of Schmutz (2007) and Molchanov and Schmutz (2008), who have shown that symmetry properties of these lift zonoids correspond to important financial parities, notably the put-call duality and put-call symmetry for multiasset options. Geometric characteristics of the lift zonoid, e.g. the curvature, are closely related to financial quantities that, for instance, determine the hedge.

Partially identified models One of the basic assumptions in the classical statistical studies is that the observed probability distribution perfectly identifies the unknown parameter θ. It has been recently appreciated (see Imbens and Manski 2004 and Manski 2003) that in many natural situations statistical observations can in the best case only identify a certain set, to which θ belongs.

These *partially identified* statistical models appear, for instance, when outcomes are missing or when the unknown parameter satisfies some inequalities instead of equalities, see Chernozhukov, Hong and Tamer (2007). Another important source of partially identified models are regression models with interval-valued response or explanatory variables, see Manski and Tamer (2002).

For instance, the relevant situation appears when respondents report only some income bracket instead of the exact value of their income or the exact data are transformed to intervals for anonymity purposes. Linear models with interval-valued response have been studied by Diamond (1990) and Gil, López-García, Lubiano and Montenegro (2001) by applying techniques based on least squares fitting.

Section 17.5 describes an alternative approach suggested by Beresteanu and Molinari (2008), which aims to treat the relevant statistical problem directly by interpreting the estimated parameters of the linear model as a set and using limit theorems for Minkowski sums of random sets in order to derive properties of the corresponding estimator.

Section 17.5.2 deals with another setting, where the aim is to estimate a parameter of a random set distribution using observations of a selection of this random set. Since a selection does not suffice to describe the full distribution of a random set, this setting also leads to partially identified statistical problems. This situation appears, for instance, in estimation problems for games with multiple equilibria, see Beresteanu, Molchanov and Molinari (2008).

17.2 Transaction costs

17.2.1 *Cone models of transaction costs*

Transaction costs in a currency market with d traded currencies can be described by means of a *bid-ask matrix* $\Pi = (\pi^{ij})_{ij=1,d}$ where π^{ij} is the number of units of currency i (or, more generally, an asset number i) needed as a payment for one unit of currency j, see Kabanov (1999) and Schachermayer (2004). It is always assumed that Π has all positive entries, the diagonal entries are all 1 and $\pi^{ij} \leq \pi^{ik}\pi^{kj}$, the latter means that a chain of exchanges never beats the result of a direct transaction.

Fix a probability space $(\Omega, \mathfrak{F}, \mathbf{P})$ and a discrete time filtration $(\mathfrak{F}_t)_{t\geq 0}$. In this time-dependent stochastic setting the bid-ask matrix Π_t is a random adapted matrix-valued process that depends on (discrete) time and is called a *bid-ask process*, see Campi and Schachermayer (2006) for the continuous time case. By discounting prices, it is always assumed that all interest rates are zero.

A currency portfolio is a vector $x \in \mathbb{R}^d$ that represents the number of physical units of the currencies held by an investor. The set of portfolios available at price zero is the cone $-\hat{K}(\Pi)$ in $\mathbb{R}^d$ spanned by the negative basis vectors $-e_i$, $1 \leq i \leq d$, and the vectors $-\pi^{ij}e_i + e_j$ for all i and j. The latter portfolios can be realised by borrowing π^{ij} units of asset i and buying with this one unit of j. This portfolio can be liquidated by buying π^{ij} units of asset i with at least $\pi^{ji}\pi^{ij} \geq \pi^{jj} = 1$ unit of asset j. The cone of portfolios available at price zero is centrally symmetric to the *solvency cone* $\hat{K}(\Pi)$, see Grigoriev (2005) and Schachermayer (2004). The hats in these notation indicate that the elements of the cones represent physical units of the assets.

Consider any x from the cone $-\hat{K}(\Pi)$. If the assets were priced at $s = (s_1, \dots, s_d)$, then this portfolio would cost $\langle x, s \rangle = \sum x_i s_i$, which should be at most zero. Thus, s might be used as a price system if and only if $\langle x, s \rangle \le 0$ for all $x \in -\hat{K}(\Pi)$. In other words, all *consistent price systems* form a cone $K^*(\Pi)$, which is polar to $-\hat{K}(\Pi)$. The cone $K^*(\Pi)$ is also called the positive dual cone to the solvency cone $\hat{K}(\Pi)$. Note that $K^*(\Pi) \setminus \{0\}$ is a subset of the interior of $\mathbb{R}^d_+$. Later on we abbreviate $\hat{K}(\Pi_t)$ as $\hat{K}_t$ with a similar agreement for other related cones.

Having started with zero initial endowment, by trading in one time unit it is possible to arrive at a portfolio which belongs to $-\hat{K}_0$. If we write $L^0(G, \mathfrak{F})$ for the family of all $\mathfrak{F}$-measurable random vectors with values in a set $G \subset \mathbb{R}^d$, then in T time units one can arrive at any portfolio from

$$\hat{A}_T = \sum_{t=0}^{T} L^0(-\hat{K}_t, \mathfrak{F}_t),$$

which is called the set of *attainable portfolios*. The sum in the right-hand side is the Minkowski sum of subsets of $L^0(\mathbb{R}^d, \mathfrak{F}_T)$, i.e. the set of all sums formed by elements of these subsets, see Chapter 1. The major issue then is to determine whether $\hat{A}_T$ might contain portfolios with a positive value, which then would lead to a sure profit without investment. The typical way to prove such results is to show the closedness of the cone $\hat{A}_T$ in L^0 and then apply the separation argument. Note that the Minkowski sum of closed but non-compact sets in a linear space is not necessarily closed.

The bid-ask process $(\Pi_t)_{t=0}^T$ is said to satisfy the *no-arbitrage* property (NA) if the intersection of $\hat{A}_T$ and the family $L^0(\mathbb{R}^d_+, \mathfrak{F}_T)$ of non-negative $\mathfrak{F}_T$-measurable random vectors is exactly $\{0\}$. In order to avoid arbitrage for the agents, who might profit from transaction costs by cancelling the effects of opposite operations and pocketing the transaction costs on the both, the sequence of portfolios is sometimes assumed to be increasing, see Jouini (2000) and Kaval and Molchanov (2006), i.e. both the short and long positions are assumed to be non-decreasing in t. However, this assumption is generally superfluous.

It is apparent that attainable portfolios in this model of transaction costs are random vectors that take values in certain polyhedral cones. These cones are determined by (possibly random) bid-ask matrices and so can be treated as random polyhedral closed cones in $\mathbb{R}^d$. Grigoriev (2005) showed that for two assets (i.e. for $d = 2$) the no-arbitrage property is equivalent to the existence of an $(\mathfrak{F}_t)$-martingale ζ_t that a.s. takes values in $K_t^* \setminus \{0\}$ for all $t = 0, \dots, T$. This martingale $(\zeta_t)_{t=0}^T$ is called a *consistent price process* associated with $(\Pi_t)_{t=0}^T$. Counterexamples confirm that this result does not hold for $d \ge 4$, see Schachermayer (2004), and for all $d \ge 3$ in case of general polyhedral cones K_t^* not necessarily associated with bid-ask matrices, see Grigoriev (2005, Ex. 5.1). The case of bid-ask matrices and $d = 3$ is still open. In general, an equivalent interpretation of the no-arbitrage condition using consistent price processes in the space of arbitrary dimension $d \geq 3$ is not yet known.

In view of this, Schachermayer (2004) has shown the equivalence of two stronger statements: the robust no-arbitrage property and the existence of a strictly consistent price process. The *robust no-arbitrage* means the existence of a bid-ask process $\tilde{\Pi}$ with smaller bid-ask spreads

$$\frac{1}{\pi^{ji}} < \frac{1}{\tilde{\pi}^{ji}} < \tilde{\pi}^{ij} < \pi^{ij}, \quad 1 \leq i \neq j \leq d,$$

which also satisfies the no-arbitrage property, meaning that the agent might offer some non-zero discounts on transaction costs while still ensuring no-arbitrage. A consistent price process ζ_t is said to be *strict* if ζ_t a.s. belongs to the relative interior of K_t^* for all t, where the *relative interior* of a set is the interior taken in the smallest affine subspace containing this set. The consistent price system $\tilde{K}_t^*$ generated by $\tilde{\Pi}$ is a subset of the relative interior of K_t^*. The appearance of the relative interior in this context is quite natural, since if K_t^* lies in a hyperplane, certain assets (or some combinations of them) can be freely exchanged. One says that the model exhibits *efficient friction* if K_t^* has non-empty interior for all t. In this case all exchanges involve positive transaction costs, the interior of K_t^* coincides with the relative interior and so each strictly consistent price process automatically belongs to the interior of K_t^*. The proof by Schachermayer (2004) first aims to show that the robust no-arbitrage implies the closedness of $\hat{A}_T$ in L^0, which in turn by a separation argument yields the existence of a strictly consistent price process. The opposite implication does not involve the closedness argument. Indeed, $\hat{A}_T$ can be arbitrage-free but not closed, see Jacka, Berkaoui and Warren (2008). They showed that it is possible to define an adjusted bid-ask process $\tilde{\Pi}$ such that the corresponding cone $\tilde{A}_T$ contains $\hat{A}_T$ and is contained in the closure of $\hat{A}_T$. Furthermore, either $\tilde{A}_T$ contains an arbitrage or is arbitrage-free and closed. The entries $\tilde{\pi}_t^{ij}$ of the adjusted bid-ask process equal either π_t^{ij} or $1/\pi_t^{ji}$ depending on the occurrence of some events.

Since a random vector that a.s. belongs to a random set is called a selection, a consistent price process is naturally called a *martingale selection* of the set-valued process $(K_t^* \setminus \{0\})_{t=0}^T$. Set-valued processes (or processes of random sets) are discussed by Hu and Papageorgiou (1997) and Molchanov (2005, Ch. 5). In view of applications to finance it is also sensible to explore strict martingale selections, i.e. martingales that take values from the relative interiors of the sets from the set-valued process. This concept has not been yet systematically studied in the literature on set-valued martingales outside finance.

17.2.2 *Models with a money account*

Assume now that the first asset represents a money account. Since this asset is traded without transaction costs, the prices of other assets can be represented as bid and ask prices in relation to the money account. Let $S_t^{a,i}$ and $S_t^{b,i}$ be the ask and bid prices (so that $[S_t^b, S_t^a]$ is the *bid-ask spread*) for asset $i = 2, \dots, d$ at time t. If we set $S_t^{a,1} = S_t^{b,1} = 1$ for all t, then it is possible to define a bid-ask matrix by setting $\pi_t^{ij} = S_t^{a,j}/S_t^{b,i}$. In the opposite direction, a bid-ask matrix admits a bid-ask spread representation with a money account if $\pi_t^{ij} = \pi_t^{i1}\pi_t^{1j}$ for all time moments t and every assets i and j. While this is always possible if $d = 2$, this assumption restricts the possible family of solvency cones in dimension $d \geq 3$.

If $d = 2$, i.e. for the bid-ask spread $[S_t^b, S_t^a]$ on a single asset, Grigoriev (2005, Cor. 2.9) showed that the no-arbitrage condition is equivalent to the existence of a martingale S_t with respect to a probability measure equivalent to the original probability measure $\mathbf{P}$, such that $S_t \in [S_t^b, S_t^a]$ a.s. for all $t = 0, \dots, T$. This means that the interval-valued process $[S_t^b, S_t^a]$ possesses a martingale selection with respect to an equivalent measure. The earlier results in this direction go back to Jouini and Kallal (1995). Although (for $d \geq 3$) this setting is more restrictive than the bid-ask matrix formulation, it corresponds to an intuitive perception of a price that belongs to some interval and is not simply rescalable as in the case of price systems given by cones. Furthermore, the family of all martingale selections can be used to determine the upper and lower prices of claims. These prices are obtained by taking supremum and infimum of the expectations of claim payoffs obtained by substituting in the payoff function all possible martingale selections.

For multiple assets with a money account, Jouini (2000) considered *set-valued price processes* being rectangles or parallelepipeds. While this setting perfectly fits into the idea of bid-ask matrices, it does not take into account possible discounts for simultaneous transactions on several related assets. This *link-save* effect has been noticed by Kaval and Molchanov (2006). There it was assumed that prices of several assets at time t are described by a convex set Z_t in the first quadrant, so that the price of a combination of assets $u = (u_1, \dots, u_d) \in \mathbb{R}^d$ (expressed in physical units) is given by the *support function* $h(Z_t, u)$ of the set Z_t. The sublinearity of support functions corresponds again to the sublinearity property of prices, since the price of the combination $u' + u''$ does not exceed the sum of the individual prices of u' and u''. It is shown in Kaval and Molchanov (2006) that the existence of a martingale selection of the set-valued process $(Z_t)_{t=0}^T$ in the square integrable case is equivalent to the no-arbitrage property. If we add a money account to this model, then the cone with base $\{1\} \times Z_t$ becomes an analogue of the cone K_t^* in Kabanov's model of transaction costs. It should be noted however that the corresponding cone is not a polyhedral cone any longer.

For (not necessarily polyhedral) cones $\hat{K}_t$ that represent solvency sets, Jacka, Berkaoui and Warren (2008, Th. 4.10) showed that the closure of $\hat{A}_T$ is arbitrage-free if and only if there exists a martingale selection for the sequence $K_t^* \setminus \{0\}$, $t = 0, \dots, T$. This martingale selection can be chosen to be bounded in norm by $c\phi$ for each $\mathfrak{F}_T$-measurable positive ϕ and some constant $c > 0$.

Rásonyi (2008) deals directly with solvency sets that are general (not necessarily polyhedral) random convex cones $\hat{K}_t$ in $\mathbb{R}^d$. Assuming that $\hat{K}_t$, $t = 0, \dots, T$, are proper (i.e. under the efficient friction condition, where the interior of $\hat{K}_t$ coincides with the relative interior) he proved that the existence of a strict martingale selection is equivalent to the strict no-arbitrage (NA^s). The latter means that the intersection of the cone $\hat{A}_t$ of portfolios available at time t starting from zero initial endowment and the family of random vectors with values in K_t^* is $\{0\}$ for all $t = 0, \dots, T$, thereby strengthening the NA condition which only requires no-arbitrage at the terminal time T.

17.2.3 *Existence of martingale selections*

An integrable set-valued process Z_t with discrete time $t = 0, \dots, T$ or continuous time $t \in [0, T]$ is said to be a set-valued $(\mathfrak{F}_t)$-martingale if $\mathbf{E}(Z_t|\mathfrak{F}_s) = Z_s$ a.s. for all $0 \leq s \leq t \leq T$. The conditional expectation of Z_t is defined as the convex set whose support function is the conditional expectation of the support function of Z_t, see Molchanov (2005, Sec. 2.1.6) and Hess (1991) for the case of unbounded Z_t. The unconditional expectation of random sets is discussed in Chapter 1. It is well known (see Hess 1999 and Molchanov 2005, Sec. 5.1.1) that a *set-valued martingale* admits at least one martingale selection and, moreover, it has a countable dense family of martingale selections, the so-called *Castaing representation*, see Molchanov (2005, Th. 5.1.12).

The specific nature of the financial setting calls for the study of set-valued processes that are not necessarily martingales themselves or are martingales with respect to another (risk-neutral) probability measure $\mathbf{Q}$ equivalent to $\mathbf{P}$. In many cases such processes are *set-valued submartingales*, i.e. $\mathbf{E}(Z_t|\mathfrak{F}_s) \supset Z_s$ a.s. for all $s \leq t$. This situation is quite typical for set-valued process obtained as convex hulls of martingales, since the supremum of martingales naturally yields a submartingale.

For instance, consider geometric Brownian motions $\zeta_t = s_0 \exp\{\sigma W_t - \sigma^2 t/2\}$ with volatility σ that belongs to an interval $[\sigma', \sigma'']$, see Avellaneda, Levy and Parás (1995) and Mykland (2003) for a discussion of such partially specified volatility models. The union Z_t of all their paths is an interval-valued submartingale, see Kaval and Molchanov (2006, Ex. 9.2). Its upper bound $\zeta_t^a = \sup Z_t$ is a numerical submartingale, while the lower bound $\zeta_t^b = \inf Z_t$ is a numerical supermartingale. Applying the Doob decomposition theorem to the both of them it is possible to come up with an interval-valued martingale that is the largest set-valued martingale inscribed in Z_t for all t. In spaces of dimension 2 and more the problem of finding the largest set-valued martingale inscribed in a set-valued process is not yet studied.

In transaction costs models, the sets Z_t become random cones for all t. Consider their bases $\bar{Z}_t = \{x = (x_1, \dots, x_d) \in Z_t : x_1 = 1\}$ obtained by setting the first coordinate to one. If $\zeta_t = (\zeta_t^1, \dots, \zeta_t^d)$ is a $\mathbf{P}$-martingale selection of $(Z_t \setminus \{0\})_{t=0}^T$, then, by normalising with $\mathbf{E}\zeta_t^1$, it is possible to assume that ζ_t^1 is a martingale with expected value one. If $\mathbf{Q}$ is a new probability measure with density ζ_t^1, then the process ζ_t/ζ_t^1 becomes a $\mathbf{Q}$-martingale selection of $(\bar{Z}_t)_{t=0}^T$. In the opposite direction, if ζ is a $\mathbf{Q}$-martingale selection of the set-valued process $(\bar{Z}_t)_{t=0}^T$, where $\mathbf{Q}$ is an equivalent martingale measure with density f, then $f\zeta_t$ becomes a $\mathbf{P}$-martingale selection of $(Z_t \setminus \{0\})_{t=0}^T$.

Consider now the mere question of the existence of a $\mathbf{Q}$-martingale selection ζ_t of a set-valued process Z_t, where $\mathbf{Q}$ is equivalent to the underlying probability measure $\mathbf{P}$. We then call ζ an *equivalent martingale selection.* For cone-valued processes Z_t the above argument shows that it is either possible to work with martingale selections of original cone-valued processes or with equivalent martingale selections of their bases.

Rokhlin (2007*b*) studied the existence of a martingale selection for a sequence of relatively open convex random sets G_t with discrete time $t = 0, \dots, T$. In view of the above discussion, the existence of such martingale selection amounts to the existence of a strictly consistent price system, if G_t is chosen to be the relative interior of K_t^*. The main result of Rokhlin (2007*b*) establishes that the existence of martingale selection is equivalent to the a.s. non-emptiness of the recursively defined set-valued process W_t, $t = 0, \dots, T$. One sets W_T to be the closure of G_T and defines W_{t-1} to be the closure of G_{t-1} intersected with the relative interior of Y_{t-1}, the latter being the convex hull of the support of the conditional expectation of W_t given $\mathfrak{F}_{t-1}$. The *support* for a random set is defined by Rokhlin (2006) as the union of the supports for all selections of the random set. It should be noted that the process G_t is no longer assumed to be cone-valued. A variant of this problem, where martingale selections are assumed not only to belong to G_t but also have increments from some sequence of random cones is studied in Rokhlin (2007*a*).

Astic and Touzi (2007) discuss another generalization of the whole setting to the case where the solvency region $\hat{K}_t$ is convex, but no longer a cone, i.e. tx does not necessarily belong to the solvency region for all $t > 0$. The solvency region at time t is then defined by

$$\hat{K}_t = \{x \in \mathbb{R}^d : P_t(\omega, x) \geq 0\},$$

where $P_t(\omega, x)$ is the random liquidation value of portfolio x at time t. In financial applications this setting reflects possible liquidation difficulties for large positions in the assets. The approach of Astic and Touzi (2007) is to work with nearly consistent price systems that are obtained as the intersection of $\hat{K}_t$ and the cone with base $\hat{K}_t \cap B_\varepsilon^1$, where B_ε^1 is the ℓ_1-ball of radius ε. In the limiting case one obtains the tangent cone to $\hat{K}_t$ at the origin.

The continuous time case is more difficult, where in many cases the only reasonable hedge is trivial, see Levental and Skorohod (1997). Guasoni, Rásonyi and Schachermayer (2008) discuss the model with a money account and d risky assets, where the bid-ask prices for asset i are given by $(1+\varepsilon)^{-1}S_t^i$ and $(1+\varepsilon)S_t^i$ representing proportional transaction costs at level ε. They prove that the existence of a martingale selection follows from the fact that the conditional distributions of the d-dimensional process $(S_t^1, \dots, S_t^d)$ have full support for all time moments between the start and a finite time horizon. Campi and Schachermayer (2006) discuss a super-replication problem for Kabanov's bid-ask process with continuous time. The existence of a martingale selection in this model is discussed by Kabanov and Stricker (2008).

17.3 Risk measures

17.3.1 *Coherent utility functions*

Consider the family L^∞ of essentially bounded random variables defined on a probability space $(\Omega, \mathfrak{F}, \mathbf{P})$. The aim is to assess the risk associated with a random variable $X \in L^\infty$ interpreted as a financial gain. Clearly, the expectation of X does not suit the purpose of risk assessment. Indeed, random variable X equally likely taking values ± 1 and random variable Y with values $\pm 10^6$ (say US$) have both expectation zero, while the risks associated with them clearly differ.

Let us associate with each $X \in L^\infty$ its *utility* $u(X)$ that satisfies all or some of the following properties:

U1 $u(X) \geq 0$ if $X \geq 0$ a.s. and $u(0) = 0$;
U2 the utility is *monetary*, i.e. $u(X + a) = u(X) + a$ for all $a \in \mathbb{R}$;
U3 u is concave, i.e. $u(\lambda X + (1 - \lambda)Y) \geq \lambda u(X) + (1 - \lambda)u(Y)$;
U4 u is homogeneous, i.e. $u(\lambda X) = \lambda u(X)$ for all $\lambda \geq 0$ and $X \in L^\infty$;
U5 u has the *Fatou property*, i.e. if X_n converges in probability to X and $\operatorname{ess\,sup}|X_n| \leq 1$ for all n, then $u(X) \geq \limsup u(X_n)$.

A random variable X is called acceptable if $u(X) \geq 0$. The monetary property implies that if $u(X)$ is negative, we have that $u(Y) \geq 0$ for $Y = X - u(X)$. In other words, $\rho(X) = -u(X)$ determines the capital to be added to (removed from if negative) X in order to create an acceptable position out of X. The functional $\rho(X) = -u(X)$ is called a *risk measure*.

Concave and homogeneous utility functions (and the corresponding risk measures) are called *coherent*. In this case u is superadditive, i.e. $u(X + Y) \geq u(X) + u(Y)$ for all $X, Y \in L^\infty$. The coherency property describes mathematically the natural financial requirement saying that diversification reduces risk, see Artzner, Delbaen, Eber and Heath (1999) and Delbaen (2002). If the coherency is not assumed, one strengthens U1 to impose the monotonicity of u. The following theorem is proved in Delbaen (2002) by using the bipolar theorem from functional analysis.

Theorem 17.1 (see Theorems 2.3 and 3.2 (Delbaen, 2002)) *The monetary utility function u is coherent if and only if*

$$u(X) = \inf_{\mu \in \mathcal{S}} \mathbf{E}_\mu X\,, \tag{17.1}$$

where $\mathcal{S}$ is a convex closed set of finitely additive probabilities.

If u also satisfies the Fatou property, then (17.1) holds with $\mathcal{S}$ being an $L^1(\mathbf{P})$-closed convex set of probability measures absolutely continuous with respect to $\mathbf{P}$.

A generalization of this representation theorem for not necessarily homogeneous concave u can be found in Föllmer and Schied (2002). In the following we consider only coherent risk measures with the Fatou property. The latter is in turn guaranteed by imposing that u is *law invariant*, i.e. the values of u on identically distributed random variables coincide, see Jouini, Schachermayer and Touzi (2006). Then the statement of Theorem 17.1 can be expressed by saying that u is the support function of a convex closed set $\mathcal{S}$ of $\mathbf{P}$-absolutely continuous probability measures.

Example 17.2 (Worst conditional expectation) Let the representing set $\mathcal{S}$ in Theorem 17.1 consist of all probability measures μ whose density $d\mu/d\mathbf{P}$ is bounded by a constant $k > 1$. If X has a continuous distribution, then a simple optimization argument shows that the infimum in (17.1) is attained for μ that allocates its mass left to the $\frac{1}{k}$th quantile $q_{1/k} = F_X^{-1}(1/k)$ of X defined from $\mathbf{P}\left\{X \le q_{1/k}\right\} = 1/k$. Then

$$u(X) = \mathbf{E}[X|X \le q_{1/k}]$$

is called the *expected shortfall*, see Acerbi and Tasche (2002) for information about the related risk measure and the case of possible atoms.

Example 17.3 (Distortion of probability) In this case $\mathcal{S}$ is the family of probability measures $\mathbf{Q}$ absolutely continuous with respect to $\mathbf{P}$ such that $\mathbf{Q}(A) \ge \phi(\mathbf{P}(A))$ for all measurable sets A, where the distortion function $\phi : [0,1] \mapsto [0,1]$ is convex and satisfies $\phi(0) = 0$, $\phi(1) = 1$. Note that $v(A) = \phi(\mathbf{P}(A))$ is called a *distorted probability.*

Note that it is not possible to construct a law invariant coherent utility function on the family of all random variables. As a reason for this, consider a negative (one-sided) stable random variable X, so that the sum $X_1 + \cdots + X_n$ of i.i.d. copies of X has the same distribution as $n^{1/\alpha}X$ for $\alpha \in (0,1)$. Then

$$n^{1/\alpha}u(X) = u(X_1 + \cdots + X_n) \ge nu(X),$$

which is impossible unless $u(X)$ is non-negative, which makes no financial sense, since X is a.s. negative. In order to come up with a meaningful extension of the utility function for not necessarily bounded random variables, it is possible to use truncation (Delbaen 2002) or work on L^p spaces with $p \ge 1$. The representing set $\mathcal{S}$ then becomes a convex closed set in the dual space L^q.

17.3.2 *Choquet integral and risk measures*

Let us return to Example 17.3 of risk measures generated by distorted probabilities. Note that $v(A) = \phi(\mathbf{P}(A))$ is convex, i.e.

$$v(A) + v(B) \le v(A \cup B) + v(A \cap B) \tag{17.2}$$

for all measurable A and B. Non-additive set-functions are used in game theory to describe payoff $v(A)$ allocated to any coalition A of players, see Shapley (1971), Schmeidler (1989) and Molchanov (2005, Sec. 1.9.1). The *core of a game* is defined as the family $\mathcal{S}_v$ of all probability measures μ that dominate v, or, equivalently, the family $\mathcal{S}^w$ of measures dominated by the dual payoff $w(A) = 1 - v(A^c)$. Note that w is concave, i.e. satisfies the opposite inequality in (17.2). If (17.2) holds, then the game is called *convex* and has a non-empty core by the Bondareva–Shapley theorem.

The utility $u(X)$ defined as infimum of $\mathbf{E}_\mu X$ over all $\mu \in \mathcal{S}_v$ can be expressed as the *Choquet integral* with respect to the non-additive measure v as

$$u(X) = \int_0^\infty v(X > s)\mathrm{d}s \tag{17.3}$$

for all non-negative random variables X, see Denneberg (1994) and Schmeidler (1986). Indeed, if μ dominates v, then $\mathbf{E}_\mu X$, being the integral of $\mu(X > s)$, dominates $u(X)$. The monetary property of X makes it possible to extend (17.3) for all $X \in L^\infty$ as $u(X + a) - a$ for sufficiently large a. The Choquet integral with respect to w equals the supremum of $\mathbf{E}_\mu X$ over all $\mu \in \mathcal{S}_v = \mathcal{S}^w$.

The functionals v and w are sometimes called lower and upper probabilities, non-additive measures or fuzzy measures. Theory of random sets provides examples of important dual pairs of non-additive set functions. If Z is a random set, then the *capacity* (or hitting) *functional* $T(A) = \mathbf{P}\{Z \cap A \neq \emptyset\}$ is completely alternating, i.e. it satisfies the concavity property of an arbitrarily high order, see Molchanov (2005, Sec. 1.1.2) and Chapter 1. The *containment functional* $C(A) = \mathbf{P}\{X \subset A\}$ is completely monotone (Molchanov 2005, Sec. 1.1.6) and so satisfies the extended variant of (17.2). Note that T and C are dual, i.e. $C(A) = 1 - T(A^c)$. If the capacity functional T is used instead of w, then the corresponding family $\mathcal{S}^T$ becomes the family of distributions for all *selections* of the random set Z. If Z is non-empty and closed, then it has at least one selection, see Molchanov (2005, Sec. 1.2.2). The Choquet integrals with respect to C and T then acquire simple probabilistic interpretations as

$$\int_0^\infty C(X > s)\mathrm{d}s = \mathbf{E}\inf\{X(z) : \ z \in Z\}, \tag{17.4}$$

$$\int_0^\infty T(X > s)\mathrm{d}s = \mathbf{E}\sup\{X(z) : \ z \in Z\}. \tag{17.5}$$

It should be noted that Z is now a random set whose values are subsets of Ω, so that it is useful to endow the space Ω of elementary events with a topological structure. Because of this, it is sensible to define a shadow probability space $(\Omega', \mathfrak{F}', \mathbf{P}')$ on which Z is defined, so that the expectation in the right-hand side of (17.5) is taken with respect to $\mathbf{P}'$. The monetary property can then be easily incorporated in the right-hand sides of (17.4) and (17.5), so that

$$u(X) = \mathbf{E}_{\mathbf{P}'} \inf\{X(z) : z \in Z\} = \mathbf{E}_{\mathbf{P}'} \inf X(Z) \tag{17.6}$$

becomes a utility function on L^∞, where $X(Z)$ is the image of Z under X. The corresponding risk measures have been studied by Sikharulidze (2006) and are called there the *Choquet risk measures.* The definition itself immediately implies that u defined by (17.6) is coherent. Since u is generated by a random closed set, operations with sets induce operations with risk measures, see Sikharulidze (2006).

Since $\inf X(Z)$ can be approximated by a monotone sequence of selections of $X(Z)$, the expectation of $\inf X(Z)$ equals the infimum of expectations for all selections of $X(Z)$, i.e.

$$u(X) = \inf \mathbf{E}_{\mathbf{P}'} X(Z)\,,$$

where the expectation in the right-hand side is the selection expectation of $X(Z)$, see Molchanov (2005, Sec. 2.1) and Chapter 1.

One particular feature of the utility function (17.6) is its dependence on a particular representation of the random variable X, in particular, which values X takes on Z. Thus, this utility function is really far from being law invariant. It is possible to come up with a law invariant modification of the Choquet risk measure if we fix the particular representation of the random variable X, e.g. as the probability integral transform $F_X^{-1}(U)$, where F_X is the cumulative distribution function of X and U uniformly distributed on $[0, 1]$. Then Ω can be identified with $[0, 1]$ and the utility function $u(X)$ becomes the expectation of $F_X^{-1}(\sup Z)$, i.e. the integral of the weighted quantiles $F_X^{-1}(t)$ over $t \in [0, 1]$. Such integrals define so-called *spectral risk measures*, see Acerbi (2002).

The properties of the Choquet integral (Denneberg 1994) imply that $u(X)$ given by (17.3) is comonotonic additive, i.e. $u(X+Y) = u(X)+u(Y)$ for comonotonic random variables X and Y. Recall that X and Y are called *comonotonic* if

$$(X(\omega') - X(\omega''))(Y(\omega') - T(\omega'')) \geq 0$$

for almost all (ω', ω'') from the product space $\Omega \times \Omega$ equipped with the product measure. Indeed, in this case the supremum of $(X+Y)(Z)$ is the sum of suprema of $X(Z)$ and $Y(Z)$. The result of Schmeidler (1986) implies that a coherent utility function admits representation (17.3) if and only if u is comonotonic with the Fatou property. This result however does not imply that v can be represented as the containment functional C of a certain random set. If u is also assumed to be law invariant, then u admits a spectral representation as integrals of the quantiles for the underlying random variable, see Tasche (2002). Such u can be obtained from (17.4) using a random set in $[0, 1]$ and the representation of X using the probability integral transform.

The representation (17.1) of a coherent utility can be written as $\inf X(\zeta)$, where ζ are Ω-valued random elements with distributions $\mu \in \mathcal{S}$. It is however not always possible to interpret all these ζ's as the family of selections of a certain random set. The decomposition theorem from Molchanov (2005, Th. 2.1.6) says that a closed subset of L^p, $p \in [1, \infty]$, is a family of selections if and only if, for each ζ_1, ζ_2 from this family and each event A, the random element

$$\zeta = \mathbf{1}_A \zeta_1 + \mathbf{1}_{A^c} \zeta_2$$

also belongs to the same family.

17.3.3 *Multivariate risk measures*

The utility function (17.6) in the univariate case can be associated with the expectation of the supremum of the random set $X(Z)$. If X is a random vector, the supremum is no longer well defined. Jouini, Meddeb and Touzi (2004) introduced and studied vector-valued risk measures for multiasset portfolios. However, it is not possible to construct a non-trivial vector-valued coherent utility function (or risk measure) for d-dimensional random vectors. It follows from Cascos and Molchanov (2007, Ex. 9.1) that if $u(X)$ is $\mathbb{R}^d$-valued coherent utility function which respects the coordinate-wise maximum order and such that $u(x) = x$ for $X = x$ a.s., then u is marginalized, i.e. $u(X)$ is a vector composed of some utility functions for the marginals of X. In view of this, it is sensible to consider set-valued risk measures and utility functions, for instance the whole set-valued expectation of $X(Z)$ is a possible candidate for a risk measure or utility function. Since the ordering of sets is defined by inclusion or reversed inclusion, the central symmetry transform does not transform risk measures to utility functions. The distinction between them relies on whether the risk of a deterministic portfolio $X = x$ is expressed as a certain set translated by x (utility) or by $-x$ (risk).

Let us fix a proper polyhedral cone K in $\mathbb{R}^d$, which is used to define a partial order $\leq_K$. Although Cascos and Molchanov (2007) assume that K does not contain any line, this condition can be easily dropped, cf Hamel (2006). We interpret K as the set of acceptable risk values or non-negative utilities. Without much loss of generality it is possible to regard K as being the first quadrant, i.e. $K = \mathbb{R}^d_+$.

It is possible to construct a rich family of multivariate set-valued utility functions by replacing X with a certain set in $\mathbb{R}^d$ that describes how the probability mass concentrates around the 'central value' of X. In multivariate statistics such sets are called *depth-trimmed* or *central* regions, see Zuo and Serfling (2000) and Chapter 13. The most well-known example of central regions are *half-space* trimmed regions defined as

$$\mathrm{HD}^{\alpha}(X) = \bigcap \{H : H \text{ closed halfspace with } \mathbf{P}\{X \in H\} \geq 1 - \alpha\},$$

see Massé and Theodorescu (1994). The value of α controls the extent of trimming. It is sensible to assume that general *depth-trimmed regions* $\mathrm{D}^{\alpha}(X)$, $\alpha \in [0, 1]$, satisfy the following properties.

Translation equivariance: $\mathrm{D}^{\alpha}(tX + y) = t\mathrm{D}^{\alpha}(X) + y$.
Nesting: $\mathrm{D}^{\alpha}(X) \subset \mathrm{D}^{\beta}(X)$ for $\alpha \geq \beta$.
$\mathrm{D}^{\alpha}(X)$ connected and closed.

These properties stem from Zuo and Serfling (2000), where the translation equivariance is strengthened to require the affine equivariance. In view of application to utility and risk measures, the following extra properties are needed.

Monotonicity: $\mathrm{D}^{\alpha}(X) \subset \mathrm{D}^{\alpha}(Y) + K$ if $Y \leq_K X$ a.s.
Subadditivity: $\mathrm{D}^{\alpha}(X + Y) \subset \mathrm{D}^{\alpha}(X) + \mathrm{D}^{\alpha}(Y)$.

The monotonicity condition reflects the fact that a utility function has a clear direction preference: larger values of the coordinates clearly increase the utility. In view of this, the *one-sided* (or monotonic) variant of the half-space trimmed regions is defined as follows

$$\mathrm{HD}^{\alpha}_{K}(X) = \bigcap \{H_u : \mathbf{P}\{X \in H_u\} \geq 1 - \alpha,$$
$$H_u \text{ halfspace with inner normal } u \in K^*\},$$

where $K^* = \{u : \langle u, v\rangle \geq 0 \ \forall v \in K\}$ is the positive dual cone to K. The one-sided half-space trimmed regions satisfy all above mentioned properties but lack the subadditivity. As a result, the corresponding set-valued utility $u(X) = \mathrm{HD}^{\alpha}_{K}(X)$ is not coherent.

It is emphasized in Cascos and Molchanov (2007) that zonoid-trimmed regions studied by Koshevoy and Mosler (1997) and convex hull trimmed regions introduced by Cascos (2007) are subadditive, and so can be used to construct coherent set-valued utility functions and risk measures for random vectors. The key idea is to consider all translations of the depth-trimmed region and compare the translated sets with either K or its origin symmetric variant $\check{K}$. If the translated set is a subset of K, then we speak about a definite acceptance, while if the translation hits $\check{K}$ we mention a possible rejection. Each translation is given by a point in $\mathbb{R}^d$ and the sets of the corresponding points (or translations) define the risk measure $\rho(X)$ or utility $u(X)$.

For instance, we may consider all translations that do not result in the definite acceptance. Then the utility function is defined by

$$u(X) = \{z \in \mathbb{R}^d : \mathrm{D}^{\alpha}(X) - z \not\subset K\} = z_0 + K,$$

where z_0 is the coordinate-wise minimum of $\mathrm{D}^\alpha(X)$, so in this case the risk measure marginalizes. The translations of $\mathrm{D}^\alpha(X)$ that intersect $\check{K}$ (and so lead to a possible rejection) yield a more interesting utility function defined as

$$u(X) = \{z \in \mathbb{R}^d : \mathrm{D}^\alpha(X) - z \cap \check{K} \neq \emptyset\} = \mathrm{D}^\alpha(X) + K\,,$$

which is a dilation of K by $\mathrm{D}^\alpha(X)$, see Matheron (1975) and Molchanov (2005, p. 396).

For illustration, consider the *convex hull* trimmed region defined as

$$\mathrm{CD}^{1/n}(X) = \mathbf{E}\,\mathrm{conv}\{X_1, \dots, X_n\}\,,$$

where $X_1, \dots, X_n$ are independent copies of X and $\mathbf{E}$ is the (selection) expectation. In the univariate case $\inf \mathrm{CD}^{1/n}(X)$ is the expected minimum $\mathbf{E}\min(X_1, \dots, X_n)$. The corresponding utility function is then given by

$$u(X) = \mathbf{E}\,\mathrm{conv}\{X_1, X_2, \dots, X_n\} + K\,.$$

17.4 Lift zonoids of asset prices

If S_T is the asset price at the time T, when a *European option* matures, then $\mathbf{E}_\mathbf{Q}(S_T - k)_+$ and $\mathbf{E}_\mathbf{Q}(k - S_T)_+$ are non-discounted prices of a call and put option with the strike price k, where the expectation is taken with respect to the martingale measure $\mathbf{Q}$ and $(x)_+ = \max(0, x)$ for real x. Note that $\mathbf{E}_\mathbf{Q} S_T = F$ is the forward price of the asset. In the simplest case this forward price is given by $F = \mathrm{e}^{(r-q)T} S_0$, where S_0 is the spot price (price at time zero), r is the risk-free interest rate and q is the dividend rate. If $r = q$, then $F = S_0$, i.e. the cost-of-carry for the asset vanishes. In this section we often assume that $r = q = 0$, so that the above expectations do not need to be discounted and become the prices of the European call and put options.

If ξ is a random variable, then the expectation of the random set X being the segment in the plane with end-points at the origin and at $(1, \xi)$ is called the *lift zonoid* of ξ and denoted by Z_ξ. This concept can also be easily formulated for ξ being a random vector, see Mosler (2002). This expectation Z_ξ can be also defined from the following identity for the support functions

$$h(Z_\xi, u) = \mathbf{E} h(X, u) = \mathbf{E}(u_0 + u_1 \xi)_+\,, \quad u = (u_0, u_1) \in \mathbb{R}^2\,.$$

Now define $\xi = S_T = F\eta$, where F is the forward price of the asset, so that $F = S_0$ for the case of vanishing cost-of-carry. Then the support function of Z_η in direction $u = (-k, F)$ becomes the non-discounted price of the call option, while the put price corresponds to the support function in direction $v = (k, -F)$.

It is well known that the lift zonoid of a random vector identically determines its distribution, see Mosler (2002). Therefore, lift zonoids carry sufficient information to determine all European option prices, also in case of options on several underlyings, the so-called *basket options*. This observation together with the above representation of the vanilla option prices was used as a starting point in Schmutz (2007) to study relationships between lift zonoids and prices of derivatives.

In particular the symmetry properties of the lift zonoid express geometrically several important financial parities. Since $\mathbf{E}_{\mathbf{Q}}\eta = 1$, the lift zonoid Z_η is always centrally symmetrical with respect to $(\frac{1}{2}, \frac{1}{2})$, which is nothing else than the *call-put parity*

$$c + k = p + S_0$$

written for the case $r = q = 0$. If η is log-normally distributed with mean 1, then Z_η is also symmetric with respect to the line $u_0 = u_1$ bisecting the fist quadrant. This property appears in the financial literature under the name of the *put-call symmetry*, see Bates (1997) and Carr, Ellis and Gupta (1998). Assuming that η has density p_η, Schmutz (2007) proved that this symmetry holds if and only if $p_\eta(z)z^3 = p_\eta(z^{-1})$ for almost all $z > 0$. This and other characterizations of symmetry properties have been generalised for the multiasset case by Molchanov and Schmutz (2008).

17.5 Partially identified models in econometrics

17.5.1 *Linear models with interval response*

Consider fitting a *linear model* for observations, where the response is interval-valued. Let (η_l, η_u, ξ) be a random vector in $\mathbb{R} \times \mathbb{R} \times \mathbb{R}$ such that $\eta_l \leq \eta_u$ almost surely. Observations (y_{il}, y_{iu}, x_i) of this random vector describe interval-valued responses $Y_i = [y_{il}, y_{iu}]$ for particular values x_i of the explanatory variable. Define by $Y = [\eta_l, \eta_u]$ the corresponding random *interval response*. This scheme gives rise to the random set (actually random segment) G in $\mathbb{R}^2$ with end-points $\xi\eta_l$ and $\xi\eta_u$ and its empirical observations $G_1, \dots, G_n$.

The conventional machinery of linear models yields that the set

$$\Theta = \begin{bmatrix} 1 & \mathbf{E}\xi \\ \mathbf{E}\xi & \mathbf{E}(\xi^2) \end{bmatrix}^{-1} \mathbf{E}G$$

is the identification region for the parameters of the linear model. Namely, the points from Θ are exactly those which (used as the intercept and slope) can be derived from the available observed intervals, see Beresteanu and Molinari (2008, Sec. 5). The next step is to replace $\mathbf{E}G$ with its estimator obtained as the Minkowski average of n segments $G_1, \dots, G_n$. Beresteanu and Molinari (2008, Th. 5.2) proved that the obtained estimator $\hat{\Theta}_n$ is strongly consistent in the Hausdorff metric ρ_H and the rate of convergence in probability is $\mathcal{O}(n^{-1/2})$. The limiting distribution of $\sqrt{n}\rho_H(\hat{\Theta}_n, \Theta)$ can be expressed using the central limit theorem for Minkowski sums of random compact sets, see Molchanov (2005, Sec. 3.2) and Chapter 1 of this volume. This limit theorem implies that the difference between the support functions of the Minkowski average $\bar{G}_n = n^{-1}(G_1 + \cdots + G_n)$ and of the expectation $\mathbf{E}G$ considered as a random element in the space of continuous functions on the unit sphere weakly converges to a Gaussian random function whose covariance is determined by the distribution of G. A variant of this result by Beresteanu and Molinari (2008, Th. 5.3) takes into account the fact that the Minkowski average $\bar{G}_n$ is transformed by a matrix whose distribution also depends on the sample. In order to circumvent the dependence of the limiting law on the estimated parameters one can use bootstrap methods. Beresteanu and Molinari (2008) also consider the case of binary responses and the corresponding interval-valued logistic regression models.

17.5.2 *Selection identification problems*

Consider a family of random closed sets $Z(\theta)$ which depend on parameter $\theta \in \Theta$. One case of this problem was considered in Chapter 12 of this volume; it concerns the case when θ itself is a set and $Z(\theta)$ is a random point whose distribution has, e.g. the support θ. The general parametric inference for random sets is complicated because of the lack of likelihood based methods. The special case, where θ is a parameter that determines the distribution of a point process, leads to statistical inference for point processes described in Chapter 9.

Here we consider another example of this situation that originates from random games. The basic example is a two players *static game of entry* analysed by Bresnahan and Reiss (1991) and Berry (1992). In the simplest case, each player's strategy y_1, y_2 is either 1 if he decides to enter the market or 0 otherwise. The profits of the players are $\pi_1 = y_1(y_2\theta_1 + \varepsilon_1)$ and $\pi_2 = y_2(y_1\theta_2 + \varepsilon_2)$, where θ_1, θ_2 are negative parameters of the game and $\varepsilon_1, \varepsilon_2$ are random profit shifters unobserved by the econometrician, but whose distribution is known up to some unknown parameters that are also estimated and so included in θ. The Nash equilibrium condition identifies possible pairs of strategies which are optimal for the both players. As a result one arrives at a random set Z which is a subset of the 4-point space $\mathcal{Y} = \{00, 01, 10, 11\}$, see Beresteanu, Molchanov and Molinari (2008). The particular nature of this game is that it might lead to multiple equilibria, i.e. the set Z may contain two or more strategies. In the above example the random set Z may take each of the four single values from $\mathcal{Y}$ and also the value $\{01, 10\}$, meaning that in some cases the entry of either player leads to an equilibrium. The econometrician observes the outcomes of the game, i.e. one of points from $\mathcal{Y}$. In case of multiple equilibria the observed point simply belongs to Z; one says that this point is sampled from Z by a random selection mechanism, see Berry and Tamer (2007, p. 68).

The corresponding statistical problem lies in the estimation of θ based on the empirically derived distribution of a selection of $Z(\theta)$. It is well known (see

Artstein 1983, Norberg 1992 and Molchanov 2005, Sec. 2.2) that ζ is a *selection* of Z if and only if $\mathbf{P}\{\zeta \in K\} \le \mathbf{P}\{Z \cap K \neq \emptyset\} = T(K)$ for all compact sets K. If $Z(\theta)$ is a subset of a finite space $\mathcal{Y}$, these inequalities need to be checked for every strict subset $K \subset \mathcal{Y}$. It is shown in Beresteanu, Molchanov and Molinari (2008) that this system of inequalities results in a *sharp* identification region for the parameter Θ. It is also possible to reduce the number of these inequalities if $Z(\theta)$ is a singleton or has a controllable cardinality. For instance, if $Z(\theta)$ is a singleton, then one only need to check them for singletons K, cf Andrews, Berry and Jia (2004) and Ciliberto and Tamer (2006).

An estimator for the set of all possible values of θ can be then obtained as the set of all θ's which solve the system of inequalities after an appropriate correction for the statistical errors that arise from replacing the theoretical distribution of a selection with its empirical variant. For this one can use the tools based on solutions of problems of moment inequalities type (Chernozhukov, Hong and Tamer 2007), estimation of sets determined by inequalities (Molchanov 1998) or general theory of set estimation from Chapter 12 of this book. A variant that allows for mixed (i.e. randomized) strategies is also considered in Beresteanu, Molchanov and Molinari (2008).

Acknowledgements

This work has been partially supported by Swiss National Science Foundation, Grants Nr. 200020-109217 and 200021-117606.

The author is grateful to Arie Beresteanu, Ignacio Cascos, Jasmin Furrer, Pavel Grigoriev, Christian Hess, Katya Kaval, Alexander Kohler, Francesca Molinari, Miklos Rásonyi, Michael Schmutz, Irina Sikharulidze for discussions and/or collaboration related to various parts of this work.

References

Acerbi, C. (2002). Spectral measures of risk: A coherent representation of subjective risk aversion. *J. Banking Finance*, **26**, 1505–1518.

Acerbi, C. and Tasche, D. (2002). On the coherence of expected shortfall. *J. Banking Finance*, **26**, 1487–1503.

Andrews, D. W. K., Berry, S., and Jia, P. (2004). Confidence regions for parameters in discrete games with multiple equilibria, with an application to discount chain store location. Working paper, Cowles Foundation, Yale.

Artstein, Z. (1983). Distributions of random sets and random selections. *Israel J. Math.*, **46**, 313–324.

Artzner, Ph., Delbaen, F., Eber, J.-M., and Heath, D. (1999). Coherent measures of risk. *Math. Finance*, **9**, 203–228.

Astic, F. and Touzi, N. (2007). No arbitrage conditions and liquidity. *J. Math. Econ.*, **43**, 692–708.

Avellaneda, M., Levy, A., and Parás, A. (1995). Pricing and hedging derivative securities in markets with uncertain volatilities. *Appl. Math. Finance*, **2**, 73–88.

Bates, D. S. (1997). The skewness premium: Option pricing under asymmetric processes. *Advances in Futures and Options Research*, **9**, 51–82.

Beresteanu, A., Molchanov, I., and Molinari, F. (2008). Sharp identification regions in games. Working Paper. Cornell University, `http://cemmap.ifs.org.uk/wps/cwp1508.pdf`.

Beresteanu, A. and Molinari, F. (2008). Asymptotic properties for a class of partially identified models. *Econometrica*, **76**, 763–814.

Berry, S. T. (1992). Estimation of a model of entry in the airline industry. *Econometrica*, **60**, 889–917.

Berry, S. T. and Tamer, E. (2007). Identification in models of oligopoly entry. In *Advances in Economics and Econometrics: Theory and Applications* (ed. R. Blundell, W. K. Newey, and T. Persson), Volume 2, Chapter 2, pp. 46–85. Cambridge Univ. Press, Cambridge.

Bresnahan, T. F. and Reiss, P. C. (1991). Entry and competition in concentrated markets. *J. Political Economy*, **99**, 977–1009.

Campi, L. and Schachermayer, W. (2006). A super-replication theorem in Kabanov's model of transaction costs. *Finance and Stochastics*, **10**, 579–596.

Carr, P., Ellis, P., and Gupta, V. (1998). Static hedging of exotic options. *J. Finance*, **53**, 1165–1190.

Cascos, I. (2007). Depth functions based on a number of observations of a random vector. Technical Report Working Paper 07-29. Statistics and Econometrics Series, Universidad Carlos III de Madrid. `http://econpapers.repec.org/paper/ctewsrepe`.

Cascos, I. and Molchanov, I. (2007). Multivariate risks and depth-trimmed regions. *Finance and Stochastics*, **11**, 373–397.

Chernozhukov, V., Hong, H., and Tamer, E. (2007). Estimation and confidence regions for parameter sets in econometric models. *Econometrica*, **75**, 1243–1284.

Ciliberto, F. and Tamer, E. (2006). Market structure and multiple equilibrium in airline markets. Working paper, University of Virginia.

Ciliberto, F. and Tamer, E. (2009). Market structure and multiple equilibrium in airline markets. *Econometrica*. To appear.

Delbaen, F. (2002). Coherent risk measures on general probability spaces. In *Advances in Finance and Stochastics* (ed. K. Sandmann and P. J. Schönbucher), pp. 1–37. Springer, Berlin.

Denneberg, D. (1994). *Non-Additive Measure and Integral*. Kluwer, Dordrecht.

Diamond, P. (1990). Least square fitting of compact set-valued data. *J. Math. Anal. Appl.*, **147**, 351–362.

Föllmer, H. and Schied, A. (2002). Convex measures of risk and trading constraints. *Finance and Stochastics*, **6**, 429–447.

Gil, M. A., López-García, M. T., Lubiano, M. A., and Montenegro, M. (2001). Regression and correlation analyses of a linear relation between random intervals. *Test*, **10**, 183–201.

Grigoriev, P. G. (2005). On low dimensional case in the fundamental asset pricing theorem with transaction costs. *Statist. Decision*, **23**, 33–48.

Guasoni, P., Rásonyi, M., and Schachermayer, W. (2008). Cosnsistent price systems and face-lifting pricing under transcation costs. *Ann. Appl. Probab.*, **18**, 491–520.

Hamel, A. H. (2006). Translative sets and functions and their applications to risk measure theory and nonlinear separation. Working paper.

Hess, Ch. (1991). On multivalued martingales whose values may be unbounded: martingale selectors and Mosco convergence. *J. Multivariate Anal.*, **39**, 175–201.

Hess, Ch. (1999). Conditional expectation and martingales of random sets. *Pattern Recognition*, **32**, 1543–1567.

Hu, S. and Papageorgiou, N. S. (1997). *Handbook of Multivalued Analysis*, Volume 1. Kluwer, Dordrecht.

Imbens, G. W. and Manski, C. F. (2004). Confidence intervals for partially identified parameters. *Econometrica*, **72**, 1845–1857.

Jacka, S., Berkaoui, A., and Warren, J. (2008). No-arbitrage and closure results for trading cones with transaction costs. *Finance and Stochastics*, **12**, 583–600.

Jouini, E. (2000). Price functionals with bid-ask spreads: an axiomatic approach. *J. Math. Econ.*, **34**, 547–558.

Jouini, E. and Kallal, H. (1995). Martingales and arbitrage in securities markets with transaction costs. *J. Econ. Th.*, **66**, 178–197.

Jouini, E., Meddeb, M., and Touzi, N. (2004). Vector-valued coherent risk measures. *Finance and Stochastics*, **8**, 531–552.

Jouini, E., Schachermayer, W., and Touzi, N. (2006). Law invariant risk measure have the Fatou property. *Adv. Math. Econ.*, **9**, 49–71.

Kabanov, Y. and Stricker, C. (2008). On martingale selectors of cone-valued processes. In *Séminaire de Probabilités XLI* (ed. C. Donati-Martin, M. Émery, A. Rouault, and C. Stricker), Volume 1934 of *Lect. Notes Math.*, pp. 437–442. Springer, Berlin.

Kabanov, Yu. M. (1999). Hedging and liquidation under transaction costs in currency markets. *Finance and Stochastics*, **3**, 237–248.

Kaval, K. and Molchanov, I. (2006). Link-save trading. *J. Math. Econ.*, **42**, 710–728.

Koshevoy, G. A. and Mosler, K. (1997). Zonoid trimming for multivariate distributions. *Ann. Statist.*, **25**, 1998–2017.

Levental, S. and Skorohod, A. V. (1997). On the possibility of hedging options in the presence of transaction costs. *Ann. Appl. Probab.*, **7**, 410–443.

Manski, C. F. (2003). *Partial Identification of Probability Distributions.* Springer, New York.

Manski, C. F. and Tamer, E. (2002). Inference on regressions with interval data on a regressor or outcome. *Econometrica*, **70**, 519–546.

Massé, J.-C. and Theodorescu, R. (1994). Halfplane trimming for bivariate distribution. *J. Multivariate Anal.*, **48**, 188–202.

Matheron, G. (1975). *Random Sets and Integral Geometry.* Wiley, New York.

Molchanov, I. (2005). *Theory of Random Sets.* Springer, London.

Molchanov, I. and Schmutz, M. (2008). Geometric extension of put-call symmetry in the multiasset setting. Technical report, University of Bern, Bern. ArXiv math.PR/0806.4506.

Molchanov, I. S. (1998). A limit theorem for solutions of inequalities. *Scand. J. Statist.*, **25**, 235–242.

Mosler, K. (2002). *Multivariate Dispersion, Central Regions and Depth. The Lift Zonoid Approach*, Volume 165 of *Lect. Notes Statist.* Springer, Berlin.

Mykland, P. A. (2000). Conservative delta hedging. *Ann. Appl. Probab.*, **10**, 664–683.

Mykland, P. A. (2003). Financial options and statistical prediction intervals. *Ann. Statist.*, **31**, 1413–1438.

Norberg, T. (1992). On the existence of ordered couplings of random sets – with applications. *Israel J. Math.*, **77**, 241–264.

Rásonyi, M. (2008). New methods in the arbitrage theory in financial markets with transaction costs. In *Séminaire de Probabilités XLI* (ed. C. Donati-Martin, M. Émery, A. Rouault, and C. Stricker), Volume 1934 of *Lect. Notes Math.*, pp. 455–462. Springer, Berlin.

Rokhlin, D. B. (2006). A martingale selection problem in the finite discrete-time case. *Theory Probab. Appl.*, **50**, 420–435.

Rokhlin, D. B. (2007*a*). Martingale selection problem and asset pricing in finite discrete time. *Electron. Commun. in Probab.*, **12**, 1–8.

Rokhlin, D. B. (2007*b*). A theorem on martingale selection for relatively open convex set-valued random sequences. *Math. Notes*, **81**, 543–548.

Schachermayer, W. (2004). The fundamental theorem of asset pricing under proportional transaction costs in finite discrete time. *Math. Finance*, **14**, 19–48.

Schmeidler, D. (1986). Integral representation without additivity. *Trans. Amer. Math. Soc.*, **97**, 255–261.

Schmeidler, D. (1989). Subjective probability and expected utility without additivity. *Econometrica*, **57**, 571–587.

Schmutz, M. (2007). Zonoid options. Master's thesis, Institute of Mathematical Statistics and Actuarial Science, University of Bern, Bern.

Shapley, L. S. (1971). Cores of convex games. *Internat. J. Game Theory*, **1**, 12–26.

Sikharulidze, I. (2006). Risk measures and random sets. Master's thesis, Institute of Mathematical Statistics and Actuarial Science, University of Bern, Bern.

Stoyan, D., Kendall, W. S., and Mecke, J. (1995). *Stochastic Geometry and its Applications* (Second edn). Wiley, Chichester.

Tasche, D. (2002). Expected shortfall and beyond. *J. Banking Finance*, **26**, 1519–1533.

Zuo, Y. and Serfling, R. (2000). General notions of statistical depth function. *Ann. Statist.*, **28**, 461–482.

INDEX

Printed and bound by CPI Group (UK) Ltd, Croydon, CR0 4YY

The manufacturer's authorised representative in the EU for product
safety is Oxford University Press España S.A. of El Parque Empresarial
San Fernando de Henares, Avenida de Castilla, 2 - 28830 Madrid
(www.oup.es/en or product.safety@oup.com). OUP España S.A. also acts
as importer into Spain of products made by the manufacturer.
Printed and bound by CPI Group (UK) Ltd, Croydon, CR0 4YY
14/09/2026
02216307-0018